# ALGAL
# CULTURING
# TECHNIQUES

Edited by

ROBERT A. ANDERSEN

ELSEVIER
ACADEMIC
PRESS

Amsterdam • Boston • Heidelberg • London • New York • Oxford
Paris • San Diego • San Francisco • Singapore • Sydney • Tokyo
Academic Press is an imprint of Elsevier

Acquisitions Editor: Frank Cynar
Project Manager: Justin Palmeiro
Editorial Coordinator: Jennifer Hele
Marketing Manager: Linda Beattie
Cover Design: Suzanne Rogers
Composition: SNP Best-set Typesetter Ltd., Hong Kong
Printer: Hing Yip Co.

Elsevier Academic Press
30 Corporate Drive, Suite 400, Burlington, MA 01803, USA
525 B Street, Suite 1900, San Diego, California 92101-4495, USA
84 Theobald's Road, London WC1X 8RR, UK

This book is printed on acid-free paper. ∞

**Library of Congress: Application submitted.**

**British Library Cataloguing in Publication Data**
A catalogue record for this book is available from the British Library

ISBN: 0-12-088426-7

For all information on all Academic Press publications
visit our Web site at www.books.elsevier.com

Printed and bound by CPI Group (UK) Ltd, Croydon, CR0 4YY

Transferred to Digital Print 2011

# Contents

**Robert A. Andersen,** Bigelow Laboratory for Ocean Sciences, West Boothbay Harbor, Maine

**Paul Behrens,** Martek Biosciences Corporation, Columbia, Maryland

**John Berges,** Biological Sciences, University of Wisconsin-Milwaukee, Milwaukee, Wisconsin

**Robert Bidigare,** Oceanography, University of Hawaii, Honolulu, Hawaii

**Susan Blackburn,** Marine Research, Commonwealth Scientific and Industrial Research Organization, Hobart, Australia

**Michael A. Borowitzka,** Algae Research Group, School of Biological Sciences and Biotechnology, Murdoch University, Murdoch, Australia

**Jerry Brand,** Molecular Cell and Developmental Biology, University of Texas, Austin, Texas

**Annette W. Coleman,** BioMed, Brown University, Providence, Rhode Island

**John J. Cullen,** Department of Oceanography, Dalhousie University, Halifax, Canada

**John G. Day,** Dunstaffnage Marine Laboratory, Culture Collection of Algae and Protozoa, Scottish Association for Marine Science, Dunbeg, United Kingdom

**R. Craig Everroad,** Center for Ecology and Evolutionary Biology, University of Oregon, Eugene, Oregon

**Thomas Friedl,** Experimental Phycology and Culture Collection of Algae, University of Göttingen, Göttingen, Germany

**Robert R.L. Guillard,** Bigelow Laboratory for Ocean Sciences, West Boothbay Harbor, Maine

**Paul J. Harrison,** AMCE Program, Hong Kong University of Science and Technology, Clear Water Bay, Hong Kong

**Masanobu Kawachi,** Environmental Biology Division, National Institute for Environmental Studies, Tsukuba, Japan

**Hiroshi Kawai,** Research Center for Inland Seas, Kobe University, Kobe, Japan

**Janice Lawrence,** Biology Department, University of New Brunswick, Fredericton, Canada

**Maike Lorenz,** Experimental Phycology and Culture Collection of Algae, University of Göttingen, Göttingen, Germany

**Klaus Lüning,** Wadden Sea Station Sylt, Alfred Wegener Institute for Polar and Marine Research, List, Germany

**Hugh MacIntyre,** Dauphin Island Sea Lab, Dauphin Island, Alabama

**Dominique Marie,** Station Biologique, Roscoff, France

**Francois Morel,** Department of Geosciences, Princeton University, Princeton, New Jersey

**Mary-Hélène Noël,** Environmental Biology Division, National Institute for Environmental Studies, Tsukuba, Japan

**Naomi Parker,** Invasive Marine Species Program, Department of Agriculture, Fisheries and Forestry, Canberra, Australia

**Hans R. Preisig,** Institute of Systematic Botany, University of Zurich, Zurich, Switzerland

**Neil Price,** Department of Biology, McGill University, Montreal, Canada

**Thomas Pröschold,** University of Cologne, Cologne, Germany

**Dinabandhu Sahoo,** Marine Biotechnology Laboratory, Department of Botany, University of Delhi, Delhi, India

**Michael Sieracki,** Bigelow Laboratory for Ocean Sciences, West Boothbay Harbor, Maine

**Nathalie Simon,** Station Biologique, Roscoff, France

**William Sunda,** National Ocean Service, National Oceanic and Atmospheric Administration, Beaufort, North Carolina

**Charles Trees,** Center for Hydro-Optics and Remote Sensing, San Diego State University, San Diego, California

**Laurie Van Heukelem,** Horn Point Laboratory, University of Maryland Center for Environmental Science, Cambridge, Maryland

**Daniel Vaulot,** Station Biologique, Roscoff, France

**Makoto M. Watanabe,** Environmental Biology Division, National Institute for Environmental Studies, Tsukuba, Japan

**John West,** School of Botany, University of Melbourne, Parkville, Australia

**Lauren Wingard,** Center for Ecology and Evolutionary Biology, University of Oregon, Eugene, Oregon

**A. Michelle Wood,** Center for Ecology and Evolutionary Biology, University of Oregon, Eugene, Oregon

**Charles Yarish,** Department of Ecology and Evolutionary Biology, University of Connecticut, Stamford, Connecticut

Why are algae important? They contribute approximately 40 to 50% of the oxygen in the atmosphere, or the oxygen in every other breath we breathe. Algae are the original source of fossil carbon found in crude oil and natural gas. Algae, for practical purposes, are the only primary producers in the oceans—an area that covers 71% of the Earth's surface. There are no grasslands or forests beneath the waves, and microscopic algae and seaweeds directly or indirectly support most life in the seas. Indeed, there are approximately $6.25 \times 10^{25}$ algal cells in the oceans at any one time. And, assuming an average diameter of 2 μm, these cells could be packed into a plank-sized volume with dimensions 7 cm thick, 30 cm wide, and long enough to extend from the earth to the moon (386,000 km)! Assuming that the cells divide once per day, the oceans produce another plank each day. Unlike the terrestrial environment, where biomass accumulates, the consumers of our oceans eat one plank each day. Therefore, while algae may seem insignificant in terms of accumulated biomass (no forests), they are very significant in terms of global productivity. Algae are also important economically. Seaweed sales account for approximately 22% of the 39.4 million metric tons of aquaculture products sold worldwide.

More than 30 years ago, the Phycological Society of America (PSA) sponsored the first of several volumes in the series *Handbook of Phycological Methods*. The first volume, *Culture Methods and Growth Measure-ments*, edited by Janet R. Stein (1973), brought together a comprehensive collection of laboratory-based techniques for working with microalgae and macroalgae. This current volume, also sponsored by the PSA, has the same goal: To provide a comprehensive resource for all aspects of algal culturing and related research. Following the tradition of previous PSA sponsored books, the editor and authors of *Algal Culturing Techniques* will generously direct all royalties to the PSA Endowment Fund. The Endowment Fund is used to promote phycology by sponsoring publications, providing various awards, and fostering other algal projects.

Many new methods are detailed in this volume. Robotic automation of flow cytometers for single cell isolation means that machines will mostly replace laborious hand isolation (see Chapter 7). Current models can isolate hundreds of cells per minute, and these numbers will increase with new technology.

The maintenance of algal cultures has been a bottleneck, and many scientists limit their isolation efforts because they are unable to maintain the growing numbers of cultures. Now that machines can isolate algae at high speeds, who will take care of all the new isolates? Fortunately, a second field, algal cryopreservation, has been developing rapidly during the past few years (see Chapter 12). Several cryopreservation protocols are currently being used with great success, and developing techniques hold promise for even better freezing methods in the near future. Together, these two advances mean that we will soon have many more algal cultures, both in public collections and in private research laboratories. These new cultures will help advance our basic knowledge in many areas, from genomics and physiology to biodiversity and systematics. With regard to applied research, we should expect increased discovery of natural products from algae, and those novel compounds will advance biotechnology and biomedicine.

Several new culture media have been developed in recent years, and some older media have been improved. We have learned that richly nutrified media work well for algae from non-oligotrophic regions, but for oligotrophic environments and so-called uncultivable organisms, low nutrient media are imperative. We include a wide range of media, new and old, in Appendix A, and we dedicate three chapters to the preparation of media (see Chapters 2–4). While traditional areas have not advanced dramatically in recent years, we add a substantial number of new and improved techniques (see Chapters 5-6, 8-11).

This book is targeted to the laboratory investigator, with additional chapters on large-scale closed systems for microalgae, outdoor systems for microalgae, and outdoor farming for macroalgae (see Chapters 13-15). Counting cells and calculating cell growth rates are an important part of microalgal culturing, and three chapters discuss these topics in detail (Chapters 16-18). General physiological experimentation (Chapter 19), special experiments involving trace metals (Chapter 4), measurements of chlorophyll pigments (Chapter 20), and measurements of endogenous rhythms (Chapter 21) provide the experimentalist with extensive background knowledge and techniques. Algal viruses (Chapter 22) are discussed both from the aspect of studying algal viruses and from the perspective of keeping algal cultures free of viruses. Sexual reproduction of algae almost always requires culturing techniques, and this topic is thoroughly addressed in Chapters 23 and 24. Finally, the last chapter introduces what may be a new area for many scientists, the *ex situ* preservation of endangered algae. It is now evident that some algae have become extinct and others are on the verge of extinction. Culture isolates of endangered algae can preserve species from extinction, as we learn in Chapter 25 from the person who helped pioneer this area of algal science.

Any edited book depends on the effort and dedication of the contributors, and this volume is no exception. The authors worked very hard, often on short notice, to write, revise, copyedit, and proofread their chapters. I am very grateful for their outstanding efforts. I also thank numerous reviewers, at least two for each chapter, who offered constructive criticism to the authors and myself. And I would like to thank Brian Palenik, who, in a review capacity, read the entire book. I thank Frank Cynar and Jennifer Helé at Academic Press and Keith Roberts at Graphic World for their long hours and hard work putting this tome together. I thank Dennis Hanisak, David Millie, and Robert Sheath, who represented the PSA and worked many hours on the sponsorship of the book. I thank my wife, Theresa, for her patience and encouragement; she has seen little of me during the past two years, and she has carried a heavy load at home. Finally, I thank Stacy Edgar, Vi Lee, Doug Phinney, Tracey Riggens, Khadidja Romari, and Julie Sexton at the CCMP for help and understanding.

Robert A. Andersen
Provasoli-Guillard National Center for
Culture of Marine Phytoplankton
October 2004

CHAPTER 1

# HISTORICAL REVIEW OF ALGAL CULTURING TECHNIQUES

HANS R. PREISIG
*Institute of Systematic Botany, University of Zurich*

ROBERT A. ANDERSEN
*Provasoli–Guillard National Center for Culture of Marine Phytoplankton, Bigelow Laboratory for Ocean Sciences*

## CONTENTS

Key Index Words: Culturing Methods, History, Microalgae, Macroalgae, Mass Culturing, Mariculture, Cryopreservation

## 1.0. INTRODUCTION

Many of the methods and basic culture medium concepts that are used today were developed in the late 1800s and early 1900s. Algal culture techniques have been described in several earlier books and articles (Moore 1903, Küster 1907, Chodat 1913, Richter 1913, Pringsheim 1924, Kufferath 1928/29, Lwoff 1932, Meier 1932, Vischer 1937, Bold 1942, Chu 1942, Pringsheim 1946, Brunel et al. 1950, Lewin 1959, Fogg 1965, Venkataraman 1969, Stein 1973, Guillard 1975, Richmond 1986), and historical information is included in many of these. In this chapter we review the main developments in algal culturing, from its origins through the 1950s.

## 2.0. NINETEENTH CENTURY ALGAL CULTURING

The German-born Ferdinand Cohn (1850), a founder of bacteriology, succeeded in keeping the unicellular flagellate *Haematococcus* (Chlorophyceae) in his laboratory in Breslau (now Wroclaw, Poland) for some time and called this procedure "cultivation." This is the first published report of algal "culture." However, Cohn did not isolate *Haematococcus* from other organisms, he did not use a culture medium, and he did not establish an indefinitely maintained culture. It was the plant physiologist Famintzin (1871), in St. Petersburg, one of the founders of this discipline in Russia (Kuznetsov and Strogonov 1995, Krasnovsky 2003), who made the first attempts to culture algae by using a solution of a few inorganic salts. He grew several green algae, especially two species that he identified as *Chlorococcum infusionum* (Schrank) Meneghini and *Protococcus viridis* C. Agardh. The solution that he used was originally devised by

**FIGURE 1.1.** **(a)** Martinus Willem Beijerinck (1851–1931); **(b)** Georg Klebs (1857–1918); **(c)** Robert Chodat (1865–1934); **(d)** Edgar Johnson Allen (1866–1942); **(e)** Ernst Georg Pringsheim (1881–1970); **(f)** Otto Heinrich Warburg (1883–1970).

Knop in 1865 for studies on vascular plants (for a recipe of Knop's solution, see Bold 1942).

The first report of pure (axenic) cultures of algae stems from the Dutch microbiologist Beijerinck (1890) (Fig. 1.1a), although Klebs (1896) later questioned this achievement. Beijerinck adopted the bacteriological technique introduced by Robert Koch 10 years earlier, and he mixed the sampled water, or the medium, with gelatin. Beijerinck (1890, 1893) was the first to isolate free-living *Chlorella* and *Scenedesmus* in allegedly bacteria-free cultures, and he also successfully isolated symbiotic green algae from *Hydra* ("*Zoochlorella*") and lichens (the green alga he identified as *Cystococcus humicola* Naegeli is now considered to be a species of *Trebouxia*). Later he also established allegedly pure cultures of other algae, including cyanobacteria (Beijerinck 1901) and diatoms (Beijerinck 1904), and he recognized that cyanobacteria such as *Anabaena* could be cultured in a medium to which no nitrogen compounds have been added (Beijerinck 1901).

Of equal importance to Beijerinck's studies of green algae was that of Miquel's (1890/92, 1892, 1892/93/98) work on diatoms, conducted in France at Paris Montsouris Observatory. Miquel, a microbiologist who was also a great pioneer in the field of aerobiology (Comtois 1997), was the first to isolate and establish pure (axenic) cultures of freshwater and marine diatoms. In addition, he introduced several new methods, such as

using a micropipette for isolating algal cells and using organic maceration as organic source supplementing the mineral medium (additions of organic nutritive material in the form of bran, straw, fragments of grasses, mosses, etc.). With a micropipette and a microscope, he isolated single cells and placed them into individual vessels containing culture medium. Miquel also used a dilution isolation method; he added a diatom-containing sample to prepared water (culture medium), and then he subdivided this mixture into a number of tubes. Miquel developed two solutions (A and B) containing mineral salts that he used to enrich seawater, and his famous A and B solutions have been widely used for algal culturing (see Allen and Nelson 1910, Provasoli et al. 1957). A method for getting pure cultures of diatoms was also described by Macchiati (1892a,b,c) in Italy.

Many other scientists also made contributions to our knowledge of algal culturing during the last decade of the 19th century. The Germans Noll (1892) and Oltmanns (1892) published papers discussing the cultivation of marine algae, but these dealt largely with maintaining the algae in good condition rather than isolating pure cultures or effecting growth and reproduction. The Swiss-born botanist Naegeli (1893) discovered that copper had a very harmful effect on growth of freshwater algae, and he introduced *Spirogyra* for testing the nontoxicity of the water used for culturing. In Germany, Krüger (1894) established pure cultures of colorless and saccharophilous coccoid green algae (*Prototheca*, *Chlorella* spp.). Molisch (1895, 1896), at the German University in Prague, and Benecke (1898), at the University of Strasbourg (Strassburg), experimented on the mineral requirements of algae. Bouilhac (1897), in France, used an organic medium to culture *Nostoc* (cyanobacteria). Perhaps the most influential studies of that time on culture work were the intensive investigations which Klebs (1896) (Fig. 1.1b) made at the University of Basel (Switzerland) (from 1898 on in Halle an der Saale and later in Heidelberg, Germany). He attempted to establish pure (axenic) cultures of filamentous and siphonous algae by placing isolated zoospores onto agar. He was successful in growing the algae, but he was unable to keep his cultures bacteria-free. He used petri dishes for culturing, and he was the first to isolate algae on agar. Gelatin, used in early microbiological studies, proved unsatisfactory because bacteria digested the gelatin, converting the solid substrate into a liquified mixture. Agar was also used by Tischutkin (1897) in Belarus, who made the first claim for pure cultures of cyanobacteria, but the purity of his cultures was never beyond all doubt (Harder 1917).

Ward (1899), in Cambridge, recommended swelling agar with dilute acetic acid, followed by thorough rinsing to remove all the salts (see Fellers 1916a,b, Bold 1942, for organic contaminants in agar). Ward also described several novel isolation methods. He mixed algae with nutrient-enriched and sterilized solution of plaster of Paris, which he dispensed into a dish where it hardened, and eventually some hardy algae exhibited growth. He mixed algae with a nutrient-enriched and sterilized silica jelly solution, and similarly he used a large quantity of lime-water into which he bubbled carbon dioxide gas to create an algal/chemical precipitate (calcium carbonate) that he poured into dishes for cultivation. Ward may also have been the first to use a stencil to create patterns (made by algal growth) on a solid substrate. That is, he covered a portion of the dish with a nontransparent sheet, but he cut out a clear area in the shape of an alphabetic letter (e.g., A) so that light reached the agar surface only through the clear area in the shape of the letter. After a period of time (unspecified), an unidentified green alga grew in the illuminated area so that when the sheet was removed, a faint green alphabetic letter was visible on the agar.

## 3.0. TWENTIETH CENTURY ALGAL CULTURING

### 3.1. Conventional Algal Culturing

Zumstein (1900), a student of Klebs and Benecke in Basel, established bacteria-free cultures of *Euglena gracilis* Klebs. He isolated single cells with a capillary pipette, and to remove contaminating bacteria he made the medium as acidic as possible without killing the alga. The investigations of Chodat (Fig. 1.1c) and his collaborators in Geneva (Switzerland) (Chodat and Grintzesco 1900, Chodat 1913) have been instrumental in extending the knowledge of algal culturing. However, the conditions in Chodat's cultures were often very different from those in nature, and he found that morphologically abnormal cells developed. During more than 30 years of research he established a collection of over 300 algal species in pure culture (Chodat 1928). Richter (1903), a Prague-born Austrian botanist, continued with Miquel's methods to grow diatoms axenically. Richter also extended his work to other algae, and in 1911 he presented a detailed publication summarizing all previous work on the nutrition of algae (Richter 1911; see also Richter 1913). In the United States,

Moore (1903) published an early summary of algal culturing. In Great Britain, Harriette Chick (1903) presented her work on *Chlorella pyrenoidosa* Chick. She established axenic cultures, which she repeatedly tested for purity, and she concluded that the alga preferred ammonia to nitrate. In Germany, Küster published his elaborate manual on the culturing of microorganisms (first edition in 1907; further editions in 1913 and 1921), and he gave special consideration to algal culturing. In 1908, Küster made the first successful attempts to grow a dinoflagellate, although he did not obtain pure cultures of the colorless marine species he tentatively described as *Gymnodinium fucorum* Küster (nom. prov.). Jacobsen (1910) was probably the first to isolate a colorless chlamydomonadalean alga (*Polytoma uvella* Ehrenberg), and he also tighten line isolated *Carteria*, *Chlamydomonas*, *Chlorogonium*, and *Spondylomorum* (Chlorophyceae). For organic enrichment, he used various sugars and peptone. Charlotte Ternetz (1912) continued Zumstein's (1900) study of *Euglena gracilis* in Basel and investigated the organic nutrition. She discovered that green forms of this organism became colorless when grown in the dark but turned green again in the light. On the other hand, constantly colorless but less vital forms of *Euglena gracilis* were also produced.

In 1910, Allen (Fig. 1.1d), Director of the Marine Biological Association of the United Kingdom, and his collaborator Nelson made significant contributions to algal culture, including some of the earliest attempts to cultivate algae for rearing marine animals. They isolated and cultured *Chaetoceros*, *Skeletonema*, and *Thalassiosira*, among other genera, to feed marine larvae, and these phytoplankters remain major food sources in aquaculture hatcheries. Allen and Nelson produced an artificial seawater based on the molecular concentrations of seawater determined by Van't Hoff (1905), and they also recognized the importance of iron as a trace metal. Nevertheless, they achieved good growth of algae only by adding small quantities of natural seawater (less than 1% to 4%) to the artificial seawater. Allen (1914) remarked that these effects may be due to products of the metabolism of bacteria and suggested an organic micronutrient similar to the first vitamin, which had just been discovered by Casimir Funk.

Because Allen and Nelson were growing mass cultures of algae, not just test tube or flask cultures, they encountered new problems. They quickly realized that light became a limiting factor in large culture vessels, and to this day, light limitation continues to plague mass culturing efforts. Because of requirements for large-volume cultures, they abandoned artificial seawater and used enriched natural seawater. They found that harbor seawater ("tank water") was polluted, but they found that seawater from the English Channel ("outside water") was much cleaner. To purify large volumes of seawater, they heated or boiled the seawater and then used animal-charcoal filters or hydrogen peroxide. They even tried ozone for purifying seawater, reporting success on a small scale with the use of an "imperfect apparatus." Allen and Nelson recognized that both "nutritive" and "protective" treatment must be added for successful algal growth. They suggested that the action of Miquel's solution B may be a "protective" one, that is, it removed or neutralized harmful substances (e.g., toxins), similar to that of animal-charcoal (which contains large quantities of calcium and magnesium phosphates) and hydrogen peroxide, which had a similar protective effect. Allen and Nelson found that potassium nitrate was the primary "nutritive" ingredient in Miquel's solution A and found that phosphate addition was also required in some cases. They grew many diatoms in "Miquel sea water," but this medium also supported growth of several species of unidentified red algae, cyanobacteria, green algae (e.g., *Enteromorpha*), *Vaucheria* (Xanthophyceae), and even young plants of *Laminaria* (Phaeophyceae). Their observations led G. H. Drew (1910) to artificially cultivate *Laminaria digitata* (Hudson) Lamouroux and to discover part of its early life history. Thus, in many regards, Allen may be considered the father of algal mariculture, having first used marine phytoplankton to feed animals and having laid the foundations for seaweed culture.

Although Drew (1910) was able to culture *Laminaria*, he missed the fact that there is a microscopic gametophyte and that the macroscopic plant is the sporophyte. The first discovery of a heteromorphic life history in kelps was made a few years later by Sauvageau (1915) in France, who cultured *Saccorhiza bulbosa* J. Agardh, another member of the Laminariales. This very significant discovery by Sauvageau aroused a great deal of interest in culturing brown algae. It was evident that the cycle of development of many of these algae could not be ascertained unless they were grown in culture.

In 1912, Pringsheim (Fig. 1.1e), at that time working in Halle an der Saale, Germany, published the first of a series of important papers on algal culture methods (of the numerous contributions which followed over a long period until 1970, his book *Pure Cultures of Algae* in 1946 and its German translation of 1954 were most highly influential). In his first paper, Pringsheim (1912) showed that chlorine of tap water was detrimental for freshwater algal media. Instead of using tap or spring water, as was often used in earlier times, he used glass-distilled water (distilled water from metal condensers

proved to be bad in general). Pringsheim also refined the technique of picking up single cells or filaments with a capillary pipette to reduce the number of contaminating bacteria. In 1912 he also started to use extracts of soil and later also of peat as supplements in purely mineral media to obtain better growth. This has been widely used since, although its mode of action, which presumably depends largely on the weak chelating properties of humic substances, has not been fully explained.

Biphasic cultures with pasteurized soil covered with water not only supported more prolific growth of inoculated material but also allowed forms that could not be grown in culture in the ordinary media to be brought into culture. The observation that inclusion of a source of organic carbon, such as cheese, allowed the development of a rich and varied flora of colorless forms led to basic investigations of algal heterotrophy. According to Harder (1917), Pringsheim (1913) was the first who unequivocally succeeded in culturing cyanobacteria without contaminating bacteria. In 1921, Pringsheim realized that acetate was an excellent substrate for heterotrophic growth of algae. He showed that a variety of species of Volvocales, Euglenophyceae, Cryptophyceae, and diatoms are able to grow in dark on acetate but not on glucose. With the years, Pringsheim built up a large culture collection, first at Halle an der Saale and later at the German University in Prague, where by 1928 it included almost 50 species (and in 1929, already more than 100 species). Later he moved to Cambridge and established the famous Culture Centre of Algae and Protozoa (CCAP). In 1953, when he left Cambridge, he took subcultures with him and set up another great collection in Germany (Sammlung von Algenkulturen Göttingen [SAG]). In total, he established approximately 2,000 cultures representing some 400 species of algae.

Warburg[1] (1919) (Fig. 1.1f), a renowned cell physiologist/biochemist in Berlin, Germany, discovered that fast-growing green microalgae (like *Chlorella*) are ideal experimental materials in biochemical and physiological research and used these cultures for his pioneering work on photosynthesis. He bubbled the liquid cultures of algae with carbon dioxide–enriched air and used an artificial light source, consisting of a 300-Watt metal-filament lamp in a glass beaker surrounded with cooling water, which acted as a screen absorbing the infrared radiation.

1. Otto Heinrich Warburg, Nobel Laureate in Medicine in 1931, received the award in recognition of his research into respiratory enzymes. In 1944 he was offered a second Nobel Prize but, being of Jewish origin, was prevented from accepting the award by the Hitler regime.

A similar artificial light source was described by Hartmann (1921) in Berlin, Germany, in his successful experiments for culturing volvocalean algae (e.g., *Eudorina*, *Gonium*), a group of algae which were formerly considered to be among the most difficult ones to be cultured (see also Hartmann 1924). The first to establish pure cultures of *Volvox* were the Russians Uspenski and Uspenskaja (1925). They used a medium containing a mixture of mineral salts including iron, and they added citrate to prevent precipitation. Wettstein (1921), in Berlin, contrived to obtain unialgal but not axenic cultures of several other groups of algal flagellates not cultured before (e.g., *Cryptomonas*, *Synura*, *Uroglena*), by growing them on agar containing peat extract.

Lwoff[2] (Fig. 1.2a), working at the Pasteur Institute in Paris, France, was a contemporary of Pringsheim. Obviously, Lwoff was more interested in protozoa, fungi, bacteria, and viruses, but he made several contributions to the growth of algae using organic compounds, especially amino acids, and he published a major text on microbial culturing techniques (Lwoff 1923, 1929, 1932). The requirement of highly specific organic compounds by certain algae is exemplified by Droop's later pursuit and discovery of ubiquinone for the growth of the dinoflagellate *Oxyrrhis marina* Dujardin (Droop 1959, Droop and Doyle 1966).

Schreiber (1925, 1927), who first worked in Würzburg and Berlin and later at the Biologische Anstalt Helgoland, Germany, was successful in preparing special nutrient combinations for culturing volvocalean freshwater algae but also marine phytoplankton. His famous marine "Schreiber solution" consisted of a mixture of nitrate and phosphate that he devised, based on the minimum requirements for the two elements shown by a diatom culture. Hämmerling (1931), a student of Hartmann in Berlin, extended these studies by adding soil extract to Schreiber's medium for growing the green dasycladalean *Acetabularia*. This so-called "Erd-Schreiber medium" (*Erde* means *soil* in German) has been used successfully for many years for growing both unicellular and benthic marine algae, which could not be grown in other media (see also Føyn 1934).

Mainx (1927, 1929, 1931), a collaborator of Pringsheim at the German University in Prague, also contributed considerably to the knowledge of pure algal

**FIGURE 1.2.** **(a)** André Lwoff (1902–1994); **(b)** Wilhelm Vischer (1890–1960); **(c)** Harold Charles Bold (1909–1987); **(d)** Luigi Provasoli (1908–1992); **(e)** Richard Cawthron Starr (1924–1998); **(f)** Hiroshi Tamiya (1903–1984).

culturing. He introduced centrifugation techniques to isolate algae and was also one of the first using phototaxis of motile stages for establishing pure cultures. Skinner, at the University of Minnesota, modified a method by Bristol-Roach (in Waksman[3] et al. 1927) to produce an innovative agar-based dilution technique for isolating soil algae (Skinner 1932). He prepared a series of nutrient agar test tubes that were cooled to nearly the gelling point. To the first tube, he vigorously mixed a few drops of soil suspended in water with the nearly gelled agar; he removed a small amount from the first tube and vigorously mixed the subsample in the second tube; he repeated this 10 more times. After incubation, he broke the glass test tube and placed the cylindrical agar onto sterile paper. He broke (not cut) the agar repeatedly, and with a small wire loop, he removed small colonies of algae growing in the embedded agar. The colonies, representing single-cell isolates, were inoculated into liquid medium, grown up, and then mixed with nearly gelled agar again. After a second round of breaking test tubes and isolating single-cell colonies, he found that approximately half of the resultant cultures were axenic.

2. In 1965 André Lwoff shared the Nobel Prize in Medicine with Jacques Monod and François Jacob for his discovery that the genetic material of a virus can be assimilated by bacteria and passed on to succeeding generations.

3. Selman A. Waksman, Nobel Laureate in Medicine in 1952, awarded for his discovery of streptomycin, the first antibiotic effective against tuberculosis.

Vischer (Fig. 1.2b), influenced in his early years by Chodat, became a great specialist in the culture of terrestrial algae, especially with regard to Chlorophyceae and Heterokontae (Xanthophyceae/Eustigmatophyceae) (Vischer 1926, 1937, 1960). In 1975, his large culture collection at the University of Basel (Switzerland), including many type strains, was transferred to the ASIB culture collection at Innsbruck (Austria), where it still forms a main part of the present collection (Gärtner 2004).

Three of the most prominent phycologists of the last century—Geitler in Vienna (Austria), Kornmann at Helgoland (Germany), and von Stosch in Marburg (Germany)—made major use of cultures in their studies of algal life histories and systematics. Their most active scientific careers started in the 1920s (Geitler) and 1930s (Kornmann and von Stosch) and lasted until the 1980s (for a detailed appreciation of their work, see corresponding chapters in Garbary and Wynne [1996]). In the United States, Bold (Fig. 1.2c) also made pioneering studies on algae using culture techniques (Bold 1936, 1942, 1974), and his excellent review "The Cultivation of Algae" (1942) established him as one of the trailblazers in this field of phycology.

Chu, who came to Great Britain from China in 1938 and first worked under Fritsch in London and later in Millport and Plymouth (in 1945 he went to America and then back to China), was also a pioneer among those who set out to devise media having some resemblance to those in which algae grow naturally. His highly successful medium No. 10 was comparable in composition and degree of dilution to the water of a eutrophic lake (Chu 1942).

Provasoli (Fig. 1.2d), first in Italy (1937/38), and later in the United States, together with Hutner, Pintner, and other associates (Hutner et al. 1950, Provasoli and Pintner 1953, Provasoli et al. 1957), was in the frontier of developing artificial culture media for algae for more than 40 years (1930s to 1980s). He and his colleagues were among the first using antibiotics for obtaining bacteria-free cultures (Provasoli et al. 1948). Although vitamin requirements were known earlier, they conducted extensive investigations to determine vitamin requirements by a wide range of algae (Hutner et al. 1949, 1950, Provasoli 1956, 1958b) and the inclusion of vitamins or organic extracts containing them in marine media has greatly increased the number of algae to be grown axenically. Another innovation that made Provasoli's medium recipes so successful was the introduction of EDTA (ethylenediaminetetraacetic acid), a metabolically inert chelator, to replace organic chelators such as citrate (Hutner et al. 1950, Hutner and Prova-

soli 1951). EDTA permitted the development of both enriched artificial media and enriched natural seawater media that were more reproducible than those depending on additions of soil extract (Provasoli and Pintner 1953, 1960, Provasoli et al. 1954, Provasoli et al. 1957, Provasoli 1958b). The increasing need to add trace metals was summarized succinctly by Provasoli and Pintner (1960), as they explained why early media required only iron and later media required cobalt, copper, manganese, molybdenum, vanadium, and zinc: "It is worth noting that the industrial methods of purification of the 'chemically pure' salts have undergone many changes since Knop's time, resulting in ever-changing sets of impurities." Other innovations by Provasoli included the use of physiologically inert pH buffers and use of sodium glycerophosphate as a soluble source of P that resisted precipitation with iron. He was the first to establish axenic cultures of species of the green foliaceous seaweed *Ulva*, which, in the absence of bacteria, he found to require plant hormones for the normal development of the thallus (Provasoli 1958a). Provasoli's legacy also exists in part in his large culture collections of marine algae that were amalgamated with those of Robert Guillard and which are now maintained at the Provasoli-Guillard National Center for Culture of Marine Phytoplankton (CCMP) at the Bigelow Laboratory for Ocean Sciences in Maine.

The discovery of penicillin, streptomycin, and other antibiotics quickly led to their application against bacteria in algal cultures. Provasoli et al. (1948), while attempting to produce axenic strains using antibiotics, discovered that streptomycin could be used to produce colorless mutants of *Euglena*. Early reports of axenic cultures produced by using antibiotics include that of Goldzweig-Shelubsky (1951), who obtained bacteria-free cultures of *Scenedesmus*, *Navicula*, and *Euglena* following penicillin treatment; Spencer (1952), who succeeded in purifying *Phaeodactylum*; and Reich and Kahn (1954), who purified *Prymnesium parvum* Carter. Droop (1967) provided a new method, as well as a summary for antibiotic treatment of algae.

Starr (Fig. 1.2e), a student of Bold, started in 1953 to establish a major culture collection of algae at Indiana University (Starr 1956), which, in 1976, he transferred to the University of Texas at Austin (UTEX Culture Collection of Algae) (Starr and Zeikus 1993). At first it contained mainly strains that he used for his research on green algae (especially Volvocales, Chlorococcales, and Desmidiales), as well as some 200 strains that he obtained from E. G. Pringsheim. This collection was greatly enlarged during the years (c. 2,000 strains in 1976) and now contains nearly 2,300 strains (approximately 200 different genera), representing today one of the largest and most diverse assemblies of living algae in the world.

## 3.2. Microalgal Mass Culturing

In addition to the achievements of Allen and Nelson (1910) and others who grew microalgae for aquaculture, scientists developed new methods to mass produce microalgae for other purposes. Pioneering work to grow microalgae (especially *Chlorella*) in dense laboratory cultures was achieved by Warburg (1919) in Berlin, Germany (see previous text). At Woods Hole Oceanographic Institution in the United States, Ketchum and Redfield (1938) described a method for maintaining continuous cultures of marine diatoms in large supplies for chemical analyses. The procedure included a periodic harvesting of a fixed portion (i.e., up to a kilogram or more of dry material) at a crucial point in a batch growth curve, while the remaining population continued to replicate and grow until the desired population was again reached and harvested. With this technique Ketchum et al. (1949) also succeeded to grow and optimally harvest cells of several other unicellular algae. This semicontinuous batch culture method is still used in aquaculture as a means of rapid production of phytoplankton to feed marine animals. In Göttingen (Germany), Harder and Witsch (1942) also started experiments in mass culturing of diatoms to determine if fat production from these cultures was possible.

A larger apparatus for growing *Chlorella* in continuous culture was constructed by Myers and Clark (1944) at the University of Texas at Austin. They devised it to hold their culture at some chosen point on its growth curve by dilution of the suspension controlled by a photometric system. In its original form, the culture vessel was a vertical sleeve-shaped chamber illuminated by vertical tubular lamps in such a way that the effective illumination was independent of the total volume of the suspension. The cells were manually harvested at intervals, leaving a small volume of suspension as inoculum.

Microalgae like *Chlorella* were suggested as a source of potential commercial application (e.g., food production) by Spoehr and Milner (1947, 1949) from the Carnegie Institution of Washington's Division of Plant Biology at Stanford University, in California. Further studies for translating laboratory methods into engineering specifications for continuous large-scale culturing of *Chlorella* were made by Cook (1950, 1951) at the Stanford Research Institute, who built a small pilot plant. Interest in algal mass culturing was continued by the Carnegie Institution through a contract with Arthur D. Little, Inc., of Cambridge, Massachusetts, who constructed and ran several pilot plant culture units on the roof of an industrial building (see Burlew 1953). Short-term yields of *Chlorella* were as high as 11 g dry weight $\cdot$ m$^{-2}$ $\cdot$ day$^{-1}$, although the average yield for 52 days of operation was only 2 g $\cdot$ m$^{-2}$ $\cdot$ day$^{-1}$. It was concluded, however, that yields up to 20–25 g $\cdot$ m$^{-2}$ $\cdot$ day$^{-1}$ were possible once culture technology improved and cultivation was performed in more suitable geographic locations.

Also in the late 1940s and early 1950s, important work on large-scale production of *Chlorella* was made in Germany by Witsch (1948). Gummert et al. (1953) started a research and development program in Essen, Germany, for large-scale production in greenhouse and open air culture (see also Meffert and Stratmann 1954).

At about the same time, another series of laboratory and pilot plant studies for growing *Chlorella* followed in Japan under the guidance of Tamiya (Fig. 1.2f) at the Tokugawa Institute in Tokyo (Mituya et al. 1953, Tamiya 1956, 1957). The same group of scientists was also successful in introducing the technique of synchronous culture (Tamiya et al. 1953). Synchronization, an experimentally obtained coordination of individual life cycles in a population of cells, was a great advance for experimental work in algal physiology and has subsequently been followed by many others using modified techniques and different species of algae (see Tamiya 1966, Pirson and Lorenzen 1966).

The results of the first boom of microalgal mass culturing were published in the classic and still fascinating report edited by Burlew (1953). For an historical outline of later developments in microalgal culturing, see Soeder (1986).

## 3.3. Mariculture of Seaweeds

Until the 1950s almost all economically important seaweeds had been harvested from natural habitats only. *Porphyra*, known as *nori* in Japan, *zicai* in China, and *laver* in the western world, is the only macroalga that has a long history of being cultivated. This is the most extensively eaten seaweed by coastal peoples in southeast Asia and the Pacific Ocean basin, and its cultivation reportedly began in the seventeenth century in Tokyo Bay (Miura 1975). Enhancement of wild stocks was originally achieved by pushing tree branches or leafless bamboo shoots into the sea bottom or by clearing rock surfaces, which served as attachment sites for the small blades (the portion of the plant which is used for human consumption). From the late 1920s on, large-meshed nets (first made of coco-fiber but later replaced by other materials) were strung out horizontally between rows of bamboo poles. These nets were more easily transported from the collecting ground to the cultivation areas, and net-farming soon became the basis of nori production in Japan. In 1949, the British botanist Kathleen M. Drew discovered the complete life history of *Porphyra*, and in particular the finding of the microscopic conchocelis stage has led to refinements of culture techniques and initiated a rapid development of the *Porphyra* industry from the 1960s onward (for a detailed appreciation of Drew Baker's work, see corresponding chapter in Garbary and Wynne 1996). Further developments in the cultivation of *Porphyra* and other economically important seaweeds, which were hardly cultivated before the late 1950s, are described by Tseng (1981) (see also Chapter 15 of the present volume).

### 3.4. Cryopreservation

Cryobiology gained real impetus only in 1949, when it was reported that glycerol protected fowl spermatozoa against freezing injury (Polge et al. 1949). Since this fortuitous discovery, storage under liquid nitrogen has also become a standard method for the long-term maintenance of algae, but the first fully documented reports on the freezing of algal cells were not published until the early 1960s (Terumoto 1961, Holm-Hansen 1963, Leibo and Jones 1963). For more information on cryopreservation, see Chapter 12.

## 4.0. REFERENCES

Allen, E. J. 1914. On the culture of the plankton diatom *Thalassiosira gravida* Cleve, in artificial sea-water. *J. Mar. Biol. Assoc. U. K.* 10:417–39.

Allen, E. J., and Nelson, E. W. 1910. On the artificial culture of marine plankton organisms. *J. Mar. Biol. Assoc. U. K.* 8:421–74.

Beijerinck, M. W. 1890. Culturversuche mit Zoochlorellen, Lichenengonidien und anderen niederen Algen. *Bot. Zeitung* 48:725–39, 741–54, 757–68, 781–85.

Beijerinck, M. W. 1893. Bericht über meine Kulturen niederer Algen auf Nährgelatine. *Zentralbl. Bakt.* 13:781–86.

Beijerinck, M. W. 1901. Über oligonitrophile Mikroben. *Zentralbl. Bakt. Ser. II* 7:561–82.

Beijerinck, M. W. 1904. Das Assimilationsprodukt der Kohlensäure in den Chromatophoren der Diatomeen. *Rec. Trav. Bot. Néerl.* 1:28–32.

Benecke, W. 1898. Über Kulturbedingungen einiger Algen. *Bot. Zeitung* 56:83–96.

Bold, H. C. 1936. Notes on the culture of some common algae. *J. Tenn. Acad. Sci.* 11:205–12.

Bold, H. C. 1942. The cultivation of algae. *Bot. Rev.* 8:69–138.

Bold, H. C. 1974. Twenty-five years of phycology (1947–1972). *Ann. Missouri Bot. Gard.* 61:14–44.

Bouilhac, R. 1897. Sur la culture du *Nostoc punctiforme* en présence du glucose. *C. R. Acad. Sci. Paris* 125:880–2.

Brunel, J., Prescott, G. W., and Tiffany, L. N., eds. 1950. *The Culturing of Algae.* Charles F. Kettering Foundation, Antioch Press, Yellow Springs, Ohio, 114 pp.

Burlew, J. S., ed. 1953. Algal culture: from laboratory to pilot plant. Carnegie Institution of Washington, Washington, D. C., 600:1–357.

Chick, H. 1903. A study of a unicellular green alga, occurring in polluted water, with especial reference to its nitrogenous metabolism. *Proc. Roy. Soc.* 71:458–76.

Chodat, R. 1913. Monographie d'algues en culture pure. *Beitr. Kryptogamenfl. Schweiz* 4(2):1–266.

Chodat, R. 1928. Les clones chez les algues inférieures. *Zeitschr. Indukt. Abst.-Vererb. Suppl. (Verhandl V. Internat. Kongr. Vererbungswiss., Berlin 1927)* 1:522–30. Verlag Borntraeger, Leipzig, Germany.

Chodat, R., and Grintzesco, J. 1900. Sur les méthodes de culture pure des algues vertes. *Congrès International de Botanique, Paris. Extrait du Compte-rendu.* Imprimerie Lucien Declume, Lons-le-Saunier, France, pp. 157–162.

Chu, S. P. 1942. The influence of the mineral composition of the medium on the growth of planktonic algae. *J. Ecol.* 30:284–325.

Cohn, F. 1850. Nachträge zur Naturgeschichte des *Protococcus pluvialis* Kützing (*Haematococccus pluvialis* Flotow). *Nov. Act. Leop. Carol.* 22(2):605–764.

Comtois, P. 1997. Pierre Miquel: the first professional aerobiologist. *Aerobiologia* 13:75–82.

Cook, P. M. 1950. Some problems in the large-scale culture of *Chlorella*. In: Brunel, J., Prescott, G. W., and Tiffany, L. N., eds. *The Culturing of Algae.* Charles F. Kettering Foundation, Antioch Press, Yellow Springs, Ohio, pp. 53–75.

Cook, P. M. 1951. Chemical engineering problems in large-scale culture of algae. *Industr. Eng. Chem.* 43:2385–9.

Drew, G. H. 1910. The reproduction and early development of *Laminaria digitata* and *Laminaria saccharina*. *Ann. Bot. London* 24:177–90.

Drew, K. M. 1949. Conchocelis-phase in the life-history of *Porphyra umbilicalis* (L.) Kütz. *Nature* 164(4174):748–9.

Droop, M. R. 1959. Water-soluble factors in the nutrition of *Oxyrrhis marina*. *J. Mar. Biol. Assoc. U. K.* 38:605–20.

Droop, M. R. 1967. A procedure for routine purification of algal cultures with antibiotics. *Brit. Phycol. Bull.* 3:295–7.

Droop, M. R., and Doyle, J. 1966. Ubiquinone as a protozoan growth factor. *Nature* 212(5069):1474–5.

Famintzin, A. 1871. Die anorganischen Salze als ausgezeichnetes Hilfsmittel zum Studium der Entwicklung niederer chlorophyllhaltiger Organismen. *Bull. Acad. Sci. St. Petersb.* 17:31–70.

Fellers, C. R. 1916a. The analysis, purification and some chemical properties of agar agar. *J. Indust. Engin. Chem.* 8:1128–33.

Fellers, C. R. 1916b. Some bacteriological studies on agar agar. *Soil Sci.* 2:255–90.

Fogg, G. E. 1965. *Algal cultures and phytoplankton ecology.* University of Wisconsin Press, Madison, 126 pp. [ed. 2, 1975, 175 pp.; ed. 3 (Fogg, G. E., and Thake, B.) 1987, 269 pp.].

Føyn, B. 1934. Lebenszyklus, Cytologie und Sexualität der Chlorophycee *Cladophora suhriana* Kützing. *Arch. Protistenk.* 83:1–56.

Garbary, D. J., and Wynne, M. J., eds. 1996. *Prominent Phycologists of the 20th Century.* Lancelot Press, Hantsport, Nova Scotia, 360 pp.

Gärtner, G. 2004. ASIB: The Culture Collection of Algae at the Botanical Institute, Innsbruck. *Nova Hedwigia* 79:71–6.

Goldzweig-Shelubsky, M. 1951. The use of antibiotic substances for obtaining monoalgal bacteria-free cultures. *Palestine J. Bot. (Jerusalem)* 5:129–31.

Guillard, R. R. L. 1975. Culture of phytoplankton for feeding marine invertebrates. In: Smith, W. L., and Chanley, M. H., eds. *Culture of Marine Invertebrate Animals.* Plenum Press, New York, pp. 29–60.

Gummert, F., Meffert, M. E., and Stratmann, H. 1953. Non-sterile large-scale culture of *Chlorella* in greenhouse and open air. In: Burlew, J. S., ed. *Algal Culture: From Laboratory to Pilot Plant.* Carnegie Institution of Washington, D. C., 600:166–76.

Hämmerling, J. 1931. Entwicklung und Formbildungsvermögen von *Acetabularia mediterranea*. *Biol. Zentralbl.* 51:633–47.

Harder, R. 1917. Ernährungsphysiologische Untersuchungen an Cyanophyceen, hauptsächlich dem endophytischen *Nostoc punctiforme*. *Zeitschr. Bot.* 9:145–242.

Harder, R., and Witsch H. von 1942. Über Massenkultur von Diatomeen. *Ber. Deutsch. Bot. Ges.* 60:146–52.

Hartmann, M. 1921. Untersuchungen über die Morphologie und Physiologie des Formwechsels der Phytomonadinen (Volvocales). III. *Arch. Protistenk.* 43:223–86.

Hartmann, M. 1924. Untersuchungen über die Morphologie und Physiologie des Formwechsels der Phytomonadinen (Volvocales). IV. *Arch. Protistenk.* 49:375–95.

Holm-Hansen, O. 1963. Viability of blue-green and green algae after freezing. *Physiol. Plant.* 16:530–40.

Hutner, S. H., and Provasoli, L. 1951. The phytoflagellates. In: Lwoff, A., ed. *Biochemistry and Physiology of Protozoa.* Academic Press, New York, pp. 27–128.

Hutner, S. H., Provasoli, L., Stokstad, E. L. R., Hoffmann, C. E., Belt, M., Franklin, A. L., and Jukes, T. H. 1949. Assay of anti-pernicious anemia factor with *Euglena*. *Proc. Soc. Exp. Biol. Med.* 70:118–20.

Hutner, S. H., Provasoli, L., Schatz, A., and Haskins, C. P. 1950. Some approaches to the study of the role of metals in the metabolism of microorganisms. *Proc. Amer. Phil. Soc.* 94:152–70.

Jacobsen, H. C. 1910. Kulturversuche mit einigen niederen Volvocaceen. *Zeitschr. Bot.* 2:145–88.

Ketchum, B. H., and Redfield, A. C. 1938. A method for maintaining a continuous supply of marine diatoms by culture. *Biol. Bull.* 75:165–9.

Ketchum, B. H., Lillick, L., and Redfield, A. C. 1949. The growth and optimum yields of unicellular algae in mass culture. *J. Cell. Comp. Physiol.* 33:267–79.

Klebs, G. 1896. *Die Bedingungen der Fortpflanzung bei einigen Algen und Pilzen.* G. Fischer, Jena, Germany.

Krasnovsky, A. A. Jr. 2003. Chlorophyll isolation, structure and function: major landmarks of the early history of research in the Russian Empire and the Soviet Union. *Photosynth. Res.* 76:389–403.

Krüger, W. 1894. Beiträge zur Kenntnis der Organismen des Saftflusses (sog. Schleimflusses) der Laubbäume. I. Über einen neuen Pilztypus, repräsentiert durch die Gattung *Prototheca*; II. Über zwei aus Saftflüssen rein gezüchtete Algen. In: Zopf, W. [ed.] *Beiträge zur Physiologie und Morphologie niederer Organismen.* Arthur Felix, Leipzig, Germany, 4:69–116.

Kufferath, H. 1928/9. La culture des algues. *Revue Algol.* 4:127–346.

Küster, E. 1907. Anleitung zur Kultur der Mikroorganismen (1. Auflage). Verlag B. G. Teubner, Leipzig, Germany, 201 pp. (2nd ed. 1913, 3rd ed. 1921).

Küster, E. 1908. Eine kultivierbare Peridinee. *Arch. Protistenk.* 2:351–62.

Kuznetsov, V. V., and Strogonov, B. P. 1995. The patriarch of Russian plant physiology (On the 160th birthday of academician A. S. Faminzin). *Russ. J. Plant Physiol.* 42:297–302.

Leibo, S. P., and Jones, R. F. 1963. Effects of subzero temperatures on the unicellular red alga *Porphyridium cruentum*. *J. Cell Comp. Physiol.* 622:295–302.

Lewin, R. A. 1959. The isolation of algae. *Revue Algol. N. S.* 3:181–97.

Lwoff, A. 1923. Sur la nutrition des Infusoires. *C. R. Acad. Sci. Paris* 176:928–30.

Lwoff, A. 1929. La nutrition de *Polytoma uvella* Ehrenberg (Flagellé Chlamydomonadinae) et le pouvoir de synthèse des protistes hétérotrophes. Les protistes mésotrophes. *C. R. Acad. Sci. Paris* 188:114–16.

Lwoff, A. 1932. *Recherches biochimiques sur la nutrition des protozoaires. Le pouvoir de synthèse. Monographies de l'Institut Pasteur.* Masson, Paris, 158 pp.

Macchiati, L. 1892a. Sur la culture des Diatomées. *Journal de Micrographie* 16:116–20.

Macchiati, L. 1892b. Comunicazione preventiva sulla coltura delle Diatomee. *Atti Soc. Naturalisti Modena, Ser.* 3, 11:53–8.

Macchiati, L. 1892c. Seconda comunicazione sulla coltura delle Diatomee. *Bull. Soc. Bot. Italiana* 1892 (7):329–34.

Mainx, F. 1927. Beiträge zur Morphologie und Physiologie der Eugleninen. I. Morphologische Beobachtungen, Methoden und Erfolge der Reinkultur. *Arch. Protistenk.* 60:305–54.

Mainx, F. 1929. Biologie der Algen. *Tab. Biol.* 5:1–23.

Mainx, F. 1931. Physiologische und genetische Untersuchungen an Oedogonien. I. *Zeitschr. Bot.* 24:481–527.

Meffert, M. E., and Stratmann, H. 1954. Algen-Grosskulturen im Sommer 1951. *Forschungsberichte des Landes Nordrhein-Westfalen* 8:1–43. Westdeutscher Verlag, Köln, Germany.

Meier, F. 1932. Cultivating algae for scientific research. *Ann. Rep. Board Regents Smithsonian Inst.* 1932:373–83.

Miquel, P. 1890/92. De la culture artificielle des Diatomées. *Le Diatomiste* 1:73–5, 93–9, 121–8, 149–56, 165–72. [Identical article was published also in *Micrographe Préparateur* 5:69–76, 104–8, 159–63, 206–11 and 6:34–8, 83–5,127–33, 226–33 (1897/98). For French, German and English summaries, see *Zeitschrift für angewandte Mikroskopie* 3:193–98 and 225–36 (1897), *American Monthly Microscopical Journal* 14:116 (1893), *Journal of the Royal Microscopical Society* 1893:111 and 1898:128–30].

Miquel, P. 1892. De la culture artificielle des Diatomées. *C. R. Acad. Sci. Paris* 114:780–2.

Miquel, P. 1892/93/98. Recherches expérimentales sur la physiologie, la morphologie et la pathologie des Diatomées. *Annales de Micrographie* 4:273–87, 321–49, 408–31, 529–58; 5:437–61, 521–47; and 10:49–59, 177–91.

Mituya, A., Nyunoya, T., and Tamiya, H. 1953. Pre-pilot-plant experiments on algal mass culture. In: Burlew, J. S., ed. *Algal Culture: From Laboratory to Pilot Plant.* Carnegie Institution of Washington, D. C., 600:273–81.

Miura, A. 1975. *Porphyra* cultivation in Japan. In: Tokida, J., and Hirose, H., eds. *Advance of Phycology in Japan.* VEB G. Fischer Verlag, Jena, Germany, pp. 273–304.

Molisch, H. 1895. Die Ernährung der Algen. I. Süsswasseralgen. *Sitzungsber. Akad. Wiss. Wien, Math.-Nat. Kl. Abt. I* 104:783–800.

Molisch, H. 1896. Die Ernährung der Algen. II. Süsswasseralgen. *Sitzungsber. Akad. Wiss. Wien, Math.-Nat. Kl. Abt. I* 105:633–48.

Moore, G. T. 1903. Methods for growing pure cultures of algae. *J. Appl. Microsc. Labor. Meth.* 6:2309–14.

Myers, J., and Clark, L. B. 1944. Culture conditions and the development of the photosynthetic mechanism. II. An apparatus for the continuous culture of *Chlorella. J. Gen. Physiol.* 28:103–12.

Naegeli, C. von 1893. Über oligodynamische Erscheinungen in lebenden Zellen. *Denkschr. Schweiz. Naturf. Ges.* 33:1–52.

Noll, F. 1892. Ueber die Cultur von Meeresalgen in Aquarien. *Flora* 2:281–301.

Oltmanns, F. 1892. Ueber die Cultur und Lebensbedingungen der Meeresalgen. *Jahrb. Wiss. Bot.* 23:349–440.

Pirson, A., and Lorenzen, H. 1966. Synchronized dividing algae. *Annual Rev. Plant Physiol.* 17:439–58.

Polge, C., Smith, A. U., and Parkes, A. S. 1949. Revival of spermatozoa after vitrification and dehydration at low temperatures. *Nature* 164 (4172):666.

Pringsheim, E. G. 1912. Kulturversuche mit chlorophyllführenden Mikroorganismen. Mitt. I. Die Kultur von Algen in Agar. *Beitr. Biol. Pfl.* 11:305–34.

Pringsheim, E. G. 1913. Zur Physiologie der Schizophyceen. *Beitr. Biol. Pfl.* 12:49–108.

Pringsheim, E. G. 1921. Zur Physiologie saprophytischer Flagellaten (*Polytoma, Astasia* und *Chilomonas*). *Beitr. Allg. Bot.* 2:88–137.

Pringsheim, E. G. 1924. Algenkultur. In: Abderhalden, E., ed. *Handbuch der biologischen Arbeitsmethoden, Abt.* XI (2/1). Urban und Schwarzenberg, Berlin, pp. 377–406.

Pringsheim, E. G. 1928. Algen-Reinkulturen. *Ber. Deutsch. Bot. Ges.* 46:216–19.

Pringsheim, E. G. 1946. *Pure Cultures of Algae. Their Preparation and Maintenance.* Cambridge University Press, Cambridge, 119 pp.

Pringsheim, E. G. 1954. *Algenkulturen, ihre Herstellung und Erhaltung.* VEB G. Fischer, Jena, Germany, 109 pp.

Pringsheim, E. G. 1970. Contribution toward the development of general microbiology. *Ann. Rev. Microbiol.* 24:1–16.

Provasoli, L. 1937/38. Studi sulla nutrizione dei Protozoi. *Boll. Zool. Agrar. Bachicolt., Milano* 8:1–121.

Provasoli, L. 1956. Alcune considerazioni sui caratteri morfologici e fisiologici delle Alghe. *Boll. Zool. Agrar. Bachicolt. Milano* 22:143–88.

Provasoli, L. 1958a. Effect of plant hormones on *Ulva. Biol. Bull. Mar. Biol. Lab. Woods Hole* 114:375–84.

Provasoli, L. 1958b. Nutrition and ecology of protozoa and algae. *Ann. Rev. Microbiol.* 12:279–308.

Provasoli, L., and Pintner, I. J. 1953. Ecological implications of in vitro nutritional requirements of algal flagellates. *Ann. N. Y. Acad. Sci.* 56:839–51.

Provasoli, L., and Pintner, I. J. 1960. Artificial media for fresh-water algae: Problems and suggestions. In: Tryon, C. A. Jr., and Hartman, R. T., eds. *The Ecology of Algae*. Pymatuning Symposia in Ecology. Spec. Pub. No. 2, University of Pittsburgh, Pittsburgh, pp. 84–96.

Provasoli, L., Hutner, S. H., and Schatz, A. 1948. Streptomycin-induced chlorophyll-less races of Euglena. *Proc. Soc. Exp. Biol. Med.* 69:279–82.

Provasoli, L., McLaughlin, J. J. A. and Pintner, I. J. 1954. Relative and limiting concentrations of major mineral constituents for the growth of algal flagellates. *Trans. New York Acad. Sci. Ser. II.* 16:412–7.

Provasoli, L., McLaughlin, J. J. A., and Droop, M. R. 1957. The development of artificial media for marine algae. *Arch. Mikrobiol.* 25:392–428.

Reich, K., and Kahn, J. 1954. A bacteria-free culture of *Prymnesium parvum* (Chrysomonadina). *Bull. Res. Council Israel.* 4:144–9.

Richmond, A. [ed.] 1986. *CRC Handbook of Microalgal Mass Culture*. CRC Press, Boca Raton, Florida, 528 pp.

Richter, O. 1903. Reinkultur von Diatomeen. *Ber. Deutsch. Bot. Ges.* 21:493–506.

Richter, O. 1911. Die Ernährung der Algen. *Monogr. Abhandl. Int. Rev. Ges. Hydrobiol. Hydrogr.* 2:1–193.

Richter, O. 1913. Die Reinkultur und die durch sie erzielten Fortschritte vornehmlich auf botanischem Gebiete. *Progressus rei Botanicae* 4:303–60.

Sauvageau, C. 1915. Sur la sexualité hétérogamique d'une Laminaire (*Saccorhiza bulbosa*). *C. R. Acad. Sci., Paris* 161:796–9.

Schreiber, E. 1925. Zur Kenntnis der Physiologie und Sexualität höherer Volvocales. *Zeitschr. Bot.* 17:336–76.

Schreiber, E. 1927. Die Reinkultur von marinem Phytoplankton und deren Bedeutung für die Erforschung der Produktionsfähigkeit des Meereswassers. *Wiss. Meeresuntersuch. Abt. Helgoland N. F.* 16(10):1–34.

Skinner, C. E. 1932. Isolation in pure culture of green algae from soil by a simple technique. *Plant Physiol.* 7: 533–7.

Soeder, C. J. 1986. An historical outline of applied algology. In: Richmond, A., ed. *CRC Handbook of Microalgal Mass Culture*. CRC Press, Boca Raton, Florida, pp. 25–41.

Spencer, C. P. 1952. On the use of antibiotics for isolating bacteria-free cultures of marine phytoplankton organisms. *J. Mar. Biol. Assoc. U. K.* 31:97–106.

Spoehr, H. A., and Milner H. W. 1947. *Chlorella* as a source of food. *Yearb. Carnegie Inst. Wash.* 47:100–3.

Spoehr, H. A., and Milner H. W. 1949. The chemical composition of *Chlorella*; effect of environmental conditions. *Plant Physiol.* 24:120–49.

Starr, R. C. 1956. Culture Collection of Algae at Indiana University. *Lloydia* 19:129–56.

Starr, R. C., and Zeikus J. A. 1993. UTEX: the culture collection of algae at the University of Texas at Austin. 1993 List of cultures. *J. Phycol.* 29 (*Suppl.*):1–106.

Stein, J. R., ed. 1973. *Handbook of Phycological Methods. Culture Methods and Growth Measurements*. Cambridge University Press, Cambridge, 448 pp.

Tamiya, H. 1956. Growing *Chlorella* for food and feed. *Proceedings of the World Symposium on Applied Solar Energy*. Stanford Research Institute, Menlo Park, California, pp. 231–41.

Tamiya, H. 1957. Mass culture of algae. *Annual Rev. Plant Physiol.* 8:309–34.

Tamiya, H. 1966. Synchronous cultures of algae. *Annual Rev. Plant Physiol.* 17:1–26.

Tamiya, H., Iwamura. T., Shibata, K., Hase, E., and Nihei, T. 1953. Correlation between photosynthesis and light-independent metabolism in the growth of *Chlorella*. *Biochim. Biophys. Acta* 12:23–40.

Ternetz, C. 1912. Beiträge zur Morphologie und Physiologie der *Euglena gracilis* Klebs. *Jahrb. Wiss. Bot.* 51:435–514.

Terumoto, I. 1961. Frost resistance in the marine alga *Enteromorpha intestinalis* (L.) Link. *Low Temp. Sci. Ser. B* 19:23–8.

Tischutkin, N. 1897. Über Agar-Agarkulturen einiger Algen und Amoeben. *Zentralbl. Bakt. Abt. 2,* 3:183–8.

Tseng, C. K. 1981. Commercial cultivation. In: Lobban, C. S., and Wynne, M. J., eds. *The Biology of Seaweeds*. Blackwell Scientific Publications, Oxford, pp. 680–725.

Uspenski, E. E., and Uspenskaja, W. J. 1925. Reinkultur und ungeschlechtliche Fortpflanzung des *Volvox minor* und *Volvox globator* in einer synthetischen Nährlösung. *Zeitschr. Bot.* 17:273–308.

Van't Hoff, J. H. 1905. *Zur Bildung der ozeanischen Salzablagerungen*. F. Vieweg, Braunschweig, Germany.

Venkataraman, G. S. 1969. *The Cultivation of Algae*. Indian Council of Agricultural Research, New Delhi, 319 pp.

Vischer, W. 1926. Études d'algologie expérimentale. Formation des stades unicellulaires, cénobiaux et pluricellulaires chez les genres *Chlamydomonas, Scenedesmus, Coelastrum, Stichococcus* et *Pseudendoclonium*. *Bull. Soc. Bot. Genève Sér. 2,* 18:184–245.

Vischer, W. 1937. Die Kultur der Heterokonten. In: *L. Rabenhorst's Kryptogamenflora von Deutschland, Österreich und der Schweiz*, vol. 11 (ed. 2). *Heterokonten*. Akademische Verlagsgesellschaft, Leipzig, Germany, pp. 190–201.

Vischer, W. 1960. Reproduktion und systematische Stellung einiger Rinden- und Bodenalgen. *Schweiz. Zeitschr. Hydrol.* 22:330–49.

Waksman, S. A., Barthel, C., Cutler, D. W., and Bristol-Roach, B. M. 1927. Methoden der mikrobiologischen

Bodenforschung. In: Abderhalden, E., ed. *Handbuch der biologischen Arbeitsmethoden, Abt.* XI (3/5). Urban und Schwarzenberg, Berlin, pp. 715–864.

Warburg, O. 1919. Über die Geschwindigkeit der photochemischen Kohlensäurezersetzung in lebenden Zellen. *Biochem. Zeitschr.* 100:230–70.

Ward, H. M. 1899. Some methods for use in the culture of algae. *Ann. Bot. London* 13:563–6.

Wettstein, F. von 1921. Zur Bedeutung und Technik der Reinkultur für Systematik und Floristik der Algen. *Österr. Bot. Zeitschr.* 70:23–9.

Witsch, H. von 1948. Beobachtungen zur Physiologie des Wachstums von *Chlorella* in Massenkulturen. *Biol. Zentralbl.* 67:95–100.

Zumstein H. 1900. Zur Morphologie und Physiologie der *Euglena gracilis* Klebs. *Jahrb. Wiss. Bot.* 34:149–98.

# CHAPTER 2

# FRESHWATER CULTURE MEDIA

MAKOTO M. WATANABE
*National Institute for Environmental Studies*

---

## CONTENTS

Key Index Words: Freshwater Culture Media, Stock Solutions, Macronutrients, Trace Metals, Vitamins, Synthetic Media, Enrichment Media, Soil Water Media

---

## 1.0. INTRODUCTION

It is well known that freshwaters display a wealth of environments and algal flora. The distribution of algal species in freshwaters depends not only on the selective action of the chemophysical environment but also on the organism's ability to colonize a particular environment. Therefore, various culture media have been developed and used for isolation and cultivation of freshwater algae. Some of them are modifications of previous recipes to meet a particular purpose, some are derived from analysis of the water in the native habitat, some are formulated after detailed study on the nutrient requirement of the organism, and some are established after consideration of ecological parameters.

This chapter provides a summary of freshwater culture medium methods that have been used during the past 30 years (see Nichols 1973). The media described here (see Appendix A) are not the full range of possible media, but they are selected to represent a variety of needs. This synopsis includes media widely used during the last 30 years, those utilized at major algal culture collections, and those newly designed. According to *World Catalog of Algae* published by Komagata et al. (1989), there are about 11,000 strains, classified into 3,000 species, that are maintained in 40 culture collections representing 16 countries. The number of cultured algal species is less than 10% of those of described algal species (ca. 40,000). Various permutations, or entirely new attempts, may be necessary to establish suitable media for uncultured algal species, and these are often based on habitat chemistry or specific nutrient requirements of the alga.

## 2.0. Materials

### 2.1. *Chemicals*

The chemical constituents necessary for the preparation of media should generally be of the highest quality available to the investigator. Quality is determined by the manufacturer, and each manufacturer uses its own code for designation of grade. Company catalogs and Web sites should be consulted for code designation and purity of constituents to be used in growth media. Recently, it was recognized that most reagent grade salts and nutrients contain levels of trace metals or other contaminants that may inhibit oligotrophic species. In addition, trace metal impurities in major salt solutions prepared from reagent grade chemicals may exceed the nominal metal concentrations of some media. In these cases, it is necessary to remove impurities from the chemical reagents by the passing of macronutrients and of some micronutrients through a resin Chelex 100 column (Bio-Rad Laboratories) (Morel et al. 1975). These methods are fully described in Chapter 4.

## 2.2. Equipment

Certain minimal equipment (e.g., glassware, plasticware, analytical balance with 1-mg sensitivity, pH meter, and magnetic stirrer) is necessary for preparing stock solutions and culture media. An autoclave is usually essential for sterilization (see Chapter 5). Filtration equipment (e.g., a vacuum source or filter syringe, filter holder, membrane filters) are also needed for sterilization of heat-sensitive substances. Disposable membrane filters, for example the Millex-GV for 1- to 100-mL filtration and the Sterivex GV for 100- to 3,000-mL filtration (Millipore Corp.), are now available and extremely useful for the sterilization of heat-sensitive substances. Sterile water should be used when preparing filter-sterilized solutions, because filters of 0.22-$\mu$m pore size are not sufficient to eliminate viruses (see Chapter 22) and some small-sized bacteria. An ultrasonic washer is useful for cleaning glassware and plasticware coated with stubborn dirt. A refrigerator with a freezer compartment is necessary for maintaining stock solutions and culture media.

## 2.3. Glassware

A variety of items of glassware (e.g., beakers, Erlenmeyer flasks, reagent bottles, pipettes, flasks, tubes, ampoules, cylinders, petri dishes, spatulas, funnels, filter holders, syringe, stirring rods, and burets) is used for preparation of culture media, including disposable glassware. Alternatively, many of these items are available in disposable plastic or reusable plastic, including Teflon-coated plasticware. For maintenance of algal cultures, test tubes with screw caps or Erlenmeyer flasks with silicon plugs are sufficient and widely used. Cotton plugs, which have been used traditionally, are also useful, but they require more labor and time to prepare. Silicon plugs (S Type, Shinetu Polymer Co. Ltd.) can be used repeatedly, and they are superior to cotton plugs for air exchange.

There are a variety of types of glassware available, and not all are suitable for culture medium preparation and culturing. Heat-resistant, hard glassware prepared from borosilicate, such as Pyrex (Corning Co. Ltd.), DURAN (Schott Co. Ltd.), and HARIO (Hario Co. Ltd.), is most satisfactory for preparing stock solutions and growth media. Borosilicate glass does not effect the pH of the contents and it is not readily corroded.

Many investigators emphasize the importance of using only chemically clean glassware. Various cleaning solutions, available in most laboratories, may be used. After cleaning, the glassware should be soaked in 1 N HCl or $HNO_3$ and then thoroughly rinsed with tap water, followed by a distilled water rinse. The glassware should then be dried and stored free from dust (see Chapter 5).

Polyethylene, polycarbonate, or Teflon-lined plastic vessels should be used in place of glass vessels for storing stock solutions of individual trace metals, combined trace metals solutions, and silicate stock solutions. Small amounts of metals will adsorb onto the walls of a glass bottle and silicic acid will dissolute from a glass bottle. These actions will change the concentration of solution, effectively making the concentration unknown.

## 2.4. Water

The earliest investigators used spring water because tap water and distilled water were heavily contaminated with metals (see Chapter 1). Pringsheim (1912) introduced the glass distilling apparatus to algal culturing because metal stills were too toxic. Today, quality water is obtained from a double-distilling apparatus with a Pyrex or quartz glass condenser or from a deionized water system further purified with carbon and membrane filters (e.g., Milli-Q, Millipore Corp.). Once-distilled water and deionized water are also used. The level of quality is dictated by the sensitivity of both the alga and the application, i.e., more critical experimental studies usually require more critical water quality.

## 2.5. Agar

Ordinary agars consist of agarose and agaropectin that are contaminated with various impurities (Krieg and Gerhardt 1981), and some agars contain water-soluble lytic agents against cyanobacteria (Allen and Gorham 1981). For most currently cultured algae, general purpose agar can be used without purification. For sensitive algae, washing the agar is necessary to remove the impurities (Carmichael and Gorham 1974, Waterbury et al. 1986). Two methods for washing agar follow:

1. Heat and dissolve a 2× concentration of agar in deionized water, and then cool to solidify. Cut the agar into pieces and soak the agar blocks in deionized water (1 to 2 times the volume of the agar). The deionized water should be changed daily for 6–8 days. The washed 2× agar blocks and 2× culture medium should be autoclaved in separate vessels and then mixed together after each has been cooled to near the gelling temperature.
2. Place the agar powder in a large beaker with double-distilled water (e.g., 100 g in 3 liters dH$_2$O) and stir for 30 minutes (Waterbury et al. 1986). Allow the agar to settle and siphon off the water; repeat until the water is clear. Remove the water (by filtration if necessary) and wash agar with 3 liters of ethanol. Separate the ethanol and agar by filtration (e.g., Whatman F4 filter and buchner funnel), and then rinse the agar with (e.g., 3 liters) analytical grade acetone. Remove the acetone and dry the agar at 50°C for 2–3 days; store the purified agar in a tightly covered container.

It is probable that even washed agar still contains traces of toxic materials for some species of cyanobacteria (Shirai et al. 1989) as well as other sensitive algae. The more expensive but more highly purified molecular biology grade agarose can often be used (usually at 0.6–1%) for sensitive organisms.

## 2.6. Soil

A liquid soil extract or solid particulate soil is used for culturing algal species when precise knowledge of nutritional requirements is not necessary and when maintaining normal morphology is critical. Success with a soil extract solution or a soil water medium depends on selection of a suitable soil. However, finding a good soil is not simple. For example, of over 40 different soil types tried for growing chrysophytes, only two were found to be reliable for general use (Robert Andersen, personal communication). Recommended sources for good soils are gardens or greenhouses where the soil has not been exposed to chemicals (e.g., fertilizers, pesticides), undisturbed deciduous forests, and grasslands that have not been tilled or grazed (Starr and Zeikus 1993, Tompkins et al. 1995). The soil should be of the loam type; soils with large amounts of clay are not suitable. For certain rare or sensitive algae, it is sometimes possible to obtain good soil near the lake or pond where they grow naturally; however, the soil must be taken from above the water level, because underwater lake and pond sediments are often anoxic and contain accumulated toxic materials.

After removal of any obvious extraneous materials (e.g., rocks, leaves, roots, worms), the soil should be dried, either at room temperature or in a drying oven at low temperature (<60°C). Once dry, the soil should easily crumble into a fine dust. A clean, chemical-free mortar and pestle can be used to grind the soil. The finely ground soil should be passed through a sieve to remove any larger particles or remaining extraneous materials, and then the soil should be stored in a dry environment.

To prepare a soil extract, add 1 part soil to 2 parts dH$_2$O and pasteurize or autoclave for about 2 hours (Tompkins et al. 1995). Allow the particulate matter to settle, and then filter (e.g., Whatman #1 filter) the liquid. The extract should be pasteurized or autoclaved again to establish a sterile solution. The solution should be tightly capped and stored at 4°C.

An alternative, alkaline extraction method is described by Provasoli et al. (1957). To prepare the alkaline soil extract solution, combine 2 parts dH$_2$O with 1 part rich organic garden soil. Add 2–3 g L$^{-1}$ NaOH. Autoclave for 2 hours, cool, and filter. The concentrated extract is then diluted 50 : 1 with dH$_2$O to make the final working stock solution.

Glazer et al. (1997) provide a method for preparing a peat extract solution. The peat extract is prepared by adding 30 g commercial dry peat moss to 1 liter of glass-distilled water; the mixture is pasteurized for 30 minutes. After cooling, the mixture is vacuum-filtered with a double layer of Whatman #1 filter paper and the filtrate is retained. The filtrate is then pasteurized again and stored at 4°C. For preparing media with soil or soil extracts, see Section 4.3.

## 3.0. STOCK SOLUTIONS

### 3.1. General Comments

Media are generally composed of three components: macronutrients, trace elements, and vitamins; all three are often prepared as stock solutions. Stock solutions in quantities of 100 mL to 1 liter are typically prepared at a nutrient concentration of 100 to 1,000 times that required. For use, some quantity (e.g., 1 mL) is removed aseptically and used. Stock solutions are useful for several reasons. Repeated individual weighing of chemicals is time-consuming and errors in weighing (e.g., mistaking mg for µg) may occur. The stock solution is made only occasionally, and once made, it provides an easy and consistent source. That is, if a liter stock solution is prepared, where 1 mL is used for each liter of final medium, then (theoretically) the stock solution can be used to make 1,000 liters of medium. Stock solutions are generally prepared as follows:

1. Add approximately 80–90% of the required volume of distilled or deionized water to a beaker.
2. Dissolve the appropriate quantity of the weighed nutrient while stirring continuously. If multiple components are included in the stock solution (e.g., a trace metals solution), completely dissolve the first component before adding the second component. Most nutrients dissolve easily with stirring; however, for some compounds, heat or pH alteration is necessary to quickly dissolve the substance.
3. Dilute to final volume with distilled or deionized water.
4. Stock solutions should be stored in tightly sealed glassware or plasticware to avoid alteration of the initial concentration due to evaporation.

Refrigerate (4°C) or freeze stock solutions when not in use.

Stock solutions containing substrates that encourage bacterial or fungal growth must be sterilized, and many other stock solutions should be sterilized as well. If the stock solution has a furry fungus growing in it or cloudy bacterial growth, it must be discarded and reprepared. Care must be given to avoid evaporation of water; if water evaporates, then the concentration will increase to some unknown concentration. When crucial and accurate experiments are to be performed, it is best to mix fresh stock solutions of both macronutrients and micronutrients. The practical protocols generally employed for stock solutions of macronutrients, trace elements, and vitamins are shown in the following.

### 3.2. Macronutrients

Separate stock solutions of each macronutrient should be prepared at a concentration of 100- to 1,000-fold of the final concentration (unless solubility problems exist) so as to use 10 mL and 1 mL, respectively, per 1,000 mL of medium. Phosphate stock solutions should never be stored in the polyethylene bottles because phosphate ions are strongly adsorbed onto polyethylene (Hassenteufel et al. 1963). Silicate stock solutions should be stored in nonvitreous material (e.g., Teflon-lined, polyethylene, or polycarbonate) because of dissolution of silicic acid from glass vessels. If silica-free experimentation is required, none of the stock solutions can be held in glass.

### 3.3. Trace Elements

These elements are usually prepared either as separate stock solutions or mixed stock solutions (see also Chapter 4). In a few cases, they are directly added to the media at concentrations of 0.1 mg to 20 mg per liter. In many, but not all, freshwater media, Na$_2$EDTA (disodium ethylenediaminetetraacidic acid) is used as a chelator. When EDTA is used, it should be dissolved first, followed by the addition of metal(s). The practical steps recommended are as follows.

#### 3.3.1. Separate Stock Solutions

In the following example, a separate (e.g., single metal) stock solution is prepared at a concentration of 1,000-fold the final medium concentration, so as to use 1 mL per 1,000 mL of stock solution for each liter of medium.

Separate stock solutions are referred to as primary stock solutions in Appendix A.

1. Into distilled and/or deionized water (i.e., 800–950 mL), dissolve the required amount of the trace element. In the case of metals such as iron, cobalt, copper, manganese, and zinc, boil for 5 to 10 minutes to hasten the process.
2. Dilute to final volume (1 liter) with distilled and/or deionized water.
3. Store in plastic, wrap tightly to avoid evaporation, and keep in a refrigerator (4°C) when not in use.

When an extremely low concentration of the trace element is required (e.g., copper for N-HS-Ca medium; see Appendix A), prepare a preliminary stock solution at even greater concentration in order to weigh the amounts accurately. To prepare the final separate stock solution, dilute the solution (e.g., 1 mL of preliminary stock solution into 1 liter of separate stock solution). Seal and store as described previously.

### 3.3.2. Mixed Stock Solution (Working Stock Solution)

1. Add approximately 80% of the required volume of distilled and/or deionized water to a beaker (e.g., 800 mL for 1 liter stock).
2. If $Na_2EDTA$ or other chelator is used, first dissolve the chelator.
3. Dispense the required volume of each trace metal from the separate stock solution, stirring continuously.
4. Dilute to final volume with distilled and/or deionized water and store in a refrigerator.

For convenience, dispense the stock into small aliquots. For example, if final medium preparation requires 1 mL of stock, 1 mL can be dispensed into Eppendorf tubes or cryovials and frozen. Alternatively, 10-mL aliquots or similar small volumes can be prepared.

### 3.4. Vitamins

Three vitamins—vitamin $B_1$ (thiamine · HCl), vitamin $B_{12}$ (cyanocobalamin), and vitamin H (biotin)—are usually used for culture of microalgae. Many algae need only one or two of the vitamins, but there seems to be no harm caused by adding a nonessential vitamin (Provasoli and Carlucci 1974). In addition to the three common vitamins, some recipes call for other vitamins.

For example, nicotinamide (nicotinic acid amide, niacinamide) is added to the culture medium for *Phacotus lenticularis* (Ehrenberg) Stein (see N-Hs-Ca medium, Schlegel et al. 2000).

Vitamins are frequently autoclaved with the final medium, and this undoubtedly results in some decomposition, but the moieties in many instances are apparently equally effective. Strictly speaking, the vitamins should be added aseptically to the final medium after autoclaving.

### 3.4.1. Separate Stock Solutions

For biotin and cyanocobalamin, it is necessary to prepare separate stock solutions, and for convenience, especially if preparing several media with different vitamin requirements, a primary stock of thiamine · HCl should be prepared. The concentration of the primary stocks should be 100-fold to 10,000-fold that necessary for use in combined (working) stock solutions (see next section). The exact concentrations depend upon the intended use, for example, one or several media, dispensing by pipette or pipetteman.

The separate stock solutions may be filter sterilized (see Chapter 5) and aseptically dispensed in small volumes (e.g., 1–10 mL) into sterile containers (e.g., Eppendorf tubes, cryovials, polycarbonate tubes). Alternatively, the primary stocks may be dispensed into small tubes and then autoclaved as acidified solutions (pH, 4.5–5.0), but keep in mind that some plastics (e.g., polyethylene) melt in an autoclave and that glass tubes may break when the liquid freezes.

### 3.4.2. Mixed Stock Solutions

To prepare a mixed stock solution, which usually contains all vitamins, thaw and dilute an aliquot (usually 1 mL) of each separate stock solution into 100 or 1,000 mL of distilled and/or deionized water. The final volume of the mixed stock solution varies with need, and the volume of separate stock solution is diluted to obtain the correct concentration. The mixed stock solution is sterilized and dispensed into small volumes as described previously; it should be stored frozen.

### 4.0. GENERAL METHODS OF PREPARATION OF MEDIA

Freshwater media are divided broadly into three categories: synthetic, enriched, and soil water. Synthetic (artificial) media are designed primarily to provide simplified, defined media, for both careful experimental studies and routine maintenance of strains. Common examples are Bold's Basal Medium, BG-11 medium, Chu #10 medium, WC medium, and V medium (see Appendix A). They can be prepared in both liquid and solid (agar) forms. Within practical limits, the media are defined, although it should be remembered that distilled and deionized water have trace contaminants, and even ultrapure chemicals have nanogram or picogram quantities of contaminating elements.

Enriched media are prepared by adding nutrients to natural lake or stream waters or by enriching a synthetic medium with soil extract, plant extracts (e.g., peat moss), yeast extract, etc. Enriched media are not defined, because the lake and stream waters have various inorganic and organic molecules; the chemical compositions of the extracts are also unknown. In general, enriched media are not used for physiological experiments, but algae often grow with normal morphology in these media. The natural water must be relatively clean and free of pollution. If the natural water contains substantial amounts of humic compounds or other organic molecules, these may cause interference for molecular biology studies, especially when nucleic acids are extracted without the cells being rinsed. Common examples are Alga-Gro lake water medium (Carolina Biological Supply Co.), *Audouinella* medium, Diatom medium, VS medium, Modified *Porphyridium* medium, Malt medium, and *Polytoma* medium (see Appendix A).

Soil water media are prepared by placing 1–2 cm of dried and sifted garden soil in the bottom of a test tube (or bottle), onto which water is added. This mimics a lake or pond, where nutritional substances are generally replenished from bottom sediments by bacterial activity and water mixing. In a soil water medium, diffusion, as well as biochemical activity of bacteria (xenic cultures) and algae at the soil/water interface, imitate nature's example. The culture medium composition is determined by the soil (e.g., pH, conductivity, nutrients, organic buffers, vitamins), and therefore it is important to find both good and appropriate soil (see previous discussion). Algae grown in soil water media usually have normal morphology, and the algae can be reliably maintained.

## 4.1. Synthetic Media

1. Add approximately 80–90% of the required volume of distilled water to a beaker.

2. If a buffer is required, dissolve appropriate quantities of weighed buffer (e.g., Tris, glycylglycine, HEPES, TAPS, Bicine, or MES) while stirring continuously.

3. Individually, add the appropriate nutrients from previously prepared stock solutions or weighed quantities, stirring continuously.

4. Dilute to final volume with distilled water.

5. Adjust the pH (if necessary) with either 1 N NaOH or 1 N HCl (when buffers are present) *or* 0.1 N HCl or 0.1 N NaOH (when no buffers are present).

6. Dispense the medium into the culture vessels. For example, add 10 mL to an $18 \times 150$-mm test tube. Autoclave or filter sterilize (see Chapter 5).

7. If autoclaved, cool the medium and let it stand for 24 hours after autoclaving to allow for re-equilibration of inorganic carbon species (especially for unbuffered media). For large flasks, filtered air can be bubbled into the medium to hasten the equilibration.

## 4.2. Enriched Media

1. Add approximately 80–90% of the required volume of distilled water or natural water to a beaker. If using natural water, you should collect it from the site where algae were collected and autoclave or pasteurize it. It is possible to filter sterilize (0.22 µm pore size) the natural water, but this will not remove any viruses in the water.

2. Individually dissolve the components (stock solution additions or weighed quantities): macronutrients, microelements, vitamins, and extracts or organic nutrients (e.g., tryptone and extracts from yeast, malt, peat moss or soil).

3. Dilute to final volume with distilled or natural water, as appropriate.

4. Adjust the pH (if necessary) with either 1 N NaOH or 1 N HCl (when buffers are present) *or* 0.1 N HCl or 0.1 N NaOH (when no buffers are present).

5. Dispense the medium into the culture vessels. For example, add 10 mL to an $18 \times 150$-mm test tube. Autoclave, pasteurize, or filter sterilize (see Chapter 5).

6. If autoclaved or pasteurized, cool the medium and let it stand for 24 hours after heating to allow for re-equilibration of inorganic carbon species.

## 4.3. Soil Water Media

1. Place a layer of 1–2 cm of garden soil (dried and sifted) in the bottom of a test tube (or flask, bottle, etc.)
2. Add distilled or deionized water until the container is 3/4 full; cover with a cotton plug or screw cap.
3. Steam for 1 hour on 2 consecutive days (see Chapter 5). Typically, one does not autoclave the soil water because true sterilization is not intended. However, for many algae (but certainly not all), autoclaved soil water will work fine.
4. Cool for 24 hours and store refrigerated until ready for use.

Several variations of soil water medium can be made by adding additional materials into the bottom of the tube before soil is added:

1. a small pinch of powdered $CaCO_3$ for many phototrophic freshwater algae (Starr and Zeikus 1993),
2. a small pinch of $NH_4MgPO_4$ for *Botryococcus*, *Synechococcus*, and some euglenoids (Starr and Zeikus 1993),
3. 1/8 of a garden pea, soaked 12 hours before use, for some euglenoids, the green flagellate *Astrephomene*, and other mixotrophic algae (Starr and Zeikus 1993, Schlösser 1994), and
4. 1/2 teaspoon of organic peat and half the amount of soil for most acidophilic algae (Starr and Zeikus 1993).

## 4.4. Solidified Media: Agar

### 4.4.1. Standard Nutrified Agar

Klebs (1896) and Tischutkin (1897) were among the first to use culture media solidified with agar for cultivating algae (see Chapter 1), and nutrified agar is still very useful for growing most freshwater algae. A small inoculum of cells is spread onto the surface of the agar (see Chapter 10), and generally the cells grow slowly on the agar surface. Agar is usually used at concentrations of 1–2%. The general procedures for preparing agar medium are given next, but see also Allen's BG-11 Medium (Appendix A) for an alternative method.

1. Heat a 2-liter Erlenmeyer flask with 1 liter of culture medium to ca. 95°C, either directly on a heating source or in a heated water bath.

2. Slowly, add the desired quantity of agar while stirring continuously so that all the agar is dispersed and in solution.
3. To make agar plates in petri dishes, first sterilize the nutrified agar by autoclaving (120°C, 20 minutes). Remove the container from the autoclave, and when the temperature is about 50°C–60°C, pour the mixture into sterile petri dishes. The mixture can also be held at the appropriate temperature (e.g., 50°C–60°C) by placing it in a heated water bath. If the temperature of mixture is too high when the mixture is poured into the petri dishes, condensation will occur. If the petri dishes are stacked after pouring and a flask with hot tap water is placed on the top of the stack, this will reduce condensation. Once the agar has gelled, the plates should be inverted (agar up) and stored in air-tight containers (e.g., plastic bags, covered containers). Store at 4°C.
4. To make agar slants in test tubes, dispense into test tubes and then sterilize by autoclaving (120°C, 20 minutes). After removing the test tubes from the autoclave, place them at the appropriate angle (for the slant) and cool. Various companies make test tube racks that are designed to be tilted while the agar cools.

### 4.4.2. Nutrified Agar Pour Plates

Some algae will not grow on the surface of agar, but they will grow if embedded in the agar (Skinner 1932). To embed the cells, one usually prepares 1× to 2× concentration agar, and then as the agar reaches its gelling temperature, a liquid suspension of cells is mixed aseptically with the sterile agar. The mixture is swirled to distribute the algae, and then the agar is poured into petri plates. For algae that are sensitive to higher temperatures, low-gelling-temperature agarose (26 ± 2°C) or ultra-low-gelling-temperature agarose (Type IX, Sigma; gelling temperature of 8°C–17°C) can be used (Shirai et al 1989, Watanabe et al. 1998).

## 5.0. RECIPES

Freshwater culture medium recipes are listed in Appendix A. I have included notes regarding preparation of the media and taxa that have been successfully grown in the media. Additional remarks can be found in the catalogs of strains from culture collections (Starr and

Zeikus 1993, Schlösser 1994, Tompkins et al. 1995, Andersen et al. 1997, Watanabe et al. 2000) and on the Web sites for these collections (see Appendix A).

## 6.0. REFERENCES

Allen, E. A. D., and Gorham, P. R. 1981. Culture of planktonic cyanophytes on agar. In: Carmichael, W. W., ed. *The Water Environment: Algal Toxins and Health*. Plenum Publishing Corp., New York, pp. 185–92.

Andersen, R. A., Morton, S. L., and Sexton, J. P. 1997. CCMP-Provasoli-Guillard National Center for Culture of Marine Phytoplankton. 1997 List of strains. *J. Phycol.* 33(Suppl.):1–75.

Carmichael, W. W., and Gorham, P. R. 1974. An improved method for obtaining axenic clones of planktonic bluegreen algae. *J. Phycol.* 10:238–40.

Glazer, A. N., Chan, C. F., and West, J. A. 1997. An unusual phycocyanobilin-containing phycoerythrin of several bluish-colored, acrochaetioid, freshwater red algal species. *J. Phycol.* 33:617–24.

Hassenteufel, W., Jagitsch, R., and Koczy, F. F. 1963. Impregnation of glass surface against sorption of phosphate traces. *Limnol. Oceanogr.* 8:152–6.

Klebs, G. 1896. *Die Bedingungen der Fortpflanzung bei einigen Algen und Pilzen*. G. Fischer, Jena.

Komagata, K., Sugawara, H., and Ugawa, Y. 1989. *World Catalog of Algae, Second Edition*. WFCC World Data Center on Microorganisms. Life Science Research Information Section, RIKEN, Wako, Saitama, Japan. 315 pp.

Krieg, N. R., and Gerhardt, P. 1981. Solid culture. In: Gerhardt, P., Murray, R. G. E., Costilow, R. N. et al., eds. *Manual of Methods for General Bacteriology*. American Society for Microbiology, Washington, DC, pp. 143–4.

Morel, F. M. M., Westall, J. C., Reuter, J. G., and Chaplick, J. P. 1975. *Description of the Algal Growth Media 'Aquil' and 'Fraquil'*. Technical report 16. Water Quality Laboratory, Ralph Parsons Laboratory for Water Resources and Hydrodynamics, Massachusetts Institute of Technology, Cambridge, Massachusetts, 33 pp.

Nichols, H. W. 1973. Growth media—freshwater. In: Stein, R., ed. *Handbook of Phycological Methods: Culture Methods and Growth Measurements*. Cambridge University Press, New York, pp. 7–24.

Pringsheim, E. G. 1912. Kulturversuche mit chlorophyllführenden Mikroorganismen. Mitt. I. Die Kultur von Algen in Agar. *Beitr. Biol. Pfl.* 11:305–34.

Provasoli, L., and Carlucci, A. F. 1974. Vitamins and growth regulators. In: Stewart, W. D. P., ed. *Algal Physiology and Biochemistry*. Blackwell Scientific, London, pp. 741–87.

Provasoli, L., McLaughlin, J. J. A., and Droop, M. R. 1957. The development of artificial media for marine algae. *Arch. Mikrobiol.* 25:392–428.

Schlegel, I., Krienitz, L., and Hepperle, D. 2000. Variability of calcification of *Phacotus lenticularis* (Chlorophyta, Chlamydomonadales) in nature and culture. *Phycologia* 39:318–22.

Schlösser, U. G. 1994. SAG—Sammlung von Algenkulturen at the University of Göttingen catalogue of strains 1994. *Bot. Acta* 107:111–86.

Shirai, M., Matsumaru, K., Ohtake, A., Takamura, Y., Aida, T., and Nakano, M. 1989. Development of a solid medium for growth and isolation of axenic *Microcystis* strains (cyanobacteria). *Appl. Environ. Microbiol.* 55:2569–71.

Skinner, C. E. 1932. Isolation in pure culture of green algae from soil by a simple technique. *Plant Physiol.* 7:533–7.

Starr, R. C., and Zeikus, J. A. 1993. UTEX—The culture collection of algae at the University of Texas at Austin. *J. Phycol.* 29(2, Suppl), 106 pp.

Tischutkin, U. 1897. Über Agar-Agarkulturen einiger Algen und Amoben. *Centr. Bakt. Par.* 3:183–8.

Tompkins, J., DeVille, M. M., Day, J. G., and Turner, M. F. 1995. *Culture Collection of Algae and Protozoa. Catalog of Strains*. Ambleside, UK, 204 pp.

Watanabe, M. M., Nakagawa, M., Katagiri, M., Aizawa, K., Hiroki, M., and Nozaki, H. 1998. Purification of freshwater picoplanktonic cyanobacteria by pour-plating in "ultra-low-gelling-temperature agarose." *Phycol. Res.* 46 (Suppl.):71–5

Watanabe, M. M., Kawachi, M., Hiroki, M., and Kasai, F. 2000. *NIES—Collection List of Strains, Sixth Edition, 2000, Microalgae and Protozoa*. Microbial Culture Collections, National Institute for Environmental Studies, Tsukuba, Japan. 159 pp.

Waterbury, J. B., Watson, S. W., Valois, F. W., and Franks, D. G. 1986. Biological and ecological characterization of the marine unicellular cyanobacterium *Synechococcus*. In: Platt, T., and Li, W. K. I., eds. *Photosynthetic Picoplankton. Can. Bull. Fish. Aquatic Sci.* 214:71–120.

# CHAPTER 3

# MARINE CULTURE MEDIA

PAUL J. HARRISON
*AMCE Program, Hong Kong University of Science and Technology*

JOHN A. BERGES
*Department of Biological Sciences, University of Wisconsin–Milwaukee*

## CONTENTS

Key Index Words: Marine Phytoplankton, Culture Media, Artificial Seawater, Nutrient Enrichments, Trace Metals, Vitamins

## 1.0. INTRODUCTION

Natural seawater (NW) is a complex medium containing more than 50 known elements and a large and variable number of organic compounds. For algal culture, direct use of NW is seldom acceptable. Without the addition of further nutrients and trace metals, the yield of algae is usually too low for culture maintenance or laboratory experiments, and thus enrichment is normally required. In addition, variations in the quality of NW throughout the year, the need to control nutrient and trace element concentrations, and the limited availability of seawater at inland locations make the option of artificial seawater (AW) attractive (see Section 4.2).

The preparation of NW with an enrichment solution of nutrients, trace metals, and vitamins and the preparation of AW are described in this chapter. For clarity, we use the term *natural seawater* to refer to unenriched NW, and *artificial seawater* for unenriched AW. The term *enrichment solution* refers to the macronutrients, trace elements, and vitamins that must be added to both NW and AW to produce a substantial algal yield. The materials required for preparation of seawater media and the main recipes of stock solutions of macronutrients, trace elements, and vitamins are described. Methods and precautions that are required in media preparation are covered. Recipes for certain algal groups and various natural and synthetic seawater recipes are compared. The recipes have been tested for planktonic algae, but reports indicate that the commonly used recipes are also suitable for benthic diatoms and some seaweeds.

Comparing today's marine culture media with those of 30 years ago (McLachlan 1973), it is interesting to see the progress that has been made and that several media are quite broad-spectrum, indicating that most culturable algae can be grown by using only a few different media. The most challenging phytoplankton to culture are still the oceanic species.

## 1.1. Historical Perspective on Previous Media and Recent Advances

Many basic media concepts that are used today were developed in the late 1800s and early 1900s (for a review see Allen and Nelson 1910; Allen 1914). Early workers soon learned that the ratios of chemicals were not always critical, and subsequently various media recipes were developed with only slight modifications. It was well known that chemicals that were added to seawater contained impurities such as trace elements, and these often improved growth (Allen and Nelson 1910). The importance of culture pH, iron, vitamins for culture growth, and the avoidance of metal toxicity and impurities in distilled water also were established very early (Allen and Nelson 1910; Allen 1914).

NW may be the preferred seawater base if large quantities are required, if a good source is really available, or if open ocean species are being cultured in the laboratory. However, for near-shore sources, the salinity may vary seasonally and large phytoplankton blooms may alter the organic compounds in the seawater. To enhance the algal yield, various additions must be made to the NW. In the early 1900s, boiling water was used to extract unknown, variable amounts of inorganic and organic compounds from soil, and when it was added it produced good growth with few morphological changes during the long-term maintenance of algal cultures. Allen (1914) concluded that organic substances were required in trace amounts. Soil extract, originally introduced by Pringsheim (1912), was established in the marine culture methodology by Foyn's (1934) now famous "Erdschreiber" medium (note that this name was originally written as Erd-Schreiber and that the original recipe was derived from Schreiber [1927]). We now know that soil extract performs numerous functions in culture media, and it has largely been replaced by specific compounds. Soil extract provides various elements and vitamins needed for plant growth, metal complexing by organic compounds that sequester potentially toxic metals, and organic compounds that keep iron in solution. In replacing soil extracts, numerous trace elements and vitamins are usually added to culture media. These include iron, manganese, zinc, cobalt, copper, molybdenum, vitamin $B_{12}$, thiamine, and biotin. Artificial chelators such as EDTA (ethylenediaminetetraacetic acid) are added to keep iron in solution and to keep free ionic metal concentrations at nontoxic levels.

The history of the development of defined media was largely learning how to dispense with soil extract. The historical development of artificial media has been thoroughly reviewed (Provasoli et al. 1957, Kinne 1976). One of the first attempts to design an AW medium for algae began about 90 years ago (Allen and Nelson 1910). More extensive analyses of NW (Lyman and Fleming 1940) stimulated the development of new recipes (Chu 1946, Levring 1946) that attempted to imitate NW precisely, but they were frequently considered too complex. Because these recipes were similar to NW, they had the same defect, namely the formation of a precipitate during autoclaving when they were enriched. Autoclaving drives carbon dioxide out of the seawater, causing a shift in the carbonate buffer system. The resulting pH of around 10 causes the precipitation of ferric phosphate and ferric hydroxides. The amount and composition of the precipitate varies, often leading to inconsistent growth in the medium. Preoccupation with making a complete medium autoclavable without precipitation led to the following extensive modifications in recipes:

1. Addition of synthetic metal chelators such as EDTA or nitrilotriacetic acid (NTA) to decrease metal precipitates
2. Addition of a pH buffer such as Tris or glycylglycine (7.0–8.5 range), because the amount of precipitate increased as the pH rose during autoclaving
3. Reduction in salinity, thereby reducing the amount of salts available for precipitation
4. Replacement of $Mg^{2+}$ and $Ca^{2+}$ with more soluble univalent salts
5. Replacement of inorganic phosphorus with an organic source (e.g., sodium glycerophosphate) to avoid the precipitation of $Ca_3(PO_4)_2$ (Droop 1969)
6. Introduction of weak solubilizers, which are acids (e.g., citric acid) having highly soluble salts with calcium

As a result of these extensive chemical modifications in ion ratios (Provasoli et al. 1957), few early recipes bore much resemblance to NW. Various simplified AW recipes were developed (Provasoli et al. 1957, McLachlan 1959, 1964), but many of these recipes could grow only a few species or favored a particular group of algae (Provasoli et al. 1957, Kinne 1976). More recently, several recipes such as the one by Kester et al. (1967) and one commercial one, Instant Ocean (King and Spotte 1974), have ion ratios of the major constituents that are very similar to NW.

Although one would expect artificial media to be much more chemically defined than NW, in some respects, artificial media are not completely defined due to the contaminants in reagent grade salts. Because of the large amounts of major salts that must be added, trace contaminants (e.g., copper, zinc, iron) in these salts can result in higher concentrations of some metals in AW than those naturally present in oceanic surface water. The use of ion exchange columns to remove these contaminants, as described by Morel et al. (1979), can greatly reduce this problem (note, however, that Chelex 100 removes only cations), but it is a time-consuming process (see Chapter 4). With the interest in trace metal availability and trace metal toxicity, the development of Aquil by Morel et al. (1979) permitted, for the first time, a complete definition of chemical speciation of various components, as calculated from thermodynamic equilibria, by controlling trace element contamination and precipitate formation. The medium Aquil is useful for trace metal studies of copper, zinc, nickel, cobalt, lead, and cadmium, because they remain in cationic form in seawater. More recent changes in the preparation of Aquil have been the purification of the Chelex column to avoid contamination by the chelating agents, use of alternative sterilization procedures, and an increase in the concentration of trace metal buffers (Price et al. 1988/89).

Medium K (Keller et al. 1987) was developed from culturing fastidious oceanic phytoplankton, and it has been tested on 200 ultraplankton clones representing seven algal classes. It is not recommended for coastal species. This medium includes selenium, both nitrate and ammonium, increased chelation, reduced copper, and a moderate level of pH buffering. The chelation to total metal ratio is 10 : 1, and the EDTA concentration is $10^{-4}$ M. There is also a synthetic counterpart. The recently formulated MNK medium and Pro99 medium (see later discussion and Appendix A) are also formulated for open ocean phytoplankton, and preliminary results suggest that they are superior to K medium for certain algae (coccolithophores and *Prochlorococcus*, respectively).

A broad-spectrum AW medium (enrichment solution with artificial water [ESAW]) was developed and tested on 83 strains (Harrison et al. 1980). The AW base was taken from Kester et al. (1967), and the ratios of the major ions closely match those found within NW. The enrichment solution was a modified enrichment solution originally developed by Provasoli (1968). The modifications were the omission of Tris (the pH buffer) and the addition of silicate. The omission of Tris was compensated for by adding equimolar amounts of $NaHCO_3$ and HCl to prevent precipitation during autoclaving. During autoclaving $CO_2$ is lost, and $CO_2$ can be added indirectly before autoclaving by adding $NaHCO_3$ or by directly bubbling $CO_2$ through the medium. This medium has a 2.3 : 1 chelator to trace metal ratio, compared to 1 : 1 in 'f' medium (Guillard and Ryther 1962), which may reduce the tendency to form metal precipitates and metal toxicity (Harrison et al. 1980). Over the past 2 decades, further changes were made to ESAW that significantly improved the medium. The forms of phosphate, iron, and silicate were changed and the trace element mixture was altered to include nickel, molybdenum, and selenium (Berges et al. 2001).

## 2.0. MATERIALS REQUIRED

### 2.1. Chemicals

Most chemicals required to make marine media are available from various chemical suppliers. Reagent grade salts (e.g., American Chemical Society grade) should be used if possible. The organic chemicals such as vitamins, buffer, and chelators are available from Sigma Chemical Company. If brands of a chemical are changed, this should be noted, because different brands are likely to have different amounts of contaminants or impurities.

### 2.2. Equipment, Glassware, and Tubing

Most required equipment comprises standard items in laboratories: analytical and top-loading balances, pH meter, hot-plate-magnetic stirrer, and so on. Borosilicate glassware should be used exclusively for all glassware, including stock bottles, beakers, and cultural tubes and flasks (examples of brand names are Pyrex and Kimax). Teflon or plastics are recommended, because they reduce breakage. Check the manufacturer's specifications for usage, such as autoclaving and storage of concentrated chemicals.

Keep the glassware and plasticware to be used in media preparation separate from general purpose laboratory use. Washing protocols vary, depending on the experiments planned, but in general it is important to be aware that tap water often contains high amounts of nutrients, trace metals, and heavy metals. Therefore, if tap water is used for washing and rinsing, then make sure that deionized water is used for the final rinse. Furthermore, domestic detergents leave a residual film on glassware. The detergents from most large chemical supply companies are satisfactory, but labels should be read to determine if the contents meet your requirements. New glassware and plasticware should be degreased in dilute NaOH, soaked in dilute HCl, and then soaked in deionized water for several days before use. Glassware should not be cleaned in chromic acid, because chromium is toxic to many phytoplankters (McLachlan 1973). Teflon is useful only for stock bottles and is not suitable for culture vessels because of its reduced light-transmission properties. Polycarbonate is good for culture vessels, especially for experiments involving trace-metal limitation. Polypropylene may yield toxins from stocks, notably silicate stocks (Brand et al. 1981). More information on culture vessels is provided in Chapters 2, 4, and 5. Glassware should be autoclaved, clean glassware and plasticware should be stored in closed cupboards, and open vessels should be covered.

One should also be aware of potential toxicity from tubing and other materials. Bernhard was one of the first to call attention to the inhibitory effects of some culture materials by testing more than 50 types on phytoplankton and zooplankton (Bernhard and Zattera 1970, Bernhard 1977). A later study examined latex tubings of rubber, polyvinyl chloride (Tygon), and silicone, and Price et al. (1986) found that latex tubing was surprisingly toxic to phytoplankton, zooplankton, and bacteria. Even using latex tubing to siphon water from one bottle to another one rendered the water toxic for phytoplankton growth. The toxic compound was not identified, but preliminary results indicated that it may be pentachlorophenols and tetrachlorophenols used to preserve the latex tubing (Price et al. 1986). Tygon tubing was generally safe, provided that the powder inside the new tubing was carefully removed by rinsing before use. Silicone tubing was completely safe to use. Colored or black rubber stoppers may be toxic, and therefore silicone stoppers are recommended, especially the ones from Cole-Palmer that are made by injecting small air-bubbles into the polymer and are thus lighter and easier to work with than the solid silicone stoppers.

All containers and tubing used for cultures and media stocks should be carefully selected to avoid toxic compounds. For general purpose culturing, we recommend flasks and test tubes made of borosilicate glass and tissue culture-grade polycarbonate or polystyrene plasticware. Teflon-lined caps are recommended for screw-top glass test tubes, and black caps should be autoclaved several times in changes of seawater, because new caps may release toxic phenolics when heated (McLachlan 1973). For studies on silicon limitation, polycarbonate is recommended. However, borosilicate glassware may be used as long as it has not been rinsed with any acid that causes severe leaching of silicate from the glass.

Likewise, rubber stoppers (or anything that releases volatile compounds when heated) should be autoclaved separately from media. Older autoclaves with copper tubing should be avoided, because excess copper is toxic to algae. The autoclave steam may be contaminated with metals or chemicals used to inhibit corrosion of the autoclave. See Chapter 5 for further information on sterilization procedures.

## 2.3. Water Sources, Treatment, and Storage

The source of seawater may determine one's success in culturing certain species. To obtain NW free from pollution, it may be necessary to collect offshore water. Oligotrophic open ocean water is ideal, because it is low in nutrients and trace metals, and these components can be added in required amounts in an enrichment solution. In addition, this water contains less sediment and possibly less phytoplankton, making it easier to filter.

Nearshore water may be seasonally variable due to rainfall and runoff inputs, which may have elevated nutrients and sediments and decreased salinity. If inshore water is used, then water below the photic zone or pycnocline is likely to have less sediments and algal biomass to remove by filtration. Water should not be collected during blooms, especially when noxious species are present. Various pumps or large water bottles may be used to obtain the water. Water should not be collected with a water bottle that uses latex tubing for the rubber spring that closes off the ends of the water bottle, because the latex rubber tubing renders the water toxic for some algae (Price et al. 1986). Because filtration may be a slow process, plastic containers or carboys are usually filled and brought back to the laboratory for filtration. Usually a large-scale filtration apparatus is used, such as 147- or 293-mm-diameter membrane filters contained in a plastic holder (e.g., Millipore or Pall-Gelman). A prefilter may be placed on top of the membrane filter to slow down the clogging of the membrane filter. Alternatively, filter cartridges (e.g., Pall-Gelman Acropak capsules) are relatively inexpensive, do not require any special filter holders, and do not clog as readily as membrane filters. The choice of filter pore size is determined by the source of the seawater, its intended use, and the volume needed. Normally, water should be filtered to 0.45 μm with membrane filters or, in special cases, down to 0.2 μm. If glass fiber filters are used, which are much faster and clog less quickly, then a GF/F filter (e.g., Whatman) with a nominal pore size of 0.7 μm is recommended.

Occasionally, dissolved organic matter removal may be required for special projects or because of suspected contamination. Dissolved organics may be removed by adsorption onto activated charcoal. The charcoal is prepared by washing with benzene, methanol, or 50% ethanol and distilled water (Craigie and McLachlan 1964). Dissolved organics in small volumes of seawater may be removed by adding 2 g of powdered, washed charcoal per liter of seawater, stirring for 1 hour and then filtering. Large volumes of seawater are passed through charcoal in a glass column. The charcoal may be washed as described previously, but it must not be allowed to dry before the addition of seawater. There is no "best method" for cleaning charcoal, and often a simple washing with seawater or distilled water may be acceptable (Guillard, personal communication). One should adjust the cleaning to suit the organism and purpose of the experiments planned. Dissolved organics may also be removed by exposing the seawater to high-intensity ultraviolet light (Armstrong et al. 1966). This destroys most dissolved organics and sterilizes the seawater.

Depending on where the seawater is collected, the salinity varies, especially with different seasons. The salinity should be noted for each collection. Offshore seawater salinity normally ranges 32 to 35 psu, whereas inshore water may often be <30 psu. Most algae grow well between 30 and 35 psu, but some species do not tolerate reduced salinities. If a lower salinity is desired, the salinity must be decreased by adding deionized water before any nutrients, trace metals, or vitamins are added, to avoid dilution of these components.

Filtered seawater can be stored in either glass or plastic carboys, often 20 liters for ease of handling. Rectangular containers require minimal storage and can be stacked. New containers should be leached for several days with diluted (i.e., 10%) HCl and then rinsed thoroughly. The seawater should be kept cool (refrigerated if possible) and in the dark (or covered with black plastic).

Filtered seawater is traditionally sterilized by steam autoclaving for 15 minutes at 121°C and 15 lb in$^{-2}$ or longer, depending on the volume. After autoclaving, leave the media for 24 hours to equilibrate, so that gases such as $CO_2$ are allowed to diffuse into the medium. To avoid formation of a precipitate during autoclaving, the following treatments are helpful:

1. Adding 1.44 mL of 1N HCl and 0.12 g of $NaHCO_3$ per liter. These additions indirectly add $CO_2$ and lower the pH, which helps to reduce the formation of a precipitate during autoclaving (Harrison et al. 1980). Carbon dioxide may be added directly by bubbling the medium before autoclaving (Morel, personal communication).
2. Cooling the seawater quickly after autoclaving by standing it in cold tap water in a sink helps to prevent precipitation.
3. Sterilizing by filtration with a 0.22-μm membrane filter.
4. Pasteurizing by heating the seawater to 90°C–95°C for 24 hours. Sometimes this heating is done for a shorter time but repeated twice or three times, with cooling between heating periods (tyndallization).
5. Adding pH buffers such as 4–5 mM Tris or glycylglycine. Note, however, that these buffers are organic compounds and may encourage bacterial growth (Fabregas et al. 1993). In addition, Tris may be toxic to some phytoplankton species (McLachlan 1973, Blankley 1973).
6. Adding high concentrations of EDTA (e.g., $10^{-4}$ M in medium K; Keller et al. 1988) for algae that tolerate it.
7. Sterilizing smaller volumes of seawater with a microwave (Keller et al. 1988).
8. Lowering the concentration of iron.

---

## 3.0. STOCK SOLUTIONS

### 3.1. Macronutrients (Nitrogen, Phosphorus, and Silicon)

Macronutrients are generally considered to be nitrogen, phosphorus, and silicon. However, silicon is required only for diatoms, silicoflagellates, and some chrysophytes. These macronutrients are generally required in a ratio of 16N:16Si:1P (Parsons et al. 1984, Brzezinski 1985), and the ambient ratio in NW is often similar to the ratio required by the algae, except in some estuaries where there are large inputs of nitrogen and phosphorus. Most media do not balance the relative concentrations of macronutrients needed for algal growth. Several popular media (e.g., f/2 medium) have nitrogen : phosphorus ratios >16 : 1, indicating that the phytoplankton would be phosphorus-limited in senescent phase (Berges et al. 2001).

Unfortunately, experimentalists usually pay little attention to the nitrogen : phosphorus or nitrogen : silicon ratios in the medium that they are using, which will ultimately determine which nutrient limits growth and influences the chemical composition and physiological rates when the cells become senescent. Similarly, carbon concentrations and carbon : nitrogen ratios are rarely considered. Many media have a bicarbonate concentration of about 2 mM and nitrogen (nitrate) of about 500 µM or higher, which yields a carbon : nitrogen ratio of about 4 : 1. According to the Redfield ratio, the chemical composition of the average phytoplankter is 106C : 16N : 1P, or 6.7C : 1N. Therefore, most media are nitrogen-rich relative to carbon, and carbon could become limiting, depending on the growth rate of the phytoplankton and the surface area of the medium through which atmospheric $CO_2$ can diffuse (Riebesell et al. 1993). When culture pH rises quickly to 9 or higher, this may be an indication that carbon may be limiting. Species that can readily use bicarbonate may be able to grow, whereas other species that are more dependent on $CO_2$ may exhibit a reduced growth rate or cell yield. Depending on the use of the seawater, one may consider bubbling with $CO_2$ or adding more bicarbonate in late exponential phase to ensure that carbon is not limiting algal growth. This is especially important in physiological experiments where it is essential to know which nutrient is limiting during senescent phase.

To simplify routine media preparation, recipes are usually divided into working stock solutions. Direct combinations of several stock solutions without dilution in water may result in undesirable precipitation. To make a working solution of several stocks, add one stock to a certain volume of water and mix thoroughly before adding the next stock solution. Nitrate and phosphate are normally added as $NaNO_3$ and $NaHPO_4 \cdot H_2O$. In some media, phosphate is added as sodium glycerophosphate to make the trace metal salts less prone to precipitation; however, sodium glycerophosphate may precipitate as a calcium salt at elevated temperatures (Provasoli 1971).

Ammonium may be an alternative nitrogen source and may be added as $NH_4Cl$. At the typical pH of seawater (8.2), there is about 90% $NH_4$ and 10% $NH_3$ (ammonia). Because considerable quantities of ammonia may be lost from the medium through volatilization during autoclaving, ammonium should be added aseptically after autoclaving. As the pH of the culture medium increases during algal growth, the ratio of $NH_4 : NH_3$ increases and reaches 1 : 1 at a pH 9.3. Therefore, substantial amounts of ammonium may be lost from the culture if the algal culture is kept mixed by bubbling with air. Ammonium, at concentrations of 100 to 250 µM, may be inhibitory to some coastal species, but most coastal species tolerate concentrations as high as 1,000 µM (McLachlan 1973). Similarly, some oceanic species show toxicity to ammonium at only 25 µM (Keller et al. 1987), whereas others (e.g., *Prochlorococcus, Bolidomonas*) tolerate concentrations >500 µM. In some special cases, urea is another form of nitrogen to consider, but it decomposes when heated (McLachlan 1973). In experiments in which different forms of nitrogen are compared, it is important to note that 1 µM urea provides 2 µM $N \cdot L^{-1}$, because the urea molecule contains two atoms of nitrogen.

Silicate is added as $Na_2SiO_3 \cdot 9H_2O$. Because silicic acid enhances precipitation, it is useful to omit it from the medium if one is culturing species that do not require silicate (e.g., most flagellates). If concentrated stock solutions (e.g., 100 mM) are prepared in deionized water and acidified to pH 2 (McLachlan 1973), silicate polymerizes. When this stock is added to seawater, it may take several days before the entire concentration of silicon that was added is available for uptake and growth (Suttle et al. 1986). It is recommended to store the 100-mM stock solution of $Na_2SiO_3$ at its pH of dissolution in deionized water (pH = 12.6) at 4°C in the dark (Suttle et al. 1986). When the silicon stock is added to seawater, it should be added slowly with rapid stirring. Autoclaving the $Na_2SiO_3$ stock solution in a glass container may result in etching of the glass and precipitation, and therefore it is recommended to prepare it in a Teflon-coated bottle (Suttle et al. 1986).

To help control contamination with bacteria or fungi, autoclave all nutrient stocks and then use sterile techniques subsequently. Another possibility is to sterilize the stocks through a 0.2-µm filter. Generally, nutrient enrichments should be added aseptically after autoclaving, but they may also be added before autoclaving, except in the case of ammonium.

For concentrations of the various macronutrient stock solutions for the various media, see Appendix A.

## 3.2. Trace Metals

Trace metals and vitamins are usually prepared as "primary" stocks of high concentrations to permit weighing of reasonable amounts. These are used to make "working" solutions from which the final medium is made (see Appendix A for examples). Because some primary or working solutions are kept for very long periods, evaporation through ground glass stoppers, screw caps, or plastic bottles may be significant. Mark the liquid level on the bottle and keep it cold or wrapped in laboratory film.

Typical trace metal stock solutions may consist of chloride or sulphate salts of zinc, cobalt, manganese, selenium, and nickel, and they are kept in a solution containing the chelator EDTA. Iron is usually kept as a separate solution, and it should be chelated or kept in $10^{-2}$ M HCl to avoid precipitation. It may be added as ferric chloride, ferrous sulphate, or ferrous ammonium sulphate, but the latter compound contains ammonium, and this may be a problem if one is conducting nitrogen uptake studies. There is sufficient boron in NW and therefore it is not necessary to add it, but boron should be added to AW. The stock solutions for various recipes are given in Appendix A. For a more detailed discussion of trace metals, see Chapter 4.

## 3.3. Vitamins

Usually three vitamins—vitamin $B_{12}$ (cyanocobalamin), thiamine, and biotin—are added, but very few algae need all three vitamins (Provasoli and Carlucci 1974). The general order of vitamin requirements for algae is vitamin $B_{12}$ > thiamine > biotin. If large-scale culturing of a single species is the goal, check what vitamins this species requires. It may be possible to omit the addition of two of the three vitamins, because most species require only one or two. Vitamins are normally added aseptically (through a 0.2-μm filter) after the medium has been autoclaved. Vitamins maintain maximum potency if they are filter sterilized (0.2-μm filtration) rather than autoclaved. Autoclaving may cause decomposition of some vitamins, but it is thought that some algae may be able to use some of these decomposition products (Provasoli and Carlucci 1974). Vitamin stocks may be frozen for long periods without noticeable degradation, and the stocks may be refrozen after each use. These three vitamin stocks may be combined into a single working solution for a 1,000-fold dilution.

## 3.4. pH Buffers

Two common pH buffers are used to prevent or reduce precipitation: Tris (2-amino-2-[hydroxymethyl]-1-3-propanediol) and glycylglycine (McLachlan 1973). Mix Tris base and Tris : HCl as per Sigma instructions and make a stock to be added as desired; for example, 1 mg $L^{-1}$ for a Tris concentration of $10^{-3}$M. Adjust the pH with concentrated HCl to obtain the desired pH of the medium. Glycylglycine is readily soluble in water, and the powder can be added directly to seawater. It is slightly acidic, and it may be necessary to make a small adjustment to the pH with several drops of 1N NaOH. Tris may be toxic to some species (McLachlan 1973; use no more than 1–5 mM), whereas glycylglycine is nontoxic. Apparently, neither Tris nor glycylglycine can serve as a nitrogen source for algal growth, but they can serve as a carbon source for bacteria (Fabregas et al. 1993). They may also interfere with the analysis of dissolved organic nitrogen and ammonium.

HEPES (N-[2-hydroxyethyl] piperazine-N′-[2-ethanesulphonic acid]) and MOPS (3-N-morpholino propane sulfonic acid) are used extensively in freshwater media (McFadden and Melkonian 1986), but they are not commonly used in marine media. Loeblich (1975) compared the growth of a marine dinoflagellate in several buffers (including MOPS, HEPES, Tris, glycylglycine, and TAPS N-Tris [hydroxymethyl] methyl-3-aminopropanesulfonic acid), and concluded that Tris and TAPS provided maximal growth with minimal pH change.

## 3.5. Chelators

Chelate : metal ratios of 1.5 : 1 to 3 : 1 are commonly used. EDTA is the most common chelator and is usually purchased as the disodium salt ($Na_2EDTA.2H_2O$) that is readily soluble in water. However, EDTA has been noted to inhibit the growth of some oceanic species (Muggli and Harrison 1996). For routine use in culture media, nitrilotriacetic acid (NTA) and citric acid are less effective than EDTA, but they are used sometimes in experimental work (Brand et al. 1986). Chelators are discussed extensively in Chapter 4.

## 3.6. Soil Extracts

In the simplest media, only nitrogen and phosphorus and soil extract are added to NW. Erdschreiber is an example of such a medium, and it has been used successfully to grow various planktonic and benthic species (McLachlan 1973). Some culture collections still use soil extract to maintain some species (see Plymouth Erdscheiber in Appendix A), but it is seldom used in physiological experiments when a defined medium is preferred.

## 3.7. Germanium Dioxide

In special cases, germanium dioxide may be added to prevent the growth of diatoms (Lewin 1966), but other algae may also be affected. For example, the addition of 100 mg $L^{-1}$ of $GeO_2$ prevents diatom growth in macro-algal cultures (Markham and Hagmeier 1982), but McLachlan et al. (1971) found that $GeO_2$ inhibited brown algal growth. Thomas et al. (1978) added germanium dioxide to natural phytoplankton samples in an attempt to separate flagellate productivity from diatom productivity and noted that dinoflagellate photosynthesis was inhibited as well. Therefore, it seems unwise to add germanium dioxide during physiological experiments.

## 4.0. GENERAL METHODS OF MEDIA PREPARATION

The various aspects and precautions involved in media preparation have already been covered in the sections on materials required (Section 2) and stock solutions (Section 3).

## 4.1. Natural Seawater

The general steps are as follows:

1. Obtain good–quality NW with salinity >30, if possible.
2. Filter as soon as possible, and store the seawater in the dark and cold (4°C).
3. Sterilize as needed by filtration (0.2 μm), autoclaving, ultraviolet light treatment, or pasteurization.
4. Add macronutrients, Fe-EDTA, trace metals, and vitamins aseptically after autoclaving. Mix thoroughly after the addition of each stock.

## 4.2. Artificial Seawater

Artificial or synthetic seawater consists of two parts: the basal (main) salts that form the "basal seawater" and the enrichment solution (often the same as the enrichment solution that is added to NW). In the early AW recipes, calcium and magnesium salts were reduced to avoid precipitation. Because precipitation can now be avoided (Harrison et al. 1980), the more recent recipes ensure that the ratio of the major ions is identical to NW. For example, in ESAW (see Appendix A), the basic salts (usually 10 or 11) are divided into the anhydrous sodium or potassium salts (i.e., NaCl, $Na_2SO_4$, KCl, $NaHCO_3$, KBr, NaF) and the hydrated chloride or sulphate salts (i.e., $MgCl_2 \cdot 6H_2O$, $CaCl_2 \cdot H_2O$, and $SrCl_2 \cdot 6H_2O$). The anhydrous salts must be dissolved separately from the hydrated salts, and then the two solutions can be mixed together and the enrichment solution components (macronutrient, trace metals, and vitamins) can be added after the salt solution is autoclaved.

Studies of the physiology of marine phytoplankton have been greatly facilitated by the ability to culture these species in a defined medium (i.e., AW). One of the great advantages of AW is that the composition of the seawater is relatively constant over several decades, unless the impurities in the major salts (e.g., NaCl, $Na_2SO_4$, and KCl) change either by a change in brands or a change in the manufacturing process. In 1985, we suddenly were unable to grow the common diatom, *Thalassiosira pseudonana* (Hust.) Hasle et Heimdal, despite the fact that we were using the same ESAW recipe that we used for the previous 7 years. We later discovered that selenium was previously an impurity in sufficient quantity in our basal salts, and therefore it had not been necessary to add it (Price et al. 1987, Harrison et al. 1988). We assume that a change in the process of producing one of the basal salts reduced selenium contamination. When we added selenium to the new enrichment solution, *T. pseudonana* grew well. Therefore, selenium is routinely added to our medium now (Berges et al. 2001).

Another advantage of AW for nutrient limitation studies is that one can control the amount of the limiting nutrient because there is little or none in the AW, unlike NW. Similarly, one can precisely control the nutrient ratios. However, cleanly collected oligotrophic ocean water may also be used, because nutrients and trace metals are also very low.

Commercial preparations are synthetic mixes and can be purchased in various amounts. One of the best is Instant Ocean, produced by Aquarium Systems, Inc. (King and Spotte 1974). The ion ratio of the major salts is very close to NW. Macronutrients, trace metals, and vitamins must be added. If large-scale preparation is required, this may be an economical alternative to buying individual salts and weighing and mixing the salts.

For all the AW basal salt recipes, if contamination is a concern, then we recommend that you analyze the basal medium for copper, zinc, cobalt, lead, manganese, nickel, chromium, etc. For example, we measured 2.5, 4.0, and 0.13 µM lead, copper, and cadmium, respectively, in ESAW, which uses reagent grade salts. This concentration of copper is sufficient for growth, depending on the concentration of the other nutrients, trace metals, and chelation. Brand et al. (1986) found that 4.0 µM copper is toxic in the absence of a chelator for some species.

## 5.0. MEDIA

Comparisons between media are complicated by many factors. McLachlan (1973) observed: "Numerous enriched and synthetic media have been formulated, which together with generally trivial modifications, almost equal the number of investigators." This has certainly remained true. Many modifications result from a desire to increase the flexibility of a medium (i.e., creating multiple nutrient stocks so that individual macronutrient and micronutrient concentrations may be manipulated) or to reduce the number of stocks necessary in cases where various different media are being used in a single laboratory. (In Appendix A note the different recipes that incorporate f/2 trace metals stock.) In many cases, minor modifications to the original recipes have been made (e.g., changing a nitrogen source or adding a single trace metal) and an entirely new name has been given to the medium with this minor modification. In other cases, fairly extensive modifications have been made, yet a medium name has been retained or simply designated as "modified." In still other cases, subsequent modifications have been made in addition to a name change to honor the originator (e.g., "Grund" versus "von Stosch"; von Stosch 1963, Guiry and Cunningham 1984). More commonly used media such as f/2 and ESAW/ESNW have not only evolved over time (Berges et al. 2001), but they can change substantially in different publications (e.g., compare the recipe for f medium given in Guillard and Ryther [1962] with that for f/2 medium in Guillard [1975] and with those given on the Web sites of major

culture collections). When compiling the recipes in Appendix A, we used both original recipes and those provided by major culture collection Web sites. The reader is cautioned that the Web site recipes are subject to change; indeed, Berges et al. (2001) recommended that Web-based recipes may have real advantages for phycological research.

In both AW and NW recipes, there is some variation in salinity. McLachlan (1973) considered that salinity was not an "inherent feature" of media recipes. Although salinities of 35 are often considered normal, most artificial media tend to produce somewhat lower salinities (27 to 30). Many coastal phytoplankton species can be cultured at a much lower salinity (cf. McLachlan 1973).

Sometimes it is necessary to grow axenic cultures. Details for preparing test media for bacteria and fungal contamination are provided in Chapter 8.

### 5.1. Artificial Seawater Media

Relatively few artificial water media are commonly used. Most are based on the ASM and ASP formulations of Provasoli and McLachlan, with minor variations (the AWs described by Goldman and McCarthy [1978] and Keller et al. [1987] both fall into this category). There are numerous species-specific artificial media that have been developed; for example, YBC-II, designed for nitrogen-fixing strains of *Trichodesmium* (Chen et al. 1996), was derived from an earlier medium by Ohki et al. (1992). As noted previously, the ion ratios of these media have diverged from that of NW (Table 3.1), particularly with respect to sulfate and magnesium (Kester et al. 1967). ESAW was formulated on the basis of Kester et al. (1967) and is thus quite similar to NW.

**TABLE 3.1** Comparison of the ratio (normalized to K) of selected major ions in different artificial seawater recipes and natural seawater.

| Recipe | Na$^+$ | Cl$^-$ | Mg$^{2+}$ | SO$_4^{2-}$ | K$^+$ |
|--------|------|-----|-----|------|-----|
| Aquil | 60 | 51 | 7.0 | 3.1 | 1.0 |
| ASNIII | 64 | 69 | 3.6 | 2.0 | 1.0 |
| ASP-M | 40 | 47 | 4.0 | 2.0 | 1.0 |
| ESAW | 52 | 59 | 5.1 | 3.1 | 1.0 |
| YBC-II | 42 | 45 | 5.0 | 2.5 | 1.0 |
| NW | 47 | 55 | 5.3 | 2.8 | 1.0 |

Several commercially available AWs are also available. Instant Ocean is one of the oldest and best-evaluated (discussed previously), but Tropic Marin Sea Salt (Tropic Marin) is also well-regarded by European aquarists, and Aus Aqua Pty Ltd (www.algaboost.com) sells various AW formulations, including ESAW, and enrichment solutions such as f/2. In published research, it appears that relatively few phycologists use ready-made salts, but at least one recipe available on the Culture Collection of Algae and Protozoa (CCAP) Web site uses a commercial sea salt preparation (Ultramarine Synthetica, Waterlife Research Industries Ltd.). More recently, Sigma-Aldrich has begun producing a dry sea salt mixture (S9883). When using these preparations, it is important to note that the dry mixture may not be homogenous; subsamples from a large package may vary considerably, especially if the salts have hydrated during storage.

## 5.2. Natural Seawater Media

Among the major NW enrichment media recommended for a broad spectrum of algae, f/2 medium and variations, different versions of Erdschreiber medium (e.g., Plymouth Erdschreiber medium), and ESNW appear to dominate citations (Berges et al. 2001, see Appendix A). The K-medium family has been specifically developed for oceanic species (Keller et al. 1987), and two new oceanic media, MNK and Pro99 media, offer promising improvements. More species-specific media that are frequently used include the L-medium family (Guillard and Hargraves 1993; Guillard 1995), GPM and variants (Sweeney et al. 1959; modified by Loeblich 1975 and Blackburn et al. 1989), and numerous media for oceanic cyanobacteria, including SN (Waterbury et al. 1986), PC (Keller in Andersen et al. 1997), PRO99 (Chisholm, unpublished), and ASNIII (Rippka et al. 1979); BG medium (Rippka et al. 1979) and variations are also recommended for marine cyanobacteria but are less commonly used with oceanic species. For macroalgal cultures, enrichments based on the "Grund" medium of von Stosch (1963) are most commonly used. Provasoli's ES medium and its modifications are also used (McLachlin 1973, West and McBride 1999).

It is difficult to provide a comprehensive set of recommendations as to which media are best for certain species. A good starting point is the media in which the species are being maintained in the culture collections from which they are obtained (see culture collection Web sites). A variety of detailed recipes are now available on culture collection Web sites, but the reader is cautioned that there is considerable variation in recipe details and even in the names used to describe recipes between sites. However, some generalizations about media recipes can be made.

In terms of vitamins there is remarkable consistency in media: thiamine, biotin, and $B_{12}$ all are normally added, at quite comparable concentrations, although they may not be required (see Appendix A). Some media specify that autoclaving should be avoided, because it can lead to decomposition of vitamins, but species that require vitamins are usually able to grow well on the decomposition products (see McLachlan 1973).

Trace metals stock solutions are more variable, but it appears that most recipes have basic similarities (see Appendix A). Elements such as iron, zinc, manganese, cobalt, copper, and molybdenum are almost always included. Less commonly found, but critical to certain species, is selenium (Harrison et al. 1988). Nickel is required by most algae if urea is the nitrogen source, because nickel is a component of the enzyme unease (Syrett 1981). For oceanic species, a widely variable set of metals has been added, including the potentially toxic vanadium and chromium. There are also numerous metals found in some of Provasoli's early media, such as aluminum, ruthenium, lithium, and iodine (Provasoli et al. 1957), that are rarely found in media currently used. As noted earlier, the trace metal mixture is more critical when AW is used, particularly if water of very high purity (i.e., >18.2 MΩ water, such as that provided by MilliQ systems) is used.

Concentrations and forms of macronutrients (e.g., f, f/2, and f/50) in media commonly vary. Typically, macronutrients are in great excess in comparison with natural concentrations, particularly in the case of media intended for aquaculture. For example, Walne's medium (Appendix A) contains >1 mM nitrate, or about 40 times the maximum levels found in coastal waters. As noted above, the ratios of nutrients relative to algal requirements often appear to make little sense (Berges et al. 2001). Lower nutrients have been found to support growth of oceanic species, but often it is only the major nutrients that have been reduced in the recipe. McLachlan (1973) noted that ammonium can become toxic to some species of phytoplankton at concentrations of 100 µM or greater, and Berges et al. (2001) speculated that this may be the reason that dilutions of media in which ammonium is included may improve growth of some species.

It is also worth noting that there are commercially available premade enrichment stocks. Some major culture collections sell nutrient stock solutions for various culture media as well as premixed culture media and seawater (e.g., CCAP, at www.ife.ac.uk/ccap, and CCMP, at http://ccmp.bigelow.org). Sigma-Aldrich sells f/2 medium (dry salt mixture G1775 or liquid G9903), and AusAqua Pty. Ltd. sells f/2 medium under the name AlgaBoost and ES enrichment stocks.

## 6.0. COMPARING RECIPES BASED ON ALGAL RESPONSE

Growing phytoplankton remains an art as well as a science, and in our experience, most investigators tend to settle on media that "work" for the species they are growing, rather than engage in wholesale comparisons to determine which media are the best. McLachlan (1973) observed that media preferences were "rarely supported by comprehensive, qualitative comparisons."

Any media comparison should be based on the use of the medium by the investigator. There are three broad categories of general use: culture maintenance, algal biomass yield, and physiological (growth rate) experiments. It is strongly advised that the growth rate in NW and the medium be compared, because growth rate is a general index of algal health in the medium. Some species grow better if there is a large surface area/volume ratio to allow the diffusion of gases into the medium.

In principle, the simplest comparison to make is whether the algae grow. Consult the culture collection Web sites for the most recent information on recipes and also to find the most appropriate medium for a particular strain. In practice, this is considerably complicated by the culture methods. For example, some investigators may maintain their culture collections quite conscientiously, transferring cultures weekly or more often. In this case, species that grow quickly (i.e., a high exponential growth rate, $\mu$) could be favored. Under these conditions, a medium that was relatively poorly buffered or contained relatively low levels of macronutrients might be regarded as superior. On the other hand, cultures are often transferred far less frequently, and consequently they can persist in stationary phase for especially long periods if they are kept in dim light. In these cases, a medium that supports a higher growth rate might not be regarded as superior, but media that are more strongly buffered or have nutrients in excess may perform much better. Berges et al. (2001) compared several variations on ESAW with ESNW and

found that assessment based on growth rate and final biomass did not always agree. For example, *Phaeocystis pouchetti* (Hariot) Lagerh. and *Karlodinium micrum* (Leadbeater et Dodge) Larsen (*Gymnodinium galatheanum* [Lohm.] Kofoid & Swezy) both grew at equal rates ($\mu$) in ESAW and ESNW, but whereas *P. pouchetti* achieved a threefold higher biomass in ESAW versus ESNW, *K. micrum* cultures were almost twice as dense in stationary phase in ESNW as in ESAW.

Growth rates are relatively straightforward to measure by performing cell counts, but this is quite labor-intensive, even when using electronic particle counters (see Chapters 16–18). Measurements of fluorescence are much quicker and agree very well with cell numbers in most species, so long as cultures remain in exponential growth phase and are measured at the same time of day, if grown on a light : dark cycle. If culturing is done in 25 × 150–mm screw-capped culture tubes, then repeated measurements of cultures can easily be made with a Turner Designs fluorometer. For further information on measuring growth rates, see Chapter 18.

Determining the biomass of stationary phase cultures requires cell counts, and it is probably wise to measure cell volumes as well; in this case, fluorescence is unsuitable (see Berges et al. 2001). Although such comparisons are rarely done quantitatively, assessing media on the basis of the length of time a culture can remain at a stationary biomass is a criterion that is probably quite relevant to culture collections.

Another criterion for media evaluation is whether the original morphology of the cells is maintained, but this is seldom evaluated (Harrison et al. 1980) and can be impractical for small species. Many recipes that include soil water extract justify the addition on the basis that it maintains the original cell morphology during long-term culturing.

## 7.0. REFERENCES

Allen, E. J. 1914. On the culture of the plankton diatom *Thalassiosira gravida* Cleve, in artificial sea-water. *J. Mar. Biol. Assoc. U.K.* 10:417–39.

Allen, E. J., and Nelson, E. W. 1910. On the artificial culture of marine plankton organisms. *J. Mar. Biol. Assoc. U.K.* 8: 421–74.

Armstrong, F. A. J., Williams, P. M., and Strickland, J. D. H. 1966. Photo-oxidation of organic matter in seawater by ultraviolet radiation, analytical and other applications. *Nature* 211:481–3.

Berges, J. A., Franklin, D. J., and Harrison, P. J. 2001. Evolution of an artificial seawater medium: Improvements in enriched seawater, artificial water over the last two decades. *J. Phycol.* 37:1138–45.

Bernhard, M., and Zattera, A. 1970. The importance of avoiding chemical contamination for a successful cultivation of marine organisms. *Helgolander wiss Meeresunters* 20: 655–75.

Bernhard, M. 1977. Chemical contamination of culture media: Assessment, avoidance and control. In: Kinne, O., ed. *Marine Ecology. Vol III. Cultivation, Part 3.* Wiley & Sons, Chichester, UK, pp. 1459–99.

Blackburn, S., Hallegraeff, G., and Bolch, C. J. S. 1989. Vegetative reproduction and sexual life cycle of the toxic dinoflagellate *Gymnodinium catenatum* from Tasmania, Australia. *J. Phycol.* 25:577–90.

Blankley, W. 1973. Toxic and inhibitory materials associated with culturing. In: Stein, J., ed. *Handbook of Phycological Methods: Culture Methods and Growth Measurements.* Cambridge University Press, Cambridge, UK, pp. 207–29.

Brand, L. E., Guillard, R. R. L., and Murphy, L. S. 1981. A method for a rapid and precise determination of acclimated phytoplankton reproduction rates. *J. Plankton Res.* 3:93–101.

Brand, L. E., Sunda, W. G., and Guillard, R. R. L. 1986. Reduction of marine phytoplankton reproduction rates by copper and cadmium. *J. Exp. Mar. Biol. Ecol.* 96:225–50.

Brzezinski, M. A. 1985. The Si-C-N ratio of marine diatoms: Interspecific variability and the effect of some environmental variables. *J. Phycol.* 21:347–57.

Chen, Y.-B., Zehr, J. P., and Mellon, M. 1996. Growth and nitrogen fixation of the diazotrophic filamentous nonheterocystous cyanobacterium *Trichodesmium* sp. IMS 101 in defined media: Evidence for a circadian rhythm. *J. Phycol.* 32:916–23.

Chu, S. P. 1946. Note on the technique of making bacteria-free cultures of marine diatoms. *J. Mar. Biol. Assoc. U.K.* 26:296–302.

Craigie, J. S., and McLachlan, J. 1964. Excretion of colored ultraviolet-absorbing substances by marine algae. *Can. J. Bot.* 42:23–3.

Droop, M. R. 1969. Algae. In: Norris, J. R., and Ribbon, D. W., eds. *Methods in Microbiology* (Vol. 3B). Academic Press, New York, pp. 1–324.

Fabregas, J., Vazquez, V., Cabezas, B., and Otero, A. 1993. TRIS not only controls the pH in microalgal cultures, but also feeds bacteria. *J. Appl. Phycol.* 5:543–5.

Foyn, B. 1934. Lebenszyklus, cytology und sexualitat der Chlorophyceae *Cladophora suhriana* Kützing. *Arch. Protistenk* 83:1–56.

Goldman, J. C., and McCarthy, J. J. 1978. Steady state growth and ammonium uptake of a fast growing marine diatom. *Limnol. Oceanogr.* 23:695–703.

Guillard, R. R. L., and Ryther, J. H. 1962. Studies of marine planktonic diatoms. I. *Cyclotella nana* Hustedt and *Detonula confervacea* (Cleve) Gran. *Can. J. Microbiol.* 8:229–39.

Guillard, R. R. L. 1975. Culture of phytoplankton for feeding marine invertebrates. In: Smith, W. L., and Chanley M. H., eds. *Culture of Marine Invertebrate Animals.* Plenum Press, New York, pp. 26–60.

Guillard, R. R. L. 1995. Culture methods. In: Hallegraeff, G. M., Anderson, D. M., and Cembella, A. D., eds. *Manual on Harmful Marine Microalgae. IOC Manual and Guides No. 33.* UNESCO, Paris, pp. 45–62.

Guillard, R. R. L., and Hargraves, P. E. 1993. *Stichochrysis immobilis* is a diatom, not a chrysophyte. *Phycologia* 32:234–36.

Guiry, M., and Cunningham, E. 1984. Photoperiodic and temperature responses in the reproduction of the northeastern Atlantic *Gigartina acicularis* (Rhodophyta: Gigartinales). *Phycologia* 23:357–67.

Harrison, P. J., Waters, R. E., and Taylor, F. J. R. 1980. A broad spectrum artificial seawater medium for coastal and open ocean phytoplankon. *J. Phycol.* 16:28–35.

Harrison, P. J., Yu, P. W., Thompson, P. A., Price, N. M., and Phillips, D. J. 1988. Survey of selenium requirements in marine phytoplankton. *Mar. Ecol. Prog. Ser.* 47:89–96.

Keller, M. D., Bellows, W. K., and Guillard, R. R. L. 1988. Microwave treatment for sterilization of phytoplankton culture media. *J. Exp. Mar. Biol. Ecol.* 117:279–83.

Keller, M. D., Selvin, R. C., Claus, W., and Guillard, R. R. L. 1987. Media for the culture of oceanic ultraphytoplankton. *J. Phycol.* 23:633–8.

Kester, D., Duedall, I., Connors, D., and Pytkowicz, R. 1967. Preparation of artificial seawater. *Limnol. Oceanogr.* 12:176–9.

King, J. M., and Spotte, S. H. 1974. *Marine Aquariums in the Research Laboratory.* Aquarium Systems, Inc., Eastlake, Ohio.

Kinne, O. 1976. Cultivation of marine organisms. In: Kinne, O., ed. *Marine Ecology, Vol III. Cultivation.* John Wiley & Sons, Toronto, pp. 29–37.

Levring, T. 1946. Some culture experiments with *Ulva lactuca.* *Kgl. Fysiogr. Sallsk. Hdl.* 16:45–56.

Lewin, J. 1966. Silicon metabolism in diatoms. V. Germanium dioxide, a specific inhibitor of diatom growth. *Phycologia* 6:1–12.

Loeblich, A. 1975. A seawater medium for dinoflagellates and the nutrition of *Cachonina niei.* *J. Phycol.* 11:80–86.

Lyman, J., and Fleming, R. H. 1940. Composition of seawater. *J. Mar. Res.* 3:134–46.

Markham, J. W., and Hagmeier. 1982. Observations on the effects of germanium dioxide on the growth of macro-algae and diatoms. *Phycologia* 21:125–30.

McFadden, G. I., and Melkonian, M. 1986. Use of HEPES buffer for microalgal culture media and fixation for electron microscopy. *Phycologia* 25:551–7.

McLachlan, J. 1959. The growth of unicellular algae in artificial and enriched seawater media. *Can. J. Microbiol.* 5:9–15.

McLachlan, J. 1964. Some considerations of the growth of marine algae in artificial media. *Can. J. Microbiol.* 10:769–82.

McLachlan, J. 1973. Growth media—marine. In: Stein, J., ed. *Handbook of Phycological Methods: Culture Methods and Growth Measurements.* Cambridge University Press, Cambridge, UK, pp. 25–51.

McLachlan, J., Chen, L. C.-M., and Edelstein, T. 1971. The culture of four species of *Fucus* under laboratory conditions. *Can. J. Bot.* 49:1463–69.

Morel, F. M. M., Rueter, J. G., Anderson, D. M., and Guillard, R. R. L. 1979. Aquil: A chemically defined phytoplankton culture medium for trace metal studies. *J. Phycol.* 15:135–41.

Muggli, D. L., and Harrison, P. J. 1996. EDTA suppresses the growth of oceanic phytoplankton from the northeast subarctic Pacific. *J. Exp. Mar. Biol. Ecol.* 205: 221–7.

Ohki, K., Zehr, J. P., and Fujita, Y. 1992. *Trichodesmium*: establishment of culture and characteristics of N-fixation. In: Carpenter, E. J., Capone, D. G., and Rueter, J. G., eds. *Marine Pelagic Cyanobacteria: Trichodesmium and other diazotrophs.* Kluwer Academic Publishers, Dordrecht, Germany, pp. 307–18.

Parsons, T. R., Takahashi, M., and Hargrave, B. 1984. *Biological Oceanographic Processes.* 3rd ed. Pergamon Press, New York, 342 pp.

Price, N. M., Harrison, G. I., Hering, J. G., Hudson, R. J., Nirel, P. M. V., Palenik, B., and Morel, F. M. M. 1988/89. Preparation and chemistry of the artificial culture medium Aquil. *Biol. Oceanogr.* 6:443–61.

Price, N. M., Harrison, P., Landry, M. R., Azam, F., and Hall, K. 1986. Toxic effects of latex and Tygon tubing on marine phytoplankton, zooplankton and bacteria. *Mar. Ecol. Prog. Ser.* 34:41–9.

Price, N. M., Thompson, P. A., and Harrison, P. J. 1987. Selenium: An essential element for growth of the coastal marine diatom *Thalassiosira pseudonana* (Bacillariophyceae). *J. Phycol.* 23:1–9.

Pringsheim, E. C. 1912. Die kultur von algen in agar. *Beitr. Biol. Pflanz.* 11:305–33.

Pringsheim, E. G. 1946. *Pure Cultures of Algae.* Cambridge Univ. Press, London, 119 pp.

Provasoli, L. 1958. Nutrition and ecology of protozoa and algae. *Annu. Rev. Microbiol.* 12:279–308.

Provasoli, L. 1968. Media and prospects for the cultivation of marine algae. In: Watanabe, H., and Hattori, A., eds. *Culture and Collection of Algae. Proceedings, U. S.–Japan Conference.* Japanese Society of Plant Physiology, Hakone, Japan, pp. 63–75.

Provasoli, L. 1971. Media and prospects for the cultivation of algae. In: Rosowski, J. R., and Parker, B. C., eds. *Selected Papers in Phycology.* Department of Botany, University of Nebraska, Lincoln, Nebraska, 876 pp.

Provasoli, L., and Carlucci, A. F. 1974. Vitamins and growth regulators. In: Stewart, W. D. P., ed. *Algal Physiology and Biochemistry.* Blackwell Scientific, UK, pp. 741–87.

Provasoli, L., McLaughlin, J. J. A., and Droop, M. R. 1957. The development of artificial media for marine algae. *Arch. Mikrobiol.* 25:392–428.

Riebesell, U., Wolfgladrow, D. A., and Smetacek, V. 1993. Carbon dioxide limitation of marine phytoplankton growth rates. *Nature* 361:249–51.

Rippka, R., Derulles, J., Waterbury, J. B., Herdman, M., and Stainer, R. Y. 1979. Generic assignments, strain histories and properties of pure cultures of cyanobacteria. *J. Gen. Microbiol.* 111:1–61.

Schreiber, E. 1927. Die Reinkultur von marinen Phytoplankton und deren Bedeutung für die Erforschung der Produktionsfähigkeit des Meerwassers. *Wiss. Meeresuntersuch., N.F.* 10:1–34.

Suttle, C., Price, N., Harrison, P., and Thompson, P. 1986. Polymerization of silica in acidic solutions: A note of caution to phycologists. *J. Phycol.* 22:234–7.

Sweeney, B., Haxo, F., and Hastings, J. 1959. Action spectra for two effects of light on luminescence in *Gonyaulax polyedra. J. Gen. Physiol.* 43:285–99.

Syrett, P.J 1981. Nitrogen metabolism of microalgae. *Can. Bull Fish. Aquatic Sci.* 210:182–210.

Thomas, W. H., Dodson, A. W., and Reid, F. M. H. 1978. Diatom productivity compared to other algae in natural marine assemblages. *J. Phycol.* 14:250–3.

Turner, M. F. 1979. Nutrition of some marine microalgae with special reference to vitamin requirements and utilization of nitrogen and carbon sources. *J. Mar. Biol. Assoc. U.K.* 59:535–52.

von Stosch, H. 1963. Wirkungen von jod un arsenit auf meeresalgen in kultur. *Proc. Int. Seaweed Symp.* 4: 142–50.

Waterbury, J. B., Watson, S. W., Valois, F. W., and Franks, D. G. 1986. Biological and ecological characterization of the marine unicellular cyanobacterium *Synechococcus.* In: Platt, T., and Li, W. K. W., eds. *Photosynthetic Picoplankton. Can. Bull. Fish. Aquatic Sci.,* 214. Department of Fisheries and Oceans, Ottawa, Canada, pp. 71–120.

West, J. A., and McBride, D. L. 1999. Long-term and diurnal carpospore discharge patterns in the *Ceramiaceae, Rhodomelaceae* and *Delesseriaceae* (Rhodophyta). *Hydrobiologia* 298/299:101–13.

# TRACE METAL
# ION BUFFERS AND
# THEIR USE IN
# CULTURE STUDIES

William G. Sunda
*National Ocean Service, National Oceanic and Atmospheric Association*

Neil M. Price
*Department of Biology, McGill University*

Francois M. M. Morel
*Department of Geosciences, Princeton University*

---

## CONTENTS

1.0. Introduction
1.1. Trace Elements as Limiting Nutrients
1.2. Trace Metal Toxicity
1.3. Trace Metal Availability to Phytoplankton
2.0. Chemically Defined Trace Metal Ion Buffer Systems
2.1. Theory of Metal Ion Buffer Systems
2.2. Computation of Concentrations of Aquated Free Metal Ions and Inorganic Species
2.3. Regulation of Dissolved Inorganic Iron Species in Fe–EDTA Buffer Systems in Seawater
3.0. Effects of Cellular Processes on Medium Chemistry
3.1. Changes in pH
3.2. Trace Metal Uptake
3.3. Release of Chelators
3.4. Redox Reactions
4.0. Preparation of Culture Media
4.1. Aquil* and Other Seawater Culture Media
    4.1.1. Chelex Preparation
    4.1.2. Medium Preparation and Composition
    4.1.3. Media Prepared from Natural Seawater
    4.1.4. Manipulation and Sterilization
4.2. Freshwater Media
5.0. Culture Experiments in Metal Ion Buffer Systems
5.1. Growth Rate Measurement
5.2. Measurement of Cell Quotas
5.3. Steady State Metal Uptake Rates
6.0. Acknowledgments
7.0. References

Key Index Words: Trace Metals, Metal Buffers, Chelators, Phytoplankton, Metal Nutrition, Algal Cultures, Culture Media, Culture Experiments

---

## 1.0. INTRODUCTION

Phytoplankton growth in culture depends not only on an adequate supply of essential macronutrient elements (carbon, nitrogen, phosphorus, silicon) and major ions ($Na^+$, $K^+$, $Mg^{2+}$, $Ca^{2+}$, $Cl^-$, and $SO_4^{2-}$) but also on a number of micronutrient metals (iron, manganese, zinc, cobalt, copper, and molybdenum) and the metalloid selenium. The macronutrients and major ions are generally highly soluble and nontoxic (with the notable exception of ammonium) and either are present at high concentrations, as in the case of $HCO_3^-$, $Na^+$, $Mg^{2+}$, $Ca^{2+}$, $K^+$, $Cl^-$, and $SO_4^{2-}$ in seawater and most freshwater media, or can be added at sufficiently high levels (e.g., for nitrate, phosphate, and silicic acid) to support vigorous algal

*Algal Culturing Techniques*

35

Copyright © 2005 by Academic Press
All rights of reproduction in any form reserved.

growth. In contrast, many essential trace metals, such as copper, zinc, and cobalt, are toxic at high concentrations, and one, iron, forms insoluble hydrous ferric oxide precipitates that are largely unavailable to aquatic algae (Rich and Morel 1990). In addition, these ferric precipitates adsorb other essential metals and lower their availability. Because of these difficulties, providing an adequate, nontoxic supply of essential trace metals to algae in batch cultures of phytoplankton historically has presented a considerable challenge to algal culturists. Initially, this problem was solved through the addition of aqueous extracts of soil, which provided a suite of micronutrient elements along with complex mixtures of high-molecular-mass organic acids, commonly referred to as humic compounds. The humic compounds chelated iron, greatly increasing its solubility, and formed organic complexes (chelates) with reactive micronutrient metals such as copper, thereby reducing their toxicity. However, the unknown and varied composition of various soil extract preparations often led to poor reproducibility in the growth rates and yields of algal cultures and prevented any serious study of the micronutrient growth requirements of aquatic algae.

A major advance in algal culturing came in the 1950s with the introduction of synthetic chelators, notably ethylenediaminetetraacetic acid (EDTA), which was first introduced as a replacement for soil extract in freshwater culture media by Hutner et al. (1950). EDTA was subsequently introduced as a metal buffering agent in seawater media (Provasoli et al. 1957, Guillard and Ryther 1962) and is now widely used in most freshwater and marine culture media (see Chapters 2 and 3). Like humic compounds, EDTA complexes trace metals, as in the reaction

$$Cu^{2+} + EDTA \leftrightarrow CuEDTA \qquad (1)$$

in which EDTA reacts with cupric ions to form the copper chelate CuEDTA. The resulting metal chelates are neither susceptible to formation of insoluble precipitates (as in the case of iron precipitation as hydrous oxides) nor directly available for cellular uptake by phytoplankton (Anderson et al. 1978; Sunda and Huntsman 1984). The nonchelated forms of the metal, including free ions (e.g., $Zn^{2+}$) and inorganic metal complexes (e.g., $ZnCl^+$) are available to phytoplankton. As these are removed by algal uptake in an exponentially growing batch culture, they are readily replaced by dissociation of an equivalent concentration of the metal chelate. The combination of EDTA and lower concentrations of cationic trace metal nutrients thus acts as a metal ion buffer and regulates the availability of metal ions in culture, much as a pH buffer regulates the avail-

ability of hydrogen ions. The use of such metal ion buffer systems has allowed the development of culture media that support reproducible growth of freshwater and marine algal species and has provided precise regulation of trace metal availability in studies of trace metal uptake, limitation, and toxicity (Sunda 1988/89; Price et al. 1988/89).

This chapter is primarily intended to help design experiments for such studies of trace metal–phytoplankton interactions. Because most of the work in this field has been focused on marine species, this chapter emphasizes the use of trace metal ion buffers in seawater cultures. The readers who are primarily interested only in preparing culture media of defined chemistry for routine culturing or for non–trace metal experimental studies will find the necessary protocols, calculations, and practical considerations in Section 4.0.

## 1.1. Trace Elements as Limiting Nutrients

We begin with a brief discussion of the metabolic roles of trace metals as limiting nutrients and toxicants and their interactions with other limiting resources such as light, nitrogen, and inorganic carbon. We have included selenium in this discussion, which, although not a trace metal, is nonetheless an important micronutrient element in freshwater and marine environments and culture media.

Trace elements, including metals (iron, manganese, zinc, cobalt, copper, molybdenum, nickel, and cadmium) and the metalloid selenium, influence phytoplankton growth and community composition largely because of their roles as limiting micronutrients. Iron is quantitatively the most important of these micronutrients and limits phytoplankton growth in many regions of the ocean: the subarctic and equatorial Pacific, the Southern Ocean, and some coastal upwelling systems along the eastern margin of the Pacific (Martin and Fitzwater 1988, Coale et al. 1996, Hutchins et al. 1998, Boyd et al. 2000). Other metals, such as zinc, cobalt, and manganese, can occasionally limit phytoplankton growth in the ocean (Buma et al. 1991, Coale 1991) but may play a more important role in regulating the composition of phytoplankton communities because of large differences in trace metal requirements among species (Brand et al. 1983, Sunda and Huntsman 1995a, Crawford et al. 2003). For example, high ratios of cobalt to zinc may favor the growth of marine cyanobacteria and coccolithophores (Sunda and Huntsman 1995a, Saito et al. 2002). Likewise, iron can have a critical influence on the composition and structure of algal communities because of differences in requirements among species, particularly coastal and oceanic ones (Brand et al. 1983, Maldonado and Price 1996) (Table 4.1) and large- and small-celled phytoplankton (Price et al. 1994, Sunda and Huntsman 1997). Trace metals are believed to be less important in regulating the growth of freshwater phytoplankton, although variations in iron availability have been found to control nitrogen fixation and algal growth in some high-phosphorus lakes (Wurtsbaugh and Horne 1983, Evans and Prepas 1997).

Trace elements play critical roles in a variety of metabolic pathways involving the utilization of essential algal resources (light, nitrogen, phosphorus, and $CO_2$); thus, their cellular requirements can be influenced by the availability of these resources. Iron is needed for the growth of all phytoplankton. It serves essential metabolic functions in photosynthetic electron transport, respiratory electron transport, nitrate and nitrite reduction, sulfate reduction, dinitrogen ($N_2$) fixation, and detoxification of reactive oxygen species (e.g., superoxide radicals and hydrogen peroxide). As a consequence of its involvement in photosynthetic electron transport (in both cytochromes and iron/sulfur centers), the cellular iron requirement for growth increases with decreasing light intensity (Raven 1990, Sunda and Huntsman 1997) and decreasing photoperiod duration

(Sunda and Huntsman 2004). The cellular requirement for iron varies with the nitrogen source and is higher for cells growing on nitrate than on ammonium (Raven 1988, Maldonado and Price 1996). Because of its large requirement in $N_2$ fixation, it is also needed in much higher amounts by cyanobacteria growing diazotrophically (i.e., growing on $N_2$) than those growing on ammonia (Kustka et al. 2003).

Manganese is an essential component of the water oxidizing centers of photosynthesis, and consequently, it is also essential for the growth of all phytoplankton. Because of its photosynthetic involvement, manganese, like iron, is needed in higher amounts for growth under low light conditions (Sunda and Huntsman 1998a). Manganese is also present in superoxide dismutase, an enzyme that removes toxic superoxide radicals (Raven et al. 1999, Peers and Price 2004). Because manganese is involved in fewer metabolic components, its requirement for growth is predicted to be much less than that of iron (Raven 1990), a prediction confirmed by results of algal culture experiments conducted in trace metal ion buffer systems (see Table 4.1).

Zinc, whose cellular growth requirement is similar to that for manganese (see Table 4.1), is needed for a variety of metabolic functions. A major use for zinc is in carbonic anhydrase, an enzyme critical to $CO_2$ trans-

**TABLE 4.1** Cellular metal growth requirements for some coastal and oceanic phytoplankton cultured in seawater at 20°C under 14 hours per day of saturating light (500 μmol photons $m^{-2} \cdot s^{-1}$).

| Algal group | Species | $\mu_{max}$ $(d^{-1})$ | Metal | Metal: C ($\mu mol \cdot mol^{-1}$) | | Reference[a] |
|---|---|---|---|---|---|---|
| | | | | $\mu = 0.5$ | $\mu = 1.2$ | |
| Coastal diatom | Thalassiosira pseudonana | 1.8 ± 0.2 | Fe | 15 ± 2 | 28 | (1,2) |
| | | | Zn[b] | 2–3 | 4 ± 1 | (3,4) |
| | | | Mn | 1 | 4 | (5,6) |
| | Thalassiosira weissflogii | 0.89 | Fe | 13 | | (1,2) |
| Coastal dinoflagellate | Prorocentrum minimum | 0.60 | Fe | 14 | | (2) |
| Coastal | Synechococcus bacillaris | 0.62 | Fe | 100 | | (4) |
| cyanobacterium | | 0.53 | Co[b] | 0.12 | | (3) |
| Oceanic diatom | Thalassiosira oceanica | 1.55 | Fe | ~3 | 5 | (1) |
| | | 1.31 | Zn[b] | | ~1 | (2) |
| | | 1.48 | Mn | | ~1 | (5) |
| Oceanic coccolithophore | Emiliania huxleyi | 1.1 | Fe | ~3 | | (1) |
| | | 1.2 | Co[b] | 1.4 | | (2) |

[a](1) Sunda and Huntsman 1995b; (2) Sunda and Huntsman 1997; (3) Sunda and Huntsman 1995a; (4) Sunda and Huntsman, unpublished data; (5) Sunda and Huntsman 1986; (6) Sunda and Huntsman 1998a.
[b]Growth requirements for Zn were measured in the absence of added Co, while those for Co were determined without added Zn.

port and fixation (Morel et al. 1994). Higher amounts of this enzyme are needed under $CO_2$ limiting conditions, and consequently the algal requirement for zinc also increases under $CO_2$ limitation. Zinc is also found in zinc finger proteins, which are involved in the transcription of DNA, and in alkaline phosphatase, which is needed to acquire organic forms of phosphorus. Cobalt, and sometimes cadmium, can substitute for zinc in zinc enzymes such as carbonic anhydrase, leading to complex interactions among the three metals in marine algae (Price and Morel 1990, Morel et al. 1994, Sunda and Huntsman 1995a). These interactions need to be heeded in the design of any algal studies of zinc, cobalt, or cadmium limitation. In addition to substituting for zinc in metalloproteins, cobalt also has a unique requirement in vitamin $B_{12}$, but the need for this cofactor is usually quite small in phytoplankton (Sunda and Huntsman 1995a). There is a specific growth requirement of cobalt in some marine cyanobacteria and prymnesiophytes, but the biochemical basis for this has not yet been established (Sunda and Huntsman 1995a, Saito et al. 2002).

Copper should be essential for all phytoplankton because of its function in cytochrome oxidase, an essential protein in the respiratory electron transport chain. Copper also serves in plastocyanin in photosynthesis, which can substitute for the iron protein cytochrome $c_6$ in some algal species (Raven et al. 1999). It also serves in a multicopper oxidase, a component of the iron transport system of *Chlamydomonas reinhardtii* Dangeard (La Fontaine et al. 2002) and likely other algae as well. These latter two molecular functions lead to colimitation interactions between copper and iron nutrition, as recently demonstrated for marine diatoms (Maldonado, unpublished data; Price, unpublished).

Molybdenum and nickel, like iron, play important roles in nitrogen assimilation. Nickel is present in the enzyme urease and thus is required for phytoplankton grown on urea as a nitrogen source (Price and Morel 1991). Molybdenum occurs with iron in the enzymes nitrate reductase and nitrogenase, and consequently it is needed for nitrate assimilation and $N_2$ fixation (Raven 1988). Unlike the previously mentioned metals, which all exist as aquated cations and metal complexes, molybdenum occurs in seawater as the oxyanion molybdate ($MoO_4^{2-}$), whose molecular structure and chemical behavior are similar to those of sulfate ($SO_4^{2-}$). Because of its high concentration in seawater (ca. 100 nM) and its conservative distribution relative to major ions such as sulfate, molybdenum is believed not to be limiting in the ocean and need not be added to culture media made with natural seawater. However, it needs to be added to synthetic seawater media such as Aquil and synthetic freshwater media such as Fraquil (for definitions of Aquil and Aquil* see Section 4.1; for definitions of Fraquil and Fraquil* see Section 4.2). Because it is an oxyanion, it cannot be buffered by addition of metal ion chelating agents such as EDTA. However, the low cellular content of this metal in cultured plankton (mean molybdenum:carbon = 0.27 µmol · mol$^{-1}$ in 16 marine algal species; Ho et al. 2003) relative to its high seawater concentration obviates the need for such buffering in marine culture media.

The metalloid selenium is also an essential element for growth of freshwater and marine phytoplankton (Wehr and Brown 1985, Price et al. 1987, Harrison et al. 1988). It occurs in glutathione peroxidase, an antioxidant enzyme that degrades hydrogen peroxide and organic peroxides (Price and Harrison 1988). Marine phytoplankton have cellular requirements for selenium (selenium : carbon = 1.6 µmol · mol$^{-1}$) similar to those for other trace elements, suggesting that selenium may have other, as yet unidentified, metabolic roles. Selenium exists in three oxidation states in natural waters: Se(–II) in organic selenides and Se(IV) (selenite) and Se(VI) (selenate), which are oxyanions like molybdate (Cutter and Bruland 1984). Both selenite and selenate exist at nanomolar concentrations in the ocean and show pronounced sea surface depletion, like other algal nutrients. Low concentrations are routinely added as selenite to freshwater and marine culture media. At present we have no way to buffer the concentration of selenite or selenate ions.

It is clear from the previous discussion that trace element requirements are influenced by a number of factors that need to be quantified and controlled in algal trace element limitation experiments. These factors include light intensity and photoperiod duration, which are particularly important for iron and manganese studies; nitrogen source (ammonium, nitrate, dinitrogen, and urea) for studies involving iron, molybdenum, or nickel; $CO_2$ availability for studies of zinc/cobalt/cadmium limitation; and phosphorus source in studies of zinc limitation. In addition, several micronutrient elements can interact with one another in cases where more than one serve the same metabolic function, by substitution either in the same enzyme (e.g., cobalt substitution in zinc carbonic anhydrase) or in separate enzymes, which serve the same metabolic function (e.g., manganese and iron in superoxide dismutase or selenium and iron in peroxidases). These interactions among micronutrients and between micronutrients and other limiting factors need close attention in the design and execution of trace element limitation experiments.

## 1.2. Trace Metal Toxicity

Many reactive micronutrient metals, such as copper and zinc, are toxic at elevated concentrations (Brand et al. 1986; Sunda and Huntsman 1992). Cadmium, which in some species can metabolically substitute for zinc (e.g., in carbonic anhydrase), is also toxic at high concentrations (Lee and Morel 1995). For these metals, optimal growth occurs at intermediate concentrations, below which growth rate is limited and above which growth is inhibited. In addition, other metals with no known metabolic function, such as lead and mercury, are toxic at high concentrations.

Toxic metals often gain entry into the cell via the transport systems of essential nutrient metals (Sunda 2000). In addition, a common mode of toxic action is the inhibition of nutrient metal uptake and intracellular interference with nutrient metal metabolism (Harrison and Morel 1983, Sunda and Huntsman 1983). As a consequence, antagonistic interactions often exist between toxic and nutrient metals; and in any laboratory study of metal toxicity, close attention must be paid to the availability of interacting nutrient metals. A good case in point is the interaction between toxic levels of copper, cadmium, or zinc and the nutrient metal manganese, which have been observed in both freshwater and marine chlorophytes (Hart et al. 1979, Sunda and Huntsman 1998b) and in coastal and oceanic diatoms (Sunda and Huntsman 1983, 1996). In these interactions, growth rate inhibition by these metals is related to an inhibition of manganese uptake by the cells and the induction of a manganese deficiency (Sunda and Huntsman 1983, 1996, 1998a). The deficiency is alleviated either by increasing the manganese ion concentration or by decreasing the ionic concentration of the toxic metal. In studies with cadmium and zinc, both metals were taken up by the cell's manganese transport system, and thus cellular uptake of the metals was inversely related to the availability of manganese in the culture medium (Hart et al. 1979, Sunda and Huntsman 1996). Similar toxic metal/nutrient metal antagonisms are observed between copper and zinc (Reuter et al. 1982, Sunda and Huntsman 1998b).

## 1.3. Trace Metal Availability to Phytoplankton

Trace metals form various complexes with both inorganic and organic ligands. Some metals, such as iron, also readily precipitate as oxyhydroxides. The formation of these different "chemical species" profoundly influences the reactivity of trace metals and their uptake by phytoplankton. Early experiments with marine algal cultures demonstrated that the cellular uptake and toxicity of copper were related to the free ion concentration (or activity) of the metal and not to the concentration of total metal or of copper chelates with organic ligands (Sunda and Guillard 1976, Anderson and Morel 1978). Subsequent experiments showed similar results for growth limitation by zinc in a marine diatom (Anderson et al. 1978), uptake of zinc in freshwater chlorophytes (Bates et al. 1982), and iron and manganese uptake in marine diatoms (Anderson and Morel 1982, Sunda and Huntsman 1985). All these studies pointed to the free metal ion concentration as a core factor that regulates the uptake, intracellular utilization, and toxicity of trace metals.

Subsequent studies showed that in certain instances, such as cellular uptake of iron from seawater, the actual controlling parameter was the concentration of kinetically labile inorganic species (aquated free ions and dissolved inorganic complexes), which is proportional to the free ion concentration in media with a constant composition of major ions and a constant pH (Hudson and Morel 1990, 1993). Most of the previously mentioned studies examining the influence of metal speciation were conducted in such constant ionic media, and thus it was usually impossible to discern the true controlling chemical variable.

Whether algal uptake is controlled by the free metal ion concentration or the concentration of dissolved labile inorganic species is determined by the interaction of the metal with uptake sites on the outer cell membrane, which are responsible for intracellular metal transport (Sunda 1988/1989, Morel and Hering 1993). A metal ion ($M^{n+}$) is taken up into the cell by first forming a coordination complex with a receptor site (X) on a membrane transport protein, as represented by the reaction

$$M^{n+} + X \leftrightarrow MX \rightarrow M(cell) + X \qquad (2)$$

Once bound, the metal can either dissociate back into medium or be transported across the membrane into the cell, as indicated in Equation 2. The cellular uptake rate of the metal equals the concentration of metal bound to the uptake site ([MX]) times the rate constant for metal internalization ($k_{in}$):

$$V_{MCell} = k_{in} [MX] \qquad (3)$$

The rate of formation of MX is given by

$$d[MX]/dt = [M'][X]k_f \qquad (4)$$

where $k_f$ is the rate constant for the binding of metal ions to the uptake site, [X] is the concentration of unbound uptake sites on the outer membrane, and [M′] is the concentration of dissolved inorganic metal species. These species, which include both free metal ions and dissolved complexes with inorganic ligands (OH⁻, Cl⁻, and $CO_3^{2-}$), have sufficiently rapid exchange kinetics to donate metal ions to the uptake site (Hudson and Morel 1993). The rate of loss of MX equals the dissociation rate plus the rate of metal uptake into the cell

$$-d[MX]/dt = [MX](k_d + k_{in}) \qquad (5)$$

where $k_d$ and $k_{in}$ are rate constants, respectively, for dissociation and internalization. At steady state the forward and reverse rates equal one another, and thus from Equations 4 and 5 we derive

$$[MX] = [M'][X]\, k_f/(k_d + k_{in}) \qquad (6)$$

Further, by combining Equations 3 and 6, we derive

$$V_{MCell} = k_{in}[MX] = k_{in}[M'][X]\, k_f/(k_d + k_{in}) \qquad (7)$$

The term $k_f/(k_d + k_{in})$ is the inverse of the half-saturation constant, the M′ concentration at which half of the sites are bound to the metal. If the dissociation rate constant for the uptake site is much larger than that for internalization (i.e., $k_d \gg k_{in}$), a pseudo-equilibrium is established between the metal in the medium and that bound to the transport site; and at equilibrium, the amount of metal bound to the site, and thus the transport rate, is determined by the free metal ion concentration or activity (Sunda 1988/1989, Morel and Hering 1993). However, if the opposite occurs and $k_d \ll k_{in}$, then the uptake rate equals [M′] [X] $k_f$. Here the rate of uptake is under kinetic control by the rate of metal reaction with the transport site and will be determined by the concentration of kinetically labile inorganic metal species (Hudson and Morel 1993). If $k_d \sim k_{in}$, then the transport system is under mixed control, and metal uptake is determined by some combination of the free metal ion concentration and that of M′. A notable example of kinetic control is the uptake of iron by eucaryotic marine algae (Hudson and Morel 1990). For copper toxicity to phytoplankton, however, the inhibition of growth rate was related to the free ion concentration rather than that of Cu′ (Sunda and Guillard 1976), and thus copper uptake and resulting toxicity were likely under equilibrium control. In most cases, however, the relative extent of kinetic versus equilibrium control of uptake is simply not known (Hudson 1998). As a rule of thumb, kinetic control should be favored for high-affinity uptake systems, which transport metals into cells at low, limiting concentrations in the medium (Hudson and Morel 1993). The high affinity for metal binding in these systems is achieved through a low dissociation rate constant ($k_d$) for the uptake site, greatly increasing the likelihood that $k_d \ll k_{in}$, the necessary condition for kinetic control (Hudson 1998).

There is one other special case where uptake is under kinetic control. For uptake of zinc and cobalt by marine phytoplankton at very low concentrations of dissolved-inorganic species (i.e., generally <10 pM), the rate of uptake becomes limited by the rate of diffusion of these labile species (Zn′ and Co′) to the cell surface (Sunda and Huntsman 1992, 1995, Hudson and Morel 1993). Under these conditions the rate of uptake is simply proportional to that of diffusion of Zn′ and Co′ species, which is proportional to their concentration in the culture medium. Such diffusion control has also been noted for iron uptake by larger-celled marine algae at low growth-limiting concentrations (Hudson and Morel 1993; Sunda 2001).

In practice, whether the controlling parameter is the free ion concentration or the concentration of dissolved inorganic species may be unimportant so long as the inorganic complexation is minor or remains constant. The former occurs for $Mn^{2+}$, $Co^{2+}$, $Zn^{2+}$, and $Ni^{2+}$, for which the free metal ion is the dominant dissolved inorganic species in both seawater and freshwater media (Byrne et al. 1988). For $Fe^{3+}$ or $Cu^{2+}$ there is substantial complexation by hydroxide and carbonate ions, respectively, and this complexation remains constant so long as the pH does not change. Likewise, cadmium is highly complexed by chloride ions in seawater, and the degree of complexation is constant, provided the salinity remains constant. The ratio between the total concentration of inorganic species (free ions plus inorganic complexes) to that of free metal ions can readily be computed from inorganic speciation models (Byrne et al. 1988), with the notable exception of iron, whose degree of inorganic complexation by hydroxide ions is still the subject of much uncertainty (Waite 2001, Sunda and Huntsman 2003). These ratios are commonly called inorganic side reaction coefficients ($\alpha_M$). They can be used to interconvert concentrations of dissolved inorganic species and free metal ions:

$$[M'] = \alpha_M[M^{n+}] \qquad (8)$$

Values for $\alpha_M$ in seawater at a salinity of 36, 20°C, and pH of 8.2 are shown in Table 4.2.

**TABLE 4.2** Conditional stability constants and computed complexation of metals in EDTA and NTA metal ion buffer systems in seawater at 20°C, a pH of 8.2, and a salinity of 36. The conditional constants are defined both in terms of the free metal ion concentration ($K^*$) and the concentration of dissolved inorganic metal species ($K'$)

| Chelator | Metal | Log $K^*$ | Log $K'$ | Log $[M_{tot}]/[M']$ | $\alpha_M = [M']/[M^{n+}]$ |
|---|---|---|---|---|---|
| EDTA (0.1 mM) | $Fe^{3+}$ | 17.35[a] | 6.94[a] | 2.94 | $2.6 \times 10^{10}$ |
| | $Mn^{2+}$ | 5.19 | 5.05 | 1.09 | 1.38 |
| | $Cu^{2+}$ | 10.12 | 8.91 | 4.91 | 16.1 |
| | $Zn^{2+}$ | 7.99 | 7.81 | 3.81 | 1.51 |
| | $Co^{2+}$ | 7.63 | 7.46 | 3.46 | 1.49 |
| | $Ni^{2+}$ | 9.89 | 9.64 | 5.64 | 1.78 |
| | $Cd^{2+}$ | 7.78 | 6.25 | 2.25 | 34 |
| NTA (0.5 mM) | $Fe^{3+}$ | — | — | — | $1.3 \times 10^{10}$ |
| | $Mn^{2+}$ | 2.88 | 2.74 | 0.11 | 1.38 |
| | $Cu^{2+}$ | 8.40 | 7.19 | 3.89 | 16.1 |
| | $Zn^{2+}$ | 6.11 | 5.93 | 2.63 | 1.51 |
| | $Co^{2+}$ | 5.82 | 5.65 | 2.35 | 1.49 |
| | $Ni^{2+}$ | 6.97 | 6.72 | 3.41 | 1.78 |
| | $Cd^{2+}$ | 5.27 | 3.74 | 0.58 | 34 |

For the ZnEDTA chelate, $K'$ was measured directly in seawater (Sunda and Huntsman 1992) and the value for $K^*$ was then computed from the inorganic side reaction coefficient, $\alpha_M$. Values for $\alpha_M$ are taken from Byrne et al. (1988), with the exception of $\alpha_{Fe}$, which taken from Sunda and Huntsman (2003). The $K'$ value for iron was also empirically measured and represents the equilibrium value in the dark (see Fe-EDTA section). All other constants were first computed as $K^*$ values and then converted to values of $K'$ with use of Equation 15. Thermodynamic constants used in these calculations were taken from Martell and Smith (1974). These calculations assumed total Ca and Mg concentrations of 0.011 and 0.057 M for the 36 salinity seawater. The free Ca and Mg ion concentrations were assumed to be 84% and 90% of the total Ca' and Mg' concentrations (Thompson 1964, Thompson and Ross 1966).

[a]Values in the dark. Because of photo-redox cycling, the effective constants will be higher in the light.

## 2.0. CHEMICALLY DEFINED TRACE METAL ION BUFFER SYSTEMS

The discovery of the importance of free metal ions in controlling algal metal uptake led to the development of defined metal ion buffer systems to quantify and control concentrations of free metal ions and dissolved inorganic metal species in seawater culture media (Sunda and Guillard 1976, Anderson and Morel 1978, Brand et al. 1983) and in freshwater media (Morel et al. 1975, Xue et al. 1988). Typically, these buffers consist of a high concentration of a well-defined organic ligand (chelator) and lower concentrations of biologically active trace metals. However, as mentioned previously, trace metal ion buffer systems based on EDTA had been developed empirically in the 1950s for use in culture media (Hutner et al. 1950, Provasoli et al. 1957). One of the most widely used early seawater media, the half-strength f medium (or f/2, Guillard and Ryther 1962), contained 11.7 µM EDTA and an equal concentration of ferric iron. This medium, although quite satisfactory for culturing marine algae, had ill-defined trace metal speciation and thus was poorly suited for controlled studies of algal trace metal uptake, nutritional limitation, or toxicity. The lack of chemical definition in the f/2 medium resulted from the precipitation of most of the high concentration of added iron as hydrous ferric oxide, which adsorbed trace metals and further reduced their availability. This precipitation varies with the chemical and physical conditions of medium preparation and sterilization, leading to variable and ill-defined chemical speciation of trace metals.

The earliest of the more chemically defined buffer systems used high concentrations of trishydroxymethylamino methane (Tris) and lower concentrations of copper to quantify and control free cupric ion concentrations in culture studies of uptake and toxicity of copper in a marine diatom (Sunda and Guillard 1976). This medium also contained a low concentration of iron (1 μM, added as FeEDTA) to minimize iron hydroxide precipitation. Subsequent chemically defined trace metal buffers employed low concentrations of added iron and used the synthetic chelators EDTA (e.g., the marine medium Aquil, Morel et al. 1979) or NTA (nitrilotriacetic acid) (Brand et al. 1986, Xue et al. 1988) to complex trace metals. These ligands were chosen because they are relatively nontoxic, are not subject to biological uptake or degradation, and are well-characterized with respect to metal complexation equilibria. The latter attribute allowed the development of metal ion buffer systems, in which all biologically relevant metals (e.g., iron, copper, manganese, zinc, cobalt, cadmium, and nickel) can be controlled simultaneously by metal complexation with a single chelator. This ability for multiple control of metals is of considerable experimental importance in trace metal culture studies, given the propensity for metal/metal interactions and the need to ensure that all metals other than the ones of immediate experimental interest are maintained at optimal levels for growth.

## 2.1. Theory of Metal Ion Buffer Systems

Metal ion buffers function similarly to pH buffers. In these buffers, the metal is complexed (bound) by the chelator to form a metal complex (or chelate) with the ligand. We can write the reaction of zinc with EDTA:

$$Zn' + EDTA^* \leftrightarrow ZnEDTA^{2-}. \qquad (9)$$

The term Zn′ represents the labile pool of dissolved inorganic zinc species and EDTA* represents the nominally free ligand. In reality, the ligand is not free but is complexed in seawater to calcium ions and, to a lesser extent, magnesium ions. Thus, EDTA* is the combined concentrations of these Ca- and Mg-EDTA complexes. The same is true of freshwater media, provided a typical river or lake water calcium concentration is used (0.1 to 1 mM), and this concentration well exceeds that of EDTA. In practice, [EDTA*] is the total EDTA concentration minus the combined concentration of EDTA complexes with trace metals.

As zinc binds with EDTA to form Zn-EDTA chelates, these chelates begin to dissociate back into the free metal and ligand. Eventually, the rates of chelate formation and dissociation equal one another, and the concentrations of the reacting species no longer change with time. Under this equilibrium condition, the concentrations of the various reacting species ([Zn′], [EDTA*], and [ZnEDTA$^{2-}$]) are related to one another by a mass action equation:

$$[ZnEDTA^{2-}]/([Zn'][EDTA^*]) = K'_{ZnEDTA} \qquad (10)$$

where $K'_{ZnEDTA}$ is the conditional stability constant for the formation of the Zn-EDTA chelate, based on reaction 9. Conditional constants apply for only a given set of critical chemical variables, such as salinity, calcium concentration, pH, and temperature and vary as these variables change. A set of these constants valid in seawater media under typical experimental conditions (pH 8.2, salinity 36, and 20°C) and typical EDTA concentration (0.1 mM) is shown in Table 4.2. Also shown in this table are equilibrium ratios of total metal (metal-EDTA chelates plus dissolved inorganic species) to dissolved inorganic metal species. At equilibrium, virtually all the zinc is present as biologically unreactive Zn-EDTA chelates, and only ~1 in 6,500 zinc atoms is present as reactive Zn′ species (see Table 4.2). For example, at a total zinc concentration of 50 nM, the equilibrium concentration of biologically available inorganic zinc species is only 8 pM, about the concentration found in surface ocean water in the North Pacific (Bruland 1989). If all the inorganic zinc species were removed by algal uptake, the system would automatically readjust to its original state by a dissociation of an equivalent amount (8 pM) of Zn-EDTA chelate. This dissociation would remove only 1/6,500 of the chelated zinc and thus would not affect the equilibrium Zn′ concentration. Thus, the system is well buffered with respect to changes in dissolved inorganic zinc species. As is readily apparent from Equation 10, the ratio of [ZnEDTA$^{2-}$]/[Zn′]—and thus the strength of the zinc buffer—increases in direct proportion to the EDTA concentration. This proportional relationship should be remembered in the design and use of metal buffer systems in algal culture experiments.

Equilibrium relationships similar to that shown in Equation 10 apply for other trace metals. The conditional stability constants differ among metals, which varies the ratio of chelated to free metal. As shown in Table 4.2, nickel and copper have the highest conditional stability constants in seawater and thus very high ratios of metal-EDTA chelates to dissolved inorganic metal species (~440,000 and 80,000, respectively) at an EDTA concentration of 0.1 mM. They are thus the most heavily buffered of the metals. Manganese, on the other hand, is the least heavily chelated of the seven trace metals in Table 4.2, and at 0.1 mM EDTA, the ratio of total manganese (MnEDTA + Mn′) to Mn′ is only 12.5, 0.016% of that for copper. Thus, this metal is poorly buffered by EDTA in seawater.

In freshwater media, EDTA is a stronger effective chelator for trace metals because of generally lower concentrations of calcium and thus a lower level of competition between calcium ions and trace metals for binding EDTA. As a result of this competition, conditional stability constants for EDTA chelation of trace metals are inversely proportional to the calcium ion concentration. This concentration is typically 10- to 100-fold lower in freshwater culture media than it is in seawater, and consequently, conditional stability constants are 10- to 100-fold higher. For example, the calcium ion concentration in the synthetic freshwater medium Fraquil* is 0.25 mM, or 1/44 of that (11 mM) in seawater at a salinity of 36. Thus, conditional stability constants for EDTA ($K^*$) are ~40-fold higher in this medium (Table 4.3) than in seawater after correcting for the slight complexation of calcium in seawater (16%) by inorganic ions (Thompson and Ross 1966). Conse-

quently, at the same concentration of EDTA and trace metal, the free trace metal concentration will be 40-fold lower in the Fraquil* medium. To compensate for the higher effective binding strength of EDTA in freshwater media, EDTA concentrations are usually reduced substantially (i.e., to 5–10 μM) relative to those used in seawater media.

In seawater media, the 0.1 mM EDTA buffer system is excellent for reproducing, in a controlled fashion, the free metal ion concentrations found in surface open ocean waters and for studying trace metal limitation of marine algal growth rates; and there are numerous examples where this system has been used for this purpose (Brand et al. 1983; Sunda and Huntsman 1983, 1992, 1995a, 1995b; Price and Morel 1990, 1991; Maldonado and Price 1996; Ho et al. 2003). In principle, one can achieve even higher levels of chelation—and thus even lower free metal ion concentrations—by increasing the EDTA concentration. But this may not always be feasible, because EDTA is toxic to some species at high

**TABLE 4.3** Conditional stability constants and computed complexation of metals in freshwater media Fraquil* and Fraquil-NTA at 20°C and pH 8. These media are based on the synthetic freshwater medium WC and have a free calcium ion concentration of 0.25 mM. Fraquil* contains 10 μM of "free EDTA" (EDTA*), while Fraquil-NTA contains 50 μM of "free NTA" (NTA*). The conditional constants ($K^*$) are defined in terms of the free metal ion concentration. Trace metal-calcium ion exchange constants ($K_{Ca,M}$) are also given to allow for calculation of trace metal ion concentrations in media with different $Ca^{2+}$ concentrations (see Equation 32).

| Chelator | Metal | Log $K^*$ | Log $K_{Ca,M}$ | Log $[M_{tot}]/[M^{n+}]$ |
|----------|-------|-----------|----------------|--------------------------|
| EDTA (10 μM) | $Fe^{3+}$ | a | a | |
| | $Mn^{2+}$ | 6.78 | 3.18 | 1.78 |
| | $Cu^{2+}$ | 11.71 | 8.11 | 6.71 |
| | $Zn^{2+}$ | 9.58 | 5.98 | 4.58 |
| | $Co^{2+}$ | 9.22 | 5.62 | 4.22 |
| | $Ni^{2+}$ | 11.53 | 7.93 | 6.53 |
| | $Cd^{2+}$ | 9.37 | 5.77 | 4.37 |
| NTA (50 μM) | $Fe^{3+}$ | a | a | |
| | $Mn^{2+}$ | 4.61 | 1.03 | 0.48 |
| | $Cu^{2+}$ | 10.13 | 6.55 | 5.83 |
| | $Zn^{2+}$ | 7.84 | 4.26 | 3.54 |
| | $Co^{2+}$ | 7.55 | 3.97 | 3.25 |
| | $Ni^{2+}$ | 8.70 | 5.12 | 4.40 |
| | $Cd^{2+}$ | 7.00 | 3.42 | 2.70 |

[a]Constants for iron are not given because of uncertainties relating to iron chelation, inorganic complexation, and photo-redox cycling.

concentrations. We found that a fivefold higher concentration of EDTA (0.5 mM) had no effect on the growth rate of coastal and oceanic diatoms (*Thalassiosira weissflogii* [Grunow] Fryxell et Hasle, *T. pseudonana* [Hustedt] Hasle et Heimdal, and *T. oceanica* [Hustedt] Hasle et Heimdal) but reduced the growth rate of the prymnesiophyte *Emiliania huxleyi* (Lohman) Hay et Mohler by 14% (Price et al. 1988/89, Sunda and Huntsman, unpublished data). Likewise, in experi-ments examining iron limitation of the $N_2$–fixing cyanobac-terium *Trichodesmium*, an EDTA concentration of 100 μM was inhibitory, and the chelator concentration had to be decreased to 10 μM (Kustka et al. 2003). In freshwa-ter media less EDTA needs to be added to achieve the same level of trace metal chelation, and consequently EDTA toxicity should be less of an issue.

For studies of trace metal toxicity, EDTA may be too strong a chelator to buffer free metal ion concentrations at the high levels known to inhibit the growth of many algal species. For effective buffering of trace metal ions, the concentration of the metal chelate should not exceed the concentration of the "free" ligand. This is because the slope of the relationship between concentrations of free metal ions (or inorganic species) and total metal con-centration steepens considerably as the total metal con-centration approaches that of the ligand, leading to poor buffering of free metal ions (Fig. 4.1). For copper, the most toxic of the more abundant trace metals (Brand et al. 1986), the equilibrium relationship for EDTA in sea-water at 20°C, a pH of 8.2, and salinity of 36 is

$$[CuEDTA^{2-}]/([Cu'][EDTA^*]) = 10^{8.91} \quad (11)$$

written in terms of the Cu' concentration and

$$[CuEDTA^{2-}]/([Cu^{2+}][EDTA^*]) = 10^{10.21} \quad (12)$$

when expressed in terms of the free cupric ion concen-tration (Table 4.2). Thus, to maintain [CuEDTA$^{2-}$] ≤ [EDTA*], [Cu'] and [Cu$^{2+}$] cannot exceed values of $10^{-8.91}$ M (1.23 nM) and $10^{-10.12}$ M (76 pM), respectively (Fig. 4.1). Because copper inhibition of the growth rate of marine algae often occurs at free cupric ion concen-trations above $10^{-10}$ M (0.1 nM) (Brand et al. 1986), this buffer is not well suited for many marine copper toxi-city studies. This situation only worsens in freshwater media because of the higher conditional stability constants (see Table 4.3).

For studies of metal toxicity, metal chelate buffers based on NTA are generally a better choice. NTA binds metals more weakly than EDTA, as indicated by its lower conditional stability constants (see Table 4.2), and thus effectively buffers metals over a higher range of metal ion concentrations. For example, its conditional

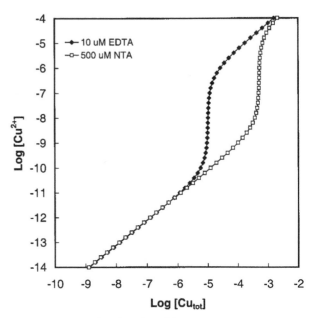

**FIGURE 4.1.** Log-log plot of free cupric ion concentration ([Cu$^{2+}$]) versus total copper concentration for two metal ion buffers (10 μM EDTA and 500 μM NTA) in seawater at 20°C, pH of 8.2, and salinity of 36. Both buffers show similar linear relationships between free cupric ion concentration and total copper at low copper concentrations, indicating similar levels of copper chelation. But as the total copper concentration approaches that of the chelator, the curves steepen consider-ably, leading to large changes in free copper with only small changes in total copper. This condition results in poor buffering of [Cu$^{2+}$] and thus is to be avoided in the design of culture exper-iments. Note that the stronger chelator, EDTA, allows effective buffering at [Cu$^{2+}$] up to $10^{-10.1}$ M (0.08 nM) (the inverse of the conditional stability constant), whereas the weaker chelator, NTA, which has a 50-fold lower conditional constant, allows effective buffering over a higher range of free cupric ion con-centrations (up to $10^{-8.4}$ M [4 nM]).

stability constant for binding copper in seawater is 1/50 of that for EDTA. Consequently, it can effectively buffer copper at free cupric ion concentrations from ~$10^{-14}$ M to $10^{-8.4}$ M (see Fig. 4.1), covering the entire range from no effect to complete inhibition of growth rate for most marine algal species (Brand et al. 1986). Similar arguments can be made for other toxic metals such as zinc and cadmium, whose stability constants for chelation by NTA are two orders of magnitude lower than those for EDTA.

Although NTA is generally preferred for most metal toxicity studies, this chelator does have some drawbacks. The much lower binding constants result in poorer chelation of micronutrient metals, which can be prob-lematic in studies of toxic metal–nutrient metal antagonisms. This problem can be lessened by using a high NTA concentration (0.5 mM), such as that employed in the seawater studies of Sunda and Huntsman (1996, 1998). However, even at this high concentration, only an estimated 22% of the manganese in seawater is bound to NTA, which means that the chelator provides almost no metal ion buffering for this metal (see Table 4.2). Thus, in studies of copper/manganese antagonisms in marine diatoms, EDTA buffer systems were used to control metal availability (Sunda and Huntsman 1983, 1998a). Clearly, the choice of chelator in metal toxicity studies is not always clear-cut. To circumvent this problem, one can, in principle, use a buffer system employing two or more chelators with differing metal binding affinities, but such mixed systems must be carefully designed to maintain proper buffering of toxic metals and constant free ion concentrations of interacting nutrient metals.

Another important issue in metal toxicity studies is the effect of the added metal on the availability of other bioactive metals, some of which may interact competitively with the toxic metal. A common mistake is to maintain the concentration of the chelator constant and to incrementally add the toxic metal at a range of concentrations up to those that totally inhibit growth rate. If the concentration of a metal approaches that of the chelator and complexes a significant fraction of the ligand, the concentration of "free" chelator (e.g., EDTA* or NTA*) decreases, causing an increase in the free ion concentrations of all other trace metals regulated by that chelator. Such metal-induced decreases in "free" chelator concentration can result in a poorly controlled experiment, in which a number of free ion variables that can influence growth rate all change simultaneously. To avoid this situation, metals should be added to the culture media as metal chelates in all situations where they would complex a significant fraction (i.e., >1%) of the existing chelator. In this way there is no further reaction of added metal with the chelator already present in the medium, and the free ion concentrations of other metals remain unchanged.

## 2.2. Computation of Concentrations of Aquated Free Metal Ions and Inorganic Species

Trace metal ion buffer systems not only allow one to maintain constant free metal ion concentrations in culture experiments, but also allow the quantification of the concentrations of aquated free metal ions and dissolved inorganic metal species. This in turn allows

one to measure relationships among these controlling chemical variables and the response of the algae in terms of cellular metal uptake rates, cellular metal concentration, and variations in growth rate. Quantification of metal speciation is often effected through the use of chemical equilibrium computer models such as MINEQL (Westall et al. 1976). These models allow calculation of trace metal speciation under varying experimental conditions of pH, calcium and magnesium ion concentrations, salinity, and temperature. They also can be used for speciation calculations in complex buffer systems containing two or more chelating ligands. Because such models are only as good as the input data for reaction equilibria and stability constants, it is advised that these be checked for correctness and for consistency with stability constant data used in previous culture studies in metal ion buffer systems.

Alternatively, calculations can be done manually with a hand-held calculator or a computer spread sheet using conditional stability constants, provided known constant conditions of salinity, calcium and magnesium concentrations, temperature, and pH are maintained. Constants for reaction of micronutrient metals (iron, zinc, manganese, cobalt, copper, and nickel) with EDTA and NTA in seawater media at a salinity of 36, pH of 8.2, and temperature of 20°C are listed in Table 4.2. Conditional constants for reaction of trace metals with EDTA in the freshwater medium Fraquil* at 20°C are listed in Table 4.3. Two different conditional constants, $K^*$ and $K'$, are shown in the tables. $K^*$ is the conditional constant for the equilibrium mass action equation for the chelation of metal M by ligand Y to form the chelate MY, written in terms of the free metal ion concentration, $[M^{n+}]$:

$$[MY]/([M^{n+}][Y*]) = K_{MY}^* \qquad (13)$$

An equivalent equation can be written in terms of the concentration of dissolved inorganic metal species, $[M']$:

$$[MY]/([M'][Y*]) = K_{MY}' \qquad (14)$$

[MY] is the concentration of metal ligand chelate. [Y*] is the nominal "free ligand" concentration, equal to the total concentration of the ligand minus the summed concentration of metal chelates with all trace metals. Comparing the two equations, we readily see that the two stability constants are related to one another by the equation

$$K_{MY}' = K_{MY}^* [M']/[M^{n+}] = K_{MY}^* \alpha_M \qquad (15)$$

where $\alpha_M$ is the side reaction coefficient for inorganic complexation (Equation 8), equal to the ratio $[M']/[M^{n+}]$. In Table 4.2, $\alpha_M$ values for divalent metals in seawater range from 1.38 for manganese to 34 for cadmium. However, in most freshwater media, $\alpha_M$ values for divalent metals approach unity, with the exception of copper at pH > 7. Consequently, for these metals, $K_{MY}' \sim K_{MY}^*$, and thus for the freshwater media Fraquil* and Fraquil-NTA in Table 4.3, values for $K_{MY}^*$ only are listed.

Equilibrium concentrations of free aquated metal ions or of dissolved inorganic species are computed from Equations 13 and 14 and the mass balance equation for the total concentration trace metal, $[M_{tot}]$:

$$[M_{tot}] = [M'] + [MY] \qquad (16)$$

Combining this equation with Equation 14, we derive the expression

$$[M_{tot}] = [M'] + [M'][Y^*]K_{MY}' = [M'](1 + [Y^*]K_{MY}') \quad (17)$$

Rearrangement of this equation yields an expression for calculating $[M']$:

$$[M'] = [M_{tot}]/(1 + [Y^*]K_{MY}') \qquad (18)$$

Likewise, from Equations 15 and 18, we derive an equation for computing the free metal ion concentration from values of $K^*$:

$$[M^{n+}] = [M_{tot}]/\left(\alpha_M + [Y^*]K_{MY}^*\right) \qquad (19)$$

Where there is heavy chelation of the metal ($[MY]/[M'] > 100$), we can ignore the terms for the inorganic metal species on the right, and Equations 18 and 19 can be simplified to

$$[M'] \sim [M_{tot}]/([Y^*]K_{MY}' \quad \text{and}$$

$$[M^{n+}] \sim [M_{tot}]/\left([Y^*]K_{MY}^*\right) \qquad (20)$$

These equations represent simple rearrangements of Equations 13 and 14. For the seawater buffer systems containing 0.1 mM EDTA and 0.5 mM NTA shown in Table 4.2, the previously simplified equations can be used in all situations except for manganese in both buffer systems and cadmium in the NTA buffer. In these latter instances, dissolved inorganic species represent a significant fraction of the total metal, and thus the full equations (18 and 19) must be used to compute trace metal speciation.

The previous approach to computing trace metal speciation in NTA and EDTA buffer systems works well, provided that the culture experiment is conducted under conditions for which the conditional constants are valid; that is, 20°C, salinity of 36, and pH of 8.2 for the constants in Table 4.2 and 20°C and a free calcium ion concentration of 0.25 mM for those in Table 4.3. In seawater, both "free EDTA" and "free NTA" are

present primarily as calcium chelates and, to a lesser extent, magnesium chelates. Consequently, calcium and magnesium ions compete with trace metals in the formation of trace metal chelates, and the conditional constants in Table 4.2 are inversely related to the free calcium and magnesium ion concentrations, which are proportional to salinity, provided the salinity variation is not too large. Thus, to convert a conditional constant in Table 4.2 defined for seawater at a salinity of 36 to one valid at a salinity of 33, one multiplies the former constant by the ratio of the two salinities (36/33):

$$K^*(S = 33) = K^*(S = 36) \times 36/33 \qquad (21)$$

Salinity adjustments in $K^*$ values are particularly useful when culture media are prepared from natural open ocean water (see Section 4.1.3), which typically varies in salinity between 33 and 37. Similar adjustments in $K^*$ values in response to variations in $Ca^{2+}$ concentrations in freshwater media will be discussed in Section 4.2.

Because concentrations of free aquated calcium and magnesium ions vary little with pH, conditional stability constants ($K^*$) for EDTA and NTA in seawater and freshwater media are largely insensitive to pH variations, with the notable exception of iron. However, pH can influence $K'$ values because of its effect on inorganic complexation to carbonate and hydroxide ligands and resulting effects on inorganic side reaction coefficients ($\alpha_M$), particularly for iron and copper. These variations in $\alpha_M$ can be calculated from inorganic speciation models for seawater (Byrne et al. 1988) and freshwater media (Morel and Hering 1993).

## 2.3. Regulation of Dissolved Inorganic Iron Species in Fe–EDTA Buffer Systems in Seawater

Of all the micronutrient metals, iron is the most important limiting nutrient in the ocean. It also has the most complex speciation chemistry, which often can be difficult to quantify and control in culture experiments. Iron uptake by phytoplankton in EDTA and NTA buffer systems appears to be related to the concentration of dissolved inorganic ferric hydrolysis species (Fe (III)′), which include ($Fe(OH)_2^+$, $Fe(OH)_3^0$, and $Fe(OH)_4^-$) (Anderson and Morel 1982, Hudson and Morel 1990). The hydrolysis speciation of iron is poorly known (Waite 2001), and consequently, it is not feasible to accurately compute Fe(III)′ concentrations in metal chelate buffer systems in culture experiments from existing thermodynamic data. Three additional factors further complicate quantification of Fe(III)′ concentrations in Fe-EDTA buffer systems. First, Fe(III)′ has limited solubility and precipitates as hydrous ferric oxides at high Fe(III)′ concentrations (i.e., above ~0.7 nM in seawater; Sunda and Huntsman 1995). Second, iron forms more than one chelate with EDTA (i.e., FeEDTA⁻ and Fe(EDTA)OH²⁻), and the proportions of the two chelates vary with pH. Finally, and most important, Fe-EDTA chelates undergo photolytic intramolecular charge transfer reactions, which result in their photoreductive dissociation (Fig. 4.2). In these photolysis reactions, the iron in the chelate is reduced to Fe(II) and the EDTA is oxidized. The Fe(II) dissociates from the oxidized ligand, forming dissolved inorganic ferrous species (Fe(II)′), which are rapidly reoxidized to Fe(III)′. This is subsequently rechelated to re-form Fe-EDTA chelates, thereby completing the photo-redox cycle. The photo-redox cycling of Fe-EDTA chelates substantially increases the steady-state concentration of Fe(II)′ and Fe(III)′ species and thus increases the uptake of iron by marine phytoplankton (Fig. 4.2). The photo-redox cycling of Fe-EDTA chelates is directly proportional to light intensity, but it also depends on light quality (spectral distribution) and thus is likely to differ among various sources of fluorescent light, such as Vita-Lite and Cool White fluorescent bulbs.

To deal with the complexity of iron speciation in Fe-EDTA buffer systems in seawater, conditional equilibrium or steady state constants have been measured for Fe-EDTA chelates as a function of pH, light, and temperature (Hudson et al. 1992, Sunda and Huntsman 2003). For ease of computation, these constants are best represented as conditional steady-state dissociation constants:

$$K_d' = 1/K_{Fe}' = ([Fe(III)'][EDTA^*])/[FeEDTA^*] \quad (22)$$

where FeEDTA* is the combined total concentration of Fe-EDTA chelates and $K_d'$ is a conditional dissociation constant for Fe-EDTA chelates, equal to the inverse of the conditional equilibrium constant for chelate formation. Values for $K_d'$ in the dark increase substantially with pH (Fig. 4.3). At 20°C and a pH of 8.0–8.9, they conform to the equation

$$\mathrm{Log}\,K_d'(\mathrm{dark}) = 2.427\,\mathrm{pH} - 2684 \quad (R^2 = 0.993) \quad (23)$$

Values for $K_d'$ also increase in the presence of light due to photo-redox cycling of iron (Fig. 4.3). This increase is much larger at low pH and low temperature. A kinetic analysis indicates that the $K_d'$ values in the light equal the sum of two intrinsic constants:

$$K_d'(\mathrm{light}) = K_d'(\mathrm{dark}) + I_{hv}K_{hv} \quad (24)$$

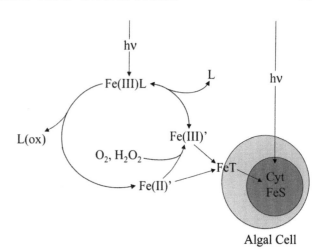

**FIGURE 4.2.** Photo-redox cycling of ferric chelates, such as those with EDTA. The cycle is initiated by the adsorption of light by the ferric chelate and a subsequent photolytic reaction in which the iron is reduced to Fe(II) and the ligand is oxidized. The Fe(II) dissociates from the degraded chelate to give dissolved inorganic ferrous species (Fe(II)′) which are then rapidly oxidized to dissolved inorganic ferric hydrolysis species (Fe(III)′) by molecular oxygen and hydrogen peroxide. These, in turn, are rechelated by the ligand to re-form the ferric chelate. The cycle increases the uptake rate of iron by algal cells by increasing the steady state concentrations of biologically available Fe(II)′ and Fe(III)′. These species have sufficiently rapid kinetics to react with iron uptake sites on the outer membrane of algal cells, the limiting step in algal uptake of iron. Once inside the cell, much of the iron is used for synthesis of cytochromes (Cyt) and FeS redox centers, needed in high amounts in photosynthesis. This model is based largely on the work of Anderson and Morel (1982) and Hudson and Morel (1990).

$K_d'$ (dark) is the equilibrium constant in the dark and equals $k_d/k_f$, the ratio of the rate constants for the dissociation and formation of Fe-EDTA chelates. $K_{hv}$ is a conditional photo-dissociation constant and equals $k_{hv}/k_f$. The term $k_{hv}$ is the rate constant for Fe-EDTA photolysis under a given set of light conditions (intensity and spectral composition), and $I_{hv}$ is the intensity of light relative to that at which $K_{hv}$ was measured. This intensity can be controlled by the distance from the light source, the number of fluorescent bulbs used, or the use of neutral-density light filters. Values for $K_{hv}$ have been measured at 10° and 20°C in Fe-EDTA seawater buffer systems exposed to light from Vita-lite fluorescent bulbs at a photosynthetically active radiation (PAR) intensity of 500 µmol photon · m⁻² · s⁻¹ (Fig. 4.3). Values for $K_{hv}$ at 20°C were linearly related to pH in the analytical range (pH, 7.8–8.4) and fit the equation

$$\mathrm{Log}\,K_{hy} = 0.776\,\mathrm{pH} - 12.92 \quad (R^2 = 0.990) \quad (25)$$

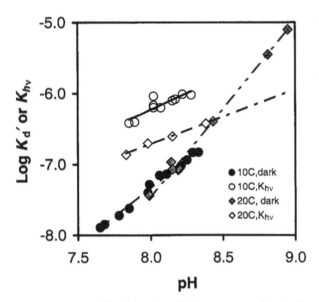

**FIGURE 4.3.** Conditional equilibrium constants for the dissociation of Fe-EDTA chelates in the dark ($K_d'$) and steady state constants for dissociation of Fe-EDTA chelates in the light ($K_{hv}$). Constants were measured in 36 ppt salinity seawater at 10° and 20°C. Light was provided from Vita-Lite fluorescent bulbs at an intensity of photosynthetically active radiation (PAR, 400–700 nm wavelength range) of 500 µmol photons · m$^{-2}$ · s$^{-1}$. $K_d'$ is the ratio of dissociation and formation rate constants for Fe-EDTA chelates ($K_d' = k_d'/k_f'$), whereas $K_{hv}$ is the ratio of the rate constants for photolysis and formation of Fe-EDTA chelates ($K_{hv} = k_{hv}/k_f$). Data are from Sunda and Huntsman (2003).

Values for $K_{hv}$ increase with decreasing temperature and are twofold to threefold higher at 10°C than at 20°C.

It is clear from the data plotted in Figure 4.3 that photo-redox dissociation of Fe-EDTA chelates is much less pH-dependent than dark thermal dissociation. For example, an increase in pH from 8.1 to 8.5 leads to an increase of only twofold in values of $K_{hv}$ at 20°C but to a 16-fold increase in $K_d'$ in the dark. Because of the much lower pH dependence of $K_{hv}$, seawater culture systems with high values of $K_{hv}$ relative to $K_d'$ (i.e., those with high light intensities, low pH, or low temperature) will exhibit much lower pH-dependent variations in [Fe'] and in cellular iron uptake rates.

Because of photo-redox cycling of Fe-EDTA chelates, Fe' concentrations and thus iron uptake rates will be higher during the day than during night in experiments using a diel light : dark cycle. Based on experimental data (Fig. 4.3), this effect will be greater at lower temperatures and at lower pH values. For such experiments, one can relate biological response parameters (e.g., specific growth rate) to the mean daily [Fe'], time averaged over the light and dark periods. To

do this, the following expression is derived from Equations 22 and 24:

$$\text{Average}\left[\text{Fe(III)}'\right] =$$
$$[\text{FeEDTA}^*](K_d'(\text{dark}) + I_{hv}K_{hv}\, h/24)/[\text{EDTA}^*] \quad (26)$$

where h is the daily hours of light and $I_{hv}$, as before, is the light intensity relative to that at which $K_{hv}$ was measured. Note that this equation can also be used to compute the steady state Fe(III)' concentration under continuous light (h = 24) or the average [Fe'] at differing light intensities. Values for $K_d'$ (dark) in seawater at a salinity of 36 and temperature of 20°C can be computed from Equation 23. Values for $K_{hv}$ at the same temperature and salinity can be computed from Equation 25, provided a Vita-Lite fluorescent light source is used and the experiments are conducted in polycarbonate bottles. The type of bottle used may influence iron photolysis rates, because different bottles exhibit different spectral absorbances, particularly in the ultraviolet wavelength range.

Iron and light also interact physiologically in phytoplankton, such that cells growing at subsaturating light intensities or a shortened photoperiod need higher external Fe(III)' concentrations and higher intracellular iron : carbon ratios to achieve a given growth rate. These interactions can be examined in Fe-EDTA buffer systems (Sunda and Huntsman 1997), but in doing so, one must consider the effect of decreasing light intensity or photoperiod on Fe' concentrations. Alternatively, one can conduct experiments in NTA buffer systems, where [Fe'] concentrations—and thus iron uptake by phytoplankton—are not influenced by light (Anderson and Morel 1982, Sunda and Huntsman, unpublished data). The lack of photo-enhancement of Fe' in NTA buffer systems may be due to a lack of photolysis of Fe-NTA chelates; however, it is more likely due to the orders of magnitude higher rates of thermal (dark) dissociation, which directly compete with photolysis for dissociation of the iron chelates. Unfortunately, conditional constants have not been measured for Fe-NTA chelation in either seawater or freshwater media. Nonetheless, culture experiments indicate a similar degree of iron limitation of growth rate in a 0.5-mM NTA buffer system in seawater and a 0.1-mM EDTA system for cultures grown at 20°C and pH 8.2 on a 14 : 10 light : dark cycle under Vita-Lite fluorescent lights (500 µmol · m$^{-2}$ · s$^{-1}$ of PAR; Sunda and Huntsman 2004). Such an NTA buffer would be a good choice if one wanted to quantify the effect of light on iron limitation of algal growth without having to worry about the influence of light on iron availability.

## 3.0. Effects of Cellular Processes on Medium Chemistry

The purpose of a chemically defined medium is to maintain known constant bioavailable concentrations of some key trace metals in batch cultures. But growing cells continuously take up major and trace nutrients from the medium, release various compounds into it, and catalyze some redox reactions on their surfaces. These cell-induced changes in the composition of the medium can in some instances overwhelm the metal ion buffers, leading to changes in available metal ion concentrations. One should thus be aware of the conditions that may lead to unintended changes in the concentrations of the key trace metal ion species.

### 3.1. Changes in pH

When the cell concentration in a batch culture becomes sufficiently high, the rate of inorganic carbon uptake and fixation exceeds the rate of $CO_2$ supply from the atmosphere. The $pCO_2$ of the medium then decreases and the pH goes up. This typically occurs at total cellular carbon concentrations above $\sim 0.5 \ mg \cdot L^{-1}$ in seawater media (i.e., $40 \ \mu mol \cdot L^{-1}$ cell carbon or ca. $10^4$–$10^5$ cells/mL, depending on cell size). Freshwater media are somewhat more susceptible to pH increases because of their lower concentrations of bicarbonate and carbonate ions, the major natural buffering agents in culture media. An increase in the pH of the medium has a host of effects on the chemistry of trace metals (some subtle, some large):

1. The complexation of metals by inorganic ligands such as $CO_3^{2-}$ or $OH^-$ increases (i.e., the inorganic side reaction coefficient $\alpha_M$ increases).
2. The effective affinity of the organic chelating agent for various metals can increase as a result of the decreasing degree of hydrogen ion binding to the ligand (i.e., $K^*$ increases).
3. The solubility of metals that form hydroxide or carbonate solids may change.

The trace metals of interest whose inorganic speciation in seawater and freshwater media depends most on pH are iron ($OH^-$ complexation) and copper ($CO_3^{2-}$ complexation) (Byrne et al. 1988). In EDTA- and NTA-buffered seawater media, a change in pH has relatively little effect on the effective affinity of the chelator for most metals, with the notable exception of Fe(III), as discussed previously. This lack of pH effect occurs because EDTA and NTA are present chiefly as calcium and magnesium complexes, and thus their free ligand concentrations vary little with changing pH. In contrast to EDTA and NTA, many chelators such as Tris are heavily protonated, and thus their effective affinities for all trace metals are very sensitive to pH. It should be noted that, in addition to affecting the speciation of nutrient metals, a change in pH may result in a number of poorly known physiological effects, thus leading to uncertainties in data interpretation for cultures with high cell concentrations.

### 3.2. Trace Metal Uptake

Buffer systems establish known constant concentrations of aquated free metal ions, $[M^{n+}]$, and dissolved inorganic species, $[M']$, provided an equilibrium is maintained. At equilibrium the rate of dissociation of the metal-ligand complex, MY, is equal to the rate of reaction of the unchelated metal $M'$ with the unbound ligand, $Y^*$, to form MY. Many metal–EDTA chelates have slow reaction rates, however, making reaction kinetics an important issue in maintaining the viability of EDTA-based trace metal buffers. A critical issue is the rate of metal uptake by the algae relative to the rate of formation/dissociation of the metal–EDTA chelate. As long as the algal metal uptake rate is small relative to the metal-EDTA reaction rate, a pseudo-equilibrium will prevail, and the concentration of dissolved inorganic metal species will remain constant with time. However, at sufficiently high cell concentrations, algal metal uptake rates may approach the rate of chelate formation and dissociation, and the inorganic metal concentrations will then decrease. When this occurs, the ability of the buffer to maintain a known, constant concentration of inorganic metal species is lost.

The metal for which this problem has been best examined is iron, the trace metal nutrient required in greatest quantity by phytoplankton. The reaction kinetics of Fe–EDTA chelates in seawater and of iron uptake by phytoplankton have been studied in detail (Hudson and Morel 1990, Hudson et al. 1992, Sunda and Huntsman 1995, 1997, 2003), so that we can examine quantitatively the problem of iron supply to marine algae growing in an Fe-EDTA buffer. For iron-limited cells, the rate of uptake of iron in such a buffered culture is roughly proportional to the unchelated iron concentration, $[Fe']$ (Sunda 2001). Because the rate of formation of the Fe–EDTA chelates is also proportional to

[Fe'] and its rate of dissociation is proportional to the concentration of Fe-EDTA chelates ([FeEDTA*]), we can write a simple differential equation for the rate of change in [Fe'] resulting from cellular uptake and the formation and dissociation of the EDTA chelates:

$$d[Fe']/dt = -k_u N[Fe'][EDTA*]$$
$$- k_f[Fe'][EDTA*] + k_d'[FeEDTA*] \quad (27)$$

where $k_u$ is a first order constant for cellular uptake, N is the concentration of cells, and $k_f$ is the effective second order rate constant for the formation of the ferric chelates; $k_d'$ is the effective first order rate constant for ferric chelate dissociation and equals the normal thermal (or dark) dissociation rate constant ($k_d$) plus the rate constant for photolysis of Fe-EDTA chelates (see Fig. 4.2). Thus, as discussed, the value of $k_d'$ increases with light intensity.

At steady state, [Fe'] is invariant and can be calculated by equating Equation 27 to zero:

$$[Fe'] = k_d'[FeEDTA*]/(k_u N - k_f[EDTA*]) \quad (28)$$

For diatoms and dinoflagellates ≤32 μm in diameter, the cellular uptake rate constant per cell is proportional to the square of the cell diameter (Sunda and Huntsman 1995, 1997):

$$k_u = k_u' \pi d^2 \quad (29)$$

assuming a spherical cell geometry. Thus the uptake per unit cell volume or biomass is inversely proportional to the cell diameter. On the basis of Equations 28 and 29, we can calculate the steady state Fe' concentration in marine algal cultures on the basis of the rate constants $k_u' = 29$ cm · s⁻¹, $k_f = 17$ M⁻¹ · s⁻¹, and $k_d' = 6 \times 10^{-6}$ s⁻¹. The results of such calculations are shown in Fig. 4.4 as a function of algal biomass [cell carbon per liter of culture = N × (πd³/6) × (15 mol carbon per liter of cell volume)] for three combinations of EDTA concentration and cell diameter. As can be seen, the ability of the Fe-EDTA buffer to maintain a constant [Fe'] is excellent for cultures of large cells (d = 32 μm) containing a high EDTA concentration. But in cultures of small cells (d = 4 μm), the steady state [Fe'] decreases appreciably as soon as the biomass reaches even modest values (>20 μmol C · L⁻¹ ≈ 0.5 mg L⁻¹ dry weight), particularly at low EDTA concentrations. Clearly, if one wants to maintain constant [Fe'] and thus constant specific rates of iron uptake and growth in culture experiments, the algal biomass must be maintained below critical levels. These levels, however, vary with the EDTA concentration, cell size, light intensity, and pH, which all influ-

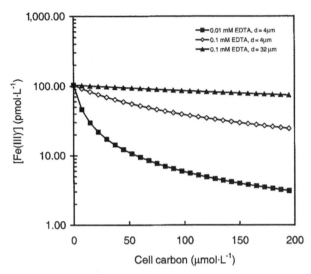

**FIGURE 4.4.** Estimated steady-state concentrations of dissolved inorganic Fe(III) species ([Fe(III)']) as a function of the total cellular fixed carbon per liter of culture medium, as based on Equations 28 and 29. Three situations are considered: (1) 4-μm-diameter (d) cells growing in media containing 0.01 mM EDTA and 3 nM iron; (2) 4-μm-diameter cells growing in seawater culture media containing 0.1 mM EDTA and 30 nM total iron; and (3) 32-μm cells growing in media containing 0.1 mM EDTA and 30 nM iron. The calculations are for 20°C, pH of 8.15, and 500 μmol quanta m⁻² s⁻¹ of PAR provided from Vita-Lite fluorescent bulbs, as well as a cell carbon concentration of 15 mol · L_cell⁻¹. Note that the decrease in [Fe(III)'] is directly related to cell carbon per liter of culture and inversely related to cell size and EDTA concentration.

ence Fe-EDTA dynamics. They likely also differ between seawater and freshwater media, owing to differences in calcium ion concentrations.

The kinetics of trace metal supply from a metal buffer depend on the nature of the metal and of the chelating agent. Nickel is the only nutrient metal that reacts at slower rates than iron (Morel and Hering 1993), but nickel is necessary only for cultures growing on urea as a nitrogen source (because urease is a nickel enzyme); the problem of nickel buffering is thus rarely encountered (Price and Morel 1991). Other commonly used chelating agents such as NTA have much more rapid chelate dissociation and association kinetics in both seawater and freshwater media than those observed for EDTA. These more rapid kinetics make metal-NTA buffer systems more stable and allow much higher biomass levels to be reached without appreciable decreases in the concentrations of dissolved inorganic metal species, including Fe'. As mentioned previously, these buffer systems have the additional benefit of not being subject to light-induced changes in Fe′ concentrations resulting from photo-redox cycling. But the lower affinity of such chelators for most trace metals does not allow practical buffering of the metal at the very low unchelated metal concentrations that are typically necessary to achieve growth rate limitation. This is particularly true for seawater media (Table 4.2), where conditional constants are generally 10–100 times lower than those in freshwater media.

### 3.3. Release of Chelators

Most of the compounds released by phytoplankton into their culture medium—typically various metabolites—have little influence on metal speciation. However, if an organism releases strong metal-chelating agents into the medium, there may be a substantial effect on the speciation and bioavailability of particular metals. There are indeed known examples of the release of such metal-complexing agents by phytoplankton. Strong iron-complexing agents, known as siderophores, are produced by many bacteria (Nielands 1989), including some cyanobacteria (Wilhelm 1995), but no firm evidence has been obtained for the production of siderophores by eukaryotic phytoplankton. Because the iron in the siderophore complex normally can be taken up by the organism that produces the siderophore, we should expect, in general, that the production of such siderophores will make iron more available. Information on the affinity of the siderophore for iron, on the kinetics of exchange between the artificial chelating agent in the culture and the siderophore, and on the rate of release of the siderophore into the medium are necessary to quantify the change in iron speciation in the culture.

Other examples of the release of metal-complexing agents by phytoplankton include the release of cadmium chelates with phytochelatins (small, thiol-containing polypeptides) by marine diatoms at high cadmium ion concentrations (Lee et al. 1996); the release of uncharacterized copper-chelating agents by marine cyanobacteria at high copper ion concentrations (Moffett and Brand 1996); and the release of cobalt-complexing agents by the same organisms, possibly for the purpose of cobalt uptake (such cobalt chelators may thus be considered "cobalophores") (Saito et al. 2002). Again, quantifying the effects of these complexing agents on the speciation of trace metals in the culture medium awaits a better characterization of their thermodynamic and kinetic properties and of their rate of release. One should note that, in the case of metal detoxification, it may be the metal chelate that is exported from the cells rather than the free chelating agent (Lee et al. 1996, Croot et al. 2003).

### 3.4. Redox Reactions

Phytoplankton are known to promote the redox transformations of some trace metals in the culture medium, directly through the activity of enzymes at their surface and indirectly through the release of reducing and oxidizing agents. The nutrient metals that can be affected in this way are iron, copper, cobalt, and manganese. Reducing and oxidizing agents that are released by the cells in the bulk medium include various metabolites including sugars (many of which can serve as electron donors under appropriate conditions), $O_2$, and transient species such as $O_2^-$ and $H_2O_2$ (Palenik et al. 1988). In most cases, however, the concentrations of these reducing and oxidizing agents are thought to be too small and the redox reactions too slow to markedly affect the speciation of metals in the bulk medium.

The situation is more complicated at the surface of the cells. High activities of metal reductases able to reduce various iron and copper complexes have been measured in marine diatom cultures (Jones et al. 1987, Maldonado and Price 2000, 2001). Furthermore, it seems likely that a direct or indirect reduction of Fe(III) to Fe(II) at the cell surface is a normal part of the iron uptake process in eukaryotic phytoplankton (La Fontaine et al. 2002), as it is in yeast (Eide 1998). As a result of such surface redox reactions, the trace metal chemistry in the immediate vicinity of the cell may be quite different from that in the bulk medium. Thus, the bioavailability to phytoplankton of Fe(III) bound to a particular chelator may depend on the susceptibility of the chelated Fe(III) to be reduced to Fe(II) at the surface of a cell and on the relative affinities of the chelating agent for Fe(III) and Fe(II), rather than on the equilibrium Fe′ concentration. For example, the surprisingly high rate of uptake by diatoms of iron bound to the fungal siderophore desferal has been shown to result from the surface reduction of the iron-desferal complex (Maldonado and Price 2000, 2001). Likewise, the reductive dissociation of Cu-EDTA chelates may account for the unexpectedly high rates of copper uptake by marine algae at low copper concentrations, which exceed limiting rates for diffusion of Cu(II)′ species to the cell surface by an order of magnitude (Hudson 1998). Clearly, further progress in elucidating the kinetics of acquisition of trace metals by phytoplankton will require a more thorough understanding of the chemical processes occurring at cell surfaces.

## 4.0. PREPARATION OF CULTURE MEDIA

### 4.1. Aquil* and Other Seawater Culture Media

Several similar seawater media formulations containing strong (0.1 mM) EDTA buffers have been used to study trace metal phytoplankton interactions in the past 2 decades. One of the most widely used of these is the synthetic medium Aquil, described by Price et al. (1988/89) as modified from an original formulation described by Morel et al. (1979). Here we describe a modification of Aquil that contains a 0.1-mM concentration of EDTA (see Table 3 of Price et al. 1988/89) and an eightfold lower iron concentration (1 μM) to minimize precipitation of hydrous ferric oxides. This modification of Aquil is called Aquil*. Aquil is a chemically well-defined artificial seawater medium that has been widely used (with various modifications) and supports the growth of many coastal and oceanic taxa (see Chapter 3).

Aquil and its various modifications employ a Chelex cation exchange treatment, initially developed by Davey et al. (1970), to remove metal contaminants present in low but significant concentrations in reagent grade salts used in its preparation. All metals are subsequently added back to Aquil in an EDTA-buffered solution that controls their speciation and thus their availability to phytoplankton. The Chelex 100 resin is composed of a styrene–divinyl benzene matrix derivatized with iminodiacetate functional groups that preferentially chelate trace metals (e.g., iron, copper, zinc, nickel, cadmium, cobalt) and remove them from solution. The medium is passed through a chromatography column packed with Chelex 100 and is collected in acid-washed polycarbonate vessels. It is then enriched with nutrients and an EDTA trace metal buffer. The Chelex procedure is time-consuming but not difficult. However, it is absolutely necessary to produce artificial seawater with low concentrations of trace metal contaminants. Preparation of Aquil medium is divided into three parts: Chelex 100 purification, medium preparation, and sterilization and manipulation. These parts have been described in detail by Price et al. (1989/1989), so only the essential elements are reproduced here.

### 4.1.1. Chelex Preparation

The purification of Chelex 100 follows a stringent procedure that lasts about 1 week (Table 4.4). It effectively removes free iminodiacetate and other strong metal-binding ligands that are soluble and leach from the resin as the medium flows over it. These dissolved ligands compete with EDTA for metal complexation. Thus, if they are not removed, they can lead to uncontrolled levels of trace metal chelation. The synthetic ocean water (SOW) component of Aquil* and all other nutrient additions, with the exceptions of the vitamin and EDTA-trace metal enrichments, are passed through Chelex 100 columns. Separate columns are used for each component to avoid the possible cross-contamination of solutions, particularly when the types of nitrogen and phosphorus compounds are varied in the medium. The 3 M $NH_4OH$ wash in step 3 invariably introduces high concentrations of $NH_4^+$ to the resin. If the resin is thoroughly rinsed, however, $NH_4^+$ concentrations in the eluant can be reduced to background levels (<0.1 μM).

---

**TABLE 4.4** Chelex 100 purification (from Price et al. 1988/1989).

Materials
1. Chelex 100 (200–400 mesh: Bio-Rad Laboratories, Richmond, CA)
2. Scintered glass filter funnel and side-arm flask for rinsing resin
3. One 500-mL polyproplyene wide-mouth bottle for soaking resin
4. Analytical reagent–grade methanol, HCL and $NH_4OH$
5. Ultrapure water (e.g., Milli-Q, 18.2 Mohms)

Methods
1. Soak Chelex 100 resin in methanol for 3–4h at room temperature (1:5 w/v); rinse with 750mL Milli-Q
2. Soak in 1 M HCl overnight at room temperature; rinse with 1 L Milli-Q
3. Soak in 3 M $NH_4OH$ for 1 week at room temperature; rinse with 1 L Milli-Q
4. Soak in 0.1 M HCl for 10min; rinse with 2 L Milli-Q; rinse with 200mL medium solution (SOW or nutrients)
5. Resuspend resin in 200mL of medium solution and titrate *slowly* to pH 8.1 with 1 M NaOH
6. Rinse resin with medium and transfer as a slurry to the chromatography column
7. Discard the first 500mL of medium passed through the Chelex column; check pH and $NH_4^+$ concentration of Chelexed medium before use

---

### 4.1.2. Medium Preparation and Composition

Full-strength synthetic ocean water (SOW) has a salinity of 35 and is prepared by mixing separate solutions of anhydrous and hydrated salts (Table 4.5). The salts should be dissolved in water from a Milli-Q RO and UF water system (Millipore Corporation) or equivalent water supply with very low metal impurities. Concentrated stocks of the minor constituents (NaF, $6 \cdot L^{-1}$; $SrCl_2 \cdot 6H_2O$, $34g \cdot L^{-1}$) are made up in advance and then diluted to their final concentrations. Major nutrients may also be added to the SOW at this time, prior to Chelex treatment. Alternatively, they may be prepared as described later and added separately if the type and amount of nutrient are varied in the experiment.

The trace metal enrichment for Aquil* or for similar EDTA–trace metal ion buffer systems (e.g., see Sunda and Huntsman 1992, 1995a–c) is prepared as a 1,000-fold concentrated stock containing EDTA. In all trace metal limitation studies it is imperative to use reagents of the highest quality. In a 100-μM EDTA buffer system, such as that in Aquil*, the EDTA is roughly 100 times more concentrated than any of the metal additions and may itself be a source of contaminating metals. It is available in various grades, from technical to analytical, which contain small amounts of some of the essential metals. The level of these impurities may range from a maximum of 0.01% to 0.0005% and, in the case of iron, may be biologically relevant. The amount of iron added as a contaminant with the EDTA, for example, may be as great as 67 nM. Most oceanic species would be able to grow at their maximum rates at this concentration, so iron limitation would be impossible to achieve. A number of biologically important trace elements, including selenium, arsenic, chromium, vanadium, and molybdenum, exist as oxyanions in seawater. Because of their negative charge, they are not removed by Chelex treatment. An exception is vanadium, which can be removed by Chelex resin (Bruland 1983).

The stock solution for each trace metal (iron, copper, zinc, cobalt, molybdenum, and manganese) is made up individually rather than as a mixture, because this allows easier modification of the medium trace metal composition. The recipe reported in Table 4.6 is for trace metal–replete Aquil*. Note that we often use a lower iron concentration than the 1-μM value reported here. At moderate light intensities (200–500 μmol photons $\cdot$ m$^{-2}$ $\cdot$ s$^{-1}$), coastal algal species are usually able to grow at maximum rates at total iron concentrations of 1 to 10 mM, whereas oceanic species are able to grow maximally at lower concentrations (0.03–0.1 μM) (Sunda and Huntsman 1995b). The stock metal solutions are prepared in 0.01 M HCl (Merck, Suprapure) at concentrations of $10^{-2}$ and $10^{-3}$ M and stored in acid-washed Teflon or polypropylene bottles. The iron stock solution (0.1 M) is stored similarly; the acidity of this stock solution prevents formation of insoluble hydrous ferric oxides.

A 1,000-fold concentrated EDTA stock solution (0.1 M) is made by dissolving EDTA (acid) or Na$_2$EDTA in Milli-Q water. EDTA added to seawater or freshwater medium will undergo reaction with calcium (or other metals, e.g., magnesium or trace metals) and release four equivalents of hydrogen ions for each mole of added EDTA:

$$H_4EDTA + Ca^{2+} \rightarrow CaEDTA^{2-} + 4H^+ \quad (30)$$

Thus, the addition of 0.1 mM EDTA (pure acid) to seawater in the preparation of Aquil* or similar seawater media will release 0.4 mM of hydrogen ions, which will decrease the final pH of the medium. To avoid this effect, four moles of base (NaOH) needs to be added to the stock EDTA solution for each mole of EDTA. The addition of base also facilitates the dissolution of the acid form of EDTA. If the disodium salt of EDTA is used, then only two equivalents of NaOH should be added for each mole of Na$_2$EDTA. Because reagent grade NaOH typically contains high levels of trace metal impurities, high-purity NaOH should be used. This can be purchased from a number of vendors (e.g., Suprapur, EM Science).

**TABLE 4.5** Composition of synthetic ocean water (SOW) used in Aquil*.

| Salt | Mass g·L$^{-1}$ of SOW | Final concentration (M) |
|---|---|---|
| **Anhydrous salts** | | |
| NaCl | 24.54 | $4.20 \times 10^{-1}$ |
| Na$_2$SO$_4$ | 4.09 | $2.88 \times 10^{-2}$ |
| KCl | 0.70 | $9.39 \times 10^{-3}$ |
| NaHCO$_3$ | 0.20 | $2.38 \times 10^{-3}$ |
| KBr | 0.10 | $8.40 \times 10^{-4}$ |
| H$_3$BO$_3$ | 0.0030 | $4.85 \times 10^{-4}$ |
| NaF | 0.0030 | $7.14 \times 10^{-5}$ |
| **Hydrous salts** | | |
| MgCl$_2$·6H$_2$O | 11.09 | $5.46 \times 10^{-2}$ |
| CaCl$_2$·2H$_2$O | 1.54 | $1.05 \times 10^{-2}$ |
| SrCl$_2$·6H$_2$O | 0.0170 | $6.38 \times 10^{-5}$ |

**TABLE 4.6** Composition of major and minor nutrient enrichments in metal-replete Aquil*. The metal nutrients and vitamins are prepared as separate stock solutions. The major nutrients may be prepared separately or mixed with the SOW salts.

| Nutrient | Final concentration (M) | Log [M'] |
|---|---|---|
| Major nutrients | | |
| P—NaH$_2$PO$_4$·H$_2$O | 1 × 10$^{-5}$ | |
| N—NaNO$_3$ | 1 × 10$^{-4}$ | |
| Si—Na$_2$SiO$_3$·9H$_2$O | 1 × 10$^{-4}$ | |
| Metal/metalloid nutrients* | | |
| Fe—FeCl$_3$·6H$_2$O | 1.00 × 10$^{-6}$ | — |
| Zn—ZnSO$_4$·7H$_2$O | 7.97 × 10$^{-8}$ | −10.93 |
| Mn—MnCl$_2$·4H$_2$O | 1.21 × 10$^{-7}$ | −8.03 |
| Co—CoCl$_2$·6H$_2$O | 5.03 × 10$^{-8}$ | −10.77 |
| Cu—CuSO$_4$·5H$_2$O | 1.96 × 10$^{-8}$ | −12.63 |
| Na$_2$MoO$_4$·2H$_2$O | 1.00 × 10$^{-7}$ | −7.00 |
| Na$_2$SeO$_3$ | 1.00 × 10$^{-8}$ | −8.00 |
| Vitamins | | |
| B$_{12}$ | 3.96 × 10$^{-10}$ | |
| Biotin | 2.50 × 10$^{-9}$ | |
| Thiamine | 2.96 × 10$^{-7}$ | |

Metal/metalloid nutrients are prepared with EDTA (see text).

A 1,000× concentrated EDTA trace metal stock solution can be prepared by adding appropriate concentrations of individual metals to the EDTA stock solution. Iron should be added first and then the remaining trace metals. Preparing the EDTA trace metal solution in this manner prevents precipitation of hydrous ferric oxides.

The best way to vary the free ion concentration of an individual metal in the medium is to add the metal complexed with EDTA (1:1). Adding the metal as an EDTA complex avoids potential problems of slow complexation kinetics and prevents re-equilibration of the added metal with the existing metal-EDTA complexes, as discussed previously.

The major nutrients (PO$_4^{3-}$, NO$_3^-$, and SiO$_3^{2-}$) are prepared as 1,000-fold concentrated stock solutions of 0.01, 0.1, and 0.1 M in Milli-Q water and are treated with Chelex, as described earlier, to remove contaminant trace metals. The nutrient stock solutions are added before sterilization of the medium. Inorganic carbon is added as bicarbonate with the SOW salts. It may be desirable to alter or supplement the nitrogen and phosphorus sources in the medium either to change the yield-limiting nutrient or to examine the metal requirements for major nutrient assimilation (Price and Morel 1991). To vary the nitrogen and phosphorus sources, the stock solutions (e.g., NH$_4^+$, urea, glycerol-2-phosphate) are prepared as 1,000-fold concentrates; they are then Chelex-treated as noted earlier and added as required. These alternative nitrogen and phosphorus–containing substrates thus replace the NO$_3^-$ or PO$_4^{3-}$ additions described in Table 4.6.

The vitamin addition follows the f/2 formulation (Guillard and Ryther 1962). Vitamin B$_{12}$, a porphyrin-like nutritional factor containing cobalt, contributes 0.37 nM cobalt to the medium and is unlikely to contribute significantly to the available cobalt in the medium. None of the vitamins are Chelex-treated. They are prepared as a 1,000-fold concentrated solution, which is sterilized by filtration to avoid bacterial degradation.

### 4.1.3. Media Prepared from Natural Seawater

Natural ocean water obtained from various locations, including the Gulf Stream and Sargasso Sea, may be used in place of SOW in the preparation of Aquil* or similarly defined seawater media. Trace metal concentrations in these waters are low enough to not significantly affect medium metal speciation in the presence of a high concentration of EDTA. A notable exception is the oxyanion molybdate, which is present in natural seawater at a concentration of ~100 nM (the same as in Aquil*) and thus need not be added to media prepared with natural seawater. One major advantage of using natural ocean water is that it eliminates the need for Chelex treatment in media prepared for routine maintenance of cultures and most non–trace metal culture studies. For culture studies examining limitation by micronutrient metals, the stock solutions of nutrients (nitrogen, phosphorus, and silicon) may introduce unwanted contaminant metals and must either be purified by Chelex treatment or be prepared from high-purity reagent salts. Culture media prepared from natural ocean water have been used successfully to study manganese, zinc, cobalt, and iron limitation of a number species of marine phytoplankton (Brand et al. 1983, Sunda and Huntsman 1983, 1986, 1992, 1995a, 1995b). Such studies typically benefit from the consistently low background concentrations of micronutrient metals in natural ocean water, something that is often difficult to achieve in Chelex-treated synthetic ocean water without an extreme level of care.

Great care also must be taken to avoid trace metal contamination during the collection of natural ocean water used to prepare culture media. A practical method is to pump the seawater through acid-cleaned polyethylene or Teflon tubing into clean polyethylene carboys with a noncontaminating pump. Peristaltic pumps (e.g., Masterflex, Cole Parmer Instrument Co.) have been used for this purpose (Flegal et al. 1991), as have Teflon diaphragm pumps (Wilden Pump and Engineering). In using a peristaltic pump, a short section of acid-washed flexible tubing (e.g., C-flex [Cole Parmer] or Teflon lined Tygon tubing) is passed through the pump head and attached by polyethylene fittings to the upstream and downstream working sections of tubing. An in-line acid-cleaned 0.45-μm-pore cartridge filter can be used to remove plankton and other particulate matter from the seawater during collection. Care must be taken to avoid contamination from the ship's hull or engine exhaust. To do this the tubing inlet can be positioned on the upwind side of the ship with an aluminum or polyvinyl chloride (PVC) plastic boom while the ship steams slowly through the water. A water vane can be used to position the incurrent end of the tubing upstream of the current to further minimize contamination. Alternatively, the tubing inlet can be positioned well below the ship's hull by attaching it to a plastic rope or cable with a noncontaminating weight on its end. We have used a well-sealed polyethylene or Teflon bottle filled with lead shot for this purpose. To minimize atmospheric contamination, carboy caps with inflow and outflow fittings can be purchased and attached directly to the outflow end of the tubing. When not in use, the ends of the tubing should be protected by enclosing them in clean plastic bags (e.g., Ziplock bags) that are tightly attached to the tubing with rubber bands.

### 4.1.4. Manipulation and Sterilization

Care also must be taken to avoid introducing trace metal contaminants during the preparation of chemically defined seawater and freshwater culture media. Media preparation and all culture transfers are performed in a dust-free, sterile laminar flow hood (HEPA filter), and all plasticware or other materials coming in contact with the medium must be thoroughly cleaned and acid-washed with trace metal clean procedures (Fitzwater et al. 1982). Even Pasteur pipettes used to transfer cells should be used cautiously; the ones we used contain millimoles of $NO_2^-$ and unknown quantities of trace metals. Sterile, autoclaved pipette tips for automatic pipettors are rinsed before use with sterile 10% HCl (Baker, Suprapure) and Milli-Q water that

has been sterilized by microwaving (Keller et al. 1988). All prepackaged sterile plastic containers, Swinex filters, and syringes are similarly acid-cleaned. The black rubber (neoprene) plugs in syringes are likely to be a source of zinc and should be avoided. Attention should be paid to all materials used in culturing and algal experimentation (Price et al. 1986: see Chapters 2 and 3 and Appendix A).

Several methods for medium sterilization may be used (Price et al. 1988/89), but the microwave method is preferred, because it is rapid and avoids the problems of trace metal contamination and pH increases that accompany sterilization by autoclaving (Keller et al. 1988). We follow the protocol outlined by Keller et al. (1988), using a 700-W microwave oven. The medium is dispensed into metal-clean Teflon or polycarbonate bottles and is microwaved for 10 minutes in intervals of 3, 2, 3, and 2 minutes and is mixed between heating cycles. Boiling of the medium is to be avoided, because this can cause increases in pH, which can alter metal speciation chemistry. We periodically test the effectiveness of our microwave to sterilize media, using cultures of marine bacteria. Regardless of the method used, it is important to add the trace metal solution as a filter-sterilized solution after medium sterilization to avoid the formation and heat-aging of metal precipitates, particularly hydrous ferric oxide. In our experience, failure to do so results in lower than maximum growth rates of *Thalassiosira weissflogii* in nutrient-sufficient medium and cell yields reduced by about one-half.

We use only polycarbonate containers for culturing phytoplankton. These are initially cleaned with a commercially available soap (Micro) and then soaked overnight in deionized, distilled water, followed by overnight soaks in 1M HCl and Mill-Q water. The polycarbonate containers are filled with Mill-Q water and then sterilized by microwaving. Sterile flasks are then rinsed once with culture medium immediately before use. Polycarbonate is largely free of contaminating trace metals and adsorbs only small amounts of metal from solution (Fitzwater et al. 1982). Prolonged treatment with acid or microwaving may cause the flasks to become discolored or cloudy. Such flasks obviously should be discarded.

### 4.2. Freshwater Media

In contrast to seawater, which has a remarkably constant composition of major ions ($Na^+$, $K^+$, $Ca^{2+}$, $Mg^{2+}$, $Cl^-$, $SO_4^{2-}$, $HCO_3^-$, $CO_3^{2-}$) and pH, freshwaters have highly variable compositions. The freshwater phytoplankton species that grow in these waters are usually adapted to their ambient chemistry. As a result, one cannot depend on a single recipe for successfully culturing and studying freshwater phytoplankton. To the extent that information is available, it is thus recommended to use a medium recipe that has the same major ion composition and the same pH as media that have been shown previously to support good growth of the organism of interest. Even changes that are seemingly minor, such as the $[Na^+]/[K^+]$ ratio, may inhibit the growth of some species.

The trace element composition of freshwaters is also more variable than that of seawater. For example, some freshwaters contain high concentrations of particular metals such as copper that are toxic to most phytoplankton. The species that grow in those waters have developed particular detoxification or tolerance mechanisms, but the trace metal buffers based on EDTA or NTA described previously permit one to vary free trace metal ion concentrations over several orders of magnitude, so that it is usually possible to design an appropriate trace metal buffer in practically all recipes for freshwater media.

Although finding an appropriate major ion composition to support optimal algal growth in freshwater media can present difficulties, the low salt content of such media often permits a major shortcut to achieve a well-defined trace metal composition. The total salt concentration of freshwater media is usually at least one hundred times lower than in seawater. As a result, the trace impurities in reagent salts are often low enough to not affect the final trace metal composition of the medium. Usually the trace metal buffers designed for optimal algal growth contain total concentrations of trace metals 0.01 to 1.0 $\mu$M (as in the Fraquil* recipe in Table 4.7) so that contaminating concentrations of trace elements in the nM range and below (in the final medium) have minimal effect on the final free metal ion concentrations. Thus, it is often not necessary to clean the major salt solutions with a Chelex ion exchange column. Because some batches of reagent grade salts have higher concentrations of some trace elements than others, it is recommended to measure these concentrations and to use only batches that have appropriately low levels of contaminants. Often the added concentrations of trace metals can be adjusted to compensate for the known background concentrations of metal contaminants. Chelex treatment may still be necessary for media used in trace metal limitation studies; however, because of the low concentrations of major ions, stock solutions of major ion salts can be made up as 100- or

**TABLE 4.7** Composition of the freshwater metal ion buffer system Fraquil* at pH 8.0[a] and 20°C. The major ion composition of this medium is that of WC (Appendix A); nutrients and vitamins are the same as in Aquil*.

| Constituent | Final concentration (M) | Log $[M^{n+}]$ |
|---|---|---|
| EDTA | $1.30 \times 10^{-5}$ | $-5.00^b$ |
| Ca | $2.60 \times 10^{-4}$ | $-3.60$ |
| Fe | $1.00 \times 10^{-6}$ | $—^c$ |
| Zn | $1.20 \times 10^{-6}$ | $-10.50$ |
| Mn | $6.03 \times 10^{-7}$ | $-8.00$ |
| Co | $5.03 \times 10^{-8}$ | $-11.00$ |
| Cu | $1.62 \times 10^{-7}$ | $-13.50$ |
| $Na_2MoO_4 \cdot 2H_2O$ | $1.00 \times 10^{-8}$ | $-8.00^d$ |
| $Na_2SeO_3$ | $1.00 \times 10^{-8}$ | $-8.00^d$ |

[a]A pH of 8.0 can be achieved through the addition of 0.5 mM $NaHCO_3$ to the medium. At moderate to high algal biomass, an additional pH buffer (e.g., 1–5 mM HEPES) may be necessary to maintain constant pH.
[b]log [EDTA*], the concentration of Ca-EDTA chelates.
[c]'Fe' concentration cannot be accurately computed.
[d]log $[MoO_4^{2-}]$ or log $[SeO_3^{2-}]$.

1,000-fold concentrates, which can then be cleaned with Chelex as described previously for nutrient stock solutions.

As an example of a recipe for a defined freshwater medium, Table 4.7 gives the composition of the medium Fraquil, which was designed in parallel with Aquil and shares its methods of preparation and its composition for major nutrients and vitamins. The major ion composition of Fraquil is that of the WC medium (Appendix A), and the chelating ligand is EDTA. The Fraquil recipe that is described here (which we will refer to as Fraquil*) has concentrations of free aquated trace metal ions that are similar to those of the Aquil* medium described previously and about an order of magnitude higher than in the original Fraquil described by Morel et al. (1975). As in Aquil*, the formation of a CaEDTA complex dominates the speciation of EDTA in Fraquil*, and as described previously, the 40 times lower calcium ion concentration results in proportionally higher conditional constants for the formation of trace metal complexes (see Tables 4.2 and 4.3). To partially compensate for this effect, the concentration of nominally "free" EDTA (EDTA*) in Fraquil* (10 $\mu$M) is 10 times lower than in Aquil*. The large effect of the calcium concentration on the free ion concentrations of trace metals in EDTA-buffered systems occurs over practically the whole range of calcium concentrations (0.2 to 2 mM) and freshwater compositions encountered in nature. In all cases, the speciation of EDTA* is dominated by the formation of a complex with calcium, and the free chelator concentration (EDTA$^{4-}$) varies inversely proportionally to that of Ca$^{2+}$. The only exception to this rule is in very-low-calcium ("soft") waters that are also acidic (pH $\leq$ 5) in which the various protonated forms of the chelating agent are dominant. In exceptional cases, the concentration of magnesium may be much larger than that of calcium and dominate EDTA speciation. In addition to the conditional binding constants applicable to the Fraquil* medium, Table 4.3 gives constants ($K_{Ca,M}$) for the reactions of exchange between CaEDTA$^{2-}$ and trace metals:

$$CaEDTA^{2-} + M^{n+} \leftrightarrow MEDTA^{(4-n)-} + Ca^{2+} \quad (31)$$

These constants permit a simple calculation of free trace metal ion concentrations in media of various Ca$^{2+}$ concentrations.

$$\log[M^{n+}] = \log[M_{tot}] - \log K_{Ca,M}$$
$$- \log[EDTA^*] + \log[Ca^{2+}] \quad (32)$$

The greater trace metal affinity of EDTA in freshwater media compared with seawater media can result in a less effective metal buffering for similar free metal ion concentrations, because the chelator concentration must be reduced to a value not much in excess of the trace metal concentrations. This effect is seen in the Fraquil* recipe, where nearly a quarter of the EDTA is complexed to the trace metals. Thus, these metals begin to titrate the EDTA as shown in Fig. 4.1 (see earlier discussion). In some situations it may then be necessary to use weaker chelating agents such as NTA or even citrate to buffer the trace metals at the appropriate free ion concentrations. An example of such a medium, which we will refer to as Fraquil-NTA, is shown in Table 4.8. This medium contains 50 μM NTA and the same calculated free ion concentrations of copper, zinc, manganese, and cobalt as the Fraquil* medium. However, because of the much lower conditional stability constants for NTA relative to those for EDTA (see Table 4.3), the total concentrations of these metals are ~10-fold lower than in Fraquil* and are ~1/100 of the total chelator concentration. Thus, one would be able to substantially increase free trace metal ion concentrations in the Fraquil-NTA medium without titrating the chelator.

In addition to the applicable EDTA constants, Table 4.3 also provides appropriate constants for NTA. It

**TABLE 4.8** Composition of the freshwater metal ion buffer system Fraquil-NTA at pH 8.0 and 20°C. The major ion composition of this medium is that of WC medium (see Appendix A); nutrients and vitamins are the same as in Aquil*.

| Constituent | Final concentration (M) | Log [M$^{n+}$] |
|---|---|---|
| NTA | $5.12 \times 10^{-5}$ | $-4.30^a$ |
| Ca | $3.00 \times 10^{-4}$ | $-3.60$ |
| Fe | $1.00 \times 10^{-6}$ | $—^b$ |
| Zn | $1.10 \times 10^{-7}$ | $-10.50$ |
| Mn | $3.00 \times 10^{-8}$ | $-8.00$ |
| Co | $1.78 \times 10^{-8}$ | $-11.00$ |
| Cu | $2.14 \times 10^{-8}$ | $-13.50$ |
| Na$_2$MoO$_4$·2H$_2$O | $1.00 \times 10^{-8}$ | $-8.00^c$ |
| Na$_2$SeO$_3$ | $1.00 \times 10^{-8}$ | $-8.00^c$ |

$^a$log [NTA*], the concentration of Ca, Mg, and protonated forms of NTA.
$^b$Fe' concentration cannot be accurately computed.
$^c$log [MoO$_4^{2-}$] or log [SeO$_3^{2-}$].

must be noted that the constants for the reactions of exchange between calcium and trace metals for NTA, K$_{Ca,M}$, are applicable only in cases where the calcium complex dominates the speciation of NTA. This occurs for media with sufficiently high pH and calcium ion concentration (log [Ca$^{2+}$] + pH > 3.4), as is the case for the Fraquil-NTA medium.

There are two other factors that need special attention in freshwater media. One is the low bicarbonate ion concentration in most freshwaters, which leads to poor buffering of pH and thus poor pH control. To alleviate this problem it may be necessary to add an additional pH buffer, and the use of nonchelating buffers such as HEPES is advised. Some commonly used pH buffers such as Tris and glycylglycine readily chelate trace metals, particularly at high pH, and thus can alter the trace metal speciation in the culture media.

A second factor is the low ratio of calcium to chelator (EDTA or NTA) in many freshwater metal ion buffer systems, which may cause unanticipated changes in the free calcium ion concentration and thus in free metal ion concentrations. In the Aquil* seawater medium, the EDTA concentration is 1% of the total calcium, and thus its addition to seawater does not appreciably change the free calcium ion concentration, which, as pointed out previously, regulates the binding strength of the EDTA. However, in the Fraquil* medium, the EDTA concentration is 4% of the calcium concentration, and in Fraquil-NTA, the NTA is 20% of the calcium. To prevent unintended changes in free calcium and resultant changes in the binding strength of the chelator, the EDTA or NTA should be added as a 1 : 1 calcium chelate, as advised earlier for trace metal additions.

## 5.0. CULTURE EXPERIMENTS IN METAL ION BUFFER SYSTEMS

Metal ion buffer systems allow one to conduct growth experiments in batch culture under constant concentrations of trace metal ions and dissolved inorganic species. These buffers thus allow one to determine relationships between the concentration of available metal species and important dependent biological variables, including the cellular metal concentration, the specific growth rate, and the steady-state metal uptake rate. In these studies the specific growth rate is determined from the exponential portion of growth curves (see Chapter 18), and the cellular metal is determined either by conventional chemical analysis or by the use of radiotracers.

In culture experiments, the algae should first be acclimated by growing them for at least 10 cell generations in the experimental culture medium at the same temperature and light conditions as will be used in the experiment. For metal toxicity experiments, the acclimation culture should be grown at a noninhibitory metal concentration, whereas for limitation experiments, the cells should be acclimated at a nutrient metal concentration sufficiently low to limit growth rate but high enough to permit the growth of a moderately dense inoculum culture. This culture should be growing exponentially and should not be allowed to become so dense as to substantially deplete major nutrients, $CO_2$, or free trace metal ion concentrations. The experimental cultures are initiated by pipetting the inoculum culture into a series of culture media containing a range of concentrations of the metal to be tested. The growth of the cells in the experimental media is then followed over time by daily measurements of *in vivo* chlorophyll fluorescence (Brand et al. 1981) or measurements of total cell concentrations or cell volume with a Coulter electronic particle counter.

Growth curves for such an experiment, examining growth limitation by zinc in the coastal diatom *Thalas-*

*siosira pseudonana*, are shown in Fig. 4.5. The cells were acclimated by first growing them in a 100-μM EDTA seawater medium at a growth-limiting Zn' concentration of 1.5 pM. They were then inoculated into a fresh aliquot of the same medium and two others containing ninefold and 100-fold higher Zn' concentrations (14 and 150 pM). The cells inoculated into the low Zn' medium continued to grow at a constant zinc-limited rate of 0.64 d⁻¹ for nine to 10 cell generations (a 700-fold increase in total cell volume), as indicated by the linear relationship between log cell volume and time. The cells inoculated into media with the two higher Zn' concentrations exhibited an increasing specific growth rate for the first 3 days and a constant specific rate of 1.6 d⁻¹ for the final 3 days. This latter rate represented the maximum zinc-sufficient rate for the algae. The specific growth rate observed at a Zn' concentration of

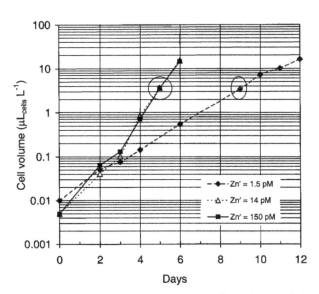

**FIGURE 4.5.** Semi-log growth curves for cultures of the marine diatom *Thalassiosira pseudonana* grown in 0.1 mM EDTA buffer systems at a salinity of 36, a temperature of 20°C, and 14 hours per day of saturating light (500 μmol · m⁻² · s⁻¹). The cells were preacclimated at a Zn' concentration of 1.5 pM and then were transferred to the same medium and to parallel culture media containing two higher Zn' levels (14 and 150 pM). Linear regression of ln cell volume versus time for the 1.5 pM Zn' medium over the initial 10 days of growth yielded a highly significant correlation ($R^2 = 0.998$). The slope of this regression line (0.64 d⁻¹) gives the specific growth rate. This rate is only 40% of the specific rate (1.6 d⁻¹) at the two higher [Zn'] over days 3 to 6, indicating substantial zinc-limitation of growth at the lowest [Zn']. The circles denote the point in the growth curves where cellular zinc quotas were measured. The data shown are from Sunda and Huntsman (unpublished).

1.5 pM ($0.64$ d$^{-1}$) was thus reduced to 40% of the maximum rate by an insufficient supply of zinc. The fact that this specific growth rate remained constant over the entire initial 10 days of the experiment indicated that the cells were acclimated to the zinc-limiting conditions and that the Zn′ concentration, which regulates the supply of zinc to the cells, remained constant during this period. On day 11, the growth rate of the zinc-limited culture declined due either to a decrease in Zn′ concentration (owing to slow dissociation kinetics of Zn-EDTA chelates; see previous section on kinetics) or to an adverse change in the chemistry of the medium, such as a decrease in $CO_2$ concentration (which exacerbates zinc limitation; Morel et al. 1994). Thus, to maintain constant growth conditions and thus constant algal growth rates in batch cultures, the cell density cannot be allowed to become too high. This invariably necessitates the use of sensitive methods for following culture growth, such as electronic particle counting of cell concentrations or of total cell volume.

## 5.1. Growth Rate Measurement

The growth of algal cells follows a simple first order rate law:

$$d\,CV/dt = \mu\,CV \qquad (33)$$

where $\mu$ is the specific growth rate and CV is the total cell volume per liter of culture or some other cellular parameter that is proportional to algal biomass. Integration over times $t_1$ and $t_2$ yields the well-known log-linear growth equation

$$\ln CV_{t2} - \ln CV_{t1} = \mu(t_2 - t_1) \qquad (34)$$

where ln CV is the natural log of the volume of cells per liter of culture medium. Thus, if a cell is growing at a constant specific rate, a plot of ln CV versus time will be linear, and the specific growth rate equals the slope of that line. Under these conditions, specific growth rates can be computed from a linear regression of ln CV versus time (see Chapter 18).

## 5.2. Measurement of Cell Quotas

Growth rate limitation by micronutrient metals is ultimately related to their intracellular concentration or their metal to biomass ratio, which is often expressed as a metal to carbon ratio. Thus, relationships between spe-

cific growth rate and cell quota are of central interest in defining the trace metal requirements of aquatic algae. Measurements of cell quotas ideally are made in cells growing at a constant acclimated specific growth rate within the log-linear portion of the growth curve. A second requirement is that a sufficient number of cells are sampled to obtain enough biomass for chemical analysis. This requirement is most easily achieved by sampling during the latter portion of exponential growth (e.g., see Fig. 4.5). Because the cell biomass cannot be allowed to reach too-high levels, sensitive analytical methods are needed such as the use of radiotracers (e.g., $^{59}$Fe, $^{65}$Zn, $^{57}$Co, and $^{109}$Cd) (Sunda and Huntsman 1983, 1992, 1995a, 1995b, Maldonado and Price 1996). Alternatively, sensitive chemical analytical methods may be used, such as atomic absorption spectrometry (Sunda and Huntsman 1995c) or ICPMS (inductively coupled plasma mass spectrometry) (Ho et al. 2003). The latter method is not only very sensitive but also allows for multi-element analysis of all important trace metals (iron, manganese, zinc, cobalt, copper, nickel, molybdenum, and cadmium), most major metals (magnesium, calcium, and potassium), and several nonmetallic nutrients such as phosphorus and sulfur.

To measure cellular metal quotas (normalized per cell, cell volume, or cell carbon), phytoplankton are typically grown in medium enriched with a radioisotope of the metal of interest. The cells are harvested by filtration, and the filters are analyzed by liquid scintillation counting or gamma ray spectroscopy. Cell densities and volumes are determined by microscopic cell counts with a counting chamber or by electronic particle counting with a Coulter counter. Cell carbon can be measured by radiotracer labeling with $^{14}$C (Sunda and Huntsman 1995b) or by elemental measurement with a C/H/N analyzer (Maldonado and Price 1996). The amount of radioactive tracer added varies, depending on the isotope used, the total metal concentration, the fraction of total metal taken up by the cells, and the efficiency of isotope counting. The specific activity of the medium is determined from direct measurement of the isotopic activity of the medium (in radioactive counts per minute) and knowledge of its total metal concentration. Phytoplankton are acclimated to the test media until steady state growth rates are achieved, which typically requires one to two culture transfers, corresponding to roughly 15 to 20 cell divisions. At this time, the cells are in isotopic equilibrium with the medium, so metal quotas are calculated from the cellular radioactivity retained by the filter and the specific activity of the medium.

Two caveats should be noted. The first is that the filters may adsorb some of the radioactive isotope. Filter blanks may be run, either by filtering similar volumes of sterile culture medium or by filtering the culture through two filters, one placed on top of the other. The second filter is used as a blank, because it contacts only the filtered medium. Normally the filter blank is negligible, but it may be significant if cell biomass and activity are low. The second caveat is that some of the metal may be bound to cell surfaces and may not be a metabolically relevant part of the phytoplankton metal quota. In metal-limiting media, most (and in many cases, virtually all) of the metal quota is intracellular. In media containing high concentrations of metal, such as might be encountered during metal-sufficient growth or short-term metal-uptake experiments, some fraction of the metal may bind to the surface of cells or be bound within or to the surface of metal precipitates (e.g., hydrous ferric oxide). Intracellular metal is typically determined in these situations by washing the cells with a nontoxic chemical reagent to remove extracellular metal. Extracellular iron is typically removed with the Ti-citrate-EDTA reagent of Hudson and Morel (1989) that removes ferric oxides and iron bound to cell surfaces via reduction to Fe(II). Manganese oxides are reduced and dissolved by a dilute (0.1 mM) ascorbate treatment at the ambient pH of the medium (Sunda and Huntsman 1987), and nonspecifically bound metals (e.g., zinc, cadmium, nickel, or copper) may be removed by rinsing them with seawater or freshwater medium containing 0.01 M DTPA (diethylenepentaacetic acid), a strong chelator (Lee et al. 1996). The Ti-EDTA-citrate wash contains high concentrations of iron and other metals and thus may not be suitable for use when the cellular metals are analyzed by ICPMS or atomic absorption spectroscopy.

## 5.3. Steady State Metal Uptake Rates

Steady state metal uptake rates ($V_{ss}$) can be computed from the product of the specific growth rate $\mu$ and the mean cellular metal concentration:

$$V_{ss} = \mu[\text{Cell metal}] \qquad (35)$$

If the specific growth rate has units of $d^{-1}$ and the cellular metal concentration is in units of $\mu$mol metal per liter of cell volume, then the units for steady state uptake are $\mu\text{mol} \cdot L_{cell}^{-1} \cdot d^{-1}$. Thus, it is a specific rate of cellular uptake, normalized to cell volume. Specific cellular uptake rates normalized to cell carbon [mol metal $\cdot$ (mol C)$^{-1} \cdot$ d$^{-1}$] can also be determined by multi-

plying the metal:carbon ratio by the specific growth rate. The rates determined in this fashion represent net rates of uptake, equal to the specific uptake rate minus the efflux rate. For nutrient metals under normal physiologic or growth-limiting ranges, efflux is usually negligible. Efflux does occur, however, for many toxic metals at high concentrations, in association with metal detoxification. Such efflux has been observed for cadmium (Lee et al. 1996, Sunda and Huntsman 1996) and copper (Croot et al. 2003) and likely occurs for other toxic metals, as found in bacteria (Silver et al. 1993).

For acclimated cells growing at a constant specific rate under continuous light and constant chemical and physical conditions, a true steady state can exist, and thus, Equation 35 can apply exactly. However, for cells grown under a diel light/dark cycle, as typically occurs in nature, photosynthetic carbon fixation, on which growth depends, occurs only during the day, leading to diel variations in both carbon-specific and biovolume-specific growth rates and associated diel oscillations in cellular metal concentrations and metal:carbon ratios (Sunda and Huntsman 2004). The cellular metal concentration usually decreases during the day because of increased daytime rates of growth and biodilution. For cultures grown under daily light : dark cycles, Equation 35 can still be applied, but average daily values for growth rate and cellular metal concentration (or metal : carbon ratio) must be used, and average daily metal uptake rates are computed. The normal methods for growth rate determination described above or in Chapter 18 provide average daily values, but determination of average daily cellular metal concentrations can require time-consuming and cumbersome multiple measurements over the entire diel cycle. In practice, the average daily cellular metal concentrations usually occur near the middle of the light period; therefore, measurements made during this time should provide a good estimate of average daily values (Sunda and Huntsman 2004). Thus, for culture studies employing daily light:dark cycles, we recommend that cellular metal concentrations be determined near the midpoint of the light period.

---

## 6.0. Acknowledgments

The writing of this chapter was funded by the National Center for Coastal Ocean Science, National Ocean Service, National Oceanic and Atmospheric Association and by the Center for Environmental Bioinorganic Chemistry, National Science Foundation.

## 7.0. REFERENCES

Anderson, D. M., and Morel, F. M. M. 1978. Copper sensitivity of *Gonyaulax tamarensis. Limnol. Oceanogr.* 23:283–95.

Anderson, M. A., Morel, F. M. M., and Guillard, R. R. L. 1978. Growth limitation of a coastal diatom by low zinc ion activity. *Nature.* 276:70–71.

Anderson, M. A., and Morel, F. M. M. 1982. The influence of aqueous iron chemistry on the uptake of iron by the coastal diatom *Thalassiosira weissflogii. Limnol. Oceanogr.* 27:789–813.

Bates, S. S., Tessier, A., Campbell, P. G. C., and Buffle, J. 1982. Zinc adsorption and transport by *Chlamydomonas variabilis* and *Scenedesmus subspicatus* (Chlorophyeae) grown in semi-continuous culture. *J. Phycol.* 18:521–29.

Boyd, P. W. et al. 2000. A mesoscale phytoplankton bloom in the polar Southern Ocean stimulated by iron fertilization. *Nature.* 407:695–702.

Brand, L. E., Guillard, R. R., and Murphy, L. S. 1981. A method for the rapid and precise determination of acclimated phytoplankton reproduction rates. *J. Plankton Res.* 3:193–201.

Brand, L. E., Sunda, W. G., and Guillard, R. R. L. 1983. Limitation of marine phytoplankton reproductive rates by zinc, manganese and iron. *Limnol. Oceanogr.* 28:1182–98.

Brand, L. E., Sunda, W. G., and Guillard, R. R. L. 1986. Reduction of marine phytoplankton reproduction rates by copper and cadmium. *J. Exp. Mar. Biol. Ecol.* 96:225–50.

Bruland, K. W. 1983. Trace elements in seawater. In: Riley, J. P., and Skirrow, G., eds. *Chemical Oceanography.* Academic Press, New York, pp. 157–220.

Bruland, K. W. 1989. Complexation of zinc by natural organic ligands in the central North Pacific. *Limnol. Oceanogr.* 34:269–85.

Buma, A. G. J., de Baar, H. J. W., Nolting, R. F., and van Bennekom, A. J. 1991. Metal enrichment experiments in the Weddell-Scotia Seas: Effects of Fe and Mn on various plankton communities. *Limnol. Oceanogr.* 36:1865–78.

Byrne, R. H., Kump, L. R., and Cantrell, K. J. 1988. The influence of temperature and pH on trace metal speciation in seawater. *Mar. Chem.* 35:163–81.

Coale, K. H. 1991. Effects of iron, manganese, copper, and zinc enrichments on productivity and biomass in the subarctic Pacific. *Limnol. Oceanogr.* 36:1851–64.

Coale, K. H., et al. 1996. A massive phytoplankton bloom induced by an ecosystem-scale iron fertilization experiment in the equatorial Pacific Ocean. *Nature.* 383: 495–501.

Crawford, D. W., et al. 2003. Influence of zinc and iron enrichments on phytoplankton growth in the northeastern subarctic Pacific. *Limnol. Oceanogr.* 48:1583–1600.

Croot, P. L., Karlson, B., Elteren, J. T., and Kroon, J. J. 2003. Uptake and efflux of copper in Synechococcus. *Limnol. Oceanogr.* 48:179–88.

Cutter, G. A., and Bruland, K. W. 1984. The marine biogeochemistry of selenium: A reevaluation. *Limnol. Oceanogr.* 29:1179–92.

Eide, D. J. 1998. Molecular biology of metal transport in *Saccharomyces cerevisiae. Annu. Rev. Nutr.* 18:441–69.

Evans, J. C., and Prepas, E. E. 1997. Relative importance of iron and molybdenum in restricting phytoplankton growth in high phosphorus saline lakes. *Limnol. Oceanogr.* 42:461–72.

Davey, E. W., Gentile, J. H., Erickson, S. J., and Betzer, P. 1970. Removal of trace metals from marine culture media. *Limnol. Oceanogr.* 15:486–88.

Fitzwater, S. E., Knauer, G. A., and Martin, J. H. 1982. Metal contamination and its effect on primary production measurements. *Limnol. Oceanogr.* 27:544–51.

Flegal, A. R., Smith, G. J., Gill, G. A., Sanudo-Wilhelmy, S., and Anderson, L. C. D. 1991. Dissolved trace element cycles in the San Francisco Bay estuary. *Mar. Chem.* 36:329–63.

Guillard, R. R. L., and Ryther, J. H. 1992. Studies of marine planktonic diatoms. I. *Cyclotella nana* (Hustedt) and *Detonula confervacea* (Cleve). *Gran. Can. J. Microbiol.* 8: 229–39.

Hart, B. A., Bertram, P. E., and Scaife, B. D. 1979. Cadmium transport by *Chlorella pyrenoidosa. Environ. Res.* 18:327–35.

Harrison, G. I., and Morel, F. M. M. 1983. Antagonism between cadmium and iron in the marine diatom *Thalassiosira weissflogii. J. Phycol.* 19:495–507.

Harrison, P. J., Yu, P. W., Thompson, P. A., Price, N. M., and Phillips, D. J. 1988. Survey of selenium requirements in marine phytoplankton. *Mar. Ecol. Prog. Ser.* 47:89–96.

Ho, T. Y., Quigg, A. Finkel, Z. V., Milligan, A. J., Wyman, K., Falkowski, P. G., and Morel, F. M. 2003. The elemental composition of phytoplankton. *J. Phycol.* 39:1145–59.

Hudson, R. J. M., and Morel, F. M. M. 1989. Distinguishing between extra- and intracellular iron in marine phytoplankton. *Limnol. Oceanogr.* 34:1113–20.

Hudson, R. J. M., and Morel, F. M. M. 1990. Iron transport in marine phytoplankton: Kinetics of cellular and medium coordination reactions. *Limnol. Oceanogr.* 35:1002–20.

Hudson, R. J. M., Covault, D. T., and Morel, F. M. M. 1992. Investigations of iron coordination and redox reactions in seawater using $^{59}$Fe radiometry and ion-pair solvent extraction of amphiphilic iron complexes. *Mar. Chem.* 38:209–35.

Hudson, R. J. M., and Morel, F. M. M. 1993. Trace metal transport by marine microorganisms: Implications of metal coordination kinetics. *Deep-Sea Res.* 40:129–51.

Hudson, R. J. 1998. Which aqueous species control the rates of trace metal uptake by aquatic Biota? Observations and predictions of non-equilibrium effects. *Sci. Total Environ.* 1998:95–115.

Hutchins, D. A., DiTullio, G. R., Zhang, Y., and Bruland, K. W. 1998. An iron limitation mosaic in the California upwelling regime. *Limnol. Oceanogr.* 43:1037–54.

Hutner, S. H., Provasoli, L., Schatz, A., and Haskins, C. P. 1950. Some approaches to the study of metals in the metabolism of microorganisms. *Proc. Amer. Phil. Soc.* 94:152–70.

Jones, G. J., Palenik, B. P., and Morel, F. M. M. 1987. Trace metal reduction by phytoplankton: The role of plasmalemma Redox enzymes. *J. Phycol.* 23:237–44.

Keller, M. D., Bellows, W. K., and Guillard, R. L. 1988. Microwave treatment for sterilization of phytoplankton culture media. *J. Exp. Mar. Biol. Ecol.* 117:279–83.

Kustka, A. B., Sanudo-Wilhelmy, S. A., Carpenter, E. J., Capone, D., Burns, J., and Sunda, W. G. 2003. Iron requirements for dinitrogen and ammonium supported growth in cultures of *Trichodesmium* (IMS 101): Comparison with nitrogen fixation rates and iron:carbon ratios of field populations. *Limnol. Oceanogr.* 48:1869–84.

LaFontaine, S., Quinn, J. M., Nakamoto, S. S., Page, M. D., Göhre, V., Moseley, J. L., Kropat, J., and Merchant, S. 2002. Copper dependent iron assimilation pathway in the model photosynthetic eukaryote *Chlamydomonas reinhardtii*. *Eukaryotic Cell* 1:736–57.

Lee, J. G., and Morel, F. M. M. 1995. Replacement of zinc by cadmium in marine phytoplankton. *Mar. Ecol. Prog. Ser.* 127:305–9.

Lee, J. G., Ahner, B. A., and Morel, F. M. M. 1996. Export of cadmium and phytochelatin by the marine diatom Thalassiosira weissflogii. *Environ. Sci. Tech.* 30:1814–21.

Maldonado, M. T., and Price, N. M. 1996. Influence of N substrate on Fe requirements of marine centric diatoms. *Mar. Ecol. Prog. Ser.* 141:161–72.

Maldonado, M. T., and Price, N. M. 2000. Nitrate regulation of iron reduction and transport by Fe-limited *Thalassiosira oceanica*. *Limnol. Oceanogr.* 45:814–26.

Maldonado, M. T., and Price, N. M. 2001. Reduction and transport of organically bound iron by *Thalassiosira oceanica*. *J. Phycol.* 37:298–310.

Martell, A. E., and Smith, R. M. 1974. *Critical Stability Constants, Vol. 1: Amino Acids*. Plenum Press, New York, 469 pp.

Martin, J. H., and Fitzwater, S. E. 1988. Iron deficiency limits phytoplankton growth in the Northeast Pacific subarctic. *Nature* 331:341–43.

Moffett, J. W., and Brand, L. E. 1996. Production of strong, extracellular Cu chelators by marine cyanobacteria in response to Cu stress. *Limnol. Oceanogr.* 41:388–95.

Morel, F. M. M., and Hering, J. G. 1993. *Principles and Applications of Aquatic Chemistry*, Wiley, New York, 588 pp.

Morel, F. M. M., Reinfelder, J. R., Roberts, S. B., Chamberlain, C. P., Lee, J. G., and Yee, D. 1994. Zinc and carbon co-limitation of marine phytoplankton. *Nature* 369:740–42.

Morel, F. M. M., Rueter, J. G., Anderson, D. M., and Guillard, R. R. L. 1979. Aquil: A chemically defined phytoplankton culture medium for trace metal studies. *J. Phycol.* 15:135–41.

Morel, F. M. M., Westall, J. C., Rueter, J. G., and Chaplick, J. P. 1975. *Description of the Algal Growth Media Aquil and Fraquil. Technical Note #16*. R. M. Parsons Laboratory for Water Resources and Hydrodynamics, Department of Civil Engineering, Massachusetts Institute of Technology, Cambridge, Massachusetts.

Nielands, J. B. 1989. Siderophore systems in bacteria and fungi. In: Beverage T. J., and Doyle, R. J., eds. *Metal Ions and Bacteria*. Wiley, New York, pp. 141–63.

Palenik, B., and Morel, F. M. M. 1988. Dark production of $H_2O_2$ in the Sargasso Sea. *Limnol. Oceanogr.* 33:1606–11.

Peers, G. S., and Price, N. M. 2004. A role for manganese in superoxide dismutases and the growth of iron-deficient diatoms. *Limnol. Oceanogr.* 49:1174–1783.

Price, N. M., Harrison, P. J., Azam, F., Landry, M. R., and Hall, K. J. F. 1986. Toxic effects of latex and tygon tubing on marine phytoplankton, zooplankton, and bacteria. *Mar. Ecol. Prog. Ser.* 34:41–49.

Price, N. M., Thompson, P. A., and Harrison, P. J. 1987. Selenium: An essential element for growth of the coastal marine diatom *Thalassiosira pseudonana* (Bacillariophyceae). *J. Phycol.* 23:1–9.

Price, N. M., and Harrison, P. J. 1988. Specific selenium-containing macromolecules in the marine diatom *Thalassiosira pseudonana*. *Plant Physiol.* 86:192–99.

Price, N. M., Harrison, G. I., Hering, J. G., Hudson, R. J., Nirel, P. M. V., Palenik, B., and Morel, F. M. M. 1988/89. Preparation and chemistry of the artificial algal culture medium Aquil. *Biol. Oceanogr.* 6:443–61.

Price, N. M., and Morel, F. M. M. 1990. Cadmium and cobalt substitution for zinc in a marine diatom. *Nature* 344:658–60.

Price, N. M., and Morel, F. M. M. 1991. Co-limitation of phytoplankton growth by nickel and nitrogen. *Limnol. Oceanogr.* 36:1071–77.

Price, N. M., Ahner, B. A., and Morel, F. M. M. 1994. The equatorial Pacific Ocean: Grazer controlled phytoplankton populations in an iron-limited ecosystem. *Limnol. Oceanogr.* 39:520–34.

Provasoli, L., McLaughlin, J. J. A., and Droop, M. R. 1957. The development of artificial media for marine algae. *Arch. Mikrobiol.* 25:392–425.

Raven, J. A. 1988. The iron and molybdenum use efficiencies of plant growth with different energy, carbon, and nitrogen sources. *New Phytol.* 109:279–87.

Raven, J. A. 1990. Predictions of Mn and Fe use efficiencies of phototrophic growth as a function of light availability for growth and C assimilation pathway. *New Phytol.* 116:1–18.

Raven, J. A., Evans, M. C. W., and Korb, R. E. 1999. The role of trace metals in photosynthetic electron transport in $O_2$-evolving organisms. *Photosynth. Res.* 60:111–49.

Rich, H. W., and Morel, F. M. M. 1990. Availability of well-defined iron colloids to the marine diatom *Thalassiosira weissflogii*. *Limnol. Oceanogr.* 35:652–62.

Rueter, J. G., and Morel, F. M. M. 1982. The interaction between zinc deficiency and copper toxicity as it affects the silicic acid uptake mechanisms in *Thalassiosira pseudonana*. *Limnol. Oceanogr.* 26:67–73.

Saito, M. A., Moffett, J. W., Chisholm, S. W., and Waterbury, J. B. 2002. Cobalt limitation and uptake in *Prochlorococcus*. *Limnol. Oceanogr.* 47:1629–36.

Silver, S., Lee, B. O., Brown, N. L., and Cooksey, D. A. 1993. Bacterial plasmid resistances to copper, cadmium, and zinc. In: Welch, A. J., and Capman, S. K., eds. *The Chemistry of Copper and Zinc Triads*. Royal Society of Chemistry, London, pp. 38–53.

Sunda, W. G. 1988/1989. Trace metal interactions with phytoplankton. *Biol. Oceanogr.* 6:411–42.

Sunda, W. G. 2000. Trace metal-phytoplankton interactions in aquatic systems. In: Lovley, D. R., ed. *Environmental Microbe-Metal Interactions*. ASM Press. Washington, D. C., pp. 79–107.

Sunda, W. G. 2001. Bioavailability and bioaccumulation of iron in the sea. In: Turner, D. R., and Hunter, K. A., eds. *The Biogeochemistry of Iron in Seawater*. Wiley, New York, pp. 41–84.

Sunda, W. G., and Guillard, R. R. L. 1976. The relationship between free cupric ion activity and the toxicity of copper to phytoplankton. *J. Mar. Res.* 34:511–29.

Sunda, W. G., and Huntsman, S. A. 1983. The effect of competitive interactions between manganese and copper on cellular manganese and growth in estuarine and oceanic species of the diatom *Thalassiosira*. *Limnol. Oceanogr.* 28:924–34.

Sunda, W. G., and Huntsman, S. A. 1985. Regulation of cellular manganese and manganese transport rates in the unicellular alga *Chlamydomonas*. *Limnol. Oceanogr.* 30:71–80.

Sunda, W. G., and Huntsman, S. A. 1986. Relationships among growth rate, cellular manganese concentrations, and manganese transport kinetics in estuarine and oceanic species of the diatom *Thalassiosira*. *J. Phycol.* 22:259–70.

Sunda, W. G., and Huntsman, S. A. 1987. Microbial oxidation of manganese in a North Carolina estuary. *Limnol. Oceanogr.* 32:552–64.

Sunda, W. G., and Huntsman, S. A. 1992. Feedback interactions between zinc and phytoplankton in seawater. *Limnol. Oceanogr.* 37:25–40.

Sunda, W. G., and Huntsman, S. A. 1995a. Cobalt and zinc interreplacement in marine phytoplankton: Biological and geochemical implications. *Limnol. Oceanogr.* 40:1404–17.

Sunda, W. G., and Huntsman, S. A. 1995b. Iron uptake and growth limitation in oceanic and coastal phytoplankton. *Mar. Chem.* 50:189–206.

Sunda, W. G., and Huntsman, S. A. 1995c. Regulation of copper concentrations in the oceanic nutricline by phytoplankton uptake and regenerations cycles. *Limnol. Oceanogr.* 40:132–37.

Sunda, W. G., and Huntsman, S. A. 1996. Antagonisms between cadmium and zinc toxicity and manganese limitation in a coastal diatom. *Limnol. Oceanogr.* 41:373–87.

Sunda, W. G., and Huntsman, S. A. 1997. Interrelated influence of iron, light, and cell size on growth of marine phytoplankton. *Nature* 390:389–92.

Sunda, W. G., and Huntsman, S. A. 1998a. Interactive effects of external manganese, the toxic metals copper and zinc, and light in controlling cellular manganese and growth in a coastal diatom. *Limnol. Oceanogr.* 43:1467–75.

Sunda, W. G., and Huntsman, S. A. 1998b. Interactions among $Cu^{2+}$, $Zn^{2+}$, and $Mn^{2+}$ in controlling cellular Mn, Zn, and growth rate in the coastal alga *Chlamydomonas*. *Limnol. Oceanogr.* 43:1055–64.

Sunda, W. G., and Huntsman, S. A. 2003. Effect of pH, light, and temperature on Fe-EDTA chelation and Fe hydrolysis in seawater. *Mar. Chem.* 84:35–47.

Sunda, W. G., and Huntsman, S. A. 2004. Relationships among photoperiod, carbon fixation, growth, chlorophyll a, and cellular iron and zinc in a coastal diatom. *Limnol. Oceanogr.* 49:1742–1753.

Thompson, M. E. 1964. Magnesium in seawater: An electrode measurement. Science 153:866–7.

Thompson, M. E., and Ross, J. W., Jr. 1966. Calcium in seawater by electrode measurement. *Science* 154:1643–4.

Waite, T. D. 2001. Thermodynamics of the iron system in seawater. In: Turner, D. R., and Hunter, K. A., eds. *The Biogeochemistry of Iron in Seawater*. Wiley, New York, pp. 291–342.

Wehr, J. D., and Brown, L. M. 1985. Selenium requirement of a bloom-forming planktonic alga from softwater and acidified lakes. *Can J. Fish. Aquat. Sci.* 42:1783–8.

Westall, J. W., Zachary J. L., and Morel, F. M. M. 1976. *MINEQL. Technical Note #18*. R. M. Parsons Laboratory for Water Resources and Hydrodynamics, Department of Civil Engineering, Massachusetts Institute of Technology, Cambridge, Massachusetts.

Wilhelm, S. W. 1995. Ecology of iron limited cyanobacteria: a review of physiological responses in implications for aquatic systems. *Aquat. Microb. Ecol.* 9:295–303.

Wurtsbaugh, W. A., and Horne, A. J. 1983. Iron in eutrophic Clear Lake, California: Its importance for algal nitrogen fixation and growth. *Can. J. Fish. Aquat. Sci.* 40:1419–29.

Xue, H. B., Stumm, W., and Sigg, L. 1988. The binding of heavy metals to algal surfaces. *Wat. Res.* 22:917–26.

# STERILIZATION AND STERILE TECHNIQUE

MASANOBU KAWACHI
*National Institute for Environmental Studies*

MARY-HÉLÈNE NOËL
*National Institute for Environmental Studies*

CONTENTS

Key Index Words: Autoclaving, Bleaching, Cell Culture Transfer, Culture Medium, Dry-Heat Sterilization, Filtration, Microwave Sterilization, Pasteurization, Sterilization, Tyndallization, Ultraviolet Radiation

## 1.0. INTRODUCTION

Sterilization is a process for establishing an aseptic condition, that is, the removal or killing of all microorganisms. Sterilization is very important in phycological research, especially when maintaining living organisms as isolated strains in culture. The use of sterile technique, in combination with sterile equipment and supplies, minimizes contamination, resulting in more precise experiments free of potential variables caused by unwanted organisms. Sterilization is not a difficult procedure, but precautions must be taken when working with sterilized material to avoid contamination. Indeed, once sterilized materials are exposed, they soon become contaminated from the ambient air, which contains dust, spores, and microorganisms.

The term *disinfect* is sometimes confused with sterilization. Disinfection is usually defined as an operation that kills or reduces the number of pathogenic microorganisms in an environment or on a surface. For example, cleaning hands and surfaces such as tables and benches with detergents or 70% ethanol before handling cultures or sterilized equipment is a disinfecting procedure that is usually conducted as part of the sterile technique. The aim of this chapter is to describe different sterilization methods and basic sterile techniques (including disinfecting) that are commonly used in phycological studies.

There are various sterilization methods, which can be roughly classified into four categories: heat sterilization, electromagnetic wave sterilization, sterilization using filtration, and chemical sterilization (Table 5.1). Heat sterilization is the most common of the general categories and usually requires high temperatures (≥100°C), implying that the materials to be sterilized can resist high temperatures (e.g., glassware, metallic instruments, and aluminum foil). Liquids are filter-sterilized when the liquid contains fragile components that are destroyed by high temperature. Electromagnetic waves (e.g., ultraviolet [UV] rays, gamma rays, x-rays, and microwaves) are used as an alternative for materials that cannot be exposed to high temperature (e.g., many plastic products or liquids with a labile component). Today, disposable plastic supplies sterilized with gamma rays are readily available. Finally, many different types of chemicals have been used for the purpose of sterilization (Table 5.1); however, chemical traces may remain after the sterilization treatment, and those chemicals may be detrimental to living algae and to the investigator. Therefore, chemical sterilization procedures have fallen into disuse in the laboratory.

## 2.0. PRESTERILIZATION CLEANING PROCEDURES

### 2.1. New Culture Vessels

New glass and plastic vessels, except ready-to-use sterilized products, should be cleaned before the first use, because chemicals or other traces of the product manufacturing can remain in the vessels and may be harmful to living cells. The procedure consists of immersing the containers in a dilute hydrochloric acid bath (generally 1 M HCl) for 1 week, rinsing them several times with running tap water, and rinsing with distilled or deionized water before drying. For a less rigorous method, new glass and plastic vessels can be washed with neutral detergent commercialized for laboratory use (e.g., M-251L without Phosphorus, Shapu Manufacturing System; Neodisher FT with neutralizer, Dr. Weigert GmbH & Co.), followed by thorough rinsing and drying.

Several types of caps for vessels are available on the market: heat-resistant plastic caps, metal caps, and expansive silicon rubber plugs. It is recommended to check the safety of the material before use, because some products can release toxic substances during the autoclaving process. For example, black Bakelite caps should be autoclaved in water several times before using with culture medium. New black caps color the water with phenolic compounds that are released when the caps are first autoclaved. New polyethylene caps may also release harmful substances when they are first autoclaved.

### 2.2. Dirty Culture Vessels

Culture vessels that were used to grow cells should be autoclaved to kill all cells. Living cells, especially as cysts, may cause contamination of local waters if discarded improperly. After autoclaving, the liquid is cooled and discarded; when agar is used, it should be cooled to near the gelling point and then discarded. The vessels should be briefly rinsed with running water and then cleaned.

The standard cleaning method consists of immersing the vessels overnight in a neutral detergent bath (commercial detergent for laboratory use), followed by scrubbing with a brush and sponge. Vessels are then rinsed several times with running water, ensuring that all detergent is removed (i.e., no suds should be present [rinsing >10 times ensures good removal]). The final rinse is with distilled or deionized water, and the cleaned vessels are dried in an area protected from dust (Fig. 5.1).

Other glassware such as petri dishes, watch glasses, and microscope slides are also cleaned with this procedure. In cases where the algal cells are strongly attached to the inner surface of the glassware, it is often necessary to soak the glassware in hot water several hours before proceeding with the subsequent cleaning steps. For cleaning caps on screw-type culture tubes, as well as the expansive type of silicon rubber plugs for flasks (Fig. 5.2), a mild detergent is used (with scrubbing when necessary), followed by rinsing and drying as described previously.

### 2.3. Reusable Glass Pipettes

Reusable glass pipettes, used to transfer cells, should be immediately rinsed with tap water after use so that cell material does not dry onto the glass surfaces. The cotton plug should be removed before the pipettes are rinsed. When immediate rinsing is not possible, the pipettes should be placed tip down into a plastic beaker containing water and then rinsed later. For washing, the pipettes can be immersed in a detergent bath (i.e., commercial detergent for laboratory use) (Fig. 5.3) for

**TABLE 5.1.**  Summary of sterilization types, including applicability and limits

| Category | Sterilization method | Effective method | Application | Limitation |
|---|---|---|---|---|
| Heat | Flame | Direct heat with fire (Bunsen burner) | Surface sterilization (test tube openings, transfer loops, glass pipettes) | Non–heat-resistant materials (e.g., most plastics) |
| Heat | Autoclaving | 2 atm (steam pressure), 121°C; time varies (10, 20 min for small liquid vol; 1 h for large vol) | For general use: liquids and agar, glass and metal vessels, equipment | Non–heat-resistant materials; pH change; metal contamination |
| Heat | Dry heat | 250°C, 3 to 5 h; current protocol at 150°C for 3 to 4 h | Dry goods: glass and metal vessels and equipment | Non–heat-resistant materials; liquids |
| Heat | Pasteurization | 66–80°C for at least 30 min, followed by quick cooling (4–10°C) | Liquids with heat-labile components | Not complete sterilization (originally for killing food germs) |
| Heat | Tyndallization | 60–80°C, 30 min, followed by quick cooling; cycle repeated 3 times in 3 d | Liquid with heat-labile components | Requires time |
| Filtration | Filtration | ≤0.2 μm pore size filter | Liquid with heat-labile components | Small volumes, high-viscosity liquids, viruses not eliminated |
| Electromagnetic waves | Microwave | 10 min at 700 W; 5 min with intervals at 600 W. For dry goods: 20 min at 600 W with water, 45 min without water | Liquids: small volume of media; dry goods: glassware, vessels | Small liquid volumes; dry goods with water require elimination of water |
| Electromagnetic waves | Ultraviolet radiation | 260 nm, 5–10 min | Surface of materials, working area | Ultraviolet-sensitive plastics |
| Chemical | Bleach (sodium hypochlorite) | 1–5 mL for 1 L water, several hours | Large volume of water for aquaculture | Cysts may survive; neutralization required (e.g., sodium thiosulfate, $250 \text{ g} \cdot \text{L}^{-1}$ stock solution; 1 mL for 4 mL of bleach) |
| Chemical | Ethanol | 50–70% solution | Popular, general disinfection | Some resistant microorganisms |
| Chemical | Ethylene oxide | Airtight room or pressure cabin | Plastic and rubber products, non–heat-resistant products | Explosive; chemical residue is problematic or toxic |
| Chemical | Corrosive sublimate, $HgCl_2$ | 0.1%; add same amount of NaCl and dissolve with distilled water | Antiseptic and disinfectant | Poison; not for materials contacting live cells |
| Chemical | Phenol (carbolic acid) | 3% solution | Antiseptic and disinfectant | Poison; not for materials contacting live cells |
| Chemical | Saponated cresol solution | 3–5% solution | Antiseptic and disinfectant | Poison; not for materials contacting live cells |
| Chemical | Formaldehyde (formalin) | 2–5% solution | Antiseptic and disinfectant | Poison; not for materials contacting live cells |

**FIGURE 5.1.** A dust-proof drying cabinet for storing clean or sterile materials and supplies.

**FIGURE 5.2.** Erlenmeyer flasks with silicon rubber plugs. *Left,* covered with aluminum foil; *center,* a standard silicon rubber plug; *right,* a silicon rubber plug with an extended shield over the lip of the flask.

1 or 2 days. An ultrasonic bath and siphon-type pipette washer are used to completely remove the detergent (Fig. 5.4). Pipettes are then rinsed with distilled water and dried in an oven at 150°C. Cotton plugs are then inserted into the top of the pipettes, the pipette end is briefly flamed to remove loose cotton fibers, and then

**FIGURE 5.3.** Detergent baths for cleaning glass inoculation pipettes (inset: top view of bath).

the pipettes are placed into their canister and autoclaved or sterilized by dry-heat technique.

## 3.0. Sterilization Methods

### 3.1. Autoclaving

Autoclaving is the most effective and popular way to sterilize heat-resistant materials and is generally used to sterilize liquids. An autoclave is a specialized apparatus consisting of a heavy-walled closed chamber (Fig. 5.5) within which high steam pressure produces a high-sterilizing temperature (ca. 121°C; Table 5.2) without boiling liquids (under ideal conditions). Duration of the autoclaving depends on the volume of liquid to be sterilized; for example, 10 minutes of autoclaving at 121°C is sufficient to sterilize test tubes of 18 mm diameter, whereas 1 hour is required to sterilize 10 liters of liquid. When properly operated, an autoclave kills all microorganisms, even the heat-resistant spores of bacteria and fungi. After autoclaving, the surface and the inside of sterilized materials are often wet, especially materials containing paper or cotton, and in most cases a dry-heat

**FIGURE 5.4.** An ultrasonic siphon-type pipette washer.

step in an oven at 150°C is required unless the autoclave is capable of a dry cycle for vessels resistant to 150°C.

Some precautions must be taken for correct autoclaving. For liquids, it is important to leave a free space in the containers (e.g., flasks and test tubes) of at least one quarter of the total volume to guarantee space for steam and potential boiling. The caps of the containers, especially screw caps of test tubes, should be loose to prevent an excessive buildup of pressure.

Steam- and heat-resistant material is required for labeling vessels to be autoclaved. Autoclave tape that is commercially available is recommended, and the date of sterilization should be added to the label. There should be space between the individual items within the autoclave; that is, do not overfill the autoclave. Before starting the autoclave, make sure that the water level in the autoclave is sufficient to ensure steam generation throughout the autoclaving process, unless the autoclave is connected to a utility steam supply. The autoclave door should be closed without excessive tightening, because the steam pressure pushes outward on the door, forming a tight seal. If tightening is excessive, then it may be difficult to open the door.

After completion of the autoclaving process, the autoclave door should not be opened until the pressure is completely reduced and the temperature is less than

**Pasteurization**

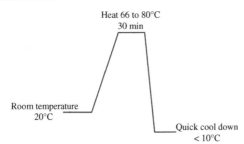

**Tyndallization**

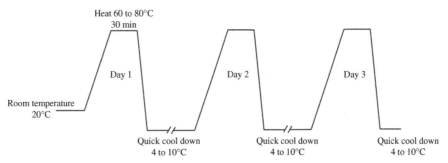

**FIGURE 5.5.** Schematic representation of the temperature cycles for pasteurization and tyndallization.

**TABLE 5.2.** Sterilization temperature and steam pressure for autoclaving.

| Temperature | | Steam pressure | |
| --- | --- | --- | --- |
| °C | °F | atm | psi |
| 100 | 212 | 1.0 | 14.7 |
| 110 | 230 | 1.4 | 20.6 |
| 115 | 239 | 1.7 | 25.0 |
| 121 | 250 | 2.0 | 29.4 |
| 134 | 273 | 3.0 | 44.1 |

100°C. Opening the autoclave prematurely may cause boiling of superheated liquids and may result in contamination from spores in the ambient room air that circulate by convection currents and enter the autoclave; however, a liquid culture medium should not be left to cool to room temperature in the autoclave, because this often results in the formation of precipitants. This is particularly important when autoclaving seawater or a culture medium containing silicate. Rapid cooling outside the autoclave minimizes the formation of precipitants. For screw-top culture tubes, the loose caps should be tightened after the liquid is cool. Hot glassware should not be placed directly into cold storage, because this may cause fine cracks to form in the glass. When trace metal contamination must be avoided, other methods of sterilization should be used (Price et al. 1989).

## 3.2. Dry-Heat Sterilization

A hot air oven is generally used for dry-heat sterilization of dry goods (i.e., not liquids). This mode of sterilization is advantageous to eliminate unacceptable autoclave residues (e.g., for certain isolation techniques; see Chapter 6) and to dry the cotton in cotton-plugged pipettes. Dry-heat sterilization requires higher temperatures and longer heating than sterilization by autoclaving. The most rigorous method involves heating to 250°C for 3 to 5 hours; however, in many instances, heating to 150°C for 3 or 4 hours is sufficient. If the oven is not equipped with a fan, then the temperature in the oven may not be uniform; caution is required, because some areas of the oven may reach a higher temperature than the preset temperature, possibly leading to damage of some materials.

The materials to be sterilized should be dry and covered (e.g., with aluminum foil or an autoclave bag) or placed inside a container (e.g., a stainless steel box or glass dish) to avoid contamination once the sterilization process is finished. The door of the oven should not be opened before the oven temperature has decreased less then 60°C, because quick cooling of the sterilized material enhances contamination by convection currents.

## 3.3. Pasteurization and Tyndallization

During the nineteenth century, Tyndall and Pasteur developed, for other purposes, procedures for steaming solutions, such as food items; hence, the procedure is called *pasteurization* and *tyndallization*. Phycologists have adopted and modified the techniques for liquids that should not be exposed to temperatures more than 100°C or that cannot be autoclaved. Heat is usually provided by unpressurized steam; thus the term *steaming* is often applied in phycology (Starr and Zeikus 1993; see also Chapter 11). Pringsheim (1946) recommended steaming for the preparation of biphasic soil-water culture medium, and steaming is often used to prepare enriched seawater culture media. Various methods are practiced, although no clear definitions are assigned to these, but they can be summarized as methods for heating a liquid solution to a high temperature and holding it at this temperature for a period, followed by a quick cooling.

For pasteurization, the liquid is traditionally raised to a temperature between 66°C and 80°C, held at this temperature for at least 30 minutes, and then rapidly cooled to a temperature less than 10°C (Fig. 5.6). However, pasteurization of seawater is usually carried out at 95°C for 1 hour. For tyndallization, a process similar to pasteurization is applied on the first day (heating to 60°C–80°C for 30 minutes, quickly cooling to ~4°C–10°C); the material is kept cool until the following day, when it is heated and cooled as before; on the third day, the cycle is again repeated (Fig. 5.6). The repeated process is designed to kill cysts that germinate after the first or second heating.

In aquaculture, where large containers of seawater or freshwater are used, flash sterilization is used. Steam is produced, often by a home furnace, and passed between the titanium plates of a heat exchanger. The seawater (or freshwater) is pumped through the titanium plates, and the water is almost instantly heated to ~70°C. This

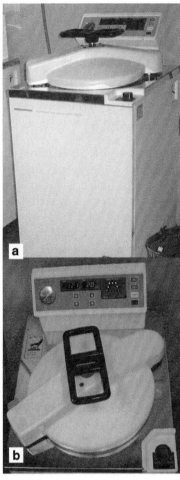

**FIGURE 5.6.** Autoclave for laboratory use. **(a)** Front view. **(b)** Top view.

**FIGURE 5.7.** A reusable filtration apparatus inside an autoclavable bag.

procedure does not kill the most resistant spores, but it does kill most life in the water.

### 3.4. Sterilization by Filtration

Sterilization by filtration is required for heat-labile components, such as vitamins, or volatile components of liquids, such as organic solvents. This method is also used for convenience and rapidity when only a small volume of liquid is to be sterilized. A variety of filters are available (i.e., different pore sizes, composition, color, and size). Membrane filters that can be autoclaved are generally used for sterilizing culture medium. The filters should have a pore size less than 0.2 μm; however, it is important to note that viruses can pass through such filters. When the solution has a high viscosity or contains suspended particles, prefiltration with a 1-μm pore size filter is required. Both disposable and reusable types of filtration equipment are commercially available. Sterile, disposable filtration sets are convenient and ready to use, but they are relatively expensive. Reusable filtration equipment (e.g., glass or polycarbonate materials) can usually be autoclaved; the membrane filter and its assembly unit are either placed in an autoclave bag (Fig. 5.7) or wrapped in aluminum foil before autoclaving. After autoclaving, the filtration set should be dried in an oven at a temperature of up to 120°C. After cooling, the filtration apparatus can be used on a bench top (i.e., without requiring a laminar flow hood), as long as the filtration reservoir is not exposed to contamination. After the liquid is filtered, the sterile solution in the reservoir bottle can be dispensed (using sterile technique) into other appropriate sterilized vessels.

For very small volumes, sterile syringes with disposable filter units (Fig. 5.8) are commonly used, although reusable filter units are also available. Disposable units are manufactured for immediate use, whereas reusable filter units must be autoclaved and dried before use as described previously.

Filter sterilization should be used with caution, however. Membrane filters, such as intertwined organic fibers or perforated polycarbonate sheets, sometimes have openings larger than the nominal size. For example, Stokner et al. (1990) found that 0.2-μm filters had numerous surface defects visible by electron microscopy and that both particles and algae that were several micrometers were identified in the filtrates. Small heterotrophic flagellates are able to squeeze through pore openings that are much smaller than their normal cell diameter.

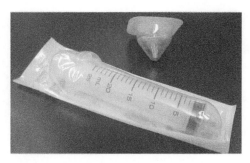

**FIGURE 5.8.**   Disposable syringe filtration unit.

## 3.5. Microwave Oven Sterilization

Sterilization with a microwave oven is quicker than steam or dry sterilization. The heat produced by microwaves is of two types: ionic polarization and dipole rotation. In practice, cells may be killed more by the heat from boiling liquids than by microwaves, and indeed, the killing mechanism is not known (Keller et al. 1988). But nevertheless, the microwave sterilization technique is effective and nontoxic. It is recommended to use a microwave oven that is equipped with a rotating table and has up to 700 watts (W) of power.

For the sterilization of liquids in previously cleaned culture vessels, there are different protocols. Keller et al. (1988) showed that, for 1 to 1.5 liters of seawater, microalgae were killed in 5 minutes, bacteria in 8 minutes, and fungi in 10 minutes. Microwave ovens manufactured today are more powerful than those produced in 1988, so the times may vary depending on the power of the microwave oven. Furthermore, the effectiveness of a microwave oven often diminishes over time, and therefore an older oven requires more time. Leal et al. (1999) describe a quick intermittent protocol, such as a 5-minute total microwave time (2, 1, and 2 minutes of treatment, separated by intervals of 30 seconds) at 600 W.

For sterilizing glassware, Boye and van den Berg (2000) described a protocol in which a small volume of distilled or deionized water was added and then microwaved for 20 minutes at 600 W. After sterilization, the water was discarded in a laminar flow hood with use of sterile technique. When water was not added to the glassware, sterilization by microwave required 45 minutes or more to destroy bacterial spores (Jeng et al. 1987), analogous to time differences between autoclave and dry-heat oven sterilization methods (see previous text).

## 3.6. Ultraviolet Radiation

Although x-rays and gamma rays (ionizing radiation) are widely used for commercial manufacturing (e.g., sterile plasticware), they are not suitable for use in laboratories for safety reasons. UV radiation is suitable for laboratory use, including sterilization of culture facility hoods and benches (see later text). UV radiation is damaging to humans (especially eyes), and care must be taken to avoid exposure. In addition, UV radiation produces ozone, and some people find this offensive (Hamilton 1973). UV radiation has its primary lethal effect at 260 nm and forms covalent bonds between adjacent thymines in DNA. These thymine-thymine dimers, in turn, cause DNA replication errors, which cause potentially lethal mutations.

UV lamps typically produce radiation with wavelengths between 240 and 280 nm. The energy varies with the bulb, ranging from 40 to 40,000 microwatt-seconds per cm$^2$, and the choice of bulb depends on the application. Low-energy bulbs are used for sterilizing hoods and benches, whereas high-energy bulbs are used for sterilizing water. UV radiation does not penetrate ordinary glass, so for sterilizing liquids, especially large volumes (e.g., carboys), a waterproof submersible bulb is usually required. Quartz test tubes allow UV radiation to penetrate, but these test tubes are too expensive for routine culture work. Furthermore, quartz glass easily scratches, which greatly reduces its efficiency (Hamilton 1973; for additional details, see Hamilton and Carlucci 1966). With regard to sterilizing large volumes of water, it must be remembered that UV light does not penetrate water uniformly—it is absorbed, and the effectiveness diminishes with distance from the bulb. Continual mixing of the water and sufficient exposure time usually circumvents this problem for the water, but if organisms are attached to the inner surface of the container, then sterilization may be ineffective unless more powerful lamps are used with longer exposure times.

## 3.7. Bleach

Bleach (sodium hypochlorite) is widely used in aquaculture hatcheries where very large volumes of water require sterilization. Although the process may not kill all cysts when smaller amounts of bleach are added, it is very effective in killing most organisms. The amount of bleach added depends on the organic matter in the water. Typically, 1 to 5 mL of commercial bleach is added per liter of water, and after gentle mixing, the water is left to stand (no mixing and no aeration) for several hours. For shorter times, add more bleach. The solution should not be exposed to direct sunlight during treatment. After the bleach treatment is finished, the solution is neutralized with sodium thiosulfate ($Na_2S_2O_3 \cdot 5H_2O$). If 250 g of sodium thiosulfate are dissolved in 1 liter of water, then 1 mL of the sodium thiosulfate solution is added for each 4 mL of bleach used.

### 3.8. Ethylene Oxide Sterilization

Historically, non-heat-tolerant materials (plastic and rubber) were sterilized with ethylene oxide (Hamilton 1973). However, problems arose after sterilization, because chemical traces remained, often killing the living cells used in the experiment. Consequently, ethylene oxide sterilization has gradually fallen into disuse, and modern heat-tolerant reusable materials or sterile disposable materials have become popular.

**FIGURE 5.9.** A stainless steel basket for holding items in an autoclave. The basket contains Pasteur pipette canisters covered with autoclavable bags. Note that the cap junction of a pipette canister is covered by aluminum foil.

---

### 4.0. STORAGE OF STERILIZED MATERIALS

Once it is sterilized, regardless of method, the main difficulty is keeping the material sterile. After cooling, sterilized materials can be stored in clean, dust-free containers, cabinets, or shelves (Fig. 5.1), and refrigerated storage is often used for culture media. It should be remembered that at least the outer surface of the container (e.g., metal box, canister, or aluminum package) is contaminated. To prevent contamination of the sterile contents, the junction of the container and its cover should be wrapped with aluminum foil before sterilization (Fig. 5.9). The body of the container should be cleaned externally with ethanol or passed through a flame if the container is small and made of glass or metal before opening to remove the sterilized contents.

---

### 5.0. STERILE TECHNIQUES

### 5.1. Sterile Rooms and General Techniques

A laminar flow hood, placed in a small, closed room, is the best choice for manipulating cultures and sterile materials. If a laminar flow hood is unavailable, then a small, closed room, isolated from the normal laboratory room, is the next best choice. These small rooms are usually equipped with UV lamps, which can sterilize the room when not in use, and a Bunsen burner (or alcohol lamp). When small rooms are not available, sterile technique can still be performed in a normal laboratory room if precautions are taken. Regardless of the setting, the primary source of contamination will be from the air, the working surface, and the external surfaces of containers, vessels, and so on. Therefore, all manipulations should be conducted in a way to avoid these risks. In the United States, Lysol is sprayed in the air 10 minutes before sterile work (R. A. Andersen, personal communication). The spray neutralizes charges on colloidal particles in the air, such as dust and spores, allowing them to fall onto surfaces, such as tables and floors. Before beginning the sterile work, the surfaces are wiped with 70% ethanol to remove the dust and spores and to ensure a good state of the environment for conducting sterile work.

The working area and materials should be arranged so that air circulation is reduced as much as possible (Fig. 5.10; order of placement of the materials is from a to f), that is, the operator should make minimum movements and any motion should be slow. The operator should avoid crossing hands or waving pipettes through the air, and pipetting action should be conducted slowly. A laminar flow hood ensures that airborne dust does not enter the working area, but

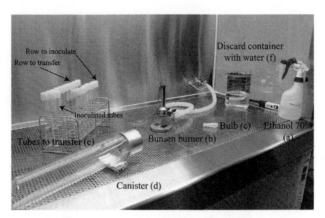

**FIGURE 5.10.** A laminar flow hood with Bunsen burner and supplies organized for use by a right-handed person. Placement and use order of the materials is indicated with the letters (a) to (f). Note the stopper for the cylindrical canister. Test tube arrangement is to facilitate handling and to avoid mistakes during manipulations.

attention must be given to maintenance of the hood filters. If a filter becomes contaminated, then the laminar flow hood becomes a very effective contaminating machine. The condition of sterile materials should be checked, and if any doubt exists, then the material should not be used. The person may wear latex gloves, which should be changed often, or alternatively the person should wipe his or her hands with 70% ethanol. Sleeves of clothing are a major source of contamination; therefore, sleeves should be rolled up and the person's arms should be wiped with 70% ethanol. Alternatively, the person may wear clean, UV light–sterilized work clothes that are not worn outside the transfer room. The work area should be wiped with 70% ethanol before the task is begun.

The Bunsen burner is usually placed in the center for subculturing manipulations or inoculations, so that all operations are conducted within a 40-cm circle around the flame (Fig. 5.10). A receptacle for used pipettes, disposable tips, and so on, should be placed on the right side for a right-handed person (or on the left side for a left-handed person) so that contamination risk is limited. Once the work area is well organized, the Bunsen burner can be ignited, and the person should once again wipe his or her hands with 70% ethanol, using care to avoid ignition of the alcohol. Thereafter, the worker should not place his or her hands outside of the hood (Fig. 5.10). If a sterile item (e.g., a pipette or disposable tip) accidentally touches the outer surface of a bottle or container, then it should be discarded and replaced with a new sterile implement. The contents of

a glass vessel should not be poured directly into another vessel unless the outer surface near the rim is passed through a flame, because the outer surface is a major source of contamination. Under no circumstances should a pipette be reused. Similarly, if a portion of sterile medium is poured into one cell culture, then the remainder should be discarded; microscopic droplets often splash from the cell culture back into the previously sterile medium.

## 5.2. Transferring Liquid Cell Cultures

The following example is a common procedure for transferring established cultures into new culture medium. Although many variations exist in different laboratories, this provides a general procedure for proper sterile technique. The example assumes that the worker is using a laminar flow hood. If you do not have a laminar flow hood, then certain steps are to be deleted. The example also assumes the worker is using reusable glass pipettes; if you use disposable plastic pipettes or pipette tips, then some minor modifications are necessary. Finally, the directions are for right-handed people; left-handed people should reverse the movements and the placement of items (see Fig. 5.10). For additional details and other manipulations, see Barker (1998).

1. Wear protective clothing that is sterilized by UV light. If clothing is maintained inside the room, then carefully do this after turning off the UV lamp and entering the room.
2. Turn off the UV sterilizing light in the room, open the door using a minimum opening, enter the room with slow motion, and gently close the door.
3. Turn on the laminar flow hood air circulator and then turn off the UV sterilizing light in the laminar flow hood. If there is an external gas valve for the hood, then open it, but only if the valve on the Bunsen burner is closed.
4. Wipe the working surface with 70% ethanol.
5. Place the Bunsen burner in the center.
6. Place the canister with sterile pipettes or disposable pipette tips on the most left side (Fig. 5.10). Spray and wipe the surface of the container with 70% ethanol.
7. Place the bottles, Erlenmeyer flasks, or test tubes on the left side (i.e., second position for

chronological order of use; see Fig. 5.10, with order of materials from a to f) at the periphery of the 40-cm circle from the Bunsen burner.

8. Place the pipette bulb on the right side of the Bunsen burner. Regularly clean the inner part of the bulb with 70% ethanol and wipe carefully the inner rim; loose cotton fibers in its opening prevent a seal from forming, causing a pipette to drip. Never use a bulb that has been used for manipulation of chemicals (e.g., osmium, glutaraldehyde).

9. Place the container for discarding used pipettes outside the 40-cm circle, on the most right side of the working area.

10. Make sure that all inoculation vessels are correctly labeled.

11. After organizing all materials, turn on the (secondary) gas valve and light the Bunsen burner. Spray hands with 70% ethanol (or put on sterile disposable gloves). Sterile manipulation can now begin; make sure to stay within the 40-cm circle around the Bunsen burner.

The following steps describe the procedure for transferring liquid cultures with a glass Pasteur pipette that will be cleaned and reused.

12. With your left hand, pick up the pipette canister and bring it close to the Bunsen burner. Remaining in the 20-cm circle of the Bunsen burner, remove the cap with your right hand by putting pressure on the cap in between your palm and two outer fingers (Fig. 5.11a). When the canister cap is too large to be held as described, place the cap at the base of the Bunsen burner and flame it before replacing it onto the pipette canister.

13. With your left hand, gently shake the canister so that one pipette is smoothly extruded a few centimeters from the canister opening (Fig. 5.11b).

14. With your right hand, catch the pipette with your thumb and index and middle fingers, and then remove it from the canister with a slow, uniform motion (Fig. 5.11c). The pipette tip should never contact the container opening.

15. Keep your right hand near the Bunsen burner; a 20-cm circle is the preferred distance to maintain pipettes and to open vessels.

16. Replace the cap onto the canister, being very careful that the pipette never touches the container surface or anything else (Fig. 5.11d). To avoid manipulations and risks during repeated culture transfers, it is recommended to keep the pipette canister open within the 40-cm circle and to extract the pipettes with a stainless forceps flamed before each extraction. In this case, the canister content should be completely used or, if not, should be resterilized before the next use.

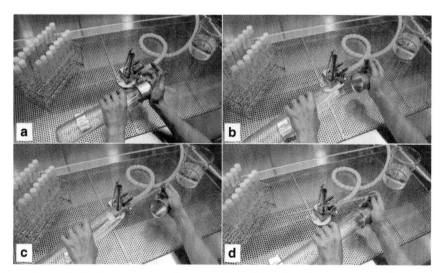

**FIGURE 5.11.** Removal of a Pasteur pipette from a canister. Note that the canister cap, though large, is held in the right hand between the palm and two outer fingers. All work is close to the Bunsen burner flame. **(a)**, Opening the pipette canister; **(b)**, Shaking a pipette partially out of the canister; **(c)**, Extraction of a pipette; **(d)**, Replacing the cap on the canister.

17. With your left hand, place the canister in its normal position on the working surface.
18. With your left hand, pick up the pipette with your thumb and index finger at the cotton-plugged end of the pipette and remain close to the flame (Fig. 5.12a).
19. With your right hand, pick up the bulb.
20. Insert the pipette bulb onto the pipette (Fig. 5.12b).
21. It should not be necessary to flame the sterile pipette; however, if you choose to flame the pipette before use, then cool the pipette with a small amount of sterile medium.
22. Handle the pipette in your right hand near the bulb (Fig. 5.12c) and slowly pick up the cell culture vessel with your left hand.
23. Bring the cell culture vessel to your right hand, and using your palm and small finger of the right hand while still holding the pipette, remove the cap from the vessel (Fig. 5.13a). The pipette should remain close to the Bunsen burner and should not touch anything.
24. With the left hand, briefly flame the opening of the cell culture vessel while slowly rotating the lip of the vessel in the flame at an angle of at least 45 degrees; avoid breathing into the vessel if the laminar flow hood window is not pulled down (Fig. 5.13b; the hood window is up only for photographic purposes). Note: do not flame plasticware.

25. With the left hand, move the vessel slowly away from the flame but keep it oriented at a 45 degree angle to reduce the possibility of contamination.
26. Slowly insert the tip of the pipette into the liquid, being careful not to touch the pipette against the opening of the vessel.
27. Slowly draw a cell suspension into the pipette by carefully controlling the pressure exerted on the bulb, making certain that the liquid level does not come near the cotton plug of the pipette (Fig. 5.13c). Collect only the necessary volume of cell suspension. If extra material is removed, then it must be discarded. To avoid any risk of spilling or dripping, the bulb should be completely expanded at the end of the collection.
28. After drawing up the appropriate amount of cell suspension, the pipette is slowly removed while avoiding contact with the vessel.
29. Orient the pipette to a nearly horizontal position without exerting pressure on the bulb, and maintain the pipette within 20 cm of the Bunsen burner.
30. Flame the mouth of the vessel again, using a rotating motion (Fig. 5.13d). Do not touch the pipette against anything.
31. Using your left hand, slowly bring the vessel to the cap that has been maintained between the small finger and palm of the right hand (Fig. 5.13e). Note: for larger caps (e.g., Erlenmeyer large soft plugs, Fig. 5.2) that cannot be held during manipulations, the cap is placed on the working area and flamed before replacement.

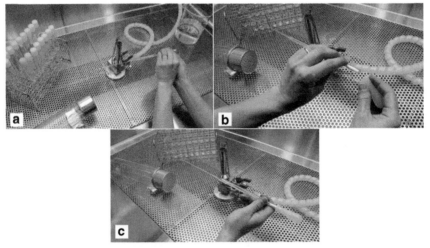

**FIGURE 5.12.** **(a)** Transferring the pipette from the right hand to the left hand. **(b)** Holding the pipette and attaching the bulb. Note that the tip of the pipette remains near the Bunsen burner. Also note the cotton plug in the pipette that retains any contaminating material from the pipette bulb. **(c)** Holding the pipette in the right hand, slowly move the left hand (not visible) to pick up the culture vessel containing cells.

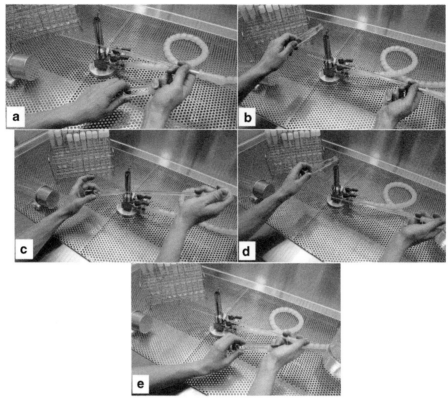

**FIGURE 5.13.** Demonstration with a test tube, in which the cap is held between the palm and small finger during aseptic work: opening the test tube cap **(a)**, flaming the test tube mouth **(b)**, drawing up a cell suspension for transfer **(c)**, the second flaming of the test tube mouth **(d)**, and replacing the test tube cap **(e)**.

32. Replace the cap, always being careful that the pipette does not touch anything. Be careful not to bring the pipette too close to the flame, because the heat may kill the cells.

33. Return the capped vessel to its previous position. Your left hand is now free. Note: it is good practice to place the drawn vessels (or, alternatively, newly inoculated vessels) in a new location so that accidental reinoculation does not occur. In the case of test tubes held in a rack, a row free of tubes can be maintained between uninoculated and inoculated tubes (See Fig. 5.10).

34. Slowly move your left hand to the vessel that will be inoculated.

35. Using the same procedure described above, open the new vessel, flame the opening, and insert the pipette into the new vessel without touching the mouth.

36. Slowly discharge the cell suspension into the vessel and carefully remove the pipette. If the tip of the pipette is placed below the surface, then avoid bubbling into the liquid.

37. Flame the mouth of the vessel and replace the cap as described above, using precautions as before.

38. The newly inoculated vessel, with cap attached, should be placed onto the bench again.

39. Discard the used pipette in the storage container, using caution to avoid any spillage of cell suspension onto the bench. If any spill has occurred, then wipe up the liquid with a Kimwipe or towel, spray the area with 70% ethanol, and wipe the surface again with a new tissue paper. If wearing gloves, then change to a new pair of sterile gloves; otherwise, spray hands with 70% ethanol.

40. Bring the pipette to the discard container (Fig. 5.14a). With the free left hand, pick up the pipette with your thumb and index finger at the cotton-plugged end of the pipette, and carefully remove the pipette bulb with your right hand.

41. With your left hand, place the pipette in the container, making sure that the tip is well immersed in the water (Fig. 5.14b). Return the bulb to its normal position. The working area is

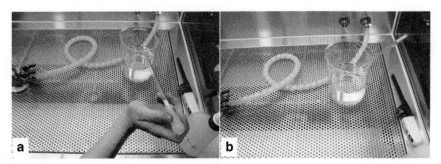

**FIGURE 5.14.** After transfer, the pipette is directed to the water-filled discard container **(a)**. The bulb is removed once the pipette is above the discard container to avoid spilling on the working area (not shown). **(b)** The discard container should be stable enough to store the appropriate number of pipettes without risk of falling. A sufficient amount of water stabilizes the container and ensures that the pipette tips remain wet.

back to its initial stage and ready for the next transfer process.

42. Once all transfers are completed, turn off the Bunsen burner, remove all the materials from the working place, and wipe the surface with 70% ethanol.

43. Close the front window of the hood and close the external gas valve, if present. Switch off the ventilation system and turn on the UV lamp of the hood. Exit the room by carefully opening and closing the door. Turn off the room light and turn on the UV lamp for the room.

44. Remove your protective clothing and place it in the UV cabinet for sterilization. Exit the airlock room carefully and switch on the UV sterilization of this room.

45. In most cases, clean shoes or slippers are recommended for the clean area, and these should be changed when entering and exiting the clean area. A sticky pad or carpet placed on the floor at the door location is very helpful to keep dirt from clean areas; dirt on shoes is perhaps the primary source of contamination entering a room, where later it may become airborne.

46. Appropriate signs should be in place to ensure a safe work area. Custodians should be generally prohibited from servicing sterile facilities unless informed and instructed.

## 5.3. Transfer of Agar Culture

Refer to Section 5.2, with modifications as follows:

1–5. Follow steps 1 through 5, described previously.

6. Place the transfer loop at the base of the Bunsen burner.

7. Place the test tubes or petri dishes on the left side at the periphery of the 40-cm circle from the Bunsen burner.

8. After organizing all materials, turn on the (secondary) gas valve and light the Bunsen burner. Spray hands with 70% ethanol, or put on sterile disposable gloves. Sterile manipulation can now begin, but make sure to stay within the 40-cm circle around the Bunsen burner.

9. Flame the transfer loop in the blue part of the Bunsen burner flame until the metal is red (Fig. 5.15a).

10. Cool the loop while maintaining it within the 20-cm circle around the Bunsen burner (Fig. 5.15b).

11. Using the left hand, bring the cell culture vessel toward the Bunsen burner within the 20-cm circle.

12. When using a test tube, follow the instructions given for transfer of liquids (steps 23 through 25 in the previous section). When using a petri dish, open the cover with your left hand, to an angle of 45 degrees; avoid touching the inner part of the cover (Fig. 5.15c).

13. Slowly insert the loop and pick up the cells without scratching the agar (Fig. 5.15d).

14. Replace the petri dish cover or the cap of the test tube (see steps 30 to 32 in Section 5.2). Keep the loop end within the 20-cm circle of the Bunsen burner without touching any surfaces and without getting too close to the flame.

15. Slowly move your left hand to the vessel that will be inoculated.

16. Using the same procedure described previously, open the new vessel and gently brush the surface of the agar (without scratching the surface). Use

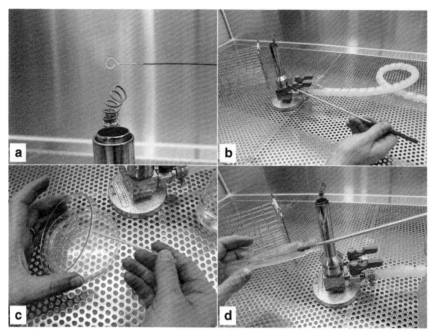

**FIGURE 5.15.** Cell culture transfer onto agar medium: flaming a platinum loop in a Bunsen burner flame **(a)**; cooling the loop in the sterile zone near the Bunsen burner **(b)**; opening a petri dish cover at a 45 degree angle **(c)**; 45 degree angle opening for spreading cells with a glass rod); and using a transfer loop to remove cells before transfer **(d)**. Note that the transfer loop is centered to avoid touching the test tube and that the test tube is oriented at a 45 degree angle.

the appropriate technique of transfer (i.e., horizontal lines in zigzag or quadrants system for a petri dish). In any case, the loop end should not touch the opening of the test tube or sides of the petri dish.

17. After a test tube transfer, flame the mouth of the test tube and replace the cap as described previously, using precautions as before.
    After a Petri dish transfer, replace the cover gently.

18. Flame the end of the loop in the blue part of the Bunsen burner until the loop is red. After cooling the loop within 20 cm of the Bunsen burner, the loop is ready for the next transfer.

19. Follow instructions given in Section 5.2., steps 42 to 46.

## 6.0. STERILIZATION OF CULTURE MEDIA

### 6.1. Stock Solutions

In most cases, stock solutions of each component of the medium are prepared in autoclavable bottles. To sterilize, each stock solution should be autoclaved or sterile filtered and stored at 4°C. (Vitamins are stored at −20°C.) To maintain sterility of stock solutions, they should be handled aseptically. If fungal or bacterial growth is discovered, then the stock should be autoclaved and the contents discarded. For details on preparing stock solutions, see Chapters 2 to 4.

### 6.2. Sterilization of Liquid Culture Media

Sterilization is typically carried out by autoclaving (121°C for 15 minutes), followed by a cooling period. The appearance of the culture medium should be checked; if the color has changed or if precipitates have formed, then the medium may not be suitable for normal use. Reasons for such changes are multiple, so the preparation procedure should be checked step by step. Heat-labile components, like vitamins, may be added after autoclaving by using a filter sterilization method. Tyndallization is used in place of autoclaving when destruction of fragile components by high temperature must be avoided (see previous section on pasteurization and tyndallization).

Using a microwave oven can also perform the sterilization of media. The temperature reached during this type of sterilization is less than 84°C (Keller et al. 1988, Hoff and Snell 2001). This allows the direct addition of some labile components before microwave treatment. Nevertheless, sterile addition of vitamins is still recommended after microwave treatment. Microwave oven sterilization of media is quick (only 10 minutes or less), and it avoids both metal contamination, which occurs during autoclaving, and carbonate precipitation (Price et al. 1989). However, microwave sterilization can be used only for small volumes of media (≤1–1.5 liters), whereas autoclave sterilization can be used for large volumes of media (e.g., 20 liters).

## 6.3. Sterilization of Agar Culture Media (See Also Chapter 2)

As for liquid culture media, sterilization is usually obtained by autoclaving (121°C for 15 minutes). For agar slants in test tubes, the tubes should be cooled at an angle so that the agar solidifies with a smooth slope that is not too close to the opening of the tube. When petri dishes are used, the agar is dispensed into sterile petri dishes under sterile conditions and then cooled. To reduce contamination, the agar should not exceed half the height of the dish.

For agar culture media containing heat-labile components, the fragile components are added aseptically after autoclaving but before the agar begins to gel (approximately 50°C–60°C for standard-melting-point agars).

## 7.0. EVALUATING STERILE CONDITIONS

Despite careful sterile technique, contamination may sometimes occur. It may not be obvious how or why the contamination occurs, or a less rigorous protocol may be necessary when suboptimal conditions exist (e.g., at a field station). Therefore, it is often helpful to employ sterility tests to evaluate the conditions. Bacteria and fungi grow in various nutrient agars and broths, although it is important to note that not all bacteria or fungi grow in any specific organic medium. For general testing purposes, a 0.1% peptone agar or broth is usually satisfactory. For best results, the peptone should be dissolved in culture medium rather than distilled water. When fungal contamination is a specific concern, malt extract is pre-

ferred, because it grows fungi better than peptone. In specific cases, for example, testing for methyl-aminotrophic bacteria, methylamine is the preferred substrate; a peptone medium tests negative, because there is insufficient substrate to detect bacterial growth.

To test for air-borne contamination over a working surface, a series of the test medium is exposed over specific periods. For example, petri dishes with sterile peptone agar can be exposed at intervals of 1 minute, 5 minutes, 15 minutes, 30 minutes, and 1 hour. To achieve this, the petri dish cover is removed for the period (e.g., 1 minute) and then replaced. The petri dish is then incubated for 1–3 days, and the number of bacterial or fungal colonies is counted. For a laminar flow hood, exposure for 1 hour should not result in any contaminating growth. For a still hood in a small room, often no contamination occurs for short periods, and only one or two colonies appear after 15–30 minutes. If numerous colonies appear after incubation on the nutrient agar plate that was exposed for a short period, then the environment is heavily contaminated, and sterile procedures are likely to fail. Similarly, a test tube of broth can be uncapped for the period, recapped, and incubated. Colonies cannot be counted when a broth test medium is used, and the only results that can be measured are positive or negative contamination. However, if one exposes 10 test tubes at intervals similar to those used for transferring a culture, if no positive growth occurs in the test medium, and if environmental conditions remain unchanged, then the worker can have confidence that at least most of the transfers will be contaminant-free. If environmental conditions change between the test and the actual sterile work, then the test results no longer predict the sterile conditions. Two examples of changing conditions are given. If the tests are conducted at a field station when there are no other people present, but the transfers are carried out when several people are active in the room, then the additional activity increases contamination above that of the tests. If the tests are conducted in a small laboratory room without activity, and then a custodian operates a vacuum cleaner a few hours before cultures are transferred, then the room contains more air-borne spores, and contamination is much higher than that of the tests.

Variations of these simple tests can be used to test culture medium, vitamins, stock solutions, pipettes, and so on. For example, if a stock solution used to prepare a culture medium has been opened several times, then the stock solution may be unsterile. By taking the stock solution and test medium into a laminar flow hood (see earlier procedures), a small amount of the stock solution can be added to the test medium. If the test medium produces bacterial growth during incubation, then the stock solution is contaminated and should be discarded. Similarly, if a culture medium is prepared with sterile filtration and there is concern about the effectiveness of the filtration (Stokner et al. 1990), then a small amount of the sterile medium can be added to a test medium. A positive bacterial growth demonstrates that bacteria have passed through the filter. If the concern is about contamination by a heterotrophic flagellate that feeds phagotrophically on bacteria, then the test medium should contain bacteria.

These simple tests, with nearly endless modifications, can be performed to help evaluate sterile conditions. The tests must be carried out under conditions that show positive results when contamination exists. Also, if the real work is to be conducted following the testing period, then the test conditions must be representative of the working conditions. The tests should not be considered as an absolute measure of sterility, but they provide an evaluation of sterility when no other means is available.

In conclusion, all sterilization protocols vary for each person and laboratory practice. Whatever protocol is used, good sterile technique is required for culturing algae. Different protocols and modifications of available materials are often required, and the goals of each individual determine the exact protocols. Control methods should be employed to verify sterile techniques, and a careful, routine work protocol should be established to minimize contamination.

---

## 8.0. REFERENCES

Barker, K. 1998. *At the Bench: A Laboratory Navigator.* Cold Spring Harbor Laboratory Press, New York, 460 pp.

Boye, M., and van den Berg, C. M. G. 2000. Iron availability and the release of iron-complexing ligands by *Emiliania huxleyi. Mar. Chem.* 70:277–87.

Hamilton, R. D. 1973. Sterilization. In: Stein, J. R., ed. *Handbook of Phycological Methods. Culture Methods and Growth Measurements.* Cambridge University Press, Cambridge, 181–93.

Hamilton, R. D., and Carlucci, A. F. 1966. Use of the ultraviolet irradiated seawater in the preparation of culture media. *Nature.* 211:483–4.

Hoff, F. H., and Snell, T. W. 2001. *Plankton Culture Manual.* 5th ed. Florida Aqua Farms, Inc., Dade City, Florida, USA, 162 pp.

Jeng, D. K. H., Kaczmarek, K. A., Woodworth, A. G., and Balasky, G. 1987. Mechanism of microwave sterilization in the dry state. *Appl. Environ. Microbiol.* 53:2133–7.

Keller, M. D., Bellows, W. K., and Guillard, R. R. L. 1988. Microwave treatment for sterilization of phytoplankton culture media. *J. Exp. Mar. Biol. Ecol.* 117:279–83.

Leal, M. F. C., Vasconcelos, M. T. S. D., and van den Berg, C. M. G. 1999. Copper-induced release of complexing ligands similar to thiols by *Emiliania huxleyi* in seawater cultures. *Limnol. Oceanogr.* 44–7:1750–62.

Price, N. M., Harrison, G. I., Hering, J. G., Hudson, R. J., Nirel, P. M., Palenik, B., and Morel, F. M. M. 1989. Preparation and chemistry of the artificial algal culture medium Aquil. *Biol. Oceanogr.* 6:443–61.

Pringsheim, E. G. 1946. *Pure Cultures of Algae. Their Preparation and Maintenance.* Cambridge University Press, Cambridge. 119 pp.

Starr, R. C., and Zeikus, J. A. 1993. UTEX—The culture collection of algae at the University of Texas at Austin. *J. Phycol.* 29(2, Suppl):1–106.

Stokner, J. G., Klut, M. E., and Cochan, W. P. 1990. Leaky filters: a warning to aquatic ecologists. *Can. J. Fish. Aquat. Sci.* 47:16–23.

# TRADITIONAL MICROALGAE ISOLATION TECHNIQUES

ROBERT A. ANDERSEN
*Provasoli-Guillard National Center for Culture of Marine
Phytoplankton, Bigelow Laboratory for Ocean Sciences*

MASANOBU KAWACHI
*National Institute for Environmental Studies*

## CONTENTS

Key Index Words: Agar, Algae, Centrifugation, Cysts, Culture, Enrichment, Isolation, Micropipette, Phototaxis, Sonication

## 1.0. INTRODUCTION

Isolation of microalgae into culture by means of traditional methods is well established, beginning with the work of Beijerinck (1890) and Miquel (1890–1893) (see Chapter 1). Some species, often called *weeds*, are easy to isolate and cultivate, whereas others are difficult or seemingly impossible to grow. Not surprisingly, organisms from extreme environments and unusual habitats are less abundant in culture collections than are those from freshwater ponds, soils, and coastal marine environments. The purpose of this chapter is to describe isolation methods regardless of natural habitat. For additional general references on algal isolation, see Küster (1907), Kufferath (1928/29), Bold (1942), Pringsheim (1946), Brunel et al. (1950), Lewin (1959),

Copyright © 2005 by Academic Press
All rights of reproduction in any form reserved.

Venkataraman (1969), Stein (1973), and Guillard (1995). For automated methods of isolating algae, see Chapter 7.

## 2.0. Important Factors Affecting Isolation

Often, the first step toward successful isolation is understanding and mimicking the naturally occurring environmental conditions. For coastal marine algae, temperature and salinity are important, and for oceanic (open ocean) phytoplankters, water quality and metal toxicity are additional concerns. Freshwater algae collected in nonwinter months are frequently less sensitive to temperature, but pH or alkalinity may be important. Polar and snow algae are very sensitive to warmer temperatures, just as protists from hot springs or hydrothermal vents are sensitive to cooler temperatures. Algae from acid environments or hypersaline environments require special culture media, but for terrestrial or soil algae, environmental factors are less important.

Taxonomic knowledge of the target species may be important. Diatoms require silica, euglenoids often require ammonia, and some genera (e.g., *Chrysochromulina*) require selenium. Mixotrophic species (e.g., certain dinoflagellates and chrysophytes) often require a bacterial food source, and colorless phagotrophic species (e.g., *Pfiesteria*) may require a eukaryotic food source.

The second step toward successful isolation involves the elimination of contaminants, especially those that can outcompete the target species. Techniques of dilution, single-cell isolation by micropipette, and agar streaking are widely used, among other methods (see Section 6). The final step requires continued growth upon subculturing. It is not uncommon for the target species to grow in the initial stages of isolation but then die after one or more transfers to fresh culture medium. This often indicates that the culture medium is lacking a particular element or organic compound, which is not immediately manifest. Unfortunately, when this is discovered, it is sometimes too late, because the original sample is gone and the original isolates are dead. Alternatively, the organism may be accumulating wastes that poison its environment, causing death. In nature, these wastes are diluted or metabolized by other organisms (e.g., bacteria).

## 3.0. Sample Collection

The collection method is sometimes crucial for success, because damaged or dead cells lead to failure. If the target species is well known and if it has been collected and studied previously, then the isolator has substantial experience to guide the collection process. For example, *Synura petersenii* Korsh. is routinely concentrated with a plankton net; the sample is placed on ice or in a cool container, and the organism survives for 24 hours or more. However, using this method with *Gonyostomum semen* (Ehr.) Diesing is difficult, because as the concentrated cells begin touching each other or other particles, mucocysts discharge and many cells die. Similarly, for marine plankton samples, *Skeletonema costatum* (Greville) Cleve is robust whereas *Akashiwo sanguinea* (Hirasaka) Hansen *et* Moestrup is not. Oceanic species are particularly sensitive to collection and concentration, and special water bottles are used, such as Teflon-coated Go-Flow Niskin bottles (General Oceanics Inc.). Samples collected at depth can be sensitive to pressure, light, or temperature changes. Thus, regardless of the collection site, whole water (not concentrated) samples, collected in clean containers and kept at a stable temperature, often provide viable cells when concentrated samples fail. The isolator must spend more time trying to find the target species in dilute samples, but once located, the cell is usually viable. Finally, if sampling an environment of which prior knowledge is lacking, or if the target organism is unknown to science (and there are many), then the collector is wise to use caution and multiple methods.

Natural samples often contain zooplankters that feed upon the algae. Once concentrated, these animals can quickly eat or otherwise kill algae. Larger animals (e.g., copepods, rotifers, ciliates) and unwanted colonial algae can be removed by gentle filtering (see next section). The filter, screen, or net is chosen so that the alga passes through the filtering device but the animal or colony does not (Fig. 6.1a). Prefiltering of samples immediately upon collection is usually effective for removing these unwanted organisms. Similarly, if the sample has an abundance of unwanted tiny organisms, such as other algae and bacteria, then gentle filtration can collect the larger, targeted species while the smaller organisms pass through the filter. However, care must be exercised to avoid damage or desiccation to the target species, and this method is usually applied only when the smaller organisms pose a problem (e.g., dilution techniques, as discussed later in the text).

**FIGURE 6.1.** Assortment of isolation supplies. **(a)** Screens for removing larger animals and algae; prepared from PVC pipe and Nitex netting. **(b)** Glass and plastic dishes. **(c)** Multiwell plates. **(d)** Syringe with 0.2-μm filter. **(e)** Pipetteman. **(f)** Small loop, fine metal probe, and forceps. **(g)** Pasteur pipettes, with tips drawn to produce micropipettes, arranged on a metal rack. **(h)** Sterile Pasteur pipettes in plastic can, resting on a simple metal support to keep the opening off the table surface. **(i)** Glass capillary tube, small tubing, and red mouthpiece. **(j)** Isolation mouthpiece (red), small tubing connected to final short piece of large tubing that forms a tight seal when a Pasteur pipette is attached. **(k)** Large isolation tubing with plastic pipette tips at both ends, one serving as a mouthpiece and the other supporting a Pasteur micropipette. **(l)** Commercial mouthpieces (red, white, yellow) and two plastic pipette tips (right one with pointed tip removed), which can be used as mouthpieces. **(m)** Glass spot plate that can be used to rinse cells during isolation. **(n)** Small sterile filter apparatus. **(o)** Glass slip caps that cover either smooth or threaded test tubes. **(p)** Smooth, unthreaded test tubes covered with slip caps. **(q)** Threaded test tubes covered with slip caps. **(r)** Sterilized test tube caps that can replace slip caps or be used when a culture tube cap is dropped or contaminated. **(s)** Plastic threaded test tubes (black cap, orange cap), which are sterilized by the manufacturer. **(t)** Two blue-capped tissue culture flasks, sterilized by the manufacturer.

Time is an important factor when isolating algae. Some organisms die quickly, even within 1 hour or a few hours after sampling. For these organisms, isolation must be conducted rapidly. When samples are enriched (see Section 6.1), some species quickly multiply but die suddenly, demonstrating a different aspect of timing. In this case, the enrichment should be monitored every 1 or 2 days. Finally, some organisms that are seemingly absent at the time of sampling appear in the sample bottle weeks or even months after collection. *Chromulina nebulosa* Cienkowski is rarely ever observed

immediately after collection. However, if the sample, which is obtained from slightly acid ponds, is left without enrichment in an incubator at 10 to 15°C for 6 months, then in many cases this alga grows and becomes abundant. For isolating this type of organism, patience is important.

Another aspect of time is the varying health or condition of the species in nature. Successful growth of isolated cells depends on the condition or state of the cells at the time of collection. Peter Miller (personal communication) has found that when repeatedly isolating cells of *Pseudo-nitzschia*, almost every isolated organism grows on some occasions, but on other occasions, most isolates fail to grow in culture. Miller concludes that this is due to the condition or health of the alga at the time of collection, not to isolation techniques, because he uses the same techniques to isolate hundreds of *Pseudo-nitzschia* cultures. One likely factor may be virus infections (see Chapter 22).

Application of good sterile technique also is important when collecting samples (see Chapter 5). Dirty sample vessels, nets, filters, and other sampling devices that have resistant stages of other organisms, especially cysts, can lead to unwanted contamination. This is particularly important when attempting to collect samples with unknown organisms, because the contaminants may be mistaken for native species.

## 4.0. EQUIPMENT AND SUPPLIES

### 4.1. Microscopes

Dissecting and inverted microscopes are the most widely used (Fig. 6.2), although compound microscopes can be used also. For the well-equipped isolator, all three microscopes can be used, each to its advantage. Dissecting microscopes should have good optical lenses, and it is often advantageous to have magnification up to at least 80×. The stage should be clear glass, devoid of scratches (plastic scratches easily). Lighting is perhaps the most crucial aspect for dissecting scopes, and both transmitted and dark-field illumination work well; reflected light is extremely difficult to use except for the largest of organisms. Dark-field illumination is superior to transmitted light, and although it is more expensive, this added expense is justified when isolating smaller cells. A Schott illuminator (Schott Corp.) with a fiber optic light source and a dark-field illumination dispers-

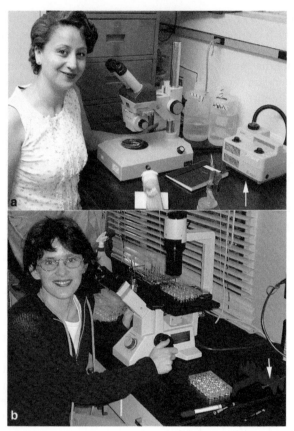

**FIGURE 6.2.** **(a)** A Zeiss dissecting microscope (magnification up to 80×), with dark-field illumination delivered by a Schott illuminator (*arrow*; light via fiber optic cable). Dark-field illumination makes tiny cells more obvious than transmitted light. Khadidja Romari. **(b)** An Olympus inverted microscope (magnification usually up to 400×), with the stage designed for multiwell plates. Note large isolation tube around the neck and the Pasteur pipette that is drawn out and ready (*arrow*). Mary-Helene Noël.

ing system is excellent (see Fig. 6.2a, *arrow*). Even tiny cells appear brightly lit against the dark background, and dust and bacteria are seen at higher magnifications. Transmitted light from a fluorescent lamp or a focused incandescent illuminator is satisfactory, provided that the stage mirror can be adjusted in all directions to focus the light. With transmitted light, the mirror is adjusted until the target cell is clearly visible.

The inverted microscope should have a long-working distance condenser to provide easy access when using a micropipette (see Fig. 6.2b). Inverted light microscopes, especially recent models, have advantages for easy isolation and observation while providing a detailed image of the target cell and contaminants. (Older models built before 1990 are less suitable.) Objective lenses of

4×, 10×, 20×, and 40× magnification are commonly used. Various sample dishes or microscope slides can be used with an inverted microscope, but this microscope excels in its easy use of multiwell plates. Therefore, a mechanical stage designed for plates is highly desirable. Enrichment cultures in multiwell plates and single-cell isolates made into multiwell plates are efficiently monitored with an inverted microscope; cells can be isolated directly from wells.

An inverted microscope with epifluorescence illumination is usually a good tool to detect and isolate cysts from sediment samples, because both the cell wall and chlorophyll may provide a fluorescent signal. Many bloom-forming species (e.g., raphidophytes, diatoms, and dinoflagellates) have been cultivated by the isolation of their cysts from sediments (Anderson et al. 1995).

Compound microscopes can be much more difficult to use than dissecting and inverted microscopes. The working distance between the objective lens and the sample is small, making it difficult to pick cells with a micropipette. Also, most compound microscopes reverse the image, making isolation procedures more difficult. Reversing prisms can be installed in some models. Although many different types of compound microscopes are available, cheap microscopes used for educational purposes have enough working distance to facilitate isolation work (ca. 30 mm for a 4× objective lens, 8 mm for a 10× lens, and 3 mm for a 20× lens; Olympus CH-2, Olympus Corp.). The resolution is the same as or better than that for an inverted light microscope. Moreover, simple compound microscopes are more popular and less expensive than inverted microscopes. The simple structure of the compound microscope makes setup and takedown easy, which is a real advantage for field-sampling trips. If a 40× water immersion lens is used, then a sample on a microscope slide can be examined at high magnification without a coverslip. Once the cell is located, a lower-magnification dry lens is used for the actual isolation. Compound microscopes are also very useful for checking single-cell picks. For example, if a tiny cell is isolated with a dissecting microscope and the cell is placed in a sterile droplet on a microscope slide, then the slide can be examined at much higher magnification with the compound microscope. This is useful when trying to distinguish several similar cell types, and it is also beneficial when checking for the presence of contaminating cells. A phase contrast microscope can also be useful to detect contaminates. For example, the 10× and 20× objectives of an Olympus CH-2 microscope produce a pseudo-dark-field image when the condenser's light annulus corresponds to that designed for the 100× objective.

## 4.2. Filters and Sieves

A sieve, strainer, net, or filter is used to separate the particulate contents of a liquid into two fractions according to size, although the shape of particles also has an influence. Woven screens of stainless steel or nylon are available with square openings as small as single-μm dimensions, but not all stainless steels resist seawater or acid water conditions. Nylon netting (Nitex) sieves are easily made by attaching the netting to short sections of polycarbonate or PVC pipe with contact cement (see Fig. 6.1a).

Membrane filters are of two basic types. One has intertwined organic fibers (e.g., cellulose acetate), leaving many passages of different effective diameters ("tortuous path" filters). They are given nominal pore sizes as low as 0.01 μm, but the cutoff is not sharp (Sheldon 1972). They are available generally with 0.45-, 0.2-, or 0.1-μm porosity. Filters of this type are also available in electrode-deposited silver (Flotronics, Flotronics, Inc.) and anodized aluminum (Anopore, Whatman, Inc.). The latter is listed with porosities down to 0.02 μm, which is the size of small viruses (see Chapter 22).

The other type of membrane filter, which is the more generally useful one in phycology, is a thin, flat sheet of polycarbonate perforated by a random distribution of circular holes of fairly uniform diameter. Occasionally, two or three of the holes overlap. This second type is designed to separate two size fractions more or less abruptly at the designated porosity, provided that the filters are not loaded too heavily. The median size of particles retained is close to the nominal pore size, so they are better approximations of an ideal sieve than are the intertwined-fiber filter type (Sheldon 1972). Stokner et al. (1990) found that commonly used filters of both types, of nominal porosity of 0.2 μm, had numerous surface defects that were visible with an electron microscope. Particles and algae several micrometers were identified in the filtrates. Particle retention (>1.0 μm) on average was 92.5% but was presumably less for smaller particles, which would include heterotrophic bacteria and picophytoplankton.

When using filtration to separate an alga for isolation, filter only a small volume, stopping when the flow rate begins to decrease or sooner. The filter should not be allowed to dry, and the liquid above the filter should be transferred to a sterile vessel. A larger-diameter filter may be used if the first filtration effort yields too few cells, or one may harvest from several primary filtrations and combine them. If the cell concentrate contains only one alga, then the cells can be transferred directly into culture medium. However, in most cases, further isolation is required with use of the techniques described later in the text.

Differential filtration can also be used to remove yeast or other fungi from samples or crude cultures. Fungal filaments are held well by filters, but their spores usually pass through with the algae. To reduce or eliminate fungal spores, add a little malt extract or other organic enrichment to the filtrate. The fungal spores will germinate in 1 or 2 days, forming filaments that can then be collected on filters before they sporulate again (see Chapter 8).

## 4.3. Isolation Glassware, Plasticware, and Utensils

Borosilicate glassware was used almost exclusively in the past, but in recent years plasticware has been rapidly replacing glass. One advantage of plasticware is that it is presterilized in ready-to-use packages. A second advantage is that tissue culture-grade plasticware is coated with growth substances that can enhance growth of many algae. The common glassware items used in isolating algae are the test tube, slip cap, quadrant dish, spot plate, Pasteur micropipette, capillary tube (see Fig. 6.1), petri dish, watch glass, microscope slide, and other items. These should be sterile before use (see Chapter 5). When isolating larger algal cells, one can sterilize these glassware items by autoclaving; however, for tiny or fastidious cells (see next section), baking is necessary. Common plasticware items are a petri dish, multiwell plate, tissue culture flask, test tube, and filter apparatus (see Fig. 6.1).

Regardless of whether they are glass or plastic, the necessary materials should be available before the isolation procedure begins. All materials should be wrapped to preserve sterility, and they should be stored in a dust-proof cabinet or clean containers (see Chapter 5). In addition to the necessary items, it is wise to have extra sterile test tube caps or slip caps available in case a cap is dropped. A sterile spatula is necessary for soil samples, and a sterile filtrating apparatus is often useful. Pens and labels are also necessary, and Parafilm is frequently used to seal petri dishes and multiwell plates, from which evaporation may occur. Finally, in an emergency, a piece of paper towel or

Kimwipe can be used as a clean surface if it is necessary to set down a sterile culture utensil, test tube cap, etc. The inner surface of these paper products is sterile or nearly sterile.

## 4.4. Growth Chambers

Growth chambers are widely used for incubating isolated cells (see Chapters 9 and 10). The light source and intensity should be considered carefully (see Chapter 19). Cool-white fluorescent lights are widely used, but new full-spectrum fluorescent bulbs provide light that more nearly matches natural light. Incandescent lighting should be avoided. When illuminated culture boxes are not available, a north-facing window (in the Northern Hemisphere) can be used if the alga will grow at room temperature. Increased light intensity often does not increase the growth rate for single-cell isolates, and in many cases it is deleterious (e.g., for many benthic diatoms). Most cultures are illuminated with 30–60 $\mu mol \cdot m^{-2} \cdot s^{-1}$, but conditions can vary for organisms from more extreme environments. Some algae require a light-dark cycle, that is, they will not grow under continuous light conditions. Therefore, for newly isolated cells, a light-dark cycle should be used until continuous light growth is established. For most algae, a light-dark cycle is used between 12:12 and 16:8 hours, although some winter strains require a longer night period (e.g., *Haslea*).

## 5.0. CULTURE MEDIUM

Selection of the proper culture medium is important (see Chapters 2 and 3 and Appendix A), and the nutrient concentration of the culture medium can be crucial. Therefore, before isolation of cells begins, serious attention should be paid to the culture medium or media that will be used. Single cells of common weedy organisms (e.g., *Chlorella*-like organisms, *Tetraselmis*, and many diatoms) grow well when placed directly into full-strength medium. However, many other algae die when a single cell is inoculated into full-strength culture medium. This single factor, perhaps more than any other, produces failed attempts to isolate common but nonweedy species, and even weedy species show growth in dilute medium. Therefore, it is a good precaution to use a very dilute culture medium when isolating organisms. To achieve more dense growth after isolation into

weak culture medium, additional nutrients or culture medium are added incrementally to improve growth. The increment (e.g., days to weeks) varies with the growth rate of the isolated alga, and careful monitoring is necessary. Once a dense culture is established, then, in most cases, subculturing can be made into full-strength medium, provided that sufficient numbers of cells are used as an inoculum.

## 6.0. STANDARD ISOLATION METHODS

### 6.1. Enrichment Cultures

Enrichment cultures have long been used as a preliminary step toward single-cell isolations. They are established by adding nutrients to the natural sample, which enrich the sample so that algal growth occurs. Common enriching substances include culture medium, soil-water extract, or macronutrients (i.e., nitrate, ammonium, and phosphate), but in some cases the limiting factor is a trace metal (e.g., Boothbay Harbor, Maine, seawater in May 2003). Soil-water extract is perhaps the simplest and most successful, provided that the original soil is of good quality (Pringsheim 1912, 1950; see Chapter 2). Peat moss can be substituted for soil-water extract when enriching for desmids and some other algae from acid habitats. Organic substances, such as yeast extract (primarily for vitamins), casein (for amino acids), or urea (for nitrogen), can be added when trying to isolate osmotrophic algae, but the amount should be small, because these organic compounds almost always result in rapid bacterial growth. If bacterial growth is too high, then the enrichment culture can become anoxic or toxic, and the alga will die. For species not previously brought into culture, it sometimes pays to be inventive. Droop (1959) purchased various fruits and vegetables and added tiny amounts when trying to isolate *Oxyrrhis* into culture with phytoplankton for prey. He discovered that unfiltered lemon juice, and subsequently extracts from lemon rind, led to success. Ultimately, from analysis of the lemon rind, he was able to determine that the requirement was satisfied with ubiquinone or plastiquinone (Droop 1966, 1971). Conversely, this "grocery cart" approach failed when trying to isolate *Dinophysis accuminata* Claparède et Lachmann (Andersen, unpublished observations).

For mixotrophs requiring bacteria, a dry grain of rice may be added, which releases organic matter very slowly. Controlled bacterial growth, in turn, provides nutrition for the phagotrophic alga (e.g., *Ochromonas*, Chrysophyceae). The rice grain should not be autoclaved in the culture medium, because too much dissolved organic matter is released as the heat cooks the rice grain, making it soft and diffuse. Once the culture is established, some organisms, especially heterotrophic algae, grow well on autoclaved rice (Lee and Soldo 1992). Dissolved organic matter can also be added (e.g., glucose, acetate, or yeast extract); however, as Pringsheim (1950) states, the amount of dissolved organic matter (e.g., acetate or glucose) should be low when culturing chrysophyte flagellates and other slow-growing mixotrophs. Other seeds (e.g., 1/4 pea seed) are sometimes added to soil-water medium when trying to isolate euglenoids, but in this case, the addition is for ammonia or organic compounds, not for bacterial growth (Nichols 1973).

Natural samples are often deficient in one or more nutrients (i.e., a limiting factor), but in nature the algae survive, because bacterial action, grazing, and death of organisms recycle those nutrients. Once the sample is collected, recycling may be reduced or altered, and nutrient stress can cause death to the target species. Thus, for some species, weak enrichment of field samples can extend the life of healthy algal cells needed for isolation. Enrichment of field samples may be effective for oceanic species where collections are made from ships, especially on long cruises where isolation is not attempted. In this case, minute amounts (e.g., 1–10 μL · L$^{-1}$ of culture medium or a specific nutrient solution) are added to the samples to avoid poisoning, and the samples are incubated in culture chambers aboard the ship. However, enrichments can also be detrimental, regardless of the collection site. For example, if the target species is rare and unable to compete with weedy species, then enrichment of samples can diminish the target organisms, and in that case, unenriched samples may be more appropriate. The enrichment culture is incubated in a culture chamber, and the culture is examined every few days for growth of the target species. Once healthy target cells are abundant, individual cells are isolated. When the target species responds favorably to enrichment, the ease and success of isolation is greatly improved.

The enriching substance, although often nitrate, phosphate, or soil-water extract, can be varied. If 10 mL of the collection sample is placed into each of 10 different test tubes, then 10 different enrichments can be attempted (i.e., nitrate may be added to the first, phosphate to the second, silica to the third, ammonia to the fourth, iron to the fifth, and so on). Also, combinations of nutrients (e.g., nitrate and silica for diatom growth) or complete culture medium additions can be attempted. A different organism often dominates each tube when various enrichments are established. Furthermore, organisms not observed in the initial sample occasionally appear 7 to 10 days later, because the addition favors their rapid growth; they establish a competitive advantage over other species, even if initially rare.

How much enriching material should be added? Although the answer varies with the sample and the substance, the general answer is "not very much." At most, the addition equals that found in a common culture medium; at a minimum, only one-thousandth of that in a culture medium is added. For example, approximately 800 μM nitrate is equivalent to f/2 medium (Guillard 1975), which is appropriate for *Skeletonema* growth, but for isolating single cells of oceanic species such as coccolithophores, 800 nM nitrate is more appropriate. Second, the enrichment may be staged (e.g., 800 nM nitrate on day 1, an additional 800 nM on day 10, 1.6 μM on day 20, 3.2 μM on day 25, and so on). Note that as the biomass doubles, the nutrient addition must be doubled. For r-selected species (e.g., *Skeletonema*), greater amounts of enrichment are not only beneficial but often necessary. Conversely, k-selected species (e.g., oceanic coccolithophores) grow very slowly and require little addition of nutrients; under these conditions the r-selected species die, and over time the k-selected species gradually dominate.

The addition of ammonium is particularly useful in enriching for species with an absolute ammonium requirement (e.g., *Aureoumbra*). Understandably, they will not flourish with nitrate addition, but they are adapted for rapid growth when ammonium is available. In some cases (e.g., *Aureococcus*), ammonium is lethal at concentrations above about 20 μM. Thus, these two brown-tide organisms, so similar in many respects, require different enrichment strategies.

Selective culturing is a type of enrichment culturing with a special purpose. If the goal is to isolate microalgae that has the ability to grow at a high $CO_2$ concentration, then an enrichment culture that is aerated with 1–5% $CO_2$ is an effective way to select for species with high $CO_2$ tolerance. Similarly, many different types of selective cultures can be designed (e.g., high or low temperature, light, salinity, or pH). The addition of specific physical conditions can have a dramatic effect on species selection and growth of enrichment cultures. For example, certain benthic raphid diatoms, especially large *Pinnularia*, may require a sediment environment through which they are free to migrate and grow.

## 6.2. Single-Cell Isolation by Micropipette

Perhaps the most common method is single-cell isolation by micropipette, although automation may replace this technique in popularity in the future (see Chapter 7). Micropipette isolation is usually performed with a Pasteur pipette or a glass capillary. A Pasteur pipette can be heated in a flame, extended, and broken (Fig. 6.3). With minimal practice, this technique becomes quick and easy, but the beginner must spend some time practicing before reliable production of micropipettes is achieved. The pipette is held in one hand, and a forceps held in the other hand supports the tip. The pipette is rotated to provide even softening as the pipette warms to the melting point (see Fig. 6.3a). When the heated area is sufficiently soft, the pipette is removed from the flame and simultaneously pulled to produce a thin tube (see Fig 6.3b). If drawn out too quickly, or if drawn out in the flame, then the thin extension breaks or burns through, and the resulting product is unsatisfactory. Some people like a very straight micropipette tip and

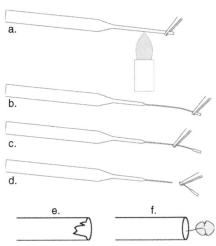

**FIGURE 6.3.** Preparation of a micropipette from a Pasteur pipette. **(a)** The Pasteur pipette is held in the hottest region of the flame, supported on the left by a hand and on the right by forceps. The pipette should be rotated as the glass is heated to a soft, pliable condition. **(b)** When the glass is soft, the pipette is quickly removed from the flame with a gentle pull to produce a thin tube. **(c)** The forceps is then relocated to the appropriate region of the thin tube. **(d)** The forceps is used to gently bend the thin area so that it breaks, forming a micropipette. **(e)** An enlarged tip of a micropipette, showing a jagged break; this tip is not suitable for use. **(f)** An enlarged tip with a very smooth break; this tip is suitable for use. Note that the diameter of the tip is larger than the flagellate cell (bearing microscopic scales), thus reducing the probability of shearing as the cell enters the micropipette during isolation.

others produce a bent or curved tip. The curved tip is advantageous when picking cells from a deep dish, but the straight tip is easier to use when discharging the captured cell into a sterile rinsing droplet. Once the tip is drawn, the glass is allowed to cool for a couple of seconds. Next, the forceps is repositioned to the thin area, approximately where the weight of the tip bends downward (see Fig. 6.3c). With a slight tugging and bending motion, the end is removed and discarded (see Fig. 6.3d). Properly done, the broken end of the pipette is smooth and round (see Fig. 6.3f). If the end is jagged or broken (see Fig. 6.3e), then the pipette should be discarded, because it will not draw up the target cell properly. Both discarded ends and used micropipettes should be carefully discarded, because they are extremely sharp. These fine tips can be pushed into a flame, where they will quickly melt into a rather irregular, dull piece of glass. Cans or bottles make good waste containers, but a boat prepared from aluminum foil is better, because at the end of the isolation session, the foil can be folded to enclose the small glass remains.

Some people prepare several micropipettes in advance, whereas others prepare a micropipette immediately before use. A previously used micropipette can be redrawn to form a new tip, and the heat required to melt the glass is sufficient to sterilize it, assuming that no contaminating liquid is further up in the pipette. Redrawn pipettes used in seawater often form a small salt crust when any remaining seawater is evaporated.

The goal of micropipette isolation is to pick up a cell from the sample, deposit the cell without damage into a sterile droplet, pick up the cell again, and transfer it to a second sterile droplet (Fig. 6.4). This process is repeated until a single algal cell, free of all other protists, can be confidently placed into culture medium. The process balances two factors: cell damage by excessive handling, which is bad, and clean isolation of a single cell, which is good. For robust organisms, repeated handling can be achieved without damage; however, for delicate organisms, cell damage is an important concern.

A microscope is necessary for observing and isolating the cell (see Section 4.1). The sample containing the target species can be placed in a glass or plastic dish, in a multiwell plate, on a microscope slide, or in similar containers. Second, sterile droplets should be prepared before isolation begins. Lewin (1959) recommends placing the droplets on agar to reduce evaporation, but in our experience this isn't necessary if the isolation proceeds without delays. Also, the agar is not as transpar-

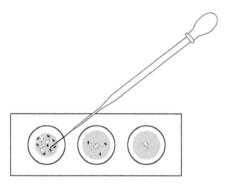

**FIGURE 6.4.** A pipette is used to remove other small cells (*left, middle*), leaving the target organism free of contamination (*right*). This procedure limits the handling of the target organism.

ent as glass or plastic, and for small cells it is more difficult to see them on agar. The droplets can be sterile seawater, sterile pond water, culture medium (diluted or not), etc. Cells must be able to survive in the droplet, preferably without stress. For example, a freshwater alga will likely die if placed in seawater, and a sensitive species will likely die if placed in full-strength culture medium. Typically, one prepares several sterile droplets, covering these with a petri dish cover or similar cover when not in use.

There are two common methods for picking up single cells with a micropipette. For one method, a flexible, latex tube is attached to a mouthpiece on one end and a micropipette or capillary tube at the other end (see Fig. 6.1i–k). Small tubing is advantageous, because it is lightweight and easy to store, but it may be necessary to use a reducing connector to connect the tubing to the micropipette or capillary tube, depending on its size. Reducing connectors can be purchased, but sections cut from plastic pipette tips (see Fig. 6.1l, right side) or glass Pasteur pipettes work quite well. A short piece of tubing attached to the reducing connector provides a quick and simple seat for the micropipette. Larger tubing can be used (see Fig. 6.1k), but its weight makes it more cumbersome to use. Large tubing is usually cut to a longer length, so that one can pass the tubing over the shoulders and around the neck, providing support for comfortable use (see Fig. 6.2b). Although various mouthpieces and micropipette connectors can be fitted to the larger tubing, 1000-μL plastic pipette tips work well at both ends. Regardless of tubing size, the operator places a small amount of sterile water into the micropipette to act as a cushion. The operator places the tongue over the mouth piece, places the micropipette tip near the target organism and then removes the tongue to gently allow the capillary

action to draw the cell up and into the micropipette tip or capillary tube. After successful capturing of the cell, the micropipette tip is removed from the sample or droplet; the tip is immersed into the next droplet, test tube, or multiwell; and then by means of gentle blowing into the mouthpiece, the captured cell is discharged into a second, sterile droplet. The drawing or expelling pressure should be slight, because excessive pressure or rapid movement can damage the cell.

Alternatively, the micropipette tip can be touched into a sterile droplet so that capillary action pulls water up into the micropipette. If a dry micropipette is immersed in the sample (i.e., without first touching the tip to the sterile droplet), then violent capillary action results, drawing substantial unwanted material into the micropipette. After capillary loading with sterile liquid, the micropipette is then directed to the selected cell, and residual capillary action gently draws the cell into the micropipette. The micropipette tip containing the captured cell is then moved and submersed in a second sterile droplet, and the cell is discharged by means of gentle blowing on the micropipette.

The sterile droplet containing the target cell, and possibly other cells, is then examined microscopically. With the same technique, a clean micropipette is then used to pick up the cell and transfer it to a third sterile droplet. This procedure is repeated only until the single cell is isolated from other cells; unnecessary additional isolation often leads to cell damage. Therefore, after the final capture, the cell is discharged into the final isolation vessel (test tube, multiwell plate, etc.).

The diameter of the micropipette opening should be at least twice that of the cell, and often several times the cell size. If the opening is too small, then fluid shearing forces can damage the cell as it passes into the micropipette, especially if it is a naked, scaled, or flagellate cell. If the opening is too large, then it becomes more difficult to pick up the cell, and there is also an increase in the amount of unwanted material that is captured. When isolating filaments, chains of cells, or long single cells (e.g., certain pennate diatoms), the micropipette should be directed to one end of the filament, chain, or cell; the micropipette should be held at an angle so that the filament or cell slides up into the micropipette tip without severe bending.

Some cells adhere to the bottom of the dish or multiwell plate. Efforts to pry a cell loose often result in damage or death. Rapid cell-handling can circumvent this problem (i.e., immediately after the sample is added to the dish, the cell should be picked up before it can sink to the bottom and adhere). Quickly, the cell should be discharged into the sterile rinsing droplet; cells can also adhere to the inside of the micropipette. Immediately after the cell is discharged into the sterile rinsing droplet and before the cell can settle, it should be picked up again and transferred. By quick action, the isolation can proceed before the cell adheres to a surface.

Although one approach is to focus on a single cell—from original sample to final isolation vessel—other techniques can be employed. A skilled technician can pick several target cells from the original sample, and with each rinse only viable cells are moved to the next stage. With experience, one can assess the viability of cells. For flagellates, cessation of swimming sometimes indicates damage. For diatoms, broken frustules can refract light differently than for intact cells. Leakage of protoplasm is an obvious sign of severe damage. Another approach is to isolate several target cells into larger sterile droplets of weakly nutrified liquid, and with attention given to avoiding evaporation (e.g., sealing with Parafilm), these cells can be left for minutes to days. Subsequently, viable cells can be processed, leaving damaged or dead cells. Finally, in many cases, it is easier to remove contaminating cells from around the target cell. This method reduces the handling of the target cell, because effort is directed to the nontarget contaminating material. When most or all contamination is removed, then the target cell is picked up and placed into the isolation vessel.

Another micropipette technique may also be employed to induce cell wall rupture in some diatoms, especially larger ones, and may be desirable for existing clonal strains that are presumably dioecious and near the end of their cell size minimum. This purposeful damage serves to remove the physical constraints of size regeneration. Rogerson et al. (1986) employed repeated introduction and ejection of cells, suspended in a 1% crude papain solution, into and from a micropipette to generate ca. 10% naked cells of *Coscinodiscus asteromphalus*. David Czarnecki (with and without the papain treatment) and Anne-Marie Schmid (without the papain treatment) have both had some success using this technique to generate naked cells of *Campylodiscus clypeus*; these naked cells are reisolated via micropipette into fresh medium, and some successfully regenerate normal larger cells (D. Czarnecki, personal communication).

## 6.3. Isolation with Use of Agar

### 6.3.1. Streaking Cells Across Agar Plates

Isolation of cells on agar plates is also an old and common method. (For preparation of agar plates, see

Chapter 2.) It is the preferred isolation method for many coccoid algae and most soil algae, not only for ease of use but also because axenic cultures can often be directly established without further treatment (see Chapter 8). For successful isolation onto agar, the alga must be able to grow on agar. Some flagellates (e.g., *Heterosigma*, *Pelagomonas*, and *Peridinium*) do not grow on agar, but others (e.g., *Chlamydomonas*, *Pavlova*, *Synura*, and *Tetraselmis*) grow very well on agar. Coccoid cells frequently grow well on or in agar, but some (e.g., *Aureococcus*, *Aureoumbra*) do not. Most diatoms and chlorarachniophytes grow very well on agar; some cryptophytes do, whereas others do not, and dinoflagellates rarely grow on agar.

In most cases, the concentration of agar is not an important factor, assuming the agar is between 0.8% and 1.5 to 2.0% (see Chapter 2). A few algae grow on "sloppy" agar (i.e., preparations with between 0.3 and 0.6% agar), but the algae are probably growing in liquid pockets rather than on the "solid" substrate.

Agar also is a good medium for fungal and bacterial growth. Field samples with substantial fungal contamination can prove frustrating, because the fungus often grows quickly, producing sporangia and spores that contaminate efforts to isolate the alga. Filters and organic substrates can be used to remove filamentous fungi (see previous text). When fungal growth appears on Parafilm-sealed agar plates, it is almost always better to discard the plate without opening it. Conversely, bacteria usually produce small, limited colonies, and unialgal cultures can be obtained if the plate has been properly streaked. One exception is the bacteria from benthic tropical samples, because they often contain agar-digesting bacteria that will "dissolve" regions of the agar plate.

Isolation is accomplished by streaking the natural sample across the agar surface, identical to the technique used for isolating bacteria. A bacterial loop is "loaded" with a small amount of sample, and then the sample is spread with the loop across the agar, with use of one of several techniques or patterns (Fig. 6.5). The origin of the streak typically has too many cells that are not separated, but as the distance from the origin increases, single cells begin to separate. After streaking, the agar plate is incubated until colonies of cells appear. The incubation time varies from a few days for soil and freshwater algae up to several months for oceanic species. When colonies, each originating from a single isolated cell, are present, these are isolated from the agar plate. The isolated colony may be removed from the agar plate with a drawn-out micropipette or with a nichrome or platinum bacterial loop. The micropipette

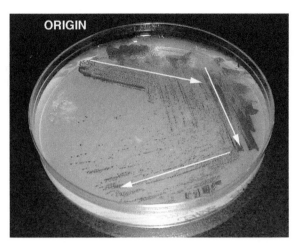

**FIGURE 6.5.** An agar plate streaked with a small, green coccoid alga, showing small isolated colonies arising from a single cell.

is usually used in the dry condition, because if the micropipette is "loaded" with sterile liquid, some seepage of liquid may disperse the colony. When the micropipette tip touches the colony, some cells are drawn into the pipette, whereas others adhere to the sides of the tip. The micropipette is then quickly immersed into an isolation vessel or onto another agar plate. The cells are then discharged by gentle blowing. If placed onto an agar plate or slant, then they are restreaked. Cells also may be picked up from the original plate with a loop and then either restreaked on a new agar plate or rinsed in liquid culture medium to free the cells. If isolated colonies from the single cell are axenic and if good sterile technique is exercised, then an axenic isolate and a single-cell isolate are obtained simultaneously (see Chapter 8).

### 6.3.2. Agar Pour Plates

Some algae do not grow on the surface of agar plates, but they do grow embedded in agar. Beijerinck (1890) first applied the pour technique to isolate cells, but he used gelatin rather than agar (see Chapter 1). Skinner (1932) used an interesting variation of the method when he poured agar and cells into test tubes and then, after incubation, broke the test tubes and removed the isolated colonies. Pringsheim (1946) summarized the history of the traditional agar pour method. More recently, the agar pour method was used to isolate oceanic picoplankters (Brahamsha 1996, Toledo and Palenik 1997). To prepare agar pour plates, cells from the field sample or enrichment culture are mixed with the nonsolidified agar and then poured into petri dishes.

The nonsolidified agar must be cool—only slightly warmer than the gelling point—to avoid overheating and killing the algae. Low-melting-point agar is preferred over normal agar, for obvious reasons. After the nonsolidified, cell mixture is poured into plates, the plates are allowed to cool until the agar solidifies. The agar plate is then incubated at the proper temperature and light conditions until colony formation occurs. To select isolated algal colonies, a micropipette tip is driven into the agar to pick up the selected cells, and the cells are then discharged into dilute liquid culture medium. Although it is possible to indefinitely maintain cells by perpetual transfer with repeated use of the agar pour technique, this technique is almost always used only to isolate cells, with subsequent maintenance by another method (see Chapter 10).

### 6.3.3. Atomized Cell Spray Technique

A fine or atomized spray of cells can be used to inoculate agar plates. The technique can vary, but in general a liquid cell suspension is atomized with forced sterile air so that cells are scattered onto the plate (Pringsheim 1946). For best results, the spray should be administered in a sterile hood or clean environment so that airborne bacteria and fungi are not dispersed onto the plate. The plates are incubated, and after colony formation, selected cells are removed and inoculated as described previously.

### 6.3.4. Isolation After Dragging Through Agar

Agar plates can also be used to remove epiphytes from filamentous algae, as a step in the isolation process. Although various methods can be employed, typically a hook is prepared on the end of a Pasteur pipette, and with the hook the filament is dragged into and through the agar. A metal hook can also be used for larger filaments (see Fig. 6.1f). This process often shears the epiphytes from the filament, and the cleaned filament can then be placed in liquid culture medium or placed onto an agar plate.

## 6.4. Dilution Techniques

The dilution technique has been used for many years (see Chapter 1), and it is effective for organisms that are rather abundant in the sample but largely ineffective for rare organisms. The goal of the dilution method is to deposit only one cell into a test tube, flask, or well of a multiwell plate, thereby establishing a single-cell isolate

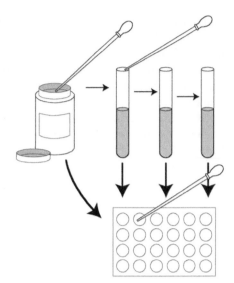

**FIGURE 6.6.** An illustration of the dilution technique. An aliquot is removed from the sample jar (*left*) and placed into a test tube containing sterile medium. After mixing, one aliquot is removed from the test tube and dispensed into multiwells containing sterile medium, and a second aliquot is removed and added to the middle test tube. After mixing, the process is repeated (i.e., dispensed into multiwells and added to the test tube on the right). Each cycle dilutes the original sample and increases the probability of single-cell isolation; the cycle stops when it is probable that no cell will be transferred.

(Fig. 6.6) (Kufferath 1928/29, Droop 1954, Throndsen 1978). If the approximate cell concentration is known, then it is easy to calculate the necessary dilution so that, on the basis of probability, a small volume (e.g., one droplet, 50 µL, 1 mL) contains a single cell. In practice, some contain more than one cell and others contain no cells at all. If the approximate cell number is unknown and cannot be immediately calculated (e.g., at a pond or aboard ship), then one can make repeated serial dilutions of 1 : 10, and between five and six repetitions is enough in most cases (i.e., six repeated 1 : 10 dilutions theoretically places a single cell into the final tube if the original sample had $10^6$ cells · $mL^{-1}$).

The dilution technique is perhaps most often used when attempting to culture random algal species from field samples, often with the goal of discovering new species. The technique can be altered in several ways. First, the dilution can be made with culture medium, distilled water (freshwater organisms), seawater (marine organisms), filtered water from the sample site, or some combination of these. The culture medium can be full-strength for weedy organisms or very dilute for fastidious organisms. Second, an intensive effort can be made at the concentration at which single-cell isolation

is expected (e.g., 10, 50, or 100 attempts). Commonly, one inoculates a large number of tubes from the last dilution, assuming that some single cells will die, some will contain two or more species, and some will receive no cells at all. Third, ammonium, selenium, or another element can be added to some isolation tubes or cell-wells to specifically select for species that require these nutrients. Similarly, one can incubate some isolates at one temperature or light regimen and others with different regimens. Axenic isolates are not often obtained with the dilution technique, because bacteria are usually more abundant than algae (see Chapter 8 for strategies to reduce bacterial contamination).

## 6.5. Gravity Separation: Centrifugation, Settling

Gravity separation can be effective for partitioning larger and smaller organisms, and the two primary methods are centrifugation and settling. Gravity is perhaps more frequently used to concentrate the target organisms rather than to establish unialgal cultures, but the latter can also be achieved. Usually, the goal is to separate the larger and heavier cells from smaller algae and bacteria. Gentle centrifugation for a short duration brings large dinoflagellates and diatoms to a loose pellet, and the smaller cells can be decanted. The target cells are then resuspended, and if necessary, the process may be repeated. The process requires quick action for strong-swimming dinoflagellates, because they will quickly swim up into the liquid again. Also, excessive centrifugation can damage cells, especially delicate organisms and those with ejectile organelles. The centrifugation speed or force and the time vary with each organism and are determined empirically. Centrifugation with use of density gradients (e.g., silica sol, Percoll) has been used to separate mixed laboratory cultures where each species was separated into a sharp band (Price et al. 1974, Reardon et al. 1979). However, to our knowledge, application of the method hasn't been reported for field samples and isolating species.

Settling is effective for nonswimming large or heavy cells. The sample is gently mixed and placed into a tube, flask, cylinder, or other vessel to settle. The sample is settled until many of the larger target cells have reached the bottom, but the smaller cells are still suspended. The upper portion of the sample is decanted or removed by a siphon, leaving a small amount of liquid over the larger cells. The large cells are resuspended, and the process can be repeated if necessary.

Both centrifugation and settling techniques are effective for concentrating larger cells, but it is difficult to obtain a unialgal culture; a single-cell isolation is nearly impossible. Therefore, it is common to employ other techniques in combination with centrifugation and settling (e.g., filtration, screening, or single-cell isolation by micropipette).

## 6.6. Isolation with Use of Phototaxis

Phototaxis has been used to isolate flagellates (Bold 1942, Meeuse 1963, Paasche 1971, Guillard 1973; see also Chapter 1). This method is effective when the sample contains one dominant flagellate that is a strong swimmer and exhibits phototaxis. When several species with similar swimming capability are present, it is more difficult to establish unialgal cultures by this method. The technique involves a focused light source at one end, and, depending upon intensity, either positive or negative phototaxis can be employed (Fig. 6.7a). In its simplest form, a channel of sterile culture medium or water is created, the sample is added to one end, a light source is employed, and the cells swim through the sterile liquid toward or away from the light source. Once the cells have moved a sufficient distance into the

sterile liquid, they are removed by micropipette and placed into a separate vessel. Meeuse (1963), who used a flask with a slender arm, has described a novel approach (Fig. 6.7b). The sample and a sterile glass bead are added to the flask, and a light source is employed. After the phototactic cells swim into the arm, the flask is tilted so that the bead closes the arm while the remaining liquid is decanted. A more sophisticated method involves a specially prepared micropipette. The flagellates swim up the special micropipette, separating themselves from the other organisms. A section of the pipette tip is broken and discarded, and the cells that swam further up the pipette are then discharged into a culture vessel (Paasche 1971, Guillard 1973). Finally, some nonflagellate cells (e.g., *Myxosarcina*; Waterbury and Stanier 1978) migrate across agar in response to light. After migration has separated the cells from the remaining material, the isolated cells are removed and inoculated into culture medium.

---

## 7.0. SPECIFIC ISOLATION METHODS

### 7.1. Isolating Picoplankton

Ultraplankton ($\leq 5$ μm) and picoplankton ($\leq 2$ μm) present special problems for traditional isolation because of their small size. Agar streaks are the simplest method if the alga grow on agar. Tiny freshwater species can often be cultivated on agar, but oceanic species frequently do not; however, for special techniques see Waterbury et al. 1986. For oceanic species, isolation by the dilution technique can be successful if weakly nutrified medium is used, but one cannot be certain about single-cell isolations. Also, new automated instruments can be successfully employed to isolate tiny cells (see Chapter 7).

The traditional method of micropipette isolation can be successfully employed if certain precautions are taken. Ultraclean droplets for rinsing are necessary, because the tiny cells cannot be easily distinguished from particles, especially when working with seawater, in which particulates such as precipitating compounds are common (Jones 1967). Although dilute soil-water extract, recommended by Pringsheim (1950), is satisfactory for larger freshwater organisms, it is not suitable for picoplankton unless it is carefully processed to remove particulates. Particulates also form on and from glass surfaces when the glass has been autoclaved. Silica shards are particularly problematic when isolating

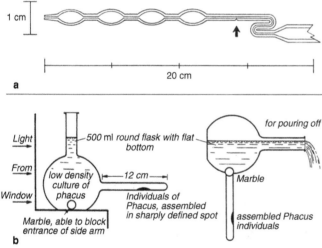

**FIGURE 6.7.** Phototaxis apparatuses. **(a)** Phototactic flagellates drawn into the tip of the pipette and then allowed to migrate through the cavities and into the pipette itself. The tip is broken at the arrow and discarded, and the cells in the pipette are then discharged into sterile culture medium (e.g., test tube, flask, or multiwell) (Paasche 1971, modified from Guillard 1973). **(b)** Negatively phototactic flagellates concentrated with bright light in the narrow arm of the flask (*left*) and then retained by decanting the original sample while the target cells are trapped in the arm by a glass bead (from Meeuse 1963).

picoplankters. To avoid particulates in liquids (e.g., seawater, soil water, or culture medium), two easy methods can be used. One method is to filter the liquid through a 0.2-µm filter. Alternatively, one can allow the seawater or medium to settle for 1–3 days and then carefully immerse a plastic pipetteman tip into the liquid without resuspending the settled material. Both methods can be used in combination by first drawing up liquid with the plastic pipetteman tip, discharging that liquid into a syringe with an attached filter, and finally dispensing filtered droplets for isolation.

One must also be careful to use only very clean, dust-free isolation materials and vessels. If glassware is used, then it should be cleaned thoroughly with acid and rinsed with high-quality doubly distilled or deionized water. Sterilization of glass should be by dry heating in an oven (i.e., 250°C for 3 hours); autoclaves leave a fine material on surfaces that release as particulates into isolation droplets of liquids. It is preferable to use presterilized plasticware, because the surfaces are usually free of particulates; sterile plastic petri dishes work very well. Particulate matter is also created on heated and drawn micropipettes if the flame is sooty, and rust or dirt that is present on the forceps used to draw micropipettes also leave particulate matter in the isolation droplets when the micropipette touches the droplet. In summary, every effort must be taken to remove particulates otherwise it is exceedingly difficult to distinguish the picoplankters from dirt. This is true when using a compound or inverted microscope, but it is especially true when using a dissecting microscope.

Lighting is critical. When using a dissecting microscope, dark-field illumination is far superior to transmitted light, the latter providing suboptimal conditions at best. With dark-field illumination, the tiny cells reflect light so that they can be observed even at a magnification of 20× (higher magnifications are recommended). The algal cell occasionally rotates as it is suspended in the liquid, and a tiny flash of light is emitted from the chloroplast as it turns, clearly distinguishing the algal cell from bacteria or other materials. Using a very finely drawn micropipette, the cell can be picked and dispensed into a sterile rinsing droplet. It is important to use very small droplets for rinsing, because the tiny cell is difficult to find again in a large droplet. Because small droplets can evaporate rather quickly, one must work more quickly. Tiny cells are more easily isolated using an inverted microscope. The higher magnification provided by the inverted microscope makes identification of the algal cells easier, and one has more confidence that the cell actually has been picked rather than just relocated.

Tiny cells are often very sensitive to compounds in the culture medium, and even if single cells are successfully selected, they can die after being dispensed into the culture medium. In the case of oceanic picoplankton, isolation of a single cell into oligotrophic seawater is often sufficient for the first few days. Subsequently, a tiny amount of culture medium can be added (e.g., 10 µL of medium for each 25 mL of isolation solution), and over a period of weeks, successively larger amounts of nutrients can be added, especially once visible cell growth is apparent.

## 7.2. Isolating Epiphytes After Sonication

Epiphytes on larger algae or other organisms are difficult to isolate by direct picking from the host with a micropipette. When the epiphyte is scraped from the host, damage often occurs. Sonication can be employed to vibrate epiphytes from the host, and once suspended, the epiphyte can be picked and rinsed in the usual way (Brown and Bischoff 1962, Hoshaw and Rosowski 1973; see Chapter 8).

## 7.3. Isolating Sand-Dwelling Organisms

Sand is a well-known substrate for the epipelon community (Round 1981). Two common types of algae are present: flagellates that swim freely among the sand grains and algae that grow attached to the sand grains. The former is much easier to isolate than the latter, especially if entire sand grains (often with other attached organisms) are removed and inoculated into culture vessels. For sampling, the upper layer of sand is collected with a glass tube, wide-mouthed bottle, or small shovel. Round (1981) suggests placing the sand and water in a petri dish, and the cells that move up into the overlaying water are either collected with a micropipette or allowed to adhere to floating coverslips. The coverslips can be removed using a forceps, with one arm on each side of the coverslip, and placed onto a microscope slide for observation. For actual isolation of cells, the coverslip is inverted and cells are removed with a micropipette. For enrichment cultures, the coverslip can be placed into a petri dish with sterile medium, wrapped with Parafilm, and incubated. Takeo Horiguchi (personal communication) provides a second method. He places a small amount of sand in a transparent plastic cup with a lid (e.g., an ice cream cup), and the appropriate amount of culture medium is added.

(Germanium dioxide is also added if diatoms are not desired.) Then the cup, with the lid in place, is tilted so that the sand grains move to one side. When the sand settles, the cup is returned to the upright position; alternatively, a bar (e.g., a chopstick) is placed at one side of the cup so that the cup is tilted all the time in the culture cabinet. In either case, half of the bottom of the cup should be devoid of sand so that observation is possible. An inverted microscope works well for examining the bottom of the cup, but a dissecting microscope is necessary for examining cells on the sides of the cup or in the upper water layer. Cells are isolated from the enrichment with a micropipette. Hoppenrath (2000) separated sand-dwelling dinoflagellates by placing a filter (i.e., 0.45 μm) above the sand and by placing frozen seawater ice on top of the filter. As the seawater melts, the filtered liquid passes through the sand and washes out the algae. This method is adapted from that of Uhling (1964).

The isolation of algae from mud is more difficult because the mud does not settle to the bottom, leaving a clear liquid. If the organisms attach to a coverslip, then this provides one means for separating them from the original sample. Also, dilution of the muddy water can provide a means for establishing enrichment cultures or even dilution cultures. Isolation by micropipette is possible only when the mud is diluted enough so that observation is possible.

## 7.4. Isolating Strongly Adhering Organisms

Some organisms adhere tightly to surfaces, and quick work is necessary to avoid problems (see previous discussion). However, sometimes the only cell of interest is already adhered to the surface. Sonication or more violent mixing may not be appropriate because that one cell may be lost among other materials in the sample before it can be found again and isolated. In this case, the cell can sometimes be washed loose with a stream of water gently blown from a pipette (with a tube and mouthpiece). Once the stream detaches the cell from the substrate and becomes free in the liquid, the cell can be quickly picked, rinsed, and inoculated into the culture medium.

Some attached algae (e.g., attached dinoflagellates) adhere so strongly that a stream of water does not detach the cell. Horiguchi (personal communication) suggests an alternative method that involves motile cell formation for *Amphidinium testudo* Herdman and *Spiniferodinium*. The enrichment culture is examined several times within 1 day, and if the cells are growing, then motile cells will be released. (The period of motile cell release can be quite short.) Some benthic dinoflagellates release motile cells a few hours after the onset of the light period. The motile cells are then isolated by micropipette.

When these methods fail, a more drastic approach can be attempted if the alga is attached to a plastic substrate (e.g., a petri dish or plastic dish). First, the unwanted material is removed around the target alga; a fine paint brush can be used to sweep away larger particles and a micropipette can be used for smaller particles. Using a razor blade or scalpel, cut out the area around the cell and transfer the plastic piece containing the cell to a new culture vessel. This can be messy, because water leaks once the plastic is cut. A petri dish cover or another cup should be placed under the original container to capture the leaking water.

## 7.5. Isolating Subaerial and Aerial Organisms

Subaerial or aerophilous algae grow on substrates that are exposed to air (e.g., bark, leaves, and stems of plants; animal surfaces; stones and soil; various man-made structures). The diversity from these substrates is surprising to some. For example, Foerster (1971) found 77 genera in a Puerto Rican forest and Schlichting and Milliger (1969) cultured 91 genera of microorganisms (mostly algae) from the giant water bug *Lethocerus* collected at night over land with a mercury vapor light trap. Soil algae are perhaps the best known (Ettl and Gaertner 1995), but other substrates have also been studied extensively (Brook 1968; Schlichting 1969, 1975). Samples are collected by scraping, chipping, or otherwise removing the algae from the substrate. In some cases, enrichment cultures are established, but direct isolation is often possible. One interesting enrichment protocol, used by Schlichting and Milliger (1969), consisted of placing the entire water bug into the culture medium. For direct isolation, cells can be streaked on agar plates, and colonies formed from single cells can be removed. Atomized spray of a scraping suspension can also be directed onto an agar plate (see Section 6.3.3), with isolates removed after growth.

Seafoam occupies a position somewhere between subaerial and aerial algae. Schlichting (1971) collected seafoam with a spoon or knife, placed the foam into sterile plastic bags, and then inoculated freshwater and marine culture media upon return to the laboratory. These enrichment cultures produced numerous algae, although they were often freshwater algae rather than marine algae.

Aerial algae have been studied less than subaerial algae, but several studies by different investigators have been published (Brown et al. 1964; Schlichting 1961, 1969; Smith 1973). Smith (1973) cultured 44 species from the air of Raleigh, North Carolina, but found that there was always less than one culturable alga per liter of air. Sampling aerial algae without contamination is described by Schlichting et al. (1971), and enrichment or agar plate cultures and isolations are carried out with the collected particulate matter.

## 7.6. Isolating Cysts from Sediments

Sediment samples have substantial amounts of debris that hinder isolation. For flagellates, the sediments can be enriched with nutrients and incubated in the light so that the cysts germinate. Once germinated, flagellate cells can be isolated with the techniques described previously. Alternatively, cysts can be separated from sediments by several methods (Wall et al. 1967, Matsuoka and Fukuyo 2000), and the concentrated cysts can then be collected and rinsed and individual cysts can be placed into multiwells or other isolation containers.

In some cases, cyst germination involves a ploidy change (e.g., a diploid cyst undergoes meiosis, producing four haploid cells) (see Chapters 23, 24). If it is important to isolate each haploid cell individually (e.g., for sexual crossing studies), then the enrichment method is inappropriate. By some means, an individual cyst must be isolated into a small isolation vessel (e.g., multiwell plate) so that the germination process can be monitored every few hours. After germination of the cyst, the four haploid cells should be isolated individually and brought into culture. The culture strains obtained from these single-cell isolates should represent male/female (+/−) cell lines.

## 7.7. Removing Diatoms with Germanium Dioxide

Although diatomists love diatoms, they can be a problem for other scientists who are attempting to isolate nondiatom algae. Diatoms grow very quickly, and they can often outcompete other algae. In most cases, the addition of germanium dioxide to the sample, an enrichment culture, or isolation culture with a diatom contaminant leads to the death of most or all diatoms (Lewin 1966). Uptake of germanium dioxide occurs readily in diatoms, and the germanium is substituted in the place of silica in biochemical reactions; this

becomes lethal. For many nondiatomeaous algae, a high concentration of germanium dioxide usually is not toxic; however, see McLachlan et al. 1971 for toxicity to brown seaweeds. A typical germanium addition results in a final concentration of 1 to 10 mg · L$^{-1}$ germanium dioxide. The time at which the germanium dioxide is added can be important. When establishing enrichment cultures, it is best to immediately add the germanium dioxide; if it is added after significant diatom growth has occurred, 5 to 10 times as much is required to inhibit diatom growth.

## 7.8. Removing Cyanobacteria with Antibiotics

Like Aesop's fable of the tortoise and the hare, contaminating cyanobacteria slowly but persistently overcome other algae. Fortunately, antibiotics are quite effective against cyanobacteria, because, like bacteria, they are prokaryotes. Some antibiotics are more effective than others, and some eukaryotic algae are sensitive to antibiotics. However, in most cases, one can find a combination of an effective antibiotic at an appropriate concentration (see Chapters 8 and 9).

## 7.9. Isolating Organisms Requiring Water Movement

Some species, especially those growing in flowing water, require water movement for successful growth (i.e., they die when isolated into stationary vessels). Stirring (spinning) culture vessels (e.g., those from Bellco, Inc.) are an easy means of providing water motion, and roller culture equipment is available as well. Similarly, a culture vessel can be placed on an orbital mixing device that provides enough motion to move the water.

---

## 8.0. SUMMARY

The isolation of algae may be daunting to the beginner, but with practice and persistence, sufficient skills will be quickly developed. In many cases, ingenuity can be applied to make the isolation easier, whether in an exorbitantly equipped lab or a remote, ill-equipped field station. A well-prepared isolation area makes isolation less frustrating, easier, and more successful. The area should have all the usual (and unusual) isolation items (see Fig. 6.1) ready to use. For isolation of algae, this equipment is often needed immediately. The items should be stored or kept in the same location at all times, so that they are immediately accessible. A well prepared, well-practiced, patient, and persistent isolator is ready for unexpected circumstances, and such a person likely succeeds when faced with adversity.

## 9.0. Acknowledgments

The authors thank Drs. David Czarnecki, Takeo Horiguchi, and Peter Miller for providing specific techniques for isolating diatoms and benthic dinoflagellates. R. A. A. was supported by National Science Foundation grant DBI-9910676.

## 10.0. References

Anderson, D. M., Fukuyo, Y., and Matsuoka, K. 1995. Cyst Methodologies. In: Hallegraeff, G. M., Anderson, D. M., and Cembella, A. D., eds. *Manual on Harmful Marine Microalgae.* IOC Manuals and Guides No. 33. UNESCO, Paris, pp. 229–49.

Beijerinck, M. W. 1890. Culturversuche mit Zoochlorellen, Lichengonidien und anderen niederen Algen. *Bot. Zeitung* 48:725–39, 741–54, 757–68, 781–85.

Bold, H. C. 1942. The cultivation of algae. *Bot. Rev.* 8:69–138.

Brahamsha, B. 1996. A genetic manipulation system for oceanic cyanobacteria of the genus *Synechococcus. Appl. Environ. Microbiol.* 62:1747–51.

Brook, A. J. 1968. The discoloration of roofs in the United States and Canada by algae. *J. Phycol.* 4:250.

Brown, R. M., Jr., and Bischoff, H. W. 1962. A new and useful method for obtaining axenic cultures of algae. *Phycol. Soc. Amer. News Bull.* 15:43–44.

Brown, R. M., Jr., Larson, D. A., and Bold, H. E. 1964. Airborne algae: Their abundance and heterogeneity. *Science* 143:583–5.

Brunel, J., Prescott, G. W., and Tiffany, L. N., eds. 1950. *The Culturing of Algae.* Charles F. Kettering Foundation, Antioch Press, Yellow Springs, Ohio, 114 pp.

Droop, M. R. 1954. A note on the isolation of small marine algae and flagellates for pure culture. *J. Mar. Biol. Assoc. U. K.* 33:511–41.

Droop, M. R. 1959. Water-soluble factors in the nutrition of *Oxyrrhis marina. J. Mar. Biol. Assoc. U. K.* 38:605–20.

Droop, M. R. 1966. Ubiquinone as a protozoan growth factor. *Nature* 212:1474–75.

Droop, M. R. 1971. Terpenoid quinones and steroids in the nutrition of *Oxyrrhis marina. J. Mar. Biol. Assoc. U. K.* 51:455–70.

Ettl, H., and Gaertner, G. 1995. *Syllabus der Boden-, Luft- und Flechtenalgen.* Gustav Fischer Verlag, Stuttgart.

Foerster, J. W. 1971. The ecology of an elfin forest in Puerto Rico. 14. The algae of Pico Del Oeste. *J. Arnold Arbor.* 52:86–109.

Guillard, R. R. L. 1973. Methods for microflagellates and nannoplankton. In: Stein, J. R., ed. *Handbook of Phycological Methods: Culture Methods and Growth Measurements.* Cambridge University Press, Cambridge, pp. 69–85.

Guillard, R. R. L. 1975. Culture of phytoplankton for feeding marine invertebrates. In: Smith, W. L., and Chanley, M. H., eds. *Culture of Marine Invertebrate Animals.* Plenum Press, New York, pp. 26–60.

Guillard, R. R. L. 1995. Culture methods. In: Hallegraeff, G. M., Anderson, D. M., and Cembella, A. D., eds. *Manual on Harmful Marine Microalgae.* IOC Manuals and Guides No. 33. UNESCO, Paris, pp. 45–62.

Hoppenrath, M. 2000. Morphology and taxonomy of six marine sand-dwelling *Amphidiniopsis* species (Dinophyceae, Peridiniales), four of them new, from the German Bight, North Sea. *Phycologia* 39:482–97.

Hoshaw, R. W., and Rosowski, J. R. 1973. Methods for microscopic algae. In: Stein, J. R., ed. *Culture Methods and Growth Measurements.* Cambridge University Press, Cambridge, pp. 53–68.

Jones, G. E. 1967. Precipitates from autoclaved seawater. *Limnol. Oceanogr.* 12:165–7.

Kufferath, H. 1928/29. La culture des algues. *Revue Algol.* 4:127–346.

Küster, E. 1913. *Anleitung zur Kultur der Mikroorganismen.* (1. Auflage). Vertlag B. G. Teubner, Leipzig, 201 pp.

Lee, J. J., and Soldo, A. T. 1992. *Protocols in Protozoology.* Society for Protozoology, Lawrence, Kansas (unpaginated).

Lewin, J. 1966. Silicon metabolism in diatoms. v. germanium dioxide, a specific inhibitor of diatom growth. *Phycologia* 6:1–12.

Lewin, R. A. 1959. The isolation of algae. *Rev. Algol.* (new series) 3:181–97.

Matsuoka, K., and Fukuyo, Y. 2000. *Technical Guide for Modern Dinoflagellate Cyst Study.* Westpac-HAB, Nagasaki University, Nagasaki, Japan, 29 pp.

McLachlan, J., Chen, L. C.-M., and Edelstein, T. 1971. The culture of four species of *Fucus* under laboratory conditions. *Can. J. Bot.* 49:1463–69.

Meeuse, B. J. D. 1963. A simple method for concentrating phototactic flagellates and separating them from debris. *Arch. Mikrobiol.* 45:423–4.

Miquel, P. 1890–3. De la culture artificielle des diatomées. *Le Diatomiste* 1: Introduction, 73–75; Cultures ordinaires des diatomées, 93–99; Culture artificielle des diatomées marines, 121–28; Cultures pures des diatomées, 149–56; Culture des diatomées sous le microscope. De l'avenir des cultures de diatomées, 165–72.

Nichols, H. W. 1973. Growth media: freshwater. In: Stein, J. R., ed. *Handbook of Phycological Methods: Culture Methods and Growth Measurements.* Cambridge University Press, Cambridge, pp. 7–24.

Paasche, E. 1971. A simple method for establishing bacteria-free cultures of phototactic flagellates. *J. Cons. Int. Explor. Mer.* 33:509–11.

Price, C. A., Mendiola-Morgenthaler, L. R., Goldstein, M., Breden, E. N., and Guillard, R. R. L. 1974. Harvest of planktonic marine algae by centrifugation into gradients of silica in the CF-6 continuous-flow zonal rotor. *Biol. Bull.* 147:136–45.

Pringsheim, E. G. 1912. Die Kultur von Algen in Agar. *Beitr. Biol. Pfl.* 11:305–32.

Pringsheim, E. G. 1946. *Pure Cultures of Algae.* Cambridge University Press, London, 119 pp.

Pringsheim, E. G. 1950. The soil-water culture technique for growing algae. In: Brunel, J., Prescott, G. W., and Tiffany, L. H., eds. *The Culturing of Algae.* Charles F. Kettering Foundation, Dayton, Ohio, pp. 19–26.

Reardon, E. M., Price, C. A., and Guillard, R. R. L. 1979. Harvest of marine microalgae by centrifugation in density gradients of 'Percoll.' In: Reid, E., ed. *Cell Populations. Methodological Surveys (B) Biochemistry.* Vol. 8. John Wiley & Sons, New York, pp. 171–5.

Rogerson, A., DeFreitas, A. S. W., and McInnes, A. C. 1986. Observations on wall morphogenesis in *Coscinodiscus asteromphalus* (Bacillariophyceae). *Trans. Am. Microsc. Soc.* 105:59–67.

Round, F. E. 1981. *The Ecology of Algae.* Cambridge University Press, Cambridge, 653 pp.

Schlichting, H. E., Jr. 1961. Viable species of algae and protozoa in the atmosphere. *Lloydia* 24:81–8.

Schlichting, H. E., Jr. 1969. The importance of airborne algae and protozoa. *J. Air Poll. Control Assoc.* 19:946–51.

Schlichting, H. E., Jr. 1971. A preliminary study of the algae and protozoa in seafoam. *Bot. Mar.* 14:24–28.

Schlichting, H. E., Jr. 1975. Some subaerial algae from Ireland. *Br. Phycol. J.* 10:257–61.

Schlichting, H. E., Jr., and Milliger, L. E. 1969. The dispersal of microorganisms by a hemipteran, *Lethocerus uhleri* (Montandon). *Trans. Amer. Microsc. Soc.* 88:452–4.

Schlichting, H. E., Jr., Raynor, G. A., and Solomon, W. R. 1971. *Recommendations for Aerobiology Sampling in a Coherent Monitoring System: Algae and Protozoa in the Atmosphere. U.S./IBP Aerobiology Handbook No. 3.* University of Michigan, Ann Arbor, Michigan, pp. 60–1.

Sheldon, R. W. 1972. Size separation of marine seston by membrane and glass-fiber filters. *Limnol. Oceanogr.* 23:1256–63.

Skinner, C. E. 1932. Isolation in pure culture of green algae from soil by a simple technique. *Plant Physiol.* 7: 533–7.

Smith, P. E. 1973. The effects of some air pollutants and meteorological conditions on airborne algae and protozoa. *J. Air Poll. Control Assoc.* 23:876–80.

Stein, J. R., ed. 1973. *Handbook of Phycological Methods.* Cambridge University Press, Cambridge, 448 pp.

Stokner, J. G., Klut, M. E., and Cochan, W. P. 1990. Leaky filters: a warning to aquatic ecologists. *Can. J. Fish. Aquat. Sci.* 47:16–23.

Throndsen, J. 1978. The dilution-culture method. In: Sournia, A., ed. *Phytoplankton Manual.* UNESCO, Paris, pp. 218–24.

Toledo, G., and Palenik, B. 1997. *Synechococcus* diversity in the California current as seen by RNA polymerase (rpoC1) gene sequences of isolated strains. *Appl. Environ. Microbiol.* 63:4298–303.

Uhling, G. 1964. Eine einfach Methode zur Extraktion der vagilen, mesopsammalen Mikrofauna. *Helgoländer wiss. Meeres.* 11:178–85.

Venkataraman, G. S. 1969. *The Cultivation of Algae.* Indian Council of Agricultural Research, New Delhi, 319 pp.

Wall, D., Guillard, R. R. L., and Dale, B. 1967. Marine dinoflagellate cultures from resting spores. *Phycologia* 6:83–6.

Waterbury, J. V., and Stanier, R. Y. 1978. Patterns of growth and development in pleurcapsalean cyanobacteria. *Microbiol. Rev.* 42:2–44.

Waterbury, J. V., Watson, S. W., Valois, F. W., and Franks, D. G. 1986. Biological and ecological characterization of the marine unicellular cyanobacterium *Synechococcus.* In: Platt, T., and Li, W. K. W., eds. *Photosynthetic Picoplankton. Can. Bull. Fish. Aquat. Sci.* 214:71–120.

# CHAPTER 7

# AUTOMATED ISOLATION TECHNIQUES FOR MICROALGAE

MICHAEL SIERACKI
*Bigelow Laboratory for Ocean Sciences*

NICOLE POULTON
*Bigelow Laboratory for Ocean Sciences*

NICHOLAS CROSBIE
*Ocean Genome Legacy Foundation*

## CONTENTS

Key Index Words: Automated Isolation, Axenic Isolation, Cell Sorting, Flow Cytometry, Fluorescence Plate Reader, Multiwell Plates, Optical Trapping, Optical Tweezer

## 1.0. INTRODUCTION

The isolation of phytoplankton from nature has a long history (see Chapter 1). Traditional methods require skill, patience, and a good microscope. Using manual and visual methods with low-magnification optical microscopy, microplankton (i.e., cells larger than 20 µm) are not too difficult to collect with a plankton net and to isolate from the complex mixtures of cells typically seen in natural samples (see Chapter 6). Smaller cells, especially those smaller than about 5 µm, are much more difficult to isolate by traditional manual and visual methods, and the importance of smaller phytoplankton has become widely recognized (Li 2002). Newer, automated technologies, like flow cytometric cell sorting, are well suited to these smaller cells, and the potential of these instruments is only beginning to be exploited. Flow cytometry is ideal for initially isolating the smallest microalgal cells, and single cells can be rapidly sorted into multiwell plates for establishing new algal cultures. The separation of algal cells from co-occurring contaminating cells (e.g., bacteria) is an important advantage as well.

The purpose of this chapter is to consider some of the critical issues regarding isolating microalgae by automated methods. The focus is on single-cell sorting, because this technology has advanced far enough to be applied for routine isolations. Optical trapping, or the "optical tweezer," is a developing technology and is briefly considered.

## 1.1. Flow Cytometry with Cell Sorting

Flow cytometry was developed in the 1960s and 1970s as a way of counting and analyzing optical properties of single cells suspended in a fluid. Details of the principles and operation of a flow cytometer are given elsewhere (Melamed et al. 1994, Shapiro 2003). An overview of the basic aspects of flow cytometry is given in Chapter 17, in the context of cell counting by flow cytometry. Reckermann (2000) has written a general review of sorting applications in aquatic ecology, and we only briefly describe the basic principles; specific details of isolation of microalgae are provided.

When a sample is analyzed by flow cytometry, distinct populations may appear as clusters in the two-dimensional cytograms of scatter and/or fluorescence parameters. With rapid electronic sorting logic, it is possible to identify single cells within these populations and make sorting decisions as the cells leave the optical interrogation point. There are two major types of sorting mechanisms now commonly available: fluid flow sorting and droplet sorting.

In fluid flow sorting, the flow stream is physically diverted to collect the portion of the stream that contains the target cell. A variation of this idea is to move a sample collection device into the sample stream at the precise moment a target cell is present. One commercial application of this technology has been implemented by Becton-Dickinson (BD Biosciences), starting with the FACSort instrument, followed by subsequent models (e.g., FACSCalibur). These systems typically

sort at rates of hundreds of cells per second. They collect sheath fluid when they are not sorting, so generally a sort sample is highly diluted with sheath fluid. They can be combined with a filtration system to remove sheath fluid and yield a more concentrated sort sample. An advantage of this instrument is its ease of use relative to droplet sorters (Shapiro 2003).

In droplet sorting (Fig. 7.1) the flow stream is oscillated so that it undulates, breaking into small drops in a controlled way. Typically in these "jet-in-air" systems the fluid flow stream leaves the flow cell through an orifice of defined diameter. The interrogation point, where the laser intersects the flow stream, is located just below the flow cell. A piezoelectric device on the flow cell typically generates the oscillation force. The stream can be charged (positive or negative) prior to droplet formation, and the drop retains the given charge. If the droplets pass near charging plates, then they can be drawn toward or away from the plates, directing the droplet into collecting devices. The sort logic can therefore charge droplets containing target cells so that they will be deflected in the proper direction for collection.

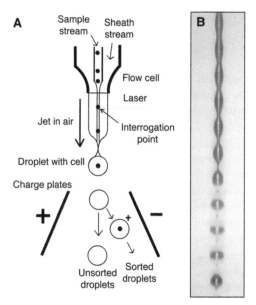

FIGURE 7.1. (a) Schematic diagram of droplet cell sorting, showing the joining of the sample stream and sheath fluid stream in the flow cell, forming the jet-in-air flow stream with the cells in single file. The laser intercepts the cells at the interrogation point, where measurements are made. If a cell meets the criteria of the sort logic, then the flow stream is charged just before the break-off point of the droplet containing the target cell. The droplet retains the charge and is deflected by the charge plates toward the collection tube or plate. (b) Image of flow stream at the droplet break-off point with use of a 70-μm tip.

Modern electronics allow these cell measurements, sort logic decisions, and plate charging to occur very rapidly (ca. $5 \times 10^4 \cdot s^{-1}$). In practice, the speed of sorting depends more on the concentration of target cells in the sample. The cells can be collected in test tubes, onto microscope slides, or into multiwell plates. The multiwell plate can be moved manually under the sort stream by an automated robotic arm so that the proper number and type of cells are put into user-designated wells.

The pressures applied to the sheath and sample fluids control fluid flow. The differential pressure between the two determines the sample flow rate. This control is called *hydrodynamic focusing*, because changing the differential pressure changes the diameter of the sample stream flowing within the sheath stream. In contrast to fluid flow sorters, droplet sorters can sort at high rates ($>10^4$ cells $\cdot s^{-1}$, under ideal conditions), and they can sort cells into very small liquid volumes or onto slides and solid media. Our experience has been with droplet sorters, and our discussion focuses on this type of instrument. Examples of such instruments currently in use include the Beckman-Coulter EPICS V and EPICS Altra, the Becton-Dickinson FACSVantage, and the DakoCytomation MoFlo (Dako).

The cell velocity in droplet sorters can be quite high. The fluid flow stream in a flow cytometer is primarily laminar once it enters the sample tubing. Cells may experience high shear forces in the sample tubing leading to the flow cell and where the droplets impact the fluid in the collection tube, breaking up fragile cells, removing flagella, etc. When sorting for clonal isolation, however, it is not necessary to sort a large number of cells, so sheath pressure can be reduced to the lowest level at which stable flow is accomplished (Durack 2000). Furthermore, droplets sorted during a single-cell cloning experiment experience a relatively low terminal velocity because aerodynamic forces act to decelerate "isolated" droplets to a greater extent than during bulk sorting (where droplets experience an aerodynamic slipstream) (Tyrer and Kunkel-Berkley 1984). A negative consequence of this greater aerodynamic deceleration, however, is that the trajectory of each droplet is more easily disturbed by surrounding air currents. Droplet impact is lower during sorting into liquid than during sorting onto slides and other solid or semisolid materials.

Experiments to determine the effects of sorting on phytoplankton cells have shown that the optical effects of the high laser power are stronger than any detrimental effects of fluid flow or droplet charging. Haugen et al. (1987) tested four fragile algae (*Chroomonas*, *Micromonas*, *Tetraselmis*, and *Gyrodinium*) for possible damaging effects by flow cytometry, and neither flow cytometric fluidics nor laser exposure caused cell loss or inability to grow. In some cases an extended growth lag was induced following flow cytometric analysis, but all four strains showed growth after passing through the instrument. Rivkin et al. (1986) tested the photosynthetic capacity of cells after sorting and found that there was some loss attributable to exposure to the laser light. There was no effect by fluid flow, droplet charging, or sorting. The detrimental effect was stronger with increased laser power, so they recommend using the lowest laser power possible.

Several authors have reported successful isolation of microalgae by flow cytometric sorting (Sensen et al. 1993, Reckermann and Colijn 2000, Crosbie et al. 2003). Sensen et al. (1993) started with unialgal, bacterized cultures of *Cyanophora*, *Haematococcus*, *Monomastix*, *Scherffelia*, and *Spermatozopsis*. From these, single-cell isolates were created by droplet sorting into 50-mL sterile culture medium in Erlenmeyer flasks. They reported a 20 to 30% success rate, with visible growth observed after 6 to 12 weeks. This is despite their use of 0.01% sodium azide in their sheath fluid, which is not recommended (see later discussion). In addition, they found that more than 20% of the cultures were axenic, and axenic cultures were produced even when the ratio of bacteria to algal cells was as high as 300 : 1.

Sorting directly from a natural sample has also been successful (Reckermann and Colijn 2000). Reckermann (2000) was able to produce various cultures from mixed natural assemblages collected from the Wadden Sea. They included strains of *Synechococcus*, *Microcystis*, *Hemiselmis*, *Teleaulax*, *Rhodomonas*, and *Thalassiosira*, and some unidentified prymnesoid flagellates, diatoms, and pico-eukaryotes. Success rates for cryptophytes and diatoms were high, whereas *Synechococcus* and pico-eukaryotes were more difficult and required more trials before cultures were produced. Isolation of cells from natural assemblages during sorting at sea has been successful for establishing several strains of *Prochlorococcus* (Moore et al. 1998). Sorting at sea has special challenges and is considered in more detail later in this chapter.

Single-cell isolations of picoalgae have been performed from enrichment cultures of lake samples (Crosbie et al. 2003). Samples were enriched with culture media, filter-fractionated through 2-μm pore filters, and incubated under low light for 3 or 4 weeks. With a B-D FACSVantage SE and ClonCyt, picoalgae

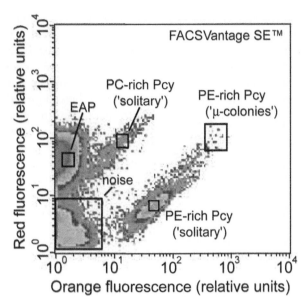

**FIGURE 7.2.** Gates used for single-cell and single-colony sorting of autotrophic picoplankton populations sourced from enrichment cultures and natural samples originating from subalpine lakes. Pcy, picocyanobacteria; PE, phycoerythrin; PC, phycocyanin; EAP, eukaryotic autotrophic picoplankton. (Crosbie et al. 2003, used with permission).

**TABLE 7.1** Percentage of sorted wells showing phycoerythrin-rich picocyanobacteria (PE), phycocyanin-rich picocyanobacteria (PC), or eukaryotic autotrophic picoplankton (EAP) growth for *a priori* sort gate classfications PE, PC, and EAP.

| Presort classification | N sorted wells | Post-sort APP classification | | |
| --- | --- | --- | --- | --- |
| | | PE | PC | EAP |
| PE | 1,193 | 14 | 0 | 1 |
| PC | 96 | 1 | 17 | 0 |
| EAP | 288 | 0 | 0 | 5 |

were sorted by single-cell or single-colony sorting (Fig. 7.2) into BG11 medium (Stanier et al. 1971). Target cell (autotrophic picoplankton) growth was observed in 15% of microplate wells putatively inoculated with single target cells (Table 7.1). Overall recoveries ranged from 5% for cells preclassified as eucaryotic autotrophic picoplankton to 17% for cells preclassified as phyco-cyanin-rich picocyanobacteria. Misclassifications resulting in growth of the "wrong" cell type were rare (1–5%).

## 2.0. Instrument Setup for Sorting

Many different issues must be considered prior to sorting, beginning with the target alga to be sorted. What is the ultimate goal for sorting the algal species of interest: enrichment, unialgal isolates, or axenic isolates? The cell sorter setup should be optimized to meet your goal. Different sorting protocols are needed for producing nonclonal, clonal, axenic, or fungus-free cultures. The details of setting up a particular sorting instrument depend on the flow cytometer model. However, a few criteria are standard for most instruments. Here we consider general factors for setting up a sorting instrument to isolate and work with algal cultures.

### 2.1. Nozzle Tip Size

The nozzle tips used with most jet-in-air flow cytometers are available with orifices of various sizes (50 to 400 μm). Larger-nozzle orifices produce larger droplets. As the drop volume increases, the potential for contamination by coincident particles or cells also increases. For larger cells (>10–20 μm), the nozzle orifice diameter should be at least three to five times larger than that of the target algal cell. The stability of the sorting jet and the ability to discriminate the target population also influence nozzle choice. For example, the isolation of freshwater autotrophic picoplankton (<2 μm) was optimized by single-cell sorting with a 100-μm nozzle mounted on a FACSVantage SE (Crosbie et al. 2003). The jets produced by smaller nozzles were more difficult to align and resulted in pure laser noise when overlays were not added to the forward scatter obscuration bar. Overlays increase the effective size of the obscuration bar to block more of the forward angle light. Adding overlays reduced the cell scatter and fluorescence signals to an extent that individual autotrophic picoplankton populations could not be readily distinguished from each other and from the noise. Optimal signals occurred when a 100-m nozzle was used in combination with a custom-made forward scatter obscuration bar overlay and a broad bandpass red filter (695 ± 40 nm).

### 2.2. Sheath Fluid

Various sheath fluids have been described in the aquatic cytometry literature, including distilled water, phosphate buffered saline, and filtered seawater. For analysis of natural aquatic samples, the sheath fluid salinity should match the sample salinity for accurate light scatter measurements because of the effect of mismatched refractive indices (Cucci and Sieracki 2001). For effective sorting, the sheath fluid must be an ionic solution (i.e., distilled water cannot be used). The solutes in the sheath allow the droplets to maintain charge so that they can be sorted. A low-salinity sheath (0.1% NaCl) should be used when sorting freshwater species, because there can be a significant diffusive exchange between the (outer) sheath stream and the sample core (Pinkel and Stovel 1985). The physical mixing of sheath and core components within sorted droplets, although brief, may cause further osmotic or toxic stress. As a general rule, when sorting freshwater phytoplankton, use the lowest possible salt concentration that results in effective sorting. Richard Stovel (personal communication) reports using salt concentrations at 1/50 the concentration of "normal" saline (i.e., 1/50 of 0.9%). Anything potentially toxic should be reduced to a minimum (e.g., use high-grade salts when making phosphate-buffered saline or relevant ionic solution).

## 2.3. Sheath Pressure and Pressure Differential

Current flow cytometers are capable of sorting at very high speeds because of faster electronic processing and increased sheath pressures. Increasing the pressure differential increases the sample flow rate, because the sample pressure is greater than the sheath pressure. Operating sheath pressures between 60 and 100 psi (414–689 kPa) are commonly used for mammalian tissue culture cells that are sorted at very high densities ($>10^7$ cells·mL$^{-1}$). For establishing an algal culture, lower sheath pressures (10–30 psi, or 69–207 kPa) are recommended because of the lower cell densities in culture and environmental samples. When sorting algal cells, from either a culture or mixed field population, reducing the sample flow rate decreases potential contamination but results in a lower event rate.

## 2.4. Laser

Usually a blue laser (488 nm) is used to detect chlorophyll autofluorescence of individual algal cells. Fluorochromes are not required for algal detection, but certain nonlethal fluorochromes such as LysoTracker

(Molecular Probes, Inc.) may be used to detect and isolate protozoa (Rose et al., 2004; see Section 5.0). It is important not to photodamage the cell with excessive laser light, because this extends the growth lag and reduces photosynthesis (Rivkin et al. 1986, Haugen et al. 1987). Therefore, use the lowest laser power that allows clear detection of the target algal population. An air-cooled argon laser, with power ranging from 15 to 120 mW, is usually sufficient to detect algal cells containing chlorophyll and phycoerythrin. Only in some cases are higher laser intensities (150 mW to 1 W) of a water-cooled laser required. This occurs when the pigment concentration of small algal cells is very low (e.g., with *Prochlorococcus*). Low pigment concentrations are usually observed in samples in the surface layers of oligotrophic water bodies because of high light levels (Olson et al. 1990).

## 2.5. Data Acquisition

To collect flow cytometric data and sort cells, one of the signals, scatter or fluorescence, must be used as the trigger signal. A scatter trigger detects all particles in a sample. A fluorescence trigger detects only the fluorescing particles within a sample (e.g., algae). When establishing an axenic culture (see Section 3.0), a scatter trigger should be used so that all the nonfluorescing particles and cells (i.e., bacteria) are detected and excluded.

Flow cytometric data can be acquired with the detectors in linear or logarithmic mode. Typically, linear data collection is used during instrument calibration with a suspension of uniform size and fluorescence beads. Logarithmic data collection is recommended for most samples in aquatic cytometry, including both algal cultures and field assemblages, because it provides a wider dynamic range.

## 2.6. Sorting Criteria

The sorting criteria determine how the drop(s) containing the target cell(s) of interest are collected and, consequently, defines the yield and purity. Sorting criteria depend on two related factors: the sort envelope and the sort mode. The "sort envelope" is the number of sequential drops that are sorted per sort decision, ranging from one to three or more drops (Fig. 7.3). The more drops collected for each sort decision, the greater the chance the target organism is collected, thereby

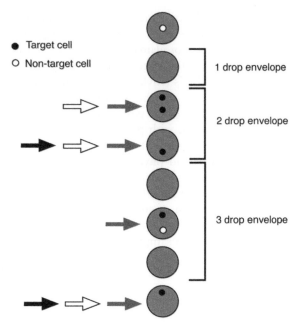

**FIGURE 7.3.** Droplet schematic representing three sort modes and envelopes that are possible when choosing criteria. The gray, white, and black arrows are drops that would be collected (containing black target cells) if an "enrich," "purify," or "single" sort mode were chosen, respectively. In all three cases a 1-drop sort envelope would be required (only one drop collected per sort decision). See text (Sorting Criteria) for further explanation.

increasing the yield. However, by increasing the sort envelope there is also a greater risk that contaminants are sorted, thereby decreasing the purity.

The "sort mode" considers the position of the target cells in the flow stream relative to other target and non-target cells. For example, a DakoCytomation MoFlo allows three different sort modes (enrich, purify, and single mode; see Fig. 7.3), for each sort envelope (Anonymous 2001). Enrich mode is used when recovery is the most important criterion; all positive target cells are collected, including any nontarget cells. Purify mode is used when purity is most important; positive target cells are sorted only when there are no negative events. Single mode can also be thought of as "single cell" mode; for each sort decision, only one positive event is collected. If two positive cells, or a positive and a nontarget cell, are within the sort envelope, then the positive target cell is not sorted. Single mode is most useful for clonal sorting. The operator must consider the criterion that best suits the purpose of the sort. For example, to establish an algal culture from a natural assemblage or enrichment culture, where neither axenic

nor single-cell isolation is necessary, the purify mode is best. In this case numerous cells (e.g., 10–100) can be sorted per tube or cell well.

## 2.7. Clonal Isolation

Configuring a cell sorter for single-cell sorting *sensu stricto* is similar to that for bulk sorting but with a few additional requirements (Battye et al. 2000; see Section 5.0.). Cell clumping should be reduced by gentle agitation and sample dilution with culture medium. Because doublets and larger clumps orient lengthwise along the flow (Gray et al. 1979, Lucas and Pinkel 1986), marked by a greater pulse width signal, the probability of sorting two or more cells can be reduced by using "pulse processing" (available on many modern sorters). Additionally, some flow sorters (e.g., B-D FACS Vantage CloneCyt system [BD Biosciences]) have the option of "index sorting," in which the data values of each putatively sorted cell are mapped to the sorting well, providing a quality-control map.

## 2.8. Precision Sorting

In some instances, cells traveling at or close to the droplet boundaries may snap the stream prematurely, and the charging pulse is only partially (or not at all) applied to the targeted droplets. This can be avoided by "phase gating," in which only cells occupying a certain portion of a droplet period (e.g., the central portion) are eligible for sorting (Merrill et al. 1979). Setup requires extra precision by the operator when setting the droplet break-off timing. Cells are typically sorted into a very small target (e.g., 3–4 mm in the case of 384-well plates). Therefore, aiming accuracy is particularly important. Minimizing air currents and sorting at or closer to a vertical angle can result in more accurate positioning (Stovel and Sweet 1979).

## 2.9. Determining the Volume of a Drop

A sorted droplet contains the target cell, sample fluid, and sheath fluid. For some applications it is useful or necessary to know the volume of a sorted droplet and the volume of the sample fluid (e.g., minus the cell volume) within a sorted drop. With this information you can determine the dilution of the sample medium in the collection vessel, or you can estimate the expected number of contaminating bacteria. Total drop volumes vary with different flow cell orifices and can vary slightly with the size of the sorted cells, especially large cells. The volume of sample fluid within a drop depends on flow cell orifice size as well. For a given flow cell orifice, the difference between the sheath and sample pressures control the sample fluid volume in the drop. Note that for large cells and high differentials, the cell itself may displace most of the sample fluid in a drop.

Measuring the total volume of a droplet ($V_d$) is not difficult with a high-speed sorter. Then, with a good estimate of the proportion (R) of sample fluid to $V_d$, one can calculate the volume of sample in one drop ($V_s$) as

$$V_s = (V_d)(R) \qquad (1)$$

The volume of sample run is determined gravimetrically with a microbalance accurate to 0.1 mg. The $V_d$ is determined by sorting a known (large) number of drops into a preweighed tube, measuring the total volume sorted by weight and dividing by the number of drops. Sample particles are not required. Set the trigger photomultiplier tube (PMT) voltage high so the sort region can be set on instrument noise. This is necessary, because there are no particles in the sample fluid, and there must be events on which to trigger the sorter.

The proportion of sample in a drop (R), for a given instrument setting, can be determined several ways. Using a dye in the sample solution, one can spectrophotometrically determine the proportion of dye in the sample tube relative to the waste stream. This gives the dilution of sample fluid with sheath fluid. A simpler method is to measure the volumes of total fluid (i.e., sample + sheath) produced and the sample consumed

over a given interval of sorting. The sample fluid should not contain cells (it can be sheath fluid), and the drop drive does not need to be on for this. Again, the volumes are determined by weighing sample and collection tubes before and after the run. Start collecting the fluid stream coming out of the flow cell (sample + sheath, otherwise called the *waste* stream) at the same moment that the sample flow is started. Because the ratios of sheath to sample fluid is high (Table 7.2), a large volume (ca. 3–4 mL) must be collected. When enough fluid has been collected, the tube is moved out of the flow stream and the sample supply is stopped simultaneously. Then both tubes are weighed and the volumes determined by difference. The ratio of these volumes is R, and $V_s$ can be calculated as in Equation 1. Because fluid is conserved, the diameter of the sample stream within the overall flow stream can be calculated from the ratio of sample to sheath fluid in the drops, assuming an overall flow stream diameter and using the formula for the volume of a cylinder.

Table 7.2 shows the drop volumes obtained with a DakoCytomation MoFlo instrument with different differential pressures. There are several interesting aspects. First, the drop diameter (equivalent spherical diameter, or EDS; 177 μm) is greater than the flow cell orifice diameter, indicating that the stream bulges out as drops are formed (as seen in Fig. 7.1B). The diameter of the sample stream is only 5 to 10% of the flow stream diameter (assumed to be the orifice diameter, 100 μm). If the sample fluid within the drop is spherical, then its diameter ranges from 26 to 38 μm. This is significantly smaller than the overall drop diameter but similar to the size of larger cells. Large cells could distort the flow stream and prevent accurate sorting,

**TABLE 7.2** Determination of sample volume in a drop ($V_s$ in units of picoliters) on the DakoCytomation MoFlo flow cytometer. A nozzle orifice of 100 μm was used, and sample pressure was about 25 psi (172 kPa). Total drop volume was determined to be 2.785 nanoliters, with an equivalent spherical diameter (ESD) of 177 μm. These results were confirmed by independently determining R by dilution of a dye measured spectrophotometrically.

| Pressure differential | Ratio of $V_s$ to $V_d$ (R) | Sample stream diameter (μm) | Sample droplet volume, $V_s$ (pL) | ESD of $V_s$ (μm) |
|---|---|---|---|---|
| 0.2 | $3.13 \times 10^{-3}$ | 5.6 | 8.7 | 25.9 |
| 0.4 | $4.98 \times 10^{-3}$ | 7.0 | 13.9 | 30.3 |
| 0.6 | $7.93 \times 10^{-3}$ | 8.9 | 22.1 | 35.3 |
| 0.8 | $9.71 \times 10^{-3}$ | 9.8 | 27.0 | 37.8 |

and therefore a larger nozzle orifice diameter is needed to sort larger cells. If $10^6 \cdot mL^{-1}$ bacterial cells were present in the sample, then there would be one bacterium every 40 to 110 drops over this range of differential pressures. In this case, sorting to produce axenic cultures would be practical (see later text).

## 3.0. FLOW SORTING UNDER STERILE CONDITIONS

### 3.1. Fluidics Preparation

Sterile sorting conditions are necessary when establishing axenic cultures. Both the sheath fluid and sheath lines (tubing) of the flow cytometer must be sterilized. The sheath fluid should be passed through a 0.2-μm filter and autoclaved to remove and kill both bacterial and viral contaminants. 0.01% Sodium azide has been used to prevent bacterial growth in the sheath fluid (Sensen et al. 1993); however, it is not recommended, because it can inhibit algal growth. Flow cytometers with an in-line 0.2-μm sheath filter should have the filter replaced periodically to maintain in-line sheath sterility. The sheath tank should also be thoroughly cleaned with ethanol and pressurized with either an inert gas (e.g., nitrogen) or compressed air that has been passed through a sterile 0.2-μm filter.

The fluidic system should be thoroughly disinfected with one of the following solutions:

1. 6 to 10% hydrogen peroxide solution (effective against vegetative bacteria, bacterial spores, fungi, and viruses).
2. 7 to 10% sodium hypochlorite (bleach) (effective, but contact with metal parts should be avoided because of its corrosive nature).
3. 70% ethanol (less virucidal activity, but useful for cleaning noncritical areas and metal parts because it evaporates quickly).

It is critical that all the disinfectant is removed from the fluidic system prior to sorting. Residual bleach or ethanol in the sheath fluid, in-line filter, or sample lines is likely to inhibit growth of target algal cells. After using $H_2O_2$, bleach, or ethanol, sterile deionized water should be passed through the fluidics system for several hours. More detailed procedures depend on the sorter and the cell sample (Merlin 2000).

Once the sheath lines are sterile, we recommend sterilizing the sample lines with 70% ethanol or 10% beach. Also, rinse the sample lines with deionized water between algal samples to eliminate cross-contamination.

### 3.2. Drop Size and Sort Trigger

The smaller the volume of the drop, the less likely there is bacterial contamination (see Section 2.9). Using a smaller nozzle size reduces the drop volume; however, minimum nozzle size depends on target cell size. Likewise, a smaller sample volume reduces the possibility of contamination, and this is achieved by reducing the sample flow rate.

To effectively detect and separate nonfluorescing bacteria, forward or side scatter is used to trigger the sort event. The sort mode is set to high stringency, such as single drop mode, where only one positive event is collected with the sort envelope chosen. In this case a small sort envelope (e.g., 1 drop) is recommended.

### 3.3. Experimental Controls

As a control, it is important to demonstrate sterile sorting methods and the ability to obtain bacteria-free drops containing algae. In our own laboratory, we have used the following method. A natural marine bacterial assemblage is enriched in marine broth. Bacteria are mixed with an axenic algal culture (*Isochrysis galbana*) at various bacteria to alga cell ratios (1:1, 10:1, and 100:1). With a 100-μm nozzle tip, a single algal cell is sorted into each well of a 96-well plate containing marine broth and incubated. If a bacterial cell is inadvertently sorted, then we assume it grows in the broth. Our results (unpublished data) show that even with increasing ratios, we were able to sort most algal cells without accompanying bacteria. At the highest level of contamination (100:1), 54% of the wells had bacterial growth after 10 days.

## 4.0. ISOLATION TECHNIQUES

A single-cell isolate is preferred, but in some cases, the establishment of a nonclonal unialgal culture may be an intermediate step in the isolation process. The best growth conditions (e.g., culture medium or light intensity) are often unknown for uncultured species, and the establishment of a nonclonal culture may be necessary before subsequent single-cell isolations are successful (i.e., optimizing the growth conditions of the nonclonal culture may improve later single-cell isolation attempts conducted under optimized conditions). Furthermore, traditional isolation methods have demonstrated that several factors can affect the success of single-cell isolation attempts. For example, dilute medium is often necessary when isolating more fastidious species, because single cells, which do not benefit from the partitioning of resources among multiple cells, are potentially subjected to relatively higher doses of growth-inhibiting substances (see Chapter 6). As shown in Tables 7.3 and 7.4, the percentage of successful isolation events increases when more than one cell is isolated, and these strains are not fastidious species. Other factors, such as culture medium formulation, culture vessel size, and deficiencies of vitamins or other growth factors, may need careful study before a single-cell isolation is successful. Bacteria may secrete essential growth factors or compounds that protect the alga from

otherwise toxic material in the medium (Gonzalez and Bashan 2000, Fukami et al. 1996).

Unlike tedious traditional methods, flow cytometric sorting is relatively easy and rapid. With little additional expense, numerous sorting trials can be attempted (e.g., different macronutrients or micronutrients, vitamins, light intensities, salinities, or temperatures), more than reasonably possible in the same period with traditional micropipette techniques.

## 4.1. Reisolation of Existing Cultures

Reisolation of an alga is sometimes necessary (i.e., when the culture is not unialgal, when a single-cell isolation must be established from a nonclonal culture, or when a culture becomes contaminated with bacteria or fungi). Flow cytometry has been used successfully to establish clonal cultures for several marine and freshwater algal species (Crosbie et al. 2003, Sensen et al. 1993). In our laboratory, we purified algal cultures contaminated with bacteria or fungi by means of a DakoCytomation high speed MoFlo cell sorter. For example, *Synechococcus* and

**TABLE 7.3** Percentages of post-sorting viability of phytoplankton cultures (larger eukaryotes) sorted into both 96-well and 24-well plate formats. Note that two culture media were used (f/2 and f/20). The number of cells per well ranged from 1 to 125. For experiments using the 96-well plate format, 200 μL of medium was added; for 1, 5, and 10 cells per well, 44 wells were inoculated; and for 25 and 125 cells per well, 22 wells were inoculated. For experiments using the 24-well plate format, 2 mL of medium was added to each well, and for each experiment, 10 wells were inoculated. A 1-month incubation was required prior to determining positive or negative growth via microscopy.

| Phytoplankton (strain no.) | Plate format (no. of wells) | Culture medium | Percent of wells with growth | | | | |
| | | | 1 Cell | 5 Cells | 10 Cells | 25 Cells | 125 Cells |
|---|---|---|---|---|---|---|---|
| *Isochrysis galbana* (CCMP 1324) | 96 | f/2 | 22 | 64 | N/A | 95 | 100 |
| | 96 | f/2 | 9 | 22 | N/A | 82 | 100 |
| *Rhodomonas salina* (CCMP 1319) | 96 | f/2 | 2 | 50 | 68 | 100 | 100 |
| | 96 | f/20 | 2 | 48 | 68 | 100 | 100 |
| | 24 | f/2 | 90 | 100 | N/A | 100 | 100 |
| | 24 | f/20 | 100 | 100 | N/A | 100 | 100 |
| *Heterocapsa triquetra* (CCMP 448) | 96 | f/2 | 0 | 18 | 55 | 100 | 100 |
| | 96 | f/20 | 0 | 30 | 100 | 100 | 100 |
| | 24 | f/2 | 60 | 100 | N/A | 100 | 100 |
| | 24 | f/20 | 30 | 100 | N/A | 100 | 100 |
| *Thalassiosira pseudonana* (CCMP 1335) | 96 | f/2 | 5 | 77 | 100 | 100 | 100 |
| | 96 | f/20 | 27 | 100 | 100 | 100 | 100 |

**TABLE 7.4** Percentages of post-sorting viability of *Micromonas* and *Synechococcus* cultures sorted into 96-and 24-multiwell plates. Viability was determined 24 days after inoculation by either microscopy or flow cytometry. The number of cells isolated into a well ranged from 1 to 1,000. Note that two culture media were used (f/2, f/20). For experiments using the 96-well plate format, 200 μL of medium was added and 22 wells were inoculated. For experiments using the 24-well plate format, 2 mL of medium was added to each well and 5 wells were inoculated.

| Phytoplankton (strain no.) | Plate format (no. of wells) | Culture medium | Percent of wells with growth | | | |
| --- | --- | --- | --- | --- | --- | --- |
| | | | 1 Cell | 10 Cells | 100 Cells | 1,000 Cells |
| *Micromonas pusilla* (Butcher) | 96 | f/2 | 0 | 5 | 55 | 100 |
| Manton et Parke (CCMP 494) | 96 | f/20 | 0 | 2 | 68 | 100 |
| | 24 | f/2 | 0 | 60 | 100 | 100 |
| | 24 | f/20 | 0 | 80 | 100 | 100 |
| *Synechococcus* sp. (S8c, 1k) | 96 | f/2 | 0 | 5 | 100 | 100 |
| | 96 | f/20 | 9 | 41 | 100 | 100 |
| | 24 | f/2 | 60 | 100 | 100 | 100 |
| | 24 | f/20 | 20 | 100 | 100 | 100 |

*Prochlorococcus* (cyanobacteria) were rendered axenic by sorting single cells into glass test tubes containing ~4 mL of sterile conditioned culture medium. In collaboration with the Provasoli-Guillard National Center for Culture of Marine Phytoplankton (CCMP), we successfully reisolated *Scrippsiella* cells (culture contained a fungal contaminant) and produced an axenic *Aureococcus* culture.

## 4.2. Viability of Sorted Cells

Using multiwell plates, we tested post-sorting viability, using six algal strains (*Isochrysis*, *Thalassiosira*, *Rhodomonas*, *Micromonas*, *Heterocapsa*, and *Synechococcus*) with well-established growth conditions. Between 1 and 1,000 cells were sorted into each well of 24- and 96-well plates (Falcon, Becton Dickinson Labware) (see Tables 7.3 and 7.4). Cells isolated into 24-well plates had a higher viability than those isolated into 96-well plates. This suggested, at least for the plate formats we tested, that cells sorted into 24-well plates had higher viability levels. It is not obvious why this result occurred, although plastic residues may be relatively higher in smaller wells. Overall, the viability of flow cytometrically sorted single-cell isolates was slightly less or comparable to those expected by traditional micropipette technique. However, with use of

flow cytometric sorting, the speed and number of possible isolations greatly outweighed any reduction in viability.

## 4.3. Test Tubes and Well Plates

Multiwell plates are gaining popularity in algal culture. Larger algae that settle to the bottom of a test tube often grow better in multiwell plates. This may be due to the flat bottom surface and the ratio of culture medium volume to bottom surface area, both of which reduce cell packing on the bottom. Also, surface-to-volume ratios are higher in multiwell plates than in test tubes. Multiwell plates are also commonly used for traditional micropipette isolation attempts, because they can be quickly scanned with an inverted microscope.

Flow cytometers can sort cells into both test tubes and multiwell plates. Historically, before automated systems were available, direct sorting into test tubes was the standard method for cell isolation. With newer instruments, cells can be isolated easily by sorting automatically into multiwell plates. These plates come in various formats, with 6 to 384 or more wells per plate, and the well volumes range from 20 μL to 12 mL. Because of the speed at which a sorted multiwell plate can be generated, a large number of plates can be inoculated quickly and easily. The choice between sorting into test tubes or multiwell plates may depend on the algal species and the amount of replication desired. In our experience, some algal species do not grow in small-volume wells, because of either the well size or sensitivity to plasticizers. An extensive survey of cell-well plates from available manufacturers may reveal that some types of well plates are better than others. If algae are maintained in well plates for longer periods, then they should be sealed with Parafilm or plate-sealing film to reduce evaporation. For choice of culture medium, see Chapters 2 through 4, Chapter 6, and Appendix A.

## 4.4. Sorting from an Enrichment Culture

The initial process of establishing a unialgal culture from a natural assemblage can be challenging. Many species or groups of algae, such as the picoeukaryotes, are difficult to isolate into culture. One method of improving the chance of algal growth is to first establish an enrichment culture. This allows cells in the natural assemblage to adapt to the higher nutrient content and establish themselves, making it easier to identify dominant populations. If culture conditions (especially the light regimen) are standardized, then it is possible to identify the same (or similar) populations in repeated sorting experiments, making it easier to recognize emergent populations over the course of the enrichment culture. By sorting at appropriate intervals (i.e., short intervals to avoid overgrowth) from the same enrichment culture(s), both common and rare populations can be isolated. However, enrichment cultures have the possible drawback of selecting for faster-growing types.

## 4.5. Sorting from a Field Sample

To obtain ecophysiological data of different plankton species, it is often necessary to isolate cells into culture from the natural environment. Because of the high diversity in a natural field sample, it is important to sort at low event rates by diluting the field sample or reducing the pressure differential to reduce nontarget cell contamination. We have successfully sorted both large eukaryotic phytoplankton and smaller picoeukaryotes from natural field samples. A relatively high degree of species diversity was observed for the large eukaryotic cell types after 2 to 3 weeks of growth, indicating broad applicability. In the case of picoeukaryotes, between one

and 1,000 cells were sorted per well. Growth was observed in both 96- and 24-well plates, in wells containing more than one sorted cell.

## 4.6. Sorting at Sea

To date, a limited number of cultures have been established using cell sorters at sea. These include multiple strains of *Prochlorococcus* from the Sargasso Sea that were established for genetic analysis and physiological studies (Moore et al. 1998). Until recently, operating a cell sorter aboard a research ship was limited by the requirements of power and cooling of large lasers. New developments in solid-state lasers, however, have reduced the size of cell sorters to near that of current benchtop analyzers, and they have much lower power and cooling requirements, making them much easier to use at sea. Currently, there are various benchtop cell sorters with solid state lasers that use violet (412 nm), blue (488 nm), green (514 nm), and red (635 nm) wavelengths. One disadvantage of using solid-state lasers is the lower power output (5–100 mW) when compared with water-cooled lasers, which provide up to several watts of power. The availability of these small, more stable lasers will revolutionize the way cell sorting is conducted in the field.

## 4.7. Sorting Controls

When isolating single cells, the possibility of multiple cell isolation is a concern. We have developed a "split-well" protocol to serve as a control (Crosbie et al. 2003). Using a multiwell plate, single cells are sorted into every other row, leaving the alternate rows of cells filled with sterile culture medium. After sorting, a sorted well (i.e., inoculated with a cell) is thoroughly mixed with a pipette using good sterile technique. Half the contents of the mixed well are then transferred to the adjacent well containing only sterile medium. The amount of medium that was removed is then replaced with culture medium. After incubation, if more than one cell was originally sorted, then growth should occur in both the sorted and transferred wells. Although post-sorting viability is a complicating factor (see Tables 7.3 and 7.4), this simple method offers a measure of confidence that single cells were sorted. Also, although it is rarely discussed, one should recall that traditional single-cell isolation by micropipette has no rigorous control regarding the inoculation of a single cell. More than one investigator has been surprised to find species "x" growing despite being confident that only a single cell of species "y" entered the isolation tube.

## 4.8. Screening Multiwell Plates

Numerous multiwell plates can be generated quickly when establishing algal cultures via automated sorting. An efficient way of screening and monitoring cell growth in these plates becomes a problem. Multiwell plates can be monitored with an inverted microscope, but this is time-consuming and tedious. Alternatively, newly available fluorescence plate readers with enhanced sensitivity can be used to monitor growth by measuring increases in chlorophyll fluorescence over time. This screening process is quite rapid; a 96-well plate can be scanned in 1 to 2 minutes. Plate readers can also measure light absorbance in the wells, implying that growth rates can be calculated, but we have not tested this aspect.

Preliminary tests with a TECAN SpectrofluorPlus fluorescence plate reader showed that the reader successfully monitored culture growth (Fig. 7.4). Six cultures, representing different cell sizes and algal classes, were diluted with culture medium and placed in multiwell plates (200 μL per well). Each plate was incubated at optimum temperature and light for 1 week. Both opaque black and clear 96-well plates were used. The opaque black optical plates provide greater sensitivity for fluorescence measurements by minimizing optical crossover between wells. The bottoms of the wells in the black plates are transparent, allowing light for cell growth and fluorescence measurements.

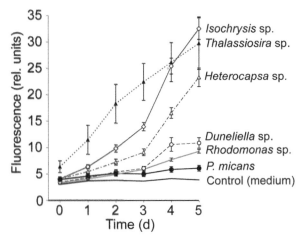

**FIGURE 7.4.** Changes in chlorophyll fluorescence (675 nm) for six cultures of marine phytoplankton in clear 96-well plates over 7 days, as determined with the TECAN SpectrofluorPlus fluorescence plate reader.

The fluorescent plate reader detected positive growth in all wells. The highest growth rates were observed in the clear 96-well plates (see Fig. 7.4). Cells in the clear plates received higher light levels, and this is most likely the cause of the better growth. To evaluate the sensitivity of the fluorescent plate reader, measurements were also made in glass tubes with a Turner fluorometer (Turner Designs, Inc.), a highly sensitive instrument used to measure chlorophyll fluorescence. Algae were cultured in glass test tubes with use of the same culture medium, inoculum concentration, and growth conditions. The fluorometer results compared favorably with the results with the clear 96-well plates, suggesting that the differences among strains were most likely due to differences in growth rates by the strains.

To further test the sensitivity of the fluorescence plate reader, each culture was diluted to extinction to determine the minimum cell number detected by the plate reader (Table 7.5). Overall, higher sensitivity was observed in the black optical plates than the clear plates, and the sensitivity in the black plates was similar to that of the Turner fluorometer. Although growth was faster in the clear plates, the black plates supported cell growth and provided a more sensitive measure of fluorescence. Illumination of the black plates could be optimized to promote faster growth, if desired. Finally, we have not yet tested the fluorescence sensitivity using plates with larger wells.

Analytical flow cytometry is an alternative or complement to using a plate reader to screen multiwell plates (see Chapter 17). Small volumes are withdrawn aseptically from the wells and stained with fluorescent nucleic acid stains, for example, Syto 13 or SybrGreen I (Molecular Probes). When the sample is analyzed, the presence of targeted phytoplankton and the presence or absence of nontarget phytoplankton, heterotrophic bacteria or fungi, and (potentially) viral infection are determined. Large numbers of subsamples can be run on an analytical flow cytometer equipped with an autoloader (e.g., B-D FACSCalibur [BD Biosciences]). These automated features decrease screening time dramatically, especially when one is trying to verify potentially clonal or axenic algal cultures. Furthermore, the analytical flow cytometry approach is intrinsically more sensitive than current microplate reader techniques. Provided that the flow-cytometric "signature" of the target cell does not overlap greatly with the signature produced by sheath fluid contaminants, "growth positive" wells are readily distinguished from those exhibiting no growth, even when the absolute cell numbers are small (Crosbie et al. 2003).

**TABLE 7.5** Cell detection limits for a variety of marine algal cultures for the TECAN fluorescence microplate reader using a 96-multiwell plate and the Turner fluorometer. Detection limits were determined by comparing the results of each dilution to the filtered seawater blank with a pairwise $t$ test ($P < 0.05$). The sensitivity is the first dilution found to be significantly different from the blank.

| Culture | Cell size (μm) | Sensitivity (cells · mL⁻¹) TECAN microplate | | Turner fluorometer |
| | | Black plate | Clear plate | |
|---|---|---|---|---|
| *Thalassiosira pseudonana* (Hustedt) Hasle et Heimdal | 3–5 | 1,800 | 3,500 | 500 |
| *Prorocentrum micans* Ehr. | 25–30 | 40 | NT | 40 |
| *Heterocapsa triquetra* (Ehr.) Stein | 15 | 100 | 400 | 100 |
| *Dunaliella sp.* | 3–5 | 500 | 500 | 500 |
| *Rhodomonas salina* (Wislouch) Hill et Wetherbee | 10–15 | 1,000 | NT | 1,000 |
| *Isochrysis galbana* Parke | 3–5 | 2,500 | 10,000 | 2,500 |

NT, Not tested.

## 5.0. ISOLATING HETEROTROPHIC ALGAE AND OTHER HETEROTROPHS

Unlike photosynthetic algae, heterotrophs usually produce no fluorescence signal to aid in sorting. Heterotrophic bacteria are near the limit of detection by light scatter but can be detected after the addition of fluorochromes (Lemarchand et al. 2001; see Chapter 17). Antibodies with fluorescent labels are used to sort bacteria (Porter et al. 1993), and some bacteria reduce 5-cyano-2,3-ditoyl tetrazolium chloride (CTC) to produce fluorescence (Bernard et al. 2000). However, these procedures kill the bacteria, so they cannot be used to cultivate new isolates.

Gel microencapsulation, where single cells are encapsulated in agarose gel microdrops (GMDs) (Katsuragi et al. 2000), was successful for isolating marine bacteria from the Sargasso Sea (Zengler et al. 2002). The GMDs are permeable to solutes, so cells can grow into colonies while encapsulated. GMDs have distinctive scatter properties that can be used to identify and sort them, for example, separating fast and slow-growing bacteria. Zengler et al. (2002) found the most diversity when the encapsulated cells were cultured in unamended, filter-sterilized Sargasso Sea water.

Because algae can be grown by gel encapsulation, both for maintenance cultures (see Chapter 10) and cryopreservation (see Chapter 12), this heterotrophic technique may have application for certain photosynthetic algae as well.

Eukaryotic heterotrophs maintained in simple culture can be detected on the basis of forward and side scatter alone. In a natural sample, or complex enrichment culture, these protists overlap with many other particles in their scatter signals. Perhaps, therefore, flow cytometric analysis of heterotrophic protists from natural samples is rarely reported in the literature. Keller et al. (1994) demonstrated that phagotrophic protists could be detected by the fluorescence of food vacuole contents, suggesting vacuole fluorescence as a means for selectively isolating heterotrophic protists on the basis of their feeding preferences. For example, bacteria labeled with fluorochromes (Sherr et al. 1987) or live bacteria induced to express fluorescing proteins (Parry et al. 2001, Fu et al. 2003) could be added to a natural or enrichment sample to identify grazing protists. Alternatively, the heterotrophs themselves could be labeled with a nontoxic fluorochrome LysoTracker (Rose et al. 2004) and sorted on the basis of the fluorochrome's signal. Isolation of protistan heterotrophs by means of cell sorting has not been described fully but only in reports of inadvertent cultures established when isolating phytoplankton. Our preliminary results show that marine heterotrophic flagellates and amoebae survive sorting and grow in multiwell plates (unpublished data), and isolating these should become routine as methods are refined.

## 6.0. OPTICAL TWEEZERS

A technique called *optical trapping* can be used to manipulate cells under a microscope (Ashkin et al. 1987). A focused light beam, typically an infrared laser emitting radiation at 1064 nm, interacts with a dialectric medium, causing the medium to become polarized. If the beam forms an intensity gradient in cross-section, then the light pulls the medium toward the center of the beam. The force produced can be strong enough to hold cells and particles within the beam, and cell absorption (i.e., heating) is minimal at 1064 nm. The instrument, combined with a micropipette, can be used to isolate and remove an individual target cell from other co-occurring cells. The captured cell is dispensed into sterile culture medium to produce a single cell isolate.

To our knowledge, optical trapping has not been used for isolating microalgae, but it has been used for photophysiological studies of phytoplankton (Sonek et al. 1995) and the isolation of bacteria and yeast (Grimbergen et al. 1993). Mitchell et al. (1993) used a flat microcapillary tube that was filled with sterile medium. They inoculated one end with a mixed bacterial culture and then sealed the tube. An individual target cell was captured in the "tractor" laser beam, and by means of moving the microscope stage, the cell was dragged 2 to 3 cm into the sterile medium. The tube was cut and the single cell was dispensed into sterile culture medium. They were able to successfully drag bacteria at high speeds using 180 mW of laser power, but at lower laser powers they had to reduce the speed for dragging cells. Mitchell et al. (1993) suggest that more critical optical alignment is required to produce enough force to trap larger protistan cells. A similar procedure was used by Huber et al. (2000) to isolate hyperthermophilic bacteria.

Sonek et al. (1995) found that optical trapping could be used to hold cells of two pico-algae (*Synechococcus* and *Nanochloris*) for photophysiological studies. They reported that increasing laser power (0–500 mW) caused a significant reduction in the fluorescence spectra after a 5-hour exposure in the trapping laser. Isolating cells with optical tweezers takes only a few

seconds, but the possible photophysiological damage during that time is unknown.

## 7.0. MICROFLUIDICS

Progress in nanotechnology may revolutionize the tools available for working at the very small scales of individual cells. A system with an air stream sheath may lead to disposable cell sorters (Huh et al. 2002). Bacteria and viruses have been successfully sorted with devices using microfluidics (Fu et al. 1999, Chou et al. 1999). Optical trapping has been combined with microfluidics to create a particle sorter, with a 100-μm flow cell, that sorts macromolecules, colloids, and small cells (MacDonald et al. 2003). The system uses an optical lattice to sort particles on the basis of refractive index, and it could be incorporated into "lab-on-a-chip" designs. Aqueous microdrops suspended in a liquid microfluidic chip can be manipulated by an electric field (Velev et al. 2003). These new flow cytometry instruments and components have much lower flow rates (ca. tens of μm or mm $\cdot s^{-1}$) than conventional cell sorters (ca. 1–10 m $\cdot s^{-1}$), but they are smaller and less complex. In the future, an optical trapping system consisting of numerous small cytometric units working in parallel, rather than serially in a flow stream, may be used to rapidly sort a sample. The optical tweezers would replace the droplet formation and charging steps of current flow cytometers, and the systems may eventually be deployed remotely in lakes and oceans to automatically analyze and sort cells.

## 8.0. ACKNOWLEDGMENTS

The authors thank T. Cucci and L. Moore for helpful discussions. This work was partly funded by National Science Foundation grants OCE-9986331 and DBI-9907566 and Office of Naval Research grant N0014-99-1-0514.

## 9.0. REFERENCES

Anonymous. 2001. Module 7: Sorting. In: *MoFlo Hands-On Training Course Manual*. DakoCytomation, Inc., Fort Collins, Colorado, pp. 117–59.

Ashkin, A., Dziedzic, J. M., and Yamane. T. 1987. Optical trapping and manipulation of single cells using infrared laser beams. *Nature* 330:769–71.

Battye, F. L., Light, A., and Tarlington, D. M. 2000. Single cell sorting and cloning. *J. Immunol. Methods* 243:25–32.

Bernard, L., Schaler, H., Joux, F., Courties, C., Muyzer, G., and Lebaron, P. 2000. Genetic diversity of total, active and culturable marine bacteria in coastal seawater. *Aquat. Microb. Ecol.* 23:1–11.

Chou H. P., Spence, C., Scherer, A., Arnold, F. H., and Quake, S. 1999. A microfabricated device for sizing and sorting DNA molecules. *Proc. Natl. Acad. Sci. USA* 96:11–13.

Crosbie, N., Pöckl, M., and Weiss, T. 2003. Rapid establishment of clonal isolates of freshwater autotrophic picoplankton by single-cell and single-colony sorting. *J. Microbiol. Methods* 55:361–70.

Cucci, T. L., and Sieracki. M. E. 2001. Effects of mismatched refractive indices in aquatic flow cytometry. *Cytometry* 44:173–8.

Durack, G. 2000. Cell-sorting technology. In: Durack, G., and Robinson, J. P., eds. *Emerging Tools for Single-Cell Analysis.* Wiley-Liss, New York, pp. 1–19.

Fu, A. Y., Spence, C., Scherer, A., Arnold, F. H., and Quake, S. R. 1999. A microfabricated fluorescence-activated cell sorter. *Nature Biotech.* 17:1109–11.

Fu, Y., O'Kelly, C., Sieracki, M., and Distal, D. L. 2003. Protistan grazing analysis by flow cytometry using prey labeled by *in vivo* expression of fluorescent proteins. *Appl. Environ. Microbiol.* 69:6848–55.

Fukami, K., Sakaguchi, K., Kanou. M., and Nishijima, T. 1996. Effect of bacterial assemblages on the succession of blooming phytoplankton from *Skeletonema costatum* to *Heterosigma akashiwo.* In: Yasumoto, T., Oshima, T., and Fukuya, Y., eds. *Harmful and Toxic Algal Blooms.* Intergovernmental Oceanography Committee of UNESCO, Paris, pp. 335–8.

Gray, J. W., Peters, D., Merrill, J. T., Martin, R., and Van Dilla, M. A. 1979. Slit-scan flow cytometry of mammalian chromosomes. *J. Histochem. Cytochem.* 27:441–4.

Gonzalez, L. E., and Bashan, Y. 2000. Increased growth of microalga *Chlorella vulgaris* when coimmobilized and cocultured in alginate beads with the plant-growth-promoting bacterium *Azospirillum brasilense. Appl. Environ. Microbiol.* 66:1527–31.

Grimbergen, J. A., Visscher, K., Gomez de Mesquita, D. S., and Brakenhoff, G. J. 1993. Isolation of single yeast-cells by optical trapping. *Yeast* 9:723–32.

Haugen, E. M., Cucci, T. L., Yentsch, C. M., and Shapiro, L. P. 1987. Effects of flow cytometric analysis on morphology and viability of fragile phytoplankton. *Appl. Environ. Microbiol.* 53:2677–9.

Huber, R., Huber, H., Stetter, K. O. 2000. Towards the ecology of hyperthermophiles: Biotopes, new isolation strategies and novel metabolic properties. *FEMS Microbiol. Rev.* 24:615–23.

Huh, D., Tung, Y. C., Wei, H. H., Grotberg, J. B., Skerlos, S. J., Kurabayashi, K., Takayama, S. 2002. Use of air-liquid two-phase flow in hydrophobic microfluidic channels for disposable flow cytometers. *Biomed. Microdev.* 4:141–9.

Katsuragi, T., Tanaka, S., Nagahiro, S., and Tani, Y. 2000. Gel microdroplet technique leaving microorganisms alive for sorting by flow cytometry. *J. Microbiol. Methods* 42:81–6.

Keller, M. D., Shapiro, L. P., Haugen, E. M., Cucci, T. L., Sherr, E. B., and Sherr, B. F. 1994. Phagotrophy of fluorescently labeled bacteria by an oceanic phytoplankter. *Microb. Ecol.* 28:39–52.

Lemarchand, K., Parthuisot, N., Catala, P., and Lebaron, P. 2001. Comparative assessment of epifluorescence microscopy, flow cytometry and solid-phase cytometry used in the enumeration of specific bacteria in water. *Aquat. Microb. Ecol.* 25:301–9.

Li, W. K. W. 2002. Macroecological pattern of phytoplankton in the northwestern North Atlantic Ocean. *Nature* 419:154–7.

Lucas, J. N., and Pinkel, D. 1986. Orientation measurements of microsphere doublets and metaphase chromosomes in flow. *Cytometry* 7:575–81.

MacDonald, M. P., Spalding, G. C., and Dholakia, K. 2003. Microfluidic sorting in an optical lattice. *Nature* 426:421–4.

Melamed, M. R., Lindmo, T., and Mendelsohn, M. L., eds. 1994. *Flow Cytometry and Cell Sorting.* 2nd ed. Wiley-Liss, New York. 820 pp.

Merlin, S. 2000. Sterilization for sorting. In: Diamond R. A., DeMaggio S., eds. *Living Color: Protocols in Flow Cytometry and Cell Sorting.* Springer, Heidelberg, pp. 572–6.

Merrill, J. T., Dean, P. N., and Gray, P. N. 1979. Investigations in high-precision sorting. *J. Histochem. Cytochem.* 27:280–3.

Mitchell, J. G., Weller, R., Beconi, M., Sell, J., and Holland, J. 1993. A practical optical trap for manipulating and isolating bacteria from complex microbial communities. *Microb. Ecol.* 25:113–19.

Moore, L. R., Rocap, G., and Chisholm, S. W. 1998. Physiology and molecular phylogeny of coexisting *Prochlorococcus* ecotypes. *Nature* 393:464–7.

Olson, R., Chisholm, S. W., Zettler, E. R., Altabet, M. A., and Dusenberry, J. A. 1990. Spatial and temporal distributions of prochlorophyte picoplankton in the North Atlantic ocean. *Deep-Sea Res.* 37:1033–51.

Parry, J. D., Heaton, K., Drinkall, J., and Jones, H. L. J. 2001. Feasibility of using gfp-expressing *Escherichia coli*, coupled with fluorimetry, to determine protozoan ingestion rates. *FEMS Microbiol. Ecol.* 35:11–17.

Pinkel, D., and Stovel, R. 1985. Flow chambers and sample handling. In: Dean, P. N., ed. *Flow cytometry: Instrumentation and data analysis*. Academic Press, New York, pp. 77–128.

Porter, J., Edwards, C., Morgan, J. A. W., and Pickup, R. W. 1993. Rapid, automated separation of specific bacteria from lake water and sewage by flow cytometry and cell sorting. *Appl. Environ. Microbiol.* 59:3327–33.

Reckermann, M. 2000. Flow sorting in aquatic ecology. *Sci. Mar.* 64:235–46.

Reckermann, M., and Colijn, F., eds. 2000. *Aquatic Flow Cytometry: Achievements and Prospects, Sci. Mar.* 64(2). 152 pp.

Rivkin, R. B., Phinney, D. A., and Yentsch, C. M. 1986. Effects of flow cytometric analysis and cell sorting on phytosynthetic carbon uptake by phytoplankton in cultures and from natural populations. *Appl. Environ. Microbiol.* 52:935–8.

Rose, J. M., Caron, D. A., Sieracki, M., and Poulton, N. 2004. Counting heterotrophic nanoplanktonic protists in cultures and aquatic communities by flow cytometry. *Aquat. Microb. Ecol.* 34:263–77.

Sensen, C. W., Heimann, K., and Melkonian, M. 1993. The production of clonal and axenic cultures of microalgae using fluorescence-activated cell sorting. *Eur. J. Phycol.* 28:93–7.

Shapiro, H. M. 2003. *Practical Flow Cytometry*. 4th ed. Wiley-Liss, New York.

Sherr, B. F., Sherr, E. B., and Fallon, R. D. 1987. Use of monodispersed, fluorescently labeled bacteria to estimate in situ protozoan bacterivory. *Appl. Environ. Microbiol.* 53:958–65.

Sonek, G. J., Liu, Y., and Iturriaga, R. H. 1995. *In situ* microparticle analysis of marine phytoplankton cells with infrared laser-based optical tweezers. *Appl. Optics* 34:7731–41.

Stanier, R. Y., Kunisawa, R., Mandel, M., and Cohen-Bazier, G. 1971. Purification and properties of unicellular blue-green algae (order Chroococcales). *Bacteriol. Rev.* 35:171–205.

Stovel, R. T., and Sweet, R. G. 1979. Individual cell sorting. *J. Histochem. Cytochem.* 27:284–8.

Tyrer, H. W., and Kunkel-Berkley, C. 1984. Multiformat electronic cell sorting system: I. Theoretical considerations. *Rev. Sci. Instr.* 55:1044–50.

Velev, O. D., Prevo, B. G., and Bhatt, K. H. 2003. On-chip manipulation of free droplets. *Nature* 426:515–16.

Zengler, K., Toledo, G., Rappé, M., Elkins, J., Mathur, E. J., Short, J. M., and Keller, M. 2002. Cultivating the uncultured. *Proc. Natl. Acad. Sci. USA* 99:15681–6.

CHAPTER 8

# PURIFICATION METHODS
# FOR MICROALGAE

ROBERT R. L. GUILLARD
*Bigelow Laboratory for Ocean Sciences*

## CONTENTS

Key Index Words: Antibiotics, Filtration, Purification, Sonication

## 1.0. INTRODUCTION

The goal of purification methods is to obtain a viable culture of a single species, free of all other species ("contaminants") whether eukaryotes, prokaryotes, or viruses. The idea of pure cultures undoubtedly started at the time of Koch and Pasteur in reference to bacteria. Extended next to eukaryotes, it first produced the term *unialgal* for single-species cultures of algae, and if cultures had no detectable contaminants, then they were called *pure* or *axenic*. The term *gnotobiotic* is usually used for individuals of larger species reared free of microorganisms or parasites. Because some algae are nutritionally partially or wholly organo-heterotrophic, bactivorous, planktivorous, or even carnivorous, it may be necessary to supply food as live or killed cultures of bacteria, other algae, or protists. Such fed cultures are sometimes called *bixenic*. Other cultures are beyond a simple definition. For example, how should a culture of *Pinnularia* with endosymbiotic bacteria be regarded (Schmid 2003a,b).

### 1.1. General Conditions

Methods for obtaining axenic cultures use all the techniques devised for getting unialgal cultures, plus additional ones, some quite new. Chapters 1 and 6 cover historical aspects and various developments of the traditional methods for obtaining unialgal cultures. More tools and compounds have come into recent use, whereas others are being reevaluated or combined in new and effective ways for purification. In particular, the automated flow cytometer-cell sorter offers entirely new prospects (Chapter 7).

Generalized schemes of possible isolation and purification methods are outlined (Figs. 8.1 and 8.2). Figure

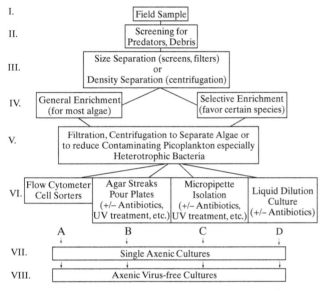

**FIGURE 8.1.** Flow diagram of purification steps of single-cell algae (after Guillard and Morton 2003).

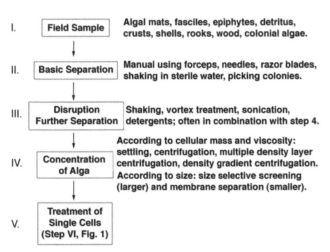

**FIGURE 8.2.** Flow diagram of purification steps for attached or colonial algae.

8.1 (after Guillard and Morton 2003) applies to field samples that can be manipulated to produce single cells or small clusters relatively free of attached contaminants. Figure 8.2 applies to specimens that occur as mats, tight fascicles, crusts, epiphytes, or in detritus or other configurations, making it difficult to separate algal units from unwanted materials or surface bacteria (see Chapter 9 for macroalgae and Starr 1973 for soil algae).

All, some, or only a few of the operations may be involved in any particular purification attempt (e.g., an axenic strain may be achieved in one step [see Fig. 8.1] by micropipette isolation or plating on agar). However, it is much easier and more usual to work at purification from a unialgal culture, as recommended by Pringsheim (1946). This amounts to starting at Level V (see Fig. 8.1) and using filtration or centrifugation to remove the bulk of the smallest contaminants, then proceeding to the processes of Level VI. For axenic isolations directly from the plankton (Level I), the flow cytometer-cell sorter offers much promise (see Chapter 7).

Some techniques applicable to special circumstances of purification are not shown explicitly in Fig. 8.1; these include the use of detergents to aid in separating cells and particles, as well as sonication, vortex mixing, or fragmentation of filaments, colonies, or thalli of certain species (see Chapters 6, 9). These are treated explicitly in Fig. 8.2, where they are of more direct application and often critical to success.

It is apparent that there are two paths toward purification. One physically separates the desired organisms (quarry) from the unwanted ones (contaminants) in one or more steps. The other kills the contaminants but not the quarry. If there are several contaminants, then it may be necessary to proceed in steps, using different methods to eliminate the various contaminants successively. This also may be needed if the contaminant has more than one stage of its life cycle in the culture. Isolation of an organism into either single-species or axenic culture should be approached as a hunting expedition; the climate, the terrain, and the nature of the quarry determine the stalk and tools employed. There is no one-size-fits-all technique, and patience, even if not a virtue, is a necessity.

## 1.2. Cultureware and Media

A first necessary condition in purification is to avoid killing the desired cell. An algal cell is extremely vulnerable to adverse chemical or physical environmental influences when it is placed alone in a tube of medium, isolated onto an agar surface, or embedded in agar. For some species and circumstances this concern is crucial. For uniform safety, see specific procedures in Brand et al. (1981) and Price et al. (1988/89). Sensitivity may result partly from the lack of protective influences of other cells, such as dilution of the toxic materials by shared absorption or chemical effects (e.g., chelation by cellular exudates). Evidence for the latter effect, "conditionings" in nature, is reviewed and discussed elsewhere (Blankley 1973, Davies 1983, Stokes 1983, Sunda 1988/89, Price et al. 1988/89).

It is remarked that a single cell is most sensitive; this is pointed out in connection with bioassay for toxic materials, but the inference for isolation and purification efforts is clear—a maintenance medium may be too concentrated or otherwise unsuited for single cells. The recommendations of Brand et al. (1981) followed many trial procedures aimed at eliminating lag phases at inoculation of successive cultures, a necessity for determining acclimated growth rates. Much of the improvement in marine culture techniques followed the discovery that trace element concentrations in unpolluted seawater were orders of magnitude lower than had been previously thought (Sunda 1988/89). Methods for collecting, handling, and storing seawater samples were improved, resulting in the techniques ("clean" or "ultraclean") that are now standard (Fitzwater et al. 1982). Cautions and ideas derived from these observations have been extended to trace element studies in culture and to the preparation of enrichments and media for isolating algae. Because of the low ionic content of many freshwaters and freshwater media, problems with leaching of elements and of chemical transformations in the media are lower, but problems with metal contamination during autoclaving and with leaching of materials into nutrient stocks remain (see Chapters 2–5).

Whereas temperatures can be assumed to be a controllable environmental factor easily chosen for best effect during purification manipulations, light may not be. This is considered by Brand (1986) in connection with marine phytoplankton species, but the same considerations apply to freshwater species as well. Guillard and Keller (1984) survey effects of light quality and quantity on dinoflagellate culture. Recipes for and details of preparation of culture media are given in Chapters 2 and 3 and the Appendix.

## 2.0. METHODS AND TECHNIQUES OF PURIFICATION

### 2.1. Size Selective Screening and Filtration

See Chapter 6 for details on the structure of various screens or filters used during separation and isolation stages prior to purification (see Fig. 8.1).

Methods of use can be at several levels but mostly apply to freeing unialgal cultures of unwanted prokaryotes or yeasts. Repeated screening with membrane filters is one of the most useful tools at the purification step, because most algae are slightly larger than most heterotrophic bacteria. When hunting a species larger than the nominal filter porosity, filter only a volume small enough not to plug the filter. The liquid above the filter, which should never be allowed to dry and which contains the quarry organism, can be transferred to another filter apparatus and washed repeatedly with sterile water or medium. For some purification processes, only a few quarry cells are needed. If more cells are needed, then use larger diameter filters or harvest from more than one primary filtration and pool harvests for repeated washing.

Algae of small or moderate size can be freed of yeast or other fungal contaminants by differential filtration. Strands of filamentous fungi are held well by filters, but their spores usually pass through with the quarry algae. If the latter occurs, add to the filtrate a little malt extract or other organic enrichment that is tolerated by the algae. The fungal spores will germinate in 1 or 2 days and form filaments that can then be trapped on filters before they sporulate again. Repeat for precocious fungal spore formation. Differential filtration can also serve to isolate the smallest photosynthetic species. Thus, for example, Waterbury et al. (1986) concentrated *Synechococcus* strains relatively free of other algae and heterotrophic bacteria by using 10- and 1-μm Nuclepore filters and made final steps of purification from the concentrate. It seems likely in retrospect that both prochlorophytes and SAR-11 type bacteria mostly washed through the 1-μm filter—an unrecognized benefit—but in any event, neither organism was likely to grow in the media employed at that time and therefore would not have interfered in the isolation of *Synechococcus*.

It must not be assumed that there is any simple definite relationship between the nominal sizes of the holes in the filters and the measured sizes of organisms being separated. Craig (1986) studied this in connection with estimates of picoplanktonic productivity made using filter fractionation of natural freshwater and marine phytoplankton populations. Samples were filtered sequentially through Nuclepore filters having pores of 3.0, 2.0, 1.0, and 0.2 μm in diameter, and the eukaryotic and prokaryotic picoplanktonic populations were enumerated by fluorescence microscopy on the filters (heterotrophic bacteria were not enumerated). The main point for productivity studies was that the 1.0-μm filter was not a universal cutoff point; it was for the eukaryotes, of which only 2% of the original population) reached the 0.2-μm filter, but approximately 30% of marine *Synechococcus* and approximately 8% of the slightly larger freshwater *Synechococcus* reached this 0.2-μm filter.

Although perhaps unfortunate for productivity estimates, this illustrates a usefulness for isolation and purification efforts. Using the same data, it can be seen that to filter a sample of seawater with ca. $10^4$ *Synechococcus* cells · $ml^{-1}$ to 0.2 μm would leave 30% of this population (3,000 cells · $ml^{-1}$) remaining *above* the 0.2-μm filter (which need not be there), but only 2% of the original eukaryote picoplankters and an unknown proportion of the original heterotrophic bacteria would remain with them. If this sample, passing through a 1.0 μm-filter, is collected in enough volume to present a reasonable cell population to another 1.0-μm filter, then the *Synechococcus* would again be reduced to 30% of the material filtered through it, or 900 cells · $ml^{-1}$, with essentially no eukaryotic competition and certainly many fewer heterotrophic bacteria (a separate consideration easily examined).

## 2.2. Differential Centrifugation

For the purposes of purifying free-living planktonic eukaryotes or an equivalent (see Fig. 8.1), centrifugation serves much the same function as differential filtration to concentrate the quarry relative to the contaminants and thus increase the chances for success in later stages of purification. In uniform density centrifugation, the timing and speed of centrifugation are critical. Centrifugation is continued until the heavier cells—usually the ones wanted—are just barely sedimented, then the supernatant is replaced by fresh sterile medium. The process is usually repeated (Hoshaw and Rosowski 1973). In density gradient centrifugation the various cells equilibrate in position in the gradient according to their density, the sedimentation rate being relatively unimportant. Sitz and Schmidt (1973) sepa-

rated *Synechococcus lividus* Copeland from heterotrophic bacteria on a Ficoll gradient and then continued purification by plating on agar. Centrifugation is extremely effective used in conjunction with sonication and other methods of separating algal cells or colonies from each other, from bacteria, or from detritus (see Fig. 8.2). Treatment of these cases, also involving centrifugation, is outlined in the section that follows. For possible application of continuous density gradient centrifugation to purification, see Reardon et al. (1979).

## 2.3. Sonication and Vortexing

Samples from large compact colonies, algal mats, damp soil, or mud can be fragmented and separated manually with forceps, needles, or scalpels (see Fig. 8.2). Further treatment can include crushing algal material between sterile glass slides, passing it through a syringe, or treating it with a glass-glass or glass-Teflon homogenizer (Hoshaw and Rosowski 1973, Allen 1973, Rippka et al. 1981). The aim is to obtain smaller and manipulable algal units. The same general procedures apply to scrapings from wood or other soft materials, including macrophytes. However, samples from rock, shells, corals, or other porous materials may require using a mechanical homogenizer (e.g., Waring blender) followed by sieving or mechanical separation. It may be necessary to make enrichment cultures in liquid or on agar to obtain separable algal units of these forms.

Two species of the aerial green alga *Trentepohlia* were obtained in axenic culture by dragging short filaments through agar using a Sherman micromanipulator (Lim et al. 1992). Tufts scraped from a rock (*T. aurea* [L.] Martius) or a wooden wall (*T. odorata* [Wiggers] Wittrock) were shaken in sterile water to produce short segments of filaments, some of which were spread in small numbers on dry agar plates. Individual filaments of 3–5 cells were picked up by the microhook and dragged through the agar to remove contaminants. In the case of *T. aurea*, which has longer cells and more rigid filaments, not all contaminants were removed by the dragging process, so a subsequent treatment with 5% lactic acid for about 30 minutes followed by more washing was used. About half the treated filaments survived, and some yielded axenic cultures. The micromanipulator described by Throndsen (1973) employs a micropipette rather than a hook for picking up single algal units; thus, it is suited for isolation from drops of liquid culture.

For material with suitably separable algal units, the favored procedure is treatment with ultrasound of suitable frequency and intensity for an appropriate time, followed by centrifugation. The technique is used to isolate and purify bacteria (Solp and Starr 1981), microalgae (Hoshaw and Rosowski 1981), and cyanobacteria, including the difficult *Microcystis* (Watanabe et al. 1985). Watanabe et al. (1985) disrupted contaminated *Microcystis* cultures for 30 seconds at 20-KHz frequency (energy not specified), which disaggregated most colonies into unicells. The cultured material was centrifuged and resuspended in fresh sterile water four times, after which single cells were isolated by micropipette (Watanabe et al. 1985). Shirai et al. (1989) plated material on an agarose medium following similar disruption and washing. Lower intensity energy ultrasonic treatment (90 KHz) for 5–20 minutes was used on eukaryotic algae (Brown and Bischoff 1962) followed by repeated centrifugation.

Sutherland (1976) employed sonication to obtain pure cultures of myxobacteria. This was done by killing the vegetative cells of the myxobacteria and contaminating heterotrophic bacteria by appropriate sonication. The microcysts of the myxobacteria were more resistant to sonic disruption and yielded pure cultures. The same technique may work with species of eukaryotic algae and cyanobacteria that have spores or other resistant stages (e.g., cysts of *Acetabularia*). Time-intensity data on differential effects of sonication on different organisms would be useful. (Note that sonication is used in this case as a killing agent rather than as a separating agent.)

Both sonication and vortexing produce shear forces acting on the surfaces of algal cells, thus serving to remove attached microorganisms. Sonication at high energy levels also produces damaging intracellular forces, and vortexing can also disrupt cells (Nelson and Brand 1979). For purposes of studying nutrient uptake dynamics it has been possible to subject phytoplankton cells to controlled shear forces in a culture apparatus (Pasciak and Gavis 1975). Shear effects were studied up to values (10 sec$^{-1}$) almost double those that occur in turbulent marine waters (ca. 6 sec$^{-1}$). Natural turbulence is of significance to phytoplankton growth (Gavis 1976) and has been considered significant in fragmenting algal colonies and disbursing cells. Controlled shear might usefully be studied in connection with algal purification.

rious methods. It is most successful when used in conjunction with washing by repeated aseptic filtration or centrifugation. The washed preparation can be incubated with antibiotics or other chemicals before the dilution series is made up (see Sections 3 and 4). The dilution method is described in Chapter 6 in general detail. To our knowledge no one has made serial dilutions of any alga directly from nature employing antibiotics in the final isolation tubes. Such use would remove the photosynthetic prokaryotes when they might be considered contaminants, as in efforts to culture eukaryotes.

In most natural samples the nonphotosynthetic bacteria outnumber all other organisms. Thus, serial dilution culture is not a method of choice for purification directly from field samples. However, Brand and Guillard (unpublished observation) and Waterbury et al. (1986) had some success getting pure cultures of *Synechococcus* from dilution cultures of marine field samples (ca. $10^5$ cells · mL$^{-1}$). These were preconcentrated and freed of heterotrophs to some extent by differential filtration (10- and 1-µm Nucleopore filters). All tubes from dilutions of 1 : $10^4$ exhibited algal growth (some with other species also); most from dilutions of 1 : $10^5$ had *Synechococcus* cultures, as well as a few from 1 : $10^6$ (Waterbury et al. 1979). About 120 tubes were used for the highest dilution. Thus, direct purification without enrichment culture is possible if the concentration of contaminants is reduced to the same order of magnitude as that of the quarry and many samples are processed.

Dilution culture is not always labor-saving. For example, a strain of *Prochlorococcus marinus* Chisholm et al. ssp. *pastoris* Rippka was purified by repeated serial dilution (Rippka et al. 2000). A first effort yielded a culture contaminated with only one large heterotrophic bacterium. A culture of this mix was centrifuged three times at 300 *g* in sterile medium, and the final supernatant was diluted in four different media from which replicate tubes were inoculated (numbers not specified). Only one tube was axenic by their extensive testing. This single success of many attempts followed previous unsuccessful efforts to grow the prochlorophytes on agar, even when prepared by their most rigorous methods.

## 2.4. Purification by Dilution Culture

Purification by serial dilution from cultures that are unialgal or nearly so is generally one of the least labo-

## 2.5. Agar Plates

Purification on agar plates is probably the oldest and the most used method of purification but, of course, is effective only for algae that can grow either on the surface of agar or embedded in it. For such species it is also the method of choice for isolation into unialgal culture; both objectives can often be met simultaneously. Including antibiotics or other selective agents (e.g., tellurite) into the agar is common. Some filamentous algae can grow or glide away from contaminants on agar surfaces, as can some diatoms or cyanobacterial spores. Some show phototactic motion, which can be exploited by locating the culture dish appropriately in unidirectional light. See Chapter 2 for preparing agar plates; see Chapter 6 for processing inoculated plates and isolating cellular colonies from them.

Preparation of the solid medium is sometimes critical. Since Pringsheim's time it has been known that it is best to autoclave agar separately from the growth medium, which is easy in the case of freshwater species (Allen 1973) but more laborious for marine species. See Chapter 2 for details of preparing culture media with agar and agarose. Low-temperature and ultra-low-temperature-gelling agaroses (with gelling as low as 17°C) can be used for temperature-sensitive algae and to prevent loss of antibiotic activity, which can be embedded with very little or no heat stress. Low-temperature setting agarose was important in purifying the difficult freshwater cyanobacterium *Microcystis*; it was used together with sonication and centrifugation (Watanabe et al. 1985). It was also employed in purifying freshwater *Synechococcus* using the pour plate technique (Watanabe et al. 1998).

Agaroses also offer the advantage of greater purity over ordinary high-gelling-temperature agars, which are variously contaminated (Krieg and Gerhardt 1981; Allen and Gorham 1981). Sigma-Aldrich Company and similar companies provide descriptive material on agarose products. Even these products may not be pure enough for very delicate marine organisms.

## 2.6. Purification by Micropipette

The process is basically the same as that used for getting unialgal cultures (see Chapter 6) with probably some extra precautions regarding sterility. Organisms have in fact been isolated directly into pure culture by micropipette. However, it is more usual to begin purification attempts from enrichment cultures (Level IV), unialgal cultures or a previously axenic culture that mysteriously acquired a contaminant. Techniques and apparatus have been described many times with emphasis on different details or on the nature of the quarry (Droop 1969, Hoshaw and Rosowski 1973, Guillard 1973, 1995,

Guillard and Morton 2003; see Chapter 6). Droop (1969) remarked "the actual manipulation requires patience but, like riding a bicycle, is not very difficult once the skill is acquired" (p. 274).

A protocol for purification by micropipette is as follows. Ensure that the culture is healthy, growing rapidly or at least well, and at less than half the maximum density it can attain (in short, is still in the log phase of growth). At this point, an enrichment culture should be size-selected by screening (II, B) using Nuclepore and possibly other filters to concentrate the quarry species to the extent possible. Then, for this screened culture and any unialgal one, wash the cells over a filter (usually Nuclepore, but not necessarily) with sterile medium prepared so as to be free of bacteria-sized particles; the filter selected should then hold the quarry but allow contaminating bacteria to pass through in the particle-free washing medium. The quarry cells must not be allowed to dry on the filter. Wash the quarry cells at least twice, moving them in some sterile medium, by pipette, into a new sterile filter apparatus if necessary. When finished with washing, move the quarry, in medium, into a sterile flask or tube. The process described should not take long if the apparatus is prepared in advance, the medium equilibrated in temperature, etc. For many kinds of cells, especially flagellates, it is advised to put the washed culture back into its usual growing conditions for a short time—up to 1 hour at most. The time can be used to add the antibiotics or set up for the micropipette isolation if that is to be done at once. If antibiotics are to be used, then they should be applied at this point, ordinarily by using a portion of the washed culture as inoculum into flasks containing the antibiotics or other chemicals. Antibiotic use is described in Section 3.0, and other chemicals are described in Section 4.0. The micropipette isolation would begin at the end of the time of exposure.

If antibiotics are not used, the micropipette isolation would begin at once (see later text and Chapter 6). The cell chosen is ordinarily transferred to a fresh washing bath at least once before it is transferred to the final tube (even if it comes from an antibiotic solution). Each washing by transfer constitutes another dilution of whatever bacterial content remains with the quarry cells. The greater the volume of washing fluid compared with that of the added droplet containing the algal cell, the better; it is also helpful if the quarry can be persuaded to move or can be stirred about with a fine jet of water (medium) from a sterile micropipette. Using 3- or 9-well depression spot plates with a suitable dissecting microscope works well (Droop 1969), because the volume of wash is comparatively large (up to 2 mL); even small cells can be tracked in it provided that the spot plate and fluid are particle free. Certain inverted microscopes are also well suited to micropipette isolation using thin depression slides, multiwell plates, or small sterile plastic dishes. The improved optical resolution is helpful with very small cells. The same advantage can be had with an upright compound microscope having a low set stage, a reversing prism in the optical path, and suitable, low power long working distance objectives. Using any of these instruments and procedures and with preliminaries done, a skillful isolator can sample material held in a growth chamber and return the first tubes of isolates to the same chamber in less than 10 minutes—this is the timing needed for fragile organisms.

If contaminants are attached to the quarry, then chances of success are poor by direct pipette isolation. The use of antibiotics or treatment to dislodge or kill the contaminants is indicated before pipette isolation.

## 3.0. ANTIBIOTIC TREATMENT

Although in principle it may be possible to destroy all bacteria in a mixed culture by means of antibiotics without killing all the algae, this is probably seldom achievable in practice. All techniques in fact reduce the viable bacteria to such small numbers that transfer of one algal cell or even a small inoculum containing many algal cells sometimes includes no viable bacteria. The lethality of the antibiotic treatment is an intensity–time relationship, the intensity being the dose (concentration) of antibiotics and the time being the period of exposure before transfer to antibiotic-free medium. A test for bacterial survival is made separately at this time. The fundamental choices are thus which antibiotics to employ, at what concentrations, and for how long. Further, if one antibiotic mix fails to get rid of all contaminating bacteria, then repeating with another selection of antibiotics may complete the task. (The action of mixtures of antibiotics is not in general the sum of the individual actions.) Three somewhat different approaches (designated I, II, and III) are described here. These may be adapted to different situations, as will be apparent from the citations given.

### 3.1. Approach I

The first approach makes antibiotics an adjunct to micropipette isolation (Guillard 1973, Hoshaw and Rosowski 1973). It uses time of exposure to antibiotics as the main variable; the numbers of viable bacteria decline to a minimum at 18–48 hours after first exposure to the antibiotics and then rise rapidly (Oppenheimer 1955). The timing of the minimum depends on temperature, the kinds of bacteria, and the antibiotics chosen. Thus, if a moderate inoculum of healthy and rapidly growing algal culture is transferred to fresh medium containing a tolerable level of antibiotics, and if, at several convenient times from 18–72 hours after exposure, one or a small number of cells is transferred by micropipette to each of several tubes of fresh medium (without antibiotics), then there is a good chance that some tubes are bacteria-free (or at worst, have fewer types of bacterial contaminants). Note that Oppenheimer's experiment was done with a mixed natural population of bacteria, which apparently had a small population of resistant cells that flourished when the others died. Unialgal cultures ordinarily have fewer strains of bacteria so that exact timing may be less critical.

The basic antibiotic solution still recommended for this purpose is made as follows: dissolve 100 mg of penicillin G (sodium or potassium salt), 25 mg of dihydrostreptomycin sulfate, and 25 mg of gentamycin sulfate in 10 mL of distilled (or deionized) water and sterilize by membrane filtration. Keep frozen until used (polycarbonate tubes are best). Thawed solutions keep at ca. 4°C for a few weeks (five days at 37°C). The "standard" dose is 0.5 mL of antibiotic mix to 50 mL of algal medium. It yields $100/25/25$ mg $\cdot$ L$^{-1}$ of penicillin, streptomycin, and gentamycin and is reasonably well tolerated by most algae. Considerably higher concentrations of gentamycin and penicillin can often be tolerated, but streptomycin rapidly becomes toxic to some species. The original antibiotic recipe (Guillard 1973) had chloramphenicol (25 mg) instead of gentamycin; chloramphenicol is somewhat harder to use and is more toxic but remains useful as needed. See additional recommendations on choices of antibiotics at the end of this section.

In a typical application to algal species growing at 20–25°C, 0.5 to 5 mL of a rapidly growing algal culture (volume depending on size, density, and fragility of the alga) would be transferred into 50 mL of fresh medium in each of five 125 mL flasks. The flasks would have added to them 0, 0.25, 0.5, 1, or 2 mL of the antibiotic solution, the antibiotic level thus varying from none to four times the "standard" dose. A very small amount (one drop, 50 µL) of a bacterial test medium (see later text) should also be added to stimulate bacterial growth, hence susceptibility to the antibiotics. The treatment should be started near the end of the working day (or in the evening) so that at approximately 15, 24, 36, and 48 hours under growing conditions single algal cells or small numbers of cells can be transferred by micropipette to sterile medium tubes. Usually it is possible to tell by microscopy that the cells are healthy or at least alive. Five to 15 tubes, and a few tubes with organic medium to test for contaminants (described later), should be started at each isolation time. A few transfers from the untreated (zero antibiotics level) flask verifies success of the pipette transfer process for the algae. Allow up to three weeks for algal growth to appear, then transfer from successful tubes to fresh medium and test for microbial contaminants, as described later. If algal cultures are growing at 15°C or lower, then the first isolations should be at approximately 48 hours, because bacterial numbers decline more slowly.

The advantage of Approach I is that the antibiotic treatment can be gentle on the algae; the disadvantage is that time and effort are required in the manipulation of cells, especially if single cells are repeatedly washed after the antibiotic treatment. This method is well suited to most flagellate cells, large diatoms, large dinoflagellates, and colonial forms.

### 3.2. Approach II

The second method is that of Droop (1967), which was invented as a convenient, routine procedure not involving much manipulation. It is good for cultures of small cells, flagellates particularly, that grow to great density. Its disadvantage is that the algae are exposed to very high antibiotic concentrations. See Droop (1967) for an exact simple procedure, or modify the protocol as appropriate. The basic idea is to dilute a strong antibiotic mix (made up in sterile algal growth medium) with a dense but still rapidly growing algal culture having a very small amount of organic matter added. Do this by means of successive twofold dilution steps so that the concentration of antibiotics is halved at each step whereas the concentration of algae is remains constant (after the first dilution of the pure antibiotic mix). Droop (1967) made six dilutions, the last thus having 1/32 the antibiotic concentration of the highest. At some point in the decreasing series of antibiotics the algae are (hopefully) still alive while the bacteria are dead. This is revealed by sterile transfer of a small inoculum from each of the six treatments to a separate tube of fresh algal growth medium (without antibiotics); a similar transfer is made to a tube of bacterial test medium. Droop recommended transfer at 24 hours and 48 hours, but not longer than 48 hours. The transfers should be examined over several weeks.

Droop (1967) used four antibiotic mixes. The highest levels to which the algae were exposed in any of these mixes were as follows: Benzyl penicillin G sulfate, 2500 mg · $L^{-1}$; chloramphenicol, 200 mg · $L^{-1}$; neomycin, 200 mg · $L^{-1}$; Actidione (for fungi), 400 mg · $L^{-1}$. Note that these concentrations are much higher than those recommended in the previous method. Further, Actidione (cycloheximide) is usually added only to inhibit fungi, although it can also be used to inhibit eukaryotic algae if one is purifying cyanobacteria (Guillard 1973). Nystatin and amphotericin B are alternative antifungal agents. Chloramphenicol, although very effective against bacteria, is likewise very toxic to algae; for example, 10–50 mg · $L^{-1}$ inhibited or killed four strains of *Alexandrium* in an incubation period of 16–17 days (Divan and Schnoes 1982) and 50 mg · $L^{-1}$ inhibited *Micromonas pusilla* (Butcher) Manton et Parke (Cottrell and Suttle 1993). Only one of five diatom species examined could tolerate chloramphenicol at 80 mg · $L^{-1}$ for 6 days; four were destroyed by 13 mg · $L^{-1}$ or less (Berland and Maestrini 1969). This last paper provides excellent quantitative data on the effects of 25 antibiotics on the five diatoms (at 6 days). All five tolerated penicillin and streptomycin well at levels of 100–400 mg · $L^{-1}$, allowing and showing growth >50% of controls. Kanamycin, an amino-glycoside (like streptomycin) active against both gram-positive and gram-negative bacteria, was also well tolerated by the diatoms. Cycloserine in the high concentration range of 1–5 g · $L^{-1}$ together with cycloheximide (Actidione) at 10–30 mg · $L^{-1}$ were used to purify two strains of marine *Synechococcus* (León et al. 1986). Cycloserine (100 mg · $L^{-1}$) also has been used on *Acetabularia* (Shephard 1970).

Note that the times of exposure to the antibiotics tested in the three papers cited (Berland and Maestrini 1969, Divan and Schnoes 1982, Cottrell and Suttle 1993) were much longer than usual, because the aims were to document differential effects on algae. Another significant point of comparison is that although Divan and Schnoes (1982) purified two dinoflagellates without single-cell isolation (by treating cultures for 16 or 17 days with an antibiotic mix followed by transfer to fresh medium), it was first necessary to work out the susceptibility of the bacteria involved to the antibiotics singly and in various combinations, as well as to determine the sensitivity of the algae (by growth inhibition). This process is what Droop (1967) described as "a last resort" (p. 297). Even so, only two of the four algal strains could be purified. The level of streptomycin necessary to kill the bacteria was apparently toxic to the other two dinoflagellates. The successful combination was dihydrostreptomycin sulfate (250 mg · $L^{-1}$), potassium penicillin G (500 mg · L$^{-1}$), neomycin (250 mg · L$^{-1}$), and the polyene antibiotic amphotericin B (5 mg · L$^{-1}$) for fungi.

Repak et al. (1982) successfully used commercial antibiotic sensitivity discs to assess the sensitivities of bacteria and marine coccoid ultraplankters prior to purification efforts. The antibiotic discs could be used because both the algae and the bacteria grow on agar, and this saved a great deal of work.

Lehman (1976) used a strong antibiotic mix to purify *Dinobryon sertularia* Ehr. by exposing a growing culture of the alga to the mix and then making small transfers daily to fresh medium without antibiotics and later testing each subculture for bacteria. (This is much like the system used by Spencer [1952]). The antibiotic mix to which *Dinobryon* cells were exposed consisted of penicillin 4,000 mg · L$^{-1}$, streptomycin 500 mg · L$^{-1}$, and 10 mg · L$^{-1}$ of each of the following sulfa drugs: Sulfisoxazole (Gantrisin), sulfamerizine, homosulfamine, sulfisomidone, and sulfisoxazolum (the last name not presently listed in the Merck Index). Sulfa drugs have been used infrequently since antibiotics of microbial origin became available. They have, however, remained in use as antibacterials in aquaculture. Exploratory studies using them in conjunction with particular antibiotics could be fruitful. A mix of penicillin, streptomycin, neomycin, tetracycline, chloramphenicol, and cephaloridine suppressed bacteria in a culture of *Aureococcus* (Dzurica et al. 1989). Jones et al. (1973) included the cephalosporin antibiotic cephaloridine (also called ceporin) in a mix that was used successfully on two diatoms. (With some other species, it was no help.)

### 3.3. Approach III

A third approach to purification is sequential transfer of the algal culture through a series of flasks of medium each containing different antibiotics at levels permitting algal survival and growth, even if at reduced rates. The aim is to lose the bacteria by attrition. There are three reasons for using a sequence of antibiotics rather than aiming for one massive dose to kill all contaminants at once. First, it can be made less toxic to the algae. Second, any one antibiotic mix may be fatal to some metabolic types of bacteria but only suppress growth of others. Another suite of antibiotics may kill some of the survivors, and so on. Last, bacterial mutations that arise resistant to one suite are not left in it to propagate. This method allows the algae to survive or increase slightly, whereas the bacteria decrease greatly in numbers and types.

This method generally requires preliminary investigation of the tolerance of the algae and the susceptibility of the bacteria ("last resort"). It was used by Cottrell and Suttle (1993) to purify the tiny flagellate *Micromonas pusilla*. The sequence of antibiotics was penicillin (1 g · L$^{-1}$), neomycin (250 mg · L$^{-1}$), gentamycin (1 g · L$^{-1}$), and kanamycin (0.5 or 1 g · L$^{-1}$). A 20% inoculum was transferred from one antibiotic treatment to the next at three-day intervals (except four days for the last). Transfers were bacteria-free after the kanamycin treatments. The results are compatible with the idea that penicillin eliminated the gram-positive bacteria, whereas the other three antibiotics were required to eliminate the gram-negative bacteria. (Note that in the past, *Micromonas* has been purified by using standard antibiotics—penicillin and streptomycin—and micropipette isolation. Success lies in the details.)

### 3.4. Comments on Antibiotics

Vance (1966) determined the sensitivities of five freshwater cyanobacteria including *Microcystis aeruginosa* (Kützing em. Elenkin) to 32 antibiotics, with the most useful being dihydrostreptomycin and neomycin. He obtained one axenic *Microcystis* culture but could not sustain it beyond two transfers; similarly maintained cultures having a small, unidentified bacterium could be subcultured indefinitely. *Microcystis* has now been established in axenic culture without the use of antibiotics (Watanabe et al. 1985, Sharai et al. 1989). Guillard and Keller (1984) listed 17 antibiotics used by various workers to purify strains of dinoflagellates. Probably at least 50 different compounds have been used in algal culture work; many of these have several different names. Some sources of general information are the following: Berland and Maestrini (1969) who group 25 antibiotics into families based on chemical structure, the Merck Index, and cell culture catalogs (e.g., Sigma-Aldrich Co.).

No realistic recommendations can currently be given regarding choice of antibiotic treatments, but some general principles and possible useful notes can be made. First, for antibacterials that inhibit cell wall synthesis, a little organic matter should be provided to stimulate cell division, because such antibacterials kill bacteria only when the cells are actively growing. The amount of organic matter necessary is estimated at 10 mg · L$^{-1}$ (ca. $10^{-4}$ M organic matter with formula weight ca. $10^2$). Second, it appears contradictory to add, simultaneously with cell wall inhibitors (e.g., penicillin), other antibiotics that slow cell growth by inhibiting protein synthesis. Nevertheless, such cocktails have often worked well; when they do not, sequential treatment is indicated. Most cell wall inhibitors used are in either the penicillin family or the cephalosporin family of β-lactams; the exception is the cyclic glycopeptide vancomycin. By far the most frequently employed is penicillin G, which like carbenicillin acts mostly against gram-positive bacteria. Ampicillin (used by Loeblich and Sherley 1979) is listed by Sigma-Aldrich as active against both gram-positive and gram-negative bacteria. Of the broad spectrum cephalosporin antibiotics, cephaloridene was included in mixes by Jones et al. (1973) and Dzurica et al. (1989). Cefotaxime, which is mainly effective against gram-negative bacteria, and hence should be especially useful against marine bacteria, was successfully used (50–100 mg · $L^{-1}$) to purify a green seaweed (Kooistra et al. 1991). They found that vancomycin, which they used as a complement to cefotaxime, because it acts chiefly against gram-positive bacteria, was too toxic to be useful in their context.

Of the other families of antibiotic substances, the available evidence suggests that with rare exceptions only the aminoglycosides (e.g., streptomycin, gentamycin, kanamycin, and neomycin) are tolerated well enough by at least some algae at bacteriostatic or bacteriocidal levels to be useful. Of these (all considered broad-spectrum antibiotics), kanamycin is reported to be active against both gram-positive and gram-negative bacteria, neomycin against principally gram-positive bacteria, and streptomycin and gentamycin mostly against gram-negative bacteria (Sigma-Aldrich). Gentamycin has been reported to be useful in animal tissue culture and was also effective in the culture of *Amyloodinium ocellatum* (Brown) Brown et Hovasse, which is an ectoparasite on marine fish (Noga 1987, Noga and Bower 1987). Of the cyclic polypeptides, bacitracin was well tolerated by some diatoms (Berland and Maestrini 1969) and the dinoflagellate *Symbiodinium microadriaticum* Freudenthal (Loeblich and Shirley 1979).

## 4.0. OTHER ANTIMICROBIAL AGENTS

Many substances differing greatly in chemical structure, mode of action, and method of use have been employed. The sulfonamides, most recently used by Lehman (1976), act as metabolic inhibitors, as do antibiotics. The purine caffeine, used at $1 - 3 \times 10^{-2}$ M to repress fungi and protozoa (Brown 1964), likely does the same. Potassium tellurite ($K_2TeO_3$) at 10 mg · $L^{-1}$ (ca. $3.9 \times 10^{-5}$ M)

has been used in agar and in liquid culture as a bacteriostat (Hoshaw and Rosowski 1973). Tellurite is toxic to many bacteria at $5.6 \times 10^{-6}$ M (Yurkov and Beatty 1998), presumably because it is a strong oxidant. The tellurite is reduced to metallic tellurium in most species of bacteria and stored as intracellular crystals. Certain aerobic photosynthetic bacteria not only accumulate large amounts of elemental tellurium but can tolerate exposure to tellurite concentrations close to $10^{-2}$ M (Yurkov and Beatty 1998). Algae are not known to survive such high concentrations. In phycological application, the use of tellurite is mostly as a strong oxidizing agent that kills heterotrophic bacteria before it does the more resistant eukaryotes and possibly large-celled cyanobacteria.

Molecular iodine as an 0.1% alcoholic solution yielding ca. $4 \times 10^{-3}$ M of $I_2$ was used to purify a marine *Chaetoceros* by a few minutes of exposure followed by repeated centrifugation (Soli 1963). Fries (1963) also used molecular iodine at ca. $6 \times 10^{-4}$ M to sterilize thalli of red algae. Cysts are relatively tough and withstand oxidants such as 1% sodium hypochlorite (ca. $7 \times 10^{-4}$ M free chlorine; Chapman 1973) and even 2% formaldehyde (ca. $2.5 \times 10^{-2}$ M) or 10% Argyrol (an organic silver preparation; Page 1973). Some ionic detergents such as sodium lauryl sulfate are used at 1% (ca. $3.5 \times 10^{-2}$ M) as bacteriocidal agents (Page 1973) or as slow-acting antifungals in algal culture. However, most detergents are used as surfactants in cultures of single-celled or colonial species to separate attached contaminants in connection with sonication and centrifugation. Long-chain ionic detergents having different pH are available, as are nonionic detergents such as Tween or Triton. Brown and Bischoff (1962) give a concise but thorough account of purification using Tween 80 at 5%, pH 5–7 for varying times with repeated sonication (not over 10 sec. at 90 KHz.), followed by repeated washing by centrifugation, then final plating.

Kim et al. (1999) employed the enzyme lysozyme alone or in combination with ampicillin to purify the cyanobacteria *Anabaena flos aquae* (L.) Fries and *Aphanotheca nidulans* Richter. Lysozyme is an enzyme (formula weight ca. 14.3 kD) that hydrolyzes particular linkages in the peptoglycans of bacterial cell walls, thus lysing the cells. Most heterotrophic bacterial contaminants are small and not protected by firm or watery external sheaths, as are *Anabaena* and *Aphanotheca*, which therefore survive much longer in a preparation of lysozyme having up to 1.0 g · $L^{-1}$ (amounting to $5 \times 10^{-6}$ M of the enzyme). Ampicillin, when employed, was at 10 mg · $L^{-1}$ (ca. $2.9 \times 10^{-5}$ M). In initial cultures, cycloheximide (Actidione) was also used, up to 100 mg · $L^{-1}$

$(3.6 \times 10^{-4}$ M), to inhibit eukaryotes in the original field samples (cycloheximide can be used to select for cyanobacteria). During exposure to lysozyme the samples were vortex mixed, washed by centrifugation four times, then spread on agar plates for isolation of single colonies. At this point, it becomes clear that the timing and concentration is more important than the choice of antibiotics or other antimicrobial agents, assuming the appropriate class of antibiotic is chosen. Furthermore, the first antibiotics used (penicillin and streptomycin) remain the most useful.

## 5.0. PURIFICATION USING ULTRAVIOLET LIGHT

Success using this process depends on differences between algae, heterotrophic bacteria, and viruses in their responses to ultraviolet (UV) light and photosynthetically active radiation (PAR). Susceptibility to UV damage (inactivation) is probably the most significant difference, whereas repair of damage (reactivation) is probably less differentiating; however, there are some bacteria that show no reactivation at all.

Most UV effects result from direct absorption of single photons by one specific molecule or portion of a molecule; in the case of damaging or lethal results, these are the nucleotide bases of the nucleic acids, which yield the chemically active pyrimidine dimers. The action spectrum for dimer formation closely parallels the dimer absorbance spectrum, with a maximum absorbance of ca. 265 nm (UV-C) or ca. 280 nm (UV-B). There are other effects of UV, a main one being damage to phytosystem I, but these are of little consequence in purification applications, which use short-term exposures aimed at the nucleic acids. Photoreactivation (nucleic acid damage repair) has two components, one of which takes place in the dark (nucleotide excision repair), whereas the other depends on absorption of light in the UV-A-PAR range (photoreactivation). Much relevant information is contained in Smith (1977) and de Mora et al. (2000). Most of the recent research on UV has been motivated by concerns over reduction in levels of atmospheric ozone, which results in increased UV-B levels at the earth's surface. (UV-B is defined as radiation from 280–320 nm.) Consequently, whereas most work, especially in aquatic systems or for aquatic organisms, has been done in the UV-B range (de Mora et al. 2000), the maximum lethal effects of UV, to the extent known, occur in the lower end of the UV-B or in the UV-C (<280 nm). Van Baalen and O'Donnell (1972) employed a monochrometer to get a killing action spectrum for the cyanobacterium *Agmenellum quadruplicatum* (Meneghini) Brébisson (= *Synechococcus*). Lethality was high from 280 nm to the lowest wavelength tried (240 nm) with a poorly defined peak at ca. 255 nm, which matched their DNA absorbance peak. A photoreactivation spectrum, similarly achieved with a monochrometer, was high between 410–450 nm. The germicidal-type lamp used for the inactivation (killing) was a mercury discharge lamp with about 85% of its energy output at the 253.7 nm line. Exposure times to UV overall were just a few minutes, three for the germicidal lamp. A Hanovia lamp (cat. 30600) used previously (Van Baalen 1962) required times of 1–6 minutes. However, the lamp employed by Gerloff et al. (1950) necessitated exposure for 20–30 minutes. It was a mercury vapor lamp but probably of the fluorescent blacklight type, emitting at ca. 275 nm.

Under ideal circumstances six action spectra would be available; for each of the three groups (algae, bacteria, viruses), an action spectrum would be available for killing, extending below the UV-B range to at least 250 nm, and another action spectrum for photoreactivation. Much information is already available for UV-B, UV-A, and PAR regions (see references). The need is for instrumentation and experimentation at the shorter wavelengths. UV treatment should be especially useful in particular cases, for example, when bacteria are attached to the exterior of larger algae. The nucleic acids in bacteria or viruses are less shielded by protoplasm in general and by certain pigments in particular, notably phycocyanin and phycoerythrin, as well as by other large molecules (de Mora et al. 2000). The opal of diatom shells may be a screen. How UV acts on intracellular bacteria (e.g., those in *Pinnularia*) or on intracellular viruses remains to be seen.

## 6.0. TESTING FOR CONTAMINANTS

Demonstrations ordinarily include microscopic examination by as many techniques as are available: bright field, dark field, phase or other interference contrast, and epi-illumination fluorescence with or without fluorochromes (e.g., DAPI). With precautions against contamination during handling, electron microscopy can reveal cryptic extracellular or intercellular contaminants. Flow cytometry is routinely used to detect and count bacteria. This can obviously be used to detect bacteria in algal cultures (see Chapters 7 and 17).

The other demonstration is inoculation from the culture under test into enrichment media designed to favor contaminants and make their presence obvious, generally by turbidity. The contaminants anticipated in early days were aerobic heterotrophic bacteria and fungi, so that test media, patterned after bacteriological media, were of the order of $10^{-1}$ M in total organic matter (i.e., mixed amino acids and carbohydrates, taking 150 as the average formula weight). Many also contained extracts of beef, liver, or yeast, often as high as $1–10 \text{ g} \cdot \text{L}^{-1}$. Some employed thioglycollate medium or brain-heart infusions. Many newer recipes are listed by Hoshaw and Rosowski (1973) and Watanabe et al. (2000). Enrichments are now usually added to the basal algal growth medium or to a dilution of it.

Tests for bacterial contamination in algal cultures should be emended or extended in two ways: (1) to include tests with organic substrates (concentrations) reduced to $10^{-4}$ M at least (ca. 15 mg $\cdot$ L$^{-1}$ peptone or 18 mg $\cdot$ L$^{-1}$ glucose), and (2) to include tests with methylamine (as the hydrochloride) added at $10^{-2}$ M to the organic enrichment or separately as methylamine $\cdot$ HCl only. Most marine bacteria countable by microscopy fail to grow with the usual media on or in agar (Jannasch and Jones 1959). This is attributed to their growth in nature under starved conditions, making them sensitive to even moderate amounts of added substrate (Jannasch and Mateles 1974). Similarly, some supposedly axenic cultures of marine microalgae contained a cryptic bacterium cf. *Caulobacter* sp. that grew very well at concentrations of $10^{-4}$ M but were killed at higher concentrations (Guillard and Watson 1962). Another point raised by chemostat studies (Jannasch and Mateles 1974) is that a lower limit of nutrient concentration could be demonstrated, which was far higher than expected on the basis of the stoichiometry of growth (e.g., carbon cell$^{-1}$) and was influenced by the physical conditions of growth. In the case of a marine *Spirillum* (from eutrophic water) the cut-off was between $2 \times 10^{-5}$ M and $1 \times 10^{-5}$ M (of asparagine). Thus, too little of an enrichment may fail to reveal bacteria characteristic of eutrophic water, whereas too much fails on the ones characteristic of oligotrophic environments (see also Chapter 5). Very low (but unspecified) concentrations were necessary to isolate and cultivate the newly discovered SAR 11 strains of bacteria (Rappé et al. 2002).

One dilemma is that bacteria do not show up easily by turbidity at very low substrate concentrations. Another is that the chemical nature of the enrichment may be critical, especially for bacteria that are commensal with the algae. The algae may provide the only successful substrate. Examination by microscopy, flow cytometry, or molecular technology is necessary.

The second recommendation follows from the finding that bacteria able to oxidize and grow on methylamine are common in the sea; many of these bacteria are obligate methylaminotrophs and are not detectable unless supplied methylamine (Sieburth and Keller 1988/89). In their study, methylamine was added at $1.6 \times 10^{-2}$ M (as the hydrochloride); however, easily visible crops of bacteria are produced by less—$10^{-3}$ M to $10^{-4}$ M (unpublished). The upper limits of bacterial tolerance to methylamine have not been reported. More to the point, methylamine at $10^{-2}$ M does not inhibit the organoheterotrophic bacteria found in contaminated algal cultures, nor does peptone at 1 g $\cdot$ L$^{-1}$ (a usual concentration) inhibit the methylaminotrophs (unpublished). Consequently, tests for axenicity can be made with the two substrates separately or together. No marine algae tested have survived methylamine at $10^{-2}$ M (unpublished), though they do at micromolar concentrations (Koike et al. 1983).

Freshwaters also have populations of obligate or facultative methylotrophic bacteria (methanotrophs, methylotrophs, or methylaminotrophs; Whittenbury and Dalton 1981), and therefore freshwater algal cultures also should be tested for methylaminotrophs. It seems likely that methylaminotrophs are more abundant in seawater than in freshwater, because seawater is an osmotically stressful environment for algae. Compounds serving as osmotica for marine algae—glycine, betaine, dimethysulfonioproprionate, choline derivatives, and proline—all yield methylated amines (Sieburth and Keller 1988/89). Therefore, methylamines should be present wherever algae are growing—not just in sediments, for example.

The basic test procedure suggested for both freshwater and marine cultures is to employ the medium suitable for the algae and add to it peptone (or a similar digest) at 1 g $\cdot$ L$^{-1}$ (ca. $6.7 \times 10^{-3}$ M) plus methylamine HCl at 0.675 g $\cdot$ L$^{-1}$ ($10^{-2}$ M). For marine test media, see Guillard and Morton 2003; Chapters 3 and 5; and Appendix A. To make dilutions of the organic matter (peptone), it is convenient to autoclave a supply of the algal medium with only the methylamine added; this can be used to dilute the peptone ($6.7 \times 10^{-3}$ M at full strength) by a factor of 10 or, for better resolution, at greater dilutions by 3.16, which is the square root of 10. If a basal seawater medium is diluted before autoclaving, then the EDTA should not be reduced below $10^{-6}$ M to prevent loss of iron (Guillard and Keller 1984).

Many of the early media for contamination testing had no trace metals or chelators, the authors possibly feeling that the high concentrations used of amino acid containing substances would provide both needs. This was the case even for the much-used STP and $ST_3$ of Tatewaki and Provasoli (1964), which were of the order of $10^{-2}$ M and $1.6 \times 10^{-3}$ M in total organics (see Appendix A); the latter medium had in addition 400 mg $\cdot$ L$^{-1}$ of glycylglycine as buffer, which is usable by bacteria. If these media are much diluted, then the absence of trace metals may become apparent. Recent versions often contain chelated trace metals (e.g., Tatewaki's MM 23 Medium [Watanabe et al. 2000]).

The situation is much the same for testing cultures of freshwater algae, except that freshwater media are less robust than most marine media and can vary more in such properties as pH, major ion ratios, trace metal availability, and total solids. The pH should be monitored and adjusted upon addition of the peptone and methylamine HCl. Mineral acids, sodium bicarbonate, or carbonate or sodium silicate can be employed to adjust pH. At this point it is not clear if organic pH buffers are without effect on the bacteria.

Tests for contaminating bacteria can be made in tubes of liquid medium, tubes of semi-solid medium ("sloppy agar"), or poured into petri plates of medium in solid agar. Incubate at the temperature used for the algae. Tubes with relatively high levels of organic additions, including methylamine, may be incubated in darkness. But at very low levels of organics (ca. $10^{-5}$ M) and even lower (but unknown) levels of methylamine, photosynthetic growth is possible and may either inhibit or stimulate growth of contaminating bacteria. Thus, incubation in both light and darkness may be useful.

Tubes and plates should be examined frequently beginning ca. 48 hours after inoculation and left for at least 21 days, or 1 month for cold-water organisms or methylaminotrophs. An occasional tube of uninoculated test medium and other tubes inoculated with raw seawater (or freshwater) serve as controls.

Fungi grow well on peptone but even better on malt extract when it is dissolved in algal culture medium (ca. 1 g $\cdot$ L$^{-1}$). Horse or bovine serum (ca. 5%) dissolved in algal culture medium favors yeasts, thraustrochytrids, and labyrinthuloids, although most other contaminants (and some algae) also grow on it.

It is not usual to test for anaerobic or microaerophilic bacteria in algal cultures, because the latter are usually saturated with oxygen, at least during periods of photosynthetic growth. However, the discovery of cryptic methylaminotrophs in cultures of marine algae raises the possibility that bacterial consorts may include oxida-tive methylotrophs that produce microzones of reduced oxygen concentration that allow growth of the more anaerobic types (Sieburth and Keller 1988/89). A suggested (but untried) technique would be to include 0.1–0.2 g of granular agar to 100 mL of test medium, swirl, dispense into screw-capped tubes, and autoclave; tighten the caps before cooling is complete, and then loosen for use. In the resulting sloppy agar, inoculated by a stab, the upper layers can remain aerobic, whereas the lower layers become progressively less aerobic. This process is recommended by Difco Laboratories for use with brain heart infusion.

Some test media reduce the concentrations of individual peptic or tryptic digests or of liver or yeast extracts or hydrolysates to the level of mg $\cdot$ L$^{-1}$ and add more of other defined carbon sources, such as glucose, sodium lactate, sodium acetate, sodium succinate, glycerol, glycine, alanine, or asparagine. A mix containing about $10^{-4}$ M of each is added to the algal culture medium and is probably an adequate test, especially if methylamine is added. Small-scale tests (without methylamine) have shown no major differences from results using peptone alone. A 10× or 20× concentrate of mixed organics in distilled or deionized water can be autoclaved and used aseptically as needed in small batches of medium or reautoclaved in larger lots of medium.

## 7.0. REFERENCES

Allen, E. A. D., and Gorham, P. R. 1981. Culture of planktonic cyanophytes on agar. In: Carmichael, W. W., ed. *The Water Environment: Algal Toxins and Health*. Plenum Publishing Company, New York, pp. 185–92.

Allen, M. M. 1973. Methods for cyanophyceae. In: Stein, J. R., ed. *Handbook of Phycological Methods: Culture Methods and Growth Measurements*. Cambridge University Press, Cambridge, pp. 127–38.

Berland, B. R., and Maestrini, S. Y. 1969. Action de quelques antibiotiques sur le developpement de cinq diatomées en culture. *J. Exp. Mar. Biol. Ecol.* 3:62–75.

Blankley, W. F. 1973. Toxic and inhibitory materials associated with culturing. In: Stein, J. R., ed. *Handbook of Phycological Methods: Culture Methods and Growth Measurements*. Cambridge University Press, Cambridge, pp. 207–29.

Brand, L. E. 1986. Nutrition and culture of autotrophic ultraplankton and picoplankton. In: Platt, T., and Li, W. K. W., eds. *Photosynthetic Picoplankton. Can. Bull. Fish. Aquatic Sciences* 214:205–33.

Brand, L. E., Guillard, R. R. L., and Murphy, L. S. 1981. A method for the rapid and precise determination of acclimated phytoplankton reproduction rates. *J. Plankton Res.* 3:93–201.

Brown, R. M., and Bischoff, H. W. 1962. A new and useful method for obtaining axenic cultures of algae. *Phycol. Soc. Amer. News Bull.* 15:43–4.

Brown, S. W. 1964. Purification of algal cultures with caffeine. *Nature* 204:801.

Chapman, A. R. O. 1973. Methods for macroscopic algae. In: Stein, J. R., ed. *Handbook of Phycological Methods: Culture Methods and Growth Measurements.* Cambridge University Press, Cambridge, pp. 87–104.

Cottrell, M. T., and Suttle, C. A. 1993. Production of axenic cultures of *Micromonas pusilla* (Prasinophyceae) using antibiotics. *J. Phycol.* 29:385–7.

Davies, A. G. 1983. The effects of heavy metals upon natural marine phytoplankton populations. In: Round, F. E., and Chapman, D. J., eds. *Progress in Phycological Research* 2:113–45.

de Mora, S., Demers, S., and Vernet, M., eds. 2000. *The Effects of UV Radiation in the Marine Environment.* Cambridge University Press, Cambridge, 324 pp.

Divan, C. L., and Schnoes, H. K. 1982. Production of axenic *Gonyaulax* cultures by treatment with antibiotics. *Appl. Environ. Microbiol.* 44:250–4.

Droop, M. R. 1967. A procedure for routine purification of algal cultures with antibiotics. *Br. Phycol. Bull.* 3:295–7.

Droop, M. R. 1969. Algae. In: Norris, J. R., and Ribbon, D. W., eds. *Methods in Microbiology.* Vol. 3B, Academic Press, New York, pp. 269–313.

Dzurica, S., Lee, C., Cosper, E. M., and Carpenter, E. J. 1989. Role of environmental variables, specifically organic compounds and micronutrients, in the growth of the chrysophyte *Aureococcus anophagefferens.* In: Cosper, E. M., Bricelj, V. M., and Carpenter, E. J., eds. *Novel Phytoplankton Blooms,* Springer-Verlag, Vienna, pp. 229–52.

Fitzwater, S. E., Knauer, G. A., and Martin, J. H. 1982. Metal contamination and its effect on primary production measurements. *Limnol. Oceanogr.* 27:544–51.

Fries, L. 1963. On the cultivation of axenic red algae. *Physiol. Plantarum* 16:695–708.

Gavis, J. 1976. Munk and Riley revisited: Nutrient diffusion transport and rates of phytoplankton growth. *J. Mar. Res.* 34:169–79.

Gerloff, G. C., Fitzgerald, G. P., and Skoog, F. 1950. The isolation, purification, and nutrient solution requirements of blue-green algae. In: Brunel, J., Prescott, G. W., and Tiffany, L. H., eds. *The Culturing of Algae.* Antioch Press, Yellow Springs, Ohio. pp. 27–44.

Guillard, R. R. L. 1973. Methods for microflagellates and nannoplankton. In: Stein, J. R., ed. *Handbook of Phycological Methods: Culture Methods and Growth Measurements.* Cambridge University Press, Cambridge, pp. 69–85.

Guillard, R. R. L. 1995. Culture methods. In: Hallegraeff, G. M., Anderson, D. M., and Cembella, A. D., eds. *Manual on Harmful Marine Algae.* UNESCO, Paris, pp. 45–62.

Guillard, R. R. L., and Keller, M. 1984. Culturing dinoflagellates. In: Spector, D. L., ed. *Dinoflagellates.* Academic Press, New York, pp. 391–442.

Guillard, R. R. L., and Morton, S. L. 2003. Culture methods. In: Hallegraeff, G. M., Anderson, D. M., and Cembella, A. D., eds. *Manual on Harmful Marine Microalgae.* UNESCO, Paris, pp. 77–97.

Guillard, R. R., and Watson, S. W. 1962. A new marine bacterium. *Oceanus* 83:22–3.

Hoshaw, R. W., and Rosowski, J. R. 1973. Methods for microscopic algae. In: Stein, J. R., ed. *Handbook of Phycological Methods: Culture Methods and Growth Measurements.* Cambridge University Press, Cambridge, pp. 53–68.

Jannasch, H. W., and Jones, G. E. 1959. Bacterial populations in sea water as determined by different methods of enumeration. *Limnol. Oceanogr.* 42:128–39.

Jannasch, H. W., and Mateles, R. J. 1974. Experimental bacterial ecology studied in continuous culture. *Adv. Microbiol. Physiol.* 11:165–212.

Jones, A. K., Rhodes, M. E., and Evans, S. C. 1973. The use of antibiotics to obtain axenic cultures of algae. *Br. Phycol. J.* 8:185–96.

Kim, J. S., Park, Y. H., Yoon, B. D., and Oh, H. M. 1999. Establishment of axenic cultures of *Anabaena flos-aquae* and *Aphanotheca nidulans* (cyanobacteria) by lysozyme treatment. *J. Phycol.* 35:865–9.

Koike, I., Redalji, D. G., Ammerman, J. W., and Holm-Hansen, O. 1983. High-affinity uptake of an ammonium analogue by two marine microflagellates from the oligotrophic Pacific. *Mar. Biol.* 74:161–8.

Koostra, W., Boele-Bos, S., and Stam, W. T. 1991. A method for obtaining axenic algal cultures using the antibiotic cefotaxime with emphasis on *Cladophoropsis membranaceae* (Chlorophyta). *J. Phycol.* 27:656–8.

Krieg, N. R., and Gerhardt, P. 1981. Solid culture. In: Gerhardt, P., Murray, R. G. E., and Costilow, R. N., eds. *Manual of Methods for General Microbiology.* American Society for Microbiology, Washington, D.C., pp. 143–4.

Lehman, J. T. 1976. Ecological and nutritional studies on *Dinobryon* Ehrenb.: Seasonal periodicity and the phosphate toxicity problem. *Limnol. Oceanogr.* 21:646–58.

León, C., Kumazawa, S., and Mitsui, A. 1986. Cyclic appearance of aerobic nitrogenase activity during synchronous growth of unicellular cyanobacteria. *Curr. Microbiol.* 13:149–53.

Lim, M., Ong, B. L., and Wee, Y. C. 1992. A method of obtaining axenic cultures of *Trentepohlia* spp. (Chlorophyta). *J. Phycol.* 28:567–9.

Loeblich, A. R. III, and Sherley, J. L. 1979. Observations on the theca of the motile phase of free-living and symbiotic isolates of *Zooxanthella microadriatica* (Freudenthal) comb. nov. *J. Mar. Biol. Assoc. U.K.* 59:195–205.

Nelson, D. M., and Brand, L. E. 1979. Cell division periodicity in 13 species of marine phytoplankton on a light : dark cycle. *J. Phycol.* 15:67–75.

Noga, E. J. 1987. Propagation in cell culture of the dinoflagellate *Amyloodinium*, an ectoparasite of marine fishes. *Science* 236:1302–4.

Noga, E. J., and Bower, C. E. 1987. Propagation of the marine dinoflagellate *Amyloodinium ocellatum* under germ-free conditions. *J. Parasit.* 73:924–8.

Oppenheimer, C. H. 1955. The effect of bacteria on the development and hatching of pelagic fish eggs, and the control of such bacteria by antibiotics. *Copeia.* 1955:43–9.

Page, J. Z. 1973. Methods for coenocytic algae. In: Stein, J. R., ed. *Handbook of Phycological Methods: Culture Methods and Growth Measurements.* Cambridge University Press, Cambridge, pp. 105–26.

Pasciak, W. J., and Gavis, J. 1975. Transport limited nutrient uptake rates in *Ditylum brightwellii.* *Limnol. Oceanogr.* 20:605–17.

Price, N. M., et al. 1988/89. Preparation and chemistry of the artificial algal culture medium Aquil. *Biol. Oceanogr.* 6:443–61.

Pringsheim, E. G. 1946. *Pure Cultures of Algae. Their Preparation and Maintenance.* Cambridge University Press, Cambridge, 119 pp.

Rappé, M. S., Connon, S. A., Vergin, K. L., and Giovannoni, S. J. 2002. Cultivation of the ubiquitous SAR 11 marine bacterioplankton clade. *Nature* 418:630–3.

Reardon, E. M., Price, C. A., and Guillard, R. R. L. 1979. Harvest of marine microalgae by centrifugation in density gradients of "Percoll," a modified silica sol. In: Reed, E., ed. *Methodological Surveys in Biochemistry*, Vol. 8. Ellis Norwood Publishing, Chichester, U.K., pp. 171–5.

Repak, A. J., Provasoli, L., and Pintner, I. J. 1982. Tailor-made antibiotic mixes for marine ultraplankton. *J. Protozool.* 29:291.

Rippka, R., et al. 2000. *Prochlorococcus marinus* Chisholm et al. 1992 subsp. *pastoris* subsp. nov. strain PCC 9511, the first axenic chlorophyll a$_2$/b$_2$-containing cyanobacterium (Oxyphotobacteria). *Int. J. Syst. Evol. Microbiol.* 50:1833–47.

Rippka, R., Waterbury, J. B., and Stanier, R. Y. 1981. Isolation and purification of cyanobacteria: some general principles. In: Starr, M. P. et al., eds. *The Prokaryotes*, Vol. I. Springer-Verlag, Vienna, pp. 212–20.

Schmid, A. M. M. 2003a. Endobacteria in the diatom *Pinnularia* (Bacillariophyceae). I. "Scattered nucleoids" explained: DAPI-DNA complexes stem from exoplastidial bacteria boring into the chloroplasts. *J. Phycol.* 39:122–38.

Schmid, A. M. M. 2003b. Endobacteria in the diatom *Pinnularia* (Bacillariophyceae). II. Host cell cycle-dependent translocation and transient chloroplast scars. *J. Phycol.* 39:139–53.

Shirai, M., et al. 1989. Development of a solid medium for growth and isolation of axenic *Microcystis* strains (cyanobacteria). *Appl. Environ. Microbiol.* 55:2569–71.

Shephard, D. C. 1970. Axenic culture of *Acetabularia* in a synthetic medium. In: Prescott, D. M., ed. *Methods in Cell Physiology.* Vol. 4. Academic Press, New York, pp. 49–69.

Sieburth, J. McN., and Keller, M. D. 1988/89. Methyaminotrophic bacteria in xenic nanoalgal cultures: Incidence, significance and role of methylated algal osmoprotectants. *Biol. Oceanogr.* 6:383–95.

Sitz, T. O., and Schmidt, R. R. 1973. Purification of *Synechococcus lividus* by equilibrium centrifugation and its synchronization by differential centrifugation. *J. Bact.* 115:43–6.

Smith, K. C. 1977. *The Science of Photobiology.* Plenum Publishing Company, New York, 430 pp.

Soli, G. 1963. Axenic cultivation of a pelagic diatom. In: Oppenheimer, C., ed. *Marine Microbiology.* Charles C. Thomas, Springfield, IL, pp. 121–6.

Solp, H., and Starr, M. P. 1981. Principles of isolation, cultivation and conservation of bacteria. In: Starr, M. P., Trüper, H. G., Balows, A., and Schlegel, H. G., eds. *The Prokaryotes.* Vol. 1. Springer-Verlag, Vienna, pp. 135–75.

Spencer, C. P. 1952. On the use of antibiotics for isolating bacteria-free cultures of marine phytoplankton organisms. *J. Mar. Biol. Ass. U.K.* 31:97–106.

Starr, R. C. 1973. Special methods—dry soil samples. In: Stein, J. R., ed. *Handbook of Phycological Methods: Culture Methods and Growth Measurements.* Cambridge University Press, Cambridge, pp. 159–67.

Stokes, P. M. 1983. Responses of freshwater algae to metals. In: Round, F. E., and Chapman, D. V., eds. *Progress in Phycological Research.* Vol. 2. Elsevier, New York, pp. 87–112.

Sunda, W. G. 1988/89. Trace metal interactions with marine phytoplankton. *Biol. Oceanogr.* 6:411–42.

Sutherland, J. W. 1976. Ultrasonication—an enrichment technique for microcyst-forming bacteria. *J. Appl. Bacteriol.* 41:185–8.

Tatewaki, M., and Provasoli, L. 1964. Vitamin requirements of three species of *Antithamnion.* *Bot. Mar.* 6:193–203.

Throndsen, J. 1973. Special methods—micromanipulators. In: Stein, J. R., ed. *Handbook of Phycological Methods. Culture Methods and Growth Measurements.* Cambridge University Press, Cambridge, pp. 139–44.

Van Baalen, C., 1962. Studies on marine blue-green algae. *Bot. Mar.* 4:129–39.

Van Baalen, C., and O'Donnell, R. 1972. Action spectra for ultraviolet killing and photoreactivation in the blue-green alga *Agmenellum quadruplicatum. Photochem. Photobiol.* 15:269–74.

Vance, B. D. 1966. Sensitivity of *Microcystis aeruginosa* and other blue-green algae and associated bacteria to selected antibiotics. *J. Phycol.* 2:125–8.

Watanabe, M. M., Kawachi, M., Hiroki, M., and Kasai, F. 2000. *NIES-Collection List of Strains Sixth Edition 2000 Microalgae and Protozoa.* Microbial Culture Collections, National Institute for Environmental Studies, Tsukuba, 159 pp.

Watanabe, M. M., Suda, S., Kasai, F., and Sawaguchi, T. 1985. Axenic cultures of three species of *Microcystis* (Cyanophyta = Cyanobacteria). *Bull. Jap. Fed. Culture Coll.* 1:57–63.

Watanabe, M. M., et al. 1998. Purification of freshwater picoplanktonic cyanobacteria by pour-plating in "ultra-low-gelling-temperature agarose." *Phycol. Res.* 46:71–5.

Waterbury, J. B., Watson, S. W., Guillard, R. R. L., and Brand, L. E. 1979. Widespread occurrence of a unicellular, marine, planktonic, cyanobacterium. *Nature* 277:293–4.

Waterbury, J. B., Watson, S. W., Valeis, F. W., and Franks, D. G. 1986. Biological and ecological characterization of the marine unicellular cyanobacterium *Synechococcus*. In: Platt, T., and Li, W. K. W., eds. *Photosynthetic picoplankton. Can. Bull. Fish. Aquat. Sci.* 214:71–120.

Whittenbury, R., and Dalton, H. 1981. The methylotrophic bacteria. In: Starr, M. P., Trüper, H. G., Balows, A., and Schlegel, H. G., eds. *The Prokaryotes.* Vol. 1. Springer-Verlag, Vienna, pp. 894–902.

Windholz, M., ed. 1983. *The Merck Index.* 10th Ed. Merck and Co., Rahway, N.J., 1463 pp + App.

Yurkov, V. V., and Beatty, J. T. 1998. Aerobic anoxygenic phototrophic bacteria. *Microbiol. Mol. Biol. Rev.* 62:695–724.

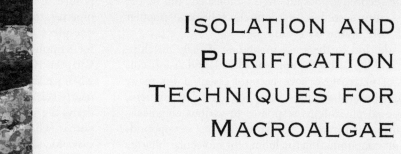

CHAPTER 9

# ISOLATION AND PURIFICATION TECHNIQUES FOR MACROALGAE

Hiroshi Kawai
*Kobe University Research Center for Inland Seas*

Taizo Motomura
*Muroran Marine Station, Field Science Center for Northern Biosphere, Hokkaido University*

Kazuo Okuda
*Graduate School of Kuroshio Science, Faculty of Science, Kochi University*

CONTENTS

Key Index Words: Antibiotics, Axenic, Clonal Culture, Crude Culture, Culture Medium, Day-length, Isolation, Life History, Macroalgae, Unialgal Culture

## 1.0. INTRODUCTION

*Macroalgae* is a general term for the algae that form a multicellular thallus at least in one stage of the life history, with the exception of siphonous ulvophycean algae that lack septa (e.g., *Caulerpa*, *Valonia*, and so on). In most cases they show differentiation between vegetative tissues and reproductive structures that release unicellular reproductive cells, as well as an alternation of generations. Therefore, to observe the development from unicellular reproductive cells to multicellular thallus, or to elucidate the whole life history that often cannot be observed from field-collected specimens, culture studies are necessary. Such studies began in the nineteenth century; however, it was often difficult to complete life histories using natural seawater and culturing at ambient temperatures. Gradually, because of the improvement of culture media and the use of controlled-temperature incubation chambers, unialgal culture techniques became well established, and by the 1960s culture studies of macroalgae became popular (Bold 1942, Tatewaki 1966).

In addition to their use in studies of early development and life histories, unialgal, clonal, and axenic cultures of macroalgae have become essential for many studies of morphogenesis, morphological development, nutritional physiology, responses to various chemicals, crossing experiments, extracting various compounds without contamination (including for molecular biology such as genomic DNA, cDNA libraries, Northern blotting, etc.), long-term strain preservation, exchange of research materials, mass culture and preparation of mariculture seed-stock, and so on.

In this chapter, a unialgal culture is a culture that includes only one species of alga (bacteria may be present). An axenic culture is unialgal and free of bacteria. A clonal culture is a culture of a single genome set (e.g., cultures derived from a single vegetative cell or tissue or from a reproductive cell) and propagated vegetatively. This chapter introduces the techniques for establishing and maintaining unialgal and axenic cultures of macroalgae.

## 2.0. Sampling

Collected specimens are transported in plastic bags, bottles, or containers suitable to their size, avoiding excess irradiation and temperature shocks relative to the prevailing habitat conditions. In general, temperature conditions 5–10°C cooler than the water temperature of the habitat (5–10°C for cold-water taxa and 20–25°C for tropical taxa) are preferable for transport. Most intertidal macroalgae (seaweeds) are tolerant of stresses such as desiccation and rapid temperature changes, compared with subtidal macroalgae. Fertile specimens collected under desiccating conditions (e.g., intertidal taxa collected during low tides) release reproductive cells (zooids, eggs, spores, etc.) as soon as they are reimmersed in seawater, such as in the containers used for transportation. Therefore, they may preferably be transported moist in plastic bags (specimens may be loosely wrapped with paper towels or newspaper) or plastic containers, instead of immersing in seawater. In contrast, subtidal macroalgae, especially those growing in deep habitats (below 5–10 m) are more sensitive to environmental changes such as desiccation and temper-

ature shock, so they should be immediately transferred to containers filled with seawater, minimizing the exposure to air or temperature fluctuations. Disposable pipettes, the kind with the bulb and pipette molded in one piece of thin polyethylene (pastette), can be used for sampling very small specimens in the field (see Chapter 10). They are cheap, unbreakable, and sterile when packaged individually (bulk-packaged pipettes are nearly sterile and are suitable for field collecting). One draws the sample into the pipette by suction. For transport it is best to fill the pipette almost completely with seawater, shake down the liquid into the bulb, and then heat-seal by carefully melting the opening at the edge of a small flame, and squeeze the molten end together with forceps. To open, just snip off the tip.

Some acidic macroalgae (e.g., some *Desmarestia* spp., *Dictyopteris* spp., *Plocamium* spp., and so on) need special care, similar to sensitive subtidal taxa. It is also important to protect other specimens from those acidic taxa at the time of collection and transportation (i.e., avoid putting both types in the same container, because even a small amount of damaged acidic algae can ruin the other specimens). The maturation of some taxa (e.g., *Dictyota*) is reported to be synchronized with lunar rhythms (Phillips et al. 1990), so special attention should be paid in scheduling collections of these taxa.

Either vegetative tissue or the cells released from reproductive structures may be used to establish cultures (see Section 4.0). In either case, cleaner plants with few epiphytes and epizoa, and fertile plants in the latter case, should be selected in the field. Fertile portions (parts of the plants bearing reproductive structures) often can be detected from the gross appearance; only those portions need be cut and transported to the laboratory.

## 3.0. Tools and Facilities

For unialgal cultures, glass and plastic petri dishes, beakers with glass covers, test tubes with screw caps, and clear thin polystyrene cups (so-called ice cream cups) are commonly used (Fig. 9.1). For small plants (less than several centimeters), plastic (polystyrene) petri dishes of 60 or 90 mm in diameter are convenient and also cheap, because their flat bottom and top are suitable for observation using an inverted microscope or stereomicroscope. They also save space, because the dishes may be stacked. For larger plants, beakers, ice cream cups, or various glass containers are used. The lids of the petri

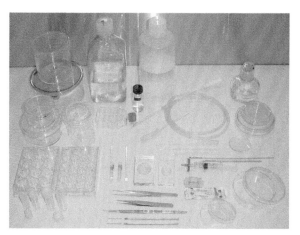

**FIGURE 9.1.** Tools used for macroalgal cultures (glassware, multiwell dishes, small vials for stocks, plastic containers for medium, Pasteur pipettes with silicone bulbs and tubing, polystyrene transfer pipettes, depression slides, glass homogenizer, forceps, scalpel, needles, paint brushes, syringe with membrane filter cartridge, razor, alcohol lamp, watch glass, and glass and plastic petri dishes).

**FIGURE 9.2.** Incubators converted from cooling incubators by adding lighting units (fluorescent tubes and a timer controlling the lighting; the ballasts should be mounted outside the chamber).

dishes should be sealed with sealing films (e.g., Parafilm, American National Can) to avoid evaporation of water and accidental leakage that contaminate other cultures. Culture media are changed every 2–4 weeks, although this interval varies depending on the material and temperature.

Cultures are normally maintained in climate-controlled culture chambers or incubators that can regulate the temperature and are illuminated by daylight-type, white fluorescent tubes. Plant growth chambers designed for higher plants generally provide illumination that is too intense, and they may not provide stable temperature control at lower ranges. Generally incubators converted from cooling incubators by adding lighting units (fluorescent tubes and a timer controlling the lighting; the ballasts should be mounted outside the chamber) are cheaper and more reliable (Fig. 9.2) (e.g., MIR-552HK, Sanyo). Because mechanical problems can occur that will cause extreme temperature fluctuations, it is important that the lights turn off when the cooling unit of the incubator fails. A temperature-gradient incubation chamber (Fig. 9.3) (e.g., TG180–5L, Nihon Ika), having several closed chambers whose temperatures can be independently controlled, is especially convenient for life history studies or to compare growth and differentiation under different temperature conditions. In many macroalgae, reproduction is controlled by a combination of temperature and daylength (photoregime) conditions (see Chapter 21 for details).

**FIGURE 9.3.** Temperature-gradient incubation chamber having several closed chambers whose temperatures can be independently controlled.

For long-day conditions, intervals of 14–16 hours lighting and 8–10 hours darkness (e.g., 16-hr light: 8-hr dark) are commonly used, and for short-day conditions, 6–8(–10) hours lighting and (14–)16–18 hours darkness. Light intensities of 10–100 μmol · m$^{-2}$ · s$^{-1}$ are commonly used for macroalgal cultures. The light intensity can be reduced by placing smoked transparent glass (or plastic panel) between the illumination and the cultures or by covering the culture containers with gray plastic window screens. Colored fluorescent tubes (e.g., FL40S-R-F, National) or photodiodes (e.g., MIL-R18, Sanyo) are available for red-light illumination (for photomorphogenesis experiments or to suppress fertilization in laminarialean gametophytes), but shading by colored plastic filters is also effective.

## 4.0. Unialgal Isolation Techniques

Most species of macroalgae have high potentials for regeneration and totipotency, so that theoretically unialgal cultures can be established for most species by cutting off vegetative tissues (cells) and cleaning them in the course of their growth. However, in practice epiphytic algae or cyanobacteria, which may be very difficult to remove from the surface of the tissues, grow faster and more vigorously than the desired algae. Therefore, isolation from vegetative tissues is usually restricted to taxa with apical meristematic growth (e.g., Dictyotales, Sphacelariales in brown algae, red algae with apical cells), siphonous green algae, and some taxa with rapid cell division (e.g., *Ectocarpus*, *Ulva*, and so on). In other cases, unialgal cultures are established from zooids such as zoospores and planogametes, or zygotes, carpospores, tetraspores, or aplanospores.

### 4.1. Crude Cultures

When the collected specimens do not immediately release reproductive cells, or as a preliminary step in isolation from vegetative tissues, the whole plant or a part of the plant may be cultured (maintained) in the laboratory. These are often called crude cultures. To clean the surface of the plants, fine paintbrushes are helpful before starting the culture (Fig. 9.4). Enriched seawater media are used, as for unialgal cultures, but to avoid overgrowth of epiphytes, plain sterilized seawater or reduced enrichment may be used. To suppress the growth of diatoms and cyanobacteria, germanium dioxide (GeO$_2$) and antibiotics may be added to the

**FIGURE 9.4.** Cleaning of algal tissue with a paintbrush.

media (see the following sections). Because the reproduction of many macroalgae is controlled by temperature and daylength conditions, to induce reproduction of the plant (formation or maturation of reproductive structures), experimentation may be required to discover the necessary temperature and daylength conditions (e.g., some brown algal taxa form reproductive structures only under low- temperature and short-day conditions).

The vegetative thalli of *Ulva* spp. (Ulvophyceae) normally do not become reproductive in crude cultures, but small pieces (a few millimeters across), cut with a razor or punched with a cork borer, become fertile within a few days (Norby and Hoxmark 1972). Various ulvophycean algae (e.g., *Bryopsis*, *Chaetomorpha*, and so on) form reproductive cells within several days to several weeks after collection and can be used for unialgal isolations; however, for some taxa, extended crude culture (up to one month or longer) is required (e.g., *Caulerpa* in Enomoto and Ohba 1987, *Polyphysa* in Berger and Kaever 1992). Field-collected vegetative plants should be cleaned and maintained in suitable culture conditions simulating the temperature and daylength of the original habitat. When the formation of reproductive cells is noticed from the change in the external appearance (especially color), the culture should be transferred into a dark box, and the release of reproductive cells can be induced the following morning by transferring the plant to a new dish filled with fresh medium and stimulating with high-intensity lighting.

### 4.2. Isolation from Vegetative Cells

If the species has obvious apical or marginal meristematic cells, isolation from these cells to establish unialgal culture is easy. In such cases, first cut out a small fragment including the apical cell(s) using a razor blade or scalpel, and place it in a 60-mm–diameter petri dish

**FIGURE 9.5.** Cutting off clean apical (distal) portion with a scalpel.

**FIGURE 9.6.** Cleaning of algal tissue by dragging the fragment through an agar plate.

**FIGURE 9.7.** Cleaning of algal tissue by dragging the fragment through an agar plate.

filled with filtered seawater. Under observation by stereomicroscopy, cut out clean, smaller pieces of tissue including intact apical cells (the smaller the better, but care is required to avoid damage to the apical cells) (Fig. 9.5). Transfer the excised pieces one by one into individual wells of a multiwell plate, petri dishes, or test tubes filled with culture medium, using a clean fine forceps, a mechanical pipettor (e.g., Pipetman, Gilson) with disposable tips, or capillary pipettes (see the following sections). Excised apical tissues may be pipetted into successive well plates of sterilized seawater, or for big apices put in a tube or vial and agitated in several changes of sterilized seawater using a vortex mixer, before putting them into growth medium. $GeO_2$ or antibiotics may be added to the medium (see the following sections). Culture the isolates for 1–2 weeks (or longer) in appropriate temperature conditions, and then observe with a stereomicroscope or an inverted microscope and select clean cultures. If contaminants are still present (epiphytes or cells on the bottom of the dishes), repeat the isolation processes (cutting off clean apical cells and isolating into new wells of a multiwell plate or individual dishes), until the culture becomes unialgal. For some taxa, culture of the primary isolate in dim light (5–10% of normal culture conditions) is effective to suppress the overgrowth of contaminants. Brief immersion in a dilute nonionic detergent (e.g., Triton X-100) or fresh water is effective for eliminating protozoa or diatoms, if the desired alga is tolerant of such treatment.

For some taxa, cleaning using agar plates is effective. Cut off a small fragment including the apical cells, and drag the fragment through an agar plate (1–2% agar in seawater) holding the proximal end with forceps or a fine needle (Figs. 9.6 and 9.7). Transfer the cleaned fragment to a petri dish filled with sterilized seawater, cut off a smaller fragment including the apex (distal fragment), and isolate it into a well of a multiwell plate filled with medium.

Certain multinucleate siphonous algae (Ulvophyceae [e.g., *Bryopsis* and *Valonia*]) show healing responses in wounded cells, forming protoplasts (Tatewaki and Nagata 1970, La Claire 1982). Protoplasts artificially induced by cutting the siphonous thallus or by puncturing the wall by a fine needle, and which are formed several hours after wounding, can be suspended in sterilized seawater by agitating the wounded thallus. Clean the protoplasts by transferring them into new petri dishes filled with sterilized seawater using fine Pasteur pipettes (or a mechanical pipettor), and then isolate them into individual wells of a multiwell plate. The protoplasts regenerate cell walls and eventually develop into thalli.

## 4.3. Isolation from Reproductive Cells

### 4.3.1. *Isolation from Swimming Zooids (Zoospores and Gametes)*

Many species release zooids or eggs/aplanospores more vigorously and synchronously on the day(s) following rather than immediately after collection, if properly stored in a dark (and cool for cold water and temperate taxa) place. Those specimens stored under cool and dark conditions release reproductive cells immediately after immersion in seawater, stimulated by the temperature rise and lighting, so they should be transferred just before isolation.

Prepare a plastic multiwell plate, test tubes, or petri dishes (see Fig. 9.1) for culturing the isolates, and fill the wells with liquid medium. Prepare a sterilized Pasteur pipette with fine tip, pulled in a flame (see Chapter 6 for the method), and attach a segment of silicone tubing with a mouthpiece (or a rubber bulb) to the wide end; a mechanical pipettor with disposable sterile tips may also be used. Fill a depression slide, small petri dish, or watch glass with liquid culture medium and place on the stage of a stereomicroscope. The temperature of the medium should be adjusted depending on the specimen (5–10°C for cold-water species, because the zooids soon settle or stop swimming in warm medium). Clean the surface of the fertile portion of the plants with a paper towel, gauze, or paintbrush, and cut off a small piece of tissue (1–3 mm in length) with reproductive structures. Gently place the tissue on the bottom of the depression slide. Expose to strong unilateral lighting, preferably with fiber-optic illuminations to avoid rapid temperature rise of the sample and medium (Figs. 9.8 and 9.9).

When swimming zooids are released, observe their phototactic behavior to determine the orientation of the taxis. Dark-field (or semi-dark-field) illumination is helpful for observing swimming zooids. If they are phototactic, adjust the direction of illumination to cause the zooids to accumulate at the surface of the medium (illumination from the upper side when they are positively phototactic, and from the lower side when negatively phototactic), and distant from the algal tissue to avoid contaminants (Fig. 9.10a,b). When the zooids are ready for isolation, dip the tip of the fine pipette into the medium of the well into which the zooids are to be isolated, and allow medium to enter by capillary action (if the tip of an empty capillary is dipped into the liquid surface of the depression slides, medium including con-

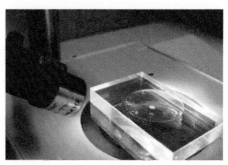

**FIGURE 9.9.** Enlargement of depression slide and illuminator.

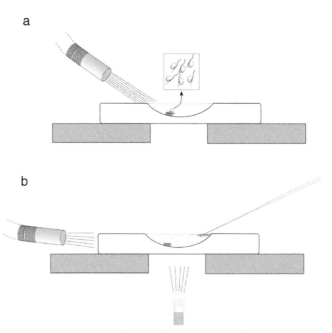

a

b

**FIGURE 9.10.** Isolation of swimming zooids using phototaxis. **(a)** Unilateral illumination from a fiber-optic light source to stimulate zooid release; **(b)** simultaneous lateral and bottom illumination to accumulate zooids at one upper edge of the depression slide well.

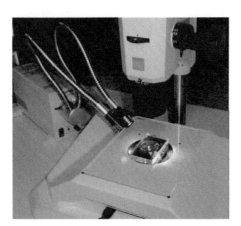

**FIGURE 9.8.** Isolation of phototactic zooids under observation with a stereomicroscope: Fertile algal tissue is placed at the bottom of a depression slide and the release of zooids is induced by intense illumination from a fiber-optic light source. Released zooids are accumulated at the surface of the medium (lighting from the upper side when they are positively phototactic, and from the lower side when negatively phototactic), and distant from the algal tissue.

taminants such as diatoms floating on the surface by surface tension would automatically flow in by capillary action). Change the lighting direction or reduce the intensity for a short time to make the zooid suspension somewhat diffuse, or let them start to swim downward. Then manipulate the tip of the fine pipette to a position just above the liquid surface over the accumulated zooids, and gently dip the tip of the pipette into the liquid and let the medium including the zooids flow underneath the liquid surface (to avoid inflow of contaminants on the surface). If pipetting by mouth, control the air pressure by using the tip of the tongue. Place the tip of the pipette in the liquid of the isolation well, test tube, or petri dish, and discharge the medium by breathing out gently until an air bubble is released. Then dip the tip of the empty pipette into the liquid of a new isolation well, and repeat this procedure several times.

### 4.3.2. Isolation from Zygotes and Aplanospores

The isolation procedure for zygotes and aplanospores (e.g., carpospores, tetraspores, monospores) of red and brown algae is similar to that for swimming zooids in the brown and green algae and can be easier, because they are larger and nonmotile. Spores of various red algae adhere to the substratum during the washing process, probably by the secretion of sticky polysaccharides and proteins. Detaching the settled spores from the substratum by force may damage the spores, and as a result, they do not develop. Tatewaki et al. (1989) found that attachment of red algal spores on the substratum is delayed in seawater conditioned by the spore-producing thallus, and they therefore recommended washing spores with filter-sterilized seawater in which the mother thallus had been immersed (10 g mother thalli in 50–100 mL seawater for 1 hour). Finally, spores are inoculated into sterilized seawater or enriched seawater medium on slides or cover glasses. After settlement of the spores, these substrates are transferred into enriched seawater medium (e.g., Provasoli's ES medium; see Chapter 3) in petri dishes or test tubes.

### 4.3.3. Isolation from Reproductive Tissues

When fertile reproductive structures (e.g., sporangia, gametangia, and so on) are found in the collected algae, but actual release of reproductive cells (spores, zooids, gametes, etc.) does not take place when observed under stereomicroscopy, the fertile reproductive structures (e.g., sporangia) or small fragments including reproductive structures may be excised and maintained (crude-cultured) in individual wells of multiwell plates

**FIGURE 9.11.** Hanging drop method: Fertile algal tissue is suspended in the drop of medium under the cover glass that is placed on the plastic spacer ring on a slide glass. The space between the cover and slide glasses and plastic spacer is sealed with petroleum jelly.

until they release reproductive cells during incubation and germlings become apparent in the culture dishes. To minimize contamination, the tissue should be removed as soon as release of reproductive cells can be detected. Transfer the clean, individual germlings attached on the bottom or walls of wells into new wells using fine forceps, capillary pipettes, or mechanical pipettors. The so-called hanging-drop method is useful to isolate algae forming flagellate reproductive cells, avoiding contamination by diatoms and other nonflagellate algae: Drop several drops of sterilized seawater onto a clean cover glass; place a small fragment of fertile tissue in the seawater; invert the cover glass quickly to leave the seawater drop, including the algal tissue, hanging, and place the cover glass on a plastic ring that is attached by petroleum jelly on a slide glass (Fig. 9.11); place the slide glass in a 90-mm–diameter petri dish and incubate in the culture chamber; observe under compound microscopy, and when the settlement of the released reproductive cells on the surface of the cover glass is noticed (often at the periphery of the drop), remove the algal tissue and contaminants by removing the drop and washing the cover glass in sterilized seawater, and then place in petri dishes filled with culture medium.

### 4.3.4. Elimination of Diatoms and Bacteria from Algal Cultures

Establishing unialgal cultures preferably starts from isolating reproductive cells, but sometimes other organisms contaminate the culture vessel. Protozoa, fungi, and diatoms are common contaminants. These contaminants can grow more rapidly than the desired alga and attach firmly; as a result, it is very difficult to eliminate these organisms. In particular, contamination by protozoa and fungi stunt the growth of unialgal cultures, although some kinds of antibiotics may be able to inhibit their growth for a time. Although diatom contamination may be eliminated by adding $GeO_2$, which inhibits silicon-metabolism in diatoms (Lewin 1966, McLachlan et al. 1971, Chapman 1973; see Chapter 11). Tatewaki and Mizuno (1979) showed that 2.5–5 mg $\cdot$ $L^{-1}$ $GeO_2$ in the medium inhibits the growth of brown algae, although it does not affect growth of green and red algae. Because diatoms could be eliminated at a $GeO_2$ concentration of 1–5 mg $\cdot$ $L^{-1}$, they recommended that the maximum concentration of $GeO_2$ should be 1 mg $\cdot$ $L^{-1}$ in the case of brown algal cultures (Markham and Hagmeier 1982).

To eliminate bacteria or inhibit the growth of bacteria in cultures, antibiotics have been used singly or in combination (Table 9.1) (Spencer 1952, Provasoli 1958, Tatewaki and Provasoli 1964). Although penicillin G (potassium or sodium salt), streptomycin sulfate, and chloramphenicol were originally used by the pioneers (Table 9.1), gentamycin is preferably used in place of chloramphenicol recently (see Chapter 8). After the antibiotic mixture is sterilized by filtration (0.22 μm), 1 mL aliquots in microtubes are stored at –20°C. The antibiotic treatment procedure is described in Section 5.0. However, it must be cautioned that antibiotics may also affect the growth of the algae, so this treatment must be kept to a minimum (a few days; see Chapter 8).

**TABLE 9.1** Antibiotic mixtures used by Provasoli (1958) and Tatewaki et al. (1989). Quantities are added to 10 mL $dH_2O$.

| Antibiotic | Provasoli | Tatewaki et al. |
|---|---|---|
| Potassium penicillin G | 120,000 U | 100,000 U |
| Chloramphenicol | 500 μg | 1 mg |
| Polymyxin B sulfate | 500 μg | |
| Neomycin | 600 μg | |
| Streptomycin sulfate | | 250 mg |
| Polymyxin B sulfate | | 25,000 U |

### 4.3.5. Clonal Cultures

Unialgal cultures established from the vegetative tissue of a single individual and propagated vegetatively are generally regarded as clones, although gene mutations might occur during the course of long-term maintenance. Cultures established from the tissues of more than one individual, or the reproductive cells formed after meiosis (e.g., zoospores), include genetic variations and are not clonal. To establish clonal cultures from these nonclonal (mixed) cultures derived from reproductive cells, at an early stage of germination of reproductive cells in petri dishes (it is recommended to prepare low-density isolates for this purpose), isolate individual germlings (or young plants) into individual petri dishes or separate wells of multiwell dishes using fine pipettes or forceps under observation by stereomicroscopy. When the cultures are already overgrown and hard to distinguish and separate from each other, homogenize a small part of the mixed culture into fragments as small as one to several cells using a glass homogenizer (1 mL volume type), spread the fragments in a petri dish filled with medium, select appropriate cells under stereomicroscopy (single, healthy cells are best, but if these cells are not available, a few-celled fragment, apparently of a single individual), and isolate them into individual wells of a multiwell dish.

### 4.4. Stock Cultures

In general, unialgal cultures of macroalgae can be maintained without changing the medium for several months under lower temperatures and light intensity conditions than normal culture conditions for the alga. Generally 5–10°C is suitable for cold-water and cool-temperate species, 15°C for warm-temperate species, and 20°C for subtropical to tropical species, illuminated by 1–10% of the light intensity of normal culture conditions. A convenient way to create dim conditions is to shade the area using a smoked transparent plastic panel or to wrap the culture containers with a plastic window screen. Small vials (Fig. 9.12), test tubes with screw caps, or plastic petri dishes tightly sealed with sealing film are used for these stocks. Disposable pipettes (e.g., pastettes) can also be used for long-term storage of isolates. Because polyethylene is sufficiently permeable to oxygen and carbon dioxide, the cultures can live indefinitely in the sealed pipette. Agar plates (0.5–1% agar in seawater medium) can also be used for stock cultures, especially for filamentous algae (e.g., *Ectocarpus*). The interval required for change of the medium depends on the

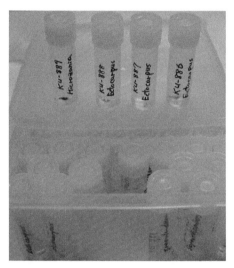

**FIGURE 9.12.** Stock cultures of unialgal strains in small vials.

species and light intensity/temperature; some brown algae can survive for more than two years without any medium change in sealed 4-mL plastic vials.

---

## 5.0. AXENIC CULTURES

### 5.1. Purification

Axenic cultures of unicellular algae are quite common; however, their use in marine macroalgae is still limited. During the 1960s–1980s, several reports were published on axenic cultures, focusing on growth regulators, effects of nutrition on morphogenesis, and the improvement of methods for establishing axenic cultures (Provasoli and Pintner 1964, 1980; Provasoli and Carlucci 1974; Fries 1975; Tatewaki et al. 1983). Obtaining axenic strains of macroalgae is difficult; however, it has merits, because the preservation of these strains is not difficult, and the interval of reinoculation into new medium can be prolonged six months or more in axenic strains compared with unialgal strains.

In this section, an outline of methods to obtain axenic cultures of macroalgae is described. To establish an axenic culture, isolation from unicellular reproductive cells (e.g., zoospores, gametes, tetraspores, and carpospores) is a standard procedure. These reproductive cells are aseptically formed in the characteristic reproductive structure, although numerous bacteria, fungi, and epiphytes are attached to the surfaces of the repro-

ductive structures and to the vegetative thalli. Although these contaminants might be killed and removed by various antibiotics, it may be impossible to completely eliminate them, because they can penetrate into the algal cell wall.

The method for isolating reproductive cells, such as zoospores and gametes of the brown and green algae, is identical to that for unialgal cultures. The liberated cells are quickly washed or diluted several times in sterilized seawater or an artificial medium. Use of glass capillary pipettes with a fine tip is convenient for isolating these cells under the stereomicroscope or inverted microscope. Then, 5–10 reproductive cells are inoculated into each screw-capped test tube containing 10 mL autoclaved artificial seawater medium. It is important to transfer several cells into each test tube, because often not all cells develop. Moreover, in some algal groups (e.g., laminarialean plants), the zoospores develop into either male or female gametophytes, and if these gametophytes grow densely, then afterward it becomes difficult to separate them. The isolation procedure for reproductive cells can be carried out in a room of normal cleanliness, so the clean bench is not always necessary during this process, because bacteria commonly found in air rarely grow in seawater.

To establish axenic cultures by washing the zoospores and gametes, a minimum of four to five washes is necessary. During sequential washes the number of swimming zooids decreases considerably, and the final isolation becomes difficult. Therefore, liberation of numerous swimming zoospores and gametes is an important condition for axenic culture. Tatewaki et al. (1989) reported a simple method using an antibiotic mixture for axenic cultures. After washing zooids 2–3 times, they are inoculated into autoclaved artificial culture medium in screw-capped glass test tubes. The next day, two to five drops of the antibiotic mixture are added to each test tube containing 10 mL culture medium, using a sterile capillary pipette in a clean bench. After antibiotic treatment for about three days, the medium in the tubes is replaced with new autoclaved artificial medium using a clean bench. Normally growing zoospores or gametes firmly attach to the glass wall, so they do not detach during the medium exchange. After about one month of culture under appropriate temperature and light conditions, several thalli of the desired algae can be detected by the naked eye. The thalli in each test tube must be checked using a sterility test for axenic cultures. When establishing axenic culture strains, it is customary to prepare 20–40 test tubes in each sample because only some of them prove axenic.

In axenic culture, many workers prefer to use artificial seawater media such as the ASP series developed by Provasoli and his co-workers (Provasoli et al. 1957, Provasoli 1958, 1963), rather than enriched seawater media (e.g., Provasoli's ES; see Chapter 3 and Appendix A).

## 5.2. Sterility Tests for Axenic Cultures

Established cultures can be tested for axenicity by several methods. The most common sterility tests use general media for marine bacteria such as ST3 medium and STP medium (Tatewaki and Provasoli 1964), commercially available Bacto Marine Broth 2216 (Difco), or DAPI (4′,6-diamidino-2-phenylindole) for directly staining bacteria on the algal thallus. Five to 10 mL of marine bacterial medium containing 0.5% agar is put into test tubes and autoclaved. One thallus or germling from each test tube obtained through the axenic culturing process is inoculated into each sterility test tube containing bacterial medium, using sterile capillary pipettes in a clean bench. The sterile capillary pipette must be changed for each sample. It is better to embed the sample into the agar medium rather than depositing it on the surface. Controls (with no inoculum, and unialgal culture samples with bacterized or nonautoclaved seawater) must be prepared simultaneously. These test tubes are maintained at 15–25°C for 2 weeks (it is noteworthy that optimum growth temperatures are different among bacterial strains). If bacteria are present, their growth is evident as a white or light yellow cloud around the algal thallus. Axenic strains of the algae are maintained in appropriate artificial culture media (e.g., $ASP_7$ or $ASP_{12}NTA$) (Provasoli 1963). Fluorescent dyes (e.g., DAPI) are useful for revealing bacteria attached on the surface of cultured thalli. Algal samples are fixed and stained in autoclaved (or preferably, sterile-filtered) seawater containing 1% formalin and 0.5 µg/mL DAPI, and observed under a fluorescence microscope.

## 6.0. Culture Media

Various types of enriched seawater media have been used for culturing macroalgae, such as Erdschreiber medium (Schreiber 1927, Føyn 1934, Gross 1937), Grund medium (von Stosch 1963, 1969), ES medium (Provasoli 1968), PESI medium (Tatewaki 1966), or SWM (McLachlan 1964, Chen et al. 1969). Artificial seawater media such as the ASP series (Provasoli et al. 1957, Provasoli 1963, Iwasaki 1967) are also frequently used. Several media are described in Chapter 3 and recipes are listed in Appendix A.

## 7.0. Acknowledgment

The authors are grateful to Eric C. Henry for helpful advice and improving the manuscript.

## 8.0. References

Berger, S., and Kaever, M. J. 1992. *Dasycladales: an Illustrated Monograph of a Fascinating Algal Order.* G. Thieme, Stuttgart, 247 pp.

Bold, H. C. 1942. The cultivation of algae. *Bot. Rev.* 8:69–138.

Chapman, A. R. O. 1973. Methods for macroscopic algae. In: Stein, J. R., ed. *Handbook of Phycological Methods. Culture Methods and Growth Measurements.* Cambridge University Press, Cambridge, pp. 87–104.

Chen, L. C. -M., Edelstein, T., and McLachlan, J. 1969. *Bonnemaisonia hamifera* Hariot in nature and in culture. *J. Phycol.* 5:211–20.

Enomoto, S., and Ohba, H. 1987. Culture studies on *Caulerpa* (Caulerpales, Chlorophyceae) I. Reproduction and development of *C. racemosa* var. *laetevirens. Jpn. J. Phycol.* 35:167–77.

Føyn, B. 1934. Lebenszyklus, Cytologie und Sexualität der Chlorophyceae *Cladophora subriana* Kützing. *Arch. Protistenk.* 83:1–56.

Fries, L. 1975. Some observations on the morphology of *Enteromorpha linza* (L.) J. Ag. and *Enteromorpha compressa* (L.) Grev. in axenic culture. *Bot. Mar.* 18:251–3.

Gross, F. 1937. Notes on the culture of some marine plankton organisms. *J. Mar. Biol. Ass. U.K.* 21:753–68.

Iwasaki, H. 1967. Nutritional studies of the edible seaweed *Porphyra tenera.* II. Nutrition of conchocelis. *J. Phycol.* 3:30–4.

La Claire, J. W., II. 1982. Cytomorphological aspects of wound healing in selected Siphonocladales (Chlorophyceae). *J. Phycol.* 18:379–84.

Lewin, C. J. 1966. Silicon metabolism in diatoms. V. Germanium dioxide, a specific inhibitor of diatom growth. *Phycologia* 6:1–12.

Markham, J. W., and Hagmeier, E. 1982. Observations on the effects of germanium dioxide on the growth of macro-algae and diatoms. *Phycologia* 21:125–30.

McLachlan, J. 1964. Some considerations of the growth of marine algae in artificial media. *Can. J. Microbiol.* 10:769–82.

McLacklan, J., Chen, L. C.-M., and Edelstein, T. 1971. The culture of four species of *Fucus* under laboratory conditions. *Can. J. Bot.* 49:1463–9.

Norby, Ø., and Hoxmark, R. C. 1972. Changes in cellular parameters during synchronous meiosis in *Ulva mutabilis* Føyn. *Exp. Cell Res.* 775:321–8.

Phillips, J. A., Clayton, M. N., Maier, I., Boland, W., and Müller, D. G. 1990. Sexual reproduction in *Dictyota diemensis* (Dictyotales, Phaeophyta). *Phycologia* 29:367–79.

Provasoli, L. 1958. Effect of plant hormones on *Ulva*. *Biol. Bull.* 114:375–84.

Provasoli, L. 1963. Growing marine seaweeds. In: De Virville, D., and Feldmann, J., eds. *Proceedings of the Fourth International Seaweed Symposium*, Pergamon Press, Oxford, pp. 9–17.

Provasoli, L. 1968. Media and prospects for the cultivation of marine algae. In: Watanabe, A., and Hattori, A., eds. *Cultures and Collection of Algae. Proc. U.S.—Japan Conf. Hakone, Sept. 1966. Jap. Soc. Plant Physiol.* pp. 63–75.

Provasoli, L., and Carlucci, A. F. 1974. Vitamins and growth regulators. In: Steward, W. D. P., ed. *Algal Physiology and Biochemistry*. Blackwell, Oxford, pp. 741–87.

Provasoli, L., and Pintner, I. J. 1964. Symbiotic relationship between microorganisms and seaweeds. *Am. J. Bot.* 51:681.

Provasoli, L., and Pintner, I. J. 1980. Bacteria induced polymorphism in an axenic laboratory strain of *Ulva lactuca* (Chlorophyceae). *J. Phycol.* 16:196–201.

Provasoli, L., McLaughlin, J. J. A., and Droop, M. R. 1957. The development of artificial media for marine algae. *Arch. Mikrobiol.* 25:392–428.

Schreiber, E. 1927. Die Reinkultur von marinem Phytoplankton und deren Bedeutung für die Erforschung der Produktinsfähigkeit des Meerwassers. *Wiss. Meeresuntersuch. Abt. Helgoland N. F.* 16(10):1–34.

Spencer, C. P. 1952. On the use of antibiotics for isolating bacteria-free cultures of marine phytoplankton organisms. *J. Mar. Biol. Ass. U.K.* 31:97–106.

Tatewaki, M. 1966. Formation of a crustaceous sporophyte with unilocular sporangia in *Scytosiphon lomentaria*. *Phycologia* 6:62–6.

Tatewaki, M., and Nagata, N. 1970. Surviving protoplasts *in vitro* and their development in *Bryopsis*. *J. Phycol.* 6:401–3.

Tatewaki, M., and Mizuno, M. 1979. Growth inhibition by germanium dioxide in various algae, especially in brown algae. *Jpn. J. Phycol.* 27:205–12 (in Japanese with English abstract).

Tatewaki, M., and Provasoli, L. 1964. Vitamin requirements of *Antithamnion*. *Bot. Mar.* 6:193–203.

Tatewaki, M., Provasoli, L., and Pintner, I. J. 1983. Morphogenesis of *Monostroma oxyspermum* (Kütz.) Doty (Chlorophyceae) in axenic culture, especially in bialgal culture. *J. Phycol.* 19:409–16.

Tatewaki, M., Wang, X. -Y., and Wakana, I. 1989. A simple method of red seaweed axenic culture by spore-washing. *Jpn. J. Phycol.* 37:150–2.

von Stosch, H. A. 1963. Wirkung von Jod und Arsenit auf Meeresalgen in Kultur. In: De Virville, D., and Feldmann, J., eds. *Proceedings of the Fourth International Seaweed Symposium*, Pergamon Press, Oxford, pp. 142–50.

von Stosch, H. A. 1969. Observations on *Corallina, Jania* and other red algae in culture. In: Margalef, R., ed. *Proceedings of the Sixth International Seaweed Symposium*, Subsecretaria de la Marina Mercante, Direccion General de Pesca Maritima, Madrid, pp. 389–99.

# PERPETUAL MAINTENANCE OF ACTIVELY METABOLIZING MICROALGAL CULTURES

MAIKE LORENZ

*Albrecht-von-Haller-Institut für Pflanzenwissenschaften, Abteilung Experimentelle Phykologie und Sammlung von Algenkulturen, Universität Göttingen*

THOMAS FRIEDL

*Albrecht-von-Haller-Institut für Pflanzenwissenschaften, Abteilung Experimentelle Phykologie und Sammlung von Algenkulturen, Universität Göttingen*

JOHN G. DAY

*Culture Collection of Algae and Protozoa, Scottish Association for Marine Science, Dunstaffnage Marine Laboratory*

## CONTENTS

Key Index Words: Microalgae, Cultures, Culture Medium, Culture Containers, Culture Rooms, Management of Culture Collections, Transfer Techniques, Maintenance Conditions

## 1.0. INTRODUCTION

The most common way to conserve microalgal cultures is perpetual maintenance under controlled environmental conditions. Routine serial subculturing is performed using aseptic microbiological technique and involves transferring an inoculum from a late log/stationary phase culture into fresh, presterilized medium.

This leads to metabolically active cultures that can be used at short notice. The objective is to retain a healthy, physiologically, morphologically, and genetically representative population. A key factor to consider is that different ages of subcultures may provide different stages of the life cycle (e.g., orange/red aplanospores in addition to green dividing and motile cells in early stationary phase cultures of *Haematococcus pluvialis* Flowtow).

The main limitations of perpetual transfer are the selective and artificial nature of the media and incubation regimens with respect to native ecological conditions. Laboratory conditions can, in extreme cases, lead to the loss of important morphological features and physiological traits. Examples of instability include the size reduction of diatom frustules (Jaworski et al. 1988), retention/loss of spines in *Micractinium pusillum* Fresenius, and loss of normal pigment composition in numerous algae (Warren et al. 2002). Further limitations include the possibility of contamination of axenic cultures and the possibility of mislabeling or other handling mistakes. Routine serial subculturing is a labor- and consumable-intensive process and certainly limits the capacity of workers to maintain large numbers of strains. To circumvent the disadvantages of routine subculturing, alternative approaches have been developed for the *ex situ* conservation of algal and cyanobacterial cultures, especially cryopreservation (see Chapter 12).

The following section discusses major considerations necessary for successful long-term maintenance, including transfer techniques, maintenance conditions, and quality control/assurance policies and procedures. The section focuses on various aspects of routine subculturing of strains and considers that there is certainly no single, best way to grow or maintain algal cultures. The basics of perpetual maintenance have not changed much in the last decades (Pringsheim 1946, Venkataraman 1969, Stein 1973, McLellan et al. 1991, Day 1999, Warren et al. 2002, Richmond 2004). Some specialized equipment is commercially available now, but in previous years it had to be homemade. However, the best results may still be obtained by employing simple equipment and facilities.

Numerous organizations, institutions, and individuals maintain collections of algal strains. These may be subdivided into the following categories:

1. Diverse collections for education and research
2. Limited-range collections of well-defined species or strains for research or practical studies
3. Collections of genetically well-defined and stable strains, often of a single or few species, for molecular studies, biotechnology development, etc.

In addition, most readers have either obtained their strains from another collection or have recently isolated them from nature.

---

## 2.0. TRANSFER TECHNIQUES

Before starting a transfer, it is extremely important to carefully check the labels to avoid mistakes. We recommend labeling the vessels of sterile culture medium prior to inoculation. For recommendations on types of culture labels, see Section 6.0. After labeling, the uninoculated vessels should be organized in parallel order with the vessels containing established cultures, and the corresponding labels should be compared.

Rigorous microbiological methods must be used, following standard guidelines for aseptic conditions and techniques (Isaac and Jennings 1995; see Chapter 5). Manipulations and transfers should be performed in a transfer-hood/laminar flow cabinet, if possible. All containers, media, caps, pipettes/loops, etc. should be sterilized before use (see Chapter 5). All containers should be opened for the shortest time possible to limit the risk of contamination. Transfer equipment includes a Bunsen burner for flaming, an inoculation wire loop for agar cultures (Fig. 10.1a), and a hook or small lancet for severing filamentous algal mats. Presterilized plastic loops and Pastettes (soft disposable polyethylene pipettes) are an efficient alternative. Transfer of liquid cultures is done either by pouring (Fig. 10.1b) or by using a sterile pipette or Pastette. Pipettes need to be plugged with nonabsorbent cotton wool at the wide end before sterilization to prevent potential contamination. This is vital, because rubber bulbs and other similar devices, which are attached to Pasteur pipettes, may contain or even have actively growing biofilms of contaminating organisms. The cotton plug prevents these from passing through the pipette to the culture. Mouth pipetting is generally discouraged and has been banned in many laboratories, although only a small number of algae have the potential to cause problems if they were inadvertently ingested.

### 2.1. Transferring Agar Cultures

To transfer algal colonies from agar, an inoculation loop is sterilized by holding it in a Bunsen flame at such an angle that the whole wire glows red, and then is allowed to cool in air or by placing it on the agar at a site where no algal colonies are growing. With petri dishes containing nutrient agar the lid is lifted at a slight angle to allow inoculations to be made with minimal risk of contamination (see Fig. 10.1a; Isaac and Jennings 1995). In case of glass test tubes, conical (Erlenmeyer) flasks, and other glass tubes and bottles in which the tops are usually sealed with cotton wool or siliconized bungs, the vessels are opened carefully in close proximity to a lit Bunsen burner to reduce the chance of contamination (see Fig. 10.1b). When several cultures are to be transferred in one step, it helps to loosen the caps of the vessels before the transfer process. The established culture is uncapped, the neck of the vessel is flamed, and a portion of a lawn of algae is removed. With a little practice, it is possible to remove and hold the cap or bung with the little finger of one hand while manipu-

lating the pipette or loop with the other fingers of the same hand (see Fig. 10.1a). The inoculum is usually 1–10% (v/v) of the original culture, but some dinoflagellates, *Synechococcus* and *Prochlorococcus*, require inocula of up to 25% (v/v).

With agar slants, the material is streaked from the bottom of the slant in loops toward the top. It is important to allow even distribution (no clumping) so that the newly inoculated cells receive even illumination. Remember that a reusable metal inoculation loop must be flamed between each streak made. With agar cultures, colonies may stick rather firmly to the agar surface (e.g., some benthic diatoms) or may even grow into the agar (e.g., some filamentous cyanobacteria). These cannot be transferred without removing agar that contains the algal material. Sometimes it is useful to

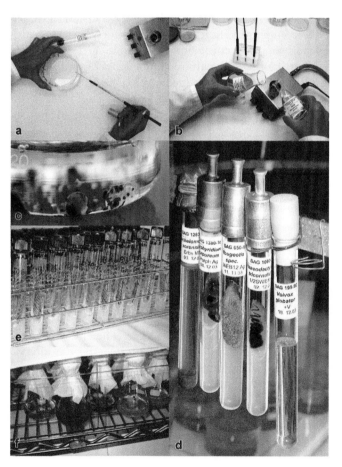

FIGURE 10.1. Transfer techniques and various culture vessels. (a) Transfer from an agar slant to a petri dish containing agarized medium. The right hand holds a platinum wire loop that has been sterilized in the flame of the gas safety burner (right upper-part) and the cap of the culture tube, and the left hand holds the lid of the petri dish and the tube with the current culture. Note that the lid of the petri dish is lifted at a slight angle to allow inoculations to be made with minimal risk of contamination. (b) Transfer of liquid cultures using Erlenmeyer flasks in close proximity to a lit gas safety burner. Note that each hand also keeps the cotton bung by which the culture vessel was capped (not visible for the right hand) and that there is a rim at the vessels to avoid drops. (c) Culture of crustaceous rhodophyte *Hildenbrandia rubra* (Summerfeldt) Meneghini SAG 18.96 where the algae also grow on glass beads as a carrier substrate to prevent damage and loss of material during transfer (SAG culture collection). (d) Glass culture tubes with different caps. The liquid culture at the left is capped with a cotton bung, covered with wax paper, and fixed with a rubber band. The three agar slants in the middle are capped with aluminum caps with handles. The liquid culture at the right is capped with a silicone bung. Note that each of the tubes is fixed with a metal wire (right end) on a stretched metal band for exposing them to the light source. The liquid culture at the left end is a biphasic culture with soil at the bottom. Note the information given on the label (SAG culture collection). (e) Cultures on agar in thick-walled glass tubes with screw caps. Note that colonies were streaked in loops over the agar surface to allow even distribution and illumination of the algal colonies (Pasteur culture collection [PCC]). (f) Cultures of cyanobacteria at different ages grown in Erlenmeyer flasks that are capped by cotton plugs, covered with greaseproof paper, and fixed with a rubber band. The cultures have been covered with filter paper, which acts as a neutral filter, to reduce light intensity. Note that the flasks are kept in polystyrene boxes to prevent contamination and stand on a shelf made of metal grid (PCC culture collection).

excise blocks of the agar using a small lancet or a sterile scalpel. The agar block is then placed upside down on the surface of the fresh agar. In addition, moving the agar block over the fresh agar surface or cutting it into smaller pieces may help to better distribute the cells. With old agar cultures, in which the algal colonies are often difficult to remove, it is recommended that the cultures be overlayered with fresh liquid medium for several hours prior to transfer. The algal cells can then be easily transferred with a wire or pipette.

To transfer from agar to liquid medium, it may be best to cut out a small piece of agar with algal colonies and drop it into the liquid medium. To transfer from liquid to agar media, two or three drops of the culture are placed on the agar surface. The cells are distributed by moving the agar plate or by placing the drops at the top of the agar slant in a tube and letting them flow down. Note that agar plates need to be stored and incubated upside down to prevent excessive moisture forming on the agar or culture surface.

## 2.2. Transferring Liquid Cultures

Detailed sterile techniques for transferring liquid cultures are provided in Chapter 5. Briefly, a few drops or milliliters of the established culture are transferred to a culture vessel with the fresh medium (see Fig. 10.1b). All tube and vessel necks should be flamed. Keep numerous sterile, empty tubes and vessels with caps in reserve should you accidentally drop a tube or cap. With liquid cultures of many algae (e.g., coccoid green algae, cyanobacteria), the culture is mixed first and then transferred. However, vigorous mixing may damage some other algae. Some fragile diatoms (e.g., *Thalassiosira* and *Rhizosolenia*) may rupture when the flask is lifted. Transfer of cells from the bottom of the vessel is sometimes necessary (e.g., *Polytoma* resting stages may be present at the bottom and develop into normal trophic cells after transfer to fresh medium). Many planktonic algae grow with an uneven distribution in the water column (e.g., gas vacuolated cyanobacteria). Many euglenophytes are best grown in biphasic soil-water medium, and they grow primarily just above the soil.

## 2.3. Transferring Filamentous Algae

Some filamentous algae can be transferred by pipette or pouring as previously described, whereas others require special treatment. It may be necessary to pour the estab-

lished culture tube or vessel into a sterile petri dish, sever the filaments with a small lancet or hook, and then transfer some material to the new culture medium. Some filamentous green algae, grown in soil-water medium, are best planted by using a wire or similar implement to push part of the tuft firmly into the soil (e.g., *Oedogonium*, *Spirogyra*). If not planted, then they may float as a scum on the liquid surface (Belcher and Swale 1982).

## 2.4. Special Problems

Some benthic diatoms and crustaceous algae adhere very firmly to the walls of the culture vessels so that damage and loss of material may occur when they are transferred. These cultures can often be grown on a carrier substrate that can be easily transferred. Small (ca. 5 mm) glass beads, which can be purchased in handcraft shops, or cover slips are suitable carrier substrates (Fig. 10.1c). The culture containers are sterilized with the beads and culture medium inside.

---

## 3.0. MAINTENANCE CONDITIONS

The maintenance of metabolically active algae usually has one of three objectives: conservation of stock cultures, achievement of a specific morphological and physiological status, or mass culture (>200 ml liquid). For the latter two objectives, optimal growth conditions are required, and these vary greatly between different microalgae. However, for stock cultures maintained by routine serial subculture, it is often desirable to use suboptimal temperature and light regimens; these factors may be similar for different algae. The aim is to minimize handling and transfers by extending the interval between subcultures. In addition, specific media can extend the transfer-interval (see Section 3.1). Freshly inoculated cultures are often incubated under optimal conditions for a short period to obtain sufficient biomass and refresh the strain, after which the strain is maintained suboptimally.

## 3.1. Choosing a Culture Medium

An alga that is poorly adapted to a particular medium is under stress and may eventually develop (sometimes irreversibly) an altered morphology. Examples include the loss of colonial habit in some volvocalean algae and *Pediastrum*, loss of functional flagella in *Chlamydomonas*, and loss of surface features of some cyanobacteria. Long-term culturing, under conditions very different from its natural environment, is likely to select for genetic variants adapted to the artificial culturing environment. For example, when cultured on medium containing high levels of inorganic nitrogen, heterocystous cyanobacteria may express poor or no heterocyst formation.

Culture conditions may dramatically change with time in continuous culture even when the external environment remains unchanged and the algal culture has not exhausted the supply of any essential nutrient. For example, pH often changes unless an appropriate buffer is present, and some nutrients are oxidized or otherwise gradually altered, especially during illumination. Intervals between transfers can sometimes be extended by identifying and compensating for these factors (e.g., by adding organic compounds [see the following section]). With long-term maintenance of cultures, effects can become obvious that are not evident during short periods. For example, some microalgae have an absolute requirement for vitamin $B_{12}$ but at such low concentrations that they can be grown for a number of generations in the absence of $B_{12}$ in the prepared medium. Some diatoms eventually become too small during continuous vegetative propagation to remain viable, so a portion of the culture must be allowed to reproduce sexually to regenerate large cells. Some strains (e.g., many Dasycladales) must go through periodic sexual reproduction to propagate indefinitely. For many brackish and marine strains, reduced salinity seawater media can be used to avoid hypersaline conditions following evaporation of some of the culture medium.

If different agar supplies are used, quality and purity may have a significant effect on the growth of some algae. Quality of water and of chemical reagents used to prepare media may also influence the vigor of some algal strains. These, and other quality-control problems, can sometimes cause a seemingly healthy culture to crash suddenly.

An important question is whether a particular strain would best be maintained for long periods in liquid medium or in agar culture. Numerous factors determine the choice. Solid media are often preferred because they are easier to handle during transfer and thus lower the risk of contamination. However, many flagellates and other planktonic species do not grow well on or in agar, whereas some edaphic and aquatic benthic microalgae do not grow well in liquid medium. This illustrates the importance of understanding the natural habitats of the organisms. For descriptions of various media, see Chapters 2 and 3 and Appendix A.

Another important issue is whether a defined mineral medium or a medium with organic supplements is most suitable for long-term culturing. It is generally desirable to provide a mineral medium for culturing nonaxenic photoautotrophic strains to minimize the culture density of heterotrophic contaminants. On the other hand, strains that must be kept rigorously axenic are sometimes best cultured in the presence of rich organic supplement so that a nonphotosynthetic contaminant can be detected soon after a culture is infected. The supplement of organic compounds or vitamins ($B_1$-thiamine · HCl or $B_{12}$-cyanocobalamin) often helps retain healthy growth of the cultures. Also, biphasic soil-water medium may be advantageous for healthy growth over long periods for nonaxenic strains, in particular for filamentous green algae and euglenoids. The addition of soil extract often helps ensure that algae retain their typical morphology (e.g., in coccoid green algae); without soil extract, cells accumulate starch or oil droplets. Some strains (e.g., some colorless euglenoids) cannot be grown in defined mineral media and require either soil-water medium (bacterized cultures) or media containing organic nitrogen and carbon sources (axenic cultures).

Large culture collections have a diversity of algal cultures that necessitate the use of several to many culture media. To reduce work, they often use only a few standard media for long-term cultivation (Starr and Zeikus 1993, Schlösser 1994, Andersen et al. 1997, Watanabe et al. 2000, Anon 2001). For example, the Sammlung von Algenkulturen at the University of Göttingen (SAG) uses ES (Erddekokt und Salze) basal medium (containing soil extract, salts, and micronutrients) for various taxonomic groups of green algae, freshwater xanthophytes, and cyanobacteria. However, one should always consider the possibility of deleterious effects to strains that are not well adapted to a standard culture medium.

## 3.2. Light and Temperature

Standard light intensities between 10–30 μmol photons $\cdot \text{m}^{-2} \cdot \text{s}^{-1}$ have proved appropriate in combination with subdued temperatures for long-term culturing of most microalgal taxa. Over-illumination is a widespread mistake in the perpetual maintenance of cultures. Not only can excessive light result in photo-oxidative stress in some algae, localized heating may also be problematic. Light and dark photoperiods are required for the maintenance of most cultures. Some algae (e.g., many tropical open-ocean coccolithophorids) may be killed by continuous light (Price et al. 1998, Graham and Wilcox 2000). In most culture collections the light : dark regimens vary between 12 : 12 to 16 : 8 hour light : dark. Inappropriate light : dark regimens may lead to unwanted photoperiodic effects (e.g., a short daylength period causes cyst formation in marine dinoflagellates like *Lingulodinium polyedrum* [Stein] Dodge [Balzer and Hardeland 1991]), and the cysts are hard to germinate under standard culture conditions.

Algae with phycobilisomes may prefer low light intensities (i.e., ~10 µmol photons · m$^{-2}$ · s$^{-1}$). Some other algal strains (e.g., most dinoflagellates) often need higher light intensities (~60–100 µmol photons · m$^{-2}$ · s$^{-1}$). Colorless algae (e.g., *Astasia, Polytomella, Prototheca*) are best kept in a closed cupboard, but otherwise they have the same maintenance regimen requirements as their photoautotrophic relatives. To culture organisms from extreme environments, specialist literature should be consulted (Elster et al. 2001).

Temperature is a major factor and should be carefully controlled; conditions may vary greatly within a culture facility or laboratory and should be periodically monitored. In general, temperature stability should be kept to ± 2°C when possible; freshwater strains are generally more tolerant of temperature variability than are marine strains. Many freshwater microalgal cultures can be maintained effectively at temperatures between 15–20°C. However, some larger service culture collections (e.g., Culture Collection of Algae at the University of Texas [UTEX]) maintain almost all strains at 20°C. Incubation temperatures higher than 20°C should be combined with increased light intensities to prevent photo-inhibition or damage. Therefore, temperatures of more than 20°C are mostly inappropriate for maintaining stocks at extended transfer cycles (exceptions include thermophilic strains and tropical marine algae). Furthermore, as maintenance temperatures increase, evaporation increases. The evaporation of the medium effectively determines the duration of their transfer cycles for many robust strains of green algae and cyanobacteria.

### 3.3. Transfer Intervals

For routine maintenance, the goal is to subculture the organism at the end of its exponential growth phase. At the large service culture collections, the shortest trans-

fer cycle is 1–2 weeks, and this is only applied to a small number of sensitive strains. Some cultures of green algae and cyanobacteria, maintained on agar slopes that are kept at low light and 10°C, are transferred only once every 6 months. A safe transfer interval can be estimated by using one quarter the time that a strain can maximally survive.

Transfers into and onto fresh medium refresh the performance of a strain. After inoculation into fresh medium many cultures require incubation under optimal conditions before they are relocated into the stock under suboptimal conditions. In larger service culture collections, the post-transfer period is an important step in regular quality-control assessment, because the cultures are visually checked for good growth and contamination before they are moved back to their suboptimal location. Optimal post-transfer conditions generally mean an elevated light and temperature regimen (e.g., 20°C or more), and light intensities of up to double the stock maintenance intensities (i.e., 20–60 µmol photons · m$^{-2}$ · s$^{-1}$). This treatment ensures that sufficient biomass of a strain is available for it to grow on subsequent transfers, because only very little growth occurs under suboptimal maintenance conditions. In addition, as perpetual maintenance over longer periods can lead to poor morphological and physiological performance of some strains, a short interval of maintenance at optimal growth conditions (a few days up to 3 weeks) may be employed to refresh the culture after inoculation. For example, some desmid strains exhibit morphological deterioration after several prolonged transfer intervals. The normal phenotype may be restored after transfer on fresh medium and maintenance under optimal conditions. In some cases, repeated transfers at shorter intervals and optimal growth conditions, often accompanied by a change of culture medium (e.g., from agarized to liquid medium) may be needed to restore the typical phenotype.

### 3.4. Establishing Maintenance Conditions for New Isolates

A newly isolated strain may be difficult to maintain, because no long-term culture strategy is yet fully known. When establishing a new, long-term maintenance protocol one should, if possible, simultaneously test different temperature and light regimens to find conditions where culture growth is reliable and subculturing is minimal (i.e., lower light and temperature conditions). Keep records of these trials, because they are a source of information for later consultation should a culture suddenly grow poorly. In some cases, a newly isolated alga grows very well for one or two subcultures but then dies. This most commonly occurs when the alga is grown in an artificial medium and some growth factor (e.g., trace metal, vitamin) is absent (see Chapters 2 and 3). It is often prudent to grow the new isolate in several culture media so that if the alga dies in one medium, then it may survive in others. Knowledge about the maintenance of closely related algae is often valuable, and catalogs or Web sites of the large service collections are a useful source of this information. For lichen photobionts, their transformation from the symbiotic to the culture state is facilitated when glucose is added to the medium (1% [w/v] final concentration).

## 3.5. Converting Cultures from Maintenance Mode to Active Use

Changes in culture conditions are required when transferring a stock-culture for subsequent use in the classroom or for research. A transfer to fresh medium and optimal conditions is required. In most cases the same medium can be used; however, to ensure typical morphology, a change of media may be required. Many flagellates must be cultured in liquid media to induce full motility (i.e., a transfer from agar slant to an Erlenmeyer flask with the liquid medium). Medium with low or no nitrogen is required to induce optimal heterocyst formation in many cyanobacteria strains. Changes in culture conditions are required to induce reproductive stages, and one needs to experiment with a particular strain for awhile to find out which conditions are most appropriate. For example, *Tribonema* filaments maintained in flasks at 12°C were transferred to a solid medium and incubated at 16°C with an increased light period. After 3 weeks, filaments were submerged with fresh liquid culture medium, and zoospore formation occurred after 1–2 days (1–3 hours into light phase) (Lokhorst 2003).

In general, higher densities and more rapid, uniform growth may be obtained when agitating liquid cultures. For some colonial flagellate and coccoid green algae, agitation and aeration may be necessary to obtain typical morphology (e.g., *Eudorina, Pediastrum*). However, other, more delicate strains (e.g., many dinoflagellates and some diatoms) cannot tolerate vigorous mixing. For long-term cultivation, aeration and agitation is generally considered inappropriate because of the increased risk of contamination and the need for short transfer intervals.

## 4.0. CULTURE CONTAINERS

Culture vessels (e.g., glass test tubes and flasks, plastic tissue culture flasks) should allow uniform illumination. Flasks provide good growth, but they occupy more space than test tubes (Fig. 10.1d–f). Test tube openings must be wide enough for easy access when transferring cultures. Reusable glassware and caps are labor-intensive, because they must be washed (see Chapter 5), and disposable plastic reduces labor but is expensive. Some Conjugatophyceae and most planktonic freshwater diatoms are cultured best in Erlenmeyer flasks (see Fig. 10.1f) and some dinoflagellates grow best in plastic tissue-culture flasks. For many algae, miniaturized cultures in 12–96 multiwell plates work well, and they can be used conveniently with an inverted microscope.

The top must fit tightly to prevent contamination and limit evaporation while allowing gas exchange. Various cotton or silicon plugs, metal or glass caps, and screw caps are available. Glass tubes used with cotton wool plugs should have a small rim, whereas rimless tubes are used for metal caps. To minimize evaporation, cotton plugs may be covered with a piece of grease-proof paper fixed with a rubber band (see Fig. 10.1d,f) or by a loose cap of aluminum foil. To reduce evaporation, *Astasia* is maintained at the Culture Collection of Algae and Protozoa (CCAP) with the plug covered with Clingfilm or Parafilm, but this sometimes leads to fungal damping or molding. Siliconized plugs are ideal, because they ensure tight closure of the tubes and minimal evaporation without extra sealing and offer better gas exchange than cotton (see Fig. 10.1d). Aluminum metal caps are used because they are unbreakable, can easily be autoclaved, and are available in different colors, which can be advantageous for labeling different culture media (see Fig. 10.1d). Metal caps usually have a spring that keeps the cap fixed on the outside of the tube. Metal caps with handles are preferred to avoid burn injuries while inoculating. Although they prevent evaporation better than cotton plugs, sealing with foils or films (e.g., Parafilm) is required to fully protect the cultures against contamination, in particular, from mites. If sealed, then evaporation can be minimal and some robust cultures may be maintained for several years without transferring. However, the growth of some strains is inhibited by the poor gas exchange. Evaporation from screw caps is also minimal, and they may be the best choice for sealing cultures (see Fig. 10.1e). For shipping cultures, tubes with screw caps are recommended. The material of any closure has to be checked carefully for toxicity and ability to withstand autoclaving. Plastic tissue-culture flasks often incorporate two-step screw caps that allow gas exchange in one position but are gas- and liquid-proof in the second position.

The use of sterile Pastettes provides another efficient and simple way for culturing and shipment. After drawing the culture liquid into the Pastette and shaking the liquid into the bulb, the opening is heat sealed (pinched after heating). The Pastette allows light penetration and gas exchange. They are nonbreakable and can be used to ship cultures. The cells are removed by snipping open the tip and expelling the cell culture.

## 5.0. EQUIPMENT AND CONDITIONS FOR PERPETUAL MAINTENANCE

Dedicated culture facilities can vary from a simple shelf at a north-facing window (in the northern hemisphere), an illuminated incubator, or a large walk-in culture room (Fig. 10.2). Normally a cabinet with temperature and light controls is adequate for a small number of cultures. For large numbers of cultures commercially available growth chambers and incubators are ideal but very expensive. Refrigerators with glass doors as used in grocery stores (e.g., for cooling beer) may be used as a cheap alternative (see Fig. 10.2a).

Racks and shelves should allow uniform illumination and temperature control and give easy access to the cultures. Varying numbers of cultures can be held together in test racks (see Fig. 10.2a,b). Shelves made of glass or metal are recommended, because they can be cleaned or sterilized easily (see Figs. 10.1f, 10.2a–c). Shelves made of a metal grid or mesh that allow air circulation are best (see Figs. 10.1f, 10.2a–f). Light sources should be above or at the side of shelves to prevent overheating (see Fig. 10.2c). Shelves without attached light sources should be fixed at right angles to the light sources to allow changing the distance of the cultures from the illumination. Suspending tubes in a row by hanging them on a stretched wire or metal band is an old but simple and effective alternative to shelves and racks (see Figs. 10.1d, 10.2d).

Air-conditioning best controls the temperature; the compressors should be removed from the culture room and ideally placed outside the building. A pre-set visual and audio-alarm system should be used to signify changes of more than 2–4°C. When overheating occurs, the alarm should turn off the lights automatically to avoid further overheating (Starr 1973). For stenother-

**FIGURE 10.2.** Various culture facilities. **(a)** Racks with culture tubes in a refrigerator with glass doors as used in grocery stores (Provasoli-Guillard National Center for Culture of Marine Phytoplankton [CCMP] culture collection). **(b)** Culture tubes in racks made from Plexiglas and other culture containers on glass shelves. Illumination is from the ceiling of the culture room (SAG culture collection). **(c)** Walk-in culture room with shelves made of metal mesh that allows air circulation. Light sources are fixed at the walls behind the shelves. Different light intensities are provided by varying distances from the fluorescent tubes (UTEX culture collection). **(d)** Culture room with a mixture of light sources. Cultures in the cupboard are illuminated by natural light through north-faced (*left*, behind the cupboard) and east-faced (*right*) windows plus artificial light from fluorescent tubes in the center of the room. Different light intensities are provided by varying distances from the light sources (SAG culture collection).

mic strains, strict temperature control usually requires commercially available culture chambers. A more simple solution is to immerse cultures in a water-bath, but this has a high risk of contaminating the culture and is not generally practicable for long-term maintenance.

Indirect natural daylight, solely artificial light, or a mixture of both can be used for the maintenance of cultures (see Fig. 10.2d). Illumination should be diffuse rather than focused. Direct sunlight should never fall directly on the cultures, because it invariably results in localized heating and other stresses that may lead to culture death. For obtaining reproducible results under artificial light, a 1 : 1 mixture of cool white and warm white fluorescent tubes works best in many culture collections. Some light sources that are designed for plant growth (e.g., xenon burners in growth cabinets for higher plants) are too bright for maintenance of algal cultures. The maximum light emission of tungsten filaments is shifted toward the red part of the spectrum in aging bulbs. This can be inappropriate for some strains and has been observed by the authors to result in premature senescence in some algal cultures. Any illumination unit produces heat and, if uncontrolled, may lead to unwanted temperature fluctuations during light : dark cycles. An optimal illumination setup allows different light intensity regimens with a single light source by varying the distance from the lights to the cultures (see Fig. 10.2a–d).

Humidity control is needed for not only preventing excessively fast evaporation of cultures but also reducing the risk of contamination by other microorganisms. In particular, fungal molds frequently grow in stock-culture rooms at condition of 60% or higher humidity. The use of air-conditioning in culture rooms may stabilize the humidity sufficiently to prevent problems. If cultures are maintained at high temperatures, then it may be necessary to humidify the air to reduce excessive evaporation of the culture medium in cultures maintained for several months.

Transfers must be performed aseptically and should be undertaken in dedicated work areas. An easy-to-clean compartment with minimal air turbulence that is separated from the rest of the lab is usually sufficient. The use of transfer hoods (commercially available laminar flow cabinets) with filter systems results in less contamination. The filter system in transfer hoods removes dust and bacteria, resulting in clean airflow across the working area and reducing the likelihood of contamination. In some transfer hoods, air circulation is needed for at least one hour before use, otherwise air turbulence can lead to an increased risk of contamination (always consult the manufacturers' recommendations). Many phycologists prefer still hoods (minimal air movement), and these can be homemade (Guillard 1995). Transfer cabinets should provide enough space to handle bigger containers or even a microscope and be equipped with a Bunsen burner for flaming of transfer hooks and needles and the openings of tubes and vessels. Cabinet surfaces should be sterilized using a solution of 70% (v/v) ethanol prior to use. In addition, a germicidal ultraviolet lamp is useful to help sterilize the surfaces of the transfer cabinet. Potentially pathogenic strains (e.g., *Prototheca*) and toxin-producing

cyanobacteria, diatoms, and dinoflagellates should also be treated with caution; in some countries, special transfer cabinets may be required. Culture workers should be aware of local health and safety legislation, and when in doubt they should consult their health or safety officer.

Cleanliness is important regardless of the size of the culturing facility. Clutter on workbenches and shelves allow the accumulation of undetected dust that harbors many kinds of spores and cysts, leading to contamination of cultures. Also, disorganization causes distractions and may result in confusion during handling of cultures, increasing the risk of errors when transferring algae from one culture vessel to another. All surfaces should be easy to clean and frequently wiped with 70% (v/v) ethanol. Field samples, soil, or other biological material must be kept tightly closed and far away from the algal cultures and culture media. Movement of personnel should be minimized in walk-in culture facilities. Adhesive foot mats at culture room entrances is used to remove dirt from footwear. In some larger collections (e.g., the National Institute for Environmental Studies [NIES] in Japan), outdoor footwear is never worn within the laboratory suite and special slippers are worn in the culture facilities. Furthermore, systems that filter the air from small particles (as commercially available for pollen-allergies) may be used, but cost implications may be prohibitive.

## 6.0. MANAGEMENT OF ALGAL STOCKS IN CULTURE COLLECTIONS

A consistent and stable labeling system is necessary for long-term maintenance of cultures, whether for a small personal collection or a large service collection. Strain numbers (e.g., SAG 3.72, 3124, 3H, CHAGRA) are the simplest and most reliable means of labeling a tube or flask. Species and genus names of an algal strain may change through improved methods of identification and progress in taxonomy, but the strain designation should always remain constant to clearly identify a strain. We recommend the following information for a label: scientific name of the organism, strain number, and culture medium (see Fig. 10.1d). Using both the strain number and species name minimizes the risk of mislabeling. For small collections of one or a few racks of tubes, transferred at different times, it may be necessary to include the transfer date to the label of each tube. However, if all tubes of the rack are transferred at the same time, then a transfer date label can be placed on the rack. Additional details (e.g., collection and isolation information, growth requirements) for each specific strain number can be maintained in a database and paper files.

The labels should be easy to read, waterproof, and, if the culture containers are reused, easy to remove. A good adhesive label stays on the tube regardless of temperature and humidity, but these are very difficult to remove, requiring substantial labor. Other labels, with adhesives designed to float away during washing, may sometimes fall off the tubes when incubated in cool, humid boxes for long periods. A label with a rubberized back (like a stamp that glues to the tube after wetting) is somewhat of a compromise between the two. For large numbers of strains, computer printers are typically used to print sheets of labels quickly and efficiently. The same types of labels can be handwritten when the number of strains does not require computer assistance. Another method is to label every tube by hand with a pencil. For reusable culture vessels, the person responsible for washing the glassware usually erases or removes each label when the cultures are discarded.

To reduce the risk of loosing strains, each strain should be maintained in more than just one culture. We recommend retaining subcultures from at least two different transfer dates. For example, at SAG a minimum of five cultures from different ages per strain is kept. Two cultures are from the last transfer date; two more represent the previous transfer date. The fifth, from a still earlier transfer, is kept as a backup under very low growth conditions and held separately from the other cultures. Some keep backup cultures at a separate location, as recommended by the World Federation of Culture Collections (Anon 1999). The transfer process should involve the inspection of all subcultures of a strain. This is an essential step to control the culture stocks, and it may be the most time-consuming stage of the perpetual maintenance procedure. A culture from the latest or previous transfer date (provided that it is still in good condition) may be used for starting a new culture, thus allowing the rapid regeneration of a vigorous healthy culture. This is advantageous not only to service culture collections for shipment but also for all applications where fresh and optimal quality cultures are needed at short notice. For easier handling it is desirable to group the stock cultures based on the same requirements (e.g., transfer interval, medium). These groups often coincide with taxonomic groups.

An important step in quality control of the holdings is to check the cultures for contamination and to purify them if contaminated. In many cases it may be sufficient to check the cultures just by using a lens or a dissection microscope. Other methods may be more efficient but rather time-consuming. Some service culture collections regularly check for contamination using test media. Methods of testing for contamination and purification of cultures are discussed in Chapters 2, 3, and 8 and Appendix A.

The more information that is available for a certain culture strain, the more valuable it is to many potential users. Information about the strains is best stored in an electronic database using appropriate applications of standard software packages. More sophisticated database programs allow the convenient printing of culture labels, storage of images, and online access of strain information to a wide audience via the Internet. Minimally, the origin (habitat and locality, name of isolator, and isolation date), culture medium, and conditions for routine serial transfer should be stored in such a database. We recommend keeping printed record sheets as well.

## 6.1. Quality Assurance

A perpetual maintenance of any algal culture collection requires a routine quality control of the stocks (i.e., a check for obvious contamination and healthy growth of the cultures). Most important is the careful selection of the one culture used for transfer. A second inspection should be undertaken when the freshly transferred subcultures are checked for good growth. A regular check for the correct identity of a strain is also desirable but time-consuming and expensive when ultrastructural or molecular methods are required. To prevent accidental loss of a strain, a secondary collection of backup cultures is invaluable. Researchers may also negotiate with the large service culture collections to maintain their strains.

## 6.2. Troubleshooting

Poor or no regeneration of the culture on transfer can usually be attributed to the inoculum, the medium, or the incubation condition. If possible it is important to try to avoid using a senescent culture as inoculum. It may be possible to use a previous subculture or a duplicate sample that is more physiologically active than the culture initially chosen. If there is no option, then a larger inoculum can be used, or, alternatively, the nutrient regimen and culture conditions can be altered (see the following paragraphs).

The most common problems associated with culture media are incorrect pH, high levels of precipitate, and incorrectly formulated media, including the omission of a vital ingredient (e.g., silicon for diatoms, vitamins). Most algae are tolerant of fairly large changes in pH; however, if the inoculum's vigor is suboptimal, then poor or no growth can result. In most cases freshwater eukaryotic algae prefer acidic environments (pH 5–7), whereas cyanobacteria prefer alkaline environments (pH 7–9). High levels of precipitate can result in nutrient limitation and osmotically stressful microenvironments. For poor inocula, improved recovery may be obtained by using a less-defined medium in combination with the standard mineral medium (e.g., *Euglena* medium : Jaworski's medium, artificial seawater medium : natural seawater). Alternatively for axenic strains, supplementation of the medium with low concentrations of proteose peptone, yeast extract, vitamins, or soil extract may assist recovery. Sometimes a transfer from agarized to liquid medium, or vice versa, improves poor growth.

Incubation conditions (e.g., temperature, light) are unlikely to be the cause of growth failure, assuming the parent culture was grown under the same conditions. It is widely recognized that transferring cultures into an incubator set at a higher, physiologically more suitable temperature stimulates the growth rate.

Incorrect light levels may cause poor or no growth. Cyanobacterial cultures are particularly susceptible to photo-inhibition, bleaching, and death from excessive light levels. The same is true for other algae that contain phycobilisomes. Light intensity should be measured at the culture's actual location. It should be noted that in dense cultures, self-shading can have a significant effect on the intensity of light reaching an individual cell, and in some cases, subcultures must be initially incubated at relatively low light levels. Low light levels can have a limiting effect on final culture density, but assuming levels were sufficiently high to support the growth of the parent culture, they are unlikely to prevent growth of the subculture. A light : dark regimen is unlikely to cause problems; however, it is worth checking the proper function of the timing or switching mechanism.

## 7.0. XENIC VERSUS AXENIC CULTURES

Ideally a culture should be derived from a single cell, and one should always attempt to purify a culture to an axenic state. However, it has been often observed that the performance of a culture was better with bacterial contamination than after purification. For example, many euglenoids and some cyanobacteria may express more typical morphology in xenic status. Also, contaminating bacteria may be providing essential vitamin or other growth factors. Therefore, with newly established axenic strains, we recommend retaining the original xenic strain until successful maintenance of the purified strain is established. Fungal contamination is generally worse than bacterial contamination, because fungi are harder to eliminate by either physical or chemical methods, and the fungus may overgrow the algae on prolonged incubation under suboptimal conditions.

## 8.0. ACKNOWLEDGMENTS

The authors are indebted to Marlis Heinemann and Ilse Kunkel for their skillful assistance with algal cultures. We also thank both anonymous reviewers and the editor for valuable comments.

## 9.0. REFERENCES

Andersen, R. A., Morton, S. L., and Sexton, J. P. 1997. CCMP–Provasoli-Guillard National Center for Culture of Marine Phytoplankton. *J. Phycol.* 33(suppl.):1–75.

Anon. 1999. *World Federation for Culture Collections: Guidelines for the Establishment and Operation of Collections of Cultures of Microorganisms.* Michael Grunenberg GmbH, Schoeppenstedt, Germany, 24 pp.

Anon. 2001. *Catalogue of the UK National Culture Collection (UKNCC). List of Algae and Protozoa.* ed. 1. UKNCC, Egham, 231 pp.

Balzer, I., and Hardeland, R. 1991. Photoperiodism and effects of indoleamines in a unicellular alga, *Gonyaulax polyedra. Science*, 253:795–97.

Belcher, H., and Swale, E. 1982. *Culturing Algae: A Guide for Schools and Colleges.* Culture Collection of Algae and Protozoa, Windermere, UK, 25 pp.

Day, J. G. 1999. Conservation strategies for algae. In: Benson, E. E., ed. *Plant Conservation Biotechnology.* Taylor and Francis Ltd., London, pp. 111–24.

Elster, J., Seckbach, J., Vincent, W. F., and Lhotský, O., eds. 2001. *Algae and Extreme Environments. Ecology and Physiology. Nova Hedwigia*, Suppl. 123, 602 pp.

Graham, L. E., and Wilcox, L. W. 2000. *Algae*. Prentice Hall, Upper Saddle River, New Jersey, 700 pp.

Guillard, R. R. L. 1995. Culture methods. In: Hallegraeff, G. M., Anderson, D. M., and Cembella, A. D., eds. *Manual on Harmful Marine Algae. IOC Manuals and Guides No. 33*, UNESCO, Paris, pp. 45–62.

Isaac, S., and Jennings, D. 1995. *Microbial Culture*. BIOS Scientific Publishers Ltd., Oxford, UK, 129 pp.

Jaworski, G. H. M., Wiseman, S. W., and Reynolds, C. S. 1988. Variability in sinking rate of the freshwater diatom *Asterionella formosa*: the influence of colony morphology. *Br. Phycol. J.* 23:167–76.

Lokhorst G. M. 2003. The genus *Tribonema* (Xanthophyceae) in the Netherlands. An integrated field and culture study. *Nova Hedwigia* 77:19–53.

McLellan, M. R., Cowling, A. J., Turner, M. F., and Day, J. G. 1991. Maintenance of algae and protozoa. In: Kirsop, B., and Doyle, A., eds. *Maintenance of Microorganisms*. Academic Press Ltd., London, pp 183–208.

Price L. L., Yin K., and Harrison P. J. 1998. Influence of continuous light and L : D cycles on the growth and chemical composition of Prymnesiophyceae including coccolithophores. *J. Exp. Mar. Biol. Ecol.* 223:223–34.

Pringsheim, E. G. 1946. *Pure Cultures of Algae*. Cambridge University Press, Cambridge, 119 pp.

Richmond A., ed. 2004. *Handbook of Microalgal Culture-Biotechnology and Applied Phycology*. Blackwell Publishing, Malden, MA, 566 pp.

Schlösser, U. G. 1994. SAG–Sammlung von Algenkulturen at the University of Göttingen. Catalogue of Strains 1994. *Bot. Acta* 107:113–86.

Starr, R. C. 1973. Apparatus and maintenance. In: Stein, J., ed. *Handbook of Phycological Methods*. Cambridge University Press, Cambridge, pp. 171–79.

Starr, R. C., and Zeikus, J. A. 1993. The culture collection of algae at the University of Texas at Austin. *J. Phycol.* 29(suppl.):1–106.

Stein, J., ed. 1973. *Handbook of Phycological Methods*. Cambridge University Press, Cambridge, 448 pp.

Venkataraman, G. S. 1969. *The Cultivation of Algae*. Indian Council of Agricultural Research, New Dehli, 319 pp.

Warren, A., Day, J. G., and Brown, S. 2002. Cultivation of protozoa and algae. In: Hurst, C. J., Crawford, R. L., Knudsen, G. R., McInerney, M. J., and Stezenbach, L. D., eds. *Manual of Environmental Microbiology*, ed. 2. ASM Press, Washington, D.C., pp 71–83.

Watanabe, M. M., Kawachi, M., Hiroki, M., and Kasai, F., eds. 2000. *NIES-Collection. List of Strains*, ed. 6. National Institute for Environmental Studies, NIES, Tsukuba, Japan, 160 pp.

# CHAPTER 11

# LONG-TERM MACROALGAL CULTURE MAINTENANCE

JOHN A. WEST
*School of Botany, University of Melbourne*

## CONTENTS

Key Index Words: Culturing; Illumination; Incubators; Macroalgae; Seawater

## 1.0. INTRODUCTION

This chapter is based largely on my personal perspective and 44 years experience in maintaining laboratory cultures of macroalgae and microalgae. I have observed culture facilities at other academic institutions and some have not been adequate for satisfactory studies on algal biology and long-term culture maintenance. In the following sections, I point out those aspects that can benefit others who wish to retain culture stocks as a gene pool for many different kinds of phycological research.

## 2.0. METHODS

### 2.1. Seawater

For routine maintenance of marine macroalgae, I prefer natural seawater-based culture medium rather than an artificial culture medium. It is best to obtain the seawater for culture from open coastal areas away from industrial and housing areas. The seawater is obtained with a 10-liter plastic bucket and then poured through a plastic funnel fitted with a Nitex mesh filter (100 μm to omit floating debris) into 20-liter black polyethylene cubical containers. Seawater can be stored for long periods at normal temperatures in these, and the black color prevents algal growth if the containers are exposed to light. For use, the seawater is siphoned off. The container should not be agitated to avoid sediment from being transferred during siphoning. Four-liter Pyrex glass flasks are the most convenient size for sterilizing seawater in large quantities. The salinity is checked when each seawater batch is collected. Most coastal seawater is about 35 psu and can be adjusted to 30 psu with either glass-distilled water or Milli-Q water (Millipore Corporation). Salinity can be checked easily with a Salinity Refractometer (Vista Model A366ATC), which is available from VWR Scientific (www.vwrsp.com/).

## 2.2. Sterilization

I do not recommend autoclaving because of the precipitation caused by the high temperature and pressure. I do not recommend sterile filtration; it is slow and breakage of the membrane filters sometimes occurs, resulting in contamination of cultures. Sterilization can be achieved with a 1 atmosphere steam sterilizer (see Chapter 5). We use a 440V electrical stainless steel sterilization unit (inside dimensions—420 mm wide × 620 mm long × 430 mm deep) (Getinge Australia Pty Ltd.) (Fig. 11.1), which was originally designed for hospital use. It is a convenient size for sterilizing four 4-liter flasks at one time for 1 hour. Similar designs should be available elsewhere or 100-liter steam boilers can be used. I use aluminum beer or soft drink cans (8 oz., 330 mL with the top removed) or 250-mL glass beakers for covers on the 4-liter flasks. Both types can be reused hundreds of times (Figs. 11.1, 11.2). Cotton plugs, silicon rubber plugs, or aluminum foil may also be used.

## 2.3. Microdissection Tools

One must have very fine microdissecting tools constructed from really good quality stainless steel for macroalgal culture work (Fig. 11.3). Roboz Surgical Instruments (www.roboz.com) produces the best I have seen and used. Less-expensive stainless steel microdissection instruments are available from Carolina Biological Supply (www.carolina.com). Fine dissecting needles can be made with 3 or 6 mm wood doweling, cut to 6–8 cm length, and a Sharps 10 sewing needle, inserted using needle-nose pliers. These instruments must be

**FIGURE 11.1.** Flasks are placed in the steam sterilizer for 50 minutes.

**FIGURE 11.2.** Seawater is collected and stored in 20-liter cubical black plastic containers. Seawater is siphoned into a 4-liter flask after adding 500-mL Milli-Q water.

carefully cleaned after each use with 70% ethanol and carefully wiped with tissue. A tall 120-mL jar of 70% or 95% ethanol with orlon wool pad on the bottom (to prevent damaging the forcep tips) is great for dipping and cleaning the forceps (Fig. 11.4). The ethanol should be replaced frequently, the container washed, and orlon wool replaced to avoid introducing foreign residues through repeated use. At no time should the implements be flame sterilized because that quickly destroys the steel quality. The forceps tips can also be protected

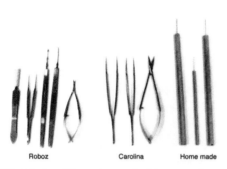

**FIGURE 11.3.** Dissecting tools. Finest quality and most expensive stainless steel by Roboz, low-cost stainless steel tools from Carolina Biological Supply, homemade needles of wood dowel and sewing needle.

**FIGURE 11.4.** 120-mL jar with 70% ethanol and orlon wool layer on bottom with forceps, Kimwipes for drying forceps.

**FIGURE 11.5.** The arrangement of equipment and materials at the bench for culture transfer and observations. From left to right are shown: Kimwipe dispenser, dissecting tools and label dispenser, compound microscope and camera system, stereomicroscope with fiber-optic illuminator and hand rest, salinity refractometer, Li-Cor 185 meter, polyethylene wash bottles of sterile seawater and Milli-Q water, storage shelf for culture record binders, open binder.

with a short piece of narrow plastic tubing (see Fig. 11.3, far-left forceps) when not in use. If the tips of forceps are damaged, they can be refinished under the stereoscope using a fine P800 waterproof carborundum paper. When rust develops on these implements, they can be refinished with the same paper or Scotch-Brite heavy-duty scouring pads.

### 2.4. Arrangement of Materials and Equipment for Efficient Transfer

To transfer large numbers of cultures most efficiently, I have arranged the equipment as shown in Fig. 11.5. The compound microscope and its photographic unit are on the left. Just behind and slightly to the left of this microscope is a small wooden rack with the 70% ethanol bottle, Kimwipe tissues in a holder, slide dispenser and coverslip box, germanium dioxide, and antibiotics. Directly in the center is the stereoscope with assorted

forceps and the label dispenser. To the right is a bookrack with the culture record binders within easy reach. The salinity meter, irradiance meter, and polyethylene squeeze-bottles with sterile seawater and distilled water are there as well. Each batch of 24 culture dishes is placed on a stainless steel cart beside the stereoscope to allow easy access to these for transfer and recording.

### 2.5. Culture Media

Chapters 3 and 4 as well as Appendix A describe the preparation of seawater culture media. The following remarks are based on my experience. The nutrient enrichment should be added to the sterile seawater just before use. If added before sterilization, cloudiness and some precipitation may develop. Mix thoroughly by swirling the flask or cover the top with Parafilm and invert several times. For many years we have used a modified version of Provasoli's ES Medium (West and McBride 1999; see Appendix A). This medium is satisfactory for most marine macroalgae and microalgae that I have grown, and it contains nitrate, phosphate, iron-EDTA, trace metals, and vitamins (thiamine · HCl, biotin, and $B_{12}$) (see Appendix A). Years ago, I eliminated the Tris buffer from Provasoli's ES medium (Provasoli 1968, McLachlan 1973) because it serves as a substrate for bacteria. Glycerophosphate also serves as a bacterial substrate and this is replaced with molar equivalent $NaH_2PO_4$. The iron originally in the PII metals mix (Provasoli 1968) is eliminated and the total iron in the Fe-EDTA is reduced by half. Because so many red algae are nitrogen limited in long-term culture, the $NaNO_3$ was increased by 25%. I normally add 10 mL of the enrichment solution to each liter of 30 psu seawater. For some brackish water or freshwater red algae, 5 psu seawater is used. This has proven successful for many years. Other culture media (e.g., f/2 medium) also work well, and they may be prepared (see Appendix A) or purchased commercially.

We spent several years working with synthetic seawater media and concluded that these are not as useful as natural seawater media for maintaining long-term cultures. Synthetic media require considerable time and expense to formulate. Invariably, we found that growth and reproduction were not as reliable with synthetic media (Woelkerling et al. 1983). Of course, synthetic media may be essential for certain critical physiological research.

## 2.6. Culture Glassware

I use glassware, not plasticware, for stock cultures to avoid waste of plastic, but plasticware can be used to culture algae. Pyrex No. 3250 storage dishes (80 × 100 mm) with Pyrex lids are very satisfactory for long-term culture. The recessed lip helps reduce spillage and evaporation. Each dish holds about 500 mL but we use just 200–250 mL to allow better gas exchange. This also minimizes spillage if the dishes are placed on a shaker for faster growth. Moreover, direct observation under the stereo dissecting microscope is quite easy with these vessels.

For smaller-scale culture, we use 50 × 70 mm Pyrex No. 3140 dishes that hold about 150 mL. These dishes are inexpensive but have limitations because the straight sides (no recessed lip) increase the risk of spillage and evaporation. A strip of Parafilm can be used to seal the dish lid to reduce water loss by evaporation. We had custom-made lids constructed of polycarbonate tubing

and sheeting because no properly fitting lids were available from commercial sources. However, the polycarbonate cannot be dry-heat sterilized at 250°C like the Pyrex glass.

All glassware is hand-washed with household detergent. Each is scrubbed thoroughly with a Scotch-Brite Dobie cleaning pad (3M Home Care Division) to remove algae, $CaCO_3$, etc., that adhere tightly to the glass. The 3M pads have proved to be very durable, giving many months of use without deteriorating. Dobie and other nylon-type pads do not scratch the glass, which will interfere with observations on the dissecting microscope. These adherent materials including calcium carbonate deposits can also be removed quickly by rinsing in 0.1 M HCl. After washing, the glassware is rinsed with warm tap water and then thoroughly rinsed with reverse-osmosis water and air-dried. Pyrex glassware is sterilized for 1 hr in a 250°C oven (see Chapter 5). Periodically, to remove brown iron deposits on the glass, the dishes require cleaning with concentrated HCl followed by thorough rinsing in tap water and reverse-osmosis water.

We do not use any plastic containers for long-term culture, but for short-term culture I use various plastic petri dishes, 6, 12, or 24 multiwell plates, and so on if these are optically suitable for microscopic observations.

## 2.7. Shakers and Aeration

For years we have used reliable and long-wearing Gyrotory G-2 Shakers (New Brunswick Scientific Co. Inc., www.nbsc.com) for enhancing growth. The best range was 60–80 rpm. Aeration also enhances the growth, but there are too many difficulties with evaporation and control of bubbling rate, etc., so we have not used this method.

## 2.8. Contamination Control

Occasionally airborne or waterborne algal contaminants occur in long-term cultures. To prevent inadvertent diatom contamination, 1–2 drops of a $GeO_2$ solution is added for each 100 mL of culture. There seems to be no adverse affect on growth of most algae from this; however, refer to Markham and Hagmeier (1982) and Tatewaki and Mizuno (1979) concerning inhibition of some algae by $GeO_2$. To prepare 100-mL stock solution (1-mg · $mL^{-1}$) of germanium dioxide (Aldrich Chemical Company), add 100-mg $GeO_2$ to 50-mL 1N NaOH.

The $GeO_2$ may take 15–20 min to dissolve with frequent stirring; if necessary, heat gently and cool. Adjust the pH to 7.8–8.2 by adding 1N $H_2SO_4$ drop by drop. Bring volume to 100 mL with glass distilled water. This can be stored in a glass bottle at 4°C. This method of preparation is different from that usually described by other authors (e.g., Lewin 1966) because it replaces HCl with $H_2SO_4$.

For cyanobacteria, the usual control is 1–5 mg · 100 mL$^{-1}$ culture of sodium penicillin G or ampicillin (e.g., Mediatech Inc., www.cellgro.com). (See Chapters 8 and 9 for additional information about antibiotics.) Either penicillin or ampicillin dissolves instantly and is stable for about three days. To be sure, add a further treatment every three days for two weeks. If this proves ineffective, because some cyanobacteria and eubacteria are resistant to penicillin, then add ciprofloxacin · HCl powder (Mediatech Inc.). Care must be taken with this DNA gyrase uncoupler because it can be lethal to some algae. Run a control first to determine its action on the cultured alga. Another antibiotic that has proven useful is Rocephin (Roche Products Pty Ltd.), a cephalosporin that blocks cell wall synthesis. Both ciprofloxacin and Rocephin are slow to dissolve and often take 2–3 days to completely dissipate. Ciprofloxacin has been especially valuable in our work with videomicroscopy of sexual reproduction in *Bostrychia*, *Murrayella*, and others because it eliminates many epiphytic bacteria. The bacteria, if abundant, can impede attachment of spermatia to the trichogyne during fertilization.

## 2.9. Culture Records

Since 1961 I have maintained records for each isolate using 8.5 × 11 inch or A4 pages in 11.5 × 10 inch clip binders (e.g., Oxford #09021 or Accohide #49259). Each page is formatted and incorporates the genus and species, culture number, and collection details at the top of the page and then lists the culture details (date, shelf or chamber number, observer, medium type, change of medium and dish, and notes on growth, reproduction, etc. for each time of observation). Each binder can hold records for about 100 culture isolates. Pages are printed on both sides to reduce the volume needed. For most isolates, pressed voucher specimens of field or culture specimens are made on herbarium paper (50 × 50 mm) placed in 70 × 70 mm glassine envelopes with an open-end flap. These are attached to the culture records.

These binders are kept in a vertical file adjacent to the dissecting microscope so they are easily accessible each time culture observations are made. Each time cultures are transferred and observations are made these are recorded by hand in the record books (Fig. 11.5). A permanent master list of cultures is also kept on computer file with a Microsoft Excel program and records are updated as new isolates are obtained. The complete list that now has more than 4,300 isolates is sent to research colleagues as an e-mail attachment.

## 2.10. Labels

The computer-generated paper labels are done with 8 font size with the genus, species, culture number, phase (e.g., ⊗, ♂, ♀, ⚥), locality, salinity, or special medium requirements. Each label is cut from A4 paper to 10–15 mm wide by 20–30 mm long and taped over with a clear 50 mm wide Bel-Art (www.bel-art.com) Lab Label Protection Tape (#F134500020) on a dispenser (#F134500000). Each label tape can be folded on one edge for easier removal and reapplication. Packs of 100 labels 12.5 × 38 mm (#F134550005) also are available for handwritten labels. It is possible to reuse the labels 20–30 times over five years and the adhesive is still functional. That alone saves hundreds of hours of time. Scotch Magic tape (3/4 inch) is also satisfactory for applying the paper labels, although it tends to tear after one or more removals and reapplications.

## 2.11. Temperature Control

For air-conditioning units, it is best to use those designed to cool large rooms. A unit with a separate external cooling system, connected to multiple ceiling suspension split systems with effective air circulation, is best. Daikin Industries, Ltd. (www.daikin.com.au) offers excellent and reliable units with long service records. Window air-conditioning units often are less suitable. Larger rooms provide multiple-use space for big culture collections, and I have found that for the many tropical/subtropical algae, a temperature range of 20–25°C is excellent. Often the temperature near the light will be 1–2°C above the room ambient level.

## 2.12. Growth Cabinets and Shelving

My years of experience with various manufacturers and models have shown that reliable units are made by

Percival Scientific Inc. (http://percivalscientific.com). The classic model I-35LL has provided 37 years continuous operation with minimal downtime. Be sure to select one with an automatic shutoff and alarm system to avoid loss of cultures due to overheating or overcooling. Earlier models did not have this feature and I lost irreplaceable cultures.

For most subtropical and tropical macroalgae, I have found that air-conditioned rooms with simply shelving are quite suitable. The materials for the shelving are either 20 mm MDF or plywood. Each unit is five shelves with overall dimensions of 1,500 mm H × 1,500 mm W × 600 mm D. Each shelf has a 1,200 mm fixture (36W) suspended at the back so the brightest light is at the back and decreases toward the front. These fixtures are wired in series so that all can operate through one connection on a single photoperiod timer. There are many reliable 24 hr timers that cost between $10 to $20.

## 2.13. Lighting

After testing a variety of light sources over 35 years of culture work, we settled with cool-white fluorescent lamps (15–36W) with fixtures 18 inches to 4 feet long. The maximum output for these is about $80 \, \mu mol$ photons $\cdot m^{-2} \cdot s^{-1}$ at a distance of 50 mm from the lamp. For all the macroalgae we have maintained, this is excessive and often causes stress or death. By contrast, the lowest irradiance level at which these algae can be maintained is about 2 $\mu mol$ photons $\cdot m^{-2} \cdot s^{-1}$. The irradiance can be effectively reduced with layers of standard gray plastic window screen (1-mm mesh size) wrapped over the lamp tube and stapled closed or simply laid over the dishes. One layer usually reduces the light level by about 40–50%, two layers by about 70–75%.

## 2.14. Minimizing Growth: Irradiance Levels and Inoculum Size

It may seem insignificant to many people, but minimizing growth is important in saving hours of work and makes the maintenance of 2,000+ cultures single-handedly a bit simpler. The exact arrangement on the shelf relative to the light is necessary to minimize growth but still not endanger the viability of the different algae. Each species and some individual isolates need to be observed through several months to determine the best shelf position. As stated previously, the fluorescent fixture is suspended at the back of the shelf in most culture units here, providing a sharp falloff in irradiance from back to front. The use of shading screen enhances that. Almost all the stock cultures of the marine macroalgae are maintained at an irradiance level of 2–7 $\mu mol$ photons $\cdot m^{-2} \cdot s^{-1}$. This slows growth considerably and I have found that the culture dishes only require transfer once every 4–5 months. Two to three shoot tips, about 5 mm in length, are inoculated at transfer. Each segment is easily excised from the plant using a shearing action of the forcep tips. This inoculum size is sufficient for most red, green, or brown algae. For unicellular red algae one drop can be removed from the dish bottom by pipette. Crustose algae such as *Erythrocladia*, various Chaetophoraceae, and *Ralfsia* can be scrapped from the dish bottom for transfer. With large numbers of dishes on each shelf it is easy to double-stack in two rows within a narrow shelf space to allow more uniform lighting and growth. Growth is monitored easily by visual inspection each week.

## 2.15. Temperature Monitoring

Classical mercury thermometers are still quite useful but, unfortunately, easily breakable. Recently, indoor-outdoor, water-resistant, max-min, digital thermometers are proving very useful, durable, and economical. The best units I have used are SW-AAA from U-LAB Instruments (ulab@bigpond.com.au).

## 2.16. Long-Term Changes in Algal Clones

It has been my experience that some isolates change after several years in culture. For example, the male isolate of *Antithamnion defectum* Kylin (now *A. densum* (Suhr) M. A. Howe) was obtained in September 1963 and deposited in UTEX (LB 2262) in 1970. The culture isolate maintained at the University of California, Berkeley was used frequently for teaching and research observations on spermatial formation and discharge for 15 years (Young and West 1979). I obtained this isolate again in 2002 from UTEX and found that the spermatangia were abnormal and no longer released spermatia. *Acrochaetium proskaueri* J. A. West was isolated in June 1964. At the time it was described as a new species, it produced monosporangia, tetrasporangia, and hair cells in culture (West 1972). It was deposited at UTEX as LB 1945 in February 1973. When I obtained this isolate from UTEX again in 2002, it easily formed many monosporangia but did not produce hair cells or tetrasporangia under the conditions in which they had formed previously.

After 2–10 years in culture some isolates of *Bostrychia* and *Caloglossa* tetrasporophytes lost the ability to form tetrasporangia, although most females and males usually continued to reproduce normally. In *Spyridia filamentosa* (Wulfen Harvey) many isolates do not reproduce sexually in culture, although one isolate (2846) has been successfully used for experimental research since first isolated in 1987 (West and Calumpong 1989). The tetrasporophyte forms normal tetrasporangia, but these often fail to release spores. It was necessary to cross the male and female stocks to reestablish a functional tetrasporophyte. I am unable to suggest ways to minimize or eliminate these long-term changes in morphology and reproduction. I am hopeful that the major genetic components remain unaltered.

## 2.17. Culture Availability

The complete list of these cultures can be provided on request (jwest@rubens.its.unimelb.edu.au) and any cultures requested can be provided for a nominal fee.

## 3.0. Acknowledgments

During much of my career, this culture collection has been maintained through the use of my own personal funds. Since 1994 the culture work has been partially supported through small grants from the Australian Research Council and the Australian Biological Resources Study program.

## 4.0. References

Lewin, J. 1966. Silicon metabolism in diatoms. V. Germanium dioxide, a specific inhibitor of diatom growth. *Phycologia* 6:1–12.

Markham, J. W., and Hagmeier, E. 1982. Observations on the effects of germanium dioxide on the growth of macroalgae and diatoms. *Phycologia* 21:125–30.

McLachlan, J. 1973. Growth media-marine. In: Stein, J. R., ed. *Handbook of Phycological Methods. Culture Methods and Growth Measurements.* Cambridge University Press, Cambridge, pp. 25–51.

Provasoli, L. 1968. Media and prospects for the cultivation of marine algae. In: Watanabe, H., and Hattori, A., eds. *Culture and Collection of Algae.* Proceedings U.S.-Japan Cont. Japanese Society of Plant Physiology, Hakone, pp. 63–75.

Tatewaki, M., and Mizuno, M. 1979. Growth inhibition by germanium dioxide in various algae, especially in brown algae. *Jpn. J. Phycol.* 27:205–12.

West, J. A. 1972. Environmental control of hair and sporangial formation in the marine red alga *Acrochaetium proskaueri* sp. nov. *Proc. Internat. Seaweed Symp.* 7:377–84.

West, J. A., and Calumpong, H. P. 1989. Reproductive biology of *Spyridia filamentosa* (Wulfen) Harvey (Rhodophyta) in culture. *Bot. Mar.* 32:379–87.

West, J. A., and McBride, D. L. 1999. Long-term and diurnal carpospore discharge patterns in the Ceramiaceae, Rhodomelaceae and Delesseriaceae (Rhodophyta). *Hydrobiologia* 298/299:101–13.

Woelkerling, W. J., Spencer, K. G., and West, J. A. 1983. Studies on selected Corallinaceae (Rhodophyta) and other algae in a defined marine culture medium. *J. Exp. Mar. Biol. Ecol.* 67:61–77.

Young, D. N., and West, J. 1979. Fine structure and histochemistry of vesicle cells of the red alga *Antithamnion defectum* (Ceramiaceae). *J. Phycol.* 15:49–57.

CHAPTER 12

# Cryopreservation Methods for Maintaining Microalgal Cultures

John G. Day
*Culture Collection of Algae and Protozoa (CCAP),*
*Scottish Association for Marine Science, Dunstaffnage Marine Laboratory*

Jerry J. Brand
*Culture Collection of Algae at University of Texas–Austin (UTEX),*
*Molecular, Cell and Developmental Biology*

## Contents

Key Index Words: Algal Cryopreservation; Cryoinjury; Cryostorage; Cyanobacteria Freezing; Microalgal Storage

## 1.0. INTRODUCTION

### 1.1. Scope

Cryopreservation may be defined as the storage of a living organism, or a portion thereof, at an ultralow temperature (typically colder than −130°C) such that it remains capable of survival upon thawing. Cryopreservation is still largely an empirical science because the underlying biological mechanisms of cell injury during freezing and thawing are not fully understood (Baust 2002). Despite this limitation, hundreds of species of cyanobacteria and eukaryotic microalgae have been successfully cryopreserved. In this chapter the term "algae" will refer to both cyanobacteria and eukaryotic algae. The term "algal unit" will refer to a single organism. Following the introduction of some basic principles of cryopreservation, protocols will be described that have the potential for successful cryopreservation of a broad range of algae. Methods that can be adapted to a variety of laboratories at modest cost will be emphasized.

### 1.2. Rationale and Minimum Requirements

Continuous maintenance of actively growing algal strains over long periods of time is often costly and time-consuming. In contrast, cultures kept alive in an arrested or retarded metabolic state generally require a minimum of attention. Resting spores or other dormant stages of some species can be maintained at ambient or cool temperatures for many years without attention. For example, living *Haematococcus pluvialis* Flotow aplanospores were recovered from air-dried soil after 27 years (Leeson et al. 1984), and the cyanobacterium *Nostoc commune* Vaucher was revived from herbarium specimens after 107 years of storage (Cameron 1962). However, the viabilities of resting stages generally decrease with time, and many aquatic algae do not exhibit any persistent dormant stage. Cryopreservation allows living algae that do not have any normal resting stage to be maintained indefinitely in an arrested state. The advantages and disadvantages of cryopreservation, with respect to continuous culturing methods, are summarized in Table 12.1 (see also Kirsop and Doyle 1991, Day and McLellan 1995).

Microalgae are cryopreserved as large populations of algal units. Success is often measured as percent viability—the percentage of frozen algal units that remain viable and resume normal physiological activities after the culture is thawed. Minimum success can be defined as the recovery of at least one living algal unit, but in practice more rigorous standards are almost always required. The percent viabilities of identical cultures recovered from separate cryogenic vials often vary, even when there are careful attempts to follow identical procedures. Variability is minimized, but not always eliminated, when each culture is in the same physiological condition immediately prior to its freezing; replicate cultures are prepared, frozen, and thawed in precisely the same way at the same time; and thawed cultures are allowed to recover and resume growth under the same set of conditions. Variability in viability is most problematic for algal strains that survive only at low (<5%) viability.

**TABLE 12.1**  Advantages and disadvantages of cryopreservation with respect to continuous culturing methods.

| Advantages | Disadvantages |
| --- | --- |
| It provides stability against changes in the genetic composition of cultures over time due to selective pressure and/or genetic drift. | It requires an initial investment in specialized equipment, supplies, and training. |
| It protects cultures against problems due to microbial contamination, handling errors and labeling errors, or mechanical failure of the culturing facility. | It requires a back-up facility for high reliability, since frozen cultures that are inadvertently thawed even briefly under noncontrolled conditions generally do not survive. |
| It requires minimal storage space per culture since cultures typically are maintained in 1-mL or 2-mL volumes. | It requires a regular (at least monthly) supply of liquid nitrogen or else a reliable ultra-low-temperature (−150°C) freezer. |
| It reduces the cost of long-term maintenance when many strains must be maintained. | It generally requires 2–3 weeks to generate substantial volumes of vigorously growing cultures from frozen vials. |

Cryopreservation should not select for freezing-tolerant genetic variants in a population of algae. Although this is a legitimate concern, there is no reported evidence that cryopreservation selects for cold-tolerant, nonrepresentative subpopulations. Potential selection is minimized for cultures that survive cryopreservation at high viability.

Viabilities higher than 60% have been recommended as the standard for cryopreserved microalgal cultures maintained in the Culture Collection of Algae and Protozoa (CCAP) (Morris 1981), although that may be unattainable for many strains. Considerably lower viabilities may be acceptable, provided that replicate samples produce consistent percent viabilities and that there is no evidence of genomic selection. For algae whose genetic content must remain constant, such as type strains, 10% may be an acceptable minimum viability, although no objective criteria are available to support a specific minimum value. Even lower viabilities may be acceptable for many purposes, such as cultures for teaching or biotechnological applications.

Some microalgae, especially cyanobacteria, can be frozen for months (occasionally years) at relatively high subzero temperatures (e.g., −20°C or −70°C), provided proper precautions are taken during the freezing and thawing processes. However, ice recrystallization and/or intracellular chemical activities cause a gradual decline in percent viability, and many algal strains lose all viability within days of storage at high subzero temperatures. Chemical activities virtually cease and ice recrystallization is prevented at sufficiently low temperatures (MacFarlance et al. 1992). Solidified water does not change states at temperatures lower than approximately −130°C, so ice recrystallization is precluded. Continuous storage below this temperature, the glass transition temperature or glass transformation temperature ($T_g$), is considered critical for long-term maintenance of cryopreserved organisms in aqueous media. It is generally believed that a population of living organisms capable of surviving the freezing and thawing processes can be stored without losing appreciable viability for an indefinite period of time, perhaps for hundreds of years, when kept below $T_g$ (Mazur 1984). Very few long-term empirical data are available for microorganisms, but bacteria and fungi retain high viability after 30 years of continuous cryostorage (D. Smith, CABI Bioscience, personal communication), and cryopreserved algae have been shown to retain a constant level of viability for more than 20 years (Day et al. 1997).

Cryopreservation is employed at major algal collections, including the American Type Culture Collection (ATCC; Lee and Soldo 1992), the Culture Collection of the Centre of Algology (CCALA; Lukavaský and Elster 2002), CCAP (Morris 1978), the Provasoli-Guillard National Center for Culture of Marine Phytoplankton (CCMP; Andersen 2002), the Culture Collection of Algae at the University of Coimbra (ACOI; Osorio et al. 2004), the Microbial Culture Collection at the National Institute for Environmental Studies (NIES; Watanabe et al. 1992), Pasteur Culture Collection (PCC; Rippka et al. 2002), the Sammlung von Algenkulturen at Göttingen University (SAG; Friedl and Lorenz 2002), and the Culture Collection of Algae at University of Texas–Austin (UTEX; Bodas et al. 1995). Protocols adapted or developed empirically at these facilities typically employ relatively simple procedures that cryopreserve a broad range of algal species. The highest success has been achieved with cyanobacteria, pennate diatoms, and unicellular green algae, most of which are small and not morphologically complex. Table 12.2 is a summary of successfully cryopreserved microalgae at the CCAP and at UTEX.

### 1.3. Physical Aspects of Freezing

Sample freezing is not uniform because the water contains solutes. Furthermore, heterogeneity of cells (e.g., water content, size, and morphology) is an important factor. Initially, ice crystals tend to form first from pure water, i.e., ice formation removes water from the solvent-solute system, concentrating solutes. Equilibrium is reached, with some ice and some solute-enriched liquid water. Before more water can be converted to ice, the solution must be cooled to a lower temperature. Thus, the process of cryopreserving a sample consists of an ever-shifting equilibrium that is driven by continuously lowering the temperature. The percentage of unfrozen water in the sample at any subzero temperature depends on the solute concentration of the original sample. Intracellular ice formation, osmotically driven cell volumetric changes, and other "solution" effects are influenced by the rate of cooling, cryoprotective additives, and the temperature of ice nucleation. Ice begins to form in the extracellular culture medium surrounding the algae, leaving most solutes in the remaining liquid. The solute concentration in the liquid phase increases as the amount of extracellular ice increases. The remaining liquid becomes hypertonic with respect to the algal cell contents, then water begins to move out through the plasma mem-

**TABLE 12.2**    Diversity of algal strains cryopreserved in the CCAP (freshwater only) and at UTEX culture collections. Taxonomic classification follows Hoek et al. (1995).

| Algal division | CCAP | | UTEX | |
|---|---|---|---|---|
| | Number of strains attempted | % Successfully cryopreserved | Number of strains attempted | % Successfully cryopreserved |
| Chlorarachniophyta | 0 | 0 | 2 | 50 |
| Chlorophyta | 700 | 68 | 1,358 | 70 |
| Cryptophyta | 20 | 0 | 13 | 8 |
| Cyanophyta | 180 | 80 | 191 | 97 |
| Dinophyta | 3 | 0 | 26 | 12 |
| Euglenophyta | 40 | 70 | 71 | 68 |
| Glaucophyta | 1 | 0 | 2 | 0 |
| Haptophyta | 0 | 0 | 19 | 16 |
| Heterokontophyta | 80 | 69 | 153 | 58 |
| Rhodophyta | 10 | 10 | 117 | 56 |

NOTE. Percent successfully cryopreserved = % of culture collection holdings with successful cryopreservation.

brane and the cells undergo plasmolysis. At any given temperature the rate of water transport increases or decreases as the difference in osmotic potential across the membrane increases or decreases. Rapidly cooled cells generally do not experience pronounced plasmolysis because they quickly reach temperatures too low for effective transport of the cryoprotective agent (CPA) and water. However, some plasmolysis may be desirable. As the volume of a cell decreases, the concentration of its intracellular solutes increases. This lowers the equilibrium temperature for intracellular ice formation and increases supercooling, thus favoring the eventual formation of intracellular vitrified water.

## 2.0. CAUSES OF CRYOINJURY

Nearly all algae are highly susceptible to damage during freezing and thawing. Thus, considerable care is required during critical stages in their cryopreservation in order to maximize viability when frozen cultures are thawed. The causes of damage vary among different algae, explaining in part why no single protocol is universally effective. However, several potential causes of cellular injury during freezing and thawing are likely of universal importance and should be considered in any cryopreservation strategy.

### 2.1. Chilling Injury

Cooling to low temperatures above the freezing point kills some algae. Species that remain in a near-constant environment in nature, such as marine tropical species, are especially susceptible to chilling injury. The cellular mechanisms of chilling injury in algae are not well known, although a temporary reduction in photosynthetic capacity has been observed in a variety of algae cooled to 0°C or supercooled to −10°C (Fleck 1998), and chilling sensitivity in cyanobacteria likely correlates with their membrane lipid composition (Nishida and Murata 1996). Even for strains that survive temporary cooling to temperatures near 0°C, the stress may sensitize them to damage during freezing and thawing. The culturing temperature prior to cooling and the rate of cooling to low nonfreezing temperatures were shown to greatly affect the chilling sensitivity of the cyanobacterium *Synechococcus leopoliensis* (Raciborski) Komárek (Rao et al. 1977). Thus, cryopreservation protocols should minimize stress while algae are cooling to their freezing temperature.

### 2.2.0. Injury at Subfreezing Temperatures

An aqueous suspension of microalgae cooled to subzero temperatures is not frozen homogenously. Solutes in the supporting medium and in intracellular compartments of cells affect both the temperature of freezing and the pattern and rate of ice formation (see Section

1.3). With continued cooling, the temperature eventually decreases to the eutectic, the temperature at which all remaining solution solidifies.

Because of the high intracellular concentration of solutes, the equilibrium temperature for ice formation within living cells is several degrees Celsius below zero. During continued cooling, intracellular ice may not form until the temperature goes well below the equilibrium temperature for ice crystallization, because there are few intracellular sites for ice nucleation. If intracellular ice first forms well below the equilibrium temperature for nucleation, then crystals may grow extremely rapidly immediately after their formation, filling much of the cell with ice almost instantaneously. This sudden appearance of highly refractile ice throughout the cell is called black flashing because it is observed in a cryomicroscope as the sudden appearance of blackened cells. Algal cells that undergo black flashing are not viable when thawed.

Both extracellular and intracellular ice formation are correlated with damage to algae during freezing and thawing, although the direct causes of damage differ. Cryopreservation protocols are designed to minimize these damaging effects by controlling the rate of ice formation and the conditions under which the ice is formed. Most successful protocols moderate the damaging effects of freezing and thawing in two ways: by the addition of a water-soluble chemical substance, called cryoprotective agent (CPA), to the algal culture prior to its freezing, and by controlling the rates of cooling and warming the algal suspension.

### 2.2.1. Mechanical Stress During Extracellular Ice Formation

As extracellular ice forms during the gradual cooling, the algae accumulate in pockets and channels of the remaining liquid. The trapped algae are subjected to increasing physical stress as the volume of liquid decreases, until they become fully immobilized when the extracellular solution has fully solidified. There is relatively little microscopic evidence of direct mechanical injury to algal cells during external freezing, although flagellate cells lose their flagella (Day et al. 1998b, Smith and Day 2000) and the cell walls of *Vaucheria* sp. are damaged (Fleck et al. 1997).

### 2.2.2. Osmotic Stress

The gradual increase in solute concentration causes a gradual decrease in osmotic potential. This may induce plasmolysis after the extracellular medium becomes hypertonic with respect to suspended algal cells. Fresh-water algae typically prefer culture media of low osmotic strength (often less than 50 mOsm). The extracellular solute concentrations may increase more than 100-fold before the solution fully solidifies, simulating a hypersaline environment. Marine algae in natural seawater (~1,050 mOsm) experience a less relative increase in osmotic concentration during freezing. Soil algae are generally able to tolerate a wide range of osmotic potentials in nature, which may explain their toleration to the osmotic stress during gradual freezing. Specific mechanisms of cellular injury in hypertonic media during freezing are not well documented, but plasmolyzed algal cells experience severe mechanical deformation, and intracellular solute concentrations may increase to toxic levels.

The rate of osmotically driven water flow across the plasma membrane decreases as temperature decreases (Franks 1985), ceasing altogether when the remaining extracellular liquid solidifies (eutectic point). The eutectic depends on the kinds of solutes present. For example, an aqueous solution of NaCl, irrespective of its initial concentration, reaches the eutectic at −21.8°C. The total amount of water driven out of a cell osmotically during freezing is minimized by rapid cooling to the eutectic.

### 2.2.3. Intracellular Ice Formation

Cell damage during freezing and thawing is often directly correlated with intracellular ice. The physical principles governing ice formation within cells and in the extracellular medium are the same. However, the intracellular environment is quite different from the surroundings, thus causing differences in the timing and the temperature of ice formation during freezing. When a cell suspension is cooled rapidly, the external medium fully solidifies before substantial osmotic dehydration occurs. The cells may remain turgid without much change in their solute concentration, encouraging intracellular ice crystal formation. However, since intracellular compartments do not contain good sites for ice nucleation, cells may cool to well below the equilibrium temperature for initial ice formation before nucleation occurs. This thermodynamically unstable state encourages very rapid freezing of the entire cell after ice crystallization is initiated. Slow cooling rates allow time for a substantial loss of intracellular water due to osmosis before the extracellular medium fully solidifies. This increases the intracellular solute concentration and decreases the equilibrium temperature for initial ice formation within cells (Mazur 1970). Slow cooling also provides more time for the initial nucleation of intracellular ice crystals. The probability of the abrupt formation of fully frozen cell contents (black flashing) is accordingly diminished as the rate of cooling is decreased during cryopreservation.

Intracellular ice may have a variety of undesirable effects on algae. There is a correlation between intracellular ice formation and ultrastructural damage (e.g., physical disruption of cellular organelles and membranes), especially in large coenocytic cells and multicellular algae (McLellan 1989, Roberts et al. 1987, Fleck 1998). Ice fronts propagate along filaments within the thallus of multicellular algae (Morris and McGrath 1981, Fleck et al. 1997). The lack of cellular compartmentalization in the coenocytic *Vaucheria sessilis* (Vaucher) De Candolle facilitates lethal propagation of intracellular ice throughout the filament (Fleck et al. 1997). Uninucleate filaments of *Spirogyra grevilleana* Hassal and Kützing (Morris and McGrath 1981) also form intracellular ice that is correlated with lack of survival following freezing and thawing at low temperatures.

### 2.2.4. Other Causes of Damage at Low Temperature

Free radicals may be generated at low temperatures. Studies of higher plants indicate that hydroxyl radicals and singlet oxygen produced during low-temperature storage may reduce storage performance (Magill et al. 1994). An antioxidant produced by *Haematococcus pluvialis* in response to low-temperature stress may contribute to its freezing tolerance. In contrast, *Euglena gracilis* Klebs has a less-regulated antioxidant response, which may explain its freeze recalcitrance (Fleck et al. 2003).

## 2.3. Minimization of Cell Damage During Freezing and Thawing

*Chlorella protothecoides* Shihira et Kraus can survive direct immersion in liquid nitrogen (cooling rate of −2,000°C min⁻¹) in the absence of a CPA (Morris et al. 1977). However, most algae do not survive cryopreservation without some control of the conditions under which they are frozen and thawed. Cryopreservation protocols are designed to minimize intracellular ice formation and the effects of excessive osmotically driven changes in cell volume, since either may cause irreversible physical and/or chemical damage (Steponkus et al. 1992). Algae generally survive freezing and thawing only when the rate of temperature change is regulated and an appropriate CPA has been added to the algal suspension prior to freezing. Rapid cooling prevents excessive dehydration

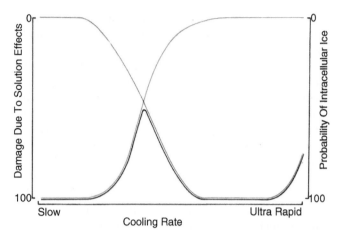

**FIGURE 12.1.** Two kinds of damaging stresses. Osmotically driven solution effects (e.g., plasmolysis) are damaging during slow cooling (*blue line*). Intracellular ice is damaging at rapid rates of cooling (*red line*). Survival is optimum when the combined effect of these two factors is minimal (*black line*). Both extracellular and intracellular water vitrify if an algal culture can be cooled to an ultralow temperature sufficiently rapidly (*red and black lines, far right*), precluding damage from either solution effects or intracellular ice. From Morris (1981).

of cells but encourages intracellular ice formation. In contrast, slow cooling minimizes damaging intracellular ice formation but causes extensive cell dehydration. Thus, an intermediate rate of cooling is usually optimal, as is illustrated in Figure 12.1.

The conditions for thawing a frozen culture are also important. New ice crystals can form and grow in a process called recrystallization, even as most ice is melting. Recrystallization generally damages cells. Virtually all cryopreservation protocols call for thawing frozen cultures as rapidly as possible to minimize recrystallization.

Cryoprotective agents are water-soluble compounds that, when added to an algal suspension prior to freezing, increase its viability as measured after thawing. The CPA displaces intracellular water if it can freely permeate the algal plasma membrane. A permeating CPA may simultaneously reduce intracellular ice formation and reduce the extent of osmotically driven changes in volume.

## 3.0. STANDARD CRYOPRESERVATION METHODOLOGIES

Most procedures can be categorized as two-step freezing protocols. Two-step protocols require the addition of a cell-permeating CPA to an algal culture prior to its freezing, and then the culture is cooled at a controlled rate to some subzero temperature (Step 1). Next, the sample is cooled rapidly to the final storage temperature (Step 2). The culture can be maintained at the storage temperature for an indefinite period of time.

Many variations are possible within this broad framework. Parameters that may affect the viability of a cryopreserved algal culture include: (1) the growth stage or physiological condition of the culture at the time of its harvesting for cryopreservation; (2) the selected CPA, its concentration, and the temperature at which it is added; (3) the exact cooling protocol, i.e., how fast it is cooled at each stage of the cooling process and the precision with which the cooling regime is controlled; (4) the intermediate temperature at which Step 1 is terminated and the culture is rapidly cooled to the final storage temperature; (5) the thawing conditions subsequent to cryostorage; and (6) treatment and incubation conditions of the culture during its recovery. These considerations will be discussed in subsequent sections.

The methods described here have been applied successfully to a broad range of microalgae, including over 700 strains in the CCAP and approximately 1,300 strains at UTEX. Much of the developmental work employed various strains of *Chlorella* at the CCAP as well as *Chlamydomonas reinhardtii* Dangeard and *Chlorococcum texanum* Archibald et Bold at UTEX.

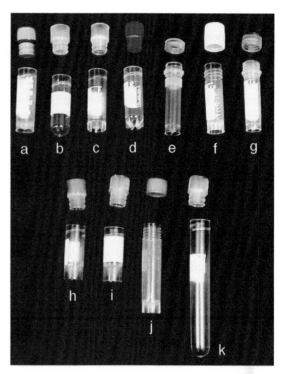

**FIGURE 12.2.** Plastic cryogenic vials: 1- to 2-mL volumes **(a–j)** and 5 mL capacity **(k)**. Vials with inside cap threads **(a–d, h, i, k)** or outer cap threads **(e–g, j)**; gasket molded into the vial **(e, g)**, molded into the lid **(f)**, or as a separate ring **(e.g., a)**; rounded bottom **(b, k)** or flat bottom and free-standing **(a, c–j)**. Outer vial diameter, labeling surface, colors of lids or lids inserts, transparency, and construction material (e.g., polypropylene, polyethylene) vary.

## 3.1. Selection and Labeling of Cryogenic Storage Vials

Presterilized thick-walled plastic (typically polypropylene) cryogenic vials, available through many major biological and medical laboratory supply companies, are used (Fig. 12.2). Vials with a flat or star-shaped base are most convenient because they are self-standing, yet can be set into a rack and opened with one hand. Although no manufacturer guarantees leak-proof vials, various design features are intended to minimize leaks while the vials are submerged in liquid nitrogen. The authors have found that silicone or natural rubber gaskets seldom allow liquid nitrogen to seep into the vial and usually maintain axenic cultures free of contamination. Shrinkable sheath tubing (e.g., Nunc Cryoflex, Nalge Nunc International) can be purchased to produce a tighter seal, although this too does not guarantee protection against liquid nitrogen seepage. The sheath increases the size of cryogenic vials and the difficulty of retrieving stored samples in tight-fitting inventory

systems (see Section 5.0). For box storage, color-coded inserts (sold by manufacturers of cryogenic vials) placed into the tops of cryogenic vials help to distinguish cultures quickly (see Fig. 12.5). Vented inserts prevent trapping of liquid nitrogen, therefore avoiding ejection of the insert during rapid thawing. For cane storage (see Section 5.1), a color tab on top of the cane is used to distinguish cultures. Cryogenic vials are best labeled with a fine-tip alcohol-resistant permanent marker. Despite limited space, the algal name or identifying number and the date of freezing should be labeled on each frozen vial. If multiple batches of vials are frozen, a batch number should also be written on each cryogenic vial and a permanent record should be kept of each batch. A rack that supports cryogenic vials is especially convenient during handling and labeling.

Although glass ampoules were once used to store cryopreserved cultures, they are now inappropriate and are avoided; any leak allows entry of liquid nitrogen, causing the glass ampoule to explode during thawing (see Section 7.0). Cryogenic straws are commonly used in mammalian cryobiology. They have better heat-transfer properties and thaw more quickly than conventional cryogenic vials, and they are available in a range of sizes (e.g., 0.25 mL, 0.5 mL, and 1.0 mL). They are available as high-security straws, which prevent liquid nitrogen leakage. However, straws have not been widely used in algal cryopreservation, primarily because they are harder to fill and to empty than are conventional cryogenic vials.

## 3.2. Pretreatment and Culture Optimization

Pre-cryopreservation culturing conditions may be selected to maximize viability subsequent to cryopreservation. Several studies indicate that growth conditions influence the susceptibility of algae to damage from freezing and thawing. For example, culture age (Morris 1978), light intensity (Beaty and Parker 1992), incubation temperature (Morris 1976b, Cañavate and Lubian 1997), osmotic potential of the medium (Cañavate and Lubian 1995), nutrient limitation (Ben-Amotz and Gilboa 1980), and nutritional mode (Morris et al. 1977) all have been correlated with viability. However, other reports show high post-cryopreservation viability without any special preadaptation conditions (e.g., Day and Fenwick 1993, Crutchfield et al. 1999). The majority of strains of microalgae in the CCAP, UTEX, and other culture collections have been successfully cryopreserved without any special preadaptation conditions.

Most protocols use actively growing cultures in late log or early stationary phase. The vigor of the culture at the time it is selected for cryopreservation may be the most important consideration. Actively growing cultures generally survive cryopreservation with higher viability than those in stationary or declining phase, or those growing under stressful conditions. Algae growing in liquid or on solidified medium (agar) may be successfully cryopreserved. Cultures do not need to be axenic. However, xenic cultures sometimes have an elevated level of contamination after thawing because many bacteria survive cryopreservation and thrive on the contents of algae that don't survive. Overgrowth by bacteria is not generally a problem in xenic marine strains (R. A. Andersen, personal communication).

## 3.3. Preparation of Cultures for Freezing

Some large filamentous and/or thalloid algae may be cut into smaller units prior to their cryopreservation in cryogenic vials. Since this is an inherently stressful treatment, it is advisable to allow at least 24 hours for damaged cells to recover. Centrifugation is frequently used to pellet microalgae when changing solutions or culture density. Viability is sometimes significantly diminished when the alga is subjected to centrifugation immediately prior to cryopreservation. Algae that contain relatively weak or flexible cell walls are especially susceptible to centrifugal stress. Thus, centrifugation should be minimized or avoided if possible. Viability may be improved if a culture that has been subjected to centrifugation is allowed to recover under normal culture conditions for at least a day prior to subsequent cryopreservation.

Algae can be cryopreserved in their normal culture medium, provided an appropriate CPA is added to the medium at a suitable concentration prior to freezing (see the following). Cultures growing on agar may first be suspended to homogeneity in liquid medium so that a uniform and controlled concentration of CPA can be added. At the CCAP, cultures grown on agar are normally transferred to liquid medium and incubated under standard conditions (see Chapter 10) for at least 2 weeks prior to cryopreservation.

Many algae can be cryopreserved directly on an agar slant. A small volume (0.3–0.5 mL) of sterile agar medium is aseptically transferred to a sterile 2-mL cryogenic vial and allowed to solidify as a slant. An algal culture is spread or streaked on the surface of the solidified agar and incubated under normal growth conditions. The culture is ready for cryopreservation after it has grown into a heavy streak or a lawn, and many cyanobacteria and unicellular chlorophytes can be cryopreserved after 2 to 3 weeks. In preparation for cryopreservation, growth medium containing an appropriate CPA is added slowly to the cryogenic vial in order to minimize the disturbance of algae growing on the agar surface. This preparation of algae can then be cryopreserved like liquid cultures. To revive the culture, rapidly thaw the vial, gently decant the liquid, and add fresh culture medium to the vial. If the alga adheres to the agar surface, then the liquid medium can again be decanted and the cryogenic vial incubated under normal culture conditions. Viable algae remaining on the agar surface typically grow into a lawn within 2 to 3 weeks. This method is convenient when a culture must be kept axenic; it requires minimum handling and is practical for cultures that are highly sensitive to mechanical stress, because centrifugation is avoided. However, it does not facilitate accurate measurements of viability and cannot be used for strains that resist growth on solid medium.

Effects of cell density on viability have not been examined extensively. However, *Chlamydomonas reinhardtii* does not survive cryopreservation with high viability unless the cell density is less than approximately $2.5 \times 10^6$ cells $\cdot$ mL$^{-1}$ (Brand and Diller 2004). This may be because algal cells killed by freezing release a substance that, when present at high concentration, kills cells that would otherwise survive cryopreservation. However, microalgal cultures cryopreserved at very low cell densities (less than $10^4$ cells $\cdot$ mL$^{-1}$) also often fail to recover when thawed. This may be due to a lack of self-shading, causing photooxidative stress from excessive light levels during recovery. The concentration of the culture medium of a frozen culture also may affect viability; cultures cryopreserved in media of half strength or less sometimes have significantly improved viabilities.

## 3.4. Cryoprotectants

Cryoprotective agents generally must be added at high concentrations to afford protection from cell damage during freezing and thawing. Two classes of CPAs may be distinguished: agents that passively move through the plasma membrane to equilibrate between the extracellular solution and the cell interior (penetrating or permeating CPAs) and those that do not pass through the plasma membrane and remain in the extracellular solution (nonpenetrating or nonpermeating CPAs). Three penetrating CPAs have been utilized quite extensively for algal cryopreservation: methanol (MeOH), dimethylsulphoxide (DMSO; Me$_2$SO), and glycerol (Taylor and Fletcher 1998). Freshwater and terrestrial strains of microalgae cryopreserved at the CCAP and UTEX have responded better to MeOH and DMSO than to glycerol, and MeOH often is the preferred CPA. Many marine phytoplankters are most effectively cryopreserved with DMSO (R. A. Andersen, personal communication), while glycerol is effective for *Tetraselmis* (Day and Fenwick 1993). A penetrating CPA lowers the temperature at which intracellular water freezes (Franks 1985) and reduces osmotically driven decreases in cell volume. In addition, permeating CPAs may confer cryoprotection by altering membrane properties such as solute permeability (Santarius 1996). Ethylene glycol and formamide (rarely used as algal CPAs), as well as DMSO, may decrease the cell membrane permeability for ions and lower the membrane potential (Chekurova et al. 1990). Penetrating cryoprotectants may also act as free radical scavengers (Benson 1990).

At 25°C, exogenously added MeOH equilibrates across the plasma membrane of the unicellular chlorophyte *Chlorococcum* (UTEX 1788) within 1 minute, which is somewhat faster than the rate of water equilibration. In contrast, DMSO requires approximately 10 minutes to fully equilibrate, which is slower than the rate of transport of water. Rates of transport of water and CPAs decrease by an order of magnitude at temperatures lowered from 25°C to near 0°C, and they continue to decline as the temperature is further lowered (Brand and Diller 2004). The relative rates of transport of different CPAs with respect to water greatly affect the direction, extent, and rate of transient changes in cell volume. The difference in transport rates of DMSO and MeOH may help explain why many freshwater and terrestrial algae with robust cell walls that can tolerate transient swelling are best cryopreserved with methanol, while many marine species, which often lack a strong wall, are more effectively cryopreserved with DMSO.

Penetrating CPAs are toxic at high concentrations. Prolonged exposure to methanol at concentrations that are used for cryoprotection (typically 5–10% v/v) is toxic to *Euglena gracilis*, and even short-term (20-minute) exposure to concentrations greater than 15% (v/v) may be damaging (Fleck 1998). Monohydric alcohols, DMSO, and ethylene glycol denature enzymes at room temperature, and DMSO destabilizes proteins (Adam et al. 1995). However, DMSO may protect isolated enzymes during freezing (Adam et al. 1995, Anchordoguy et al. 1992). This apparent paradox has been attributed to temperature-dependent, hydrophobic interactions between DMSO and nonpolar moieties of proteins. At temperatures below −22°C, low concentrations of rapidly permeating cryoprotectants may act as cryosensitizers, thereby accelerating membrane damage (Santarius 1996). In addition, DMSO has been observed to cause artificial phospholipid bilayers to become leaky due to a hydrophobic association between DMSO and the bilayer (Anchordoguy et al. 1992). Thus, a permeating CPA should be added to a culture only immediately prior to its cryopreservation and should be removed as soon as possible after thawing.

Occasionally, permeating CPAs are added after cultures are cooled to 0°C or lower, in order to minimize intracellular toxicity (Fleck 1998). However, the CPA should not be added after ice has formed in the culture. Due to slow CPA membrane transport at low temperatures, an algal culture to which CPA has been added at a low temperature should be incubated several minutes before it is further cooled, in order to ensure adequate equilibration of the CPA and water across the plasma membrane. An equilibration time as long as 30 minutes may be required if DMSO is added at 0°C or lower, while less equilibration time is required for MeOH.

If MeOH or DMSO is used as CPA, it is convenient to first prepare a 20% (v/v) CPA stock solution. Both MeOH and DMSO have high heats of hydration, generating large amounts of heat when combined with aqueous solutions. Preliminary dilution of these agents to 20% solutions releases most of the heat, and very little additional heat is produced when the 20% solution is further diluted into an algal culture. Solutions of MeOH and DMSO are sterile when purchased and will remain so if diluted with sterile medium and handled aseptically. Alternatively, 20% stock solutions of CPA can be sterilized by passage through a 0.22-μm membrane filter.

An appropriate volume of CPA stock solution is combined with an algal liquid culture to achieve the desired final concentration of CPA. For an agar slant culture grown in a cryogenic vial, a 1.0-mL volume of culture medium containing the final concentration of CPA may be added directly to the vial.

For species of algae that have no prior cryopreservation history, it is often most efficient to determine their tolerance to a range of concentrations of CPA before attempting cryopreservation. Identical cultures of the alga are prepared in normal growth medium containing several different concentrations of the CPA. The cultures are incubated in darkness at normal growth temperature for 1 to 2 hours, and then the CPA is removed or diluted to a concentration less than 0.2% (v/v) and the cultures are allowed to resume normal growth. The highest concentration of CPA that does not kill or severely retard growth of the algal culture may facilitate cryopreservation with highest viability.

Concentrations of MeOH or DMSO less than 2% (v/v) are seldom effective as CPAs, while concentrations higher than 12% (v/v) are often toxic. Within this range the most effective concentration varies greatly among species, sometimes even among closely related strains. Many strains of microalgae in the UTEX and CCAP collections have been successfully cryopreserved with 5% (v/v) MeOH or 5 to 8% (v/v) DMSO, but the most effective concentrations for individual strains often must be determined empirically. For example, concentrations of DMSO higher than 9% (v/v) are toxic to *Pfiesteria*, while concentrations lower than 5% are ineffective in preventing freezing/thawing damage. *Pfiesteria* retains high viability when cryopreserved with 7 to 8% (v/v) DMSO (R. A. Andersen, personal communication).

Nonpermeating CPAs such as polyvinylpyrrolidone (PVP), hydroxyethyl starch (HES), and polyethylene glycol (PEG) have not been utilized extensively for cryopreserving algae. Their effective usage in plant cryopreservation may result from dehydration of cells, thereby reducing the amount of intracellular water available for ice formation (Benson 1990). These agents may also have a protective effect on plant thylakoid membranes during rapid freezing (Santarius 1996). Morris (1976a) used 10% (w/v) PVP as a CPA to cryopreserve *Chlorella*, but achieved higher levels of viability using DMSO, a penetrating CPA. The cryoprotective effects of nonpenetrating CPAs, including HES, PEG, and PVP, have been tested on a variety of microalgae in CCAP and UTEX that are recalcitrant to cryopreservation. None of these compounds afforded cryoprotection in any of the strains examined.

Free radicals may be produced when primary metabolism is perturbed or electron transfer processes become uncoupled during cooling or freezing. Free radical injury occurs during storage at temperatures higher than −130°C (Fuller and Green 1986). Oxidative stress occurs in mammalian transplant organs exposed to low-temperature storage, and free radical–mediated loss of organ function can be ameliorated by applying antioxidants and free radical–scavenging agents (Whiteley et al. 1992). Fleck et al. (2000) demonstrated that incorporating $10 \, mg \cdot L^{-1}$ of the chelating agent desferrioxamine in the cryopreservation solution significantly enhances survival of *Euglena gracilis* upon thawing. Free radical scavengers have not been examined extensively in algal cryopreservation protocols, but may have wider applicability.

### 3.5.0. Cooling Protocols

For most algae, cooling too rapidly causes formation of lethal intracellular ice, while cooling too slowly causes osmotically induced cell damage. These contradictory limitations make it difficult to achieve a truly optimum protocol. However, cooling rates in the range of $0.1°C \cdot min^{-1}$ to $-10°C \cdot min^{-1}$ represent a compromise that for many strains allows a statistically high percentage of algal units in a population to survive cryopreservation. Cooling at a rate of $-1°C \cdot min^{-1}$ is a widely adapted standard. A calibrated reliable cooling device is required for reproducible rates of cooling during cryopreservation. Cooling devices for cryopreservation generally fall into two categories: passive freezing systems and controlled-cooling-rate freezers.

### 3.5.1. Passive Freezing Systems

Passive freezing systems are often employed in two-step freezing protocols. In Step 1, cryogenic vials or straws containing algal cultures prepared for cryopreservation are placed into a static system (an insulated container) that is exposed to a very low temperature (typically by insertion into a −80°C freezer). The insulation of the container retards heat transfer, so the contents of the container cool relatively slowly. The interior of a properly designed passive freezing system cools at an approximately linear rate from 0°C to less than −40°C. When the contents of the cryogenic vials reach a sufficiently low temperature (typically −30 to −80°C), they are removed from the insulated container and transferred directly to a permanent ultracold storage location such as a liquid nitrogen dewar (Step 2 of two-step cooling protocols). Step 1 cooling is terminated at −45°C to −55°C when MeOH is used as the CPA at UTEX, and at −30°C to −40°C when DMSO is used as the CPA at the CCAP. The sample must be transferred quickly from the insulated container to a permanent storage location to avoid excessive warming during the transfer time. The temperatures of the transferred samples quickly (within a few minutes) approach the internal temperature of the storage container (−196°C for liquid nitrogen).

An insulated box is a very inexpensive passive freezing device. A cubical Styrofoam box 10 to 15 cm in internal dimensions, with walls and a tight-fitting lid approximately 3.5 cm in thickness (e.g., a dry ice shipping container) often works well. The rate of temperature decrease in Step 1 cooling can be determined by placing a thermocouple into the insulated box before the box is closed and placed into the −80°C freezer, then recording the temperature at time intervals during cooling. More meaningful temperature measurements can be taken during cooling if a hole is placed into the lid of a cryogenic vial that contains a culture prepared for cyropreservation and then a small thermocouple is inserted through the hole into the culture. Highly reproducible cooling rates require that cryogenic vials always contain the same volume of solution, and vials are always placed in approximately the same position within the box. A method for varying the cooling rate in a controlled way is to place a closed container (metal or plastic with good heat transfer properties) of isopropyl alcohol along with the samples into the insulated box. The rate of cooling is then inversely related to the volume of isopropyl alcohol in the container.

Commercially available controlled-cooling canisters, such as "Mr. Frosty" (Fig. 12.3) (Nalge Nunc Interna-

**FIGURE 12.3.** Mr. Frosty, a passive freezing device. Isopropyl alcohol is placed into the clear plastic container (*left*). Cryogenic vials (1–2 mL) containing prepared algal cultures are placed into the vial holder (*right*), the lid is placed securely on the container, and the entire unit is placed into a −80°C freezer (see text for details).

tional) and "Handi-Freeze" (Taylor Wharton), are inexpensive and provide highly reproducible cooling rates. Isopropyl alcohol is placed into a reservoir adjacent to the chamber containing cryogenic vials, and the unit is then placed into a −80°C freezer. The chamber cools at approximately −1°C · min$^{-1}$ over a temperature range of 0°C to −50°C. A convenient protocol suitable for many strains of microalgae is to (1) prechill the freezing canister to 4°C, (2) place room-temperature cryogenic vials containing algal cultures to be cryopreserved into the prechilled canister, (3) immediately place the closed canister into a −80°C freezer, and (4) remove the canister after the desired temperature is reached and quickly transfer the frozen vials to an ultracold storage vessel.

### 3.5.2. Controlled-Cooling-Rate Freezers

Some algae require a more carefully controlled cooling rate or a more complex cooling pattern than can be achieved with passive freezing devices. A variety of commercial instruments (e.g., Biotronics; Planer Products; CryoMed, Thermo Electron Corporation; Gordinier Electronics CryoLogic) allow accurate control and manipulation of the cooling regime. A temperature probe is inserted into a cryogenic vial that contains solution very much like the contents of vials of cultures prepared for cryopreservation. The vials prepared for cryopreservation, along with the vial containing the temperature probe, are then inserted into the cooling chamber of the controlled-cooling-rate freezer. This probe, along with an additional probe projecting into the cooling chamber, is connected to an electronic device that regulates the entry of vapor-phase nitrogen into the chamber. The rate of cooling is determined by the rate of entry of cold nitrogen into the chamber in response to temperatures sensed by the probes according to a user-defined cooling protocol. Electronic and printed outputs that describe the cooling protocol allow the details of each cryopreservation run to be recorded automatically.

Controlled-cooling-rate freezers can be programmed to produce a wide range of customized cooling protocols. For example, they can produce user-defined cooling rates over various temperature ranges and can allow the temperature to remain constant for specified periods of time at critical points in the cooling process. As with passive freezing devices, frozen cryogenic vials must be transported rapidly from the cooling chamber of a controlled-cooling-rate freezer to the permanent storage vessel after Step 1 cooling.

A simple protocol calls for cooling the chamber cavity at $1°C$ $min^{-1}$ from $+20°C$ to $-40°C$, then dwelling at $-40°C$ for 30 minutes prior to transferring cryogenic vials to liquid nitrogen. A more complex cooling program successfully employed at CCMP for marine strains involves cooling the contents of cryogenic vials from ambient temperature to $4°C$ at $-1°C$ $min^{-1}$ then holding the temperature constant for up to 5 minutes. This dwell time is especially effective for cold polar strains and is sometimes required for adequate penetration of the CPA. Vials' contents are next cooled at $-1°C$ $min^{-1}$ until they reach $-9°C$. Seawater remains a supercooled liquid at that temperature. The cooling chamber is then cooled rapidly to $-45°C$ in order to quickly drive the contents of cryogenic vials down to $-12°C$. This causes ice nucleation and rapidly removes the latent heat of fusion. The contents of vials are then cooled at $-1°C$ $min^{-1}$ until they reach $-45°C$, which is below the eutectic. The vials are then cooled rapidly to $-90°C$ and finally transferred from the cooling chamber to a liquid nitrogen storage system (R. A. Andersen, personal communication).

### 3.5.3. Comparison of Passive Cooling Systems and Controlled-Cooling-Rate Freezers

Passive freezing systems for Step 1 cooling are inexpensive (<$100 U.S.) and are highly effective for developing protocols and cryopreserving many algal strains. However, their calibration is time-consuming and they cannot perform complex time-dependent cooling protocols. Controlled-rate cooling instruments are generally preferred for cryopreservation research because of their accuracy, reproducibility, and flexibility. However, they are expensive (often costing over $10,000), and they must be attached to a liquid nitrogen reservoir.

### 3.6. Controlled Ice Nucleation

It is possible to minimize supercooling in the extracellular solution during cooling. The cryogenic vial is allowed to cool to the equilibrium temperature for initial ice formation. A probe with good heat transfer properties such as the tip of a metal spatula is cooled to liquid nitrogen temperature and then is physically pressed against the outside surface of the cooled cryogenic vial at the level of the top of the liquid. This seeds the formation of ice in the region of the vial in closest proximity to where the probe contacts the vial.

In an alternative procedure, the cryogenic vial is cooled to the equilibrium temperature for initial ice formation and then is plunged into a bath of ethyl or isopropyl alcohol maintained at $-45°C$. Ice forms quickly within the cryogenic vial, and the heat of fusion is rapidly transferred out of the vial. Immediately after ice crystals have formed, before the temperature of the content of the cryogenic vial has dropped by more than a few degrees below the equilibrium temperature for ice nucleation, the vial is placed into a controlled-rate cooling device and cooling is resumed to complete Step 1.

Ice nucleation is standard practice for cryopreserving mammalian sperm and embryos but seldom is used for algae. However, this might improve the reproducibility of viability of replicate cultures, especially those with low viabilities.

---

## 4.0. ALTERNATIVE FREEZING PROTOCOLS

Two-step freezing protocols are widely utilized for cryopreserving algae, but many species remain recalcitrant. A variety of alternative methods that are successful with other organisms are now being examined with algae.

### 4.1. Plunge Cooling

A few small unicellular Chlorophyceae such as *Chlorella protothecoides* Krüg. can withstand direct immersion into liquid nitrogen without a cryoprotectant or any control over the rate of cooling (Morris 1981). Some strains of *Chlorella* can be preserved by incubating a culture in 5% (v/v) DMSO at ambient temperature for 5 minutes in a cryogenic vial and then plunging the vial directly into liquid nitrogen. In a slight modification of this method, the cryogenic vial may be placed into the vapor phase of liquid nitrogen temporarily in order to retard the rate of cooling, or it may be stored permanently in the vapor phase.

## 4.2. Nonlinear Cooling Systems

Although most cryopreservation protocols utilize linear cooling rates, Morris et al. (1999) demonstrated that nonlinear changes in temperature were superior for cryopreserving spermatozoa. A similar approach has been used successfully to cryopreserve *Euglena gracilis* (Morris and Day, unpublished data). The biophysical basis of these different responses to differing cooling regimes were examined using freeze substitution and a scanning electron microscope equipped with a cryostage. Surprisingly, samples in the frozen state were not osmotically dehydrated, nor did they have any visible intracellular ice. Viability subsequent to thawing did not appear to correlate with conventional theories of cellular freezing injury. Nonlinear cooling protocols show potential, but widespread application is unlikely since they require highly specialized cooling equipment.

## 4.3. Encapsulation-Dehydration

A vitrification-based technique called encapsulation-dehydration, originally developed for higher plant tissues (Fabre and Dereuddre 1990), has been applied to algae. The procedure involves (1) the encapsulation of algal units in calcium-alginate beads, (2) osmotic dehydration, usually by incubation for 24 hours in 0.5 to 1.0 M sucrose, and (3) desiccation to a moisture content of 20 to 30% (see Fig. 12.4; Harding et al. 2004). The dehydrated beads are cooled to a cryogenic temperature, whereupon they become vitrified (Benson 1990). The intracellular viscosity becomes so high during dehydration that residual intracellular water also forms a glass. The vitrified beads containing imbedded algae may be stored indefinitely at cryogenic temperatures.

The following encapsulation-dehydration procedure may be used to cryopreserve microalgae. A liquid culture of algae is subjected to centrifugation at minimum force to pellet the alga. The supernatant is decanted and the algal units are suspended in culture medium containing 2 to 5% (w/v) sodium alginate and 0.5 M sucrose (to increase the specific gravity of the solution and to initiate dehydration). The mixture is then dispensed dropwise into a container of 100 mM

$CaCl_2$ solution several cm in depth. Solid beads of calcium alginate containing trapped algae form spontaneously as the drops fall through the solution. The solution is left undisturbed for 30 to 60 minutes, during which algal cells begin to dehydrate. The beads containing encapsulated algae are then transferred to a fresh solution of sucrose (0.4–0.9 M) in culture medium for 24 hours to dehydrate. Alternatively, in a two-stage process the beads may be incubated for 24 hours in 0.5 M sucrose, followed by an additional 24 hours in 0.75 to 1.2 M sucrose. The encapsulated algae can be kept at ambient light and temperature conditions during dehydration.

The alginate beads may be dried by air flow desiccation rather than incubation in concentrated sucrose solution. In this method beads are first transferred from the calcium alginate/sucrose solution to the surface of sterile filter paper in a petri dish to blot away excessive surface moisture. The beads are then transferred to a sterile 9-cm-diameter petri dish that is left open in a laminar air-flow hood for 1 to 4 hours. Alternatively, after removal of surface moisture, the beads may be transferred to a sterile glass petri dish containing 10 to 15 grams of active (preheated at 160°C overnight) silica gel that has been covered with a sterile filter paper. The petri dish is sealed with Parafilm and placed under ambient light and temperature for 5 to 8 hours.

Regardless of the method of dehydration, the dehydrated beads are transferred with sterile forceps into 2.0-mL cryogenic vials. Often the vials containing dehydrated beads are plunged directly into a permanent ultra-cold storage facility, which causes rapid cooling. Alternatively, vials may be cooled more slowly, exactly as in two-step protocols (Day et al. 2000). An effective protocol calls for holding vials at 0°C for 5 minutes, cooling at $-0.5°C \cdot min^{-1}$ to a terminal temperature of $-40°C$ to $-60°C$, and then holding for 30 minutes at that terminal temperature before plunging them into liquid nitrogen. This protocol was marginally more effective for cryopreserving *Euglena gracilis* than conventional two-step cooling procedures (Fleck 1998). Although calcium alginate encapsulation-dehydration has only recently been attempted with algae, this method has successfully cryopreserved marine and hyper-saline algae such as *Dunaliella tertiolecta* Butcher, *Nannochloris* sp., *Brachiomonas submarina* Bohlin, and *Nannochloropsis* sp., as well as several freshwater strains, including *Chlorella emersonii* Shihiri et Klaus, *Nostoc commune*, and *Euglena gracilis* (Hirata et al. 1996, Day et al. 2000).

Encapsulation-dehydration minimizes ice formation during freezing and prevents recrystallization during thawing if the vitrified sample is warmed rapidly. The

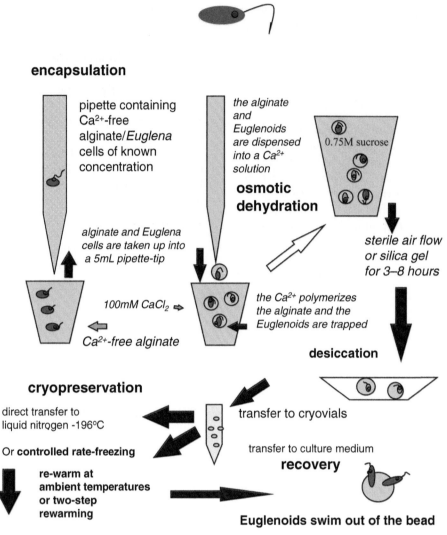

**encapsulation**

pipette containing Ca²⁺-free alginate/*Euglena* cells of known concentration

the alginate and *Euglenoids* are dispensed into a Ca²⁺ solution

0.75M sucrose

**osmotic dehydration**

alginate and *Euglena* cells are taken up into a 5mL pipette-tip

sterile air flow or silica gel for 3–8 hours

100mM CaCl₂ ⇒

Ca²⁺-free alginate

the Ca²⁺ polymerizes the alginate and the *Euglenoids* are trapped

**desiccation**

**cryopreservation**

direct transfer to liquid nitrogen -196°C

transfer to cryovials

**Or controlled rate-freezing**

transfer to culture medium

**recovery**

re-warm at ambient temperatures or two-step rewarming

**Euglenoids swim out of the bead**

**FIGURE 12.4.** Algal encapsulation-dehydration protocol for cryopreserving *Euglena* (after Harding et al. 2004).

protective environment of calcium alginate beads may extend its usefulness to other methods that utilize severe dehydration. Encapsulation-dehydration is more labor-intensive than most established methods for cryopreserving algae. Cultures are somewhat difficult to maintain axenically because a number of exposed manipulations are required and the dehydrating solutions encourage heterotrophic microbial growth.

## 4.4. Vitrification

Vitrification is a process in which water is transformed directly from liquid to a glassy state (an extremely viscous liquid), thus circumventing the formation of ice crystals. Vitreous solutions remain unchanged when maintained at temperatures below $T_g$ but convert spontaneously to crystalline ice if warmed above $T_g$. Pure water and dilute aqueous solutions form a vitreous state if cooled extremely rapidly ($>1,000°C \cdot min^{-1}$) (MacFarlane 1987), although it is seldom possible to achieve such rapid rates of cooling under normal laboratory conditions. Highly viscous aqueous solutions can be vitrified without cooling so rapidly.

Vitrification can be utilized in cryopreservation protocols by preparing a culture in a highly viscous solution such as a very high concentration of sucrose and then cooling to a temperature less than $T_g$ (Stillinger 1995). The viscous sucrose solution not only facilitates vitrification of the extracellular solution, but the low osmotic potential causes severe dehydration of cells, which also encourages intracellular vitrification. The extent of cellular dehydration can be controlled if the culture is cooled to a subzero temperature, held at that temperature just long enough to achieve the desired degree of dehydration, and then plunged into liquid nitrogen. Relatively rapid cooling is necessary because even highly viscous aqueous solutions will crystallize rather than vitrify at low temperatures above $T_g$ if given sufficient time.

Conventional cryopreservation protocols sometimes require extensive testing to determine which conditions simultaneously minimize intracellular ice formation and damaging osmotic effects. Vitrification, although not likely to be successful for cultures highly sensitive to osmotic stress, can be tested quickly without preliminary studies. Vitrification also may circumvent problems of chilling injury, and it does not require expensive or specialized equipment. However, vitrification causes some of the same stress conditions as those encountered with two-step freezing protocols because of the requirement for very low extracellular osmotic potentials (Steponkus et al. 1992). As with other protocols, cultures revived from a cold vitreous state must be warmed very rapidly in order to minimize ice formation.

In attempts to cryopreserve vitrified algae at the CCAP with the approach of Sakai et al. (1991) for cryopreserving meristematic tissue of higher plant species, only *Enteromorpha intestinalis* (L.) Nees retained viability. In other algae, lethal injury was attributed to the toxicity of the vitrification solution when added to the algal suspension (Fleck 1998).

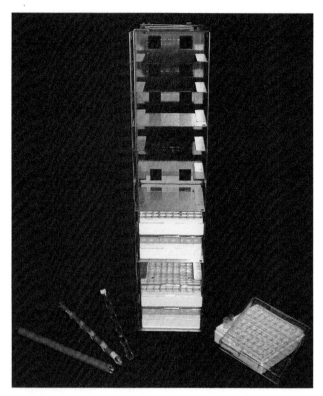

**FIGURE 12.5.** Components of a cryostorage system. Boxes (typically plastic) that contain frozen samples in cryogenic vials (*right*) are slid into position in a vertical storage rack (*center*), which is then placed into the ultracold dewar (see Fig. 12.6) for permanent storage. Alternatively, cryogenic vials may be clipped into place in a vertical column on metal canes (*left*); a sleeve (*far left*) may be used to enclose the cane.

## 5.0. STORAGE OF CRYOPRESERVED MATERIAL

Any storage location and system that continuously maintains frozen algal cultures at temperatures below $T_g$ is acceptable for long-term cryopreservation. However, different systems vary greatly in cost, reliability, and convenience. Since the storage system may be the most expensive capital investment when establishing a cryopreserved inventory of algae, it is important to carefully consider which system is best for the required application before making an initial purchase.

### 5.1. Storage Holders for Cryopreserved Samples

A system of racks such as the one shown in Figure 12.5 can be filled with a column of boxes which each hold 81 or 100 cryogenic vials, thus allowing a large number of frozen cultures to be stored in an efficient and highly organized manner while facilitating relatively easy visual inspection. However, adding or removing individual cryogenic vials to this system is somewhat cumbersome and time-consuming, and a box is easily dropped during rushed operations, thus risking the loss of a number of samples.

An alternative system utilizes a device called a cane to hold several (typically a maximum of five) cryogenic vials in a vertical column (Fig. 12.5). Although not quite as orderly as racks of boxes for permanent storage, canes can be transferred to liquid or vapor-phase nitrogen very quickly, thus minimizing uncontrolled warming during handling procedures. Since canes can be added or removed from the storage container individually, the risk to the remaining inventory is minimal. Canes are equipped with sleeves that surround the cryogenic vials, thus minimizing the risk of losing vials into the storage container.

The convenience of canes and the efficiency of storage boxes can be combined by placing cryogenic vials into canes for temporary initial storage in liquid or vapor-phase nitrogen. After several hours the vials can be transferred into boxes in racks for permanent storage.

### 5.2.0. Long-Term Storage Facilities

Cryopreserved cultures in cryogenic vials are generally maintained for long periods of time in one of three ways: submerged in liquid nitrogen (−196°C), in the cold vapor phase above liquid nitrogen (approximately −165°C), or in an ultracold freezer (−150°C). An appropriate storage system as well as a convenient system for recording and maintaining accurate records should be planned before beginning a cryopreservation program.

### 5.2.1. Liquid Nitrogen and Vapor-Phase Storage Systems

The liquid phase of nitrogen provides a low and stable storage temperature. However, pressure differences during insertion and removal of samples may cause liquid nitrogen to seep into cryogenic vials, leading to potentially explosive evaporation when the vials are warmed quickly to revive cultures. Direct contact with liquid nitrogen does not harm algal cells, but seepage of liquid nitrogen into cryogenic vials often facilitates entry of freeze-tolerant bacteria, thereby contaminating the sample. Liquid-phase storage vessels require regular (typically monthly or biweekly) filling with liquid nitrogen.

Cryogenic vials stored in the vapor phase above liquid nitrogen are less likely to experience bacterial contamination and have a lower risk of explosive rupture during thawing than vials submerged in liquid nitrogen. However, vapor-phase storage increases the risk that the contents of cryogenic vials will temporarily become warmer than $T_g$ due to convective heat transfer in the storage container while cultures are added or removed from the inventory. Temperatures as high as −55°C have been observed near the top of older storage dewars when inventory was added or removed (D. Smith, CABI Bioscience, personal communication). Modern vapor-phase dewars designed for long-term cryostorage maintain a temperature of −165 to −170°C in the highest positions where samples are stored. Vapor-phase storage systems maintain a relatively small volume of liquid nitrogen in the bottom of the storage container and must be replenished frequently (typically once per week). Some manufacturers (e.g., Custom Biogenic Systems), now produce vapor-phase storage systems

that retain liquid nitrogen in the vessel walls, improving safety and providing a superior vertical temperature profile.

A reliable regular source of liquid nitrogen is essential for all systems that rely on liquid nitrogen to maintain inventory at ultracold temperature. Thus, the cost of the liquid nitrogen and the quality of the delivery service are important considerations when planning and implementing a long-term cryogenic storage system. Liquid nitrogen–generating systems are commercially available (e.g., Cryomech, Sterling Cryogenics) and can be located on site, but are expensive to purchase, set up, and maintain.

It is important not to choose an excessively small liquid nitrogen storage unit when starting a cryopreservation program because inventories of stored cultures accumulate quickly and small systems lose nitrogen faster than larger ones in proportion to their storage capacity. It is much less expensive in initial investment and in annual liquid nitrogen cost to maintain a given storage capacity in one large container rather than in two or more smaller containers.

Liquid nitrogen storage dewars of similar capacity vary greatly in cost. Those which lose nitrogen vapor most slowly are often the most expensive but are less expensive to maintain. The static nitrogen loss (the rate at which liquid nitrogen evaporates while the tank is left closed and undisturbed) is an especially important consideration when purchasing a storage dewar for storing archival material. The most efficient systems have only a small opening for adding or removing inventory. The inconvenience of a small access port is compensated for by the lower recurring cost for liquid nitrogen if access to the inventory is seldom required. Containers with a narrow opening have the added advantage that the internal temperature of nitrogen vapor above the liquid generally remains well below $T_g$, even at positions near the top of the vessel.

Bin-type storage dewars with wide openings and large lids (Fig. 12.6) are much more convenient to access than dewars with narrow openings. However, their rate of nitrogen loss may double the cost of liquid nitrogen, and access to vapor-phase bin-type systems must be limited to very brief periods of time in order to avoid excessive warming of the bin contents.

Automated liquid nitrogen storage dewars can be equipped with many safety and convenience features. Some have an external display that shows the level of liquid in the container. Some are equipped with a safety alarm that signals malfunction or a dangerously low level of liquid in the container. They may be purchased with an automatic filling device that connects directly

**FIGURE 12.6.** Removing a vertical storage rack from an automatic-fill storage dewar (*left*) that is connected to a liquid nitrogen reservoir (*right*).

to an external source of liquid nitrogen such as a standard 160-liter transport cylinder (Fig. 12.6). The dewar fills automatically to a preset user-determined high level of liquid nitrogen whenever the level drops lower than a user-determined low level. These features add substantially to the cost of the storage dewar, but are strongly recommended (especially a safety alarm) for storage of highly valuable or irreplaceable cryopreserved cultures.

### 5.2.2. Ultracold Mechanical Freezers

Mechanical freezers that cool to −150°C are commercially available. They provide relatively maintenance-free low-temperature storage and eliminate the requirement for a constant supply of liquid nitrogen. However, they are expensive to purchase, and their long-term reliability is yet to be proven. Although they are equipped with elaborate warning systems in case of malfunction such as compressor failure, when they do fail the temperature quickly increases to values higher than $T_g$. Thus, a back-up system is necessary to avoid high risk of catastrophic loss. Mechanical freezers are available that connect to a liquid nitrogen reservoir and automatically receive nitrogen in case of failure. They are more reliable than stand-alone mechanical freezers

but offer less of an advantage over liquid nitrogen storage systems.

## 5.3. Designing a Cryostorage Facility

Storage vessels should be maintained in a dedicated room with controlled access. Vessels located in a working laboratory or other accessible area should be kept locked to prevent unauthorized access. The storage area must be well ventilated to prevent oxygen depletion during periods of high nitrogen gas release, such as during filling or in case of an accidental nitrogen spill. Storage facilities should be located at a site that allows convenient vehicular delivery since liquid nitrogen is transported in very heavy containers. Transport of liquid nitrogen through pipes over long distances is not efficient and not generally recommended. Even vacuum-sheathed pipes designed to minimize heat transfer do not prevent substantial losses. For example, a vacuum pipe 6 m long may vaporize up to 40 liters of liquid nitrogen before it cools to −196°C.

## 5.4. Inventory Stock Control and Data Management

Accurate records of stored materials are crucial for maintenance of cryopreserved stock and for managing the inventory of stored cultures. Various commercial database systems are designed for managing records, but it is important to select a data management system that is flexible over the full range of user requirements. It is best to maintain up-to-date hard-copy records in addition to electronic records. For many purposes the minimum required data are: (1) strain identification (genus and species); (2) date cryopreserved; (3) protocol employed; (4) location in the cryostorage inventory; (5) number of cryogenic vials stored; (6) removal dates; and (7) an indication of viability subsequent to thawing (ideally, % viability).

## 6.0. POST-STORAGE RECOVERY

Conditions for thawing and reviving frozen cultures seldom receive as much attention as freezing protocols. Yet, algae recovering from the frozen state are highly stressed and are susceptible to further damage.

## 6.1. Thawing Protocols

Rapid warming is critical during the interval from when melting first begins until the culture is fully thawed in order to minimize localized refreezing. Usually the cryogenic vial is plunged into a water bath immediately after its removal from the storage container. A warm (e.g., 35°C) water bath is generally preferred since the speed of thawing increases as the temperature of the water bath is increased, but to minimize stress the vial should be removed from the bath before the algal culture warms above its normal growth temperature.

For many strains of cryopreserved algae, the CPA must be removed immediately after thawing to maximize viability. Some strains remain viable when kept in the presence of high concentrations of CPA for several hours after thawing, provided they are kept away from bright light. The gentlest method for removing CPA is to add fresh medium to the thawed culture until the concentration of CPA has been diluted to an acceptable level. This may require dilution by at least 50×. Many algal strains are very fragile immediately after thawing, so minimal centrifugal force should be applied if centrifugation is needed to separate algae from the solution containing the CPA. The culture should be protected from bright light after thawing until the CPA has been removed or highly diluted.

For encapsulated-dehydrated samples, cryogenic vials containing alginate beads may be warmed to room temperature (20 to 25°C) and maintained at that temperature for approximately 30 minutes. Beads are then transferred aseptically into 5 mL of culture medium for 1 hour for preswelling and finally transferred to fresh medium and incubated under standard light and temperature conditions. For encapsulated samples cryopreserved in a two-step cooling protocol, vials are rapidly thawed in a water bath at 40°C and then transferred to fresh sterile medium before preswelling and swelling of the beads.

## 6.2. Post-Thaw Incubation Conditions

Some algae survive cryopreservation with an acceptably high viability when placed under normal growth conditions (see Chapter 10) immediately after the CPA has been removed or diluted to a nontoxic level. However, higher viabilities are sometimes achieved when thawed cultures are left in darkness or in subdued light for 12 to 24 hours after the CPA has been removed and then subjected to a normal lighting regime. The addition of a small amount of yeast extract (0.1 g · L$^{-1}$), proteose peptone (0.1 g · L$^{-1}$) or soil extract (10 mL · L$^{-1}$) sometimes enhances viability.

## 6.3. Assessment of Viability

It is usually difficult to determine the precise fraction of algal units in a culture that survive cryopreservation and resume normal growth after thawing, and no one method can be applied to all species. The most direct way to determine percent viability is to observe the fraction of cryopreserved algal units that grow and reproduce normally. This method is possible only when individual algal units can be separated and allowed to grow and/or reproduce in isolation. Each isolate can be observed at a later time to determine if it has expanded in size or in number of cells. The most common method involving this strategy is to observe the ability of individual algal units to form colonies on an agar plate. Prior to its cryopreservation, an aliquot of the algal culture is diluted and placed on replicate identical agar plates containing a suitable algal culture medium.

A sufficient number of algal units should spread on 100-mm plates to yield between 50 and 200 colonies per plate, since that provides statistically meaningful numbers yet allows convenient and accurate counting. When all viable algal units have grown into visible colonies, the total number of colonies on each plate is counted, and the averaged value for all plates is defined as the "control cell count." The same culture for which the control cell count was determined is used for cryopreservation. The culture is cryopreserved, thawed, and allowed to revive. Within 24 hours of when the culture has been thawed and the CPA removed, replicate aliquots of the culture are placed on agar plates identical to the plates used to determine control cell counts. The same amount of culture should be plated as with the control. The plates are incubated, and the number of colonies per plate is determined exactly as for the controls. Percent viability of the cryopreserved culture is then determined as

$$\% \text{ viability} = (\text{post-treatment cell count}) / (\text{control cell count}) \times 100\% \qquad (1)$$

This approach is especially effective with pour plates, where molten agar at 40°C is added to 1-mL aliquots of dilutions of algae in petri dishes. For temperature-sensitive strains, melted agarose with a lower melting point may be mixed with the culture prior to transferring to plates.

In practice, plating methods often do not provide an accurate estimate of viability. Many species do not grow on or within solid media. Strains that normally grow on agar may not grow well immediately after thawing because of the combined stress of freezing/thawing and their isolation on agar plates. Also, overgrowth by contaminants sometimes precludes accurate counts of xenic cultures, especially if the agar is prepared with an organic medium. Despite this limitation, many unicellular and small colonial algae can be plated to give a reasonable estimate of viability.

Viability immediately after thawing can be determined in *Chlamydomonas reinhardtii* and some related species by observing the ability of cells to exclude Evan's blue, a water-soluble dye. A 0.1% (w/v) solution of Evan's blue dye in water is mixed with an equal volume of cell culture immediately after the CPA has been removed from the culture. After incubation for 5 to 10 minutes, the culture is observed on a microscope slide at 250× or 400× under bright-field or phase-contrast optics. Cells that exclude the dye appear yellow-green and are viable, while nonviable cells appear blue. Viability is calculated by determining the percentage of observed cells that are yellow-green. Quantitative plating and Evan's blue dye exclusion measurements yield similar viabilities, as measured for *C. reinhardtii* (Crutchfield et al. 1999). However, this method does not work for algae with thick cell walls or other extracellular matrices because they exclude Evan's blue dye even when they are not viable.

Fragile strains of algae such as *Dunaliella acidophila* (Kalina) Massjuk often disintegrate when they are killed during cryopreservation, so only living cells can be seen in the microscope after thawing. For these strains it is necessary to know only the culture density immediately before freezing and immediately after thawing in order to calculate the percent viability.

A variety of alternative approaches has been used to determine post-thaw viability of specific algae and other microorganisms, but none has been used extensively. Biochemical measurements potentially allow rapid and accurate assessment of post-thaw viability. They have the added advantage that biochemical activities can often be averaged in nonhomogeneous algal suspensions and in populations of algae with a loose organization of cells where individual algal units are hard to define. A danger with these kinds of measurements is that biochemical functions may be arrested temporarily in recovering cells, thus yielding an underestimate of survival, or algal cells may retain biochemical activity for some time after they are rendered nonviable, thus yielding an overestimate of survival. At the CCAP an overnight recovery stage is included in biochemical viability assays, which radically reduces the incidence of false-positives and false-negatives. Studies undertaken at UTEX have demonstrated that photosynthetic oxygen evolution in bulk cultures of *C. reinhardtii* correlates well with quantitative plating and Evan's blue dye measurements (Crutchfield et al. 1999); this approach is applicable to a wide range of microalgal species (Pearson and Day, unpublished data).

Many other methods of determining viability after cryopreservation have been utilized in a wide variety of organisms. Methods that have potential but that have seldom been used with algae include: (1) microscopic observations of the fluorescence of fluorescein diacetate (a measure of biochemical function), (2) measurements related to photosynthetic function such as chlorophyll fluorescence induction kinetics and light-dependent chlorophyll bleaching (measures of biochemical or physiological function), and (3) the use of dyes such as trypan blue or naphthalene blue that may be excluded only from cells with physiologically intact plasma membranes (measures of membrane physiological integrity).

## 7.0. A SIMPLE METHOD FOR CRYOPRESERVING MICROALGAE

The method described here can be adapted to a wide range of algae without extensive training or expensive equipment. It is amenable to numerous variations, many of which are described elsewhere in this article. This method cryopreserves with high viability a variety of strains of small (less than 50 μm diameter) freshwater unicellular algae. It is less effective for colonial and filamentous strains, and seldom retains viability of large algal cells with a pronounced vacuole. Its effectiveness with marine algal strains has not been tested extensively.

### 7.1. Culture Preparation and Freezing

1. Identify a liquid nitrogen storage dewar equipped with boxes or canes where cryogenic vials may be stored. Many medical, molecular biology, and microbial facilities store samples in liquid nitrogen dewars, and they often have excess space that may be borrowed or rented. Plastic boxes with transparent tops containing 81 preidentified locations (e.g., Nalgene 5026-0929) are convenient for storing and viewing 2-mL frozen

vials and for keeping records of their exact locations. Precool the box by placing it in the storage dewar for at least 1 hour.

2. Purchase presterilized 2-mL graduated polypropylene cryogenic vials such as Nalgene 5000-0050. It is preferable to not reuse vials, since they do not seal effectively after having been used previously for cryopreservation.

3. Place a $1°C \cdot min^{-1}$ freezing container (for 2-mL cryogenic vials) in a refrigerator overnight to equilibrate at approximately 4°C. A Nalgene 5100 "Mr. Frosty" containing isopropanol as per instructions is effective for many strains of algae.

4. Prepare a 20% (v/v) solution of MeOH (e.g., Aldrich 15,490-3) in the culture medium used to grow the alga. If axenic strains are to be cryopreserved, then sterilize the 20% methanol in culture medium by passing it through a 0.45-μm filter (see Chapter 5). Millipore MF (mixed cellulose acetate and nitrate) filters are not damaged by aqueous solutions of 20% MeOH.

5. Culture the algal strain in a liquid growth medium that facilitates its vigorous growth. Harvest the culture for cryopreservation before it has reached stationary phase.

6. Pigmented algal cultures may be cryopreserved at a cell density equivalent to 5 to 15 $\mu g \cdot mL^{-1}$ of chlorophyll $a$. Dense cultures may be diluted with fresh culture medium to the desired density. Excessively dilute cultures may be concentrated by allowing them to settle, such that they can be collected at a higher culture density. Use centrifugation only at low centrifugal force and only when necessary.

7. Dispense the previously prepared 20% MeOH solution into a prelabeled 2-mL cryogenic vial (0.45 mL for vials referenced above), such that liquid algal culture to be added will dilute the MeOH to 5% (v/v).

8. In subdued light, transfer the harvested algal culture to the cryogenic vial (3 volumes of culture to 1 volume of 20% MeOH; e.g., 1.35 mL of culture to 0.45 mL of MeOH).

9. Immediately close the vial lid securely and gently invert the cryogenic vial several times to mix the MeOH uniformly. Keep the vial from bright room light.

10. Transfer the cryogenic vial to a prechilled freezing container and immediately transfer the container to a −80°C freezer. Leave the container undisturbed for 1.5 hours, by which time the temperature of the contents of the vial is less than −50°C. For some strains it may be desirable leave the container in the −80°C freezer for only 60 minutes, at which time the content of the cryogenic vial reaches approximately −40°C.

11. Remove the freezing container from the −80°C freezer and remove the permanent storage box from the liquid nitrogen storage dewar. Quickly transfer the cryogenic vial to the storage box and place the storage box back into the liquid nitrogen storage dewar. The cryogenic vial may then be left undisturbed in the box indefinitely.

12. Record the permanent storage location and other pertinent information regarding the cryopreserved culture.

## 7.2. Thawing and Revival of the Sample

1. Prepare a water bath at 35°C. Place a floating vial-holder (e.g., Nalgene 5974-4015) into the water bath.

2. Remove the storage box from the liquid nitrogen storage dewar and transfer the frozen cryogenic vial as rapidly as possible from the box to the floating holder in the water bath. Agitate the cryogenic vial by rotating the holder continuously with irregular motion to maximize the rate of heat transfer out of the vial.

3. Remove the cryogenic vial from the water bath as soon as all ice in the cryogenic vial has melted, but before the temperature of the algal culture reaches 20°C (typically less than 3 minutes). The culture may be left undisturbed briefly (no longer than 1 hour) in darkness at ambient temperature to allow time for algae that quickly settle to accumulate at the bottom of the vial. Gently remove the lid from the cryogenic vial to avoid disturbing the settled algae, and carefully remove and discard the liquid from above the settled algae. Transfer the algae to fresh culture medium. The volume of fresh medium should be large enough for the concentration of MeOH to be reduced to less than 0.2% (v/v). For cultures that do not settle rapidly, the cryogenic vial may be subjected to gentle centrifugation in order to pellet the culture. Discard the supernatant and dilute the pellet with fresh culture medium.

4. Allow the recovered culture to remain in darkness or in subdued light (normal room light is generally acceptable, but not close proximity to a source of artificial illumination or a window exposed to

bright outdoor light) for several hours, preferably overnight. Then place the culture under normal culture conditions and expect growth of viable cells to resume within 1 to 2 days.

## 8.0. HEALTH AND SAFETY CONSIDERATIONS

Liquid nitrogen handling and storage require safety precautions, including: (1) wearing safety glasses and protective gloves whenever dispensing ultracold liquids, especially liquid nitrogen, and when immersing objects into ultracold liquids; (2) dispensing liquid nitrogen only into containers designed for that purpose; (3) providing adequate ventilation where liquid nitrogen is stored or used; and (4) placing liquid nitrogen into an enclosed vessel only when the vessel is equipped with a pressure-relief valve, making sure that the valve remains fully functional and free of ice.

Valves on automatic-fill storage dewars occasionally freeze in the open position, thus overfilling the dewar. Dual solenoid valves in series reduce this risk. A compressed air–activated shut-off valve (e.g., Thames Cryogenic) and a drainable in-line ice filter further reduce the risk.

Occasionally cryogenic vials leak during storage in liquid nitrogen, allowing the liquid to enter the vial. Although this has never caused a problem in the authors' experience, vials that contain even a small volume of liquid nitrogen can explode during rapid thawing because of explosive vaporization of the liquid nitrogen trapped in the vial. To reduce this risk, cryogenic vials may be held for several minutes in the cold vapor above liquid nitrogen before they are completely removed from the storage container.

## 9.0. SUMMARY

The following successes and limitations have been observed in attempts to cryopreserve virtually all of the algal strains of UTEX and in the freshwater algal section of the CCAP. Most cyanobacteria and soil microalgae can be cryopreserved with relatively high viability. Many freshwater and marine eukaryotic algae can also be cryopreserved, but typically with a lower viability. Most dinoflagellates, cryptophytes, synurophytes, and raphi-dophytes are difficult to cryopreserve. Marine diatoms generally can be cryopreserved well and have a high viability, although freshwater diatoms have thus far proven more problematic. When large numbers of strains have been examined, most notably at CCAP, CCMP, and UTEX, chlorarachniophytes, eustigmatophytes, pelagophytes, phaeothamniophytes, ulvophytes, etc., also have very high success rates, comparable to the other green algae and cyanobacteria. Virtually all algae with a large cell size, such as macrophytic siphonous and coenocytic forms, are recalcitrant to cryopreservation by standard methods. Many multicellular eukaryotic algae, including nearly all coenobic strains and many eukaryotic filamentous strains, cannot yet be cryopreserved. There are no known fundamental reasons why large and complex algae cannot be successfully cryopreserved. Thus, it is anticipated that further research on the basic mechanisms of freezing damage and the empirical development of improved protocols will continue to expand the number and diversity of algal species that can be successfully cryopreserved.

## 10.0. ACKNOWLEDGMENTS

The authors thank the European Union (EU contract no. QLRI-CT-2001-01645) and NSF (DBI-9601026) for financial assistance. They also acknowledge all those who have been involved with cryopreservation research at the CCAP and UTEX, including Dr. G. J. Morris, Dr. K. R. Diller, Dr. R. A. Fleck, Dr. J. Walsh, Dr. E. E. Benson, Dr. K. Harding, Mr. K. J. Clarke, Dr. J. J. McGrath, Dr. M. R. McLellan, Dr. M. Engles, Prof. M. M. Watanabe, Dr. D. Smith, Dr. M. Ryan, Mr. B. Piasecki, Mrs. J. Pearson, Dr. K. Bodas, Ms. A. Crutchfield Holland, Ms. Y. Fan, and Dr. C. Fenwick. In addition, they thank the CCMP and other organizations that are currently undertaking programs of algal cryopreservation for invaluable additional information.

## 11.0. REFERENCES

Adam, M. M., Rana, K. J., and McAndrew, B. J. 1995. Effect of cryoprotectants on activity of selected enzymes in fish embryos. *Cryobiology* 32:92–104.

Anchordoguy, T. J., Carpenter, J. F., Crowe, J. H., and Crowe, L. M. 1992. Temperature-dependent perturbation of phospholipid bilayers by dimethylsulfoxide. *Biochim. Biophys. Acta.* 1104:117–22.

Andersen, R. A. 2002. The Provasoli-Guillard National Center for Culture of Marine Phytoplankton: past, present and future. In: *Abstracts of Algae 2002 Satellite Symposium, Culture Collection and Environmental Researches*. National Institute for Environmental Studies (NIES), Tsukuba, Japan, 23 July 2002.

Baust, J. M. 2002. Molecular mechanisms of cellular demise associated with cryopreservation failure. *Cell Preserv. Technol.* 1:17–31.

Beaty, M. H., and Parker, B. C. 1992. Cryopreservation of eukaryotic algae. *V. A. J. Sci.* 43:403–10.

Ben-Amotz, A., and Gilboa, A. 1980. Cryopreservation of marine unicellular algae. II. Induction of freezing tolerance. *Marine Ecology Progress Series* 2:221–4.

Benson, E. E. 1990. *Free Radical Damage in Stored Plant Germ Plasm*. International Board of Plant Genetic Resources, Rome, 50 pp.

Bodas, K., Brennig, C., Diller, K. R., and Brand, J. J. 1995. Cryopreservation of blue-green and eukaryotic algae in the culture collection at the University of Texas at Austin. *CryoLetters* 16:267–74.

Brand, J. J., and Diller, K. R. 2004. Principles and Applications of Algal Cryopreservation. *Nova Hedwigia* 79:175–89.

Cameron, R. E. 1962. Species of *Nostoc* Vauch. occurring in the Sonoran desert in Arizona. *Trans. Am. Microsc. Soc.* 81:379–84.

Cañavate, J. P., and Lubian, L. M. 1995. Relationship between cooling rates, cryoprotectant concentrations and salinities in the cryopreservation of marine microalgae. *Mar. Biol.* 124:325–34.

Cañavate, J. P., and Lubian, L. M. 1997. Effect of culture age on cryopreservation of marine microalgae. *Eur. J. Phycol.* 32:87–90.

Chekurova, N. R., Kislov, A. N., and Veprintsev, B. N. 1990. The effects of cryoprotectants on the electrical characteristics of mouse embryo cell membranes. *Kriobiologiya* 1:25–30.

Crutchfield, A. L. M., Diller, K. R., and Brand, J. J. 1999. Cryopreservation of *Chlamydomonas reinhardtii* (Chlorophyta). *Eur. J. Phycol.* 34:43–52.

Day, J. G., and Fenwick, C. 1993. Cryopreservation of members of the genus *Tetraselmis* used in aquaculture. *Aquaculture* 118:151–60.

Day, J. G., Fleck, R. A., and Benson, E. E. 1998b. Cryopreservation of multicellular algae: Problems and perspectives. *CryoLetters* 19:205–6.

Day, J. G., Fleck, R. A., and Benson E. E. 2000. Cryopreservation-recalcitrance in microalgae: novel approaches to identify and avoid cryo-injury. *J. Appl. Phycol.* 12:369–77.

Day, J. G., and McLellan, M. R. [Eds.]. 1995. Cryopreservation and freeze-drying protocols. *Methods in Molecular Biology 38*. Humana Press, Totowa, New Jersey, 254 pp.

Day, J. G., Watanabe, M. M., Morris, G. J., Fleck, R. A., and McLellan, M. R. 1997. Long-term viability of preserved eukaryotic algae. *J. Appl. Phycol.* 9:121–7.

Fabre, J., and Dereuddre, J. 1990. Encapsulation/dehydration a new approach to cryopreservation of *Solanum* shoot-tips. *CryoLetters* 11:413–26.

Fleck, R. A. 1998. *The Assessment of Cell Damage and Recovery in Cryopreserved Freshwater Protists. PhD Thesis.* University of Abertay, Dundee, 393 pp.

Fleck, R. A., Benson, E. E., Bremner, D. H., and Day, J. G. 2000. Studies of free radical-mediated cryoinjury in the unicellular green alga *Euglena gracilis* using a nondestructive hydroxyl radical assay: A new approach for developing protistan cryopreservation strategies. *Free Radical Res.* 32:157–70.

Fleck, R. A., Benson, E. E., Bremner, D. H., and Day, J. G. 2003. Studies of antioxidant protection in freeze-tolerant and freeze-sensitive microalgae: Applications in cryopreservation protocol development. *CryoLetters* 24:213–28.

Fleck, R. A., Day, J. G., Rana, K. J., and Benson, E. E. 1997. Visualisation of cryoinjury and freeze events in the coenocytic alga *Vaucheria sessilis* using cryomicroscopy. *CryoLetters* 18:343–55.

Franks, F. 1985. *Biophysics and Biochemistry at Low Temperatures*. Cambridge University Press, Cambridge, 250 pp.

Friedl, T., and Lorenz, M. 2002. The SAG culture collection: microalgal biodiversity and phylogeny research. In: *Abstracts of Culture Collections of Algae: Increasing Accessibility and Exploring Algal Biodiversity*. Sammlung von Algenkulturen at Göttingen University (SAG), Göttingen, Germany.

Fuller, B. J., and Green, C. J. 1986. Oxidative stress in organs stored at low temperatures for transplantation. In: Rice-Evans, C., ed. *Free Radicals Cell Damage and Disease*. Richelieu Press, London, pp. 223–40.

Harding, K., Day, J. G., Lorenz, M., Timmermann, H., Friedl, T., Bremner, D. H., and Benson, E. E. 2004. Introducing the concept and application of vitrification for the cryo-conservation of algae: A mini review. *Nova Hedwigia* 79:207–26.

Hirata, K., Phunchindawan, M., Tukamoto, J., Goda, S., and Miyamoto, K. 1996. Cryopreservation of microalgae using encapsulation/dehydration. *CryoLetters* 17:321–8.

Hoek, van den C., Mann, D. G., and Jahns, H. M. 1995. *Algae: An Introduction to Phycology*. Cambridge University Press, Cambridge, 621 pp.

Kirsop, B., and Doyle, A. 1991. *Maintenance of Microorganisms and Cultured Cells*. Academic Press, London, 308 pp.

Lee, J. J., and Soldo, A. T. 1992. *Protocols in Protozoology.* Society of Protozoologists, Lawrence, Kansas.

Leeson, E. A., Cann, J. P., and Morris, G. J. 1984. Maintenance of algae and protozoa. In: Kirsop, B. E., and Snell, J. J. S., eds. *Maintenance of Microorganisms.* Academic Press, London, pp. 131–60.

Lukavský, J., and Elster, J. 2002. CCALA Trebon. Culture collections of algae: increasing accessibility and exploring algal biodiversity. In: *Abstracts of Culture Collections of Algae: Increasing Accessibility and Exploring Algal Biodiversity.* Sammlung von Algenkulturen at Göttingen University (SAG), Göttingen, Germany, pp. 1913–2002.

MacFarlance, D. R., Forsythe, M., and Barton, C. A. 1992. Vitrification and devitrification in cryopreservation. In: Steponkus, P. L., ed. *Advances in Low-Temperature Biology,* Vol. 1. Jai Press, London, pp. 221–78.

MacFarlane, D. R. 1987. Physical aspects of vitrification in aqueous solutions. *Cryobiology* 24:181–95.

Magill, W., Deighton, N., Pritchard, H. W., Benson, E. E., and Goodman, B. A. 1994. Physiological and biochemical studies of seed storage parameters in *Carica papaya. Proc. Roy. Soc. Edinburgh B (Biol. Sci.)* 102B:439–42.

Mazur, P. 1970. Cryobiology: The freezing of biological systems. *Science* 168:939–49.

Mazur, P. 1984. Freezing of living cells: mechanisms and implications. *Am. J. Physiol.* 247:C125–42.

McLellan, M. R. 1989. Cryopreservation of diatoms. *Diatom Res.* 4:301–18.

Morris, G. J. 1976a. The cryopreservation of *Chlorella* 1. Interactions of rate of cooling, protective additive and warming rate. *Arch. Microbiol.* 107:57–62.

Morris, G. J. 1976b. The cryopreservation of *Chlorella* 2. Effect of growth temperature on freezing tolerance. *Arch. Microbiol.* 107:309–12.

Morris, G. J. 1978. Cryopreservation of 250 strains of Chlorococcales by the method of two step cooling. *Br. Phycol. J.* 13:15–24.

Morris, G. J. 1981. *Cryopreservation: An Introduction to Cryopreservation in Culture Collections.* Institute of Terrestrial Ecology, Cambridge, 27 pp.

Morris, G. J., Acton, E., and Avery, S. 1999. A novel approach to sperm cryopreservation. *Hum. Reprod.* 14: 1013–21.

Morris, G. J., Clarke, K. J., and Clarke, A. 1977. The cryopreservation of *Chlorella* 3. Effect of heterotrophic nutrition on freezing tolerance. *Arch. Microbiol.* 114:249–54.

Morris, G. J., and McGrath, J. J. 1981. Intracellular ice nucleation and gas bubble formation in *Spirogyra. CryoLetters* 2:341–52.

Nishida, I., and Murata, N. 1996. Chilling sensitivity in plants and cyanobacteria: the crucial contribution of membrane lipids. *Ann. Rev. Plant Physiol.* 47:541–58.

Osorio, H., Laranjeriro, N., Santos, L. M. A., and Santos, M. F. 2004. First attempts at cryopreservation of ACOI strains and use of image analysis to assess viability. *Nova Hedwigia* 79:227–35.

Rao, S. V. K., Brand, J. J., and Myers, J. 1977. Cold shock syndrome in *Anacystis nidulans. Plant Physiol.* 59: 965–9.

Rippka, R., Iteman, I., Coursin, T., Comte, K., Singer, A., Araoz, R., Laurent, T., Herdman, M., and Tandeau de Marsac, N. 2002. Recent progress in the Pasteur Culture Collection of Cyanobacteria. In: *Abstracts of Culture Collections of Algae: Increasing Accessibility and Exploring Algal Biodiversity.* Sammlung von Algenkulturen at Göttingen University (SAG), Göttingen, Germany.

Roberts, S., Grout, B. W. W., and Morris, G. J. 1987. Consecutive observations of a frozen cell sample by cryogenic light microscopy and cryogenic scanning electron microscopy. *CryoLetters* 8:122–9.

Sakai, A., Kobayashi, S., and Oiyama, I. 1991. Survival by vitrification of nucellar cells of navel orange *Citrus sinensis* var. *brasiliensis tanaka* cooled to −196°C. *J. Plant Physiol.* 137:465–70.

Santarius, K. A. 1996. Freezing of isolated thylakoid membranes in complex media. X. Interactions among various low molecular weight cryoprotectants. *Cryobiology* 33: 118–26.

Smith, D., and Day, J. G. 2000. *MAFF Pilot Study. Securing the UK National Microbial Genetic Resources: Development of Cryopreservation Protocols for Some Recalcitrant Organisms.* Ministry of Agriculture, Fisheries, and Food (MAFF), CEH, Ambleside, UK, 52 pp. (CEH Ref No. WI/C01406/1).

Steponkus, P. L., Langis, R., and Fujikawa, S. 1992. Cryopreservation of plant tissues by vitrification. In: Steponkus, P. L., ed. *Advances in Low-Temperature Biology.* Jai Press, London, pp. 1–61.

Stillinger, F. A. 1995. Topographic view of supercooled liquids and glass formation. *Science* 267:1935–9.

Taylor, R., and Fletcher, R. L. 1998. Cryopreservation of eukaryotic algae—a review of methodologies. *J. Appl. Phycol.* 10:481–501.

Watanabe, M. M., Shimizu, A., and Satake, K. 1992. Microbial Culture Collection at the National Institute of Environmental Studies: Cryopreservation and Database of Culture Strains of Microalgae. In: Watanabe, M. M., ed. *Proceedings of Symposium on Culture Collection of Algae.* National Institute for Environmental Studies (NIES), Tsukuba, Japan, pp. 33–41.

Whiteley, G. S. W., Fuller, B. J., and Hobbs, K. E. F. 1992. Deterioration of cold-stored tissue specimens due to lipid peroxidation: Modulation by antioxidants at high subzero temperatures. *Cryobiology* 29:668–73.

# CHAPTER 13

# PHOTOBIOREACTORS AND FERMENTORS: THE LIGHT AND DARK SIDES OF GROWING ALGAE

PAUL W. BEHRENS

*Martek Biosciences Corporation*

## CONTENTS

Key Index Words: Aeration, Fluorescent Lighting, Heterotrophic Growth, Mixing, Phototrophic Growth

## 1.0. INTRODUCTION

This chapter addresses large-scale, indoor algal production using highly controlled photobioreactors or fermentors. A photobioreactor is defined here as a closed (or mostly closed) vessel for phototrophic production where energy is supplied via electric lights. A fermentor is defined here as a closed (or mostly closed) vessel for heterotrophic production where energy is supplied as organic carbon. Outdoor algal growth systems are covered in Chapters 14 and 15. The perspective is the use of photobioreactors or fermentors for growing algal biomass to produce a commercial product, either cells or an extract. Production methods must produce biomass that is cost effective and consistent in quality from batch to batch. The product may be large quantities of a low value or small quantities of a high value. Quality is particularly important when the final product is subject to regulatory oversight (e.g., human food).

A multitude of photobioreactors have been designed, built, and described (Samson and Leduy 1985, Buelena et al. 1987, Chaumont et al. 1987, Radmer et al. 1987, Rathford and Fallowfield 1992, Delente et al. 1992, Iqbal et al. 1993, Lee and Palsson 1994, Eriksen et al. 1998, Hu et al. 1998, Carlozzi 2000, Csögör et al. 2001, Degen et al. 2001, Morita et al. 2001, Pulz 2001, Muller-Feuga et al. 2003, Suh and Lee 2003), and fermentation technology has a very well-established knowledge base (Bailey and Ollis 1977, Wang et al. 1979, Moo-Young and Blanch 1987, Blanch and Clark 1997, Hilton 1999). The photobioreactor and fermentor technology developed and used over the last two decades by Martek Biosciences Corporation are used as examples. Emphasis is placed on photobioreactors because more algae are phototrophs than heterotrophs.

Photobioreactor design differs depending on the ultimate goal. In general, sophisticated photobioreactors are more versatile but are more expensive to construct and more complicated to operate. Martek began designing photobioreactors two decades ago to produce stable isotopically labeled compounds ($^{13}$Carbon, $^{2}$Hydrogen, $^{15}$Nitrogen) (Behrens et al. 1989, 1994, 1996). We have also used these vessels to produce pigments, fatty acids, and bioactive molecules (Kyle et al. 1989, Behrens 1992, Radmer and Parker 1994, Behrens and Kyle 1996, Apt and Behrens 1999). In addition to producing biomass for algal products, photobioreactors have been designed for life support in outer space (Radmer et al. 1987, Godia et al. 2002), removal of various compounds from water (An and Kim 2000, Gaffney et al. 2001), production of gas vesicles in cyanobacteria (Kashyap et al. 1998, Sundararajan and Ju 2000), $CO_2$ removal (Keffer and Kleinheinz 2002), hydrogen production (Kosourov et al. 2002, Tsygankov et al. 2002), and macroalgal production (Huang and Rorrer 2002, Polzin and Rorrer 2002, Barahona and Rorrer 2003).

Our large-scale conventional fermentation of *Crypthecodinium* and *Schizochytrium* has produced biomass for aquaculture (Gladue and Maxey 1994, Barclay and Zeller 1996, Gladue and Behrens 2002, Harel et al. 2002) and for docosahexaenoic acid (DHA)–rich oils for human nutrition (Kyle et al. 1992, Barclay et al. 1994, Behrens and Kyle 1996, Kyle 1996). The overwhelming majority of algae are classified as photosynthetic, and few algae are believed capable of completely heterotrophic growth (Bold and Wynne 1985). However, this may be false because Martek found numerous species from all algal groups that are capable of heterotrophic growth. It is important to note that Martek's approach to photobioreactors and fermentors was influenced in different directions based on the status of the individual technologies.

## 1.1. Strategies for Large-Scale Growth of Algae

Several general approaches have been used to grow large quantities of phototrophic algae. These approaches include outdoor systems such as ponds and tanks where the light is supplied as sunlight (see Chapter 14), and indoor systems such as photobioreactors where the light is supplied by electric lights (Oswald 1988, Chaumont 1993, Hu and Richmond 1994, Molina Grima et al. 1995, Pushparaj et al. 1997, Tredici and Zittelli 1998, Mazzuca Sobczuk et al. 2000, Pulz 2001, Janssen et al. 2002, Ugwu et al. 2002, Barbosa et al. 2003, Hall et al. 2003, Lebeau and Robert 2003, Richmond 2004). Many indoor designs have been described for the closed culture of algae using electric lights for illumination (Delente et al. 1992, Ratchford and Fallowfield 1992, Wohlgeschaffen et al. 1992, Iqbal et al. 1993, Lee and Palsson 1994, Miroget et al. 1999, Csögör et al. 2001, Degen et al. 2001). These photobioreactors are closed systems and provide the ability to control and optimize culture parameters (Ratchford and Fallowfield 1992).

For heterotrophs, fermentors have a number of important advantages over photobioreactors, for example, a large preexisting knowledge base, highly sophisticated hardware, worldwide large-scale availability, and relatively low unit operation cost.

## 1.2. Choosing between the Photobioreactor and the Fermentor

The choice depends on several factors, including the algal growth mode, the final product, the value of the product, and the anticipated purpose for the alga or algal product. If heterotrophic growth is possible, it will generally be more economical than phototrophic growth.

It is quite simple to determine whether algae are capable of heterotrophic growth. Biolog, Inc. sells microtiter plates that easily facilitate screening algae for heterotrophic growth. These 96-well plates come with different organic carbon compounds in each well, and in one step, one can easily determine whether the alga is capable of heterotrophic growth, and if so, which carbon compounds are suitable for growth.

Some genera that we have grown heterotrophically include *Amphora, Ankistrodesmus, Chlamydomonas, Chlorella, Chlorococcum, Crypthecodinium, Cyclotella, Dunaliella, Euglena, Nannochloropsis, Nitzschia, Ochromonas,* and *Tetraselmis.* One requirement, however, is that the algae must be axenic to avoid the growth of bacteria.

Although algae possess unique and wonderful biochemistries, producing many different compounds, the means by which they are grown must be compatible with the ultimate economic value of the final product. Many good products have been developed but have never made it to the marketplace because the production process was not cost competitive.

### 1.3. Convert the Phototroph to a Heterotroph

Advancements in algal molecular biology (Dunahay et al. 1995, Apt et al. 1996, Dawson et al. 1997) have made it possible to genetically transform an obligate phototroph into a heterotroph (Zaslavskaia et al. 2001). *Phaeodactylum tricornutum* Bohlin is an obligate phototroph despite having the ability to internally metabolize glucose for cellular energy (Hellebust and Lewin 1977). Zaslavskaia et al. (2001) transformed this alga with the *Glut*1 gene from human erythrocytes, and transformed cells thrive on glucose in the absence of light. The potential now exists to transform many algae from obligate phototrophs to heterotrophs. A dual growth mode expands the options for large-scale culturing (photobioreactor or fermentor) so that the most appropriate and lowest cost scale-up option can be used. The long-term benefit should be an expansion of the number of products that can be made cost effectively from algae.

## 2.0. PHOTOBIOREACTORS

### 2.1. Design Considerations

Many configurations of photobioreactors have been devised and built. These range from tubular and cylindrical systems (Pirt et al. 1983, Radmer et al. 1984, Chaumont et al. 1987, Pohl et al. 1987, Richmond et al. 1993, Spectorova et al. 1997, Ericksen et al. 1998, Merchuk et al. 2000, Suh and Lee 2001, Babcock et al. 2002) to conical systems (Watanabe and Hall 1996, Morita et al. 2000) to flat-sided vessels (Delente et al. 1992, Iqbal et al. 1993, Hu et al. 1998), and these systems employ a wide variety of light sources (Radmer 1990, Takano et al. 1992, Lee and Palsson 1994, An and Kim 2000). As such, each is usually a highly customized system and the availability of commercial photobioreactors is limited.

Despite various configurations, several basic design features must be considered when building a photobioreactor: how to provide light, how to circulate the algae, which materials to use for construction, how to

provide $CO_2$ and remove $O_2$, and how to control pH and temperature. Sophistication is driven by purpose, and if whole algal biomass is the desired final product, a relatively unsophisticated system will suffice.

Glass and acrylic are widely used in the construction of photobioreactors, and ultraviolet (UV)-stabilized acrylic is superior because it is lighter, more flexible, stronger, and easier to machine, cut, bond, and so on. Construction requires some minimal construction skills, but certainly not professional skills. Assembly of the photobioreactor system is merely the integration of various monitor and control subsystems (e.g., pH control and temperature control).

### 2.2. Light

Light is the most important parameter in the design and construction of a photobioreactor. Despite its importance, light can be a very difficult input to measure for efficient use (Pirt et al. 1983, Kirk 1994, Ogbonna et al. 1995, Janssen et al. 2002). Light can be supplied continually or in light-dark cycles. As cell concentration changes, the light requirements change. Algal growth is limited by too little light, but too much light can be as deleterious. Phototrophs must receive sufficient light to exceed their light compensation point for their net growth; insufficient growth detracts from the net growth of the culture because of respiratory loss (Radmer et al. 1987). Increasing light beyond the compensation point results in an increase in the growth rate until the culture becomes light saturated, and higher light intensities can lead to photoinhibition (Acien Fernandez et al. 1998, Csögör et al. 2001, Yun and Park 2001, Wu and Merchuk 2002, Barbosa et al. 2003, Suh and Lee 2003).

Light is refracted and reflected when it passes from a medium of one index of refraction to another except when it enters the second medium at the normal angle (perpendicular to the surface of the second medium). Wherever possible, the surface of the photobioreactor should be designed to minimize reflection and refraction; those made with tightly curved surfaces like tubes will have less light available than those made with flat surfaces (Tredici and Zittelli 1998). Some designs have incorporated sophisticated parabolic light collection devices, fiberoptics, or light guides (Radmer 1990, Ogbonna et al. 1999, Janssen et al. 2002). We experimented with these types of configurations, and although the design is elegant, we found that the efficiency was often offset by the difficulty and cost of construction.

Chlorophyll has a very high extinction coefficient; that is, it is very good at absorbing light (Kirk 1994). For a given cell concentration, the light intensity transmitted through a culture drops very quickly with distance from the light source because of the high absorption by chlorophyll. At a high cell concentration, nearly all of the incident light may be absorbed within a small layer of cells, and the remaining cells are virtually in darkness. Consequently, photobioreactors generally have short light-path lengths. Adequate mixing and circulation "evens out" the light intensity while providing efficient gas exchange and better pH and temperature control.

Incandescent lamps are inferior to the other lamps and are seldom used for photobioreactors (Table 13.1). Both fluorescent and high-intensity discharge lamps have very good electrical efficacy and substantial lifetimes. Fluorescent lamps distribute light more or less uniformly along the length of the lamp; light emanates from a point source in high-intensity discharge lamps, requiring distance between the lamp and algae for proper light dispersion. Although high-intensity discharge lamps are slightly more efficient than fluorescent lamps, we have found that fluorescent lamps offer the best all-around choice (Radmer et al. 1987, Radmer 1990, Delente et al. 1992).

We have experimented with various lighting methods, including fluorescent lamps encased in acrylic and submerged in the photobioreactor. Ensuring watertight construction and preventing algal adhesion to the acrylic surfaces (i.e., blocking light) are significant problems. We tried external illumination on flat-sided photobioreactors using high-intensity discharge lamps, but light distribution from the point source causes problems. Parabolic surfaces and novel light distribution materials were also tried but were difficult and costly to construct (Radmer 1990). Our current systems are a hybrid of several of these ideas (Fig. 13.1). We use flat-sided vessels with fluorescent lamps through channels.

## 2.3. Circulation

Circulation is important to ensure optimum illumination of the algae, adequate gas exchange, and temperature and pH control within the culture. We use an "airlift" principle in which circulation is accomplished by creating differences in water density in different regions of the photobioreactor (Manfredini and

**TABLE 13.1** Comparison of light sources for photobioreactors.

| Feature | Incandescent | Fluorescent | High intensity discharge |
|---|---|---|---|
| Electrical efficiency | 11% | 23% | 24% |
| Life time (hours) | 1,000 | 20,000 | 24,000+ |
| Distribution pattern | Point | Linear | Point |
| Lumens/watt | 17 | 79 | 89 |

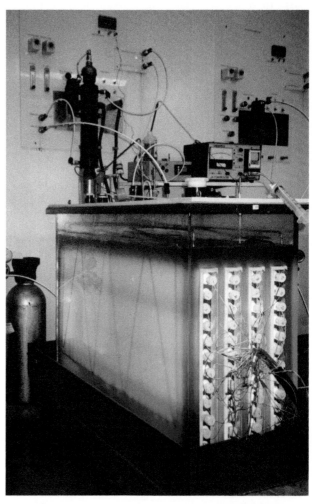

**FIGURE 13.1.** Martek 120-liter photobioreactor.

Cavellera 1983, Ratchford and Fallowfield 1992, Molina Grima et al. 1995, Blanch and Clark 1997). This principle is common in fermentors (see later discussion) and has been adapted for use in photobioreactors. Our photobioreactors use a series of channels, which are connected at the top and bottom by liquid (see Fig. 13.1), and air or an air/$CO_2$ mixture is bubbled in alternate channels. Bubbled channels have a lower density (less liquid per volume), creating a natural circulation between the channels.

## 2.4. Gas Exchange

Photobioreactors can be bubbled with air, but the low $CO_2$ concentration in air (0.033%) will often limit phototrophic growth. With an airflow of $1 \text{ L} \cdot \text{min}^{-1}$, assuming all carbon dioxide is used and the biomass is 50% carbon, there is enough carbon to support $3.54 \times 10^{-4}$ grams biomass $\cdot \text{min}^{-1}$; this is a very low productivity. The simplest approach is to blend $CO_2$ with air, for example, 0.2 to 5.0% of the total gas flow (Lee and Pirt 1984, Merchuk et al. 2000, Morita et al. 2001, Babcock et al. 2002). Care must be taken to ensure that the $CO_2$ input does not adversely lower the pH level of the culture. Also, with an open system, most of the $CO_2$ will exit the vessel unused. Small bubble size helps facilitate diffusion. Another strategy is to use $CO_2$ to control the pH of the culture (Lee and Pirt 1984, Delente et al. 1992, Babcock et al. 2002). Adding $CO_2$ acidifies the culture medium, though nitrate consumption by algae causes an alkalization of the culture medium. A system can easily be set up using a pH controller (Omega Engineering), which activates a solenoid to admit $CO_2$ in response to an alkalization of the medium (Sonnleitner et al. 1999). In practice, this system works well as long as there is a continual nitrate uptake. However, if the alkalization stops, then $CO_2$ is not added and the culture may quickly become $CO_2$ limited. For the production of secondary algal products under nitrate-deficient conditions, this system of $CO_2$ addition has limitations (Behrens et al. 1989, Hoeksema et al. 1989). A third, more flexible approach involves uncoupling pH control from $CO_2$ control. The $CO_2$ level is monitored using either a liquid submersible $CO_2$ probe or a gaseous infrared carbon dioxide analyzer (Servomex); we have found that the infrared analyzer is superior. Submersible probes generally have slow response times and are subject to fouling. Infrared gas analyzers require plumbing to direct the effluent gas from the photobioreactor to the analyzer, which is more expensive than submersible probes.

As important as it is to provide $CO_2$, it is nearly equally important to remove the $O_2$. Excess $O_2$ leads to photoxidative damage and increasing rates of photorespiration, both of which are detrimental to overall productivity. If the photobioreactor is continually bubbled with air or an air/$CO_2$ mixture, the oxygen level will remain close to ambient air; however, $CO_2$ may become limiting (see previous discussion). A more sophisticated system involves removing the oxygen with a periodic purge of nitrogen gas. This provides more efficient use of $CO_2$ and allows the $O_2$ level to be reduced below ambient air. The nitrogen purge can be periodic or controlled by an oxygen sensor, such as either a dissolved oxygen probe or a paramagnetic gaseous analyzer (Servomex). A third method uses a chemical system to remove oxygen. The $O_2$ in the gas stream is combusted with hydrogen gas to form water using a catalyst for purifying $H_2$ that is commercially available (Matheson Gas Company). $H_2$ is admitted to the catalyst when the $O_2$ concentration exceeds a predetermined set point, making it possible to very precisely regulate the $O_2$ level. There is also a certain simplicity to the approach because the oxygen that was produced from the oxidation of water is in turn combusted with $H_2$ to produce water. This method minimizes the need to vent gases to the atmosphere. This approach has been used for a decade for the production of [13]Carbon-labeled compounds, and the high degree of system closure prevents loss of expensive [13]$CO_2$ while maintaining the isotope purity of the final products.

## 2.5. Temperature and pH

After light, $CO_2$, and $O_2$, pH and temperature are the next most important parameters to measure and control. Fortunately, these are very simple parameters to control using existing off-the-shelf technology (Sonnleitner 1999). Commercially available pH controllers (Omega Engineering) are recommended. Photosynthetic systems will always generate heat because of the inefficiency of photosynthesis in converting light energy into chemical energy (Pirt 1983, Morita et al. 2001). The theoretical conversion of red light into chemical energy (NADPH) is only 31%; 69% is lost as heat. The amount of cooling depends on the incident light intensity and the cell concentration (i.e., how much light is absorbed), but regardless, cooling will be necessary. In principal, it is quite easy to control the temperature using commercially available temperature controllers (Omega Engineering). Cooling is achieved with a heat exchange system; external cool water is circulated through a good heat conducting material, which then draws heat from the photobioreactor. The more significant challenge is how to install a cooling system on a photobioreactor. We experimented with various systems. Early versions consisted of applying either aluminum or stainless steel water jackets to the external surfaces (usually the bottom to not interfere with illumination) of the photobioreactors. It was difficult to construct an external water jacket with materials that have good heat transfer while avoiding air bubbles, which can severely limit the heat transfer. Our preferred system uses stainless steel cooling coils submerged in the culture medium. There is excellent contact between the cooling coil and water, which provides good cooling, but the cooling coil sometimes interferes with the circulation within the vessel. A refrigerated water source (or tap water, depending on the cooling demands of the culture) is used, and a temperature probe, connected to a controller, operates a solenoid to regulate cooling water flow. The best strategy is to use a "normally open" solenoid so that if there is a failure of the controller, the photobioreactor will overcool, which is preferred to overheating.

## 2.6. Sterilization

For many phototrophs, gross contamination by bacteria and fungi is not a significant problem because there is generally very little free organic carbon to support their growth (see Chapter 5 on sterility). A higher concern is to prevent contamination of the photobioreactor by other phototrophs.

Photobioreactors are made of optically clear materials (e.g., glass or acrylic) which do not lend themselves to steam sterilization. Furthermore, the size of most photobioreactors exceeds what can be accommodated in an autoclave. Systems exist for sterilization with ozone (Quesnel 1987), but these systems are expensive and difficult to use. A more practical solution is sanitization rather than sterilization. Photobioreactor sanitization can be easily accomplished with bleach. Air pumps, analyzers, and other equipment can be kept free of algae and other microorganisms by the use of the appropriate prefilters.

## 2.7. Operational Strategies

With adequate measurement and control of the basic culture parameters of light, $CO_2$, $O_2$, pH, and temperature, it is possible to optimize these parameters to achieve the desired end product from the photobioreactor. The end product dictates the alga, which in turn dictates the general growth conditions.

*Yield* and *productivity* are terms that require careful definition. Yield is the production of mass per unit volume, and it is often expressed in terms of $g \cdot L^{-1}$. Productivity is yield per unit time and is often expressed in terms of $g \cdot L^{-1} \cdot hour^{-1}$ or $g \cdot L^{-1} \cdot day^{-1}$. The growth conditions that produce maximal yield are rarely the same conditions that produce maximal productivity.

For a given alga and/or a given product, it is possible to methodically test and optimize each culture parameter to determine the best value for maximizing either yield or productivity. Although this is certainly a straightforward approach, a more efficient means to optimizing culture parameters is to employ a statistically based factorial or fractional factorial approach to experimental design (Anderson and Whitcomb 2000). This method uses statistics to identify the culture parameters and the interactions between culture parameters that are most important for achieving maximal yield and/or productivity. Perhaps more importantly, this approach identifies those parameters that are not important for maximum yield and/or productivity and thus eliminates the need for much unnecessary experimentation.

## 2.8. Harvesting Methods

The algal culture is still relatively dilute in a fully optimized photobioreactor. A dry weight concentration of even 5 to $10 \, g \cdot L^{-1}$ is far below that of a fermentor system ($>50 \, g \cdot L^{-1}$). The two most common harvesting approaches for photobioreactors are flocculation and centrifugation (Sukenik and Shelef 1984, Becker 1994). Although flocculents enable a concentration of the algal biomass, centrifugation is usually still necessary to achieve a suitable volume reduction. Small-scale recovery can be easily done in large centrifuge bottles, but larger scale cultures are best done with a continuous flow centrifuge such as a Sharples (Sharples Penwalt).

## 2.9. Algal Products from Photobioreactors

Numerous reports in the literature indicate that changes in nutrient levels, light intensity, pH, and temperature can alter the growth and secondary metabolism of algae (Richardson et al. 1969, Brand and Guillard 1981). For example, many algae will channel carbon into storage products (either carbohydrate or lipid) when nitrogen is limited but carbon is still available. The specific nitrogen content of the medium can be empirically determined based on the light field for the given photobioreactor, and we have often found that there is an interaction between (1) the dry weight and cell concentration of the culture at nitrogen depletion and (2) the light intensity. We have used the interaction to produce a wide variety of stable isotopically labeled and unlabeled compounds from various algae, including *Chlorella vulgaris* Beijerinck, [13]C-glucose; *Neochloris oleoabundans* Chantanachat et Bold, [13]C-triglyceride and [2]H-lubricants; *Chlamydomonas ulvaensis* R. A. Lewin, [13]C-xylose; *Chlamydomonas sajao* R. A. Lewin, [13]C-galactose; *Navicula saprophila* Lang-Bertalot et Bonik, eicosapentaenoic acid; and *Porphyridium cruentum* (J. E. Smith) Naegeli, phycoerythrin (Behrens et al. 1989a, 1989b, 1994, Hoeksema et al. 1989, Parker 1994, Kyle 1995, Behrens and Kyle 1996, Apt and Behrens 1999). Several of these products are briefly described as examples of products from our photobioreactors.

*C. vulgaris* grown under nitrogen-limiting conditions can accumulate starch to more than 50% of its dry weight (Fig. 13.2) (Behrens et al. 1989), making it a good source for producing [13]C-glucose, a compound used to analyze the structure of bacterial macromolecules. The high degree of closure of the system ensured that the [13]Carbon content of the final product was not compromised by [12]Carbon from atmospheric carbon dioxide. *Dunaliella* was used as a source of [13]C-glycerol, a compound used to elucidate three-dimensional macromolecule structure. We found that with prolonged culture time, there was a progressive diffusion of glycerol from the algal cells into the medium (Apt and Behrens 1999, Behrens, unpublished observations). This is counterintuitive because cells produce glycerol to osmotically balance the high salinity in the medium, but "leakage" is substantial and can reach levels of $10 \text{ g} \cdot \text{L}^{-1}$. Glycerol purification is rather simple, merely a matter of separating it from the saline growth medium. *N. saprophila* is a source of the polyunsaturated fatty acid eicosapentaenoic acid (EPA), and other algae are alternative sources (Molina Grima et al. 1995, Acien Fernandez et al. 2000). A maximum fatty acid content of 50% dry biomass was obtained, and EPA made up 20% of the total fatty acid (Behrens, unpublished observations). We noted a relationship between the dry cell biomass and the fatty acid levels. With constant light, the final dry cell biomass and EPA fraction increased with higher nitrogen levels even though the total fatty acid content decreased.

## 3.0. FERMENTORS

Fermentors come in a wide range of sizes from 1 liter to more than 500,000 liters. Commercial fermentors are readily available, so it is not necessary to describe their design. For general operation of fermentation systems, see Bailey and Ollis (1977), Wang et al. (1979), Blanch and Clark (1997), and Demain and Davies (1999). Fermentor operation includes both batch and continuous modes, as well as liquid and solid systems. Batch fermentation is the most common (Fig. 13.3). There are common features between photobioreactors and fermentors (pH and temperature control, harvesting, etc.) (Table 13.2). The significant differences between a fermentor and a photobioreactor are energy source, circulation, $O_2$ supply, and sterility. Organic carbon catabolism requires adequate oxygen, often making oxygen the single largest operating constraint (Corman 1957, Brown 1970, Zeigler et al. 1980, Sinclair 1984, Bartow 1999). Glucose is the most widely used source of organic carbon and it is relatively inexpensive; however, acetate, citrate, and other organics have been used (Blanch and Clark 1997, Humphrey 1998, Kuhlmann et al. 1998).

### 3.1. Circulation

Circulation within the fermentor is necessary for uniform nutrient distribution and adequate gas

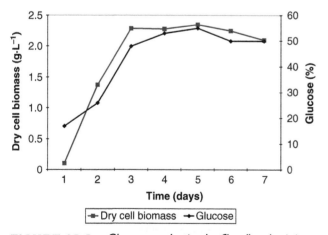

**FIGURE 13.2.** Glucose production by *Chorella vulgaris* in a photobioreactor designed by Martek Biosciences Corporation.

**FIGURE 13.3.**    One hundred liter fermentor.

exchange. The viscosity of the culture medium affects circulation, increasing with cell concentration and/or the production of viscous extracellular material. Many fermentors achieve adequate mixing with impellers and baffles (Bader 1986, Blanch and Clark 1997, Bartow 1999, Demain and Davies 1999); a wide variety of impeller configurations have been developed, including Rushton and marine impellers (Bader 1986). Shear sensitivity of the organism influences impeller choice (Bader 1986). The shear is proportional to the tip speed of the impeller, so for a shear sensitive alga, the tip speed must be reduced. A reduction in the tip speed reduces mixing, but this can be at least partially compensated for by increasing the number of impellers in the fermentor (Bader 1986, Blanch and Clark 1997). An alternative mixing approach is based on the airlift principle (Moo-Young and Blanch 1987, Blanch and Clark 1997, Demain and Davies 1999). Fermentors with an internal plenum or draught tube have true airlift. Air is supplied on one side of the plenum or draught tube, creating a density difference between the side that is aerated and the side that is not. To be effective, airlifts usually require a substantial airflow rate, because the sparging of air must supply all of the energy for mixing and mass transfer within the fermentor (Moo-Young and Blanch 1987). A bubble column is a variation on the theme of an airlift (Moo-Young and Blanch 1987). There is no internal plenum or draught tube, and the algae are circulated merely as a result of the input of air at the bottom of the chamber. Mixing in a bubble column tends to be less uniform than in an airlift, but it is a simpler system.

**TABLE 13.2**    Comparison of photobioreactor and fermentor features.

| Feature | Photobioreactor | Fermentor |
| --- | --- | --- |
| Energy source | Light | Organic carbon |
| Cell density/dry weight | Low | High |
| Limiting factor for growth | Light | Oxygen |
| Harvestability | Dilute, more difficult | Denser, less difficult |
| Vessel geometry | Dependent on light penetration | Independent of energy source |
| Control of parameters | High | High |
| Sterility | Usually sanitized | Can be completely sterilized |
| Availability of vessels | Often made in-house | Commercially available |
| Technology base | Relatively new | Centuries old |
| Construction costs | High per-unit volume | Low per-unit volume |
| Operating costs | High per-kg biomass | Low per-kg biomass |
| Applicability to algae | Photosynthetic algae | Heterotrophic algae |

Airlifts and bubble columns are not suitable for all types of algae. Importantly, both systems generate some shear despite the absence of impellers (Moo-Young and Blanch 1987, Blanch and Clark 1997). Algae that have a high specific gravity will tend to settle easily, and it may be difficult to keep the algae suspended. Conversely, algae with a low specific gravity (e.g., oleaginous algae) have a tendency to float and may not be suitable for airlift or bubble column mixing. This situation is further exacerbated if the specific gravity of the alga changes significantly over the course of the fermentor cycle (Kyle et al. 1992).

## 3.2. Oxygen

Providing sufficient $O_2$ is often the single most limiting factor for achieving a rapid growth rate and a high cell concentration. $O_2$ is generally supplied to a fermentor as compressed air (about 21% oxygen). Supplying higher levels of $O_2$ by blending pure $O_2$ with air is possible for small-scale experimentation but is generally cost prohibitive at large scale. Oxygen has a relatively low solubility in aqueous media, and the dissolved oxygen level for air-saturated water is about 250 μM (Blanch and Clark 1997). For convenience, the dissolved $O_2$ concentration in air-saturated water is referred to as 100% dissolved oxygen or 100% D.O.; the D.O. level in a fermentor will always be less than 100% because of the metabolism of the alga. Furthermore, the D.O. level changes as the growth rate and the algal concentration changes and must be empirically determined. The D.O. level is determined by several physical factors, such as total airflow rate, bubble size, bubbles residence time, temperature, and viscosity (Corman et al. 1957, Brown 1970, Wang et al. 1979, Manfredini and Cavellera 1983, Sinclair 1984, Moo-Young and Blanch, 1987, Clark et al. 1995). Bubble size is primarily determined by the turbulence and shear within the fermentor. The bubble size entering the fermentor is less important because even small bubbles can coalesce quickly in the fermentor. Impellers and internal baffles in a fermentor "break" the airstream, making smaller bubbles and increasing the D.O. in the fermentor. Finally, the longer the residence time of bubbles, the greater the D.O. concentration will be in the culture. Bubble rate rise is determined by gravity, but residence time is increased with increased height of the fermentor or with increased viscosity (Moo-Young and Blanch, 1987).

## 3.3. Sterilization

Organic carbon is an energy source for contaminating bacteria and fungi, which quickly outcompete the alga. Therefore, it is absolutely essential that the fermentor is completely sterilized before use and that the algal inoculum be axenic. Fermentor sterilization methods include heat (dry or steam), irradiation (UV rays, gamma rays, or x-rays), and chemicals (hydrogen peroxide, ethylene oxide, formaldehyde) (Quesnel 1987). Of these, steam is the most common method because it is effective, easy to use, and nontoxic to humans. Steam is preferred over dry heat because the steam is more effective for killing spores (Quesnel 1987; see Chapter 5).

The culture medium must also be sterilized. Two general methods are filter sterilization and steaming. Filter sterilization is effective for small volumes or for heat-sensitive components such as vitamins. Steam sterilization of culture medium is more common. The culture medium can be either sterilized and aseptically transferred to the vessel or sterilized in the vessel (Quesnel 1987, Demain and Davies 1999). Sterilizing the medium separately can be easily done with ultrahigh temperature systems (UHT systems) in which the medium is exposed to high temperatures for only a brief period and then rapidly cooled. This process has the advantage of not exposing the nutrients to high temperatures for a prolonged period and thereby minimizing any effects of heat on the nutrients. If the medium is sterilized in the vessel, longer sterilization times are generally required to ensure complete sterilization because of the slower heat transfer in a larger vessel. In addition, if the medium contains significant levels of chloride or other halides, sterilization in the vessel can accelerate corrosion of the stainless steel fermentor. The ideal combination of temperature/pressure and length of exposure must be empirically determined and will depend on the composition and viscosity of the medium. To determine the effectiveness of a sterilization procedure, a sterile challenge is performed by leaving the sterilized medium uninoculated for a period of time. Failure of a microorganism to grow in the medium indicates successful sterilization.

## 3.4. Algal Products from Fermentors

A number of algae have been successfully grown in large-scale fermentors, including *Crypthecodinium*, *Schizochytrium*, *Chlorella*, and *Tetraselmis*. *Crypthecodinium cohnii* (Seligo) Javornicky has been grown in fermentors exceeding 100,000 liters for the production docosahexaenoic acid (DHA) (Kyle et al. 1992, Behrens and Kyle 1996, Kyle 1996). The culture medium is composed of simple inorganic salts plus glucose and yeast extract. When optimized, *Crypthecodinium* DHA comprises more than 40% of the total fatty acids. Similarly, *Schizochytrium* has been grown in fermentors (>100,000 liters) for polyunsaturated fatty acids (Barclay et al. 1994). When cultured with glucose at low salinities, yields of 20 g · L$^{-1}$ dry biomass and productivity of approximately 1.0 g · L$^{-1}$ · day$^{-1}$ can be achieved. New strains are capable of producing more than 70% of their biomass as fatty acids, of which 35% is DHA (Nakahara et al. 1997). *Tetraselmis* is commonly used as a feed in aquaculture, and it has been grown in large-scale fermentors (Day et al. 1990, 1991).

Various strains of *Chlorella* have been grown to high cell concentrations in fermentors, for example, from a few liters to more than 100,000 liters. Some species and strains of *Chlorella* grow extremely well under heterotrophic conditions, with doubling times of a few hours (Fig. 13.4) (Behrens, unpublished observations, Doncheck et al. 1996, Running et al. 2002). Although the precise kinetics of growth will vary depending on the photobioreactor design and operation, this figure illustrates the clear advantage of the fermentor over the photobioreactor for producing algal biomass.

## 4.0. ECONOMIC COMPARISON OF PHOTOBIOREACTORS AND FERMENTORS

Although both photobioreactors and fermentors can be used for the production of algae, they are different systems (see Table 13.2), and the economics of each system are different. Several excellent references provide a detailed description of the capital and operating costs for fermentors (Sfat 1984, Van Brunt 1986, Blanch and Clark 1987, Hepner and Male 1987, Wilkinson 1998, Reisman 1999). Similar descriptions are not readily available for photobioreactors because they are much less common systems and are usually customized. Table 13.3 lists some of the major cost factors in the construction and operation of photobioreactors and fermentors. Because photobioreactors are built in-house, they are usually of relatively small scale, and this precludes the economy of scale of both construction and operating costs.

Table 13.4 compares the calculated cost of producing a kilogram of dry algal biomass from a photobioreactor and from a fermentor. For simplicity, only the cost of the energy is considered: light for the photobioreactor and glucose for the fermentor. The following assumptions are made: (1) The cost of electricity is $0.07 per kW-hr, (2) 20% of the energy of the electricity is converted into visible light (based on the efficiency of fluorescent lamps), (3) all of the light energy is absorbed by the phototroph (difficult to attain), (4) the photosynthetic efficiency of converting absorbed light into ATP and NADPH is 20% (theoretical efficiency for red light conversion into chemical energy), (5) the energy content of the algal biomass is 6.41 kW-hr per dry kilogram of algal biomass, (6) the carbon content of algae is 50%, and (7) all of the carbon of glucose is converted into algal biomass. The table also indicates the actual costs for producing 1 kg of algal biomass. For phototrophs, the actual cost is much greater than $11.22 because the assumptions are aggressive. For heterotrophs, the cost of energy per kilogram of dry algal

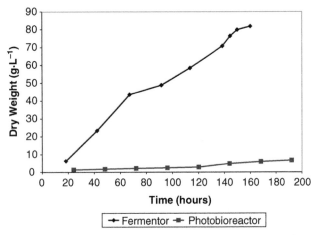

**FIGURE 13.4.** Comparison of *Chlorella* growth in a photobioreactor and a fermentor.

**TABLE 13.3** Some of the major cost factors for photobioreactors and fermentors.

| Cost factor | Photobioreactor | Fermentor |
|---|---|---|
| Construction method | Individually constructed | Mass produced by craftsmen |
| Scale | Relatively small scale | Up to 500,000 liters |
| Algal concentration | Dilute | High |
| Energy source | Light | Organic carbon |

**TABLE 13.4** Calculation of the energy cost to produce a kilogram of algal biomass. See text for assumptions.

|  | Phototroph | Heterotroph |
|---|---|---|
| Energy source | Light | Glucose |
| Energy cost | $0.07/kW-hr | $0.67/kg |
| Estimated cost/kg of dry weight | $11.22 | $0.81 |
| Actual cost/kg of dry weight | Less than $11.22 | $2.01 |
| Productivity | $0.4 \, g \cdot L^{-1} \cdot day^{-1}$ | $5.8 \, g \cdot L^{-1} \cdot day^{-1}$ |

biomass is based on experimental data. The cost of phototrophic growth greatly exceeds the cost of heterotrophic growth, largely because of the inefficiency of converting electricity into ATP and NADPH via photosynthetic electron transport (Table 13.4). These inefficiencies are not present for heterotrophs. When the cost data are coupled with the productivity data (see Fig. 13.4), it becomes clear that the economics of biomass production greatly favor fermentors over photobioreactors. Therefore, photobioreactors are economically feasible only when the value of the products are high enough to justify the costs of production.

## 5.0. References

Acien Fernandez, F. G., Garcia Camacho, F., Sanchez Perez, J. A., Fernandez Sevilla, J. M., and Molina Grima, E. 1998. Modeling of biomass productivity in tubular photobioreactors for microalgal cultures: effects of dilution rate, tube diameter, and solar irradiance. *Biotechnol. Bioengin.* 58:605–16.

Acien Fernandez, F. G., Sanchez Perez, J. A., Fernandez Sevilla, J. M., Garcia Camacho, F., and Molina Grima, E. 2000. Modeling of eicosapentaenoic acid (EPA) production from *Phaeodactylum tricornutum* cultures in tubular photobioreactors. Effects of dilution rate, tube diameter, and solar irradiance. *Biotechnol. Bioengin.* 68:173–83.

An, J.-Y., and Kim, B.-W. 2000. Biological desulfurization in an optical-fiber photobioreactor using an automatic sunlight collection system. *J. Biotechnol.* 80:35–44.

Anderson, M. J., and Whitcomb, P. J. 2000. DOE *Simplified: Practical Tools for Effective Experimentation.* Productivity, Inc., Portland, Oregon, 236 pp.

Apt, K. E., and Behrens, P. W. 1999. Commercial developments in microalgal biotechnology. *J. Phycol.* 35: 215–26.

Apt, K. E., Kroth-Pancic, P. G., and Grossman, A. R. 1996. Stable nuclear transformation of the diatom *Phaeodactylum tricornutum. Mol. Gen. Genet.* 252:572–9.

Babcock, R. W., Malda, J., and Radway, J. C. 2002. Hydrodynamics and mass transfer in a tubular air-lift photobioreactor. *J. Appl. Phycol.* 14:169–84.

Bader, F. G. 1986. Modeling mass transfer and agitator performance in multiturbine fermentor. *Biotechnol. Bioeng.* 30:37–51.

Bailey, J. E., and Ollis, D. F. 1977. *Biochemical Engineering Fundamentals.* McGraw-Hill Book Co, New York, 753pp.

Barahona, L. F., and Rorrer, G. L. 2003. Isolation of halogenated monoterpenes from bioreactor-cultured microplantlets of the macrophytic red algae *Ochtodes secundiramea* and *Portieria hornemannii. J. Nat. Prod.* 66:743–51.

Barbosa, M. J., Janssen, M., Ham, N., Tramper, J., and Wijffels, R. H. 2003. Microalgae cultivation in air-lift reactors: modeling biomass yield and growth rate as a function of mixing frequency. *Biotechnol. Bioengin.* 82:170–9.

Barclay, W. R. 1992. Process for the heterotrophic production of microbial products with high concentrations of omega-3 highly saturated fatty acids. U.S. Patent 5130242.

Barclay, W. R., Meager, K. M., and Abril, J. R. 1994. Heterotrophic production of long chain omega-3 fatty acids utilizing algae and algae-like microorganisms. *J. Appl. Phycol.* 6:123–9.

Barclay, W., and Zeller, S. 1996. Nutritional enhancement of n-3 and n-6 fatty acids in rotifers and *Artemia nauplii* by feeding spray-dried *Schizochytrium* sp. *J. World Aquaculture Soc.* 27:314–22.

Bartow, M. V. 1999. Supersizing the aerobic fermentor. *Chem. Engin.* 106:70–5.

Becker, E. W. 1994. *Microalgae: Biotechnology and Microbiology.* Cambridge University Press, New York, 293 pp.

Behrens, P. W. 1992. Microalgae as a source of bioactive products. In: Coombes, J. D., ed. *New Drugs From Natural Sources.* Information Press, Oxford, pp. 176–182.

Behrens, P. W., and Kyle, D. J. 1996. Microalgae as a source of fatty acids. *J. Food Lipids* 3:259–72.

Behrens, P. W., Sicotte, V. J., and Delente, J. J. 1994. Microalgae as a source of stable isotopically labeled compounds. *J. Appl. Phycol.* 6:113–22.

Behrens, P. W., Bingham, S. E., Hoeksema, S. D., Cohoon, D. L., and Cox, J. C. 1989a. Studies on the incorporation of $CO_2$ into starch by *Chlorella vulgaris. J. Appl. Phycol.* 1:123–30.

Behrens, P. W., Hoeksema, S. D., Arnett, K. L., Cole, M. S., Heubner, T. A., Rutten, J. M., and Kyle, D. J. 1989b. Eicosapentaenoic acid from microalgae. In: Demain, A. L., Somkuti, G. A., Hunter-Cevera, J. C., and Rossmoore, H. W., eds. *Novel Microbial Products for Medicine and Agriculture*. Elsevier, New York, 253–9.

Behrens, P. W., Piechocki, J. A., Purdon, P. A., and Delente, J. J. 1996. Microalgal production of $^{13}$C-galactose and its use as a measure of liver function. *J. Phycol.* S32:6.

Blanch, H. W., and Clark, D. S. 1997. *Biochemical Engineering*. Marcel Dekker, New York, 702 pp.

Bold, H. C., and Wynne, M. J. 1985. *Introduction to the Algae*. Prentice-Hall, Englewood Cliffs, New Jersey, 720 pp.

Brand, L. E., and Guillard, R. R. L. 1981. The effects of continuous light and light intensity on the reproduction rates of twenty two species of marine phytoplankton. *J. Exp. Mar. Biol. Ecol.* 50:119–32.

Brown, D. E. 1970. Aeration in the submerged culture of microorganisms. *Methods Microbiol.* 2:125–74.

Buelena, G., Pouliot, Y., and de la Noue, J. 1987. Performance and operating parameters of a photobioreactor. In: Stadler, T., Moolion, J., Verdus, M.-C., Karamanos, Y. Morvan, H., and Christiaen, D., eds. *Algal Biotechnology*. Elsevier Applied Science, New York, 189–98.

Carlozzi, P. 2000. Hydrodynamic aspects and *Arthrospira* growth in two outdoor tubular undulating row photobioreactors. *Appl. Microbiol. Biotechnol.* 54:14–22.

Chaumont, D. 1993. Biotechnology of algal biomass production: a review of systems for outdoor mass culture. *J. Appl. Phycol.* 5:593–604.

Chaumont, D., Thepenier, C., Gudin, C., and Junjas, C. 1987. Scaling up a tubular photobioreactor for continuous culture of *Porphyridium cruentum* from laboratory to pilot plant (1981–1987). In: Stadler, T., Moolion, J., Verdus, M.-C., Karamanos, Y. Morvan, H., and Christiaen, D., eds. *Algal Biotechnology*. Elsevier Applied Science, New York, 199–208.

Chen, F. 1997. High cell density culture of microalgae in heterotrophic growth. *Trends Biotech.* 14:421–6.

Clark, G. J., Langley, D., Bushell, M. E. 1995. Oxygen limitation can induce microbial secondary metabolite formation: investigations with miniature electrodes in shaker and bioreactor culture. *Microbiology* 141:663–9.

Corman, J., Tsuchiya, H. M., Koepsell, H. J., Benedict, R. G., Kelley, S. E., Feger, V. H., Dworschack, R. G., and Jackson, R. W. 1957. Oxygen absorption rates in laboratory and pilot plant equipment. *Appl. Microbiol.* 5:313–8.

Csögör, Z., Herrenbauer M., Schmidt, K, Posten C. 2001. Light distribution in a novel photobioreactor—modeling for optimization. *J. Appl. Phycol.* 13:325–33.

Dawson, H. N., Burlingame, R., and Cannons, A. C. 1997. Stable transformation of *Chlorella*: rescue of nitrate reductase-deficient mutants with the nitrate reductase gene. *Curr. Microbiol.* 35:356–62.

Day, J. G., Edwards A. P., and Rodgers, G. A. 1990. Large scale, heterotrophic production of microalgal biomass. *Br. Phycol. J.* 25:86.

Day, J. G., Edwards, A. P., and Rodgers, G. A. 1991. Development of an industrial-scale process for the heterotrophic production of a micro-algal mollusc feed. *Bioresource Technol.* 38:245–9.

Degen, J, Uebele, A., Retze, A., Schmid-Staiger, U., and Trösch, W. 2001. A novel airlift photobioreactor with baffles for improved light utilization through the flashing light effect. *J. Biotechnol.* 92:89–94.

Delente, J., Behrens, P. W., and Hoeksema, S. D. 1992. Closed photobioreactor and method of use. U.S. Patent 5,151,347.

Demain, A. L., and Davies, J. E., eds. 1999. *Manual of Industrial Microbiology and Biotechnology*. ASM Press, Washington, D.C., 830 pp.

Doncheck, J. A., Huss, R. J., Running, J. A., and Skatrud, T. J. 1996. L-Ascorbic acid containing biomass of *Chlorella pyrenoidosa*. U.S. Patent 5,521,090.

Droop, M. R. 1974. Heterotrophy of carbon. In: Steward, W. D. P., ed. *Algal Physiology and Biochemistry*. University of California Press, Berkeley, California, 530–59.

Dunahay, T. G., Jarvis, E. E., and Roessler, P. G. 1995. Genetic transformation of the diatoms *Cyclotella cryptica* and *Navicula saprophila*. *J. Phycol.* 31:1004–12.

Eriksen, N., Poulsen, B., Iversen, J. 1998. Dual sparging laboratory-scale photobioreactor for continuous production of microalgae. *J. Appl. Phycol.* 10:377–82.

Gaffney, A. M., Markov, S. A., and Gunasekaran, M. 2001. Utilization of cyanobacteria in photobioreactors for orthophosphate removal from water. *Appl. Biochem. Biotechnol.* 91–93:185–93.

Gladue, R. M. 1998. Heterotrophic microalgae as an inexpensive feed for rotifers. *J. Shellfish Res.* 17:325–6.

Gladue, R. M., and Maxey, J. E. 1994. Microalgal feeds for aquaculture. *J. Appl. Phycol.* 6:131–41.

Gladue, R., and Behrens, P. W. 2002. DHA-containing aquaculture feeds and methods for their production. U.S. Patent 6,372,460.

Godia, F., Albiol, J., Montesinos, J. L., Perez, J., Creus, N., Cabello, F., Mengual, X., Montras, A., and Lasseur, C. H. 2002. MELISSA: a loop of interconnected bioreactors to develop life support in space. *J. Biotechnol.* 99:319–30.

Hall, D. O., Acien Fernandez, F. G., Canizares Guerrero, E., Krishna Rao, K., and Molina Grima, E. 2003. Outdoor helical tubular photobioreactors for microalgal production: modeling of fluid-dynamics and mass transfer and assessment of biomass productivity. *Biotechnol. Bioengin.* 82: 62–73.

Harel, M., Koven, W., Lein, I., Barr, Y., Behrens, P., Stubblefield, J., Zohar, Y., and Place, A. 2002. Advanced DHA, EPA and ArA enrichment materials for marine aquaculture using single cell heterotrophs. *Aquaculture* 213:347–62.

Hellebust, J. A., and Lewin, J. 1977. In: Werner, D., ed. *The Biology of Diatoms*. University of California Press, Berkeley, California, 169–97.

Hepner, L., and Male, C. 1987. Economic aspects of fermentation processes. In: Prave, P., Faust, U., Sittig, W., and Sukatsch, D. A., eds. *Fundamentals of Biotechnology*. Weinheim, Deerfield Beach, Florida, 685–97.

Hilton, M. D. 1999. Small scale liquid fermentations. In: Demain, A. L., and Davies, J. E., eds. *Manual of Industrial Microbiology and Biotechnology*. ASM Press, Washington, D.C., 49–60.

Hoeksema, S. D., Behrens, P. W., Gladue, R. M., Arnett, K. L., Cole, M. S., Rutten, J. M., and Kyle, D. J. 1989. An EPA-containing oil from microalgae in culture. In: Chandra, R. K., ed. *Health Effects of Fish and Fish Oils*. ARTS Biomedical Publishers and Distributors, St. Johns, Newfoundland, Canada, pp. 337–47.

Hu, Q., Kurano, N., Kawachi, M., Iwasaki, I., and Miyahci, S. 1998. Ultrahigh-cell-density culture of a marine green alga *Chlorococcum littorale* in a flat-plate photobioreactor. *Appl. Microbiol. Biotechnol.* 49:655–62.

Hu, Q., and Richmond, A. 1994. Optimizing the population density in *Isochrysis galbana* grown outdoors in a glass column photobioreactor. *J. Appl. Phycol.* 6:391–6.

Huang, Y.-M., and Rorrer, G. L. 2002. Optimal temperature and photoperiod for the cultivation of *Agardhiella subulata* microplantlets in a bubble-column photobioreactor. *Biotechnol. Bioengin.* 79:135–44.

Humphrey, A. 1998. Shake flask to fermentor: what have we learned? *Biotechnol. Prog.* 14:3–7.

Iqbal, M., Grey, D., Stepan-Sarkissian, F., and Fowler, M. W. 1993. A flat sided photobioreactor for culturing microalgae. *Aquacul. Engin.* 12:183–90.

Janssen, M., Tramper, J., Mur, L. R., and Wijffels, R. H. 2002. Enclosed outdoor photobioreactors: Light regime, photosynthetic efficiency, scale-up, and future prospects. *Biotechnol. Bioengin.* 81:193–210.

Kirk, J. T. O. 1994. *Light and Photosynthesis in Aquatic Ecosystems*. Cambridge University Press, New York, 509 pp.

Kuhlmann, C., Bogle, I. D. L., and Chalabi, Z. S. 1998. Robust operation of fed batch fermentors. *Bioproc. Engin.* 19:53–9.

Kyle, D. J., Behrens, P. W., Bingham, S. E., Arnett, K. L., and Lieberman, D. 1988. Microalgae as a source of EPA-containing oil. In: Applewhite, T. H., ed. *World Conference on Biotechnology for the Fats and Oils Industry*. American Oil Chemists Society, Champaign, Illinois, 117–21.

Kyle, D. J., Sicotte, V. J., Singer, J. J., and Reeb, S. E. 1992. Bioproduction of docosahexaenoic acid DHA by microalgae. In: Kyle, D. J., and Ratledge, C., eds. *Industrial Applications of Single Cell Oils*. American Oil Chemists Society, Champaign, Illinois, 287–300.

Kyle, D. J. 1995. Method for diagnosing fatty acid metabolism and absorption disorders using labeled triglyceride oils produced by cultivation of microorganisms. U.S. Patent 5,466,434.

Kyle, D. J. 1996. Production and use of a single cell oil which is highly enriched in docosahexaenoic acid. *Lipid Technol.* 2:106–12.

Lebeau, T., and Robert, J.-M. 2003. Diatom cultivation and biotechnologically relevant products. Part I. Cultivation at various scales. *Appl. Microbiol. Biotechnol.* 60:612–23.

Lee, C.-G., and Palsson B. O. 1994. High-density algal photobioreactors using light-emitting diodes. *Biotechnol. Bioengin.* 44:1161–7.

Lee, R. E. 1989. *Phycology*. Cambridge University Press, New York, 645 pp.

Lee, Y. K., and Pirt, S. J. 1984. $CO_2$ absorption rate in an algal culture: effect of pH. *J. Chem. Tech. Biotechnol.* 34B:28–32.

Manfredini, R., and Cavellera, V. 1983. Mixing and oxygen transfer in conventional stirred fermentors. *Biotechnol. Bioeng.* 25:3115–31.

Mazzuca Sobczuk, T., Garcia Camacho, F., Camacho Rubio, F., Acien Fernandez, F. G., and Molina Grima, E. 2000. Carbon dioxide uptake efficiency by outdoor microalgal cultures in tubular airlift photobioreactors. *Biotechnol. Bioengin.* 67:465–75.

Merchuk, J. C., Gluz, M., and Mukmenev, I. 2000. Comparison of photobioreactors for cultivation of the red microalga *Porphyridium* sp. *J. Chem. Technol. Biotechnol.* 75:1119–26.

Mirón, A., Gómez, A., Camacho, F., Grima, E., and Chisti, Y. 1999. Comparative evaluation of compact photobioreactors for large-scale monoculture of microalgae. *J. Biotechnol.* 70:249–70.

Molina Grima, E., Sanchez Perez, J. A., Garcia Camacho, F., Fernandez Sevilla, J. M., Acien Fernandez, F. G., and Urda Cardona, J. 1995. Biomass and icosapentaenoic acid productivities from an outdoor batch culture of *Phaeodactylum tricornutum* UTEX 640 in an airlift tubular photobioreactor. *Appl. Microbiol. Biotechnol.* 42:658–63.

Moo-Young, M., and Blanch, H. W. 1987. Transport phenomena and bioreactor design. In: Bu'Lock, J., and Kristiansen, B., eds. *Basic Biotechnology*. Academic Press, New York, 133–72.

Morita, M., Watanabe, Y., and Saiki, H. 2000. Investigation of photobioreactor design for enhancing the photosynthetic productivity of microalgae. *Biotechnol. Bioeng.* 69: 693–98.

Morita, M., Watanabe, Y., and Saiki, H. 2001. Evaluation of photobioreactor heat balance for predicting changes in culture medium temperature due to light irradiation. *Biotechnol. Bioeng.* 74:466–75.

Morita, M., Watanabe, Y., Okawa, T., and Saiki, H. 2001. Photosynthetic productivity of conical helical tubular photobioreactors incorporating *Chlorella* sp. under various culture medium flow conditions. *Biotechnol. Bioengin.* 74:136–44.

Muller-Feuga, A., Le Guedes, R., and Pruvost, J. 2003. Benefits and limitations of modeling for optimization of *Porphyridium cruentum* cultures in an annular photobioreactor. *J. Biotechnol.* 103:153–63.

Nakahara, T., Yokochi, T., Higashihara, T., Tanaka, S., Yaguchi, T., and Honda, D. 1996. Production of docosahexaenoic and docosapentaenoic acids by *Schizochytrium* sp. isolated from Yap islands. *JAOCS* 73:1421–26.

Ogbonna, J. C., Soejima, T., and Tanaka, H. 1999. An integrated solar and artificial light system for internal illumination of photobioreactors. *J. Biotechnol.* 70:289–97.

Ogbonna, J. C., Yada, H., and Tanaka, H. 1995. Light supply coefficient: a new engineering parameter for photobioreactor design. *J. Ferment. Bioengin.* 80:369–76.

Oswald, W. J. 1988. Large scale algal culture systems (engineering aspects). In: Borowitzka, M. A., and Borowitzka, L. J., eds. *Microalgal Biotechnology.* Cambridge University Press, Cambridge, 357–410.

Parker, B. C., ed. 1994. Microalgal biotechnology and commercial applications. *J. Appl. Phycol.* 6:1–165.

Pirt, S. J. 1983. Maximum photosynthetic efficiency: a problem to be resolved. *Biotechnol. Bioeng.* 25:1915–22.

Pirt, S. J., Lee, Y. K., Walach, M. R., Pirt, M. W., Balyuzi, H. H. M., and Bazin, M. J. 1983. A tubular bioreactor for photosynthetic production of biomass from carbon dioxide: design and performance. *J. Chem. Tech. Biotechnol.* 33B:35–58.

Pohl, P., Kohlhase, M., and Martin, M. 1987. Photobioreactors for the axenic mass cultivation of microalgae. In: Stadler, T., Moolion, J., Verdus, M.-C., Karamanos, Y. Morvan, H., and Christiaen, D., eds. *Algal Biotechnology.* Elsevier Applied Science, New York, 209–17.

Polzin, J. P., and Rorrer, G. L. 2002. Halogenated monoterpene production by microplantlets of the marine red alga *Ochtodes secundiramea* within an airlift photobioreactor under nutrient medium perfusion. *Biotechnol. Bioengin.* 82:415–28.

Pulz, O. 2001. Photobioreactors: production systems for phototrophic microorganisms. *Appl. Microbiol. Biotechnol.* 57:287–93.

Pushparaj, B., Pelosi, E., Tredici, M., Pinzani, E., and Materassi, R. 1997. An integrated system for outdoor production of microalgae and cyanobacteria. *J. Appl. Phycol.* 9:113–19.

Quesnel, L. B. 1987. Sterilization and sterility. In: Bu'Lock, J., and Kristiansen, B., eds. *Basic Biotechnology.* Academic Press, New York, 197–215.

Radmer, R. J. 1990. Photobioreactor. U.S. Patent 4,952,511.

Radmer, R. J., Behrens, P. W., Fernandez, E., Ollinger, O., and Howell, C. 1984. Algal culture studies related to a closed ecological life support system. *The Physiologist* 27:25–8.

Radmer, R. J., Behrens, P. W., and Arnett, K. L. 1987. An analysis of the productivity of a continuous algal culture system. *Biotechnol. Bioeng.* 24:488–92.

Radmer, R. J., Cox, J. C., Lieberman, D., Behrens, P. W., and Arnett, K. L. 1987. Biomass recycle as a means to improve energy efficiency of CELSS algal culture systems. *Adv. Space Res.* 7:11–5.

Radmer, R. J., and Parker, B. C. 1994. Commercial applications of algae: opportunities and constraints. *J. Appl. Phycol.* 6:93–8.

Ratchford, I. A. J., and Fallowfield, H. J. 1992. Performance of a flat plate, air-lift reactor for the growth of high biomass algal cultures. *J. Appl. Phycol.* 4:1–9.

Reisman, H. B. 1999. Economics. In: Demain, A. L., and Davies, J. E., eds. *Manual of Industrial Microbiology and Biotechnology.* ASM Press, Washington, D.C., 273–88.

Richardson, B., Orcutt, D. M., Schwetner, H. A., Martinez, C. L., and Wickline, H. E. 1969. Effects of nitrogen limitation on the growth and composition of unicellular algae in continuous culture. *Appl. Micro.* 18:245–50.

Richmond, A., ed. 2004. *Handbook of Microalgal Culture—Biotechnology and Applied Phycology.* Blackwell Science, Oxford, 566 pp.

Richmond, A., Boussiba, S., Vonshak, A., and Kopel, R. 1993. A new tubular reactor for mass production of microalgae outdoors. *J. Appl. Phycol.* 5:327–32.

Running, J. 1999. Process for the production of ascorbic acid with *Prototheca*. U.S. Patent 5,900,370.

Running, J. A., Huss, R. J., and Olson, P. T. 1994. Heterotrophic production of ascorbic acid by microalgae. *J. Appl. Phycol.* 6:99–104.

Running, J. A., Severson, D. K., and Schnedider, K. J. 2002. Extracellular production of L-ascorbic acid by *Chlorella protothecoides*, *Prototheca* species, and mutants of *P. morifirmis* during aerobic culturing at low pH. *J. Industr. Microbiol. Biotechnol.* 29:93–8.

Sfat, M. R. 1984. Economics of batch microalgal heterotrophic production of chemicals. *J. Phycol.* 20:S30.

Samson, R., Leduy, A. 1985. Multistage continuous cultivation of blue-green alga Spirulina maxima in flat tank photobioreactors with recycle. *Can. J. Chem. Engin.* 63:105–12.

Sinclair, C. G. 1984. Formulation of the equations for oxygen transfer in fermentors. *Biotechnol. Lett.* 5:111.

Sonnleitner, B., 1999. Instrumentation of small scale bioreactors. In: Demain, A. L., and Davies, J. E., eds. *Manual of Industrial Microbiology and Biotechnology.* ASM Press, Washington, D.C., 221–35.

Spektorova, L., Creswell, R. L., and Vaughan, D. 1997. Closed tubular cultivators. An innovative system for commercial culture of microalgae. *World Aquaculture* 28:39–43.

Suh, I. S., and Lee, S. B. 2001. Cultivation of a cyanobacterium in an internally radiating air-lift photobioreactor. *J. Appl. Phycol.* 13:381–8.

Suh, I. S., and Lee, S. B. 2003. A light distribution model for an internally radiating photobioreactor. *Biotechnol. Bioengin.* 82:180–9.

Sukenik, A., and Shelef, G., 1984. Algal autoflocculation—verification and proposed mechanism. *Biotechnol. Bioeng.* 26:142–7.

Sundararajan, A., and Ju, L.-K. 2000. Evaluation of oxygen permeability of gas vesicles from cyanobacterium *Anabaena flos-aquae. J. Biotechnol.* 77:151–6.

Takano, H., Takeyama, H., Nakamura, N., Sode, K., Burgess, J. G., Manabe, E., Hirano, M., and Matsunaga, T. 1992. $CO_2$ removal by high density culture of a marine cyanobacterium *Synechococcus* sp. using an improved photobioreactor employing light-diffusing optical fibers. *Appl. Biochem. Biotechnol.* 34:449–58.

Tredici, M. R., and Zittelli, G. C. 1998. Efficiency of sunlight utilization: tubular versus flat photobioreactors. *Biotechnol. Bioengin.* 57:187–97.

Ugwu, C. U., Ogbonna, J. C., and Tanaka, H. 2002. Improvement of mass transfer characteristics and productivities of inclined tubular photobioreactors by installation of internal static mixers. *Appl. Microbiol. Biotechnol.* 58:600–7.

Van Brunt, J., 1986. Fermentation economics. *Biotechnology* 4:395–401.

Wang, D. I. C., Cooney, C. L., Demain, A., Dunnill, P., Humphrey, A. E., and Lilly, M. D., eds. 1979. *Fermentation and Enzyme Technology.* Wiley, New York, 374 pp.

Yun, Y.-S., and Park, J. M. 2001. Attenuation of monochromatic and polychromatic lights in *Chlorella vulgaris* suspensions. *Appl. Microbiol. Biotechnol.* 55:765–70.

Watanabe, Y., and Hall, D. O. 1996. Photosynthetic production of the filamentous cyanobacterium *Spirulina platensis* in a cone-shaped helical tubular photobioreactor. *Appl. Microbiol. Biotechnol.* 44:693–8.

Wilkinson, L. 1998. Criteria for the design and evaluation of photobioreactors for the production of microalgae. World Aquaculture Society Annual Meeting, Las Vagas, NV p. 584.

Wohlgeschaffen, G. D., Subba Rao, D. V., and Mann, K. H. 1992. Vat incubator with immersion core illumination—a new, inexpensive setup for mass phytoplankton culture. *J. Appl. Phycol.* 4:25–9.

Wu, X., and Merchuk, J. C. 2002. Simulation of algae growth in a bench-scale bubble column reactor. *Biotechnol. Bioengin.* 80:156–68.

Zaslavskaia, L. A., Lippmeier, J. C., Shih, C., Ehrhardt, D., Grossman, A. R., and Apt, K. E. 2001. Trophic conversion of an obligate photoautotrophic organism through metabolic engineering. *Science* 292:2073–5.

Ziegler, H., Dunn, I. J., and Bourine, J. R. 1980. Oxygen transfer and mycelial growth in a tubular loop fermentor. *Biotechnol. Bioeng.* 22:1613–35.

# CHAPTER 14

# CULTURING MICROALGAE IN OUTDOOR PONDS

MICHAEL A. BOROWITZKA

*Algae Research Group, School of Biological Sciences &*
*Biotechnology, Murdoch University*

## CONTENTS

Key Index Words: Raceway Pond, Pond Design, *Dunaliella*, *Spirulina*, *Chlorella*, *Haematococcus*, Productivity, Light, Temperature, Carbon Dioxide, Paddle Wheel, Scale-Up

## 1.0. INTRODUCTION

Commercial-scale culture of microalgae generally requires the ability to economically produce ton quan-

tities of algal biomass. This requires culture volumes of 10,000 to greater than 1,000,000 liters, and therefore almost all commercial-scale culture is currently in open outdoor ponds. *Chlorella* spp., *Spirulina platensis* Geitler, *S. maxima* Geitler (=*Arthrospira* [see Castenholz 1989], but here called *Spirulina* to reflect common usage), *Dunaliella salina* (Dunal) Teodoresco, *Haematococcus pluvialis* Flotow, and *Nannochloropsis* species are grown outdoors in open ponds, the latter for use in aquaculture. Several other species have also been grown successfully in outdoor ponds on a small scale: *Porphyridium* species, *Monodus* species, *Phaeodactylum tricornutum* Bohlin, and *Scenedesmus obliquus* (Turpin) Kützing. This chapter describes the key features of pond design, operation, and management for the culture of microalgae outdoors, with particular emphasis on large-scale commercial culture.

Open pond culture is cheaper than culture in closed photobioreactors (Borowitzka 1999a) but is limited to a relatively small number of algae species. Furthermore, commercial outdoor cultivation is generally restricted to tropical and subtropical zones in regions of low rainfall and low cloud cover. Although most microalgae require light and carbon dioxide, they are very diverse in their other environmental requirements. Each species has fairly specific requirements, and the various culturing systems and methods reflect this diversity (Table 14.1).

There is little literature on actual commercial culture systems because much of the fine detail of the culture process is commercially sensitive. However, details can be found in the following reports on the culture of the alga, *Spirulina* (Belay et al. 1994, Fox 1996,

**TABLE 14.1** Summary of growth conditions for typical microalgal strains cultured in outdoor ponds; variation between strains means that some strains have optima outside the limits listed.

| Species condition | Chlorella vulgaris | Dunaliella salina | Haematococcus pluvialis | Phaeodactylum tricornutum | Spirulina platensis |
|---|---|---|---|---|---|
| Natural habitat | Freshwater | Hypersaline brines | Freshwater | Marine | Alkaline soda lakes |
| Salinity, optimum (% [w/v] NaCl)[a] | 0 | 22% (growth) 35% (carotenogenesis) | 0 | 3% | 0 to 1% |
| Salinity, maximum (% [w/v] NaCl)[a] | ~1% | 35% | ~1% | ~5% (?) | <3% |
| Temperature (°C), optimum | ~25 | 30–40 | ~18–22 | ~18–24 | 30–38 |
| pH, optimum | 6.5–7.5 | ~9.0 | ~7.0 | ~8.0 | ~9.0 (–10.0) |
| Commonly used media[b] | Bolds basal | Modified Johnson's | Bolds basal | Guillard's f | Zarrouk |

[a]For practical reasons salinity is expressed here as % (w/v) NaCl rather than the usual unit of p.s.u., given the great variability in the salt composition of algal media and the fact that salinity is generally adjusted by the addition of NaCl.

[b]References for medium composition: Borowitzka 1988, Vonshak 1997a.

Belay 1997), *Dunaliella* (Borowitzka and Borowitzka 1989, Schlipalius 1991, Ben-Amotz 1999), *Chlorella* (Gummert et al. 1953, Tsukada et al. 1977, Kawaguchi 1980), and *Haematococcus* (Bubrick 1991, Olaizola 2000). Wastewater treatment systems using algae are described by Oswald (1988b), and the production of cyanobacteria for biofertilizers in India is reviewed by Kaushik (1998).

## 2.0. TYPES OF OPEN PONDS

The design of open ponds for outdoor cultivation must meet numerous basic criteria (e.g., construction and maintenance costs should be low while providing an optimal hydrodynamic and light environment for the algae). Several types of open pond systems have been developed for large-scale outdoor algal culture.

Very large (extensive), shallow, unlined, and unmixed ponds are used in Australia for the cultivation of *D. salina* and are unsuitable for most other species of algae (Fig. 14.1a). Pond areas range from 1 ha to more than 200 ha, with an average depth of 20 to 30 cm. The extreme, high-salinity environment in which *D. salina* grows means that contamination can be controlled solely by maintaining high salinity. The high amounts of salt required are obtained through the solar evaporation of seawater, and a low-cost, highly efficient harvesting process is used to separate the low-cell-density

biomass. This type of culture process also requires very low land costs because of the large area required, as well as near-optimal climatic conditions (e.g., low cloud cover and limited rainfall restricted to a short time of the year).

Deep ponds (tanks), usually mixed by aeration, are mainly used for small-scale production of marine algae such as *Nannochloropsis* for aquaculture (Fig. 14.2c). Pond areas are generally less than 10 m², with depths of 50 cm or more. Although very inefficient, these systems are easy to operate and are low in cost. Because the algae are fed directly to rotifers or other invertebrates, the cost of harvesting is not an issue. Some aquaculture hatcheries use open fiberglass cylinders approximately 1 m in diameter rather than large tanks. These cylinders are easier to mix, which reduces problems with culture "crashes." The algae also receive more light through the translucent sides of the cylinders.

Circular ponds with a centrally pivoted rotating agitator are used in Japan and Taiwan for the production of *Chlorella* (Fig. 14.2a). These circular ponds are the oldest large-scale algal culture systems and are based on similar systems used in wastewater treatment. Depth is about 30 cm. The design of these systems, however, limits pond size to about 10,000 m², because relatively even mixing by the rotating arm is no longer possible in larger ponds.

Single, rectangular ponds with a paddle wheel (raceway ponds) are the most widely used for the production of *Spirulina*, *D. salina*, and *Haematococcus* and represent the most efficient design for the large-scale

**FIGURE 14.1.** **(a)** Extensive open ponds at the Hutt Lagoon, Western Australia, plant operated by Cognis Nutrition & Health, growing *D. salina* as a source of natural beta-carotene. The largest ponds are about 200 ha in area. **(b)** Raceway ponds used for the culture of *Spirulina* by Microbio Inc. at Calipatria, California (photo courtesy of Dr. Ahma Belay).

**FIGURE 14.2.** **(a)** Center-pivot pond used for the cultivation of *Chlorella* in Taiwan. Note the variable mixing effectiveness, as illustrated by the foaming at the pond perimeter. **(b)** Biocoil, a 1,000-liter helical tubular photobioreactor. **(c)** Deep aerated tank used for the culture of *Nannochloropsis* for aquaculture.

culture of most microalgae (Fig. 14.1b). Individual ponds are up to 1 ha in area, with an average depth of about 20 to 30 cm.

Joined rectangular ponds with paddle wheels (meander ponds) are used in high-rate oxidation ponds for the treatment of wastewater (e.g., in Hollister, California) and also at several sites for the production of low-grade *Spirulina* for animal feed. The slower flow rates achieved in these ponds make them unsuitable for commercial production of high-quality algal biomass, because the areal productivity is low and contamination is harder to control. Pond areas are approximately 1 ha or larger.

A sloped meandering pond with circulating pump was used in a pilot plant in Peru but has never been used in a commercial algae plant.

A very shallow sloped pond with circulating pump (cascade pond) was developed by Setlík (1970) and is used for the production of *Chlorella* in Trebon, Czech Republic. It is very well suited for algae such as *Chlorella* and *Scenedesmus*, which can tolerate repeated pumping. Very high cell densities and productivities are achieved, because the slopes have an extremely shallow water depth (~10–20 mm). The original design had a glass bottom for the sloping channels, with very high construction costs; however, the availability of new materials has greatly reduced these.

Commercial-scale ponds range in surface area from about 0.5 to 1.0 ha for raceway or central pivot ponds and up to 200 ha or more for the extensive ponds used in the culture of *D. salina*.

## 3.0. SELECTING POND TYPES

The choice of production pond area is a compromise among numerous considerations. Several small ponds, rather than a single large pond, may be advantageous in the case of culture problems, because the whole production plant is not as affected. However, small ponds are more expensive to construct per unit area than large ponds. Pond size affects water circulation, which in turn affects the design and operating cost of the circulation/mixing system. The selection of the type of outdoor culture system requires careful evaluation of many other factors.

Climatic conditions include the daily and annual temperature range, annual rainfall and rainfall pattern, number of sunny days, and degree of cloud cover. If the local climatic conditions are suitable for only part of the year, then a more intensive culture system is preferred.

Availability and cost of land are important. If the land costs are high, then intensive systems such as raceway ponds are essential. However, if land costs are low, then extensive systems can be considered.

Source and quality of water are significant factors. The larger the pond area, the greater the amount of water lost by evaporation. A natural source of seawater or brines is desirable for marine or hypersaline algae, because the cost of salt is prohibitive. Some water sources also may contain heavy metals or other contaminants, and these are unsuitable for microalgae culture because of final product contamination.

Well-mixed, intensive systems achieve significantly higher cell densities, and this means that the volume of culture that must be processed to harvest the biomass is reduced. The type of harvesting system (e.g., centrifugation, filtration, flocculation) is therefore a critical consideration, because the economics of harvesting are greatly affected by the cell density.

The final product affects pond choice. If the biomass is used for human consumption, then the system must be capable of "cleaner" production than if the biomass is used for animal feed or if a specific product is extracted from the algae and contamination is less of a concern.

## 4.0. POND MIXING

Turbulent flow (mixing) is essential for maximum production of microalgae in open ponds; the only unmixed large-scale ponds are those used for the extensive culture of *D. salina*. These unmixed ponds achieve a maximum biomass concentration of about $0.1 \text{ g} \cdot \text{L}^{-1}$ dry weight, compared with *D. salina* cultures grown in well-mixed raceways, which reach dry weights up to $1.0 \text{ g} \cdot \text{L}^{-1}$. Mixing prevents the settling of cells and avoids thermal and oxygen stratification in the pond. Experience has shown that ponds with adequate mixing

show higher productivities and more stable cultures (Richmond and Vonshak 1978, Laws et al. 1983, Richmond 1986, Bosca et al. 1991). In raceway cultures, velocities of $5.0 \text{ cm} \cdot \text{s}^{-1}$ are sufficient to eliminate thermal stratification and maintain most species of algae in suspension. However, such a low velocity is difficult to maintain in a large pond because of frictional losses in the channel and the corners and losses due to irregularities of the pond lining material. In practice, flow velocities of at least 20 to 30 cm $\cdot$ s$^{-1}$ are required and, if experience shows that more turbulence significantly enhances growth, then extra passive turbulence-generating devices can be installed in the pond. However, higher flow rates require more energy, thus increasing operational costs (Oswald 1988a).

Aspects of pond design to maintain adequate water flow in raceway ponds are also important. The flow speed of water in sloping straight channels can be described by Manning's equation (Oswald 1988a):

$$V = \frac{1}{n} R^{2/3} S^{1/2} \qquad (1)$$

where $V$ = the mean velocity (m $\cdot$ s$^{-1}$) and $R$ = mean hydraulic radius (m), where $R$ is the area of flow ($A$) divided by the perimeter in contact with the water ($P$) (i.e., $R = A/P = dw/[w + 2d]$, where $w$ = the channel width and $d$ = the depth). For very wide, shallow channels $R$ is approximately equal to the depth $d$. The rate of energy loss ($S$) is in the channel per unit length, that is, $d \cdot L^{-1}$ (dimensionless), where $d$ is the change in water depth (head loss) and $L$ is the channel length; and n = Manning's friction coefficient (s $\cdot$ m$^{-1/3}$), a measure of channel roughness. Table 14.2 gives estimated mean

**TABLE 14.2** Estimated mean value for Manning's n in open channels (Oswald 1988a).

| Materials for channel liner | Manning's n |
|---|---|
| Smooth plastic on smooth concrete | 0.008 |
| Plastic with "scrim" on smooth earth | 0.010 |
| Smooth plastic on granular earth | 0.012 |
| Smooth cement concrete | 0.013 |
| Smooth asphalt concrete | 0.015 |
| Coarse trowelled concrete, rolled asphalt | 0.016 |
| Gunnite or sprayed membranes | 0.020 |
| Compacted smooth earth | 0.020 |
| Rolled coarse gravel, coarse asphalt | 0.025 |
| Rough earth | 0.030 |

values for Manning's $n$ in open channels constructed of various materials.

The interaction among channel length, water velocity, and water depth also must be considered in pond design. Because hydraulic energy is low due to friction and because friction increases as the square of the velocity, the depth of flowing water in the channel decreases as a complex function of the channel length (Oswald 1988a). For practical purposes it is best to determine the channel length $L$ that corresponds to a given change in depth $d$ for a given friction factor, $n$, hydraulic radius, $R$, and velocity, $V$. This is done by squaring both sides of the Manning equation (Equation 1) and solving for $s$:

$$s = \frac{V^2 n}{R^{4/3}} \quad (2)$$

and, since $s = \Delta d / L$ and $R = dw/(W + 2d)$,

$$\Delta d = \frac{LV^2 n^2}{\left(dw/w + 2d\right)^{4/3}} \quad (3)$$

and

$$L = \frac{\Delta d \left(dw/(w + 2d)\right)^{4/3}}{V^2 n^2} \quad (4)$$

The limiting pond area, $A$, therefore equals

$$A = Lw \quad (5)$$

Raceway ponds are generally mixed with paddle wheels, and experience has shown that paddle wheels are by far the most efficient for mixing the algal cultures and are the easiest to maintain. The design of the paddle wheel also affects flow rate and energy requirements (see later discussion). Earlier designs also used air lifts, propellers, and drag boards (Becker 1994). Drag boards are boards that close the pond section except for a slit of a few centimeters above the bottom and along the sides. The drag board is dragged through the culture pond to create turbulence (Valderrama et al. 1987). Although these boards are very effective, they are extremely difficult to maintain and to scale up.

## 4.1. Mixing-Depth Relationships

Algae require light for growth. In ponds the algae near the pond surface receive more light than those deeper in the water column, and if the cell density is high, then most of the light is absorbed in the upper few centimeters of the pond (Grobbelaar et al. 1990). It is therefore desirable to maintain pond depths as shallow as possi-

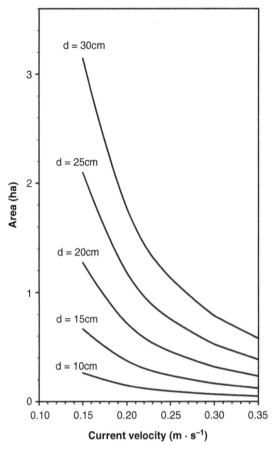

**FIGURE 14.3.** Relationship among current velocity, average pond depth, and the pond area, based on a channel width of 6 m, a Manning's $n$ value of 0.015, and a $\Delta d$ equal to $(1/2)d$.

ble to provide the maximum amount of light to the algae. However, the area that can be mixed by a paddle wheel is a function of water velocity and water depth. From Equations 4 and 5, it can be seen that the limiting pond area decreases with decreasing pond depths and increasing mixing velocities (Fig. 14.3). However, by increasing the pond depth and decreasing the mixing velocity, the limiting pond area becomes large enough so that constructional and operational, rather than hydraulic, considerations limit actual pond size.

To optimize productivity, pond depth also should be optimized. The theoretical maximum possible photosynthetic efficiency ($f_m$) that can be attained by a culture can be derived from the Bush equation (Burlew 1953):

$$f_m = (s_s/s_o)[\ln(s_o/s_s) + 1] \quad (6)$$

where $s_s$ is the photon flux density at which photosynthesis is saturated and $s_o$ is the photon flux density at any instant. This predicted efficiency cannot be attained in real cultures, because they are also limited by the quantum efficiency of various wavelengths. Furthermore, one must account for night-time respiration before a net efficiency is attained. The actual light-utilization efficiency of mixed cultures of algae ranges from less than 1% to a (rarely achieved) maximum of 5%.

For green algae, the following relationship between the concentration of algae, $C_c$ (mg $\cdot$ L$^{-1}$), and the light penetration depth, $d_p$ (cm), has been determined empirically by Oswald (1988a):

$$d_p = 6000/C_c \qquad (7)$$

Field observations of large-scale cultures further indicate that the culture concentration in light-limited, continuously mixed cultures approaches that which permits light to penetrate two-thirds of the actual culture depth, that is:

$$d_p = (2/3)d \qquad (8)$$

and

$$C_c = 9000/d \qquad (9)$$

Therefore, a continuously mixed outdoor culture of green algae at a depth of 30 cm should achieve an average light-limited algal concentration of about 300 mg $\cdot$ L$^{-1}$ dry weight. Higher photon flux densities increase this only slightly, because the penetration of light is proportional to the log of its intensity. High concentrations of algae can therefore be obtained only at shallow culture depth. Alternatively, one can control the cell density to achieve the optimum photon flux density for the algae at a particular temperature and thus maximize the photosynthetic efficiency (Richmond 1992). Any system that enhances turbulence without damaging the algal cells should increase the productivity. However, the final culture depth used is a compromise between achieving optimum photosynthetic efficiency and the hydraulic limitations of the mixing system, while obtaining the maximum productivity per unit pond area (areal productivity).

## 5.0. RACEWAY POND DESIGN AND CONSTRUCTION

Ponds can be excavated and lined with impermeable material, or they can be constructed above ground with walls of bricks or concrete blocks and floors of either concrete or some other suitable liner. Some type of pond liner is generally necessary to maintain relatively uniform flow rates throughout the pond and to avoid clouding of the water due to resuspension of sediments. Excavated ponds have several disadvantages. These include the fact that the sloping walls can lead to an increase of insect and other contamination, which can more easily be blown into the pond by wind. Furthermore, depending on the slope, a constant depth is more difficult to maintain, and this may interfere with water flow. However, they are the most cheaply constructed ponds. Above-ground concrete ponds overcome many of these problems. However, over time the concrete may wear unevenly, thus leading to water flow problems. Concrete is also fairly rough unless epoxy-coated, leading to reduced water flow due to increased "friction." Finally, concrete ponds are expensive to construct.

The choice of materials for pond construction depends on many factors, and the salinity of the medium restricts the choice available. For example, concrete ponds are quite suitable for the culture of freshwater algae such as *Chlorella* but are highly unsuitable for culture of the hypersaline *D. salina* because concrete gradually deteriorates at the high salinities. Although the concrete may be coated with various materials to protect it, this has generally been ineffective in the long term (Ben-Amotz and Avron 1989).

Ponds constructed with walls of concrete blocks appear to be the most effective in terms of both cost of construction and operation. Such ponds are constructed on levelled ground by building the external walls and the central divider of concrete blocks and then installing a suitable impermeable liner.

Several liners are available, and the most appropriate choice depends on its cost, product life, and effect on algal growth and quality. A well-manufactured polyvinyl chloride (PVC) liner of 0.75 mm lasts about 5 years in temperate desert climates. However, PVC may inhibit the growth of *Spirulina* and other algae (Dyer and Richardson 1962, Bernhard et al. 1966, Blankley 1973). PVC also contains phthalic acid esters as plasticising agents and usually contains some unreacted vinyl chloride; these compounds are potentially harmful to humans. Lead is frequently used to stabilize the PVC linings, and mercury may also be present to inhibit degradation by soil microorganisms. Algae accumulate these heavy metals efficiently, and it is therefore not recommended that PVC be used if the algae are to be used for food or feed. Special double-layered linings are available for food use; the top layer is stabilized with tin, and the bottom layer, with its heavy metal microbial inhibitor, is isolated from the culture.

An alternative to PVC is chlorinated polyethylene (CPE). CPE has little effect on algal growth (Bernhard et al. 1966, Blankley 1973) and does not appear to present a human health hazard. It should be noted, however, that black polyethylene materials can be inhibitory to algal cultures (Bernhard et al. 1966) and certain black CPE linings are not approved for food use in the United States. CPE is also more difficult to work with than PVC.

During construction it is very important to ensure that the liner has as few wrinkles as possible so that the pond bottom has the lowest possible Manning's roughness coefficient to allow the highest possible water flow rate. For large ponds the liners are usually welded or glued *in situ*. The edges of the liners must be fixed in such a way that wind cannot get below them.

## 5.1. Flow Rectifiers

Ponds with right-angled corners require flow rectifiers, which serve to prevent buildup of material in the corners by evening out the flow. Such buildups can lead to localized anaerobic regions, excessive bacterial growth, and ultimately a decline in the algal culture. The structure and operation of flow rectifiers have been described (Shimamatsu and Tominaga 1980, Shimamatsu 1987). They consist of closely spaced turning vanes placed along the diagonals from the center wall to the corners of the pond. These vanes are arranged less than 50 cm apart and produce two right-angle changes in direction. These rectifiers are most easily installed after the lining and simply sit on the plastic surface at the corners of the pond. Pond rectifiers are usually constructed of steel, and no special finish or coating is required because the steel does not degrade markedly and is nontoxic.

Small "dead" spots form between the outside vane of the rectifier and the outer wall, as well as between the inside vane and the center wall. To reduce these, the outside wall should be built up before laying the plastic liner, so that the walls curve with the same radius as the rectifier vanes. This cannot be done with the center wall because of its small width, but this does not present a significant problem.

An alternative consists of an eccentrically placed curved wall and baffles at the end of the pond furthest away from the paddle wheel (Dodd 1986). This creates a curved zone of accelerating flow, followed by a flow-expansion zone after the directional change has been made. The rate of constriction of the curved zone is sufficient to avoid eddies or velocities below that causing deposition on the back side of the center wall and baffles.

## 5.2. Paddle Wheel Design

Experience has shown that paddle wheels are the most efficient mixing devices for raceway ponds. The typical paddle wheel described here is based on that of Dodd (1986) and seems to be very effective. The paddle wheel sits in a depression or invagination (sump) on the pond bottom. This sump serves to maintain a minimum clearance between the blades and the bottom so that back-flow can be reduced. In a conventional paddle wheel over a flat bottom, the clearance varies with time, and so does the back-flow. For the clearance to be a constant minimum value, it is necessary that at least one blade be in the sump at all times. Using this criterion, we can calculate the minimum size of the sump for a given diameter of paddle wheel with a particular number of blades.

The greater the paddle wheel diameter, the greater the paddle wheel efficiency because of a lower mean back-flow. However, a larger diameter also means a higher construction cost and a slightly greater weight. A suitable diameter therefore is 1,500 mm. Similarly, the more paddle wheel blades, the higher the paddle wheel efficiency and the less the motor shock. However, a paddle wheel with more than eight blades is impractical from the construction viewpoint and does not significantly improve efficiency. Finally, the smaller the clearance between the blade and the pond floor, the more efficient the paddle wheel. In practice, a clearance of 20 mm on both sides and the bottom is feasible and allows a suitable safety margin.

Using the above parameters, we can now calculate the minimum depth required for the sump. The situation can be described by the following equations:

$$\omega = \pi/P_n \qquad (10)$$

$$D = (r+T)\cos\omega \qquad (11)$$

$$B = r+T-D \qquad (12)$$

where $\omega$ = half the angle between blades, $P_n$ = the number of blades, $r$ = the paddle wheel radius, $T$ = the clearance between blade and the pond bottom, $B$ = the depth of the sump, and $D$ = the distance from the paddle wheel axis to the normal pond bottom. If we substitute Equation 10 into Equation 11, and Equation 11 into Equation 12, then we obtain:

$$B = (r+T)(1-\cos[\pi/N]) \qquad (13)$$

If $P_n = 8$, $r = 750$, and $T = 20$, then $B = 59$. Thus, a sump of at least 59 mm in depth and 770 mm in radius is required. Because there are imperfections in construction and installation, an additional safety margin should be applied; the suggested total depth for the sump is therefore 100 mm.

The center of the paddle wheel shaft is therefore 670 mm above the flat pond bottom. The height of each blade must be at least the water depth ($d$) plus the sump depth ($B$) minus the clearance ($T$), (i.e., $d + B - T$). For a water depth of 200 mm, a 50-mm excess should be added to account for head loss changes; the resulting height of each blade is 330 mm. Adding a safety margin of 20 mm results in a final height of 350 mm.

In a typical paddle wheel, the shaft and spokes are constructed from carbon steel pipe. They should be sandblasted and coated with two-part epoxy paint. The paddle wheel blades can be constructed from ~2-mm-thick mild steel plate and the blades crimped to provide a greater resistance to bending; these are bolted onto the spokes. In highly corrosive media the paddle wheel can also be constructed of fiberglass.

The sump under the paddle wheel must be accurately constructed and must be able to withstand the scouring forces at higher water velocities. It is therefore recommended that this sump be constructed of concrete. A concrete base under the paddle wheel is also desirable even if the pond bottom is flat beneath the wheel.

## 5.3. Motor

The choice of motor is more difficult, because the exact power requirements for paddle wheels with sumps in the pond floor are not well-known. However, we can use data for ponds without a sump as a general indicator. Paddle wheels with sumps are more efficient.

The hydraulic power requirement can be calculated from the head losses, channel dimensions, and speed. Thus, the power required is

$$P = \frac{QW\Delta d}{102e} \qquad (14)$$

where $P$ = the power (kW), $Q$ = the quantity of water in motion ($m^3 \cdot s^{-1}$), $W$ = the specific weight of water ($kg \cdot m^{-3}$), and $e$ = the efficiency of the paddle wheel; 102 is the conversion factor required to convert $m \cdot kg \cdot s^{-1}$ to kW. At 15°C the specific weight of water is approximately 1,000 $kg \cdot m^{-3}$.

For a channel of width $w$ and depth $d$, the quantity of flow $Q$ is given by

$$Q = wdV \qquad (15)$$

Thus, for a 6-m-wide channel, 20 cm deep, with a velocity of 15 cm $\cdot s^{-1}$, and at a paddle wheel efficiency of 0.17 (an average value for paddle wheels operating over a flat bottom), the power requirement is 1.038 kW. Over 24 hours this would be 24.9 kW-h $\cdot$ day$^{-1}$. As a total of 1.27 ha is mixed, this is equal to 24.9/1.27 = 19.59 kW-h $\cdot$ ha$^{-2}$ $\cdot$ day$^{-1}$). If the velocity is increased to 20 cm $\cdot s^{-1}$, then the power requirement increases to 33.2 kW-h $\cdot$ day$^{-1}$ (46.43 kW-h $\cdot$ ha$^{-2}$ $\cdot$ day$^{-1}$), and at 30 cm $\cdot s^{-1}$ it would be 49.8 kW-h $\cdot$ day$^{-1}$ (156.70 kW-h $\cdot$ ha$^{-2}$ $\cdot$ day$^{-1}$); that is, a doubling of the water velocity requires an almost 100× increase in power.

With these data it can be calculated that a 2-kW motor is sufficient to produce speeds of 30 cm $\cdot s^{-1}$ in a 1,000 $m^2$ pond. If higher speeds are required, then a more powerful motor probably is needed. It should be noted that this estimated power requirement is very conservative and that experience is likely to show that a smaller motor would suffice.

Stones and other large pieces of debris that find their way into the pond can threaten the paddle wheels, especially in ponds with a sump. The stones can lodge in the sump and may jam the paddle wheel. To avoid damage to the paddle wheel or motor, a cut-off and alarm for the motor should be installed. It is also possible to use reversing motors, which reverse when an obstruction is encountered; however, if the obstruction is not pushed out of the sump, then the motor direction just oscillates back and forth. A coarse net can also be placed just upstream from the paddle wheel to trap debris, as long as the extra head loss can be tolerated.

The rotations per minute (rpm) required for the paddle wheel range from about 10 to 15 to achieve speeds of about 30 cm $\cdot s^{-1}$. A variator is therefore required. Because of the changing requirements with varying speeds, the variator should be of the constant-power type rather than the constant-torque type. Otherwise there is insufficient power at lower speeds and a subsequent mechanical degradation of the reducers.

Flexible gear couplings and roller bearings are recommended. Journal bearings have proven to be unreliable. A shock absorber can be used but is not absolutely necessary.

## 5.4. Carbonator

Most large-scale algae cultures are $CO_2$-limited (Olaizola et al. 1991). Addition of $CO_2$ to the culture results in increased growth, but its cost-effectiveness should be evaluated. In the case of *Spirulina* culture, $CO_2$ addition is usually essential to maintain the high alkalinity required. Several methods have been developed for efficient addition to algae ponds. These (Becker 1994) are (1) active gas transfer through sparging small gas bubbles into the medium or spraying the liquid through the gas phase and (2) passive transfer by creation of large contact areas between a $CO_2$ atmosphere and the surface of the culture medium.

The simplest but most effective type basically consists of a plastic sheet supported by a floating frame made from PVC pipe (Vasquez and Heussler 1985). Diffusers on the pond bottom release $CO_2$ into the water. The gas inflates the plastic dome, which prolongs its contact with the water. Spoilers across the injector produce a high turbulence in the running algal suspension for efficient gas transfer into the liquid.

The limiting factor for $CO_2$ gas transfer is at the liquid film side of the gas/liquid interface. The mass transfer, $Q$ (mM · L$^{-1}$), for gas into or from liquid is given by the equation:

$$Q = kA(C_s - C_d) \qquad (16)$$

where $k$ = the mass transfer coefficient (M · min$^{-1}$), $A$ the interface area (m$^2$), $C_s$ the equilibrium concentration of dissolved gas at the interface (mM · L$^{-1}$), and $C_d$ the concentration of gas in the liquid (mM · L$^{-1}$). If the $CO_2$ flow into the injector and the partial pressure of the gas mixture are known, then the mass transfer coefficient $k_i$ (M · min$^{-1}$) for a given injector area can be estimated from the following equation:

$$k_i = \frac{Qt}{A_i(C_{si} - C_d) \times 1000} \qquad (17)$$

where $Qt$ is the $CO_2$ inflow (mM · min$^{-1}$), $A_2$ the area of the injector (m$^2$), $C_{si}$ the saturation concentration of dissolved $CO_2$ in equilibrium with the partial pressure below the injector (mM · L$^{-1}$), and $C_d$ the actual concentration of dissolved $CO_2$ in the algal suspension (mM · L$^{-1}$). For an optimum performance, the difference $C_{si} - C_d$ should be as high as possible. This means that a high concentration of $CO_2$ must be maintained in the gas phase below the injector. According to Becker (1994), flow rates of 70 to 130 mM $CO_2$ · min$^{-1}$ · m$^{-2}$ can be expected at culture temperatures of 25°C to 30°C and saturation concentrations of the injector between 15 and 25 mM · L$^{-1}$ at the maximum concentration of 1.5 mM · L$^{-1}$ of dissolved $CO_2$. This means that 1 m$^2$ of injector area can supply to the pond 1.8 to 3.4 kg of $CO_2$ over 10 hours. At these temperatures the maximum concentration of 1.5 mM $CO_2$ · L$^{-1}$ causes losses of about 7 mM · min$^{-1}$ · m$^{-2}$. On the basis of measurements in an intensely growing pond of *Scenedesmus obliquus* (Turp.) Kütz, Becker (1994) concludes that 1 m$^2$ of injector area could supply 40 to 75 m$^2$ of pond area.

An important consideration here is the source of the $CO_2$ and its cost. If LPG gas is used to power a generator on site, then it should be possible to use the exhaust gas as a source of "free" $CO_2$. It is very unlikely, however, that the available amount of $CO_2$ is sufficient and must be supplemented with another $CO_2$ source. To optimize $CO_2$ supply and avoid excessive waste, the supply system should be controlled via a pH sensor (pH-stat) in the pond.

## 6.0. STRAIN SELECTION

Depending on the culture system used, there are several ways to select for productive strains that are well-adapted to the prevailing conditions. The general characteristics selected for are growth rate, biochemical composition, temperature tolerance, and resistance to mechanical and physiological stress. Selection for these characteristics may be done in the laboratory, with subsequent testing in outdoor ponds or in the actual production ponds by manipulating growth conditions (i.e., nutrient concentrations, pH, pond depth) to provide a competitive advantage to the desired strains.

The development of commercial production of *Spirulina* has involved extensive strain selection (Belay 1997, Vonshak 1997b). On the other hand, the use of extensive culture systems for the production of *D. salina* in Australia precludes the introduction of specific strains due to continued contamination by "wild" strains of *D. salina* and other *Dunaliella* species such as *D. parva* Lerche and the noncarotenogenic *D. bioculata* Butcher and *D. viridis* Teodoresco from adjacent natural sources. Maintenance of highly productive strains therefore relies on *in situ* selection through manipulation of the conditions in the culture ponds. *D. salina* strains show very little variation in physiology compared with other microalgae (n.b., the so-called *Dunaliella bardawil* Ben-Amotz et Avron is actually a strain of *D. salina*).

## 7.0. SCALE-UP

Scale-up is one of the most difficult tasks in outdoor mass culture. Experience has shown that this is the stage at which contamination by other algae and bacteria poses the greatest problems because of the dilute inoculum (Richmond 1990, Belay 1997). The inoculum must be sufficiently dense and growth conditions optimal to prevent overgrowth of the culture by other, contaminating algae.

Two main processes are used to scale up to the production ponds. In the first the culture is scaled up stepwise, starting with small laboratory flask cultures. The scale-up usually follows an approximately 1 : 10 dilution ratio through successive volumes up to the production pond (for *Spirulina* this ratio is usually 1 : 5). An alternative process is to derive the inoculum from existing culture ponds. This is generally possible only for algae such as *Spirulina* and *Dunaliella* that can be maintained without significant contamination because of the extreme environments they grow in. It is not possible for algae such as *Chlorella* and *Haematococcus* that require a "clean" inoculum.

Whenever possible it is preferable to use the latter method, because scale-up from a 20-mL flask culture to a 10,000-m$^3$ raceway production pond can take at least 8 to 9 weeks. As well as the long time required, there is the cost of additional ponds for the scale-up cycle.

Continuous and semicontinuous culture is preferred for economic reasons, because it makes best use of the available pond infrastructure and requires less labor. The challenge, therefore, for commercial large-scale culture is developing a continuous culture process while reliably maintaining the productivity, purity, and quality of the culture.

## 8.0. CULTURE MEDIUM

The culture media used in the large-scale culture of microalgae are the same media used in the laboratory, with a few small modifications. The choice of medium used depends on several factors: the growth requirements of the algae, how the constituents of the medium may affect final product quality, and cost. If the algae are to be sold to the health food and nutraceutical markets, then food-grade chemicals are used, whereas if the algal biomass is used as animal feed, then it may be possible to use cheaper industrial grade chemicals, although care must still be taken not to introduce contaminants such as heavy metals.

The cost of nutrients accounts for 10 to 30% of the total production costs (Borowitzka 1999b), and many producers recycle the medium that still contains nutrients to keep costs to a minimum. Recycling the medium also reduces the possible environmental problems associated with discharging large volumes of nutrient-rich

water. Some reduction in medium costs may also be achieved by using industrial grade chemicals wherever possible and by substituting cheaper alternatives (e.g., urea for nitrate). Some algal species and applications also allow for the use of wastewaters as the basis of the medium. For example, in Thailand, treated sago starch effluent is being used to produce animal feed–grade *Spirulina* (Tanticharoen et al. 1993, Bunnag et al. 1998).

## 9.0. POND MANAGEMENT

Outdoor algae ponds are exposed to the vagaries of local climate, and key environmental factors affecting culture growth such as light and temperature show diel, day-to-day, and seasonal variations. For example, in the dry semiarid Australian environments where *D. salina* is grown, pond temperature can vary by 20°C or more between day and night. This means that in the morning the pond temperature is well below the optimal for photosynthesis and growth. These suboptimal temperatures and the rapidly increasing irradiance in the morning also lead to photoinhibitory stress from which the algae never fully recover over the rest of the day, thus reducing the overall productivity of the pond (Vonshak and Guy 1988, Lu and Vonshak 1999, Masojidek et al. 1999). Heating the pond in the morning reduces this photoinhibition, but this is generally not practical. It is possible only in systems such as the very shallow, sloping, cascade system used in the Czech Republic, where the cultures are stored in deep tanks overnight, which reduces heat loss. Low temperatures in autumn and winter also limit the growing season in many parts of the world, and this is a major reason for many algae production plants being located in low-rainfall Mediterranean or subtropical climates such as Hawaii, Israel, California, and parts of Australia.

Productivity in the ponds can also be improved by maintaining a cell concentration at which productivity is greatest. This concentration changes during the year with changing light intensity, and the improvement in productivity is greatest in summer, when the main limiting factor is light rather than temperature (Vonshak 1997b).

Rainfall is also a problem, especially in the culture of *D. salina*, because it can significantly dilute the medium in the ponds, leading to reduced algal growth and even rapid invasion by predators such as ciliates and brine shrimp. In small experimental ponds this problem can be minimized by either covering the ponds when it rains or adding salt to restore the salinity rapidly. However, this is not feasible for the large commercial production ponds. At Hutt Lagoon, in Western Australia, a thick layer of NaCl is allowed to build up on the pond floor during summer by operating the plant at NaCl saturation. If rain dilutes the ponds, then this layer dissolves rapidly, restoring the salinity.

## 9.1. Control and Management of Contaminants

It is impossible to prevent contamination in open ponds. The main contaminants in "clean" algal cultures are bacteria, viruses, fungi, other algae, and zooplankton. Other contaminants include insects, leaves, and other airborne material. It is essential to control these contaminants within acceptable limits. In raceway ponds, large contaminants can be removed regularly by placing a suitably sized screen in the water flow. This can be done manually, or it can be automated. Heavy contaminants that settle to the bottom of the pond can be trapped in pits arranged at a right angle to the flow and can then be removed from these sediment traps.

Although the algal species grown commercially in outdoor ponds generally grow under highly selective conditions (e.g., *Chlorella* spp, *Nannochloropsis*, and *P. tricornutum* are grown at high nutrient concentrations, *Spirulina* is grown at high bicarbonate concentrations and a high pH, and *D. salina* is grown at very high salinities), contamination by other, unwanted algal species is common. These are controlled by effectively operating the culture system as a batch culture and restarting the culture at regular intervals with fresh, unialgal inoculum. This is the primary process used for the culture of *Chlorella* in Japan and Indonesia and for *H. pluvialis* in Hawaii. Careful culture maintenance, providing conditions that favor the desired species over the contaminating species, also allows long-term continuous culture. For example, contamination of *Spirulina* cultures with green algae can be minimized by maintaining the bicarbonate concentrations above 0.2 M and the pH above 10 and operating at high cell densities (Richmond et al. 1982). Repeated pulses of 1 to 2 mM NH$_3$, followed by a 30% dilution of the culture, is also an effective treatment, based on the differential sensitivity of *Spirulina* and *Chlorella* cells to NH$_3$. However, the green alga *Oocystis* species, which is also an alkaliophile, can still present significant problems (Belay 1997). Raceway cultures of *P. tricornutum* have also been

maintained successfully in our laboratory for more than 1 year by maintaining high concentrations of nitrogen and phosphorus. Contamination of *D. salina* ponds by other noncarotenogenic species of *Dunaliella* (e.g., *D. viridis*, *D. parva*, *D. bioculata*) is managed mainly by maintaining high salinities. Mitchell and Richmond (1987) proposed that rotifers be used to keep *Spirulina* cultures free of the unicellular green algae *Monoraphidium* and *Chlorella*; however, this approach is not used by any commercial producer because it is too difficult to manage on a large scale.

Aquatic insects are unavoidable in freshwater outdoor ponds. The major groups are the Ephydridae (brineflies), Corixidae (waterboatmen), and Chironomidae (midges). The ephydrids have been reported to be particularly common in *Spirulina* ponds (Venkataman and Kanya 1981, Belay 1997) and these are controlled mainly by netting *in situ* and during preharvest screening.

In the hypersaline *D. salina* cultures, the brine shrimps *Artemia* and *Paraartemia* can create significant problems. These zooplankters proliferate when salinity drops below approximately 15% (w/v) NaCl, at which stage the cysts can hatch successfully. The hatchlings can survive much higher salinities and consume significant quantities of the algal biomass. In raceway ponds, netting can be used to remove the animals, but this is not possible in the very large, unmixed ponds used in Australia. In this case the ponds must be maintained at 20% (w/v) NaCl or higher for several months until the animals die. This problem can be avoided, however, by a pond management regimen that never allows the salinity to fall below the critical point.

In freshwater and marine cultures, rotifers, especially *Branchionus*, have been reported to impair the growth of algal cells. Becker (1994) recommends lowering the pH to about 3.0 by the addition of acid and allowing the culture to stand at this pH for 1 to 2 hours. Following this, the pH can be readjusted to 7.5 with KOH. This treatment does not affect the algal cultures but effectively eliminates the rotifers. Amoebae can also occur in ponds of *Chlorella* or *Spirulina*, and Lincoln et al. (1983) have shown that the amoebae populations can be reduced significantly by using ammonia as the main nitrogen source. Ammonia treatment cannot be used with *D. salina* cultures, because this alga is very sensitive to ammonia, especially under conditions of high light and high temperatures (Borowitzka and Borowitzka 1988).

In *D. salina* ponds at salinities below about 20% NaCl, various protozoa, especially the flagellate amoeba *Fabrea salina* and some ciliate protozoa, can rapidly decimate the algal culture (Post et al. 1983). These can be controlled by increasing the salinity. Chemical control is also possible—10 mg · $L^{-1}$ quinine is especially effective (Moreno-Garrido and Cañavate 2001)—but prevention of infection by maintenance of sufficiently high salinity is much more cost-effective. Ciliates can also cause problems in cultures of *Spirulina* (Belay 1997), especially in colder months, when environmental conditions and mechanical stress render *Spirulina* trichomes susceptible to breakage and decomposition.

In well-mixed cultures, bacteria are generally no problem, but poor mixing and pond construction, which allows build-up of biomass in parts of the pond, can lead to localized anaerobic regions and cell death. This in turn leads to an increase in bacterial numbers and possible collapse of the culture.

If the alga is to be used for human consumption, then rodent control is also important. For example, the U.S. Food and Drug Administration (FDA) allows only less than 0.5 rodent hair per 50 g of algal powder. The entry of rodents into the culture site must therefore be prevented, and windborne contaminants must be minimized.

## 9.2. CULTURE MONITORING

Successful culture maintenance requires continuous monitoring. The most basic kind of monitoring is regular microscopic examination to detect any abnormal morphological changes and the presence of contaminating organisms such as other algae and protozoa. Early detection of these may allow control measures to be undertaken in time to be effective. Routine tests on the nutrient concentration in the pond must also be carried out to avoid unexpected nutrient deficiencies. Regular monitoring of changes in pH and $O_2$ levels in the pond over the day can also be a useful early warning system. "Healthy" cultures show a fairly regular diurnal pattern in these parameters, and any significant deviation is usually a sign of problems in the culture. Monitoring of pH and $O_2$ has the advantage of being able to be automated. Recently, a new technique, pulse amplitude modulated fluorometry (PAM), has become available and appears to be very sensitive for determining the physiological state of the algae (Torzillo et al. 1998, Lippemeier et al. 2001). Although not yet applied routinely to large-scale algal cultures, this method shows great promise as a sensitive, automated early warning system. Regular microscopic observation of the cells,

however, remains an essential part of any culture monitoring system.

## 10.0. CONCLUSIONS

Large-scale economical culture of microalgae in open ponds is very effective for a limited number of species. The success of such systems depends on a very good understanding of the physiology (and ecology) of the algae being cultured and the application of appropriate engineering principles to the design of the culture system. Although these systems have been in operation for more than 20 years, advances in the design and operation of these systems continue in light of experience gained, and "slime ranching," as it is affectionately known, continues to be a major way of producing commercially valuable algal biomass.

## 11.0. REFERENCES

Becker, E. W. 1994. *Microalgae. Biotechnology and Microbiology.* Cambridge University Press, Cambridge, 293 pp.

Belay, A., Ota, Y., Miyakawa, K., and Shimamatsu, H. 1994. Production of high quality *Spirulina* at Earthrise Farms. In: Phang, S. M., Lee, K., Borowitzka, M. A., and Whitton, B., eds. *Algal Biotechnology in the Asia-Pacific Region.* Institute of Advanced Studies, University of Malaya, Kuala Lumpur, pp. 92–102.

Belay, A. 1997. Mass culture of *Spirulina* outdoors: The Earthrise Farms experience. In: Vonshak, A., ed. *Spirulina platensis (Arthrospira): Physiology, Cell-biology and Biochemistry* Taylor & Francis, London, pp. 131–58.

Ben-Amotz, A., and Avron, M. 1989. The biotechnology of mass culturing *Dunaliella* for products of commercial interest. In: Cresswell, R. C., Rees, T. A., and Shah, N., eds. *Algal and Cyanobacterial Biotechnology.* Longman Scientific & Technical, Harlow, U. K., pp. 91–114.

Ben-Amotz, A. 1999. Production of beta-carotene from *Dunaliella.* In: Cohen, Z., ed. *Chemicals from Microalgae.* Taylor & Francis, London, pp. 196–204.

Bernhard, M., Zattera, A., and Filesi, P. 1966. Suitability of various substances for use in the culture of marine organisms. *P.S.Z.N. I: Mar. Ecol.* 35:89–104.

Blankley, W. F. 1973. Toxic and inhibitory materials associated with culturing. In: Stein, J., ed. *Handbook of Phycological Methods: Culture Methods and Growth Measurements.* Cambridge University Press, Cambridge, pp. 207–99.

Borowitzka, L. J., and Borowitzka, M. A. 1989. β-Carotene (provitamin A) production with algae. In: Vandamme, E. J., ed. *Biotechnology of Vitamins, Pigments and Growth Factors.* Elsevier Applied Science, London, pp. 15–26.

Borowitzka, M. A., and Borowitzka, L. J. 1988. Limits to growth and carotenogenesis in laboratory and large-scale outdoor cultures of *Dunaliella salina.* In: Stadler, T., Mollion, J., Verdus, M. C., Karamanos, Y., Morvan, H., and Christiaen, D., eds. *Algal Biotechnology.* Elsevier Applied Science, Barking, U. K., pp. 371–81.

Borowitzka, M. A. 1988. Algal growth media and sources of cultures. In: Borowitzka, M. A., and Borowitzka, L. J., eds. *Microalgal Biotechnology.* Cambridge University Press, Cambridge, pp. 456–65.

Borowitzka, M. A. 1999a. Commercial production of microalgae: Ponds, tanks, tubes and fermenters. *J. Biotechnol.* 70:313–21.

Borowitzka, M. A. 1999b. Economic evaluation of microalgal processes and products. In: Cohen, Z., ed. *Chemicals from Microalgae.* Taylor & Francis, London, pp. 387–409.

Bosca, C., Dauta, A., and Marvalin, O. 1991. Intensive outdoor algal cultures: How mixing enhances the photosynthetic production rate. *Bioresource Technol.* 38:185–8.

Bubrick, P. 1991. Production of astaxanthin from *Haematococcus. Bioresource Technol.* 38:237–39.

Bunnag, B., Tanticharoen, M., and Ruengjitchatchawalya, M. 1998. Present status of microalgal research and cultivation in Thailand. In: Subramanian, G., Kaushik, B. D., and Venkataraman, G. S., eds. *Cyanobacterial Biotechnology.* Oxford & IBH Publishing Co., New Delhi, pp. 325–8.

Burlew, J. S., ed. 1953. *Algae Culture. From Laboratory to Pilot Plant.* Carnegie Institution of Washington, Washington, D.C., 357 pp.

Castenholz, R. W. 1989. Subsection III. Oscillatoriales. In: Stanley, J. T., Bryant, M. P., Pfenning, N., and Holt, J. G., eds. *Bergey's Manual of Systematic Bacteriology.* Vol. 3, p. 1771.

Dodd, J. C. 1986. Elements of pond design and construction. In: Richmond, A., ed. *CRC Handbook of Microalgal Mass Culture.* CRC Press, Boca Raton, Florida, pp. 265–83.

Dyer, D. L., and Richardson, D. E. 1962. Materials of construction in algal culture. *Appl. Microbiol.* 10:129–32.

Fox, R. D. 1996. *Spirulina: Production and Potential.* Edisud, Aix-en-Provence, France, 232 pp.

Grobbelaar, J. U., Soeder, C. J., and Stengel, E. 1990. Modeling algal productivity in large outdoor cultures and waste treatment systems. *Biomass* 21:297–314.

Gummert, F., Meffert, M. E., and Stratmann, H. 1953. Non-sterile large-scale culture of *Chlorella* in greenhouse and open air. In: Burlew, J. S., ed. *Algal Culture. From Laboratory to Pilot Plant.* Carnegie Institution of Washington, Washington, D.C., pp. 166–76.

Kaushik, B. D. 1998. Use of cyanobacterial biofertilizers in rice cultivation: A technology improvement. In: Subramanian, G., Kaushik, B. D., and Venkatamaran, G. S., eds. *Cyanobacterial Biotechnology.* Oxford & IBH Publishing Co, New Delhi, pp. 211–22.

Kawaguchi, K. 1980. Microalgae production systems in Asia. In: Shelef, G., and Soeder, C. J., eds. *Algae Biomass Production and Use.* Elsevier/North Holland Biomedical Press, Amsterdam, pp. 25–33.

Laws, E. A., Terry, K. L., Wickman, J., and Chalup, M. S. 1983. A simple algal production system designed to utilize the flashing light effect. *Biotechnol. Bioeng.* 25:2319–35.

Lincoln, E. P., Hall, T. W., and Koopman, B. 1983. Zooplankton control in algal cultures. *Aquaculture* 32:331–7.

Lippemeier, S., Hintze, R., Vanselow, K. H., Hartig, P., and Colijn, F. 2001. In-line recording of PAM fluorescence of phytoplankton cultures as a new tool for studying effects of fluctuating nutrient supply on photosynthesis. *Eur. J. Phycol.* 36:89–100.

Lu, C., and Vonshak, A. 1999. Photoinhibition in outdoor *Spirulina platensis* cultures assessed by polyphasic chlorophyll fluorescence transients. *J. Appl. Phycol.* 11:355–9.

Masojidek, J., Torzillo, G., Koblížek, M., Kopecky, J., Bernardini, P., Sacchi, A., and Komenda, J. 1999. Photoadaptation of two members of the Chlorophyta (*Scenedesmus* and *Chlorella*) in laboratory and outdoor cultures: Changes in chlorophyll fluorescence quenching and the xanthophyll cycle. *Planta* 209:126–35.

Mitchell, S. A., and Richmond, A. 1987. Use of rotifers for the maintenance of monoalgal mass cultures of *Spirulina. Biotechnol. Bioeng.* 30:164–8.

Moreno-Garrido, I., and Cañavate, J. P. 2001. Assessing chemical compounds for controlling predator ciliates in outdoor mass cultures of the green algae *Dunaliella salina. Aquacult. Eng.* 24:107–14.

Olaizola, M., Duerr, E. O., and Freeman, D. W. 1991. Effect of $CO_2$ enhancement in an outdoor algal production system using *Tetraselmis. J. Appl. Phycol.* 3:363–6.

Olaizola, M. 2000. Commercial production of astaxanthin from *Haematococcus pluvialis* using 25,000-liter outdoor photobioreactors. *J. Appl. Phycol.* 12:499–506.

Oswald, W. J. 1988a. Large-scale algal culture systems (engineering aspects). In: Borowitzka, M. A., and Borowitzka, L. J., eds. *Micro-Algal Biotechnology.* Cambridge University Press, Cambridge, pp. 357–94.

Oswald, W. J. 1988b. Micro-algae and waste-water treatment. In: Borowitzka, M. A., and Borowitzka, L. J., eds. *Microalgal Biotechnology.* Cambridge University Press, Cambridge, pp. 305–28.

Post, F. J., Borowitzka, L. J., Borowitzka, M. A., Mackay, B., and Moulton, T. 1983. The protozoa of a Western Australian hypersaline lagoon. *Hydrobiologia* 105:95–113.

Richmond, A., Karg, S., and Boussiba, S. 1982. Effects of bicarbonate and carbon dioxide on the competition between *Chlorella vulgaris* and *Spirulina platensis*. *Plant Cell Physiol.* 23:1411–17.

Richmond, A., and Vonshak, A. 1978. *Spirulina* culture in Israel. *Arch. Hydrobiol.* 11:274–80.

Richmond, A. 1986. Outdoor mass culture of microalgae. In: Richmond, A., ed. *CRC Handbook of Microalgal Mass Culture*. CRC Press, Boca Raton, Florida, pp. 285–329.

Richmond, A. 1990. Large scale microalgal culture and applications. *Progr. Phycolog. Res.* 7:269–330.

Richmond, A. 1992. Open systems for the mass production of photoautotrophic microalgae outdoors: Physiological principles. *J. Appl. Phycol.* 4:281–6.

Schlipalius, L. 1991. The extensive commercial cultivation of *Dunaliella salina*. *Bioresource Technol.* 38:241–3.

Setlík, I., Veladimir, S., and Malek, I. 1970. Dual purpose open circulation units for large scale culture of algae in temperate zones. I. Basic design considerations and scheme of pilot plant. *Algolog. Studies (Trebon)* 1:111–64.

Shimamatsu, H., and Tominaga, Y. 1980. *Apparatus for Cultivating Algae*. U.S. Patent Office, patent no. 4217728.

Shimamatsu, H. 1987. A pond for edible *Spirulina* production and its hydraulic studies. *Hydrobiologia* 151/152:83–9.

Tanticharoen, M., Bunnag, B., and Vonshak, A. 1993. Cultivation of *Spirulina* using secondary treated starch wastewater. *Australasian Biotechnol.* 3:223–6.

Torzillo, G., Bernardini, P., and Masojidek, J. 1998. On-line monitoring of chlorophyll fluorescence to assess the extent of photoinhibition of photosynthesis induced by high oxygen concentration and low temperature and its effects on the productivity of outdoor cultures of *Spirulina platensis* (Cyanobacteria). *J. Phycol.* 34:504–10.

Tsukada, O., Kawahara, T., and Miyachi, S. 1977. Mass culture of *Chlorella* in Asian countries. In: Mitsui, A., Miyachi, S., San Pietro, A., and Tamura, S., eds. *Biological Solar Energy Conversion*. Academic Press, New York, pp. 363–5.

Valderrama, A., Cárdenas, A., and Markovits, A. 1987. On the economics of *Spirulina* production in Chile with details on drag-board mixing in shallow ponds. *Hydrobiologia* 151/152:71–4.

Vasquez, V., and Heussler, P. 1985. Carbon dioxide balance in open air mass culture of algae. *Arch. Hydrobiol. Ergeb. Limnol., Beiheft* 20:95–113.

Venkatamaran, L. V., and Kanya, T. C. S. 1981. Insect contamination (*Ephydra californica*) in the mass outdoor cultures of blue green, *Spirulina platensis*. *Proc. Indian Acad. Sci.* 90:665.

Vonshak, A., and Guy, R. 1988. Photoinhibition as a limiting factor in outdoor cultivation of *Spirulina platensis*. In: Stadler, T., Mollion, J., Verdus, M. C., Karamanos, Y., Morvan, H., and Christiaen, D., eds. *Algal Biotechnology*. Elsevier Applied Science, London, pp. 365–70.

Vonshak, A., ed. 1997a. *Spirulina platensis (Arthrospira): Physiology, Cell-Biology and Biotechnology*. Taylor & Francis, London, 233 pp.

Vonshak, A. 1997b. *Spirulina*: Growth, physiology and biochemisty. In: Vonshak, A., ed. *Spirulina platensis (Arthrospira): Physiology, Cell-Biology and Biochemistry*. Taylor & Francis, London, pp. 43–65.

# CHAPTER 15

# MARICULTURE OF SEAWEEDS

DINABANDHU SAHOO
*Marine Biotechnology Laboratory, Department of Botany,
University of Delhi*

CHARLES YARISH
*Department of Ecology and Evolutionary Biology,
University of Connecticut*

## CONTENTS

Key Index Words: *Eucheuma, Gracilaria, Kappaphycus, Laminaria,* Macrophyte, Nori, *Porphyra,* Seaweed Cultivation, *Undaria*

## 1.0. INTRODUCTION

Macroscopic marine algae form an important living resource of the oceans. Seaweeds are food important for humans and animals, as well as fertilizers for plants and a source of various chemicals (Lembi and Waaland 1988, Sahoo 2000). Seaweeds have been gaining momentum as a new experimental system for biological research (Sahoo et al. 2002) and as an integral part of integrated aquaculture systems (Chopin et al. 2001, Troell et al. 2003, Neori et al. 2004). We all use seaweed products in our day-to-day life in some way or other. For example, some seaweed polysaccharides are used in toothpaste, soap, shampoo, cosmetics, milk, ice cream, meat, processed food, air freshener, and many other items. In many oriental countries such as Japan, China, Korea, and others, seaweeds are diet staples.

## 1.1. Need for Mariculture

Traditionally, seaweeds were collected from natural stocks or wild populations. However, these resources were being depleted by overharvesting, so cultivation techniques have been developed (Critchley and Ohno 1998). Today, seaweed cultivation techniques are standardized, routine, and economical. Several factors, such as morphology and regeneration capacity of the thallus, as well as complex interaction of irradiance, temperature, nutrients, and water movement, are responsible for the success of large-scale seaweed cultivation.

Different taxa require different farming methods. Although some seaweeds need one-step farming through vegetative propagation, others need two-step or multistep farming. The latter must be propagated from spores and cannot survive if propagated vegetatively. Thus, *Eucheuma, Kappaphycus, Chondrus*, and *Gracilaria* are propagated vegetatively (one step), whereas *Porphyra, Enteromorpha, Laminaria, Undaria*, and others are started from spores.

Although large-scale open water cultivation of some species has been carried out in many Asian countries, other species are cultivated in tanks and ponds. For example, *Chondrus crispus* Stackhouse has been mainly cultivated in tanks in Canada and in Portugal. *Gracilaria* is being cultivated in tanks and raceways in Israel and in human-made ponds in Taiwan and Thailand. Tank cultivation of some seaweeds was tried in Florida and other parts of the world, but was not very successful. Epiphytes, fouling, and heavy nutrient requirements caused problems. Tank cultivation is generally unpopular and unsuccessful, mainly for economic reasons (Neori et al. 2004).

During the last 50 years, approximately 100 seaweed taxa have been tested in field farms, but only a dozen are being commercially cultivated today. Table 15.1 provides production data for the top taxa. Several countries are involved in seaweed cultivation, and China, Korea, Japan, and the Philippines are leaders. Because it is difficult to describe the cultivation methodologies for all the taxa, we discuss the top five cultivated genera in the world.

## 2.0. Porphyra Cultivation

*Porphyra* has an annual value of more than U.S. $1.8 billion (Yarish et al. 1999) and is considered the most valuable maricultured seaweed in the world. According to the Food and Agriculture Organization of the United Nations (FAO) (2003), nearly 1,011,000 metric tons (wet

**TABLE 15.1** Top cultivated seaweed genera in the world during 2000 (FAO 2003).

| Taxon | Value ($10^6$ U.S.$) | Raw material (metric tons) | U.S.$ per metric ton |
|---|---|---|---|
| Laminaria | 2,811 | 4,580,000 | 613 |
| Porphyra | 1,118 | 1,011,000 | 1,105 |
| Undaria | 149 | 311,105 | 480 |
| Eucheuma and Kappaphycus | 46 | 628,576 | 73 |
| Gracilaria | 11 | 12,510 | 879 |
| Total | 4,632 | 5,972,737 | |

weight) of *Porphyra* were produced through mariculture. *Porphyra* has nearly 133 species distributed all over the world, including 28 species from Japan, 30 from the North Atlantic coasts of Europe and America, and 27 species from the Pacific coast of Canada and the United States (Yoshida et al. 1997). Six species of *Porphyra*, namely *P. yezoensis* Ueda, *P. tenera* Kjellman, *P. haitensis* Chang et Zhen Baofu, *P. pseudolinearis* Ueda, *P. dentata* Kjellman, and *P. angusta* Okamura et Ueda, are usually cultivated; the first three are more common in Asia.

The plants can grow from 5 to 35 cm in length. The thalli are either one or two cells thick, and each cell has one or two stellate chloroplasts with a pyrenoid. *Porphyra* has a heteromorphic life cycle with an alternation between a macroscopic foliose thallus (gametophytic phase) and a filamentous sporophyte called the *conchocelis phase*. *Porphyra* reproduces by both sexual and asexual modes of reproduction. In sexual reproduction, certain mature vegetative cells of the thallus differentiate into carpogonia, and others on the same or different thalli differentiate into colorless spermatangia. After fertilization, the carpogonia divide to form packets of spores called *zygotospores* (carpospores). After release, the zygotospores usually germinate unipolarly to produce the filamentous "*conchocelis*" phase. The conchocelis phase can survive in adverse environmental conditions but gives rise to conchosporangia and conchospores under suitable environmental conditions. The conchospores germinate in a bipolar manner to give rise to young chimeric thalli, thus completing the life cycle.

Cultivation of *Porphyra* began in Japan, Korea, and China during the seventeenth century. Modern techniques for *Porphyra* cultivation were introduced to these countries in the 1960s. Subsequently, efforts were made in the 1980s and 1990s to expand *Porphyra* cultivation into the United States; however, they were unsuccessful because of social and economic reasons. The culture methods of *Porphyra* in all countries are basically very similar, with minor modifications, such as adaptations to local growing areas and traditional practices of local farmers.

The cultivation technique involves four major steps: (1) culture of conchocelis, (2) seeding of culture nets with conchospores, (3) nursery rearing of sporelings, and (4) harvesting.

## 2.1. Conchocelis Stage

In Asia, the culture of conchocelis starts around mid spring (March/April). In nature, the conchocelis phase grows on oyster, clam, and scallop shells. However, in Japan artificial substrata, made of transparent vinyl films covered with calcite granules, are beginning to be used as substitutes for mollusc shells. Usually, the oyster shells or artificial shells are placed on the bottom of shallow tanks filled with seawater for seeding to take place (Fig. 15.1a–c).

Seeding may be accomplished by introducing chopped pieces of fertile (ripe) *Porphyra* blades into the seeding tank, which are then removed after the release of the carpospores. The carpospore suspension can be prepared artificially by either air drying the fertile thalli overnight and then immersing them in seawater for 4–5 hours the next morning (this will induce mass shedding of carpospores) or by grinding the fertile blades and separating the suspension of carpospores by filtration. The carpospore suspension is then introduced into the seeding tank, where the spores settle on the substrata.

**FIGURE 15.1.** Culture of *Porphyra*. **(a)** Attaching oyster shells to a net before inoculation with carpospores. **(b)** Seeding oyster shells in a seeding tank. **(c)** Young conchocelis growing on oyster shells. **(d)** Net rotated in a seeding tank containing mature conchospore inoculum from the free-living conchocelis culture. **(e)** Nets raised daily to expose the young thalli to air and sun to inhibit fouling organisms; the Japanese "Ikada" system. **(f)** Floating A-frame system in China; raised and lowered to expose the young thalli, inhibiting fouling organisms. (a–c courtesy M. Notoya; e courtesy of I. Levine; f from X. G. Fei.)

The carpospores germinate and bore into the shells or the artificial substrates under very low light conditions (<5 μmol photons · m$^{-2}$ · s$^{-1}$). An alternative method, growing in popularity, is the mass cultivation of conchocelis that is then seeded directly on calcareous substrates. The advantage of this technique is the use of defined strains that give a more consistent crop. In each case, the seeded substrates are kept in large tanks (0.25–0.5 m deep). The substrates are either hung vertically or spread across the bottom. The conchocelis is allowed to develop during the rest of the summer under slightly higher light levels (25–50 μmol photons · m$^{-2}$ · s$^{-1}$), at 16 : 8 h L : D and at 23°C.

Recently, bioreactors have been developed in our laboratory for the culture of "free-living conchocelis" of native North American *Porphyra* species. In this method, clones of conchocelis are vegetatively propagated to produce enough biomass for mass production of conchospores. Large amounts of conchocelis can be produced and maintained in bioreactors under controlled temperature, light, photoperiod, and salinity. The process is less cumbersome and requires less labor. Conchosporangia formation and maturation are inhibited by reducing the salinity of the seawater to 22 psu. Mass release of conchospores can be induced as needed and seeded onto nets.

## 2.2. Control of Conchospore Release

Conchospore release is promoted by stirring, by using compressed air bubbling, or by treating cultures with low-temperature seawater (18° to 20°C). For the latter treatment, conchocelis-bearing shells are transferred to low-temperature seawater tanks 5 to 7 days before seeding. Replacing the seawater in conchocelis culture tanks and adding vitamin B$_{12}$ can also promote conchospore release. To inhibit conchospore release, conchocelis tanks may be covered by black vinyl sheets and the culture seawater maintained calm with little, if any, aeration.

## 2.3. Seeding Culture Nets with Conchospores

Although the traditional method of the cultivation of *Porphyra* (natural seeding of conchospores on bamboo sticks and nets) may still be practiced in some areas, the bulk of nori produced in Japan, China, and Korea depends on artificial seeding of conchospores from hatchery grown conchocelis. In these countries, seeding

commences in the autumn when the water temperature is 23°–24°C. The nets are 1.8 m wide and 18–45 m long, made of synthetic twine 3–5 mm in diameter. The nets have a mesh size of about 15 cm. Artificial seeding may be done outdoors in the sea or indoors in hatchery tanks. Outdoor seeding is carried out in nursery grounds by setting up layers of (12–16) nets on support systems. Hatchery-produced mature conchocelis on shells or on artificial substrata are placed in plastic bags and hung under the nets. The conchospores float in the water and are collected on the nets.

Indoor seeding may be done by fixing the net either over a rotary wheel or a belt conveyor that is rotated in a seeding tank containing mature conchospore inoculum from the free-living conchocelis culture (Fig. 15.1d). Another method, popular in China, is the indoor seeding of nets that are suspended on the surface of the water in shallow seeding tanks. The conchospores are inoculated by circulating water via a submersible pump that is in the tank.

## 2.4. Sporeling Stage

The nets that are seeded with the conchospores are stacked into bundles of four nets. They may be layered to have as many as 12 to 16 nets. These stacks of nets are transferred to the sea for nursery cultivation. In nursery culture, the nets are put into the sea and carefully monitored for blade development. During this early stage, the nets are raised out of the water daily to expose the young thalli to air and sun. This exposure is necessary to inhibit fouling organisms (e.g., other seaweed species or diatoms). In Japan, a popular system, called the "Ikada" system, is used for this process (Fig. 15.1e). In China, Professor X. G. Fei has developed a floating A-frame system that can be easily raised and lowered for supporting nursery nets and controlling the degree of exposure (Fig. 15.1f). Once blades are 2 to 3 mm, the nets can be transferred to the farm sites or frozen for later use. These nets are initially air dried to reduce the water content of *Porphyra* to 20 to 40% and then are stored at –20°C. The frozen nets can then be used to replace lost or damaged ones. When nursery-reared seedlings are 5 to 30 mm long, they are ready for outplanting. Seedlings may be separated from the bundles using one of three methods of cultivation: fixed pole, semifloating raft, or floating raft (Fig. 15.2a). The fixed pole method is used in shallow intertidal areas (Fig. 15.2b). The floating type is designed for deeper waters (Fig. 15.3b). Specially designed harvesting boats

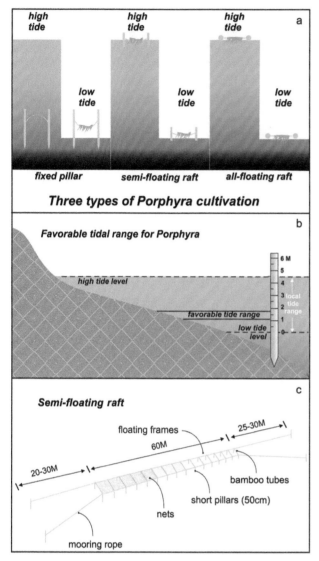

**FIGURE 15.2.** (a) *Porphyra* seedlings separated from the bundles: fixed pole, semifloating raft, and floating raft. (b) Determining of the best level for fixed pole intertidal *Porphyra* cultivation. (c) Diagrammatic illustration of a semifloating raft. (a after Tseng 1981; b–c courtesy X. G. Fei.)

lift the growing *Porphyra* nets (Fig. 15.3a,b). The semi-floating raft is a hybrid of these two methods and is most suitable for areas that have extensive intertidal zones (Fig. 15.2c, Fig. 15.3a).

The growth of *Porphyra* thalli is inhibited mostly by high temperatures during mid to late November. In these cases, nori production used to suffer heavy damage. To avoid this, the technique of frozen storage of nori nets was devised in 1969. When the thalli have grown to 1–3 cm, nets are dried to 20 to 30% moisture content, packed in polyethylene bags and kept at −20°

to −30°C. The seedlings that are stored under such conditions remain alive for 6 months and can be used whenever required.

### 2.5. Harvesting and Processing

The outplanted *Porphyra* seedlings are allowed to grow to 15–30 cm in about 40–50 days before they are mechanically harvested (Fig. 15.3c,d). The remaining thalli are allowed to grow and may be ready for a second harvest after another 15 to 20 days. Several harvests may be made from the same nets in one growing season (Merrill 1989, Yarish et al. 1999). The harvested crop is washed and transferred to an automatic nori-processing machine, which cuts the blades into small pieces. The nori is then processed by the cultivator into dried rectangular sheets or processed by the manufacturer per market requirements.

---

### 3.0. GRACILARIA CULTIVATION

The red alga *Gracilaria* contributes approximately 53% of the total agar production (Armisen 1995, FAO 2000). Although more than 150 species of *Gracilaria* have been reported from different parts of the world, the taxonomy of the genus is still chaotic (Bird 1995). *Gracilaria* is widely distributed all over the world, but most of the species are reported to be from tropical waters. Major *Gracilaria*-producing countries are Argentina, Brazil, Chile, China, Thailand, Malaysia, Indonesia, Philippines, South Africa, Taiwan, New Zealand, and so on (Critchley and Ohno 1998). Morphologically, the thallus of *Gracilaria* is cylindrical, compressed, or bladelike and irregularly branched, giving a bushy appearance. *Gracilaria* has a typical *Polysiphonia* type of triphasic life history. The male and female gametophytes in the early stages look identical, but subsequently the latter can be easily identified by the presence of cystocarps, which appear as distinct hemispherical lumps all over the thalli. The cystocarp releases a large number of carpospores (2n) that give rise to tetrasporophytic plants (2n). Each diploid tetrasporophytic plant is morphologically similar to the haploid gametophytic plants (e.g., they are isomorphic). The tetrasporophyte produces haploid tetraspores by meiotic sporogenesis within cortical sporangia, which ultimately give rise to male and female plants, thus completing the triphasic life cycle.

**FIGURE 15.3.** **(a)** Attachment of *Porphyra* nets to a semifloating raft. **(b)** Korean-style all-floating rafts. **(c)** Fleet of *Porphyra* harvesting boats. Note the tubular frames that lift the nets (see **d**). **(d)** Harvesting boat with *Porphyra* net lifted from the sea. (**a** courtesy E. Hwang; **d** courtesy M. Notoya.)

*Gracilaria* is being cultivated commercially in a number of countries such as Chile (Santelices and Ugarte 1987, Retamales 1993, Westermeir et al. 1993, Buschmann et al. 1995, 1996), China (Ren et al. 1984, Wu et al. 1993, Wu 1998, Fei et al. 2000), Taiwan (Shang 1976, Chiang 1981, Yang 1982), Israel (Friedlander and Levy 1995), Thailand (Lewmanomont 1998), the United States (Hanisak 1987, Glenn et al. 1998), and to a lesser extent in several other countries like Japan and the Philippines through a number of methodologies (Critchley 1993, Olivieria et al. 2000). Some of these methodologies are briefly discussed in the following section.

### 3.1. Site Selection

Selection of sites for *Gracilaria* cultivation: (1) It should be located near seawater sources for open water cultivation; (2) for pond cultivation, the site should be located near both seawater and fresh water sources; (3) the site should be protected from strong winds; and (4) *Gracilaria* tolerates a wide range of salinities (10–24 psu), but it is important to check other ecological conditions like temperature, light, and pH; the pH of water should be slightly alkaline (7.5–8.0).

### 3.2. Seedling Selection

Healthy branches of *Gracilaria* from natural stock should be selected for successful farming. Thalli of *Gracilaria* may be vegetatively propagated, or if the plants produce spores, they could also be used to seed farms.

### 3.3. Cultivation Methods

*Gracilaria* cultivation is mainly practiced in three different ways; open water cultivation, pond culture, and tank culture. Open water cultivation is practiced in estuaries, bays, and upwelling areas. This method can be implemented either by "bottom-stocking" or by "rope cultivation" (Fig. 15.4a–g). Bottom-stocking is the simplest method for transferring vegetative thalli that are growing in the natural field conditions and can be of two types. The direct method consists of the direct insertion of thalli into the sandy bottom using different types of tools (Fig. 15.4a–c). It is a labor-intensive method and is most effective in areas where gracilariods are growing naturally; thus, bottom-stocking increases the local density (Santelices and Doty 1989, Oliviera et al. 2000). *Gracilaria* can also be attached to

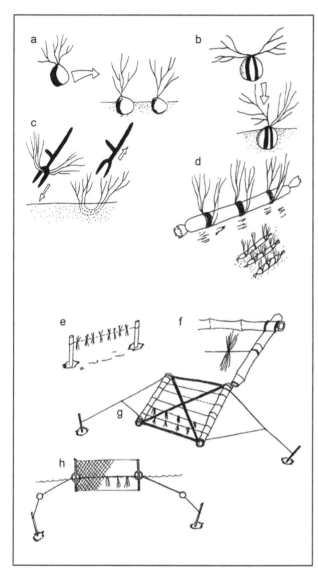

FIGURE 15.4. **(a)** Rocky substratum with attached *Gracilaria* being transplanted to a new site. **(b)** *Gracilaria* attached to rocks with rubber bands; method used for anchoring transplants in soft sediments. **(c)** Plants inserted with a fork directly into soft sediments. **(d)** Plants attached to sand-filled plastic tubes. **(e)** *Gracilaria* attached to a rope, which is stretched between two poles pushed into the sediments. **(f)** Attachment of plants for rope culture. **(g)** Bamboo floating frame with *Gracilaria* attached to ropes spanning the frame. **(h)** A floating net with *Gracilaria* attached to the meshes. (a–d modified from Santelices and Doty 1989, Oliveira and Alveal 1990, Critchley and Ohno 1997; e from Critchley and Ohno 1997.)

rocks with rubber bands, stabilizing the thalli in soft sediments. The algae start to propagate vegetatively, creating a dense seaweed culture bed in only a few months.

A second type of bottom-stocking is the plastic tube method, generally used in a subtidal region (Fig. 15.4d). The method consists of fixing bundles of *Gracilaria* thalli to plastic tubes filled with sand, which anchor the algae to the sea bottom. About 90 g of the material should be attached to each tube with rubber bands. In this method, divers place the sand-filled plastic tubes in parallel rows about 1 meter apart, perpendicular to the sea bottom, creating an underground thallus system that sustains the production in time.

Rope cultivation uses two types of outplanting (Fig. 15.4e–g). The first involves vegetative thalli, which are tied to or inserted within a rope (Fig. 15.5a,b). The rope lines may be monofilament, nylon, thin high-density polyethylene (HDPE) string, rope, or other suitable line that is locally available and cost effective. Super-rope, a plastic netting tube (20-mm mesh size), can also be used in this method. In all cases, care must be taken to keep the seed plants cool and moist while they are being attached to the line. Once the thalli are attached, ropes are suspended and stretched under tension in the sediment or they are supported at different levels by buoys or rafts. Usually, ropes are held horizontally at various depths or may be allowed to hang vertically. Transparency of water is an important limiting factor for the growth of *Gracilaria* on ropes. Such methods are used in Burma, India, Brazil, Namibia, and the United States. The second type involves a suspended spore cultivation method. It is a common method used in several countries such as Venezuela, Japan, India, and others. The material used for attachment of the spores is thin HDPE string, coir rope, or monofilament line. For spore settling, PVC frames (40 × 100 cm) are wrapped with HDPE string or nets (Santelices and Doty 1989). Healthy thalli (tetrasporic or cystocarpic) are cut into small pieces, and the seaweed may be spread evenly over the net or rope. Spore settling takes about 2 days, after which the spore-coated substrata are placed out in the ocean to produce the crop. Once spores settle on a hard substrata, they may begin to germinate within 24 hours (Glenn et al. 1996), after which, the sporelings develop into erect branches. After a few months, they attain full size. Where bamboo is available, a raft system for suspending lines may be preferred because it requires less raw material to secure a given number of lines. In the Caribbean, 2 × 4-meter bamboo frames are used to

**FIGURE 15.5.** **(a)** Rope cultivation of *Gracilaria* using monofilament long lines. **(b)** Enlarged image showing a *Gracilaria* plant held within the twists of the lone line rope. **(c)** *Eucheuma* farmers harvesting plants from submerged long lines. **(d)** Open water cultivation and harvesting of *Kappaphycus*; diver bringing a large plant to the boat. **(e)** Typical outrigger boat used for open water harvesting. (**d** and **e** courtesy T. Chopin.)

anchor lines. The system is based on a floating, kelp pilot unit (Kain 1991).

### 3.4. Pond Culture and Polyculture

*Gracilaria* has been cultivated in ponds on a large scale only in China (Wu et al. 1993) and Taiwan (Shang 1976). Ponds are generally located in areas not exposed to strong wind, situated near the sources of both freshwater and seawater. Water management for this system is an important factor. Water depth is used as a mechanism for modifying temperature; for example, when air temperatures are 32°C in summer, the pond depth should be increased to 50 to 60 cm. In winter, when air temperatures are less than 10°C, depth of the pond should be lowered to 20 to 30 cm. Usually pond depth is kept at 30 to 40 cm. Water exchanges are made after every 2 to 3 days (can be up to 7 to 15 days), which adjusts the salinity and mineral nutrient supplies for algal growth. The pH level of water should be slightly alkaline (7.3–8.0).

Several problems such as eutrophication have increased in lagoon and other areas. However, with careful management, China and Taiwan have developed a system of polyculture. Several species of economically important marine organisms are grown in the same pond at the same time. Shrimp, crabs, fish, and prawns are used in this co-culture system. The fish and crustaceans provide a useful source of income to farmers and an efficient usage of the ponds.

## 3.5. Tank Culture

The use of tanks may provide the greatest productivity per unit area and is more efficient than any other type of farming (Hanisak 1987, Oliviera et al. 2000). In this type of system, several steps can be precisely controlled and managed to reduce the labor input, although this type of system has high operational costs (Friedlander and Levy 1995, Oliviera et al. 2000, Neori et al. 2004). Tank systems also may hold promise for the processing of polluted water for specific products, the removal of extra nutrients from wastewater (Goldman et al. 1977, Ryther et al. 1979), or for energy production (Hanisak 1987). Most of the work on tank cultivation has been carried out using sporeling and juvenile plants in outdoor tanks. Edding et al. (1987) presented a schematic diagram for the flow of seawater in a tank cultivation system in Chile. Efficiency of these systems is very much dependent on the input of various types of energy (compressed air for bubbling, $CO_2$, and pumping water) and nutrients (Craigie and Shacklock 1995, Craigie et al. 1999, Friedlander and Levy 1995). Periodic aeration is required to maintain the seaweeds in circulation, bringing the biomass to the surface into the light and reducing diffusion gradients at the surface of the crop (Santelices and Doty 1989, Bird 1990, Oliveira and Alveal 1990). Carbon supply can be improved by either pumping more seawater or by adding $CO_2$. The temperature and salinity also can be manipulated by pumping more seawater. The pH of the tanks should be managed in the range of 7.9 to 8.3, and the nutrient status of the medium must be monitored (Craigie and Shacklock 1995). Tanks should be cleaned regularly and epiphytes must be controlled (Shacklock and Doyle 1983, Friedlander and Gonen 1996).

To reduce operating costs, Ugarte and Santelices (1992) investigated a polyculture system and Buschmann et al. (1996) further elaborated on an intensive salmon–*Gracilaria* integrated farm design that improved efficiency and was cost effective. In the Philippines, there is potential for co-farming *Kappaphycus* spp. with *Gracilaria lichenoides* (Turn.) Greville (Barraca 1989).

## 3.6. Harvesting and Drying

Usually after 40 to 60 days, *Gracilaria* is harvested manually or with scoop nets. After the harvest, the crop is thoroughly washed in seawater to remove the silt, sand, pieces of shells, and other extraneous materials. The cleaned *Gracilaria* is spread uniformly on bamboo screens, plastic sheets, newspapers, or the ground. After harvest, drying is an important aspect that affects the quality. Usually 2 to 3 days are necessary for complete sun drying. The dried seaweeds are then packed in plastic sacks for storage and marketing.

## 4.0. *Eucheuma* and *Kappaphycus* Cultivation

*Eucheuma* and *Kappaphycus* are important carrageenophytes (80% of world's carrageenan production), and they are abundant in the Philippines, tropical Asia, and the Western Pacific region (Critchley and Ohno 1998). Of more than two dozen species known, only *Kappaphycus alvarezii* (Doty) Doty (*Eucheuma cottoni* Weber-van Bosse) and *Eucheuma denticulatum* (Burman) Collins et Hervey (*Eucheuma spinosum* [Sonder] J. Agardh) are of major commercial importance. During the last 30 years, these species have been successfully introduced to more than 20 countries for commercial cultivation. The thalli of *Eucheuma* and *Kappaphycus* are very cartilaginous, and they may be prostrate or erect in habit and consist of cylindrical to compressed branches. Gametophytic and sporophytic thalli have been reported for many species. Fertile female thalli develop distinct cystocarps, which appear as mammillate structures. The male thalli, however, appear to be uncommon.

The life cycle is triphasic and consists of three stages: tetrasporophyte (2n), gametophyte (n), and carposporophyte (2n). The gametophyte is dioecious. The male thallus produces spermatia, and the female thallus produces few-celled carpogonial branches in the cortex, of which the tip cell acts as the carpogonium. Fertilization results in the carposporophyte within the tissue of the female gametophyte. The carposporophyte produces carpospores (2n) that develop into the tetrasporophyte. Subsequently, the tetrasporophyte produces tetrasporangia, which undergo meiotic divisions producing tetraspores (n). Upon their release, the tetraspores develop into male and female gametophytes, thus completing the life cycle.

Excellent primers have been prepared by Ask (1999) and Sahoo (2000). These handbooks were designed for the technologist to promote the cultivation of eucheumoid taxa to coastal villagers. For a more in-depth discussion of abiotic and biotic factors that affect cultivation practices, reviews have presented by Dawes (1987) and Trono (1993). The initial steps in the farming include the following: (1) site selection, (2) selection of cultivation methodology, (3) farm mainte-nance, and (4) harvesting and drying.

## 4.1. Site Selection

Site selection is very important. The site should be far from sources of freshwater such as rivers, creeks, and estuarine areas, as well as other sources of pollution. The site should be protected from strong tidal or wind-generated waves, which can destroy the farm. The pres-ence of islands or reefs will act as buffer zones for the farms by minimizing the destructive effects of strong waves, especially during the monsoon season. Water current should be between 20 and 40 meters per minute. Strong currents are destructive to the farm site. Water movement (current) influences the growth of the plant because it facilitates rapid absorption of nutrients. It also prevents extreme fluctuations in temperature, salinity, pH, dissolved gases, and so on. Water temper-ature should be between 27° and 30°C and the salinity between 30 and 33 psu. The water level should be 0.5 to 1.0 meter during low tide and not more than 2.0 to 3.0 m during high tide. Plants should be submerged in water, even at the lowest tide. Areas with coarse sandy to corally bottom substrata are good sites for cultiva-tion. Small and inexpensive tools are required, making cultivation very cost effective.

After a site has been identified, the question of whether the site will support farming is resolved by test planting. Small test plots should be identified at strate-gic locations, and seedlings should be tied to monolines and fixed to the field. The growth of test plants should be monitored weekly and the daily growth rate (DGR) should be determined. Areas supporting a DGR between 3 and 5% are potentially good sites. Usually 2 to 3 months of monitoring the growth rate is sufficient for test planting, but a 1-year trial is recommended.

## 4.2. Cultivation Methods

A fixed, off-bottom monoline method is the most popular and convenient method used. It is cheap and easy to install and maintain. In this method, mangrove or bamboo stakes are driven deep into the substratum, spaced at 10-meter intervals, in rows 1 meter apart. The end of the nylon monofilament line is tied to one stake, stretched, and tied to another stake in the oppo-site row. The monolines are constructed to form plots or units of a standard size (Fig. 15.5c). Seedlings weigh-ing between 50 and 100 g are tied to the monolines at 25- to 30-cm intervals using soft plastic tying material called *tie-tie*. The plants are allowed to grow to 1 kg before they are harvested (usually 45–60 days) (Fig. 15.5d).

Floating methods (raft or long lines) are used when space prohibits the use of the off-bottom monoline method. Grazing by bottom-associated animals is min-imized or eliminated with this method, and plants near the surface of the water column are exposed to more moderate water movement caused by waves, thereby enhancing nutrient absorption. Seedlings are tied in a similar fashion as in the monoline method, but here they are tied to rafts constructed of bamboo or wood. Styrofoam or empty plastic containers are used to keep the rafts floating, and the raft is anchored to the bottom by means of ropes. For the long-line floating method, six nylon monofilament lines (10 m or longer) are attached to bamboo (2 cm long), which are set at 5-m intervals. The nylon lines are attached to the bamboo at 30-cm intervals. Each long-line unit can be planted with 400 cuttings, which are tied to the nylon line at 15-cm intervals. The four corners of the unit are anchored to wooden stakes. One hundred long-line units are equivalent to a one-hectare farm using the fixed, off-bottom monoline method.

## 4.3. Farm Maintenance

Maintenance of the farm consists of weeding, repair to the support system, replacement of lost seedlings, and removal of benthic grazers. A number of epiphytes compete with *Eucheuma/Kappaphycus* for nutrients, light, and space. Grazers such as sea urchins, starfish, and siganid fish have been found to consume large quantities of seaweed, resulting in significant losses of biomass. Farm maintenance is a critical aspect of euchemoids cultivation.

## 4.4. Harvesting and Drying

Usually the plants are harvested after 45 days when each seedling weighs up to 1 kg (Fig. 15.5e). The best looking healthy plants are selected to serve as seedlings for the next crop. The remaining plants are sun dried or sold fresh in the market. Drying is an important post-harvest activity that will affect the quality. Drying can be done either on drying platforms made of bamboo or on the village ground. The seaweeds should be regularly turned over to facilitate drying and should be protected from rain at any cost. Usually 2 to 3 days of complete sun drying is necessary. Dried materials should not contain more than 30% moisture. The dried materials are then packed in plastic sacks for storage, marketing, or further processing for carrageenan.

## 5.0. *LAMINARIA* CULTIVATION

*Laminaria* is mainly cultivated in China, Korea, and Japan (Brinkhuis et al. 1987). According to the FAO (2000), nearly 4,055,027 tons (wet weight) of *Laminaria* was harvested globally, mainly through cultivation. China was the largest producer of *Laminaria*, contributing nearly 3.8 billion kg wet weight. *Laminaria* is a member of the family Laminariaceae, order Laminariales, class Phaeophyceae. In Japan, 13 species of *Laminaria* have been described, of which *Laminaria japonica* Areschoug is the most widely cultivated species. China leads the world in the aquaculture of *L. japonica*. The plants grow well on reefs or stones in the subtidal zone, at a depth of 2 to 15 m (sometimes up to 30 m). They prefer sheltered and calm water, rather than open seas. The seawater temperature suitable for growth of *L. japonica* ranges between 3° and 20°C.

The thallus of *L. japonica* is large, up to 2–5 m in length, but sometimes may grow up to 10 m. The life cycle of *Laminaria* is well understood. It consists of an alteration of generations between a small gametophytic phase and a dominant large sporophytic phase. In the field, the frond (sporophyte) usually matures during summer and autumn. The sporophyte releases the zoospores that settle down on the substratum and immediately germinate and grow into microscopic male and female gametophytes in equal ratios. Upon reaching maturity (about 10–14 days), the male gametophyte releases motile sperm that fertilize a large nonmotile egg that will be extruded from the oogonium. After the fusion of the gametes, zygote formation immediately follows, and within 15 to 20 days, young sporophytes develop, thus completing the life cycle.

Usually *L. japonica* is a biennial, where the frond becomes of a harvestable size about 20 months after germination in nature. Tseng et al. (1957), Tseng and Wu (1962), and Hasegawa (1971) reduced the cultivation period to 11 months through a technique called *forced cultivation*. As with *Undaria*, the cultivation of *Laminaria* consists of four phases: collection and settlement of zoospores on seed strings; production of seedlings; transplantations and outgrowing of seedlings; and harvesting.

### 5.1. Zoospores and Seedlings

Traditionally the collection of zoospores and seedlings is carried out in autumn when fertile *Laminaria* thalli are available and the water temperature decreases to about 15°C. However, in China, it is also possible to produce the seedlings during summer in hatcheries by controlling the water temperature and light. Fertile *Laminaria* thalli with abundant sporangial sori are partially air dried. The seed string (a synthetic twine of 2 to 6 mm in diameter or palm line 6 mm in diameter) is placed on the bottom of the seeding container, which is filled with cool seawater; the partially air-dried thallus with sori is placed in the container. To remove microscopic contamination from wild collected sori, they should also be rinsed with clean, filtered seawater and wiped dry. The sori are stored in a cool environment overnight, and upon re-immersion in seawater, the dehydrated thalli will release zoospores (within 15–60 minutes). The zoospores settle and germinate on the seed string (50 m) that is wrapped around plastic rectangular frames (0.5 × 0.5 m) in the seeding tank (Fig. 15.7d). The Chinese technique of intermediate nursery culture is employed to bring the plants to a size of 5 to 10 cm before transplanting to the nursery grounds (Fig. 15.6f). This nursery stage may continue for up to 8 weeks, usually in shallow raceways with running, filtered temperature-controlled seawater in a glass house. This technique is particularly useful if the temperature of the culture grounds is above good growing conditions for the juvenile sporophytes. The seed string (after 30 to 60 days) may then be transferred to the sea and fixed to rafts consisting of a series of bamboo segments anchored to the bottom by ropes or long lines (50 m). The adoption of this method has solved the problems of uniform light requirements. The culture of sporelings is kept in the open water nursery grounds until they reach 10 to 15 cm in length, when they can be transplanted from the farm to produce mature *Laminaria* plants.

Merrill and Gillingham (1991) constructed a spool technique that saves time, space, and effort and that may be a good substitution for seeding operations in pilot plant projects. Briefly, in this technique, 36- by 5-cm cylindrical sections of PVC pipe are wrapped with 50 m of seed string. After the inoculation, the culture spools are transferred to specially designed cultivation cham-

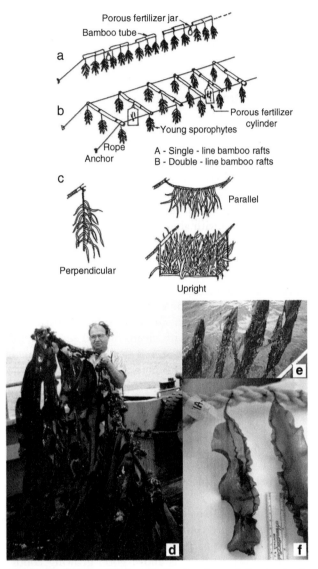

**FIGURE 15.6.** **(a)** A single-line bamboo raft for *Laminaria* sporelings. **(b)** A double-line bamboo raft used in Japan. **(c)** Perpendicular, parallel, and upright culture methods. **(d)** Long line with *Laminaria* after 8 months of growth (Yellow Sea, China). **(e)** Long line showing attached *Laminaria* plants (South Korea). **(f)** Young sporophytes growing on long line. (**a–c** from Cheng 1969; **e** courtesy E. Hwang.)

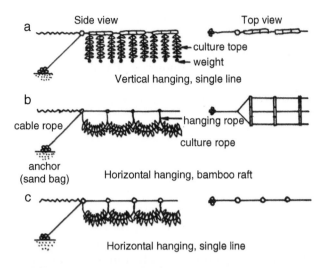

**FIGURE 15.7.** **(a–c)** A bamboo raft and vertical and horizontal single lines used for *Undaria* cultivation. **(d)** *Undaria* spore collector; frame is 50 by 50 centimeters and a synthetic seed string is wound around the frame. **(e)** Open water cultivation of *Undaria*. (**a–c** from Akiyama and Kurogi 1982; **e** courtesy E. Hwang.)

bers. The number of spools per chamber depends on the needs of the project. These tanks are constructed of either glass or transparent acrylic plastic at least 0.64 cm thick, illuminated on the side with cool white fluorescent lights at a level of 80 to 100 μmol photons · m$^{-2}$ · s$^{-1}$. Neutral density screening may be used to control the quantity of illumination, because the light requirement increases as they mature. The tanks have support rods for each of the spools and are placed vertically in the culture tanks in two rows so each spool in the tank receives illumination. This technique ensures seed stock development on the entire spool. Half-strength Provasoli's ES medium should cover the spools (Provasoli 1968; see Chapter 3 and Appendix A). Seawater exchange should be at 10- to 14-day intervals. Germanium dioxide may be used to inhibit the growth of diatoms (Lewin 1969 as cited by McLachlan 1975; see Chapter 9).

Another major enhancement in the cultivation of *L. japonica* has been raising of sporelings via the mass culture of gametophyte clones (Li et al. 1999). Female and male gametophyte clones are mass cultured, followed by the induction of gametogenesis. With reared clones, mass quantities of sporelings (with desired traits) can be successfully produced, thereby reducing the need for long-term cultivation in temperature-controlled greenhouses or environmentally controlled rooms. With this new technology, sporeling raising times may be reduced by up to 50%, with a significant reduction in production costs (estimated to be >50%).

## 5.2. Seedlings—Sporelings Transplant, Outgrowth

The sporelings are transplanted to outgrowing ropes (long lines that are usually 50–100 m long; Fig. 15.6a). Thirty or more sporelings are transplanted to a 2-meter (or shorter) growing rope. One end of the growing rope with sporelings is attached at regular intervals (50 cm) to the main support rope of a raft or long-line system so that the culture rope hangs vertically (the standard Japanese drop-line system; Fig. 15.6b). The free ends of the culture ropes (bottom ends) are weighted so they remain vertical. The culture is maintained through the growing season, which is up to late spring (Kawashima 1993). In the spool-system design described by Merrill and Gillingham (1991), once the juvenile sporophytes are about 1 to 2 mm, the spools are removed from the culture tanks. The design of the spools enables the long line to be inserted through the inside of the spool. Starting at one end of the long line, the seed string can be wrapped tightly around the long line in a spiral fashion as the spool is moved down the long line. These long lines are then "snaked" back and forth in the raceways and remain there until ready for deployment in the sea. As the plants grow, their haptera attach to the thicker long-line rope.

Raft-supported long-line systems are of two types, namely the single-line (bamboo or rubber tube) and the double-line raft system (Fig. 15.6). In the first type, pieces of bamboo or rubber tubes are tied end to end to a main support rope, forming a single line of tubes about 60 meters long. The ends of the support line are anchored to the bottom. In the double line, raft pieces of bamboo or rubber tube about 1 meter long are tied to two main support ropes at 1-meter intervals, forming a ladder-like structure (Fig. 15.6b). The ropes may be hung vertically from the support rope or positioned parallel to the surface of the water by tying one end to one of the support ropes and the other end to the other. The first type is called the *perpendicular arrangement*, and the second the parallel type (Fig. 15.6c). A third upright type is constructed by fastening 1 by 3-meter bamboo beds suspended between the two main support ropes or between two parallel single-line raft systems. The positioning of the growing ropes may vary in the double-line raft system. Modifications of these systems have been described by Brinkhuis et al. (1987) and Holt and Dawes (1989).

In the culture of "autumn sporelings," the seed string, with zoospores that are transferred to the sea for the development, is immediately overgrown by many opportunistic or weedy species. The presence of these "weeds" adversely affects the growth and development of the gametophytes, thereby hindering sporeling production. Studies on the growth requirements of the gametophytes indicate that the best time for deployment must be determined individually for any culture site. Water temperature is the primary controlling factor for sporeling growth.

Aside from light and water temperature, nutrients (especially nitrate-nitrogen) were found to be a limiting factor in the culture of *Laminaria*. Traditional culture was often limited to waters that are highly fertile, such as those found in bays. More recently, the introduction of the raft culture and application of fertilizers have significantly extended the kelp-growing area to deeper oceanic waters.

## 5.3. Harvesting

Harvesting of *L. japonica* is during late May to June (Fig. 15.6d,e). The blades are cut from the cultivation ropes and washed in seawater to remove diatoms, hydrozoans, and other attached organisms. After this, the blades are dried naturally in the sun for several days on any available surface. When weather conditions are unfavorable for natural drying, oil-powered dryers are used.

## 6.0. UNDARIA CULTIVATION

The brown algal genus *Undaria* is an important food delicacy in Japan, Korea, and China (wakame). It is sold boiled or dried and is especially appreciated as an ingredient for soybean paste soup ("misoshiru") and seaweed salad. *Undaria* belongs to the family Alariaceae in the order Laminariales. It has three species (*Undaria pinnatifida* [Harvey] Suringar, *Undaria undarioides* [Yendo] Okamura, and *Undaria peterseniana* [Kjellman] Okamura), and of these, *U. pinnatifida* is the most important (Sohn and Kain 1989, Ohno and Matsuoka 1993, Sohn 1996). According to the FAO data (2003), nearly 459,390 tons (wet weight) of *Undaria* was produced globally through culture. Experimental cultivation of *Undaria* has also started recently on the Normandy coast of France (Kaas and Pérez 1989).

In nature, *Undaria* commonly grows in open seas or within bays on the temperate coasts of Japan, Korea, China, and other areas of the northwest Pacific. The thalli grow on rocks and reefs to a depth of 1 to 8 m below the low tide level (Saito 1975). Usually, the frond grows to a length of 1 to 2 m. The life cycle of *Undaria* consists of a reduced gametophytic phase and a large dominant sporophytic phase. Young fronds of *Undaria* appear late in October to early November and grow rapidly until early spring. The best growth takes place when the seawater temperature ranges between 5° and 13°C. The mature sporophytes release zoospores that germinate on rocky substrata and develop into male and female gametophytes. After fertilization, the zygote develops into a sporophyte, thereby completing the life cycle.

The cultivation of *Undaria* consists of four stages: (1) collection of zoospores and seedlings, (2) culture of gametophyte germlings, (3) outgrowing of thalli, and (4) harvesting and processing.

## 6.1. Zoospores and Seedlings

The collection of zoospores begins around May or June in the northern parts and during April or May in the southern parts of Japan. The mature sporophylls are first partially dehydrated to induce release of zoospores and are then placed in culture tanks filled with seawater. The sporophylls are removed from the tank after the release of the zoospores, after which they are seeded by dipping a spore collector into the tank (Fig. 15.7d). Usually the spore-collector frame consists of a synthetic fiber twine (3–5 mm in diameter), wrapped uniformly at 1-centimeter intervals around a plastic frame (50 × 50 cm). Another seeding process called the *free-living technique* uses vegetative gametophytes grown in flasks (Kaas and Pérez 1990) and is similar to the *Laminaria*-rearing technology of clonal gametophyte technology (Li et al. 1999).

## 6.2. Gametophyte Germlings

The seeded twine frames are removed from the seeding tank and arranged vertically in a culture tank about 1 meter deep. During summer and early autumn, the zoospores develop into microscopic gametophytes on the seeding twines. The water temperature is maintained at less than 23°C. Salinity is maintained at 30 to 33 psu, and light intensity is controlled between 40 and 120 $\mu$mol photon $\cdot$ m$^{-2}$ $\cdot$ s$^{-1}$ at the surface of the tank. The required quantity of salts such as sodium nitrate and sodium phosphate is added to the tanks to stimulate the growth of germlings. The germlings are ready for outgrowing in the autumn when the water temperature falls to 23°C or below. The frames are removed

from the culture tanks and transferred to the nursery areas. The germlings are allowed to develop to seedlings (2–3 cm) before removal and outplanting along the open coast.

## 6.3. Outgrowing of Thalli

The outplanting of nursery-grown seedlings starts in the autumn when water temperature is about 20°C. The original seeding ropes are cut into lengths of 4 to 6 cm (about 10 seedlings each) and inserted into the twist of the main cultivation rope; they are set into the sea using rafts or long lines (Fig. 15.7a–c). The farm is supported by floating buoys and proper anchorage. The depth of the water varies from 0 to 5 m depending on site conditions. The growth of the young thalli from the seedlings is best at temperatures ranging from 15° to 17°C, whereas large thalli grow best in temperatures between 12° and 13°C. Akiyama and Kurogi (1982) developed a popular "floating rope system" for the cultivation of *Undaria* in deep and wave-exposed sea.

## 6.4. Harvesting and Processing

Generally *Undaria* reaches harvestable size about 3 months after it is transferred to the sea when they reach a length of 0.5 to 1.0 m (Fig. 15.7e). After the rope is hauled up from the sea on the boat, the cultivated thalli are cut from the rope with a sickle. Each harvested frond is approximately 5 kg in wet weight. The crop may be sold fresh, sun dried, or artificially dried.

## 7.0. ACKNOWLEDGMENTS

D. B. Sahoo thanks Ms. N. Sahu, Mr. D. Sahoo, Ms. P. Baweja, and Ms. N. Kushwa, Research students of the Marine Biotechnology Laboratory, Department of Botany, University of Delhi for their help during the preparation of this manuscript. D. B. S. also thanks Dr. M. Munshi-Sahoo for her moral support and unconditional help. C. Yarish thanks Mrs. K. J. Yarish for the moral support and understanding during the preparation of this chapter. C. Y. also wants to acknowledge the support of J. P. McVey who has spent time on many occasions supporting and working with him in China, Korea, and Japan. C. Y. also acknowledges the invaluable support of Professors X. G. Fei, C. K. Tseng, and C. Y. Wu of the Chinese Academy of Sciences and Professor P. He (Shanghai Fisheries University) for the introduction to Chinese aquaculture and philosophy. Without their gracious understanding and friendship, this chapter would not be possible. The authors want to acknowledge the invaluable efforts of Virge Kask, for assistance with the biological illustrations, and Professors M. Ohno, A. Critchley, M. Notoya, X. G. Fei, T. Chopin, I. Levine, and E. Hwang, for use of their photographs. C. Y. wants to acknowledge the support of the State of Connecticut's Critical Technology Program, the National and Connecticut Sea Grant College Programs, the Office of International Programs of the Office of Oceanic and Atmospheric Research, and the NOAA's National Marine Aquaculture Initiative.

## 8.0. REFERENCES

Abbott, I. A. 1988. Food and food products from seaweeds. In: Lembi, C. A., and Waaland, J. R., eds. *Algae and Human Affairs*. Cambridge University Press, New York, 135–48.

Abbott, I. A. 1992. *Taxonomy of Economic Seaweeds with Reference to Some Pacific and Western Atlantic Species, III.* California Sea Grant College Program, University of California, La Jolla, 241 pp.

Abbott, I. A, and Norris, J. N., eds. 1985. *Taxonomy of Economic Seaweeds with Reference to Some Pacific and Caribbean Species.* California Sea Grant College Program, University of California, La Jolla, 167 pp.

Adnan, H., and Porse, H. 1987. Culture *of Eucheuma cottonii* and *Eucheuma spinosum* in Indonesia. *Hydrobiologia* 151/152:355–8.

Akiyama, K., and Kurogi, M. 1982. Cultivation of *Undaria pinnatifida* (Harvey) Suringar, the decrease in crops from natural plants following crop increase from cultivation. *Bull. Tôhoku Res. Lab.* 44:91–100.

Armisen, R. 1995. World-wide use and importance of *Gracilaria. J. Appl. Phycol.* 7:231–43.

Ask, E. I. 1999. Cottonii and spinosum Cultivation Handbook. FMC Corporation, Philadelphia, 49 pp.

Baibaroux, O., Perez, R., and Dreno, J. P. 1984. L'algue rouge *Eucheuma spinosum* possibilités d'exploitation et de culture aux Antilles. Sci. Peche, *Bull. Inst. Peches Marit.* 348:2–9.

Barraca, R. 1989. *Eucheuma* and *Graciliaria lichenoides* co-farming possibility: *Graciliaria* production in the Bay of Bengal Region. *Bay of Bengal Program Report* 45, pp. 18–9.

Bird, C. J. 1995. A review of recent taxonomic concepts and developments in the Gracilariaceae (Rhodophyta). *J. Appl. Phycol.* 7:225–67.

Bird, K. T. 1990. Intensive seaweed cultivation. *Aquaculture Mag.* 15:29–34.

Bird, C. J., and Kain, J. M. 1995. Recommended names of included species of Gracilariaceae. *J. Appl. Phycol.* 7:335–8.

Braud, J. P., and Perez, R. 1978. Farming on a pilot scale of *Eucheuma spinosum* (Florideophyceae) in Djibouti waters. In: Jensen, A., and Stein, J., eds. *Proceedings of the Ninth International Seaweed Symposium.* Science Press, Princeton, pp. 533–9.

Braud, J. P., Perez, R., and Lacherade, G. 1974. Etude des possibilites d'adaptation de l'algue rouge *Eucheuma spinosum* aux cotes des Afais et des Issas. Sci. Peche, *Bull. Inst. Peches Marit.* 238(Juillet-Aout):1–16.

Brinkhuis, B. H., Levine, H. G., Schlenk, C. G., and Tobin, S. 1987. *Laminaria* cultivation in the far east and North America. In: Bird, K. T., and Benson, P. H., eds. *Seaweed Cultivation for Renewable Resources. Developments in Aquaculture and Fisheries Science,* Elsevier, New York, pp. 107–46.

Buschmann, A. H., Wetermeier, R., and Retamales, C. A. 1995. Cultivation of *Gracilaria* in the seabottom in southern Chile: a review. *J. Appl. Phycol.* 7:291–301.

Buschmann, A. H, Troell, M., Kautsky, N., and Kautsky, L. 1996. Integrated tank cultivation of salmonids and *Gracilaria chilensis* (Gracilariales, Rhodophyta). *Hydrobiologia* 326/327:75–82.

Carmona, R., Chanes, L., Kraemer, G., Chopin, T., Neefus, C., Zertuche, J. A., Cooper, R., and Yarish, C. 2002. Nitrogen uptake by *Porphyra purpurea.* Its role as a nutrient scrubber. In: Van Patten, P., ed. *Proceedings of the Long Island Sound Research Conference, Nov. 17–18, 2000, Stamford, CT.* Connecticut Sea Grant College Program, pp. 87–91.

Chiang, Y. M. 1981. Cultivation of *Gracilaria* (Rhodophyta, Gigartinales) in Taiwan. In: Lerving, T., ed. *Proceedings of the Tenth International Seaweed Symposium.* Walter de Gruyter, Berlin, 569–74.

Chopin, T., Yarish, C., Wilkes, R., Belyea, E., Lu, S., and Mathieson, A. 1999. Developing *Porphyra*/salmon integrated aquaculture for bioremediation and diversification of the aquaculture industry. *J. Appl. Phycol.* 11:463–72.

Chopin, T., Buschmann, A. H., Halling, C., Troell, M., Kautshy, N., Neori, A., Kraemer, G. P., Zertuche-Gonzalez, J. A., Yarish, C., and Neefus, C. 2001. Integrating seaweeds into mariculture systems: a key towards sustainability. *J. Phycol.* 37:975–86.

Chung, I., Kang, Y. H., Yarish, C., Kraemer, G., and Lee, J. 2002. Application of seaweed cultivation to the bioremediation of nutrient-rich effluent. *Algae* 17(3):187–94.

Craigie, J. S., and Shacklock, P. F. 1995. Culture of Irish moss. In: Borhen, A. D., ed. *Cold-Water Aquaculture in Atlantic*

*Canada*, ed 2. The Canadian Institute for Research on Regional Development, Université, Monction, pp. 363–90.

Craigie, J. S., Staples, L. S., and Archibald, A. F. 1999. Rapid bioassay of a red food alga: accelerated growth rates of *Chondrus crispus*. *World Aquaculture* 30:26–8.

Critchley, A. T. 1993. *Graciliaria* (Rhodophyta, Gracilariales): an economically important agarophyte. In: Ohno, M., and Critchley, A. T., eds. *Seaweed Cultivation and Marine Ranching*. Nagai: Kanagawa Int. Fisheries Training Center and JICA, Yokosuka, Japan, pp. 89–112.

Critchley, A. T., and Ohno, M. 1998. *Seaweed Resources of the World*. JICA, Yokosuka. Dalton, S. R., Longley R. E., and Bird, K. T. 1995. Hemagglutinins and immunomitogens from marine algae. *J. Mar. Biotechnol.* 2:149–55.

Critchley, A. T., and Ohno, M. 1997. *Cultivation and Farming of Marine Plants*. ETI World Biodiversity Database, CD-ROM Series. Available at: www.eti.uva.nl.

Dawes, C. J. 1987. The biology of commercially important tropical marine algae. In: *Seaweed Cultivation for Renewable Resources*. Bird, K. T., and Benson, P. H., eds. *Developments in Aquaculture and Fisheries Science*, Elsevier, New York, pp. 155–99.

Dawes, C. J., 1989. Temperature acclimation in cultured *Eucheuma isiforme* from Florida and *E. alvarezii* from the Philippines. *J. Appl. Phycol.* 1:59–69.

de Paula, E. J., Pereira, R. T. L., and Ohno, M. 1999. Strain selection in *Kappaphycus alvarezii* var. *alvarezii* (Doty) Doty ex P. Silva (Rhodophyta, Solieriaceae) using tetraspore progeny. *J. Appl. Phycol.* 11(1):111–21.

de Reviers, B., 1989. Realisation d'Une Ferme de Culture Industrielle de *Eucheuma* aux Maldives. *Oceanis* 15(5):749–52.

Doty, M. S., 1978. *Eucheuma*—current marine agronomy. In: Krauss, R. W., ed. *The Marine Plant Biomass of the Pacific Northwest Coast*. Oregon State University Press, Corvallis, pp. 203–14.

Doty, M. S. 1980. Outplanting *Eucheuma* species and *Gracilaria* species in the tropics. In: Abbott, I. A., Foster, M.S., and Ekiund, L. F., eds. *Proceedings of the Symposium sponsored by Pacifica Area Sea Grant Advisory Program and the California Sea Grant College Program California Sea Grant*, Pacific Seaweed Aquaculture, La Jolla, California, pp. 19–22.

Doty, M. S. 1985a. *Eucheuma alvarezii*, sp. nov. (Gigartinales, Rhodophyta) from Malaysia. In: Abbott, I. A., and Norris, J. N., eds. *Taxonomy of Economic Seaweeds: with Reference to Some Pacific and Caribbean Species*. California Sea Grant College Program. Rep. T-CSGCP-OII, La Jolla, California, pp. 37–45.

Doty, M. S. 1985b. *Eucheuma* species (Solieriaceae, Rhodophyta) that are major sources of carrageenan. In:

Abbott, I. A., ed. *Taxonomy of Economic Seaweeds: With Reference to Some Pacific and Caribbean Species*. California Sea Grant College Program. Rep. T-CSGCP-OII, La Jolla, California, pp. 47–61.

Doty, M. S. 1995. *Betaphycus philippinesis* gen. et. sp. nov. and related species (Solieriaceae, Gigartinales). In: Abbott, I. A., ed. *Taxonomy of Economic Seaweeds*. California Sea Grant College, University of California, vol. 5, La Jolla, 237–45.

Doty, M. S., and Alvarez, V. B. 1973. *Seaweed farms: a new approach for U.S. industry*. Proceedings of the 9th Annual Conference Proceedings, University of Hawaii, 701–8.

Doty, M. S., and Norris, J. N. 1985. *Eucheuma* species (Solieriaceae, Rhodophyta) that are major sources of carrageenan. In: Abbott, I. A., and Norris, J. N., eds. *Taxonomy of Economic Seaweeds with Reference to Some Pacific and Caribbean Species*. California Sea Grant College Program, La Jolla, pp. 47–61.

Drew, K. M. 1949. Conchocelis-phase in the life history of *Porphyra umbilicalis* (L.) Kütz. *Nature*. 164:748–9.

Edding, M., Leon, C., and Amber, R. 1987. Seaweed biotechnology: current status and further prospects. *Ann. Proc. Phytochem. Soc. Europe*. 28:335–50.

Fei, X. G. 1983. Macroalgal cultivation in California and China. In: McKay, L. B., ed. *Seaweed Raft and Farm Design in the United States and China*. New York Sea Grant Institute, Albany, pp. 31–6.

Fei, X. G., Bao, Y., and Lu, S. 2000. Seaweed cultivation—traditional way and its reformation. *Oceanologia et Limnologia sinica*. 31:375–80.

Fei, X. G., Lu, S., Bao, Y., Wilkes, R., and Yarish, C. 1998. Seaweed cultivation in China. *World Aquaculture* 29:22–24.

Food and Agriculture Organization of the United Nations. 2000. The state of world fisheries and aquaculture 2000 (electronic edition). Available at: www.fao.org/docrep/003/x8002e/x8002e00.htm.

Food and Agriculture Organization of the United Nations. 2003. Review of the state of world aquaculture. Inland Water Resources and Aquaculture Service. FAO Fisheries Circular No. 886, Rev. 2. Available at: www.fao.org/DOCREP/005/Y4490E/Y4490E00.HTM.

Fredericq, S., and Hommersand, M. H. 1989a. Proposal on the Gracilariales ord. nov. (Rhodophyta) based on an analysis of the reproductive development of *Gracilaria verrucosa*. *J. Phycol.* 25:213–27.

Fredericq, S., and Hommersand, M. H. 1989b. Comparative morphology and taxonomic status of *Gracilariopsis* (Gracilariales, Rhodophyta). *J. Phycol.* 25:228–41.

Freidlander, M., and Levy, I. 1995. Cultivation of *Gracilaria* in outdoor tanks and ponds. *J. Appl. Phycol.* 7:315–24.

Friedlander, M., and Gonen, Y. 1996. *Gracilaria conferta* and its epiphytes, III. Allelopathic inhibition of the red seaweed by *Ulva* cf. *lactuca*. *J. Appl. Phycol.* 8:21–5.

Gerwick, W. H., Roberts, M. A., Proteau, P. J., and Chen, J. L. 1994. Screening cultured marine algae for anticancer-type activity. *J. Appl. Phycol.* 6:143–9.

Glenn, E. P., Moore, D., Fitzsimmons K., and Menke, K. 1996. *Atlas of Gracilaria Spore Culture.* National Coastal Resources Institute, Portland, Oregon, 33pp.

Glenn, E. P., Moore, D., Brown, J. J., Tanner, R., Fitzsimmons, K. Akutigawa, M., and Napolean, S. 1998. A sustainable culture system for *Gracilaria parvispora* (Rhodophyta) using sporelings, reef grow out and floating cages in Hawaii. *Aquaculture* 165:221–32.

Goldman, J. C., Tenore, K. R., Ryther J. H., and Corwin, N. 1997. Inorganic nitrogen removal in a combined tertiary treatment-marine aquaculture system. I. Removal efficiencies. *Water Res.* 8:45–54.

Hanisak, M. D. 1987. Cultivation of *Gracilaria* and other macroalgae in Florida for energy production. In: Bird, K. T., and Benson, P. H., eds. *Seaweed Cultivation for Renewable Resources*, vol. 16, *Developments in Aquaculture and Fisheries Science*, Elsevier, New York, pp. 191–218.

Hasegawa, Y. 1971. Forced cultivation of *Laminaria*. *Bull. Hokkaido Reg. Fish. Res. Lab.* 37:49–52.

Holt, T. J., and Dawes, C. P. 1989. Cultivation of Laminariales in the Irish Sea. In: Kain, J., Andrews, W., and McGregor, B. J., eds. *Proceedings of the 2nd Workshop of COST 48.* Commission of the European Communities, Brussels, pp. 34–7.

Kaas, R., and Perez, R. 1990. Study of intensive culture of *Undaria* on the coast of Brittany. FAO RAS/90/002.

Kain (Jones), J. M. 1991. Cultivation of attached seaweeds. In: Guiry, M. D., and Blunden, G., eds. *Seaweed Resources in Europe: Uses and Potential.* John Wiley and Sons, Chichester, U.K., pp. 304–77.

Kawashima, S. 1992. Kombu. In: Miura, A., ed. *Farming of Sea Vegetables* [in Japanese]. Fisheries Series. 88:43–51.

Kawashima, S. 1993. Cultivation of the brown alga, Laminaria 'Kombu.' In: Ohno, M., and Critchley, A. T., eds. *Seaweed Cultivation and Marine Ranching.* Nagai: Kanagawa International Fisheries Training Center and JICA, Yokosuka, Japan, pp. 25–40.

Kraemer, G. P., Carmona, R., Chopin, T., and Yarish, C. 2002. Use of photosynthesis measurements in the choice of algal species for bioremediation. In: Van Patten, P., ed. *Proceedings of the Long Island Sound Research Conference, Nov. 17–18, 2000, Stamford, CT.* Connecticut Sea Grant College Program, pp. 113–17.

Lembi, C. A., and Waaland, J. R. 1988. *Algae and Human Affairs.* Cambridge University Press, New York, 590 pp.

Lewmanomont, K. 1998. The seaweed resources of Thailand. In: Critchley, A., and Ohno, M., eds. *Seaweed Resources of the World.* JICA, Yokosuka, Japan, pp. 70–8.

Li, D., Zhou, Z., Liu, H., and Wu, C. 199. A new method of *Laminaria japonica* strain selection and sporeling raising by the use of gametophyte clones. *Hydrobiologia* 398/399: 473–6.

Lirasan, T., and Twide, P. 1993. Farming *Eucheuma* in Zanzibar, Tanzania. *Hydrobiologia* 260–261:353–5.

Luxton, D. M., and Luxton, P. M. 1999. Development of commercial *Kappaphycus* production in the Line Islands, Central Pacific. *Hydrobiologica* 398/399:477–86.

Luxton, L. M., Robertson, M., and Kindley, M. J. 1987, Farming of *Eucheuma* in the South Pacific Islands of Fiji. *Hydrobiologia* 151/152:359–62.

Mairh, O. P., Soe-Htun, U., and Ohno, M. 1986. Culture of *Eucheuma striatum* (Rhodophyta, Solieriaceae) in subtropical waters of Shikoku, Japan. *Bot. Mar.* 29:185–91.

Mairh, O. P., Zodape, S. T., Tewari, A., and Rajyaguru, M. R. 1995. Culture of marine red alga *Kappaphycus striatum* (Schmitz) Doty on the Saurashtra region, west coast of India. *Ind. J. Mar. Sci.* 24:24–31.

McHugh, D. J. 1996. *Seaweed Production and Markets.* FAO/GLOBEFISH Research Programme, Rome, 73 pp.

Melo, R. A. 1998. *Gelidium* commercial exploitation: natural resources and cultivation. *J. Applied. Phycol.* 10:303–14.

Merrill, J. 1989. The commercial nori (*Porphyra*) sea farming in Washington State. In: Kain, J., Andrews, W., and McGregor, B. J., eds. *Aquatic Primary Biomass-Marine Macroalgae: Outdoor Seaweed Cultivation. Proceedings 2nd Workshop of COST 48.* Commission of the European Communities, Brussels, pp. 90–102.

Merrill, J., and Gillingham, D. M. 1991. *Bull Kelp Cultivation Handbook.* NCRI publication no. NCRI-T-91–011. 70 pp.

Mitman, G. G., and Van der Meer, J. P. 1994. Meiosis, blade development and sex determination in *Porphyra purpurea* (Rhodophyta). *J. Phycol.* 30:147–59.

Mumford, T. F., and Miura, A. 1988. *Porphyra* as food: cultivation and economics. In: Lembi, C. A., and Waaland, J. R. eds. Cambridge University Press, New York, pp. 87–117.

Neefus, D. D., Mathieson, A. C., Klein, A. S., Teasdale, B., Bray, T., and Yarish, C. 2002. *Porphyra birdiae* sp. nov. (Bangiales, Rhodophyta): a new species from the Northwest Atlantic. *Algae* 17(4):203–16.

Neori, A., Chopin, T., Troell, M., Buschmann, A. H., Kraemer, G., Halling, C., Shpigel, M., and Yarish, C. 2004. Integrated aquaculture: rationale, evolution and state of the art emphasizing seaweed biofiltration in modern aquaculture. *Aquaculture* 231:361–91.

Noda, H. 1993. Health benefits and nutritional properties of nori. *J. Appl. Phycol.* 5:255–8.

Ohno, M., and Matsuoka, M. 1993. *Undaria* cultivation "Wakame." In: Ohno, M., and Critchley, A. T., eds. *Seaweed Cultivation and Marine Ranching.* Nagai: Kanagawa International Fisheries Training Center and JICA, Yokosuka, Japan, pp. 41–9.

Ohno, M., Nang, H. O., and Hirase, S. 1996. Cultivation and carrageenan yield and quality of *Kappaphycus alvarezii* in the waters of Vietnam. *J. Appl. Phycol.* 8:431–7.

Ohno, M., Nang, H. O., Dinh, D. H., and Triet, V. D. 1995. On the growth of cultivated *Kappaphycus alvarezii* in Vietnam. *Jpn. J. Phycol. (Sorui)* 43:19–22.

Ohno, M., and Matsuoka, M. 1997. Cultivation of the brown alga *Laminaria* "Kombu." In: Critchley, A. T., and Ohno, M., eds. Cultivation and farming of marine plants. ETI World Biodiversity Database CD-ROM Series, ETI Information Services Ltd., and Unesco. Available at: www.eti.uva.nl.

Oliveira, E. C., and Alveal, K. 1990. The mariculture of *Gracilaria* (Rhodophyta) for the production of agar. In: Akatsuka, I., ed. *Introduction to Applied Phycology.* SPB Academic Publishing, The Hague, The Netherlands, pp. 553–64.

Oliveira, E. C., Alveal, K., and Anderson, R. J. 2000. Mariculture of the agar-producing Gracilarioid reef algae. *Rev. Fisher. Sci.* 8(4):345–77.

Parker, H. S. 1974. The culture of the red algal genus *Eucheuma* in the Philippines. *Aquaculture* 3:425–39.

Perez, R., and Braud, J. P. 1978. Possiblite d'une culture industielle de l'algau rouge *Eucheuma spinosum* dans le goife de tadjourah. Sci. Peche, *Bull. Fast. Peches Marit.* 285:1–27.

Provasoli, L. 1968. Media and prospects for the cultivation of marine algae. In: *Cultures and collections of algae.* Watanabe, A. A., Hattori, A., eds. Proceedings of the U.S.-Japan Conference Hakkone, Japanese Society of Plant Physiology, pp. 63–75.

Ren, G. Z., Wang, J. C., and Chen, M. Q. 1984. Cultivation of *Gracilaria* by means of low rafts. In: Bird, C. J., and Ragan, M. A., eds. *Proceedings of the Eleventh International Seaweed Symposium.* Walter de Gruyter, Berlin, pp. 72–6.

Retamales, C. A. 1993. Estudio de un cultivo intermareal de *Gracilaria chilensis* (Rhodophyta, Gigartinales) en Punta Pilluco, Puerto Montt, Chile [thesis]. Institute Professional de Osorno, Osorno, Chile, 115 pp.

Richards-Rajadurai, N. 1990. Production, marketing and trade of seaweeds. In: *Technical Resources Papers, Regional Workshop on the Culture and Utilization of Seaweeds,* vol. 2. Regional Seafarming Development and Demonstration Project. RAS/90/002-FAO/UNDP Seafarming Project, August 1990, Cebu City, Philippines, pp. 149–80.

Rincones, R. E., and Rubio, J. N. 1999. Introduction and commercial cultivation of the red alga *Eucheuma* in Venezuela for the production of phycocolloids. *World Aquaculture* 30(2):57–61.

Robertson, M. 1989. Growing seaweed in Fiji. In: Adams, T., and Foscarini, R., eds. *Proceedings of the Regional Workshop on Seaweed Culture and Marketing.* South Pacific Aquaculture Development Project, Food and Agriculture Organization of the United Nations, Suva, Fiji, pp. 37–41.

Rogers, D. J., and Hori, K. 1993. Marine algal lectins: new developments. *Hydrobiologia* 260/261:589–93.

Russell, D. J. 1982. Introduction of *Eucheuma* to Fanning Atoll, Kiribati, for the purpose of Mariculture. *Micronesia* 18(2):35–44.

Ryther, J. H., DeBoer, J. A., and Lapointe, B. E. 1979. Cultivation of seaweeds for hydrocolloids, waste treatment and biomass for energy conversion. In: Jensen, A., and Stein, J. R., eds. *Proceedings of the Ninth International Seaweed Symposium.* Science Press, Princeton, NJ, pp. 1–6.

Sahoo, D. 2000. *Farming the Ocean: Seaweed Cultivation and Utilization.* Aravali Publishing Corporation, New Delhi, 40 pp.

Sahoo, D., Nivedita, and Debasish. 2001. *Seaweeds of Indian Coast.* APH Publications, New Delhi, 283 pp.

Sahoo, D., and Ohno, M. 2003. Culture of *Kappaphycus alvarezii* in deep seawater and nitrogen enriched medium. *Bull. Mar. Sci. Fish. Kochi Univ.* 22:89–96.

Sahoo, D., Ohno, M., and Masanori, H. 2002. Laboratory, field and deep seawater culture of *Eucheuma serra*—a high lectin yielding red alga. *Algae* 17:127–33.

Sahoo, D., Tang, X., and Yarish, C. 2002. *Porphyra*—the economic seaweed as a new experimental system. *Curr. Sci.* 11(83):1313–16.

Saito, Y. 1975. *Undaria.* In: Tokida, J., and Hirose, T., eds. *Advances of Phycology in Japan.* Dr. W. Junk bv. Publishers, The Hague, pp. 304–21.

Santelices, B. 1996. Seaweed research and utilization in Chile: moving into a new phase. *Hydrobiologia* 326/327:1–14.

Santelices, B., and Ugarte, R. 1987. Production of Chilean *Gracilaria*: Problems and prospectives. In: Ragan, M. A., and Bird, C. J., eds. *Proceedings of the Twelfth International Seaweed Symposium.* Development in Hydrobiology. 41:295–299. *Hydrobiologia* 151/152:295–9.

Santelices, B., and Doty, M. 1989. A review of *Gracilaria* farming. *Aquaculture* 78:98–133.

Serpa-Madrigal, A., Areces, A. J., Cano, M., and Bustamante, G. 1997. Depredacion sobre las carragenofitas comerciales *Kappaphycus alarezii* (Doty) Doty and *K. striatum* (Schmitz) Doty (Rhodophyta: Gigartinales) introducidas en Cuba. *Rev. Invest. Mar.* 18(1):65–9.

Shacklock, P. F., and Doyle, R. W. 1983. Control of epiphytes in seaweed cultures using grazers. *Aquaculture* 31:141–51.

Shang, Y. C. 1976. Economic aspects of *Gracilaria* farming in Taiwan. *Aquaculture* 8:1–7.

Sohn, C. H. 1996. Historical review on seaweed cultivation of Korea. *Algae* 11:357–64.

Sohn, C. H., and Kain, J. M. 1989. *Undaria, Laminaria* and *Enteromorpha* cultivation in Korea. In: Kain, J., Andrews, W., McGregor, B. J., eds. *Aquatic Primary Biomass-Marine Macroalgae: Outdoor Seaweed Cultivation: Proceedings 2nd Workshop of COST 48*, Commission of the European Communities, Brussels, pp. 42–5.

Steentoft, M., Irvine, L. M., and Bird, C. J. 1991. Proposal to conserve the type of *Gracilaria*, nom. cons. as *G. compressa* and its lectotypification (Rhodophyta: Gracilariaceae). *Taxon* 40:663–6.

Tang, X. R., and Fei, X. G. 1999. Artificial cultures from conchocelis to conchosporelings of *Porphyra katadai* var. *hemiphylla. Ocean Limnol. Sci.* 30:180–5.

Troell, M., Halling, C., Neori, A., Chopin, T., Buschmann, A. H., Kautsky, N., and Yarish, C. 2003. Integrated mariculture: asking the right questions. *Aquaculture.* 226: 69–90.

Trono, G. C. 1993. *Eucheuma* and *Kappaphycus*: taxonomy and cultivation. In: Ohno, M., and Critchley, A. T., eds. *Seaweed Cultivation and Marine Ranching*. Nagai: Kanagawa International Fisheries Training Center and JICA, Yokosuka, Japan, pp. 75–88.

Tseng, C. K. 1962. China's kelp growing industry. In: Tseng, C. K., Wu, C. Y., and Fei, X. G., eds. *Kelp Culture*. The Science Press, Beijing, China, pp. 99–112.

Tseng, C. K. 2001. Algal biotechnology and research activities in China. *J. Appl. Phycol.* 13:375–80.

Tseng, C. K., and Wu, C. Y. 1962. *Manual of Cultivation of Haidai* (Laminaria japonica) [in Chinese]. Science Press, Beijing, 186 pp.

Tseng, C. K., Wu, C. Y., and Sun, K. Y. 1957. Effects of temperature on growth and development of kelp sporophytes. *Acta Bot. Sinica* 6:103–30.

Tsujii, K., Ichikawa, T., Matasuura, Y., and Kawamura, M. 1983. Hypercholesteremic effect of taurocyamic or taurine on the cholesterol metabolism in white rats. *Sulfur Amino Acids* 6:239–48.

Ugarte, R., and Santelices, B. 1992. Experimental tank cultivation of *Gracilaria chilensis* in central Chile. *Aquaculture* 101:7–16.

Westermeier, R., Gómez., I., and Rivera, P. 1993. Suspended farming of *Gracilaria chilensis* (Rhodophyta, Gigartinales) at Cariquilda River, Maullin, Chile. *Aquaculture* 113:215–29.

Wu, C. 1998. The seaweed resources of China. In: Critchley, A., and Ohno, M., eds. *Seaweed Resources of the World*. JICA, Yokosuka, Japan, pp. 34–45.

Wu, C. Y., Li, J. J., Xia, E. Z., Peng, Z. S., Tan, S. Z., Li, J., Wen, Z. C., Huang, X. H., Cai, Z. L., and Chen, G. J. 1988. Transplant and artificial cultivation of *Eucheuma striatum* in China. *Oceanol. Limnol. Sin.* 19:410–17.

Wu, C., Renshi, L., Guangheng, L., Wen, Z., Dong, L., Zhang, J., Huang, X., Wei, S., and Lan, G. 1993. Some aspects of the growth of *Gracilaria tenuistipitata* in pond culture. *Hydrobiologia* 260/261:339–43.

Yang, S. S. 1982. Seasonal variation of the quality of agar-agar produced in Taiwan. In: Tsuda, R. T., Chiang, Y. M., eds. *Proceedings of Republic of China—United States Cooperation Science Seminar on Cultivation and Utilization of Economic Algae*. University of Guam Marine Laboratory, Mangilao, Guam, pp. 65–80.

Yarish, C., Chopin, T., Wilkes, R., Mathieson, A. C., Fei, X. G., and Lu, S. 1999. Domestication of nori for northern America: the Asian experience. *Bull. Aquacult. Assoc. Can.* 1:11–17.

Yarish, C., Neefus, C. D., Kraemer, G. P., Chopin, T., Nardi, G., and Curtis, J. 2001. NOAAOAR's National Marine Aquaculture Initiative. Development of an Integrated Recirculating Aquaculture System for Nutrient Bioremediation in Urban Aquaculture. 9/01/01–8/31/04.

Yoshida, T., Notoya, M., Kikuch, N., and Miyata, M. 1997. Present and future on biology of *Porphyra. Nat. Hist. Res. (Special Issue)* 3:5–18.

# CHAPTER 16

# COUNTING CELLS IN CULTURES WITH THE LIGHT MICROSCOPE

ROBERT R. L. GUILLARD
*Bigelow Laboratory for Ocean Sciences*

MICHAEL S. SIERACKI
*Bigelow Laboratory for Ocean Sciences*

## CONTENTS

Key Index Words: Counting Chambers, Digital Images, Epifluorescence Microscopy, Hemocytometer, Palmer-Maloney Chamber, Sedgwick-Rafter Chamber

## 1.0. INTRODUCTION

Counting cells in cultures—by any means—has two principle applications. The first is to estimate the size of the cultured population, expressed as the total number of cells (colonies, occasionally) in the culture as a whole or, more usually, as individuals per unit volume of culture. Although there are many surrogate estimates of population size—biomass or wet or dry weight, chlorophyll content, content of organic nitrogen, phosphorus, or iron—there is a more fundamental aspect to the cell numbers; populations in nature survive or not in the form of individuals, not biomass, and predators eat cells or colonies.

Further, the important concept of the cell quota, meaning the average single cellular content of some constituent (e.g., nitrogen, phosphorus, iron, vitamin $B_{12}$), requires both cell counts and chemical (often radiochemical) determination of the constituent.

The second application of cell counting is in the estimate of the rate of culture augmentation, equivalent to the rate of population increase; often this is expressed as the rate of cell division, because the fundamental process of increase is by the division of one individual cell into two (occasionally more) in regular fashion (see Chapter 18).

## 2.0. COUNTING BY TRANSMITTED LIGHT

The practical aim is to tally all of the cells in a known volume of the cultured material when they are all settled into one plane, or nearly one plane, within the depth of focus of the objective-ocular system of the microscope. It may be necessary to immobilize or stain cells to facilitate observation or to preserve them for counting at a more convenient later time. This can be illustrated by a simple and time-honored method, one also used for preliminary survey of phytoplankton samples (Throndsen 1995, Andersen and Throndsen 2003).

Of the cultured material, place 0.05 mL (= 0.05 cm$^3$ = 50 mm$^3$) onto a slide. (Note that 0.05 mL = 1/20 mL, which is about 1 drop from many pipettes or droppers.) Cover the drop with a 20 × 20-mm-square coverslip (400 mm$^2$ area); the liquid should fill the space under the coverslip without serious overflow. Therefore, each 1 mm$^2$ of coverslip area holds cells from 1/400 × 1/20 mL = 1/8,000 mL = 1.25 × 10$^{-4}$ mL of culture. If $n$ cells are (on average) counted under 1 mm$^2$ of the coverslip, then the culture has $n$ × 8,000 cells per mL. Moreover, the depth of liquid (d) under the coverslip is 0.125 mm because the area (400 mm$^2$) times the depth (d) equals the total volume, 50 mm$^3$; i.e., 50/400 = 0.125 mm = 125 μm. Note that this depth, 0.125 mm, is close to the depth of the chambers of many hemacytometers (0.1 mm).

For hemacytometers exactly 0.1 mm deep, the factor by which $n$ (cells/mm$^2$) is multiplied is exactly 10,000 (10$^4$). For hemacytometers 0.2 mm deep, the factor is half that, 5,000 (10$^4$/2), because the 1-mm square has collected cells from a column of water twice as deep (see details in Guillard 1973).

The area of observation has to be identified in the field of view and measured with a stage micrometer, so that the number of cells per mm$^2$ can be calculated for subsequent computation of the number of cells per unit volume. Although it is possible, with some combinations of ocular and objective lenses, to use the entire ocular field of view for tallying, it is much more convenient and usual to restrict the area of observation (tallying cells) to the center of the field of view with a Whipple disc (Fig. 16.1), a subdivided square inserted into the ocular. Its area must also be measured with a stage micrometer. This is the technique used for all counting chambers that lack impressed rulings on the surface of observation. Hemacytometers have rulings of known sizes impressed on the viewing area and specify the depth of the chambers, so that conversions from area counts to volume concentrations are easily made.

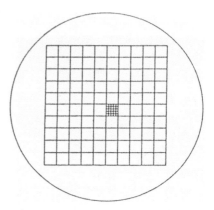

**FIGURE 16.1.** Ocular field with Whipple disc (from Throndsen 1995, Andersen and Throndsen 2003).

### 2.1. General Sampling Methods

In almost all culture studies, sampling is solved by procedural rather than statistical considerations. The culture should be well mixed (rendered homogeneous to the degree thought necessary) and collected in sufficiently accurate pipettes or volumetric measuring devices. Sampling and subsampling uncertainties are of extreme importance in studies of natural populations of planktonic organisms (Venrick 1978a, 1978b), and statistical studies have been numerous and computationally intense, but in culture studies there are only three main occasions in which problems arise.

#### 2.1.1. Sampling Attached Algae

To begin with the most difficult situation, if an alga grows largely attached to the culture vessel walls, a sample can be taken only by scraping the cells loose and then homogenizing the resulting suspension. In practice, this generally means sacrificing that particular culture. The resulting suspension may or may not be countable, depending on the separation of the cells. Some algae grow in suspension, but in such large and irregular clumps that individual cells cannot be distinguished. Sometimes the clumps can be dispersed by brief ultrasonic treatment or by the use of 3 to 5% formalin or a detergent (e.g., Tween, Sigma-Aldrich), followed by agitation or ultrasonic treatment. If not, growth must be estimated by some means other than counting devices. For cell quota studies, nets or filters are used to get a suspension of single cells or small countable clumps. It will be necessary to make the chemical or radiochemical determinations on this same preparation.

### 2.1.2. Sampling Colonial Algae

Colonial growth poses no unmanageable problems, provided the cells can be distinguished for counting and the number of cells per colony does not vary too widely. This problem is considered by Lund et al. (1958). The basic point is that colonies, not individual cells, are the units distributed over the surface of the counting area. Therefore, colonies must be tallied and the average number of cells per colony must be determined separately. In some instances—for example, that of chain-forming diatoms—ultrasonic or other treatment (see previous text) will separate colonies into such small fragments that little accuracy is lost if cells rather than colonies are tallied directly. Reducing the number of cells per colony is generally an advantage. The error involved in counting colonial forms is the product of the errors involved in counting the colonies and in determining the average number of cells per colony, which Lund et al. (1958) mention as generally negligible.

### 2.1.3. Sampling Dividing Cells

Counting dividing cells can be problematic. In the simplest situation (e.g., a flagellate dividing by fission or a centric diatom elongated and about to divide), the individual cell can be counted as two because it has about twice the amount of cell material. There are usually few enough such specimens so that the culture need not be considered to consist of "colonies" of one or two cells. If there are many such instances, it implies synchrony of division, which suggests simply changing the time of sampling. Species that divide by producing several daughter cells (e.g., autospores) in general have to be treated as a mixture of colonies and single cells. It is easier if dealing with a species that produces a definite (fixed) number of autospores per dividing cell and that tends to divide in synchrony under the experimental growth conditions.

For collecting samples from large-volume cultures, the side arm device shown in Figure 16.2 is simple and easy to use. The main culture has to be mixed thoroughly, and repeated samples can be counted to see if they yield the same estimate of population size according to the precision of the counting method employed (see Section 2.3).

Note this caution about vortex-mixing of cultured material: samples are often vortex-mixed in glass or plastic tubes after treatment with either formalin or $I_2KI$ solution (see later text). Cells of some species are broken under particular treatments. Of seven flagellate species preserved in formalin, only the cryptomonad *Chroomonas salina* (Wislouch) Butcher remained unbroken;

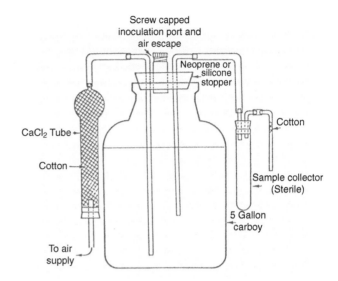

**FIGURE 16.2.** Sample collector for large culture vessel (from Guillard 1985). Suction applied to the sample tube draws the sample; relieving suction drains the siphon into the culture, permitting aseptic change of the sample tube. The sample tube must be above the surface level of the culture.

six diatom species were unbroken, but like most flagellates, the virtually unsilicified *Phaeodactylum tricornutum* Bohlin was also broken (Nelson and Brand 1979). The authors gave no further details, but our experience suggests that the exact nature of the protocol—tube used, vortex intensity and timing, pH of the preservative, especially of formalin—is involved. Vortexed samples of smaller or hyaline species should be examined for fragments by means of phase-contrast microscopy.

### 2.2. Preservation

For counting cultured cells, some version of Lugol's iodine solution is the best preservative, with either acetic acid or sodium acetate added. The acidic solution is best for preserving flagella but destroys coccoliths and other calcareous structures (e.g., plates of the dinoflagellate *Thoracosphaera*); the weakly basic acetate solution will not destroy them (Throndsen 1978). To make the acidic version (Rodhe et al. 1958), dissolve 20 g KI into 200 mL $H_2O$ and add 10 g crystalline $I_2$. Then add 20 mL glacial acetic acid. Filter before storing, and check for precipitates when using. Add to a sample to achieve a light brown (medium tea) color; usually, adding 0.05 mL (1 drop) to a 5-mL sample is adequate (1%). The recipe for the neutral or weakly alkaline solution is given by Utermöhl (1958) as the following: add

20 g KI to 40 mL H$_2$O, then add 10 g I$_2$ (crystalline) and 100 mL H$_2$O. Then add 10 g sodium acetate. Use per the acidic version. Either solution should be filtered when made and after prolonged storage.

Formalin is often available and can be used as an acidic version (some formaldehyde will decompose to formic acid) or neutralized with hexamethyl tetramine; see Throndsen (1978) or Andersen and Throndsen (2003) for preparation of stocks. The usual dose of half-strength commercial formalin is 2% in the sample, yielding 0.4% formaldehyde. Glutaraldehyde can also be used for fixing cells. Add the biological grade to yield a final concentration of approximately 1% (Murphy and Haugen 1985, Johnson and Sieburth 1982). It does not stain cells. The exceedingly volatile and dangerous acrolein (Strickland 1965, p. 138) and osmium tetroxide (OsO$_4$) can be used under a ventilated hood. A short-term fixative useful for observing and counting marine flagellate cells is a solution of 1 g uranyl acetate in 20 mL of seawater and used at great dilution. It immobilizes cells without undue distortion and may leave them intact longer than other preservatives (Andersen and Throndsen 2003).

## 2.3. Statistics of Counting: How Many Cells to Count

It is assumed (Section 2.1) that the sample drawn from the culture adequately represents the culture population, so that the problem is reduced to estimating the density (cells $\cdot$ mL$^{-1}$) of the sample itself to the degree of confidence desired. The concentration of cells in the sample can vary over some seven orders of magnitude, from about 10 cells $\cdot$ mL$^{-1}$ in the case of a starting culture of a large-celled species to over 10$^8$ cells $\cdot$ mL$^{-1}$ for a mature culture of algal cells of bacterial dimensions (e.g., *Ostreococcus*, *Micromonas*, or *Synechococcus*).

In the case of very sparse cultures, it may be difficult to get enough cells to satisfy the statistical requirements for the confidence level desired. The accuracy of an estimate increases only as the square root of the number of items counted; Fig. 16.3 and Table 16.1 show the confidence limits (the expected value [as the average, in this case] and ±2 standard deviations) for different total numbers of cells counted (based on a Poisson distribution). For example, counting all of the cells in a 5-mL sample having 10 cells $\cdot$ mL$^{-1}$ yields 95% confidence level with limits of 37 and 66 cells $\cdot$ mL$^{-1}$, about ±30% of the count (Fig. 16.3). Table 16.2 shows that the level of diminishing returns is quickly reached. To tally even

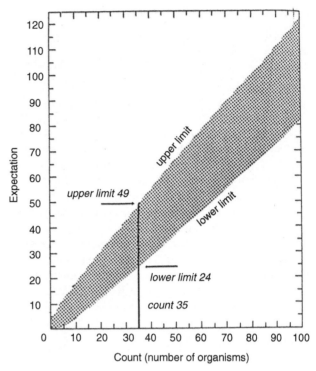

**FIGURE 16.3.** The 0.05 confidence limits for the expectation of a Poisson variable (from Lund et al. 1958).

50 cells total would require use of a 5-mL settling chamber. These chambers or ones of larger volume are available or can be made, but such large volumes of sample are not always available in experimental work. The Sedgwick-Rafter chamber (described later), which holds 1 mL, would yield only an order-of-magnitude estimate, at least at one filling. This, or small settling chambers, can be filled repeatedly, but such subsampling has its own statistical implications, which can be important (see Venrick 1978b).

Counting denser cultures presents a different problem. The number of cells in a viewing area, whether some portion of a Whipple field or of a hemacytometer grid, must not be so large that the observer cannot keep track of them confidently. Hence, choice of the counting device is important for dense cultures also (see Section 2.4.2).

Lund et al. (1958), in considering statistics applicable to counting algal populations, state that for practical purposes the important point is to test for a random distribution of cells or colonies in the counting chamber. This can be done by the χ$^2$ test (Lund et al. 1958). Randomness permits use of standard confidence limits. If the distribution is not random, then it seems reasonable to assume some degree of contagion or clumping of

**TABLE 16.1** Confidence limits for the expectation of a Poisson variable (from Lund et al. 1958).

| Confidence coefficient count | 0.95 (1 in 20) | | 0.99 (1 in 100) | | 0.998 (1 in 500) | | 0.999999 (1 in 1,000,000) | |
|---|---|---|---|---|---|---|---|---|
| | LL | UL | LL | UL | LL | UL | LL | UL |
| 10 | 5 | 18 | 4 | 21 | 3 | 24 | — | 28 |
| 50 | 37 | 66 | 34 | 71 | 31 | 76 | 20 | 88 |
| 100 | 82 | 122 | 77 | 129 | 72 | 135 | 56 | 152 |

LL = lower limit; UL = upper limit.

**TABLE 16.2** Size of count and accuracy obtained; approximate 0.95 confidence limits (from Lund et al. 1958)

| No. of organisms counted | Percentage of count | Range |
|---|---|---|
| 4 | ±100% | 0–8 |
| 16 | ±50% | 8–24 |
| 100 | ±20% | 80–120 |
| 400 | ±10% | 360–440 |
| 1,600 | ±5% | 1,520–1,680 |
| 10,000 | ±2% | 9,800–10,200 |
| 40,000 | ±1% | 39,600–40,400 |

individuals, which may reflect such factors as association of cells after division. Venrick (1978c) illustrates the use of normal probability paper to test a set of observations (e.g., counts of field subdivisions in a hemacytometer) for normality. She also discusses the use of nonparametric methods, which do not require the condition of normality, and refers to sources for this information. These methods are often simple to use and can yield efficiencies >95% if applied to normally distributed data and efficiencies of ca. 60% in other cases. They may be time-savers. The pamphlet by Wilcoxon and Wilcox (1964) illustrates useful methods.

Although it is not possible to determine by eye that a given distribution of cells in a chamber is randomly distributed, it is often possible to tell that it is not. For example, some chain-forming algae will cluster at the edges or near the entry slot of the Palmer-Maloney chamber (see later text), or along the edges of a thin hemacytometer chamber. If an alga extrudes mucilage

and the culture has not been mixed adequately, long strands of cells may be seen in the chamber fields. Cells that are too large for a counting chamber may cluster near the entry port. In these situations, a change of mixing technique, sample treatment, or counting chamber is required. However, with some colonial forms, no treatment or chamber seems to give random distributions, so that all individuals in the chamber must be counted. If the number of individuals is too large, the sample can be diluted in a "to contain" volumetric apparatus without introducing appreciable error, provided pipetting errors are minimized by the use of large-bore pipettes of suitable accuracy. The rejection of a suspect count (a deviant in a replicate series) can be made by the Chauvenet criterion. This is illustrated by Calvin et al. (1949, their Appendix II).

A final precaution should be noted regarding counting chambers. The depths are specified by the manufacturers (and marked on the glass of hemacytometers). Chamber depths are usually correct to a few percent, but a few severe inaccuracies have been found (unpublished observations). On hemacytometers and shallow counting chambers, the depth between the bottom of a coverslip and the bottom surface of the chamber itself should be checked with the calibrated fine-focus adjustment on good-quality microscopes. This depth determines the volume of the liquid counted.

## 2.4. Practical Methods

### 2.4.1. Microscopes and Slide Preparation

A compound microscope of reasonable quality is necessary. A condenser of long working distance to permit Köhler illumination with thick slides is desirable.

Phase-contrast illumination is frequently convenient, and often necessary for observation and as relief from bright-field illumination, thus aiding in preventing fatigue. Objective lenses should include achromatic lenses of 20 to 25×, with numerical aperture (NA) ≥0.5, and of 40 to 45×, with NA ≥0.6, both with the longest working distances available (see next section for counting slide chamber heights). Correct combinations of objectives and oculars permit magnification of at least 500×. Wide-field 12.5 to 15× oculars are commonly used. A magnification changer built into the microscope is convenient but entails care in its use because the dimensions subtended by ocular micrometers and Whipple discs change with changes in magnification. No. 1½ coverglasses (about 0.17 mm thick) of appropriate size can be used for all slides to improve resolution and to reduce the overall working distance over the various chambers. (Thicker covers are satisfactory for use with low power.)

Many counting slides have no marking grids; hence, a grid or other field-limiting device (Lund et al. 1958) is needed in the ocular. The commercially available Whipple disc (Fig. 16.1) provides a square field divided into 100 squares; one of the four central squares is further divided into 25 smaller squares, which are useful for estimating cell sizes. Typically, at a magnification of 125×, the side of a Whipple square subtends ca.700 μm in the image plane; the smallest squares measure ca. 14 μm. The exact area subtended by the Whipple square must be measured with a stage micrometer; in the example given, it is about 0.5 mm² at 125×.

Ethanol (90%), commercial denatured alcohol, isopropanol, or acetone can be used for the necessarily scrupulous final cleaning of counting chambers and coverglasses, following flushing of each sample by distilled water. Clean surgical gauze (cheesecloth) cut into 10-cm squares is used for wiping. Plastic squeeze bottles for water (distilled or deionized) and ethanol (or other solvent), plus a waste receptacle to receive water and ethanol used to flush the counting slides or coverglasses, are necessary, as is a small staining dish to hold coverglasses upright in ethanol.

Large-bore 1-mL pipettes are used for filling Sedgwick-Rafter chambers. Pasteur-type pipettes are selected for their uniformity and large-diameter bore, and their tips may be fire-polished or smoothed by rubbing on emery paper. These are for filling Palmer-Maloney chambers, hemacytometers, and bacteria counters. Mechanical pipettes and certain medicine droppers will also serve this purpose. Settling tubes are used for counting dilute cultures. The best are 20 × 150-mm flat-bottomed screw-capped test tubes or small-

volume graduated cylinders with stoppers. A moist chamber can be used to prevent evaporation from altering cell distributions while cells are settling in a counting slide. Put the slide into a petri dish or other closed container on a glass triangle or other support to hold it off the bottom, and add a little water to maintain a saturated environment. Commercial petri plates with bottoms divided into four sectors are best suited for this purpose. A hand tally device is sometimes useful when counting, especially when the areas being enumerated have many cells (20 or more for many observers).

### 2.4.2. Choice of Counting Device

This depends upon culture density, the size and shape of the cells or colonies being counted, and the presence and amount of extracellular threads, sheaths, or dissolved mucilage, which can influence the filling of the counting chamber. With the apparatus properly chosen, experience shows that cleanliness of the counting chamber and coverslip is the most important factor in attaining proper distribution of the cells or colonies. Care should be taken to distribute the count over a wide area of the counting apparatus or to count more than one complete chamber. Table 16.3 lists various commercially available slide-type counting devices, together with suggested applications for use under varying circumstances. Box 16.1 gives essential dimensions of each chamber, rulings if present, and magnification of microscope objectives to be used. Microscope oculars are selected to give an appropriate field of view and total magnification.

**TABLE 16.3** Commercial counting devices. The cell sizes and the culture densities are those handled easily.

| Counting device | Cell size (μm) | Culture densities (cells/mL) |
|---|---|---|
| Sedgwick-Rafter | 50–500 | 30–$10^4$ |
| Palmer-Maloney | 5–150 | $10^2$–$10^5$ |
| Speirs-Levy (0.2-mm depth) | 5–75 | $10^4$–$10^6$ |
| Hemacytometer (0.2-mm depth) | 5–75 | $10^4$–$10^6$ |
| Hemacytometer (0.1-mm depth) | 2–30 | $10^4$–$10^7$ |
| Petroff-Hausser | <1–5 | $10^6$–$10^8$ |

## BOX 16.1 Microscope Counting Devices

### The Sedgwick-Rafter Counting Slide

The chamber is without rulings and is rectangular (50 × 20 mm), 1 mm deep, of area 1,000 mm$^2$ and volume 1.0 mL (Fig. 16.4a). With no. 1$^1$/$_2$ coverslips, most ×20 N.A. 0.5 objectives can be used, permitting magnification to ×500. The largest phytoplankters can be held in this slide, and many species as small as 10 μm can be recognized in it. It is best suited to large and relatively scarce organisms, which can be detected in the concentrate at just a few per milliliter. A concentration of 10$^4$ cells/mL in the concentrate yields *ca.* 5 cells/Whipple field at ×125 magnification.

### The Palmer-Maloney Slide

The chamber is without rulings and is circular, of diameter 17.9 mm, depth 400 μm, area 250 mm$^2$, and volume 0.1 mL (Fig. 16.4b). It has a loading channel (slot) on each side. High-dry objectives can be used. Cells (of some species) as large as 150 μm will enter and be reasonably well-distributed in this chamber. Even very small phytoplankters can be detected in it. Species in the concentrate at more than 10/mL should at least be detected; 10$^4$ cells/mL yield *ca.* 2 cells/Whipple field at ×125 magnification.

### Hemacytometer, 0.2 mm Deep

The Speirs-Levy eosinophyll counter has four separate chambers, each having 10 squares that are 1 mm on a side, arranged in two rows of five squares each (Fig. 16.5a). The squares have a modified Fuchs-Rosenthal ruling (Fig. 16.5e); each 1-mm square is divided into 16 250-μm squares. Total volume in all 40 1-mm squares is 0.008 mL; thus, a sample having 10$^4$ cells · mL$^{-1}$ will yield *ca.* 80 cells total (two cells per 1-mm square). Cell concentrations in the range of 5 × 10$^3$ · mL$^{-1}$ can at least be estimated and certainly detected. Objectives of 20× can be used with thin coverslips.

There are also two chamber hemacytometers 0.2 mm deep (Fig. 16.5b). Each chamber consists of 16 squares (1 mm) on a side that are further divided into 16 squares that are 250 μm on a side (Fig. 16.5e)—the Fuchs-Rosenthal ruling. The 32 1-mm squares hold a total of 0.0064 mL of sample. A sample of 10$^4$ cells · mL$^{-1}$ will yield *ca.* 64 cells total (two per 1-mm square).

Note that some manufacturers supply slides 0.2 mm deep, but with the Neubauer ruling (Fig. 16.5f) as described next. These have only 18 1-mm squares in all, so that a sample of 10$^4$ cells · mL$^{-1}$ yields a total of only 36 cells. Thus, they are less convenient at low cell densities. With all hemacytometers of depth 0.2 mm, algal species larger than *ca.* 75 μm will seldom distribute themselves well, and long, thin species or those forming long colonies will often accumulate near the entry slit or at the chamber edges.

### Hemacytometer with Chambers 0.1 mm Deep

This has two chambers each with a tic-tac-toe arrangement of nine 1-mm squares having several layers of subdivision, including 250-μm squares, 250 × 200-μm rectangles, 200-μm squares, and 50-μm squares (Fig. 16.5c). This is the Neubauer ruling (Fig. 16.5f). The total volume in both ruled chambers (18 squares) is 0.0018 mL; thus, a sample with 10$^4$ cells · mL$^{-1}$ yields *ca.* one cell per 1-mm square, 18 cells in all. Counting is best done with 10× or 20× objectives, but standard high dry objectives (40–45×) can be used with no. 1$^1$/$_2$ coverslips.

### Petroff-Hausser Bacteria Counting Slide

This slide has but one chamber (Fig. 16.5d), 0.02 mm deep, with improved Neubauer ruling (Fig. 16.5f). High dry and oil immersion objectives can be used. Total volume of the ruled area is 0.00018 mL; thus, a sample with 5 × 10$^4$ cells · mL$^{-1}$ yields *ca.* one cell per 1-mm square. A sample with 10$^6$ cells · mL$^{-1}$ can be tallied, though with some tedium, because the 20 cells (average) within each 1-mm square have to be looked for carefully. For dense cultures, use the central 400 squares of 50-μm sides. A culture of 2 × 10$^7$ cells · mL$^{-1}$ yields one cell per 50-μm square (average).

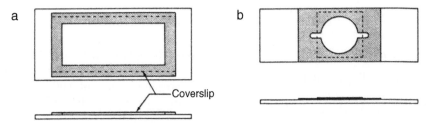

**FIGURE 16.4.** Sedgwick-Rafter chamber **(a)** and Palmer-Mahoney chamber **(b)** (from Guillard 1978).

For culture densities too low for counting in any of the slide-type devices (Table 16.3), the best technique is use of separable sedimentation cylinders (combined plate chambers) with the inverted microscope (Hasle 1978). Alternatively, the bottom of the separable chamber can be scanned with an upright microscope with use of low-power objectives or water-immersion "dipping" lenses (Guillard 1978). Throndsen (1995) describes homemade settling chambers. In the absence of an inverted microscope or separable chambers, a sample can be concentrated as follows for counting in one of the devices described in Table 16.3. Use the methods and precautions described previously to transfer a subsample to a cylinder or tube. Allow the sample to settle at least 1 hour for each 10-mm height of the sample in the cylinder (in practice, usually overnight). With tubing, attach a Pasteur-type pipette (with tip bent upwards) to a reservoir (small suction flask) and then to an aspirator. Gently aspirate 75 to 90% of the supernatant fluid, thus concentrating the sample fourfold to 10-fold. The reservoir permits both recovery of a sample lost by error and a check to be certain all cells have settled. Alternatively, a siphon can be used for very slow concentration of a sample by gravity flow. Next, homogenize the cells in the remaining liquid and continue with the counting procedure, using devices listed in Table 16.3. The concentration factor is the initial volume/final volume.

For large-celled or sparse cultures, the two counting chambers commonly used for natural phytoplankton samples serve well (Box 16.1; Fig. 16.4a,b). For denser cultures of sufficiently small size, the use of hemacytometers having chambers 0.2 millimeters deep and ruled areas is indicated (Box 16.1, Fig. 16.5a,b,e). For dense cultures of small cells (to ca. 4 μm), the hemacytometer of 0.1 millimeters depth (Box 16.1; Fig. 16.5c) with Neubauer or Improved Neubauer ruling (Fig. 16.5f) is standard. Hemacytometers made on thin slides for use with phase contrast are recommended. Otherwise, special long-working-length condensers are necessary for phase contrast and helpful to achieve Köhler

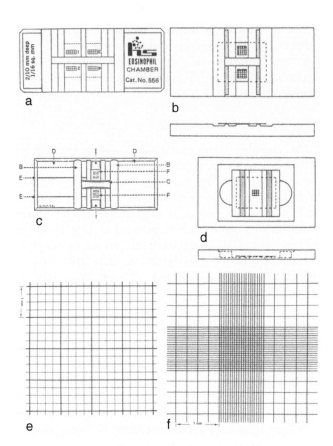

**FIGURE 16.5.** Counting chambers. **(a, b)**: Hemacytometer (blood-counting) type. Each of the 40 (or 32) small squares is 1 mm on a side and 0.2 mm deep. Each 1-mm square has the Fuchs-Rosenthal subdivision pattern (shown in **e**). **(a)** Speirs-Levy eosinophil counting slide. **(b)** Hemacytometer with Fuchs-Rosenthal pattern (from Hausser Scientific Co. leaflet). **(c)** Hemacytometer 0.1 mm deep with Improved Neubauer ruling (shown in **f**) (from A. H. Thomas leaflet). **(d)** Petroff-Hausser slide with Improved Neubauer ruling (shown in **f**) (from Guillard 1978). **(e)** Fuchs-Rosenthal ruling. **(f)** Improved Neubauer ruling. Division of the ruling into nine 1-mm squares is by double or triple lines, not shown in the figure.

illumination. The Petroff-Hausser bacteria-counting slide (Box 16.1, Fig. 16.5d) is the only one well suited to counting organisms of bacterial dimensions (ca. 1 μm) for two reasons. The first is to facilitate observations; the shallow chamber (0.02 mm) keeps the cells in the depth of focus of the microscope somewhat better. The second is that this shallowness of the chamber (one-fifth of that of a standard hemacytometer) yields one-fifth the number of cells per unit area, so that dense cultures can be tallied in the 50-μm central squares of the Neubauer ruling (Fig. 16.5f) with greater ease. Brownian motion moves the cells about significantly. The slide must be cleaned and loaded very carefully.

Occasionally it is necessary to dilute a culture for counting. If it is too dense, take a sample by pipette and dilute in a volumetric apparatus, using particle-free water or growth medium. Dilution may also be indicated for certain cultures of non–randomly distributed species. Some of these form clumps, whereas others consist of long colonies of cells (e.g., *Skeletonema*, *Chaetoceros*) that cross the ocular fields used for tallying (e.g., Whipple disc fields). In these cases, the whole chamber must be counted. The dilution factor is the final volume/initial volume.

### 2.4.3. Manipulations, Cleaning, and Filling

Wash grease off hands before working with counting chambers. A surprisingly small amount of grease can make it almost impossible to load the chamber uniformly, and the precision and accuracy of counting thus drop significantly. The chambers should be washed initially with mild detergent, water, and ethanol (or acetone) and then dried with clean cheesecloth (see Guillard 1973). Between samples the chamber need only be held over the waste container and flushed with distilled water and then ethanol or acetone and wiped dry. A square of cheesecloth three or four layers thick can be used perhaps 10 times and then should be discarded. The coverglass should be cleaned and dried the same way and then set on the counting chamber or rested on a pair of clean glass or metal rods. Examine the empty chamber occasionally to see that particles and organisms have been removed by the cleaning process.

For filling Sedgwick-Rafter slide chambers (Fig. 16.4a), set a coverglass diagonally across the chamber, leaving a space open at opposite corners. Deliver 1.0 mL of concentrate to the chamber through one opening with a large-bore pipette. Then slide the coverglass into position to seal the chamber. If the coverglass interferes

with dispersal of large species, it may be best not to use it, although visibility, especially near the edges of the chamber, will be lowered in quality and evaporation will alter the distribution of specimens (McAlice 1971).

For filling a Palmer-Maloney chamber (Fig. 16.4b), cover the chamber with a round or square 22-mm no. 1½ coverslip, making secure contact with the glass or plastic chamber ring. Tilt the slide slightly (to right or left) and add 0.1 mL of suspension via the lower of the two entry channels with a large-bore Pasteur pipette. A small piece of paraffin paper over each loading channel will prevent evaporation and does not interfere with observation.

For filling a hemacytometer (Fig. 16.5), place the coverglass in position, seating it firmly on the support pillars. Fill a smooth-tipped pipette of adequate bore diameter with algal suspension and touch off any liquid remaining on the tip. Hold the pipette at an angle of 45° for small-bore pipettes, to about half that for large-diameter ones. Place the tip on the pipette next to the entry slit or groove of the hemacytometer chamber. Release the liquid and almost simultaneously remove the pipette from contact so that liquid flows quickly and evenly into the chamber, barely filling it, with no overflow or even bulging of the liquid surface into the canals surrounding the chamber. Refill if this is not achieved. The Petroff-Hauser slide is filled in essentially the same way, with a smaller-bore pipette. It may help to remove the glass slide from its plastic holder.

### 2.4.4. Settling Cells and Counting

Allow the algal cells to settle for 3 to 5 minutes in the counting device before counting. This is chiefly important when cells are small and contained in relatively deep chambers. It is common to use two counting devices, allowing the organisms to settle in one while the other is being counted. Use a moist chamber (see previous text) for settling.

Before counting, examine the contents of the chamber under low magnification to detect any obviously unsatisfactory distribution patterns (see previous discussion). (High power of a stereomicroscope will do.) On the basis of this examination, decide on a pattern for distributing the count over the area of the chamber used, as follows. In the case of ruled chambers, decide how many areas of known size (e.g., 1-mm squares or 250-μm squares) must be surveyed to yield the total number needed for the confidence level desired. The count should be spread over all the chambers of hemacytometers and in a prearranged pattern within each chamber. (This may yield more cells than are needed, but keep all.) When using unruled devices and the Whipple disc, either the whole chamber can be counted, or randomly selected fields can be tallied to get the desired count. If the latter, do not look through the microscope while changing to another field; the eye will stop the hand when there is an attractive field to count, spoiling the randomness of choice. In either case, record the separate counts for any statistical treatment needed.

*Tallying cells.* The aim is a satisfactory estimate of the number of units—cells or colonies—per square mm of area in the counting slide used. This determines the density, as cells · mL$^{-1}$ (= cell · cm$^{-3}$), in the sample contained in the chamber of the counting slide. This density in turn is converted to density in the original culture sampled by use of any concentration or dilution factor that may have been introduced along the way. Some specific protocols are described.

To count the whole Sedgwick-Rafter chamber, position it so the Whipple square is in one corner of the counting chamber, e.g., the upper left as seen through the microscope. Slowly move the stage horizontally (left to right), tallying the organisms as they pass the leading boundary (right side) of the square, until the Whipple square is in the upper-right corner. (Count the organisms that are cut by the lower boundary; they will be ignored on the next sweep, when they will lie on the upper boundary). Move the stage vertically so that the Whipple square has (apparently) moved down by its own width. (A particle or mark on the glass of the chamber, on or near the bottom of the Whipple square, can be used as an index.). Then sweep horizontally from right to left, again tallying organisms. Ignore organisms that lie on the upper boundary of the square. Repeat until the whole chamber has been covered. The total count is the algal density in the chamber as either cells/mL or colonies/mL.

If the number of sweeps required to cover the chamber is known, then the number of algae tallied can be reduced. When this method is indicated, count only those in every other sweep or some other fraction. The concentration of organisms (cells or colonies per mL) in the sample placed in the chamber will then be the number of individuals counted, multiplied by the reciprocal of the fraction of the chamber scanned.

To count using random Whipple fields, decide first on a system for selecting the positions to be counted. (See McAlice [1971] for detailed discussion of the statistics of sampling with the Sedgwick-Rafter chamber.) Count every organism within the boundaries of the square and every other organism cut by a boundary. The area of the Sedgwick-Rafter chamber is 1,000 mm$^2$; the chamber holds 1.0 mL. The measured area of the

Whipple field is $A$ mm$^2$. Obtain the average number of cells (or colonies) per Whipple field, $N$. The density ($d$) of the algal suspension in the chamber (cells or colonies per mL) will be

$$d = \text{cells/field} \times \text{Whipple fields/}$$
$$\text{chamber} \times \text{chambers/mL} \quad (1)$$

$$d = N \times (\text{area of Sedgwick-Rafter chamber})/$$
$$(\text{area of Whipple field}) \times 1/1.0 \quad (2)$$

$$d = N \times 1,000/A \quad (3)$$

If the Whipple disc area is ca. 0.49 mm$^2$, Equation 3 becomes $d = 2040 \times N$ per mL.

For counting with a Palmer-Maloney chamber, see the general remarks and directions for the Sedgwick-Rafter device (preceding). There are two points of difference: the Palmer-Maloney chamber is round, and thus the sweeps must begin at the top or bottom inner tangent (there is no upper-right-hand corner), and the succeeding sweeps differ in length. It is therefore necessary to estimate the fraction of the area of the chamber that will be surveyed if alternate sweeps (or some other fraction) are tallied.

Assume that fields are counted, that there are $M$ organisms per field, and that the measured area of the Whipple field is $A$ mm$^2$. The area of the Palmer-Maloney chamber is 250 mm$^2$, and it holds 0.1 mL. Then the density ($d$) of the suspension in the chamber is

$$d = \text{number of cells/field} \times \text{number of}$$
$$\text{Whipple fields/chamber} \times \text{chambers/mL} \quad (4)$$

$$d = M \times 250/A \times 1/0.1 \quad (5)$$

$$d = M \times 2,500/A \quad (6)$$

For $A = 0.49$ mm$^2$, as before, $d = 5,100 \times M$ (cells or colonies) per mL.

It sometimes appears that the algal populations along the long edges of the Sedgwick-Rafter chamber or around the periphery of the Palmer-Mahoney cell are smaller than nearer the center. This edge effect is perhaps one or two Whipple fields wide. Prudence suggests that those areas be counted in approximate proportion to their area, relative to that of the entire chamber.

All hemacytometers are divided into squares, 1 mm per side, numbering 40, 32, 18, or just 9. The process is to determine the average number of organisms per 1-mm square, which can be done by counting cells in all the squares present or in selected ones, or in any fraction of the area of selected squares if the cell suspension is dense. Spread the total accumulated count desired over all chambers of the hemacytometers (4, 2, or 1) and at least two fillings of the slide (depending on density of the sample).

For the Speirs-Levy slide or others with chambers that are 0.2 mm deep, if the computed average number of cells per 1-mm square is $P$, then the density, $d$, of the suspension in the chamber is

$$d = 5 \times 10^3 \times P \text{ cells} \cdot \text{mL}^{-1} \qquad (7)$$

For hemacytometers 0.1 mm deep (most frequently used), if the computed average number of cells per 1-mm square is $Q$, then the density, $d$, of suspension in the chamber is

$$d = 10^4 \times Q \text{ cells} \cdot \text{mL}^{-1} \qquad (8)$$

In the Petroff-Hauser chamber it may not be possible to count all the cells in a 1-mm square or even in 1/20 of a square (a 250-μm square in the four corners of the array of nine). If it is possible and the average is $S$ cells per 1-mm square, then the density, $d$, in cells $\cdot$ mL$^{-1}$ in the suspension is

$$d = 5 \times 10^4 \times S \text{ cells} \cdot \text{mL}^{-1} \qquad (9)$$

For dense cultures it is usual to tally the number of cells ($V$) per 50-μm square (the smallest in the center of the tic-tac-toe array). Then the density, $d$, is $d = 2 \times 10^7 \times V$ cells $\cdot$ mL$^{-1}$ in the sample in the chamber. Concentration or dilution factors should be taken into account in all cases.

## 3.0. Counting Cells by Epifluorescence Microscopy

For picoalgae it may be convenient to count cells by epifluorescence microscopy. For some picoalgae in xenic cultures, the algal cells cannot easily be distinguished from bacteria by transmitted light microscopy. In these cases epifluorescence microscopy will provide a clear distinction based on the pigment autofluorescence of the phototroph. The very dim autofluorescence of some phototrophic prokaryotes may not be easily visible to the eye and may fade rapidly under epifluorescence excitation. Examples include the divinyl chlorophylls of prochlorophytes, and the bacteriochlorophyll of aerobic anoxygenic photoheterotrophs (Kolber et al. 2001). Many of these cells, including pico-eukaryotes, coccoid cyanobacteria, and prochlorophytes, can readily be counted by flow cytometry (see Chapter 17). Alterna-

tively, a high-sensitivity fluorescence camera can be used to visualize and count these cells (see Section 3.3).

Detailed methods for preparing and counting natural plankton samples by epifluorescence microscopy are described elsewhere (e.g., MacIsaac and Stockner 1993) and are easily adapted to counting samples of algal cells in culture. Such samples of algal cultures are much simpler than natural samples because the cultured cell type should dominate and the contaminants are usually distinctive (e.g., bacteria, protozoans). Epifluorescence microscopy could be useful in identifying and enumerating such contaminants. If the goal is simply to count the number of algal cells in a culture, then no fluorochrome is necessary because the algal cell autofluorescence should be sufficient to identify the cells. If the sample is preserved with glutaraldehyde, a slight green fluorescence will result, giving a faint outline to all cells and including any heterotrophic protists present. Another option is to use a nucleic acid stain like DAPI to visualize the cells. However, in a xenic culture this can be a problem because of a high background of fluorescing bacteria in the preparation. Concentrating the sample on a polycarbonate filter means that very low cell densities can be counted, provided there is enough sample volume available. One major advantage of epifluorescence microscopy using filters is its sensitivity. This is in terms of both the minimum number of cells that can be counted with confidence and the ability to detect small cells.

### 3.1. Sample Preparation

Filtering the sample onto a black polycarbonate filter with small pores (e.g., 0.2 or 0.4 μm) provides the best visual contrast for detecting the weak fluorescence of picoalgal cells. Fragile cells are more likely to be intact if the sample is preserved with an aldehyde fixative prior to filtering, e.g., with para-formaldehyde or glutaraldehyde at a final concentration of 0.5 to 1% (v/v). The fixed sample should be refrigerated for at least 1 hour before filtering. A 25-mm filter is placed on a glass frit filter holder, with a backing filter of cellulose acetate placed underneath to help with even cell dispersion on the filter. The volume of sample filtered depends upon the concentration of cells, so a bit of trial and error may be required to get an optimal number of cells (ca. 30 per counting field). If a sample volume of less than 2 mL is required, it is necessary to dilute the sample with several milliliters of filtered seawater or distilled water in the filter funnel so that the cells are evenly dispersed on the filter. If a fluorescent stain is to be used (see previous text), it is often convenient to stain the sample in the filter funnel. For larger volumes of sample (>5 mL) we typically filter down to a 5-mL mark on the funnel and then add stain, reducing the amount of fluorochrome used per sample. The sample is then filtered down and the filter (polycarbonate only) is removed before it can dry. The filter is placed on a prelabeled microscope slide. Slightly moistening the slide with breath helps the filter lay flat on the slide. A drop of immersion oil is placed on top of the filter and a coverslip is added. The slide is then placed horizontally in the dark, and the oil is allowed to spread. It can be examined immediately under the microscope or stored in the refrigerator for up to 24 hours or in the freezer (–20°C) indefinitely.

## 3.2. Microscopy and Counting

A standard epifluorescence microscope outfitted with a 50-W mercury lamp and a proper filter set for exciting chlorophyll is necessary for counting picoalgae. The microscope should be located in a room that can be darkened so that the operator's eyes can adapt to low light. Chlorophyll is excited with blue light and emits fluorescence in the red. This is a large Stokes shift (spectral range between excitation and emission) and requires an appropriate filter set. Two filter sets for two microscopes that work for chlorophyll are shown in Table 16.4. Long-pass barrier filters for emission, rather than a band-pass, are needed to visualize the red chlorophyll fluorescence. For picoalgae a high magnification (60–100×) oil immersion objective is necessary. Fluorescence microscopy requires objectives with high light-gathering power (i.e., numerical aperture [NA]); so

often a high-NA 60× or 63× objective is brighter than standard (lower NA) 100× objectives and yields a better image of small, dimly fluorescing cells.

Because this method depends on an even dispersion of cells on the filter, it is necessary to check this by scanning the filter at low power before counting. Sometimes filters are defective, and the cells are not evenly distributed across the filter. This is visually obvious when viewed at low power (10× or 20× objective). As cells are exposed to the high-intensity excitation light being focused on them through the objective, the chlorophyll autofluorescence will fade. There are several things that can minimize the exposure of the cells to excitation light and thereby minimize photofading. Count each field quickly after it is brought into view and focused; close the shutter on the excitation light path when not viewing the field; close the diaphragm down on the excitation light source so that only the field of view counted is illuminated.

Typically cells are counted in microscope fields until at least 200 cells and seven fields have been counted. Ideally, 7 to 10 fields are counted with 25 to 30 cells per field. The discussions of counting statistics (see above) given by Venrick (1978) and Andersen and Throndsen (2003) apply to epifluorescence counting as well as bright-field, inverted microscopy. Kirchman et al. (1993) conducted a cost-benefit analysis of variance for bacteria counting that yielded the criteria mentioned above. Randomly chosen whole fields or subfields defined by an ocular grid (e.g., a Whipple grid, Fig. 16.1) can be counted. Using a grid has the advantage of making visible the cells on the edges of the counting field. A good way to deal with cells on the edge is to count all the cells within the grid and those touching two of the edges of the counting square (e.g., top and left) but not the other two.

To calculate cells $\cdot$ mL$^{-1}$ of original sample, determine the mean cells counted per field ($M$) and apply a conversion factor ($C_f$) that is equivalent to the number of field areas on the whole filter. Cells $\cdot$ mL$^{-1}$ ($N$) is then calculated by multiplying the mean by the conversion factor and dividing by the volume filtered ($V$) in mL:

$$N = (C_f \times M)/V \qquad (10)$$

The conversion factor is the ratio of the area of the filter ($A_R$) to the area of the microscope field ($A_f$), or

$$C_f = A_R/A_f \qquad (11)$$

This factor is most easily determined by measuring the diameters of the filter and the field of view. The filter diameter, $d_R$, can be determined by measuring the inside

---

**TABLE 16.4** Spectral bands and cutoffs for the optical filters used on two different epifluorescence microscopes for visualizing chlorophyll fluorescence.

| Microscope | Filter set part number | Exciter bandpass (nm) | Dichroic beam splitter (nm) | Emitter long-pass (nm) |
|---|---|---|---|---|
| Zeiss Axioskop | 487709 | 450–490 | 510 | 520 |
| Olympus IX-70 | U-M566 | 470–490 | 500 | 515 |

diameter of the filter funnel (often ca. 17 mm). The diameter of the field of view, $d_f$, is determined with a stage micrometer and will be different for each objective used. If the field area is a circle, then $A_f = \pi d_f^2/4$, and $A_R = \pi d_R^2/4$, so the expression $C_f = A_R/A_f$ reduces to the simpler calculation of the ratio of the two diameters squared, after they have been converted to common units (μm):

$$C_f = d_R^2 / d_f^2 \qquad (12)$$

If the field area is square (such as a Whipple grid), the expression $C_f = A_R/A_f$ should be used.

## 3.3. Digital Imaging Cytometry

For cells with very dim fluorescence, it may be necessary to use a sensitive camera to image their autofluorescence. Here we consider basic issues related to using digital imaging methods for counting samples; details of digital imaging cytometry of microbes can be found in Wilkinson and Schut (1998). A standard video camera is not sensitive enough to image fluorescence, because of the fast video frame rate (30 frames · s$^{-1}$) and the high noise produced when the gain is increased. Even relatively bright fluorescence is not bright enough for these cameras. A slow-scan camera, which can integrate light over longer periods (seconds) and is cooled to reduce dark current noise, is ideal for imaging fluorescence. These cameras can produce grayscale or color images. There are several companies that manufacture such cameras, and their cost is ca. 10 to 20 times higher than that of consumer-grade video cameras. Slow-scan cameras are usually combined with a computer interface and software for basic digital image acquisition and analysis. Another advantage of digitizing fluorescence images is that cells are not exposed to the excitation light for very long, so photofading is minimized. With this software it is not difficult to automatically count cells in images, particularly when they are all similar in size, shape, and fluorescence, as would be the case with cultures. When mixed natural assemblages are imaged, there is a wide variation in cell brightness. This precludes the use of simple brightness thresholds to separate cells from background in these digital images and requires more sophisticated image analysis algorithms (Sieracki et al. 1989, Sieracki and Viles 1998). The method for digitally separating the objects of interest from the background is referred to as segmentation. Culture cell images are rather uniform, so simple threshold segmentation should be sufficient. Different

software packages will handle this process differently, but most systems now have the capability of counting segments in an image after a simple threshold has been applied. Another option is to simply count the cells visually on the computer monitor. This can be done after a series of images have been acquired and stored on the computer.

### 3.3.1. Calculations

The determination of cells per milliliter with imaging cytometry is done essentially the same way as with visual microscopy. The field area in this case is the area visualized by the camera. This area can be determined by acquiring an image of the stage micrometer. With a digital system the pixels per micrometer can be counted. Thus the length and width of the image can be converted from pixels to μm. Similarly, the field area is the length times the width, converted from pixels to μm. As with the Whipple grid, cells on the edge of only two sides of the image can be counted to prevent a bias in the count.

## 4.0. REFERENCES

Andersen, P., and Throndsen, J. 2003. Estimating cell numbers. In: Hallegraeff, G. M., Anderson, D. M., and Cembella, A. D., eds. *Manual on Harmful Marine Microalgae.* UNESCO Publishing, Paris, pp. 99–129.

Calvin, M., Heidelberger, C., Reid, J. C., Tolbert, B. M., and Yankwick, P. F. 1949. *Isotopic Carbon. Techniques in Its Measurement and Chemical Manipulation.* John Wiley and Sons, Inc., New York. 376 pp.

Guillard, R. R. L. 1973. Division rates. In: Stein, J. R., ed. *Handbook of Phycological Methods: Culture Methods and Growth Measurements.* Cambridge University Press, Cambridge, pp. 289–311.

Guillard, R. R. L. 1983. Culture of phytoplankton for feeding marine invertebrate animals. In: Berg, J. C. J., ed. *Culture of Marine Invertebrates: Selected Readings.* Hutchinson Ross Publishing Co., Stroudsberg, Pennsylvania, pp. 108–32.

Hasle, G. R. 1978. The inverted microscope method. In: Sournia, A., ed. *Phytoplankton Manual, Monographs on Oceanographic Methodology 6.* UNESCO, Paris, pp. 88–96.

Johnson, P. W., and Sieburth, J. M. 1982. In-situ morphology and occurrence of eucaryotic phototrophs of bacterial size in the picoplankton of estuarine and oceanic waters. *J. Phycol.* 18:318–27.

Kirchman, D. L. 1993. Statistical analysis of direct counts of microbial abundance. In: Kemp, P. F., Sherr, B. F., Sherr, E. B., and Cole, J. J., eds. *Handbook of Methods in Aquatic Microbial Ecology.* Lewis Publishers, Boca Raton, Florida, pp. 117–9.

Kolber, Z. S., Plumley, F. G., Lang, A. S., Beatty, J. T., Blankenship, R. E., Dover, C. L. V., Vertriani, C., Kolbizek, M., Rathgeber, C., and Falkowski, P. G. 2001. Contribution of aerobic photoheterotrophic bacteria to the carbon cycle in the ocean. *Science.* 292:2492–5.

Lund, J. W. G., Kipling, C., and LeCren, E. D. 1958. The inverted microscope method of estimating algal numbers and the statistical basis of estimations of counting. *Hydrobiologia* 11:143–70.

MacIsaac, E. A., and Stockner, J. G. 1993. Enumeration of phototrophic picoplankton by autofluorescence microscopy. In: Kemp, P. F., Sherr, B. F., Sherr, E. B., and Cole, J. J., eds. *Handbook of Methods in Aquatic Microbial Ecology.* Lewis Publishers, Boca Raton, Florida, pp. 187–97.

McAlice, B. J. 1971. Phytoplankton sampling with the Sedgwick-Rafter cell. *Limnol. Oceanogr.* 11:19–28.

Murphy, L. S., and Haugen, E. M. 1985. The distribution and abundance of phototropic ultraplankton in the North Atlantic. *Limnol. Oceanogr.* 30:47–58.

Nelson, D. M., and Brand, L. E. 1979. Cell division periodicity in 13 species of marine phytoplankton on a light:dark cycle. *J. Phycol.* 15:67–75.

Rodhe, W., Vollenweider, R. A., and Nauwerck A. 1958. The primary production and stranding crop of phytoplankton. In: Buzzati-Traverso, A. A., ed. *Perspectives in Marine Biology.* University of California Press, San Francisco, pp. 299–322.

Sieracki, M. E., Reichenbach, S. E., and Webb, K. L. 1989. Evaluation of automated threshold selection methods for accurately sizing microscopic fluorescent cells by image analysis. *Appl. Environ. Microbiol.* 55:2762–72.

Sieracki, M. E., and Viles, C. L. 1998. Enumeration and sizing of micro-organisms using digital image analysis. In: Wilkinson, M. H. F., and Schut, F., eds. *Digital Image Analysis of Microbes.* John Wiley & Sons, New York, pp. 175–98.

Strickland, J. D. H. 1965. Phytoplankton and marine primary production. *Ann. Rev. Microbiol.* 19:127–62.

Throndsen, J. 1978. Preservation and storage. In: Sournia, A. A., ed. *Phytoplankton Manual: Monographs on Oceanographic Methodology 6.* UNESCO, Paris, pp. 69–74.

Throndsen, J. 1995. Estimating cell numbers. In: Hallegraeff, G. M., Anderson, D. M., and Cembella, A. D., eds. *Manual on Harmful Marine Microalgae.* UNESCO, Paris, pp. 63–80.

Utermöhl, I. 1958. Zur Vervolkommnung der quantitativen Phytoplankton-Methodik. *Mitt. Int. Ver. Theor. Angew. Limnol.* 9:1–38.

Venrick, E. L. 1978a. Sampling strategies. In: Sournia, A., ed. *Phytoplankton Manual: Monographs on Oceanographic Methodology 6.* UNESCO, Paris, pp. 7–16.

Venrick, E. L. 1978b. The implications of subsampling. In: Sournia A., ed. *Phytoplankton Manual: Monographs on Oceanographic Methodology 6.* UNESCO, Paris, pp. 75–87.

Venrick, E. L. 1978c. Statistical considerations. In: Sournia, A., ed. *Phytoplankton Manual: Monographs on Oceanographic Methodology 6.* UNESCO, Paris, pp. 238–50.

Wilcoxon, F., and Wilcox, R. A. 1964. *Some Rapid Approximate Statistical Procedures.* Lederle Laboratories, Pearl River, New York, 60 pp.

Wilkinson, M. H. F., and Schut, F. 1998. *Digital Image Analysis of Microbes: Imaging, Morphometry, Fluorometry and Motility Techniques and Applications.* John Wiley & Sons, New York, 551 pp.

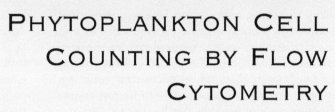

CHAPTER 17

# PHYTOPLANKTON CELL COUNTING BY FLOW CYTOMETRY

Dominique Marie
*Station Biologique de Roscoff, UMR 7127, CNRS and
Université Pierre & Marie Curie*

Nathalie Simon
*Station Biologique de Roscoff, UMR 7127, CNRS and
Université Pierre & Marie Curie*

Daniel Vaulot
*Station Biologique de Roscoff, UMR 7127, CNRS and
Université Pierre & Marie Curie*

## CONTENTS

Key Index Words: Flow Cytometry, Cell Fixation, Fluorescence, Stains, Oligonucleotide Probes, Molecular Biology, Bacteria, Microalgae, Picoplankton

## 1.0. INTRODUCTION

On many occasions, cultures can be easily monitored and counted by optical microscopy (see Chapter 16). However, researchers have recognized early on the need to automate cell counting. Automated cell counting is usually much faster than optical microscopy counting, and it minimizes errors associated with human counting. Because many more cells can be counted, the statistical significance of the data is considerably improved. Furthermore, other cellular parameters can also be determined, for example, cell volume or DNA content. Finally, the smallest phytoplankton (picoplankton, <2 to 3 μm) cannot be distinguished from bacteria when examined using a light microscope (e.g., *Prochlorococcus*), but they can be counted using automated techniques. However, automated cell counting also has its problems and limitations. Instruments are relatively expensive, starting at U.S. $20,000 for simple counters and reaching up to U.S. $300,000 for the most sophisticated flow cytometers. Some of the high-end flow cytometers can only be used by highly trained personnel. Finally, because cells are measured blindly, proper controls are necessary to ensure that the signal measured does not result from nontarget particles (detritus, contaminants).

In the 1970s, the introduction of the Coulter Counter (now marketed by Beckman Coulter) for phytoplankton counting (Sheldon and Parsons 1967, Sheldon 1978) constituted the first major advance toward automated phytoplankton counting. Particles in solution are drawn through a small aperture, separating two electrodes between which an electric current flows. As each particle passes through the aperture, it displaces its own volume of conducting liquid, briefly increasing the impedance of the aperture. This signal is converted into a voltage pulse. By counting the number of pulses for a given volume passing through the aperture, one obtains an estimate of the particle concentration. The equivalent spherical volume of each cell can also be estimated from the amplitude of the pulse. Although widely used to count large phytoplankton cells in cultures, the Coulter Counter has several limitations. First, even with the smallest aperture available, it is technically very challenging to count cells smaller than 1 to 2 μm, making this technique not applicable to picoplankton. Second, because a single cell parameter (cell volume) is determined, it is difficult to discriminate phytoplankton cells from other particles such as bacteria, detritus, and even air bubbles or to work with mixed cultures containing, for example, several phytoplankton species with overlapping sizes. Another instrument that uses a light beam to measure particle size is the HIAC counter (Pacific Scientific Instruments). It is less widespread than the Coulter Counter but is particularly well adapted to the continuous monitoring of cultures (Malara and Sciandra 1991, Sciandra et al. 2000). Because a single parameter is measured, it suffers from the same type of limitation as a Coulter Counter.

Although flow cytometry (FCM) was developed more than 30 years ago, it is only in the mid-1980s that it was first applied to phytoplankton analysis both in culture (Trask et al. 1982, Olson et al. 1983, Yentsch et al. 1983) and in the field (Olson et al. 1985). Since then, it has been increasingly used by aquatic (mostly marine) biologists for the analysis of small particles (Veldhuis and Kraay 2000) as well as phycologists (Collier 2000). Its major advantage over the Coulter Counter is that it simultaneously records several parameters for each event, allowing for the discrimination between cells and detritus. It is also particularly well adapted to the analysis of picophytoplankton, which are difficult or tedious to analyze with traditional methods such as epifluorescence microscopy. These tiny organisms are present in all aquatic environments at concentrations up to $5 \times 10^5$ cell·mL$^{-1}$. Direct analysis provides information on the abundance, cell size, and pigment content of the major photosynthetic picoplankton groups (cyanobacteria and eukaryotes). Although well suited for picophytoplankton, FCM can also be used for cultures of larger phytoplankton up to a size of 100 to 200 μm; beyond this size, custom modifications or special instruments must be used (Dubelaar et al. 1989). Moreover, some instruments are able to physically sort cells to identify them or bring them into culture (see Chapter 7). The use of benchtop instruments on board ships has helped to improve our knowledge of the geographical distribution and population dynamics of the picoplankton in relation to its physical and chemical environment (Partensky et al. 1999). FCM has been little used in freshwater research, although this trend seems to be reversing (Crosbie et al. 2003). Highly sensitive nucleic acid–specific stains such as TOTO-1, YOYO-1, and the SYBR Green family have also made possible the detection and enumeration of heterotrophic bacteria (Li et al. 1995, Marie et al. 1997) or more recently of viruses (Marie et al. 1999). The application of FCM extends also to physiological analyses (e.g., DNA analysis) (Vaulot et al. 1986) and to phylogenetic analyses with the help of fluorescent molecular probes (Simon et al. 1995).

In this chapter, we detail the use of FCM for marine algae culture work. Most of the protocols described can be applied equally well to field samples and to freshwater organisms. In the latter case, only obvious modifications, such as replacing seawater wherever mentioned by freshwater, are required (Lebaron et al. 2001, Crosbie et al. 2003).

## 2.0. PRINCIPLES OF FLOW CYTOMETRY

### 2.1. General Principles

FCM measures cells in liquid suspension. Cells are aligned hydrodynamically by an entrainment fluid (sheath fluid) into a very narrow stream, 10 to 20 µm wide, onto which one or several powerful light sources (arc mercury lamp or laser) are focused. Each time a particle passes through the beam, it scatters light; angular intensity depends on the refractive index, size, and shape of the particle. Moreover, if the particle contains a fluorescent compound whose absorption spectrum corresponds to the excitation source (e.g., blue light for chlorophyll), it emits fluorescence at a higher wavelength (e.g., red light for chlorophyll). These light pulses are detected by photodiodes or more often by photomultipliers and then are converted to digital signals that are processed by a computer. Measurement rates vary between 10 and 10,000 events per second. On the more sophisticated instruments, it is then possible to physically sort cells of interest based on any combination of the measured parameters (see Chapter 7).

### 2.2. Fluidics

Flow cytometers are equipped with a tank supplying the sheath liquid (buffer, distilled water, seawater) that carries the cells through the instrument; a second tank collects the waste fluid. Cell suspensions are injected or pushed through a capillary into a sheath fluid stream. Under laminar flow conditions, the sheath liquid aligns the cells into a narrow centered stream. The illumination of cells can be performed in the air, just outside a nozzle through which the sheath fluid exits, or in a quartz cuvette through which the sheath fluid flows. The latter solution increases the detection sensitivity, which is required for picophytoplankton. The flow rate must be adjusted depending on the cells of interest to keep laminar flow conditions and to control the number of events to be analyzed per unit time.

### 2.3. Optics

When a particle passes through the excitation beam, light can be reflected or refracted. In most flow cytometers, the light scatter detectors are located at 180° (forward scatter or FSC) and at 90° (side scatter or SSC) with respect to the light source. Both parameters are related to cell size, but the side scatter is more influenced by the cell surface and internal cellular structure (Morel 1991, Green et al. 2003).

Many fluorescent molecules can bind to a wide range of cytochemical compounds such as proteins, lipids, or nucleic acids. Each fluorescent dye is characterized by its excitation and emission spectra. Flow cytometers are usually equipped with a laser emitting at a single wavelength (488 nm). Therefore, only fluorescent molecules excited at that particular wavelength can be used. If multiple excitation wavelengths are available, then the choice of the fluorochromes is much wider.

The flow cytometer is equipped with highly sensitive photomultiplier tubes that are able to measure and amplify the brief pulse of light emitted by the cells. When a cell intersects the excitation beam, the emitted light is collected by a lens and passes through a series of filters that remove the excitation light, allowing only the emission light to be detected. With several photomultipliers, multiple wavelength emission ranges can be collected (e.g., orange and red fluorescence for algal cells).

### 2.4. Electronic and Software Processing

To be usable, analog data from the photomultipliers must be converted to digital form, that is, to a number on a scale ranging, for example, from 1 to 256 ($2^8$) corresponding to 8-bit conversion. To avoid saturation of the conversion circuitry, only events of interest must be converted. Therefore, the operator needs to select one or several signals (called *discriminators* or *triggers*) and must set thresholds for each discriminator. When the value of one of the discriminator signals is larger than the corresponding threshold, all signals from the triggering particle are converted. Choosing adequate discriminators and thresholds is critical to correctly record the cells of interest, especially when working with very small cells or particles. As an example, to record chlorophyll fluorescing microalgae, it is best to choose red fluorescence as the discriminator and to select a threshold that is high enough so optical and

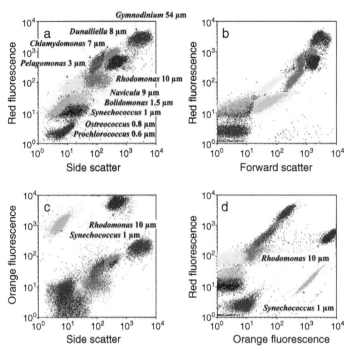

**FIGURE 17.1.** Cytograms obtained with a mixture of eight photosynthetic eukaryotes and two cyanobacteria in culture, analyzed using a FACSort flow cytometer using the natural fluorescence of phycoerythrin (orange) and chlorophyll (red). (See Table 17.1 for culture origin.)

electronic noise are left out but that is low enough so no cells are missed.

Digital data are then transmitted to a computer that displays and records the results. They will appear on the screen as mono-parametric histograms or bi-parametric cytograms (Fig. 17.1). Data can be further processed to discriminate specific cell populations and estimate their cell concentration and average cellular parameters using software provided with the instrument or public domain programs such as WinMDI (facs.scripps.edu/software.html) or CytoWin (www.sb-roscoff.fr/Phyto/cyto.html). For more sophisticated analyses, such as cell cycle deconvolution, specialized commercial software is available (e.g., from Verity Software House: www.vsh.com).

## 2.5. Available Instruments

A few companies occupy the FCM market. These include Becton Dickinson (www.bd.com), Beckman Coulter (www.beckman.com), Dako Cytomation (www.dakocytomation.com), and Partec (www.partec.de). Available instruments fall into three categories, which are covered in the following sections.

### 2.5.1. Benchtop Analyzers

Benchtop analyzers are small instruments widely used in medicine for blood analysis of antigen-responding cells. These instruments usually possess a single excitation wavelength at 488 nm, delivered by a small air-cooled laser. Possible choices are the FACS family (FACScan, FACSort, FACSCalibur) from Becton Dickinson, the EPICS XL from Beckman Coulter, or the CyAn from DakoCytomation. Some of these instruments are fitted with limited sorting capacity (FACSort, FACSCalibur). Their reduced footprint and moderate price make them perfect for ship-board analysis and small laboratories. They do not require highly trained operators, and used instruments can be easily purchased.

### 2.5.2. High-Speed Sorters

High-speed sorters are, in general, fairly large instruments that use several lasers delivering both ultraviolet (UV) and visible excitation lines. The more powerful water-cooled lasers require special open or closed water cooling circuits, as well as three phase power lines. Their capacity to excite many dyes makes them more versatile, allowing several components to be measured simultaneously within the same cell (e.g., DNA and proteins). The sorting capacity of these high-end instruments reaches up to 20,000 cells per second. These include the FACS Vantage from Becton Dickinson, the EPICS Altra from Beckman Coulter, and the MoFlo from DakoCytomation. Because of their cost, the need for a dedicated room, and the complexity of their operation (usually requiring a special technician), they are much less widely used. Moreover, their sensitivity is, paradoxically, often lower than that of the smaller benchtop analyzers, because they use a jet-in-air nozzle design.

### 2.5.3. Custom Instruments

Some researchers have devised their own instruments, for example, based on a microscope (Olson et al. 1983) or modified existing flow cytometers to improve sensitivity (Dusenberry and Frankel 1994) or measurement size range (Cavender-Bares et al. 1998). More recently, flow cytometers have been specifically designed for *in situ* applications such as the continuous monitoring of phytoplankton in marine waters. These include the CytoBuoy (www.cytobuoy.com) (Dubelaar and Gerritzen 2000) and the FlowCytobot (Olson et al. 2003).

### 2.5.4. Choice of Instrument for Algal Cultures

All commercial instruments are suitable to analyze algal cultures. For picoplanktonic organisms or for virus detection, the use of a very sensitive flow cytometer, such as a benchtop FACS (Becton Dickinson), is critical.

## 3.0. Counting Phytoplankton by Flow Cytometry Using Natural Optical Properties

Phytoplankton possess fluorescing photosynthetic pigments (see Chapter 20) that can be used to discriminate cells from heterotrophic organisms and nonliving particles. The most common fluorescing pigments are chlorophyll, phycoerythrin, and phycocyanin. The latter two are phycobiliproteins that are typical of cyanobacteria, cryptophytes, and rhodophytes. Both chlorophyll and phycoerythrin are well excited with the common 488-nm excitation line and fluoresce at 690 nm (red) and 570 nm (orange), respectively. In contrast, phycocyanin is excited at 620 nm and emits at 640 nm. Therefore, it can be detected only with a red-emitting laser. Counting phytoplankton does not require any pretreatment of the samples, but if samples cannot be counted immediately, they must be preserved by aldehydes (formaldehyde or glutaraldehyde) and stored deep frozen, either in a −80°C freezer or in liquid nitrogen.

### 3.1. Measurement of Cell Concentration

Phytoplankton abundance is best obtained using fresh unfixed samples; fixation will always result in some degree of cell disruption and cell loss (Vaulot et al. 1989). Samples can be kept, however, at 4°C in the dark for up to 12 hours with only minimal change in abundance or optical parameters (Jacquet et al. 1998).

### 3.1.1. Materials

1. Phytoplankton cultures
2. 0.95-μm fluorescent microspheres diluted at $10^5 \cdot mL^{-1}$ in distilled water (e.g., Polysciences)
3. 0.2-μm filtered seawater
4. Micropipettes and tips for 10 to 1000 μL
5. Chronometer
6. Flow cytometer equipped with a 488-nm argon laser (e.g., benchtop FACS)

### 3.1.2. Culture Dilution

If two (or more) cells pass simultaneously through the excitation beam, or if two cells are too close to each other, and the second cell arrives while the instrument is busy recording the first cell, then the two cells are recorded as a single event (see Chapter 7). This phenomenon is called *coincidence*, and it results in an underestimate of the cell concentration. Coincidence threshold is best determined empirically. For example, it is possible to analyze a range of cell concentrations at a given sample flow rate. The coincidence threshold corresponds to the maximum cell concentration beyond which the number of recorded cells is not linearly related to the sample concentration (Gasol and Del Giorgio 2000). As an example, for instruments of the benchtop FACS family, coincidence for picoplankton cells begins at more than 800 cells per second. To prevent coincidence, it is necessary either to reduce the flow rate (e.g., from high to medium on a FACS) or to dilute cultures before analysis.

### 3.1.3. Instrument Settings

Phytoplankton acclimates to changes of photon-flux densities by changing pigment content, leading to a decrease in chlorophyll fluorescence per cell as light increases (Sosik et al. 1989). Fluorescence range for a given strain is very wide for *Prochlorococcus* (up to fifty-fold) and less pronounced for eukaryotes (fivefold to tenfold only). The intensity of the other cellular parameters such as scatter can also vary with light conditions and available nutrients. For example, as nutrients become limiting, scatter typically decreases in intensity. Thus, the voltage of the photomultipliers must be adjusted accordingly, depending on the size and fluorescence of the organisms of interest.

Typical settings for picoplankton and nanoplankton cultures on a benchtop FACS flow cytometer are as follows: forward scatter (FSC) = E00, side scatter (SSC) = 400, green fluorescence (FL1) = 650, orange fluorescence (FL2) = 650, and red fluorescence (FL3) = 550. All parameters are set on logarithmic amplification and the trigger is set on the red fluorescence.

### 3.1.4. Sample Analysis

We describe here a general procedure that can be used either for fresh or preserved samples.

1. Turn on both the instrument and the computer.
2. Prepare the sheath fluid. Because cell scatter (especially FSC) is dependent on the nature of the sheath fluid (Cucci and Sieracki 2001), it is necessary to match the sheath and sample fluids. For example, for marine samples, 0.0.2-μm pore-size filtered seawater can be used as sheath fluid. In this case, it is best to remove any inline sheath filter, which becomes easily contaminated and quickly tends to release particles.
3. If samples were fixed and frozen (see Section 3.2.3), then thaw them at room temperature or at 37°C.
4. Transfer about 1 mL (minimum of 250 μL) of the culture into a flow cytometer tube.
5. Add 10 μL of the microsphere solution containing about $10^5 \cdot$ beads mL$^{-1}$ as an internal reference.
6. Select (low, medium, or high) and calibrate the flow rate (see Section 3.1.5).
7. Set the discriminator on the red (chlorophyll) fluorescence with a threshold of 50.
8. Insert the sample tube in the instrument sample holder, and after about 15 seconds (to allow the flow rate to stabilize), start data acquisition. Typical analysis of a culture lasts 2 to 3 minutes with a delivery rate of 50 to 100 μL $\cdot$ min$^{-1}$.
9. Record the duration of analysis, which is necessary to estimate the cell concentration (see Section 3.1.6.).

### 3.1.5. Flow-Rate Calibration

On most commercial flow cytometers, it is not possible to deliver a specific sample volume or to precisely set the sample flow rate. Therefore, the sample flow rate must be determined by the operator for accurate cell concentrations.

Often, a solution of fluorescent microspheres with a known concentration (determined by epifluorescence microscopy) is used for the measurement of the flow rate. Because the electrostatic charges of the beads make them stick onto the plastic tubes (particularly in seawater), this modifies their initial concentration, and we do not recommend this method (still, we always add fluorescent beads to our samples to check flow stability and to normalize cell scatter and fluorescence; see Section 3.1.6). We use instead the procedure described below that works on flow cytometers from the benchtop FACS family that can be adapted to most instruments. This calibration should be repeated several times a day.

1. Select the same sample flow rate used for analysis.
2. Fill a tube with the same liquid as samples (e.g., seawater for marine phytoplankton).
3. Measure the volume of the sample (or weigh precisely the tube containing the sample).
4. Remove the outer sleeve of the injection system. The sheath fluid will drop down the sample needle.
5. Wait until a droplet has fallen, and then before the next one forms, place the sample tube and close the sample arm in the running position.
6. Simultaneously start the chronometer.
7. Run the sample for at least 10 minutes.
8. Remove the sample tube and simultaneously stop the chronometer.
9. Measure (or weigh) the remaining volume.

The rate (R), expressed in microliters per minute, is given by the following formula:

$$R = (V_i - V_f)/T, \qquad (1)$$

where $V_i$ = initial volume (μL), $V_f$ = final volume (μL), and T = the time (minutes). The use of a scale leads to better precision for the determination of the flow rate:

$$R = (W_i - W_f)/(T \times d), \qquad (2)$$

where $W_i$ = initial weight (mg), $W_f$ = final weight (mg), T = time (minutes), and d = density of the liquid used for calibration (distilled water = 1.00, seawater = 1.03).

### 3.1.6. Data Analysis

Phytoplankton cells cover a wide range of size and fluorescence properties (see Fig. 17.1). Therefore, data are always collected using logarithmic amplifications and recorded as list-mode files, which allows detailed offline analysis. In practice, 20,000 to 100,000 events are collected for microalgae and up to 150,000 for bacteria or viruses. List-mode files are then analyzed using either the instrument software or publicly available programs.

The different populations are discriminated using a combination of parameters: scatters (FSC and SSC) and fluorescences (usually red and orange). Figure 17.1 illustrates data obtained for a mixture of 10 phytoplankton species (8 eukaryotes and 2 prokaryotes) (Table 17.1) ranging in size from 0.6 to 60 μm and how the different cultures can be discriminated by using combination of parameters. For example, cyanobacteria (*Synechococcus*) and Cryptophyceae (*Rhodomonas*) can be discriminated from other eukaryotic algae (red algae excluded) based on their high orange/red fluorescence ratio linked to the presence of phycoerythrin (see Fig. 17.1d).

Absolute cell concentration for each population is computed as follows:

$$C_{pop} = N_{pop} \times (V_{total}/V_{sample})/(R \times T), \qquad (3)$$

where $C_{pop}$ = concentration of population (cells $\cdot$ μL$^{-1}$), $N_{pop}$ = number of cells acquired, T = acquisition time (minutes), R = sample flow rate (μL $\cdot$ min$^{-1}$) as determined for the sample series (see Section 3.1.5), $V_{total}$ = volume of sample plus additions (fixatives, beads, etc.) (μL), and $V_{sample}$ = volume of sample (μL).

To compare different samples, cell parameters are normalized to parameters obtained for 0.95 μm of fluorescent microspheres added as internal reference. The mean value of each parameter (for the different populations) is divided by the mean value of the parameter for the beads:

$$X_{rel} = X_{pop}/X_{beads}, \qquad (4)$$

where $X_{pop}$ is the average value of a cell parameter (scatter or fluorescence) for a given population, and $X_{beads}$ is the same parameter for the reference beads. Both $X_{pop}$ and $X_{beads}$ must be expressed as linear values (not channels) after conversion from the logarithmic recording scale.

### 3.2. Cell Fixation and Preservation

If samples cannot be analyzed immediately, they must be fixed and then frozen in liquid nitrogen and stored at –80° C or in liquid nitrogen until analysis. Physical treatments such as centrifugation and classic or tangential filtration must be avoided because they induce vari-

**TABLE 17.1** Cultures referred to in this chapter. The RCC column corresponds the reference number of the culture in the Roscoff Culture Collection (http://www.sb-roscoff.fr/Phyto/collect.html).

| RCC | Class | Taxon | Size (μm) |
|---|---|---|---|
| 1 | Chlorophyceae | *Chlamydomonas* sp. | 7 |
| 6 | Chlorophyceae | *Dunaliella tertiolecta* Butcher | 8 |
| 22 | Chrysophyceae | *Picophagus flagellatus* Guillou et Chrétiennot-Dinet | 2 |
| 29 | Cyanophyceae | *Synechococcus* sp. | 1 |
| 80 | Bacillariophyceae | *Navicula transitans* Cleve | 9 |
| 89 | Dinophyceae | *Gymnodinium sanguineum* Hirasaka | 60 |
| 100 | Pelagophyceae | *Pelagomonas calceolata* Andersen et Saunders | 3 |
| 116 | Prasinophyceae | *Ostreococcus tauri* Courties et Chrétiennot-Dinet | 0.8 |
| 238 | Bolidophyceae | *Bolidomonas mediterranea* Guilou et Chrétiennot-Dinet | 1.5 |
| 286 | Pelagophyceae | *Ankylochrysis lutea* Billard | 6 |
| 350 | Cryptophyceae | *Rhodomonas baltica* Karsten | 10 |
| 407 | Cyanophyceae | *Prochlorococcus* sp. | 0.6 |

able losses. Because phytoplanktonic cells are discriminated on the basis of scatter and pigment fluorescence, the fixation procedure must preserve these properties. Classic methods such as formalin (a generic term that describes a solution of 37% formaldehyde gas dissolved in water usually containing 10 to 15% methanol) or Lugol's iodine fixation are generally inadequate because they modify cell shape or drastically affect fluorescence. Alcohol fixation will extract lipophilic pigments and lead to a loss of autofluorescence. Formaldehyde fixation (1% final concentration) is the best method, because in our experience, it minimizes cell loss. Moreover, solutions of formaldehyde are buffered and do not strongly modify the pH level of seawater samples. Formaldehyde is obtained by heating paraformaldehyde powder (that has no fixation properties) in distilled water or phosphate-buffered saline (PBS). Fixation with formaldehyde can be supplemented with glutaraldehyde 0.05% (final concentrations), particularly when cell cycle analysis is performed.

However, formaldehyde solutions are neither easy to prepare nor stable over time. Therefore, for inexperienced operators, or when there is any doubt on the quality of the formaldehyde to be used (e.g., during an important cruise), we recommend replacing formaldehyde with a commercial solution of glutaraldehyde at 0.1% (final concentration). This will lead to slightly higher cell loss but is clearly preferable to fixation with bad formaldehyde, which leads to a lot of background noise (cell debris, small particles), making flow cytometric analysis impossible.

### 3.2.1. Materials

1. Paraformaldehyde powder (Sigma P-6148)
2. Glutaraldehyde 25% aqueous solution (Sigma G-6257)
3. Sodium hydroxide in pellets
4. Cryovials (e.g., Nunc)
5. Pipettes and tips
6. 0.2-μm pore-size syringe filters
7. Paper filter

### 3.2.2. Preparation of 10% Formaldehyde

1. Note: If not confident with this procedure, use only glutaraldehyde.
2. Under a fume hood, add 10 g of paraformaldehyde powder to 70 mL of boiling distilled water.
3. Mix vigorously for at least 2 hours under the fume hood.

4. Add progressively small amounts of sodium hydroxide (0.1 M).
5. Agitate until the solution becomes clear.
6. Add 10 mL of 10% PBS.
7. Adjust the pH to 7.5.
8. Bring final volume up to 100 mL with distilled water.
9. Filter through paper filter.
10. Filter again through 0.2-μm pore-size syringe filters.
11. Aliquot to 15-mL tubes and store at –20°C.
12. Use unfrozen aliquoted formaldehyde solutions for not more than 1 week.

### 3.2.3. Fixation Procedure

1. Add 1% of formaldehyde or 0.1% of glutaraldehyde or a mixture of both (1% and 0.05%, respectively) to the sample.
2. Mix by vortexing rapidly.
3. Incubate for at least 15 minutes at room temperature.
4. If samples cannot be analyzed immediately, then they must be quickly frozen using liquid nitrogen. They can then be stored at –20°C for a short period (several days) but must be kept at μ80°C or in liquid nitrogen for long-term storage, because storage at –20°C beyond 1 week will result in rapid sample degradation.

## 3.3. Fluorescent Dyes

Fluorescent dyes that recognize specific molecules within cells extend considerably the application of FCM. Among these dyes, the most useful are probably nucleic acid stains. They are extremely diverse and can be used to detect contaminating bacteria or viruses (Marie et al. 1997) and to measure cell viability (Brussaard et al. 2001, Veldhuis et al. 2001) or cell cycle progression (Vaulot et al. 1986). A wide range of nucleic acid–specific dyes synthesized and manufactured by Molecular Probes (www.probes.com), such as YOYO-1, PicoGreen, or SYBR Green-I (Li et al. 1995, Marie et al. 1997), are now available and replace the UV-excited dyes, DAPI or Hoechst 33342, initially used for this purpose (Monger and Landry 1993, Button and Robertson 2001). Other markers that could be used but have received limited application for phytoplankton include protein stains such as SYPRO (Zubkov et al. 1999) or cellular activity stains such as FDA (Brookes et al. 2000).

### 3.3.1. Analysis of Heterotrophic Eukaryotes and Bacterial Contaminants

The analysis of heterotrophic cells requires a fixation step by aldehydes, as mentioned earlier, and the use of nucleic acid–specific stains. The affinity of the cyanine dyes (TOTO-1, YOYO-1) and their monomeric equivalents (TO-PRO-1, YO-PRO-1) decreases significantly with ionic strength, which makes them inappropriate for direct analysis of seawater samples (Marie et al. 1996). Other dyes such as SYBR Green-I (SYBR-I), SYBR Green-II, and SYTOX Green are less dependent on the composition of the medium and can be used for the enumeration of marine bacteria. Because SYBR-I has a very high fluorescence yield, we recommend this dye to enumerate bacteria in marine samples, although SYTO-9 may provide better results for freshwater samples (Marie et al. 1997, Lebaron et al. 1998).

### 3.3.2. Materials

1. 0.2-μm pore-size filtration units for plastic syringes
2. 0.95 μm fluorescent microspheres (see Section 3.1.1)
3. DNA-specific stains such as SYBR Green-I (all stock solutions *except* SYBR-I must be prefiltered onto 0.2 μm or less to avoid contamination)
4. Flow cytometer equipped with a 488-nm argon laser
5. Glutaraldehyde 25% aqueous solution and/or formaldehyde 10% (see Section 3.2)

### 3.3.3. Sample Preparation

1. If samples are live, add either 1% formaldehyde or 0.1% glutaraldehyde (final concentrations) and wait 20 minutes to allow a good fixation.
2. If samples have been preserved and frozen, thaw them at 37°C.
3. Dilute the sample in 0.2-μm pore-size filtered seawater if necessary (see Section 3.1.2).
4. Add the SYBR-I at a final dilution of 1 : 10,000 of the commercial solution.
5. Incubate for 15 minutes at room temperature and in the dark.
6. Add 10 μL of a suspension of 0.95 μm fluorescent microspheres at a concentration of $10^5$ beads · $mL^{-1}$ in 1 mL of sample.

### 3.3.4. Data Acquisition

1. Turn the flow cytometer and computer on.
2. Prepare the sheath fluid (distilled water can be used as sheath fluid, but for natural seawater

samples, 0.2-μm pore-size filtered seawater is preferred).
3. Calibrate the flow rate (see Section 3.1.5).
4. Set the discriminator to green (SYBR-I) fluorescence with a threshold of 150.
5. Set logarithmic amplification for all parameters.
6. Typical settings on our FACSort flow cytometer are as follows: FSC = E01, SSC = 450, FL1 = 550, FL2 = 650, and FL3 = 650.
7. Run the sample. The flow rate and the cell concentration must be adjusted to avoid coincidence. Typically, we analyze samples for 1 to 2 minutes at a delivery rate of 25 to 50 μL · $min^{-1}$ and the number of events is kept below 1,000 per second by sample dilution, so the total number of recorded events is about 100,000.

### 3.3.5. Data Analysis

The distribution of bacteria in cultures of *Ostreococcus tauri* (Fig. 17.2a,b) and *Pelagomonas calceolata* (Fig. 17.2c,d) are illustrated, as well as the detection of the heterotrophic eukaryote *Picophagus flagellatus* (Fig. 17.2e). In natural seawater samples and cultures, the use of SYBR-I allows the discrimination of two or three different bacteria clusters (Fig. 17.2a) that correspond to different taxonomic groups (Zubkov et al. 2001).

Some samples contain a lot of small particles and debris, which increase the level of background noise. This can induce coincidence or lead to the generation of large list-mode files. In such cases, the discriminator threshold must be increased to reduce the number of events seen by the flow instrument, or a "bitmap" window (nonrectangular region) can be defined that includes the population of bacteria so only the events in this area are recorded.

### 3.4. Analysis of Viral Infection

Because viruses can induce rapid decay of algal cultures, it is sometimes necessary to analyze cultures to evaluate the level of infection (see Chapter 22). The study of viroplankton initially required techniques like transmission electron microscopy (TEM), which are time consuming and allow the analysis of only a limited number of samples. During the 1990s, the use of nucleic acid–specific dyes detected by epifluorescence microscopy (EFM) improved our knowledge of viruses (Hennes and Suttle 1995) (see Chapter 22). More recently, FCM has been successfully used for the analy-

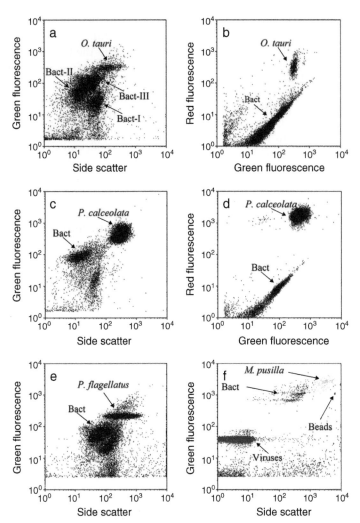

**FIGURE 17.2.** Analysis of heterotrophic eukaryotes, bacteria, and viruses in algal cultures after staining with SYBR-I (green fluorescence). (See Table 17.1 for culture origin.) **(a,b)** Bacterial contaminants of *Ostreococcus tauri* (Prasinophyceae). Three populations of bacteria are visible. **(c,d)** Bacterial contaminants of *Pelagomonas calceolata* (Pelagophyceae). **(e)** *Picophagus flagellatus*, a heterotrophic Chrysophyceae, in co-culture with bacteria. **(f)** Detection of viruses in a *Micromonas pusilla* (Butcher) Manton et Parke (Prasinophyceae) culture. (Fluorescent 0.95-μm microspheres were added in the samples.)

sis of viruses in solution, using the nucleic acid–specific dye SYBR-I (Marie et al. 1999).

The preparation of the samples for the analysis of viruses is similar to that of heterotrophs, although a certain number of precautions must be taken. No significant difference has been found for virus enumeration performed on samples fixed with formaldehyde, glutaraldehyde, or a mixture of both aldehydes. For virus samples that are freshly fixed (i.e., have not been frozen), or for recalcitrant material, it is necessary to heat the samples for 10 minutes at 80°C in the presence of a detergent such as Triton X-100 (0.1% final concentration). Because a large fraction of virus particles

can pass through 0.2 μm pore-size filters, filtered seawater cannot be used to dilute the samples. The best solution to minimize the background noise is to dilute samples in Tris-EDTA buffer (Tris 10 mM, EDTA 1 mM). Different buffers have been tested, but Tris-based buffers give the best results, probably because Tris has free amines that interact with aldehydes. Distilled water must be used as sheath fluid. Samples are then stained with SYBR-I at a dilution of 1/20,000 of the commercial solution.

Typical settings on a FACSort flow cytometer are as follows: FSC = E03, SSC = 600, FL1 = 600, FL2 = 650, and FL3 = 650. Discriminator is set on the green fluorescence (FL1) with a threshold value of about 100. Analysis must be performed with a suspension of about $2 \times 10^5$ to $2 \times 10^6$ viruses/mL (final concentration). To avoid generating large files, samples can be run for 1 or 2 minutes at a rate ranging from 10 to 30 $\mu$L $\cdot$ min$^{-1}$.

Natural viroplankton displays a wide range of sizes, and these particles are often difficult to separate from background noise. However, viruses that contaminate cultures are usually simple to analyze (see Fig. 17.2f). Viruses are too small to be discriminated only by their SSC or FSC properties. Detection must, therefore, be performed using the green SYBR-I fluorescence (see Fig. 17.2f). Because FCM was not designed for the analysis of such small particles, care must be taken to obtain reliable data. If samples are too diluted, there is a loss in the emission signal of the nucleic acid–dye complex. If they are not diluted enough, coincidence occurs or the population of viruses is overlapped by background noise. For viruses, coincidence seems to occur at more than 600 events per second on a FACSort, that is, at a lower rate than for beads, bacteria, or small algae.

## 3.5. Counting Dead vs. Live Cells

It is sometimes necessary to evaluate the percentage of living and/or dead cells in a sample. Propidium iodide (PI) or SYTOX Green can penetrate into cells that have lost membrane integrity so that dead cells exhibit fluorescence. However, PI cannot be used with phytoplankton because its red fluorescence interferes with that of chlorophyll; SYTOX induces green fluorescence, which is more suitable (Veldhuis et al. 2001). Conversely, fluorescent dyes from the SYTO family (Molecular Probes), such as SYTO-9, or calcein-AM can penetrate into intact cells and induce live cells to fluoresce green (Brussaard et al. 2001).

## 4.0. MOLECULAR PROBES

Phytoplankton can be discriminated from other particles by FCM based on their natural scattering and fluorescence properties. However, these natural properties are not sufficient to separate lower level taxa (e.g., genera, species). Antibodies labeled with fluorescent markers have been used in this context (Peperzak et al. 2000), but their use remains limited because of the lack

of specificity for polyclonal antibodies and the cost for developing monoclonal antibodies. Nucleic acid probes targeting ribosomal RNA (Amann et al. 1995) offer a much more flexible solution. Probes can be easily designed to target any phylogenetic level from the division to the species, and various probes are available for phytoplankton.

The probes used are generally oligonucleotides (15 to 30 bases). Different probe-labeling techniques are available. Probes may be directly labeled with a fluorochrome (Simon et al. 1995), or labeling may be indirect (Not et al. 2002). For indirect labeling, hybridization of the probes and labeling with the fluorochrome are realized in two steps, as in the tyramide signal amplification of fluorescent in situ hybridization (TSA-FISH) technique. Indirect labeling increases the intensity of fluorescence and thus raises the limit of detection and the signal/noise ratio (Not et al. 2002), which is critical for small cells. Recently, TSA-FISH has been successfully adapted for the identification and enumeration of phytoplankton cells by FCM (Biegala et al. 2003).

The most common fluorochrome used is fluorescein isothiocyanate (FITC) (excitation = 488 nm; emission = 525 nm), but other fluorochromes that have higher fluorescence yield, such as CY3 (excitation = 550 nm; emission = 570 nm) or CY5 (excitation = 650 nm; emission = 670 nm), are also suitable, provided that the flow cytometer can be set to the corresponding excitation and emission wavelengths. For phytoplankton, the combined use of FCM and molecular probes may be useful to assess culture identity or when cultures are not pure and it is difficult to distinguish the taxon of interest from the contaminants.

## 4.1. Probe Design and Labeling

A database of the oligonucleotide probes for cyanobacteria and protists is available at www.sb-roscoff.fr/Phyto/Databases/RNA_probes_introduction.php. Although probes have been designed against some of the major algal groups such as the Chlorophyta, Prymnesiophyceae (Simon et al. 1995, 2000), or some key genera such as *Phaeocystis* (Lange et al. 1996), considerable work remains to be done to cover all existing taxa. Probes can be designed from ribosomal DNA databases using a public domain software such as ARB (www.arb-home.de). Advice for the design of new taxa-specific probes is available in Amann et al. (1995). Probes may be purchased directly labeled, but cost may be reduced by custom labeling of oligonucleotide probes with fluorochromes such as FITC or CY3 (Amann et al. 1995).

## 4.2. Cell Labeling

The cell labeling protocol was designed for the identification of cells but was not optimized for cell counting.

### 4.2.1. Materials

1. Hybridization oven set at 46°C
2. Microcentrifuge
3. Fixatives: formaldehyde (stock at 10%) (see Section 3.2.2) and ethanol
4. Hybridization buffer: 0.9 M NaCl, 20 mM Tris HCl (pH 7.8), 0.01 sodium dodecyl sulfate (SDS), 0% to 50% formamide. For every 1% increase in the concentration of formamide, the melting temperature ($T_m$) of the hybrid is reduced by 0.7°C. The percentage of formamide must be adapted for each probe to ensure a specific labeling (Amann et al. 1995).
5. Wash buffer: 0.028 to 0.9 mM NaCl, 5 mM EDTA, 0.01% SDS, 20 mM Tris-HCl, pH 7.5. The concentration of NaCl must be adapted for each probe so the stringency of the washing buffer is equivalent to the stringency of the hybridization buffer.
6. Resuspension buffer: PBS

### 4.2.2. Procedure

1. Cell fixation and permeabilization: Samples (5 mL of cell suspension at $10^5$ cells/mL) should be fixed with formaldehyde (1% final concentration; glutaraldehyde should not be used because it prevents probe binding) and stored at 4°C for 1 hour. Cells are then pelleted (3 minutes, 4000 × g) and resuspended in a cold (−80°C) mixture (70 : 30, v/v) of ethanol and PBS (500 µL).
2. Hybridization: Cells are then pelleted again in an Eppendorf type of tube and resuspended in 20 to 100 µL of hybridization buffer. A 20-µL aliquot of the cell suspension is then incubated at 46°C for 3 hours with the oligonucleotide probe (2.5 ng · µL⁻¹). An aliquot without probe incubated in the same condition can serve as a negative control for autofluorescence. Hybridization is stopped by the addition of 1 mL of cold PBS at a pH of 9.0. Samples are then stored at 4°C until analysis with FCM.

## 4.3. Analysis of Hybridized Cells with Flow Cytometry

First calibrate the sample flow rate (see Section 3.1.5). Set the discriminator to green fluorescence (if the fluorochrome used is FITC). Set all parameters on logarithmic amplification (Fig. 17.3). Events are recorded in list mode. The flow rate and the cell concentration must be adjusted to avoid coincidence. Typically, we analyze samples for 1 to 2 minutes at a delivery rate of 25 to 50 µL · min⁻¹ and the number of events is kept at less than 1,000 per second (by diluting samples that are too concentrated).

## 4.4. Limits and Troubleshooting

The protocol used for whole-cell hybridization involves several centrifugation steps that generally lead to the

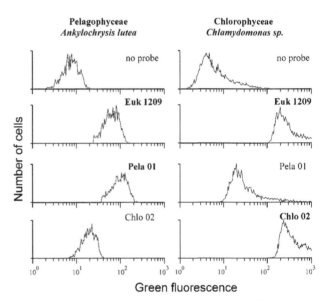

**FIGURE 17.3.** Flow cytometric analysis of fluorescence signals for whole-cell hybridization of exponentially growing *Ankylochrysis lutea* (Pelagophyceae) and *Chlamydomonas* sp. (Chlorophyceae) with fluorescein isothiocyanate (FITC)–monolabeled probes. Both species are nanoplanktonic (diameter 7 to 10 µm). For each species, the distribution of green fluorescence intensity per cell is plotted on a three-decade log scale. The intensity of green fluorescence per cell was measured for cells incubated without probe (green autofluorescence, no probe) and in the presence of a general eukaryotic probe (Euk1209), a Pelagophyceae-specific probe (Pela 01), and a Chlorophyta-specific probe (Chlo 02). (Modified from Simon et al. 2000.)

formation of cell clumps and/or to cell losses due to the adhesion of the cells to the surfaces of the tubes. Such losses can be reduced by treating the tubes with surfactants, by adding surfactants to the cells, and by sonication (Biegala et al. 2003).

The protocol based on monolabeled probes is quite short and simple. It is best suited, however, for large microplankton and nanophytoplankton cells. For smaller cells (picoplankton), the intensity of fluorescence conferred by the probes is, in general, not sufficient to distinguish target from nontarget cells, especially if cultures are not in exponential growth phase. In this case, enzymatic amplification (TSA) may be needed because it increases 20 to 40 times the fluorescence intensity of target cells (Schönhuber et al. 1997, Biegala et al. 2003).

## 5.0. CONCLUSION

FCM is now a well-established technique to analyze phytoplankton both in the field and in culture. There has been very little change in instrument design, although novel flow cytometers such as the FACSAria (released in late 2002) that combine small footprint, high sensitivity, and very fast sorting could prove ideal for both culture (in particular for the isolation of novel strains) and field work. Most progress has come from the development of novel fluorochromes such as SYBR Green that allow routine and enumeration analysis of bacteria and viruses. The application of molecular probes that permit accurate cell identification will probably develop considerably soon as the number of available algal sequences, a prerequisite for probe design, increases.

## 6.0. ACKNOWLEDGMENTS

We thank Isabelle Biegala for giving access to her work before publication and Florence Le Gall for assistance with cultures. This work has been partially funded by the following programs: PICODIV (European Union EVK2-1999-00119), PICOMANCHE (Région Bretagne), Souchothèque de Bretagne (Plan Etat-Région), Centre de Ressources Biologiques (Ministère de la Recherche), and BIOSOPE (CNRS).

## 7.0. REFERENCES

Amann, R. I., Ludwig, W., and Schleifer, K.-H. 1995. Phylogenetic identification and in situ detection of individual microbial cells without cultivation. *Microbiol. Rev.* 59:143–69.

Biegala, I., Not, F., Vaulot, D., and Simon, N. 2003. Quantitative assessment of picoeucaryotes in the natural environment using taxon specific oligonucleotide probes in association with TSA-FISH (tyramide signal amplification—fluorescent in situ hybridization) and flow cytometry. *Appl. Environ. Microbiol.* 69:5519–29.

Brookes, J. D., Geary, S. M., Ganf, G. G., and Burch, M. D. 2000. Use of FDA and flow cytometry to assess metabolic activity as an indicator of nutrient status in phytoplankton. *Mar. Freshw. Res.* 51:817–23.

Brussaard, C. P. D., Marie, D., Thyrhaug, R., and Bratbak, G. 2001. Flow cytometric analysis of phytoplankton viability following viral infection. *Aquat. Microb. Ecol.* 26:157–66.

Button, D. K., and Robertson, B. R. 2001. Determination of DNA content of aquatic bacteria by flow cytometry. *Appl. Environ. Microbiol.* 67:1636–45.

Cavender-Bares, K. K., Frankel, S. L., and Chisholm, S. W. 1998. A dual sheath flow cytometer for shipboard analyses of phytoplankton communities from the oligotrophic oceans. *Limnol. Oceanogr.* 43:1383–8.

Collier, J. L. 2000. Flow cytometry and the single cell in phycology. *J. Phycol.* 36:628–44.

Crosbie, N. D., Teubner, K., and Weisse, T. 2003. Flowcytometric mapping provides novel insights into the seasonal and vertical distributions of freshwater autotrophic picoplankton. *Aquat. Microb. Ecol.* 33:53–66.

Cucci, T. L., and Sieracki, M. E. 2001. Effects of mismatched refractive indices in aquatic flow cytometry. *Cytometry* 44:173–8.

Dubelaar, G. B., Groenewegen, A. C., Stokdijk, W., van den Engh, G. J., and Visser, J. W. 1989. Optical plankton analyser: A flow cytometer for plankton analysis, II: Specifications. *Cytometry* 10:529–39.

Dubelaar, G. B. J., and Gerritzen, P. L. 2000. CytoBuoy: A step forward towards using flow cytometry in operational oceanography. *Sci. Mar.* 64:255–65.

Dusenberry, J. A., and Frankel, S. L. 1994. Increasing the sensitivity of a FACScan flow cytometer to study oceanic picoplankton. *Limnol. Oceanogr.* 39:206–9.

Gasol, J. M., and Del Giorgio, P. A. 2000. Using flow cytometry for counting natural planktonic bacteria and understanding the structure of planktonic bacterial communities. *Sci. Mar.* 64:197–224.

Green, R. E., Sosik, H. M., Olson, R. J., and DuRand, M. D. 2003. Flow cytometric determination of size and complex refractive index for marine particles: comparison with independent and bulk estimates. *Appl. Opt.* 42:526–41.

Hennes, K. P., and Suttle, C. A. 1995. Direct counts of viruses in natural waters and laboratory cultures by epifluorescence microscopy. *Limnol. Oceanogr.* 40:1050–5.

Jacquet, S., Lennon, J. F., and Vaulot, D. 1998. Application of a compact automatic sea water sampler to high frequency picoplankton studies. *Aquat. Microb. Ecol.* 14:309–14.

Lange, M., Guillou, L., Vaulot, D., Simon, N., Amann, R. I., Ludwig, W., and Medlin, L. K. 1996. Identification of the class Prymnesiophyceae and the genus *Phaeocystis* with ribosomal RNA-targeted nucleic acid probes detected by flow cytometry. *J. Phycol.* 32:858–68.

Lebaron, P., Parthuisot, N., and Catala, P. 1998. Comparison of blue nucleic acid dyes for flow cytometric enumeration of bacteria in aquatic systems. *Appl. Environ. Microbiol.* 64:1725–30.

Lebaron, P., Servais, P., Agogue, H., Courties, C., and Joux, F. 2001. Does the high nucleic acid content of individual bacterial cells allow us to discriminate between active cells and inactive cells in aquatic systems? *Appl. Environ. Microbiol.* 67:1775–82.

Li, W. K. W., Jellett, J. F., and Dickie, P. M. 1995. The DNA distributions in planktonic bacteria stained with TOTO or TO-PRO. *Limnol. Oceanogr.* 40:1485–95.

Malara, G., and Sciandra, A. 1991. A multiparameter phytoplankton culture system driven by microcomputer. *J. Appl. Phycol.* 3:235–41.

Marie, D., Vaulot, D., and Partensky, F. 1996. Application of the novel nucleic acid dyes YOYO-1, YO-PRO-1, and PicoGreen for flow cytometric analysis of marine prokaryotes. *Appl. Environ. Microbiol.* 62:1649–55.

Marie, D., Partensky, F., Jacquet, S., and Vaulot, D. 1997. Enumeration and cell cycle analysis of natural populations of marine picoplankton by flow cytometry using the nucleic acid stain SYBR Green I. *Appl. Environ. Microbiol.* 63:186–93.

Marie, D., Brussaard, C. P. D., Thyrhaug, R., Bratbak, G., and Vaulot, D. 1999. Enumeration of marine viruses in culture and natural samples by flow cytometry. *Appl. Environ. Microbiol.* 65:45–52.

Monger, B. C., and Landry, M. R. 1993. Flow cytometric analysis of marine bacteria with Hoechst 33342. *Appl. Environ. Microbiol.* 59:905–11.

Morel, A. 1991. Optics of marine particles and marine optics. In: Demers, S., ed. *Particle Analysis in Oceanography.* Springer-Verlag, New York, pp. 142–88.

Not, F., Simon, N., Biegala, I. C., and Vaulot, D. 2002. Application of fluorescent *in situ* hybridization coupled with tyramide signal amplification (FISH-TSA) to assess eukaryotic picoplankton composition. *Aquat. Microb. Ecol.* 28:157–66.

Olson, R. J., Frankel, S. L., Chisholm, S. W., and Shapiro, H. M. 1983. An inexpensive flow cytometer for the analysis of fluorescence signals in phytoplankton: chlorophyll and DNA distributions. *J. Exp. Mar. Biol. Ecol.* 68:129–44.

Olson, R. J., Vaulot, D., and Chisholm, S. W. 1985. Marine phytoplankton distributions measured using shipboard flow cytometry. *Deep-Sea Res. A* 32:1273–80.

Olson, R. J., Shalapyonok, A., and Sosik, H. M. 2003. An automated submersible flow cytometer for analyzing pico- and nanophytoplankton: FlowCytobot. *Deep-Sea Res. I* 50:301–15.

Partensky, F., Hess, W. R., and Vaulot, D. 1999. *Prochlorococcus*, a marine photosynthetic prokaryote of global significance. *Microb. Mol. Biol. Rev.* 63:106–27.

Peperzak, L., Vrieling, E. G., Sandee, B., and Rutten, T. 2000. Immuno flow cytometry in marine phytoplankton research. *Sci. Mar.* 64:165–81.

Schönhuber, W., Fuchs, B., Juretschko, S., and Amann, R. 1997. Improved sensitivity of whole-cell hybridization by the combination of horseradish peroxidase-labeled oligonucleotides and tyramide signal amplification. *Appl. Environ. Microbiol.* 63:3268–73.

Sciandra, A., Lazzara, L., Claustre, H., and Babin, M. 2000. Responses of growth rate, pigment composition and optical properties of *Cryptomonas* sp to light and nitrogen stresses. *Mar. Ecol. Prog. Ser.* 201:107–20.

Sheldon, R., and Parsons, T. 1967. A continuous size spectrum for particulate matter in the sea. *J. Fish. Res. Bd. Canada* 24:909–15.

Sheldon, R. 1978. Electronic counting. In: Sournia, A., ed. *Phytoplankton Manual.* UNESCO, Paris, pp. 202–14.

Simon, N., Lebot, N., Marie, D., Partensky, F., and Vaulot, D. 1995. Fluorescent in situ hybridization with rRNA-targeted oligonucleotide probes to identify small phytoplankton by flow cytometry. *Appl. Environ. Microbiol.* 61:2506–13.

Simon, N., Campbell, L., Ornolfsdottir, E., Groben, R., Guillou, L., Lange, M., and Medlin, L. K. 2000. Oligonucleotide probes for the identification of three algal groups by dot blot and fluorescent whole-cell hybridization. *J. Euk. Microbiol.* 47:76–84.

Sosik, H. M., Olson, R. J., and Chisholm, S. W. 1989. Chlorophyll fluorescence from single cells: interpretation of flow cytometric signals. *Limnol. Oceanogr.* 34:1749–61.

Trask, B. J., van den Engh, G. J., and Elgerhuizen, J. H. B. W. 1982. Analysis of phytoplankton by flow cytometry. *Cytometry* 2:258–64.

Vaulot, D., Olson, R. J., and Chisholm, S. W. 1986. Light and dark control of the cell cycle in two phytoplankton species. *Exp. Cell. Res.* 167:38–52.

Vaulot, D., Courties, C., and Partensky, F. 1989. A simple method to preserve oceanic phytoplankton for flow cytometric analyses. *Cytometry* 10:629–35.

Veldhuis, M. J. W., and Kraay, G. W. 2000. Application of flow cytometry in marine phytoplankton research: current applications and future perspectives. *Sci. Mar.* 64:121–34.

Veldhuis, M. J. W., Kraay, G. W., and Timmermans, K. R. 2001. Cell death in phytoplankton: correlation between changes in membrane permeability, photosynthetic activity, pigmentation and growth. *Eur. J. Phycol.* 36:167–77.

Yentsch, C. M., Horan, P. K., Muirhead, K., Dortch, Q., Haugen, E. M., Legendre, L., Murphy, L. S., Phinney, D., Pomponi, S. A., Spinrad, R. W., Wood, A. M., Yentsch, C. S., and Zahurenec, B. J. 1983. Flow cytometry and sorting: A powerful technique with potential applications in aquatic sciences. *Limnol. Oceanogr.* 28:1275–80.

Zubkov, M. V., Fuchs, B. M., Eilers, H., Burkill, P. H., and Amann, R. 1999. Determination of total protein content of bacterial cells by SYPRO staining and flow cytometry. *Appl. Environ. Microbiol.* 65:3251–7.

Zubkov, M. V., Fuchs, B. M., Burkill, P. H., and Amann, R. 2001. Comparison of cellular and biomass specific activities of dominant bacterioplankton groups in stratified waters of the Celtic Sea. *Appl. Environ. Microbiol.* 67:5210–8.

# CHAPTER 18

# MEASURING GROWTH RATES IN MICROALGAL CULTURES

A. MICHELLE WOOD
*Center for Ecology and Evolutionary Biology, University of Oregon*

R. C. EVERROAD
*Center for Ecology and Evolutionary Biology, University of Oregon*

L. M. WINGARD
*Center for Ecology and Evolutionary Biology, University of Oregon*

## CONTENTS

Key Index Words: Chemostat, Continuous Culture, Genetics

## 1.0. INTRODUCTION

Phycologists have many reasons to measure growth rates of algae in culture. In evolutionary studies, for instance, the population growth rate of an asexually dividing culture can be viewed as a measure of inclusive fitness, and in physiological studies, there is an interest in the degree to which biochemical constituents of a culture vary as a function of growth rate. Growth can be interpreted as any form of accretion of the biomass of algae in a culture, but the methods described here specifically pertain to situations in which the parameter being used to follow growth increases as a fixed percentage of the total per unit time (exponential growth). When the parameter of interest is cell number or a proxy measure that is directly proportional to cell number, these methods also provide an estimate of the population growth rate.

## 2.0. CONTINUOUS CULTURE

There is an extensive literature concerning calculation of growth rates for populations growing in chemostats and continuous culture, but even modern investigators should start with the foundational reports by Monod

(Monod 1949, 1950). Also useful are several other early publications (Novick and Szilard 1950; Herbert et al. 1956; Luedeking 1976) and the recent treatment by Gerhardt and Drew (1994). In continuous cultures, a fresh supply of medium is added continuously at the same rate at which it is withdrawn. In principal, this technique allows cultures to remain in exponential growth indefinitely. The steady-state concentration of cells in a continuous culture can be determined either by the concentration of a single limiting nutrient (chemostats) or by the selection of a dilution rate that maintains a particular cell density (turbidostat). At steady state the specific growth rate ($\mu$) of the population within either type of continuous culture is determined by the dilution rate:

$$\mu = F/V = D \qquad (1)$$

where F is the medium flow rate to and from the culture vessel (usually liters per hour for bacterial cultures and liters per day for algal cultures), V is the volume of the culture vessel (usually liters), and D is the dilution rate. This equation assumes that the specific death rate in the steady state culture is so much lower than the specific growth rate that it can be ignored. As noted by Gerhardt and Drew (1994), this is usually a reasonable assumption, but it may not be correct in cultures grown at very low densities or under physiological stress from the environment. All approaches to studying algal cells in continuous culture assume uniform mixing in the culture vessel, an assumption that does not always hold (see Chapter 19). In addition, because algae require light for photosynthesis, the time frame for measurement must take into account any diurnal pattern to cell division and growth in the culture (Caperon and Meyer 1972, Davis et al. 1978, Chisholm and Costello 1980).

---

## 3.0. BATCH CULTURE

Few references are as valuable as Guillard (1973) for working with discrete data from batch cultures or semi-continuous batch cultures. Because batch culture has so many advantages in terms of expense, ease of manipulation, volume of media required, and various manipulations that are possible, it is important to understand how to collect growth rate information from this type of culture.

### 3.1. Basic Data Collection

Any estimate of growth rates requires a time series of measurements that allow an estimate of the rate of change in biomass. If the goal of the study is to estimate population growth rates, then cell number must be counted using one of the methods described in previous chapters. Alternatively, another parameter can be measured as a proxy for cell number if it can be shown to be linearly correlated with cell number. Typical proxy measures are *in vivo* fluorescence, biomass (as dry weight, particulate organic material), and optical density. The concentrations of chlorophyll, protein, carbohydrate, and lipid in the culture are also used as proxy measures, but only when they can be shown to be linearly correlated with either cell number (for calculation of population growth rates) or biomass (for calculation of growth rate, in the simple sense of accretion of material). To use a proxy measure it is essential to know the growth conditions under which the proxy measure and cell number or biomass are linearly correlated and the concentration range under which this can be detected. For many parameters (e.g., pigment per cell, protein per cell, carbon per cell), there is a period of acclimation to any new growth condition, during which the relationship between the per-cell value of the parameter and cell number is quite variable. This period can last for 20 or more generations (Brand 1981, Kana and Glibert 1987). In general, parameters like chlorophyll fluorescence, particulate organic carbon (POC), particulate organic nitrogen (PON), protein, or carbohydrate cannot be used to follow changes in population size until a physiological steady state is achieved. Some components, notably lipid or carbohydrate content, continue to increase after a culture enters stationary phase and may be particularly misleading when used incautiously (Fisher and Schwarzenbach 1978; Shifrin and Chisholm 1981; Reitan et al. 1994; Zhekisheva et al. 2002).

The condition of steady-state growth is called *balanced growth* (Shuter 1979, Eppley 1981, Cullen 1992). Under these conditions, the average per-cell concentration of major cell constituents remains constant and the rate of change of these constituents in the population (culture) is constant. For algal cultures in balanced growth, this constant rate of change of cell components (and cell number) is apparent when examined on a per-photoperiod basis (c.f. Cullen 1992; see Chapter 19). Balanced growth is the standard condition for cultures growing in continuous culture (Eppley 1981), but it occurs in batch cultures only during the exponential phase of growth (Cullen 1992).

In general, cell constituents, like lipid or protein, are used in specialized cases where the actual research question concerns the rate of production of these materials in a culture. Cell number is a good choice if methods for counting are easily available and can be applied to the experimental system; using cell counts generally involves removal of at least a small volume of material from the batch culture. This can introduce contamination risk and actually alter the growth conditions if the total culture volume is substantially depleted during the course of the experiment. Before using a measure like fluorescence, carbon, protein, dry weight, or chlorophyll as a proxy for cell number in an experiment that requires measurement of growth rate, it is necessary to conduct preliminary experiments to verify the existence of a linear relationship between cell number and the particular parameter of interest under the experimental conditions.

When working with photosynthetic organisms like algae, it is also necessary to take into account the diel cycle when planning the sampling strategy. The synthesis of various cell components and the timing of cell division show diel periodicity in most phytoplankton species (Chisholm and Costello 1980; Chisholm 1981; Olson and Chisolm 1983; Vaulot et al. 1996; Sherry and Wood 2001). This means that sampling should occur at about the same time(s) during each photoperiod. For example, when sampling a culture growing on a 16 : 8 light-dark cycle with a light period beginning around 6:00 AM, one may measure cell densities or fluorescence at about 11:00 AM each morning. In practice, we find that sampling at the same time every day, plus or minus about a half hour, minimizes scatter in the time series of measurements. We also find that scatter is reduced if we avoid periods when cytokinesis is happening in cultures that have highly synchronized cell division at a particular point in the light-dark cycle. Thus, it can be useful to conduct preliminary studies to determine what the pattern of cell division is in the strain or species under investigation.

Nearly all calculations of growth rate use an exponential growth equation. Therefore it is helpful to have some advance concept of how rapidly cell number or biomass is changing and how long it takes to exhaust the medium when planning a growth-rate experiment. Ideally, one should ensure that the experiment is designed so that sampling occurs on at least three time points between the time the culture enters the exponential growth phase and the time it enters the T1 transition phase (Palenik and Wood 1988). In general, this involves using a small enough initial inoculum that the

culture medium can support five or more generations of growth.

## 3.2. Caution for Working with Diatoms

In most cases, the average size of diatoms decreases over time when a population is growing asexually (Paasche 1973; Wood et al. 1987; Armbrust and Chisholm 1992). This means that the per-cell value of fluorescence, chlorophyll, and biomass decreases over time, even in a population experiencing balanced growth. If the actual research question requires only an estimate of the growth rate of the parameter being measured (e.g., rate of biomass increase as POC or PON), this is not a problem. However, if the goal is to estimate population growth rates, with an ultimate interest in the number of cells produced, the changing size of diatoms complicates the use of proxy measures. Most investigators ignore the possible effects of cell size change when culture experiments last only a short time. However, if a culture is maintained for long periods between experiments, change in cell size that occurs during the intervening time can be significant and should be investigated. For example, the average diameter of isolates of the centric diatom *Thalassiosira tumida* (Janisch) Hasle in Hasle, Heimdal & Fryxell decreased by 11% during a 6-month period of routine maintenance (Wood et al. 1987). The cell volume of *Thalassiosira weissflogii* (Grunow) Fryxell et Hasle in exponential growth routinely decreases by nearly half after slightly more than 3 months (Armbrust and Chisholm 1992). These changes simply mean that calibration curves for parameters used to estimate cell number cannot be assumed to hold constant for a given diatom culture.

## 4.0. CALCULATING GROWTH RATES

During exponential growth, the rate of increase in cells per unit time is proportional to the number of cells present in the culture at the beginning of any unit of time. In other words, population growth follows this equation (we use $r$ for the exponential growth rate of the population, equal to "$K_e$" by Guillard [1973]):

$$dn/dt = rN \qquad (2)$$

the solution of which is

$$N_t = N_0 e^{rt} \qquad (3)$$

where $N_0$ is the population size at the beginning of a time interval, $N_t$ is the population size at the end of the time interval, and $r$ is the proportional rate of change, also called the *intrinsic rate of increase*, the *Malthusian parameter*, or the *instantaneous rate of increase* (Gotelli 1995). Units for $r$ are always expressed per unit time ($t^{-1}$). Solving Equation 3 for $r$, we get

$$r = \frac{\ln(N_t/N_0)}{\Delta t} = \frac{\ln N_t - \ln N_t}{\Delta t}, \qquad (4)$$

where $\Delta t$ is the length of the time interval ($t_t - t_0$). Equation 2 is equivalent to the classic growth equation

$$r = \mu - m, \qquad (5)$$

where $r$ is equal to $\mu$, the specific growth rate, when mortality ($m$) is zero. If "$t$" is expressed in days, then the growth rate $r$ can be converted to doublings per day ($k$) by dividing $r$ by the natural log of 2.0, according to the equation:

$$k = r/0.6931 \qquad (6)$$

Doublings per day ($k$) can be calculated directly from estimates of $N$ with use of the equation

$$k = \frac{\log_2(N_t/N_0)}{\Delta t} \qquad (7)$$

These equations can also be used to calculate doublings per unit time for any increment of time. In this chapter, and in most phycological papers, the nomenclatural assumption is that "$k$" refers to doublings per day and that "$t$" is expressed in days (or fractions of days). If time increments are less than 1 day, it may not be possible to assume that the growth rate calculated from any two consecutive points applies over a 24-hour period. In instances where the growth rate is calculated from an inclusive data set that spans less than a complete photoperiod, direct extrapolation to the daily growth rate is likely to be erroneous. In other words, the best estimates for growth rates of algal cultures require measurement over at least one and preferably several daily photocycles. Even cultures grown under continuous light may show daily rhythms of cell division and should be sampled over more than one daily cycle.

Doubling time, $T_2$, for the culture, expressed in the same units of time as $r$, can be calculated from an estimate of $r$ with use of the equation:

$$T_2 = 0.6931/r \qquad (8)$$

Regardless of the approach used, it is important to bear in mind the conceptual differences that underlie the terms *doubling time*, *division rate*, and *divisions per day*.

Intuitively, it may seem that a division per day would be equivalent to a growth rate of $1.0 \cdot d^{-1}$; however, this is not true. The continuous nature of exponential growth means that a culture growing at a rate, k, of one division per day has a doubling time, $T_2$, of 1 day and a growth rate, $r$, of $0.69 \ d^{-1}$.

## 5.0. SAMPLE DATA FOR CALCULATION OF GROWTH RATE

As noted previously, calculation of growth rate using standard equations requires exponential growth. When only two data points are available, it is possible to calculate a growth rate, but confidence in the rate is only as good as the investigator's certainty that the culture was in exponential growth for the entire time that elapsed between sampling for each data point. As noted previously, it is much more satisfactory to collect at a minimum of three time points in a series while the culture is in exponential growth.

In honor of R. R. Guillard's many contributions to phytoplankton research, we use essentially the same data set he used in 1973 (Guillard 1973) to demonstrate several methods for calculating growth. These are taken from an experiment on the growth of the centric diatom *Cyclotella cryptica* Reimann, Lewin, et Guillard. The raw data are presented in Table 18.1 and plotted as a semilog plot in Figure 18.1. A very slight lag phase is apparent during the first day after inoculation. Between day 1 and day 4 there appears to be good fit to a straight line, as would be predicted for exponential growth. Fitting a curve to these points, even by eye, highlights the presence of a possibly high cell count on day 3 (line A–A''). The culture appears to enter T1 transition to stationary phase between days 4 and 5, so it appears that four data points (days 1, 2, 3, 4) were obtained during the exponential growth period.

### 5.1. Calculation of Growth Rate from Two Points

In this example, the most reasonable choice of two points to use would be the data from days 1 and 4. Because this encompasses the longest possible period of exponential growth, it also provides the largest net increase in biomass. It is easier to get a good estimate of the increase in biomass during a time interval if there has been a large proportional change. When biomass is estimated at two closely spaced time intervals, the change in biomass can be so small that it is hidden in the measurement error for a single time point. Indeed, with use of Equations 4, 6, and 7, the value of $r$ calculated for the culture between days 1 and 4 is 0.75 and k is 1.09. As indicated in Table 18.2, calculations based on 24-hour time intervals are misleading; the inclusion of data points collected after the culture entered stationary phase dramatically reduce the value of the calculated growth rate. The high data point on day 3 also leads to low estimates of $r_{2-3}$ and $k_{2-3}$ and high estimates of $r_{3-4}$ and $k_{3-4}$.

A rough estimate of k can also be obtained with use of two data points and the graph provided in Figure 18.2. Using any pair of values for $N_1$ and $N_2$, read off the number of divisions during the time interval $T_1$ and $T_2$ and calculate k as explained in the figure legend. The estimates obtained for the *Cyclotella* data are given in Table 18.2. As in the previous example, the best estimate appears to be the one obtained from the data collected over the period day 1 to day 4 (1.09).

## 5.2. Inference from Graphical Representation of Growth

It is advisable to create a semilog graph of biomass or population size by plotting cell number, fluorescence, or whatever proxy measure is being used to follow growth versus time on semilog paper or by using the appropriate settings in a graphing program. Such a plot makes it easy to see when the culture is in exponential growth because of the straight-line relationship between the variables during exponential growth. If it is possible to maintain these plots daily during an experiment, then they also provide an easy way to monitor the growth stage of the culture and to properly time manipulations that are growth-stage-dependent. For example, Figure 18.1 shows that the population did not quite double during day 1 and that there was more than one doubling between days 2 and 3. If the plot were being maintained daily, this would have alerted the investigator to the possible onset of exponential phase between day 1 and day 2. On the basis of just these data points, the investigator would anticipate seeing slightly more than one doubling per day for the next day or so, depending on initial inoculum size. By day 3, it would be clear that biomass in the culture was doubling at an exponential rate. If there were *a priori* knowledge of the carrying capacity of the medium, straight-line extrapolation of the slope by eye would also provide warning of the onset of stationary phase between days 4 and 5.

Useful estimates of doubling time during the period of exponential growth can also be obtained by inspection of these simple plots. For example, in Figure 18.2,

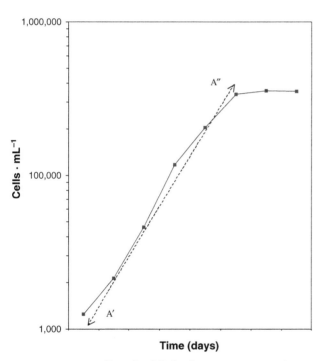

**FIGURE 18.1.** Growth of *Cyclotella cryptica* presented as a semilog plot using the data in Table 18.1. The straight line A′–A″ was fit by eye.

**TABLE 18.1** Growth curve for *Cyclotella cryptica* (modified from Guillard 1973).

| Day | 0 | 1 | 2 | 3 | 4 | 5 | 6 | 7 |
|-----|-----|------|------|------|------|------|------|------|
| Cell number | 1.25 | 2.14 | 4.62 | 11.8 | 20.5 | 33.8 | 35.7 | 35.3 |

Cell number = $n \times 10^4 \cdot mL^{-1}$.

**TABLE 18.2** Computation of population growth rate ($r$), divisions per day (k), and population doubling time ($T_2$) using equations from the text.

| Time interval (days) | 0–1 | 1–2 | 2–3 | 3–4 | 4–5 | 5–6 | 6–7 | 1–7 | 1–4 |
|---|---|---|---|---|---|---|---|---|---|
| $N_2/N_1$ | 2.14/1.25 | 4.62/2.14 | 11.8/4.62 | 20.5/11.8 | 33.8/20.5 | 35.7/33.8 | 35.3/35.7 | 35.3/2.14 | 20.5/2.14 |
| | 1.71 | 2.16 | 2.55 | 1.74 | 1.65 | 1.06 | 0.99 | 16.5 | 9.58 |
| $\ln(N_2/N_1)$ | 0.54 | 0.77 | 0.94 | 0.55 | 0.5 | 0.05 | −0.01 | 2.8 | 2.26 |
| $\log_2(N_2/N_1)$ | 0.77 | 1.11 | 1.35 | 0.80 | 0.72 | 0.08 | −0.01 | 4.04 | 3.26 |
| $\Delta_t = (t_2 - t_1)$ | 1 | 1 | 1 | 1 | 1 | 1 | 1 | 6 | 3 |
| $r$ (Eq.4)[a] | 0.54 | 0.77 | 0.94 | 0.55 | 0.5 | 0.05 | −0.01 | 0.47 | 0.75 |
| k (Eq.6) | 0.78 | 1.11 | 1.35 | 0.8 | 0.72 | 0.08 | −0.01 | 0.67 | 1.09 |
| k (Eq.7) | 0.78 | 1.11 | 1.35 | 0.8 | 0.72 | 0.08 | −0.01 | 0.67 | 1.09 |
| $T_2$ (Eq.8) | 1.28 | 0.9 | 0.74 | 1.26 | 1.39 | 13.86 | — | 1.49 | 0.92 |

[a]Assuming exponential growth and zero mortality, $r = \mu$, the intrinsic growth rate of the population.
Eq = Equation in text.

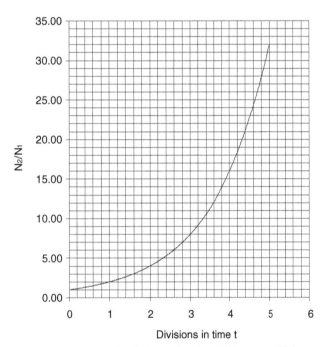

**FIGURE 18.2.** Graph for estimating division rate (k) from two counts of a culture, assumed to be in exponential growth. The ratio $N_2/N_1$ of the counts at times $t_2$ and $t_1$ ($\Delta t = t_2 - t_1$, expressed in days) is located on the ordinate. The corresponding abscissa then yields the number of divisions in the time interval $\Delta t$ needed to produce the increment in N. Divide by $\Delta t$ to obtain k as divisions per day. The function plotted is $y = 2^x$, $x \leq 5$ (from Guillard, 1973).

it is easy to see that slightly less than 1 day elapsed between the time the population size was $3 \times 10^4$ cells per mL and $6 \times 10^4$ cells per mL. From this, one could estimate that cell density would reach $10^5$ cells · mL$^{-1}$ in just 1 more day. This kind of quick estimate is often useful for planning experiments and workloads during the course of an experiment.

### 5.3. Estimating Growth Rate with Least-Squares Regression

One of the corollaries of the straight-line relationship between the log of cell number and time in an exponentially growing culture is that the growth rate $r$ corresponds to the slope of the straight line. Numerous authors explain how to calculate a linear regression for this type of data (Sokal and Rohlf 1994, Zar 1996). In the modern era, it is relatively easy to set up a spreadsheet program that allows entry of data, updates of log-linear plots, and calculation of the slope and regression coefficient for any set of points in the plot. This makes it easy to examine the data set and determine which set of consecutive time points provides the best fit to a straight line. These time points would then be viewed as bracketing the time period during which the culture was in exponential phase. The slope of the regression for the log-transformed values of N during that time interval is used as an estimate of $r$. Equations for manual calculation of growth rate by least squares regression of log-transformed data are described in Guillard (1973;

note nomenclatural differences discussed in footnote 1 if this method is used).

## 6.0. MEASURING ACCLIMATED GROWTH RATES

Culture experiments are often used to help evaluate the effect of different environmental conditions on algal growth (Doblin et al. 1999a,b; Taguchi and Kinzie 2001; Yamaguchi et al. 2001), physiology (Needoba et al. 2003), or the production of key metabolites (Keller et al. 1999a,b; Hamasaki et al. 2001). There is also a slowly growing literature that uses growth-rate experiments to demonstrate intraspecific genetic variation in growth rate (Brand 1985; Rynearson and Armbrust 2000; Doblin et al. 2000). Most of these experiments rely on measurement of the acclimated growth rate of the culture. The acclimated growth rate of a culture is the growth rate after the population has made all physiological adjustments required to maintain the same rate of balanced growth through multiple serial transfers into new media. These adjustments include those involved in acclimating to a major change in the incubation conditions (e.g., changes in light or temperature) and those involved in any transition from stationary or lag phase growth to exponential phase growth. By definition, an estimate of balanced growth made with batch culture techniques requires multiple serial transfers, all performed while the strain is still in exponential growth. This procedure of serial transfer while cultures are still in balanced growth is often called *semicontinuous batch culture* (Paasche 1973, 1977; Brand 1985). Like the continuous culture methods described earlier, semicontinuous batch culture provides a mechanism for maintaining an algal culture in exponential growth more or less indefinitely. Semicontinuous batch cultures differ from chemostats and turbidostats because cell density changes in the former, even if growth rate does not. In chemostats and turbidostats, both cell density and growth rate remain constant once the culture is in steady state.

### 6.1. Experimental Approach

1. If the culture to be used in the experiment is being removed from long-term maintenance at slow growth rates in a culture collection, inoculate one or two cultures in advance of the experiment so that the experiment can be started with a healthy inoculum containing as little contaminating material as possible (e.g., cell debris, waste metabolites, or bacteria). It may be necessary to step through several gradual transitions so that the culture does not expire from shock. For example, if the stock cultures are maintained at 12°C and the growth experiment is conducted at 23°C, it may be good to coax the stock culture through a series of transfers involving two- or three-degree increments of temperature change per transfer. It is also prudent to save some of the cultures grown at intermediate conditions as backup in case a transfer is unsuccessful.

2. Prepare an adequate amount of medium for the entire experiment, with replication. Clonal cultures of algae are known to show different acclimated growth rates when grown under the same conditions but with different batches of medium.

3. Once the culture appears stable in the new environment, estimate the time course of the growth curve for the culture under the experimental conditions. This can be done with one or two cultures, using whatever measure of cell growth that is used in the experiment. Begin by estimating the cell concentration in the source culture and inoculating the experimental culture at a very low dilution. The dilution constant can be used to estimate initial cell number in the experimental culture. Follow growth through the lag and exponential phases and into stationary phase by making cell number estimates daily. It is advisable to plot the values for N each day so that transition to stationary phase does not unexpectedly pass by.

4. Data collected in step 3 provide a good estimate of how quickly the culture will enter the transition to stationary phase. These data are used for deciding what inoculum size is required to ensure that data from at least three time points can be collected during exponential phase, before any nutrient starvation occurs in the culture. For example, the data in Figure 18.1 suggest that the *Cyclotella* culture would have stayed in exponential growth for 6 days if the inoculum size had been ~1,000 cells · mL$^{-1}$ instead of ~10,000 cells · mL$^{-1}$.

5. Steps 5–8 are the heart of an estimation of acclimated growth rate. This is the maximum growth rate that a strain will show when fully acclimated to a given set of environmental conditions. It is a method for ensuring that the

growth rate is estimated during balanced growth. Under these conditions, the estimate of growth rate should also be highly replicable. To begin, inoculate a tube of medium with the culture that recently entered stationary phase. Be sure to collect data necessary to estimate $N_0$. Collect data necessary to estimate $N_{1...t}$, sampling at approximately the same time every day. Plot the data on a semilog plot. Exponential growth will be identifiable as the region in the plot where there is a straight-line relationship between cell number and time.

6. Before the culture reaches the T1 transition to stationary phase, remove the appropriate amount of medium and inoculate another tube; follow growth as above. Plot on the same graph or a graph of similar dimensions, so that the slope of the growth curves can be followed and compared.

7. Repeat Step 6 until the slopes of three consecutive transfers are the same. Once this occurs, and while the culture is still in exponential phase, it can be considered "acclimated." It is not sufficient to go through a certain number of transfers or generations to achieve acclimation; the critical step is to verify that the exponential growth rate is constant over multiple transfers (usually at least three). Even cultures that have been maintained under constant incubation conditions will go through an acclimation phase when transferred to new media, unless they were in balanced growth at the time of transfer. If a culture goes into stationary phase before transfer, the cells in the inoculum require some time, and often several divisions, to achieve balanced growth again. Thus, the timing of transfers for this method demands that the inoculum used for each transfer is obtained from its parent culture before any cells begin to experience starvation for the limiting nutrient. Brand (1985) recommends making transfers before the culture reaches one-tenth of the density at which the T1 transition begins. We have had success with this method by making transfers before the culture reached 50% of the density at T1 transition, although it is probably safer to follow Brand's recommendation. (Note that when an exponentially growing culture reaches 50% of the carrying capacity in batch culture, it will be in transition phase after one more generation.)

8. Once the slope of three consecutive transfers is the same, inoculate three tubes or flasks with an identical amount of inoculum from the most recent

transfer while it is still in exponential phase. Use the insight gained in the preceding steps to ensure that the inoculum is small enough so that at least three measurements are collected while the culture is in the exponential phase of growth. Inoculation of three tubes ensures the ability to calculate a mean and variance for the growth rate of the culture under acclimated conditions. Although duplicate cultures are sufficient for this purpose, they provide no backup in case of culture loss. Additionally, the power of t-tests used to compare means among different estimates of acclimated growth rates increases dramatically if n = 3 rather than 2 (Sokal and Rohlf 1994). The added power of additional replication is significant, but the rate at which statistical power increases with increasing replication is greatest when moving from one to two or three replicates (Sokal and Rohlf 1994).

A shortcut that can save time in some cases is to pool the data collected in the last three transfers during Step 7 to estimate the mean and standard deviation for the acclimated growth rate. This practice is not acceptable if the data are going to be used for the analysis of variance, because the three data sets do not represent true replicates. Step 8 provides true replication and also serves as a control against premature determination that a culture was acclimated. If the cultures are genuinely acclimated by the end of Step 7, then the mean of the three growth rates observed in the last three sequential transfers in Step 7 should be within a standard deviation of the mean obtained in Step 8.

---

## 7.0. Measuring Acclimated Growth with Using *In Vivo* Fluorescence

One of the more popular proxy measures for cell number or biomass is *in vivo* fluorescence, particularly because of the development of a rapid method for monitoring growth by measuring *in vivo* fluorescence (Brand et al. 1981). The Brand approach relies on two major assumptions and the timely development of the Turner AU-10 fluorometer, a reasonably priced instrument that is ideally suited for this type of work. The basic assumptions of the method are that the fluorescence yield of a culture in balanced growth is linearly correlated with cell number and that it is detectable over a wide range of cell densities. A great many variables influence the fluorescence yield of a phytoplankton culture (see

Falkowski et al. 1986, Chapter 19). Brand (1981) noted that diatoms did not appear to maintain the constant growth rates required for the method. However, in practice, it appears that the instrumental configuration of a Turner Designs AU-10 equipped with the basic *in vivo*/extracted chlorophyll *a* kit (TD 10–037-R) has been remarkably applicable to growth rate measurements with a wide range of microalgal groups under a wide range of conditions (Brand 1985, 1991; Doblin et al. 1999a,b, 2000; Maldonado and Price 2000, 2001; Tarutani et al. 2001; Uchida et al. 1999). We have found that it worked well with the diatom *Skeletonema costatum* (Greville) Cleve and various picoplankton and dinoflagellates.

This method requires uniform suspension of the organisms in the culture, at least during the short period it takes to measure fluorescence. Different taxa respond differently to various mechanisms for mixing the culture; some are highly tolerant of agitation with a vortex mixer and others are not. Gentle agitation can be adequately effective. The method is not very well suited to highly clumped suspensions or for cultures that adhere to the walls of the culture vessel. When using this method, it is important that the culture be briefly acclimated to subsaturating irradiances for a few minutes before it is put in the fluorometer and allowed to reach steady-state fluorescence yield when exposed to the excitation beam. A blank tube of medium should be run at the beginning of each day's measurements and subtracted from the reading for each culture (Cullen and Davis 2003). Additionally, optical variations in the culture tube can influence the readings; it is often helpful to use a mark on each tube (including the blank) to ensure that it is always oriented in the same way when in the sample holder. Tubes should be clean and wiped free of fingerprints before each reading.

There are several technical considerations that investigators should bear in mind when using Turner Designs instruments in the standard Brand methodology. For growth rate measurements, the TD-10–037-R optical kit has application to a broader range of algal taxa than the specialized optical kit for measuring extracted chlorophyll without acidification (10-040-R). This is because the optical package excites a broader range of photosynthetically active molecules in living algal cells and the emission is measured over a broader range of wavelengths. Specifically, the TD-037-R uses a Daylight white lamp (F4T5D), a wide band-pass excitation filter (CS-5-60, 340–500 nm) and a >665-nm-long pass emission filter (CS-2-64). The TD-10-037R can be used to follow growth in essentially all eukaryotic algae, and as shown in the example in this chapter,

it works well even for phycoerythrin (PE)-containing cyanobacteria. In contrast, the excitation wavelengths provided by the blue mercury vapor lamp in the 10-040-R optical kit are too narrow to excite most spectral forms of phycoerythrin efficiently, and the emission band-pass (>680 nm) does not permit transmission of even the longest wavelengths emitted by phycoerythrin. It is always important to know the precise specifications of the optical kit installed in a Turner Designs fluorometer and to report this information when using a Turner Designs instrument to follow growth.

One aspect of the Turner Designs instrument that makes it particularly popular for monitoring microalgal growth is the fact that 25-millimeter culture tubes can be placed directly into the instrument for each reading. There is no subsampling of the culture, and measurement of an individual culture takes less than a minute (Brand 1981).

The Turner Designs TD700 is a relatively new, fairly inexpensive, benchtop fluorometer that measures fluorescence from samples in 25-millimeter culture tubes. Although it can be outfitted with the same optical kits as the AU-10, the reference correction present in the AU-10 has been eliminated. This may cause problems with use of the TD700 for estimating growth rate of phytoplankton cultures, because the lack of an internal reference means that there is no correction for slight variations in lamp output. Use of solid secondary standards (available from Turner Designs) would be a solution to this problem if it arises. However, it is quite difficult to maintain cultures with aeration in 25-millimeter tubes. Thus, algal cultures that require bubbling with air or $CO_2$ to prevent $CO_2$ limitation during growth may need to be maintained in flasks or other culture vessels, which means a slight loss of convenience for monitoring growth.

## 7.1. Semicontinuous Batch Culture

In Step 5 (listed previously) we discuss the semicontinuous batch culture approach as involving transfer of a small amount of exponentially growing cells to a new tube or flask of medium as a means of keeping the cells in exponential growth. This ensures that the medium remains very fresh, as well as providing the cells with a continuous supply of nutrients. An alternative approach has been suggested by John Cullen (pers. comm.). This approach achieves the same conditions of extended exponential growth by maintaining the algal cells in the same culture tube throughout the acclimation period; instead of transferring cells to a tube of new medium, the serial transfers are achieved by removing a specific volume of culture and replacing it with fresh medium. In other words, after several generations of growth, some volume of culture is removed and new medium is added. This approach minimizes use of glassware and works well as long as biofilms on the tube do not affect fluorescence measurement or the growth conditions (John Cullen, pers. comm.). When this approach is used, the initial water level in the culture tube should be marked on the tube with a felt-tip marker. Additionally, nearly all the medium (and accompanying cells) need to be removed during each instance of removal and replenishment so that the old medium and any metabolites it has accumulated do not markedly influence the chemical composition of the new medium.

## 7.2. Example of Monitoring Acclimation with *in Vivo* Fluorescence

Figure 18.3 illustrates the steps taken to estimate the acclimated growth rate of a culture of phycoerythrin containing picocyanobacteria with use of the Brand approach. Cells were grown in 25-millimeter borosilicate tubes under a 14:10 day-night cycle at 25°C. Illumination was provided by cool white fluorescent lights ($115 \ \mu E \cdot m^{-2} \cdot s^{-1}$). A Turner Designs AU-10 fluorometer equipped with a 10-037-R optical kit was used for this experiment, and the strain of *Synechococcus* is described in Wingard et al. (2002). Figure 18.3a illustrates the nonlinearity of the relationship between fluorescence and cell number that occurs at high cell densities. This probably results from changes in fluorescence yield associated with self-shading; in time, this would also affect growth rate. For these cells, under these conditions, it is clear that *in vivo* fluorescence is going to be useful only as a measure of cell number while the culture density remains between $10^5$ and $10^7$ per milliliter and fluorescence is between zero and 250 relative fluorescence units. Figure 18.3b shows the fluorescence readings obtained during the last three sequential transfers of the culture during acclimation; black arrows show the point in time at which a small volume was removed from the growing culture to inoculate the next culture. The open arrow in Figure 18.3b indicates the point at which cells were removed to inoculate the three replicate cultures from which the acclimated growth rate was calculated. Growth of these three cultures is plotted in Figure 18.3c; as is apparent,

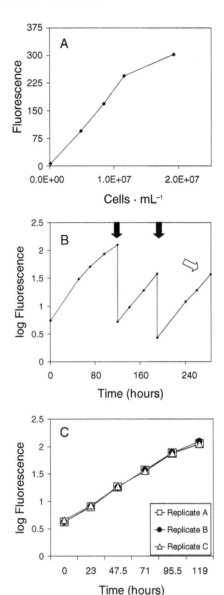

**FIGURE 18.3.** Example of experimental approach to measuring acclimated growth rate in marine *Synechococcus* using *in vivo* fluorescence and semicontinuous batch culture. **(a)**, the calibration of the fluorometer with cell number, illustrating the range of cell densities over which a linear relationship between population size and *in vivo* fluorescence can be assumed. Although linear regression of all points yields a satisfactory $R^2$ value (0.95), inspection by eye indicates that linearity cannot be assumed after fluorescence exceeds 225 relative fluorescence units in this particular combination of cell type and instrument. All cells are in exponential growth and acclimated to the growth environment. **(b)**, the time course of growth, measured by fluorescence, during the last three transfers in Step 7 of the acclimation process (see text). **(c)** the growth curve for three replicate cultures inoculated with cells collected from the culture noted with an open arrow in (b) (Step 8; see text). Both solid and open arrows show the point at which inocula were obtained from the growing culture and used to inoculate the next culture in the series.

the coefficient of variation among the three replicate cultures was very low, providing the basis for detection of even small differences in acclimated growth rate among replicates of the same strain acclimated to different environments or among different strains or species acclimated to the same environment.

## 7.3. Using Growth Rate to Demonstrate Genetic Differences

Earlier reviews have emphasized the importance of using more than one genotype or clonal isolate when trying to characterize the growth rate of a species in any given environment (Wood 1988; Brand 1991; Wood and Leatham 1992). Like many ecologically important characteristics, growth rate is a quantitative trait. In other words, it is determined by the combined action of multiple genes and can be studied by means of the statistical methods of quantitative genetics (c.f. Lynch and Walsh, 1998). Despite many advances in the application of molecular genetics to studies of phytoplankton diversity (Wood and Townsend, 1990; Stabile et al. 1992; Medlin et al. 1996; Toledo and Palenik 1997; Rynearson and Armbrust 2000, 2004; Montresor et al. 2003; Steglich et al. 2003), it is still difficult to determine what, if any, phenotypic diversity is associated with this genetic diversity. Because natural selection acts at the phenotypic level, this poses a problem for investigators interested in the potential for microalgal species to evolve in response to changing environmental conditions. One solution to the problem is to use quantitative genetic approaches to determine if there is a genetic component to the variance of any trait that is expected to be the target of selection. Because the growth rate of unialgal cultures is a direct measure of inclusive fitness, it is a particularly informative trait to examine. Differences in the acclimated growth rate of two strains or isolates provide a direct estimate of the selection differential and can be used in population genetic models (Fig. 18.4; Wood 1989, Rynearson and Armbrust, 2000).

To determine whether two (or more) strains have a genetically determined difference in their acclimated growth rate, it is necessary to eliminate all other sources of variation from the comparison. Normally, this is done by growing the strains together under identical conditions. This type of common garden experiment is widely used in plant ecology and was pioneered by Larry Brand in phytoplankton ecology (Brand 1981, 1985). It has been used recently by Rynearson and Arm-

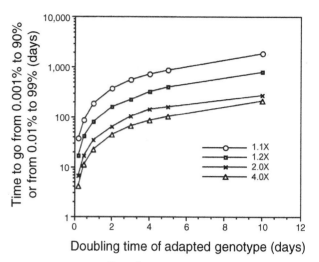

**FIGURE 18.4.** Time for a rare, newly adapted genotype to become abundant in a culture (or community) when the doubling time of the predominant genotype is 10 to 400% greater than the doubling time of the rare genotype. When considering the competitive advantage of one strain over another, it is important to remember that growth rates measured in cultures cannot be extrapolated directly to the field, even when the intention of the culture experiment was to simulate environmental conditions of direct ecological relevance. (From Wood 1989, reprinted with permission, Japan Scientific Societies Press.)

brust (2000, 2004), Doblin et al. (2000), Moore et al. (1998), and others. The underlying assumption is that statistically significant differences in the acclimated growth rate of different clonal isolates must have a genetic basis if all sources of environmental variation are eliminated. This reflects a fundamental tenet of evolutionary population genetics: phenotypic variation has only two major components: environment and genotype. In other words, regardless of the molecular basis for differences in the acclimated growth rate of two species, one can determine whether there is a statistically significant genetic component to the difference in growth rate between two strains by simply measuring their acclimated growth rates in a common garden experiment. The relationship between the genetic and nongenetic components of variation in growth rate is described by substituting $r$ for $z$ (phenotypic value for any quantitative trait) in the standard equation:

$$r = r_G + r_E + r_{G \times E} \qquad (8)$$

where $r$ is the acclimated growth rate for a particular strain in a specific environment, $r_G$ refers to the sum of the total effects of all loci on the acclimated growth rate (i.e., the genotypic value), $r_E$ to the deviation from the genotypic value caused by the environment, and $r_{G \times E}$ to the effect of any interaction between the genotypic value and the environment (Wood et al. 1987; c.f. Wood 1988; Lynch and Walsh 1998). Measurement error is a third source of observed phenotypic variation and must be taken into account in order to detect environmental and genetic effects.

In a common garden experiment, most sources of environmental variation are eliminated by incubating all the replicates of all genotypes under the same environmental conditions at the same time. The location of individual culture tubes or flasks within the incubator is repeatedly randomized during the experiment. This helps to eliminate bias caused by microenvironmental variation within the incubator and ensures that each replicate of a particular genotype has an independent positional history relative to other replicates. If the experiment is analyzed with multivariate techniques, the effects of microenvironmental variation in the growth chamber and the effects of measurement error are part of the overall error variance. Excellent discussions of the proper design and analysis of this type of experiment may be found in the statistical textbooks already mentioned (Sokal and Rohlf 1994, Zar 1996), as well as other books devoted to the design of ecological experiments (e.g., Underwood 1996, Scheiner and Gurevitch 2001). The study of intraspecific genetic variation in diatom morphology by Wood et al. (1987) also provides a model for the design and analysis of common garden experiments.

We consider that the minimum design for detection of intraspecific genetic variation includes at least six individual cultures: three replicates of two clonal isolates. If these are all incubated together under the same conditions, they usually provide enough power to detect a difference in the mean of the two genotypes that is great enough to be of interest to us with a reasonable level of statistical significance. Sokal and Rohlf (1994) provide an excellent discussion of the increase in power of an experimental design with increasing replication. It is critical to take the matter of replication into account so that the optimal balance between cost, time, and statistical power is achieved. For example, if the acclimated growth rate of four strains will be compared with a randomized block design, and the true difference between the means is 15% of the mean and the true standard error per unit $\sigma$ is 5% of the mean, it would take three replicates to detect the difference with a probability of 90% at the 0.05 confidence level. If the goal is to ensure a 90% chance of detecting the differ-

ence at the 0.01 confidence level, the number of replicates would have to be increased to six (Cochran and Cox 1957).

---

## 8.0. Final Comments

### 8.1. Other Methods for Measuring Growth Rates

The advent of new technologies like flow cytometry, fluorescent staining, and high pressure liquid chromatography have introduced a wide range of methods for examining growth rate that were not in the literature when Guillard wrote the first version of his chapter (Guillard 1973). Although the technical details of these new approaches are beyond the scope of the current chapter, it should be noted that in most cases, even the newest methods still rely on the same basic mathematical formulations for calculating exponential growth rate. For example, the equation for calculating carbon-specific growth rates, $\mu_C$, from rates of incorporation of radiolabeled $^{14}C$–bicarbonate,

$$\mu_C = -1/t \left[ \ln\left(1 - {}^*P_{Chl\,a}\right)\right] \qquad (9)$$

is a simple rearrangement of Equation 4 in which $[\ln(1 - {}^*P_{Chl\,a})]$ is substituted for $(\ln_{Nt} - \ln_{N1})$, t is equivalent to the incubation time of the experiment and thus is equivalent to $\Delta t$, and $P_{Chl\,a}$ is the specific activity of chlorophyll $a$ divided by the specific activity of the inorganic carbon (Goericke and Welschmeyer 1993). Techniques based on cell cycle analysis and the frequency of dividing cells use the theory of exponential growth described here. However, the actual calculations for these methods require discrete time equations, as discussed by numerous authors (McDuff and Chisholm 1982; Campbell and Carpenter 1986; Liu et al. 1998; Sherry and Wood 2001). Techniques for measuring growth rates of natural populations using diffusion theory (Lande et al. 1989) or measurement of rRNA content using rRNA-targeted nucleic acid probes (Binder and Liu 1998, Worden and Binder 2003) are not usually used to estimate the growth rate of algal cultures. To the contrary, the basic *in vivo* fluorescence method described here is the principal method used to demonstrate the growth-rate dependence of rRNA content by investigators who are developing the targeted rRNA approach (Binder and Liu 1998, Worden and Binder 2003). Similarly, Lin and Corstjens (2002) used growth rate calculations based on daily cell counts to examine the potential for expression of the proliferating cell nuclear antigen gene to be used as a measure of growth rate.

## 8.2. WHEN TO MEASURE GROWTH RATES

We have tried to emphasize the care with which growth rates must be measured. Any application of the formulae in this chapter implicitly assumes that the culture is in exponential growth throughout the time period over which growth rate is being calculated. Often the equations presented here are used inappropriately to estimate "growth rates" when the assumption of exponential growth doesn't apply, either because the culture was not in exponential growth during the entire interval of time used to parameterize $\Delta t$ or because the growth habit of the organism does not lead to exponential accumulation of biomass or cell number. In many of these cases, the investigator really needs to know only the culture yield, and this is what should be calculated and reported.

Yield is determined by calculating the difference between initial inoculum size and the standing stock after some interval of growth. If this value is divided by the length of time that elapsed between inoculation and harvest (or biomass measurement), a *yield rate*, $Y$, can be calculated:

$$Y = (N_1 - N_2)/\Delta t \qquad (10)$$

where $N_1$ and $N_2$ are the biomass or population size at the beginning and end, respectively, of the measurement period and $\Delta t$ represents the time elapsed between the times when $N_1$ and $N_2$ were estimated. Yield rate is often useful, but investigators should be careful not to imply that it represents a rate that was constant throughout the time interval of measurement unless additional data support such a conclusion. Additionally, exponential growth cannot be assumed when the growth pattern of the organism does not conform to models of exponential growth. In some colonial or filamentous species, cell division occurs in only particular cells or locations in the colony. In these situations, the rate of biomass accumulation is not a fixed proportion of the total biomass per unit time. In these cases, the

only way to estimate growth is by repeated measurement of yield.

## 8.3. Maternal Effects

Effects of transferring cytoplasmic material from parent to offspring are included under the genetic concept of "maternal effects." This type of maternal effect can affect growth rates, particularly if growing cells experience a change in environment. When transferring cultures that are rapidly growing, or when the suspension of cells used to inoculate the culture is markedly different from the new medium (i.e., one is complete and the other is not), the growth of the culture in the new environment or medium will be influenced by the effects of the old environment or medium on cell composition for several generations. A typical example of this would be growth of cells transferred from nutrient-replete medium to conditions of nitrogen starvation. Initially, the cells would continue to divide, transferring some of the nitrogen present in the cytoplasm of the original nutrient-replete cells to each daughter cell. However, once all the nitrogen pools that could be mobilized under starvation conditions were used up, growth would stop. Similarly, cells transferred from low light to high light, or vice versa, will respond by changing the concentration of light-harvesting pigments per cell. This can take several generations to reach steady state; during that time, there will not be a good correlation between growth rate changes and pigment per unit culture volume or *in vivo* fluorescence.

## 8.4. Evolution in Clonal Cultures

Once the acclimated growth rate of a clonal culture has been determined for a particular environment, it is reasonable to expect that this will remain constant with time. However, this assumes that the culture remains the same genetically. Unfortunately, even though a clonal culture is established from a single cell by propagation through asexual division, it cannot be assumed that the cells in the culture will remain genetically identical to the original founding cell. Mutation and sexual recombination within the culture are the two most likely mechanisms by which the genotype of a clonal culture of microalgae can change. Mitotic recombination, as occurs in somatic cells of some model organisms, is an additional, although highly speculative, possible source of genetic variation in clonal cultures.

Given the large population sizes of most microalgal cultures, it is likely that at least one spontaneous mutation occurs in nearly every generation. Thus, the most probable reason that clonal cultures maintain genetic fidelity with the founding cell is that the mutations are deleterious and thus the mutant cell and its offspring will be outcompeted by the founding genotype. However, if the mutation is favorable, the new genotype will likely replace the original genotype. This can happen even within the time frame of some growth experiments if the mutant has a substantial selective advantage (Fig. 18.4; Wood 1989, Lynch et al. 1991).

If cells in a clonal culture are diploid, then considerable genetic diversity can also arise in a "clonal" culture as a result of sexual reproduction within the culture. This can occur only if compatible gametes can be produced by the single founding genotype. Species that have mating types will not be able to mate in culture because the genotype that founded the clonal lineage will be able to produce only one of the two needed gametic types. On the other hand, if the genotype is self-compatible, diploid during vegetative growth, and heterozygous at some loci, the potential for generating a large amount of genetic diversity in a "clonal" culture is considerable. Armbrust's painstaking study of changes of cell size in a diatom culture during prolonged culture makes it clear that sex does occur in microalgal cultures (Armbrust and Chisholm 1992). Finally, given the evidence that recombination can occur spontaneously during mitosis in yeast (Freedman and Jinks-Robertson 2002) and can be induced in other eukaryotes (Beumer et al. 1998, Liu et al. 2001), mitotic recombination cannot be ruled out for algal cultures. In yeast and *Drosophila*, the recombination event does not affect subsequent generations because it occurs in somatic cell lines; however, for single-celled organisms, all cells are effectively part of the germ line. If the mechanisms that permit recombination during mitosis in these model organisms can operate in eukaryotic algae, mitotic recombination is a potential source of genetic variability.

One way to check to see if prolonged culture under new conditions has led to an evolutionary change is to reacclimate it to the original growth conditions. If directional selection had caused the culture to become better adapted to the new environment, its growth rate in the old environment is likely to have decreased. Absence of this effect is not conclusive proof that evolution did not happen in the culture, but because growth rate is a product of the combined effects of all genes in the genome, it is potentially sensitive to genetic change at any locus.

---

## 9.0. Acknowledgments

The authors thank Bob Guillard for his correspondence and encouragement over the years, particularly in connection with the development of this chapter; Bob Andersen for his heroic efforts in shepherding this volume to publication; ONR for financial support through grant N0014-99-1-0177; and John Cullen, Jim McCormick, the members of the Wood/Castenholz laboratory group, and two anonymous reviewers for comments on earlier versions of the manuscript. A. M. W. thanks all her students and postdocs for their patience and dedication to careful algal growth measurements, especially Samantha Garbush, Craig Everroad, Lauren Wingard, Cindy Thieman, Zahit Uysal, Feran Garcia-Pichel, Scott Miller, and Mark Teiser.

---

## 10.0. References

Armbrust, E. V., and Chisholm, S. W. 1992. Patterns of cell change in a marine centric diatom: variability evolving from clonal isolates. *J. Phycol.* 28:146–56.

Beumer, K. J., Piminelli, S., and Gollic, K. G. 1998. Induced chromosomal exchange directs the segregation of recombinant chromatids in mitosis of *Drosophila*. *Genetics* 150:173–88.

Binder, B. J., and Liu, Y. C. 1998. Growth rate regulation of rRNA content of a marine *Synechococcus* (Cyanobacterium) strain. *Appl. Environ. Micro.* 64:3346–51.

Brand, L. E. 1981. Genetic variability in reproduction rates in marine phytoplankton populations. *Evolution.* 35:1117–27.

Brand, L. E. 1985. Low genetic variability in reproduction rates in populations of *Prorocentrum micans* Ehrenb. (Dinophyceae) over Georges Bank. *J. Exp. Mar. Biol. Ecol.* 88:55–65.

Brand, L. E. 1991. Review of genetic variation in marine phytoplankton species and the ecological implications. *Biol. Oceanogr.* 6:397–409.

Brand, L. E., Guillard, R. R. L., and Murphy, L. S. 1981. A method for the rapid and precise determination of acclimated phytoplankton reproduction rates. *J. Plankton Res.* 3:193–201.

Campbell, L., and Carpenter, E. J. 1986. Diel patterns of cell division in marine *Synechococcus* spp. (Cyanobacteria): Use of the frequency of dividing cells technique to measure growth rate. *Mar. Ecol. Prog. Ser.* 32:139–48.

Caperon, J., and Meyer, J. 1972. Nitrogen-limited growth of marine phytoplankton. II. Uptake kinetics and their role in nutrient growth of phytoplankton. *Deep-Sea Res. Part I. Oceanogr. Res. Pap.* 19:619–32.

Chisholm, S. W., and Costello, J. C. 1980. Influence of environmental factors and population composition on the timing of cell division in *Thalassiosira fluviatilis* (Bacillariophyceae) grown on light:dark cycles. *J. Phycol.* 16:375–83.

Chisholm, S. W. 1981. Temporal patterns of cell division in unicellular algae. In: Platt, T., ed. *Physiological Bases of Phytoplankton Ecology.* Canadian Bulletin of Fisheries and Aquatic Sciences, 210:150–81.

Cochran, W. G., and Cox, G. M. 1957. *Experimental Designs.* 2nd ed. John Wiley & Sons, New York, 612 pp.

Cullen, J. 1992. Factors limiting primary productivity in the sea: In: Woodhead, A. D., and. Falkowski, P. G., eds. Nutrients: Nutrient Limitation and Marine Photosynthesis. Primary Productivity and Biogeochemical Cycles in the Sea. *Environ. Sci. Res.,* 43:69–88.

Cullen, J. J., and Davis, R. F. 2003. The blank can mean a big difference in oceanographic measurements. *Limnol. Oceanogr. Bull.* 12:29–35.

Davis, C. O., Breitner, N. F., and Harrison, P. J. 1978. Continuous culture of marine diatoms under silicon limitation. 3. A model of Si-limited diatom growth. *Limnol. Oceanogr.* 23:41–52.

Doblin, M., Blackburn, S., and Hallegraeff, G. 1999a. Comparative study of selenium requirements of three phytoplankton species: *Gymnodinium catenatum, Alexandrium minutum* (Dinophyta) and *Chaetoceros* cf. *tenuissimus* (Bacillariophyta). *J. Plankton Res.* 21:1153–69.

Doblin, M. A., Blackburn, S. I., and Hallegraeff, G. M. 1999b. Growth and biomass stimulation of the toxic dinoflagellate *Gymnodinium catenatum* (Graham) by dissolved organic substances. *J. Exp. Mar. Biol. Ecol.* 236:33–47.

Doblin, M. A., Blackburn S. I., and Hallegraeff, G. M. 2000. Intraspecific variation in the selenium requirement of different geographic strains of the toxic dinoflagellate *Gymnodinium catenatum. J. Plankton Res.* 22:421–32.

Eppley, R. W. 1981. Relations between nutrient assimilation and growth in phytoplankton with a brief review of estimates of growth rate in the ocean. In: Platt, T., ed. *Physiological Bases of Phytoplankton Ecology.* Canadian Bulletin of Fisheries and Aquatic Sciences, 210:251–63.

Falkowski, P. G., Wyman, K., Ley, A. C., and Mauzerall, D. C. 1986. Relationship of steady-state photosynthesis to fluorescence in eukaryotic algae. *Biochim. Biophys. Acta.* 849:237–40.

Fisher, N. S., and Schwarzenbach, R. P. 1978. Fatty acid dynamics in *Thalassiosira pseudonana* (Bacillariophyceae): Implications for physiological ecology. *J. Phycol.* 14:143–50.

Freedman, J. A., and Jinks-Robertson, S. 2002. Genetic requirements for spontaneous and transcription-stimulated mitotic recombination in *Saccharomyces cerevisiae. Genetics.* 162:15–27.

Gerhardt, P., and Drew, S. W. 1994. Liquid culture. In: Gerhardt, P., Murray, R. G. E., Wood, W. A., and Kreig, N. R., eds. *Methods for General and Molecular Bacteriology,* 2nd ed. American Society for Microbiology, Washington, D.C., 224–47.

Goericke, R., and Welschmeyer, N. A. 1993. The chlorophyll-labeling method: Measuring specific rates of chlorophyll *a* synthesis in cultures and in the open ocean. *Limnol. Oceanogr.* 38:80–95.

Gotelli, N. J. 1995. *A Primer of Ecology.* Sinauer, Sunderland, Massachusetts, 206 pp.

Guillard, R. R. L. 1973. Division rates. In: Stein, J. R., ed. *Handbook of Phycological Methods: Culture Methods and Growth Measurements.* Cambridge University Press, Cambridge, pp. 289–312.

Hamasaki, K., Horie, M., Tokimitsu, S., Toda, T., and Taguchi, S. 2001. Variability in toxicity of the dinoflagellate *Alexandrium tamarense* isolated from Hiroshima Bay, western Japan, as a reflection of changing environmental conditions. *J. Plankton Res.* 23:271–8.

Herbert, D., Elsworth, R., and Telling, R. C. 1956. Continuous culture of bacteria: A theoretical and experimental study. *J. Gen. Microbiol.* 14:601–22.

Kana, T. M., and Glibert, P. M. 1987. Effect of irradiances up to 2000 μE · m$^{-2}$ · s$^{-1}$ on marine *Synechococcus* WH7803. I. Growth, pigmentation, and cell composition. *Deep-Sea Res. Part I. Oceanogr. Res. Pap.* 34:479–95.

Keller, M. D., Kiene, R. P., Matrai, P. A., and Bellows, W. K. 1999a. Production of glycine betaine and dimethylsulfoniopropionate in marine phytoplankton. I. Batch cultures. *Mar. Biol.* 135:237–48.

Keller, M. D., Kiene, R. P., Matrai, P. A., and Bellows, W. K. 1999b. Production of glycine betaine and dimethylsulfoniopropionate in marine phytoplankton. II. N-limited chemostat cultures. *Mar. Biol.* 135:249–57.

Lande, R., Li, W. K. W., Horne, E. P. W., and Wood, A. M. 1989. Phytoplankton growth rates estimated from depth profiles of cell concentration and turbulent diffusion. *Deep-Sea Res.* 36:1141–59.

Liu, H., Campbell, L., Landry, M. R., Nolla, H. A., Brown, S. L., and Constantinou, J. 1998. *Prochlorococcus* and *Synechococcus* growth rates and contributions to production in the Arabian Sea during the 1995 Southwest and Northeast

Monsoon. *Deep-Sea Res. Part II. Top. Stud. Oceanogr.* 45:2327–52.

Lin, S., and Corstjens, P. 2002. Molecular cloning and expression of the proliferating cell nuclear antigen gene from the coccolithophorid *Pleurochrysis carterae* (Haptophyceae). *J. Phycol.* 38:164–73.

Liu, P., Jenkins, N. A., and Copeland, N. G. 2001. Efficient Cre-*lox*P-induced mitotic recombination in mouse embryonic stem cells. *Nat. Genet.* 30:66–72.

Luedeking, R. 1976. Fermentation process kinetics. In: Blackebrough, N., ed. *Biochemical and Biological Engineering Science*, Vol. 1. Academic Press, New York, pp. 181–243.

Lynch, M., Gabriel, W., and Wood, A. M. 1991. Adaptive and demographic responses of plankton populations to environmental change. *Limnol. Oceanogr.* 36:1301–12.

Lynch, M., and Walsh, B. 1998. *Genetics and Analysis of Quantitative Traits.* Sinauer, Sunderland, Massachusetts, 980 pp.

Maldonado, M. T., and Price, N. M. 2000. Nitrate regulation of Fe reduction and transport in Fe-limited *Thalassiosira oceanica. Limnol. Oceanogr.* 45:814–26.

Maldonado, M. T., and Price, N. M. 2001. Reduction and transport of organically bound iron by *Thalassiosira oceanica* (Bacillariophyceae). *J. Phycol.* 37:298–309.

McDuff, R. E., and Chisholm, S. W. 1982. The calculation of *in situ* growth rates of phytoplankton populations from fractions of cells undergoing mitosis: A clarification. *Limnol. Oceanogr.* 27:783–8.

Medlin, L. K., Kooistra, W. H. C. F., Gersonde, R., and Wellbrock, U. 1996. Evolution of the diatoms (Bacillariophyta). II. Nuclear-encoded small-subunit rRNA sequence comparisons confirm a paraphyletic origin for the centric diatoms. *Mol. Biol. Evol.* 13:67–76.

Monod, J. 1949. The growth of bacterial cultures. *Ann. Rev. Microbiol.* 3:371–94.

Monod, J. 1950. La technique de culture continue. Theorie et applications. *Ann Rev. Inst. Pasteur Paris.* 79:390–410.

Montresor, M., Sgrosso, S., Procaccini, G., and Kooistra, G. 2003. Intraspecific diversity in *Scrippsiella trochoidea* (Dinophyceae): Evidence for cryptic species. *Phycologia.* 42:56–70.

Moore, L. R., Rocap, G., and Chisholm, S. W. 1998. Physiology and molecular phylogeny of coexisting *Prochlorococcus* ecotypes. *Nature.* 393:464–7.

Needoba, J. A., Waser, N. A., Harrison, P. J., and Calvert, S. E. 2003. Nitrogen isotope fractionation in 12 species of marine phytoplankton during growth on nitrate. *Mar. Ecol. Prog. Ser.* 255:81–93.

Novick, A., and Szilard, L. 1950. Experiments with the chemostat on spontaneous mutations of bacteria. *Proc. Nat. Acad. Sci. U.S.A.* 36:708–14.

Olson, R. J., and Chisholm, S. W. 1983. Effects of photocycles and periodic ammonium supply on three marine phytoplankton species. I. Cell division patterns. *J. Phycol.* 19:552–8.

Paasche, E. 1973. The influence of cell size on growth rate, silica content and some other properties of four marine diatom species. *Norw. J. Bot.* 20:197–204.

Paasche, E. 1977. Growth of three plankton diatom species in Oslo Fjord water in the absence of artificial chelators. *J. Exp. Mar. Biol. Ecol.* 29:91–106.

Palenik, B., and Wood, A. M. 1998. Molecular markers of phytoplankton physiological status and their application at the level of individual cells. In: Cooksey, K. E., ed. *Molecular Approaches to the Study of the Oceans.* Chapman & Hall, London, pp. 187–205.

Reitan, K. I., Rainuzzo, J. R., and Olsen Y. 1994. Effect of nutrient limitations on fatty acid and lipid content of marine microalgae. *J. Phycol.* 30:972–9.

Rynearson, T. A., and Armbrust, E. V. 2000. DNA fingerprinting reveals extensive genetic diversity in a field population of the centric diatom *Ditylum brightwellii. Limnol. Oceanogr.* 45:1329–40.

Rynearson, T. A., and Armbrust, E. V. 2004. Genetic differentiation among populations of the planktonic marine diatom *Ditylum brightwellii* (Bacillariophyceae). *J. Phycol.* 40:34–43.

Scheiner, S. M., and Gurevitch, J. 2001. *Design and Analysis of Ecological Experiments.* 2nd ed. Oxford University Press, Oxford, 415 pp.

Sherry, N. D., and Wood, A. M. 2001. Phycoerythrin-containing picocyanobacteria in the Arabian Sea in February 1995: Diel patterns, spatial variability, and growth rates. *Deep-Sea Res. Part II. Top. Stud. Oceanogr.* 48:1263–83.

Shifrin, N. S., and Chisholm, S. W. 1981. Phytoplankton lipids: Interspecific differences and effects of nitrate, silicate and light dark cycles. *J. Phycol.* 17:374–84.

Shuter, B. 1979. A model of physiological adaptation in unicellular algae. *J. Theor. Biol.* 78:519–52.

Sokal, R. R., and Rohlf, F. J. 1994. *Biometry: the Principles and Practice of Statistics in Biological Research.* 3rd ed. Freeman, New York, 887 pp.

Stabile, J. E., Wurtzel, E. T., and Gallagher, J. C. 1992. Comparison of chloroplast DNA and allozyme variation in winter strains of the marine diatom *Skeletonema costatum* (Bacillariophyta). *J. Phycol.* 28:90–4.

Steglich, C., Post, A. F., and Hess, W. R. 2003. Analysis of natural populations of *Prochlorococcus* in the northern Red Sea using phycoerythrin gene sequences. *Environ. Microbiol.* 5:681–90.

Taguchi, S., and Kinzie III, R. A. 2001. Growth of zooxanthellae in culture with two nitrogen sources. *Mar. Biol.* 138:149–55.

Tarutani, K., Nagasaki, K., Itakura, S., and Yamaguchi, M. 2001. Isolation of a virus infecting the novel shellfish-killing dinoflagellate *Heterocapsa circularisquama*. *Aquat. Microb. Ecol.* 23:103–11.

Toledo, G., and Palenik, B. 1997. *Synechococcus* diversity in the California current as seen by RNA polymerase (rpoC1) gene sequences of isolated strains. *Appl. Environ. Microbiol.* 63:4298–303.

Uchida, T., Toda. S., Matsuyama, Y., Yamaguchi, M., Kotani, Y., and Honjo, T. 1999. Interactions between the red tide dinoflagellates *Heterocapsa circularisquama* and *Gymnodinium mikimotoi* in laboratory culture. *J. Exp. Mar. Biol. Ecol.* 241:285–99.

Underwood, A. J. 1996. *Experiments in Ecology: Their Logical Design and Interpretation Using Analysis of Variance.* Cambridge University Press, Cambridge, 504 pp.

Vaulot, D., LeBot, N., Marie, D., and Fukai, E. 1996. Effect of phosphorus on the *Synechococcus* cell cycle in surface Mediterranean waters during summer. *Appl. Environ. Microbiol.* 62:2527–33.

Wingard L., Miller, S., Selker, J., Stenn, E., Allen, M. M., and Wood, A. M. 2002. Cyanophycin production in a phycoerythrin-containing strain of marine *Synechococcus* of unusual phylogenetic affinity. *Appl. Environ. Microbiol.* 68:1772–7.

Wood, A. M., Lande, R. S., and Fryxell, G. A. 1987. Quantitative genetic analysis of morphological variation in an Antarctic diatom grown at two light intensities. *J. Phycol.* 23:42–54.

Wood, A. M. 1988. Molecular biology, single cell analysis, and quantitative genetics: New evolutionary approaches in phytoplankton ecology. In: Yentsch, C. M., Mague, F., and Horan, P. K., eds. *Immunochemical Approaches to Coastal, Estuarine, and Oceanographic Questions.* Springer-Verlag, New York, pp. 41–71.

Wood, A. M. 1989. Ultraphytoplankton community structure and evolution in the seasonal thermocline. In: Hattori, T., Ishida, Y., Maruyama, Y., Morita, R. Y., and Uchida, A., eds. *Recent Advances in Microbial Ecology. Proceedings of the 5th International Symposium on Microbial Ecology.* Japan Scientific Societies Press, Tokyo, pp. 336–40.

Wood, A. M., and Townsend, D. 1990. DNA Polymorphism within the WH7803 serogroup of marine *Synechococcus* spp. *J. Phycol.* 26:576–85.

Wood, A. M., and Leatham, T. 1992. The species concept in phytoplankton ecology. *J. Phycol.* 28:723–9.

Worden, A. Z., and Binder, B. J. 2003. Growth regulation of rRNA content in *Prochlorococcus* and *Synechococcus* (marine cyanobacteria) measured by whole-cell hybridization of rRNA-targeted peptide nucleic acids. *J. Phycol.* 39:527–34.

Yamaguchi, M., Itakura, S., and Uchida, T. 2001. Nutrition and growth kinetics in nitrogen- or phosphorus-limited cultures of the "novel red tide" dinoflagellate *Heterocapsa circularisquama* (Dinophyceae). *Phycologia.* 40:313–8.

Zar, J. H. *Biostatistical Analysis.* 1996. Prentice Hall, Upper Saddle River, New Jersey, 929 pp.

Zhekisheva, M., Boussiba, S., Khozin-Goldberg, I., Zarka, A., and Cohen, Z. 2002. Accumulation of oleic acid in *Haematococcus pluvialis* (Chlorophyceae) under nitrogen starvation or high light is correlated with that of astaxanthin esters. *J. Phycol.* 38:325–31.

# USING CULTURES TO INVESTIGATE THE PHYSIOLOGICAL ECOLOGY OF MICROALGAE

HUGH L. MacINTYRE
*Dauphin Island Sea Lab*

JOHN J. CULLEN
*Centre for Marine Environmental Prediction, Department of Oceanography,*
*Dalhousie University*

Key Index Words: Microalgae, Experiment, Pigment, Photosynthesis, Growth, Irradiance, Spectrum, Nutrient Limitation, Nutrient Starvation, Temperature

---

## 1.0. INTRODUCTION

Cyanobacteria and their eukaryotic descendants, the single-celled algae (to which we refer collectively as *microalgae*), play a vital role in the ecology of the planet, being responsible for roughly 45% of global productivity (Field et al. 1998) and supporting food webs in waters from ponds to oceans. However, trying to quantify their contributions to ecological dynamics and biogeochemical cycles remains a daunting process for various reasons. Standing crops of microalgae (expressed as the concentration of the pigment chlorophyll *a* [Chl*a*] per unit volume) vary over more than six orders of magnitude (Fig. 19.1).

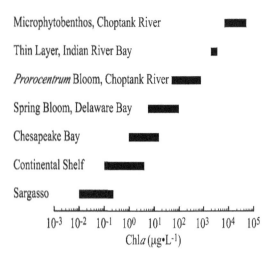

**FIGURE 19.1.** Variation in microalgal biomass, as chlorophyll *a* (Chl*a*) on a gradient from the oligotrophic Sargasso Sea (36° 20' N, 72° 20' W) to eutrophic estuaries in Delaware and Maryland. Note the log scale. All samples were collected in late spring to early summer. (From MacIntyre, Richard Geider and Jason Adolf, unpublished data.)

Chl*a* (see Table 19.1 for abbreviations) is used by field ecologists and modelers as an index of biomass because it is unique to photosynthetic organisms and relatively easy to measure or derive from remote sensing. However, translating Chl*a* abundance to the more useful currency of microalgal carbon is complicated by the fact that the organisms are physiologically plastic and alter their quotas of cellular carbon and pigments in response to environmental factors such as irradiance, nutrient availability, and temperature (Fig. 19.2a), thereby altering their Chl*a*:C ratios. Consequently, the ratio of cellular carbon to Chl*a* for a single species can vary by more than a factor of 10 in response to varying growth conditions (Geider and Osborne 1987). In addition to this phenotypic variability (acclimation), there is also considerable genotypic variability (Fig. 19.2b). This reflects adaptations (genetically determined differences) that are rooted in the polyphyletic evolution of taxonomic groups with a wide diversity in the structure and function of their organelles (Delwiche 1999) and inherent differences in cellular structures (Chan 1980). Physiological plasticity is not confined to the Chl*a*:C ratio but is also reflected in other indices of biomass such as the Chl*a*:N ratio (Fig. 19.2c,d). The abundance of microalgae cannot, therefore, be simply described by any single currency of biomass such as Chl*a* or N content. The uncertainties in defining ecological roles extend beyond finding a convenient parameter for describing biomass. Microalgal biomass-specific physi-

ological responses also vary with changes in environmental characteristics such as irradiance, temperature, and nutrient availability (Fig. 19.3). Although defining the biomass of microalgae in a water body is difficult, predicting their biomass-specific rate of photosynthesis, which is a highly desirable description of their contribution to the global carbon cycle, is even more so.

The influences of environmental factors on the growth, chemical composition, and physiological responses of microalgae, though complex and time-dependent, are determined solely by biochemical processes under genetic control. The role of the microalgal physiologist is to describe these influences and to determine the role of the environment in regulating the organisms' growth and survival. However, there can be no fixed parameterization of cause and effect that applies across all groups: There are taxonomic differences in the structure and integration of such photosynthetic components as pigments, electron carriers, and enzymes. Whereas the microalgae as a group do vary widely in terms of their biochemical composition and physiological responses, the variability is ordered, reflecting genetically determined reactions to the characteristics of the environment in which they are growing. For example, microalgae differ widely in size and shape, and within the diverse taxa, their pigmentation and photosynthetic responses will to some extent scale according to allometric laws (Finkel and Irwin 2000, Finkel 2001, Raven and Kübler 2002).

Experiments with cultures are and will remain central to our understanding of microalgal responses to environmental variability. We describe here the types of responses that can be studied, the systems to use, and some of the issues that must be addressed to relate results from laboratory experiments to the growth of microalgae at large. Much of the information is general in nature and is intended to illustrate the principles rather than the specifics of culturing. For instance, we do not provide detailed step-by-step directions for use of techniques such as fluorescence or spectrometry because of variations among different models of instruments and the innumerable possibilities for experimental design. We refer you to other sections of this volume for detailed information on the preparation of media (Chapters 2–5), maintenance of cultures (Chapter 10), and enumeration of cells and determination of growth rate (Chapters 16–18).

Although most of the data that we show are from species isolated from estuarine and marine waters, the principles of culturing for examining physiological responses apply equally to species isolated from fresh waters and/or benthic habitats. The data illustrate the

**TABLE 19.1**  Summary of the symbols and abbreviations used in the text.

| Symbol | Description | Typical Unit |
|---|---|---|
| $a^{Chl}$ | Mean value of $a^{Chl}(\lambda)$ over PAR (400–700 nm) | $m^2 \cdot [mg\ Chl a]^{-1}$ |
| $a^{Chl}(\lambda)$ | Chl$a$-specific absorption at wavelength $\lambda$ | $m^2 \cdot [mg\ Chl a]^{-1} \cdot nm^{-1}$ |
| $A_e$ | Weighting coefficient to convert PAR to PUR | dimensionless |
| $\alpha$ | Initial, light-limited rate of photosynthesis in a PE curve | $g\ C \cdot [g\ Chl a]^{-1} \cdot hr^{-1} \cdot (\mu mol\ photons \cdot m^{-2} \cdot s^{-1})^{-1}$ or mol $O_2$ or $C \cdot [g\ Chl a]^{-1} \cdot hr^{-1} \cdot (\mu mol\ photons \cdot m^{-2} \cdot s^{-1})^{-1}$ |
| $\beta$ | Parameter describing reduction in photosynthesis at high irradiance | $g\ C \cdot [g\ Chl a]^{-1} \cdot hr^{-1} \cdot (\mu mol\ photons \cdot m^{-2} \cdot s^{-1})^{-1}$ or mol $O_2$ or $C \cdot [g\ Chl a]^{-1} \cdot hr^{-1} \cdot (\mu mol\ photons \cdot m^{-2} \cdot s^{-1})^{-1}$ |
| $CFC$ | Ratio of variable to maximum fluorescence in the dark $\pm$ DCMU: $(F_d - F_0)/F_d$ | dimensionless |
| Chl$a$ | Chlorophyll $a$ | $\mu g \cdot L^{-1}$ or $pg \cdot cell^{-1}$ |
| Chl$a$:$C$ | Ratio of Chl$a$ to POC | $g \cdot g^{-1}$ |
| Chl$a$:$N$ | Ratio of Chl$a$ to PON | $g \cdot g^{-1}$ |
| $DIC$ | Dissolved inorganic carbon | $g \cdot L^{-1}$ or M |
| $DOC$ | Dissolved organic carbon | $g \cdot L^{-1}$ or M |
| $E$ | Mean value of $E(\lambda)$ over PAR (400–700 nm) | $\mu mol\ photons \cdot m^{-2} \cdot s^{-1}$ |
| $E(\lambda)$ | Irradiance at wavelength $\lambda$ | $\mu mol\ photons \cdot m^{-2} \cdot s^{-1} \cdot nm^{-1}$ |
| $E'$ | Hypothetical reference irradiance with equal intensity between 400 and 700 nm | $\mu mol\ photons \cdot m^{-2} \cdot s^{-1}$ |
| $E_a$ | Irradiance absorbed by a cell | $mol\ photons \cdot [g\ Chl a]^{-1} \cdot s^{-1}$ or $\mu mol\ photons \cdot m^{-2} \cdot s^{-1}$ |
| $E_d$ | Downwelling irradiance | $\mu mol\ photons \cdot m^{-2} \cdot s^{-1}$ |
| $E_k$ | Saturating parameter in a P vs. E curve: $= P_m/\alpha$ | $\mu mol\ photons \cdot m^{-2} \cdot s^{-1}$ |
| $F, F_s$ | Fluorescence in darkness or light, measured before saturating flash | arb. |
| $F_0$ | Dark-adapted fluorescence without DCMU | arb. |
| $F_d$ | Dark-adapted fluorescence in the presence of DCMU | arb. |
| $F_m, F_m'$ | Maximum fluorescence in darkness or light, measured after a saturating light flash | arb. |
| $F_v/F_m$ | Ratio of variable to maximum fluorescence in the dark: $(F_m - F)/F_m$ | dimensionless |
| $F_v'/F_m'$ | Ratio of variable to maximum fluorescence in the light: $(F_m' - F_s)/F_m'$ | dimensionless |
| $K_C$ | Compensation irradiance in $\mu$ vs. E curve, where $\mu = 0$ | $\mu mol\ photons \cdot m^{-2} \cdot s^{-1}$ |
| $K_E$ | Saturating parameter in a $\mu$ vs. E curve | $\mu mol\ photons \cdot m^{-2} \cdot s^{-1}$ |
| $\lambda$ | Wavelength | nm |
| $\mu$ | Specific growth rate | $d^{-1}$ |
| $\mu_{E,T}$ | Nutrient-replete specific growth rate at irradiance E and temperature T | $d^{-1}$ |
| $\mu_m$ | Maximum specific growth rate | $d^{-1}$ |
| $\mu_N$ | Nutrient-limited specific growth rate | $d^{-1}$ |
| $\phi$ | Quantum yield of carbon fixation | $mol\ C \cdot [mol\ photons]^{-1}$ |
| $\phi_m$ | Maximum quantum yield of carbon fixation | $mol\ C \cdot [mol\ photons]^{-1}$ |
| PAR | Photosynthetically active radiation, 400–700 nm | $\mu mol\ photons \cdot m^{-2} \cdot s^{-1}$ |
| PE | Photosynthesis as a function of irradiance | |
| $P_E$ | Photosynthesis at irradiance E | $g\ C \cdot [g\ Chl a]^{-1} \cdot hr^{-1}$ or mol $O_2$ or $C \cdot [g\ Chl a]^{-1} \cdot hr^{-1}$ |
| $P_m$ | Light-saturated rate of photosynthesis in a PE curve | $g\ C \cdot [g\ Chl a]^{-1} \cdot hr^{-1}$ or mol $O_2$ or $C \cdot [g\ Chl a]^{-1} \cdot hr^{-1}$ |
| POC or C | Particulate organic carbon | $\mu g \cdot L^{-1}$ or $pg \cdot cell^{-1}$ |
| PON or N | Particulate organic nitrogen | $\mu g \cdot L^{-1}$ or $pg \cdot cell^{-1}$ |
| PQ | Photosynthetic quotient; unit oxygen evolved per unit carbon assimilated | dimensionless |
| $PUR^{Chl}$ | Photosynthetically usable radiation | $mol\ photons \cdot [g\ Chl a]^{-1} \cdot s^{-1}$ |
| PUR | Photosynthetically usable radiation | $\mu mol\ photons \cdot m^{-2} \cdot s^{-1}$ |
| $R_E$ | Respiration rate at irradiance E | $g\ C \cdot [g\ Chl a]^{-1} \cdot hr^{-1}$ or mol $O_2$ or $C \cdot [g\ Chl a]^{-1} \cdot hr^{-1}$ |
| $X$ | Index of biomass (cell number, Chl$a$, POC, PON, etc.) | See above |

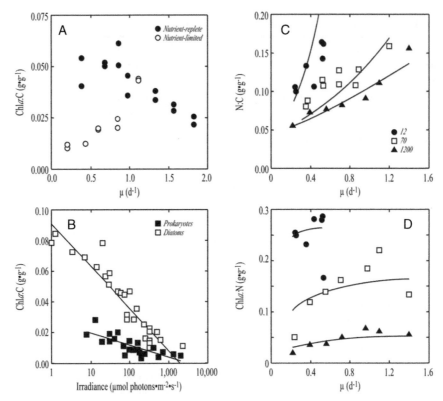

**FIGURE 19.2.**   Variation in the chemical and physiological characteristics of microalgae with growth rate (redrawn and modified from Geider 1992, Cullen et al. 1993). **(a)** Variation in the Chl$a$ : C ratio as a function of growth rate in the diatom *Thalassiosira pseudonana* (Hustedt) Hasle et Heimdal (Xiaolong Yang, MacIntyre and Cullen, unpublished data). Nutrient-replete cultures = semicontinuous culture at irradiances between 25 and 900 µmol photons $\cdot$ m$^{-2}$ $\cdot$ s$^{-1}$. Nutrient-limited cultures = chemostats at 200 µmol photons $\cdot$ m$^{-2}$ $\cdot$ s$^{-1}$. **(b)** Variation in the Chl$a$ : C ratio as a function of growth irradiance in the diatoms *Cylindrotheca fusiformis* (Ehr.) Reimann et Lewin, *Phaeodactylum tricornutum* Bohlin, *Thalassiosira eccentrica* (Ehr.) Cleve and *Thalassiosira weissflogii* (Grun.) Fryxell et Hasle, and the prokaryotes *Microcystis aeruginosa* Kütz. em. Elenkin, *Prochlorothrix hollandica* Burger-Wiersma, and *Synechococcus* spp. Nutrient-replete conditions under continuous light (Chan 1980, Raps et al. 1983, Falkowski et al. 1985, Geider et al. 1985, 1986, Kana and Glibert 1987a, 1987b, Burger-Wiersma and Post 1989). Note the 3x difference in taxonomic trends. **(c)** Variations in the N : C ratio of the diatom *Skeletonema costatum* (Greville) Cleve as a function of nutrient-limited growth rate at 12, 70, and 1200 µmol photons $\cdot$ m$^{-2}$ $\cdot$ s$^{-1}$. Data are from Sakshaug et al. (1989). Lines are relationships predicted by the dynamic balance model of Geider et al. (1998). **(d)** Variations in the Chl$a$ : N ratio of *S. costatum* as a function of nutrient-limited growth rate at 12, 70, and 1200 µmol photons $\cdot$ m$^{-2}$ $\cdot$ s$^{-1}$. Symbols and data sources as in **(c)**.

species' responses to particular growth conditions and should not be taken as representative for other species or other conditions of growth.

---

## 2.0. DESCRIBING THE DISTRIBUTIONS AND ACTIVITIES OF MICROALGAE IN NATURE

### 2.1. Measurements in the Field

Field studies of the microalgae have been central to our understanding of their ecology (e.g., Riley 1947,

Sverdrup 1953, Redfield 1958, Margalef 1978, Reynolds and Bellinger 1992) and remain vital to our efforts to describe their dynamics. Broadly, the central objective of field studies is to describe and explain the distributions and activities of microalgae in nature. However, interpretation of data collected on natural assemblages is always subject to the following major constraints:

- Many direct measures of biomass such as particulate organic carbon (POC) are compromised by the presence of contaminants (bacteria, zooplankton and their fecal pellets, other detrital material, and the occasional charismatic vertebrate) (Eppley et al. 1977).

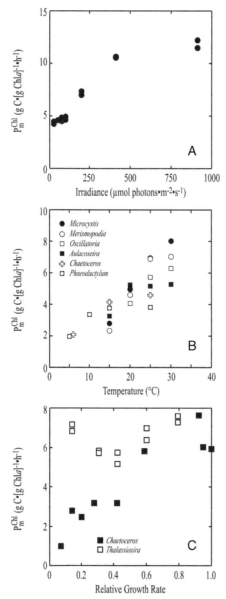

**FIGURE 19.3.** Variation in the chlorophyll *a* (Chl*a*)–specific light-saturated rate of photosynthesis, $P_m$, with irradiance, temperature, and nutrient availability. **(a)** $P_m$ as a function of growth irradiance in nutrient-replete cultures of the diatom *Thalassiosira pseudonana* at 20° C (Xiaolong Yang, MacIntyre and Cullen, unpublished data). **(b)** $P_m$ as a function of growth temperature in nutrient-replete cultures of the cyanobacteria *Microcystis aeruginosa*, *Merismopedia tenuissima* Lemmermann, and *Oscillatoria* spp. (Coles and Jones 2000) and the diatoms *Aulacoseira granulata* (Ehr.) Simonson (Coles and Jones 2000), *Chaetoceros calcitrans* Paulsen (Anning et al. 2001), and *Phaeodactylum tricornutum* (Li and Morris 1982). **(c)** $P_m$ as a function of nutrient-limited growth rate in the diatoms *Chaetoceros neogracile* (Schütt) van Landingham (formerly *C. gracilis* Schütt) and *T. pseudonana*. Both were grown in N-limited continuous culture: *C. neogracile* at 160 μmol photons · m⁻² · s⁻¹ and *T. pseudonana* at 200 μmol photons · m⁻² · s⁻¹. Data on *C. neogracile* are from Thomas and Dodson (1972). Data on *T. pseudonana* are from Xiaolong Yang, MacIntyre and Cullen; redrawn from Cullen et al. (1992b).

- The environmental history of cells before their sampling is unknown. Physiological plasticity and the time lags involved in responses to environmental influences can confound the interpretation of cause (the environmental conditions) and effect (biochemical and physiological characteristics) if the cells have been transported or have migrated from waters with different environmental conditions.
- When microalgae are collected and confined for experimental determination of rate processes, conditions in the natural turbulent environment cannot be simulated reliably (Legendre and Demers 1984, Cullen and Lewis 1995).

Thus, few measurements made in the field can be taken as accurate measures of the distributions or physiological responses of natural microalgae. The general approach is to measure all of the bulk properties of the water attributable to microalgae (e.g., the concentration of Chl*a* and other pigments, rate of photosynthesis during incubations, uptake of nutrients, nitrogen fixation, etc.) and develop models to relate the measurements that can be made (Chl*a*, irradiance, nutrient concentration, activities of enzymes) to the rates and properties of interest (e.g., microalgal biomass, instantaneous rate of photosynthesis, growth rate, or nutrient limitation). Resulting models, such as those used to calculate productivity from remotely sensed ocean color, have proved to be effective in converting large amounts of readily acquired data into estimates of a difficult-to-measure process on large scales (Platt and Sathyendranath 1993, Behrenfeld and Falkowski 1997a). The reliability of extrapolations to conditions other than those for which the models were calibrated, such as when predicting global responses to changes in atmospheric $CO_2$, cannot be easily assessed.

## 2.2. Models of Physiological Responses

An alternative approach to describing the characteristics of microalgae as a function of environmental factors is to use a first-principles model, in which the physiological response is defined in terms of light, temperature, and nutrient supply. The model can be parameterized to approximate the biochemical and physiological control points, such as the redox state of the plastoquinone pool (Escoubas et al. 1995, Maxwell et al. 1995). Extrapolating from cultures to field populations requires a detailed understanding of how environmental conditions influence microalgal growth for the species in culture and for the different algal assemblages in the natural environment (Flynn and Hipkin 1999, Flynn and Fasham 2003). One such dynamic model has proved to be robust in reproducing microalgal responses to conditions of fluctuating nutrient and irradiance availability when calibrated on the responses of algae fully acclimated to their environment (Taylor et al. 1997, Geider et al. 1998, Lefèvre et al. 2003). Such a calibration is possible only under the controlled conditions that can be obtained when populations of microalgae are grown in cultures.

## 2.3. The Need for Experiments with Cultures

The organization of pigments, end products of photosynthesis, and growth rate respond to changes in light, temperature, and nutrient availability in a dynamic way (see Figs. 19.2 and 19.3). This process must be characterized to describe the distributions, growth rates, and physiological activities of microalgae based on limited data, such as measurements of chlorophyll, irradiance, and temperature. Physiological characteristics, such as optimal temperatures for growth, maximum chlorophyll content, and capabilities to store pulses of nutrients, differ between taxa, reflecting adaptations that have evolved through natural selection. Appreciation of these adaptations can lead to a better understanding of the environmental factors that define the niches of microalgae, thereby leading the way to a description of environmental controls on community composition and ultimately on food web structure and biogeochemical cycles.

## 3.0. Approaches to Culturing

Microalgae can be cultured in many ways. The simplest classification of these depends on whether they are in balanced growth.

- If conditions are kept constant and within the bounds of survivability, the cells will eventually become fully acclimated to their environment; their physiological responses will be at equilibrium and their growth will be balanced.
- If cultures are responding to repeatable and predictable changes, they may be in dynamic equilibrium and their growth will be balanced, when considered over the cycle of variability.

- If cultures are subjected to step-changes in conditions or stochastic variability, they will not be at equilibrium, and thus, their responses will not be acclimated and their growth will be unbalanced.

Because microalgae respond to perturbations in their environment, and because their physiology takes a finite time to respond, fully acclimated growth occurs only after long exposure to highly stable conditions. However, long-term stability is atypical of natural environments and microalgal growth that is fully acclimated to constant conditions is a completely artificial state. Its relevance, and therefore the relevance of culturing under conditions that do not reflect the full range of natural variability, is open to some debate. The rationale for studying acclimated growth has much in common with Plato's Theory of Forms, which holds that an individual Thing is the product of the interaction between an eternal, ideal, and abstract Form and the Thing's surroundings. As a result, the abstract Form embodies a higher level of reality than its temporal imitation, the Thing. By analogy, the state of a cell (the Thing) can be viewed as the deviation from its Form (the fully acclimated state) in response to its surroundings (the environment). A critical difference from Plato's theory is that each set of environmental conditions would have its own fully acclimated condition so that the Form would be a continuum rather than a single state. However, the central relationship between Thing and Form is analogous so that the fully acclimated state remains the expression of an idealized condition from whose contours we can discern the underlying elements of the cell's expression in response to a mutable world. This approach is inherently reductionistic and, in our opinion, pragmatic: Much can be learned by observing and describing the mechanisms by which microalgae regulate their physiology to attain the acclimated state. An alternative, and incompatible, view of the cell's ecology is holistic and holds that the whole cannot be derived simply from the sum of its parts. To a holist, a description of acclimated growth is an abstraction made meaningless by its detachment from the complexity of the real world. The two opposing views have been explored extensively by Ernest Nagel and Robert Pirsig, among others, neither of whom found a resolution between them.

The approach to modeling and culturing that we describe in the following is fundamentally reductionistic, holding that the interactive effects of light, temperature, and nutrient availability on microalgal physiology can be described meaningfully in experimental studies, and that a description of physiology under artificial controlled conditions can inform our understanding of dynamics in natural assemblages. We argue that acclimation can be described with response functions that are general (MacIntyre et al. 2002) and consistent with dynamic models that are parameterized with observations of cultures in balanced growth (Geider et al. 1998, Flynn and Hipkin 1999). We also argue that both the chemical composition and growth rate under acclimated conditions and the metabolic means by which they are achieved are adaptations that define in part the niche of each microalga (Cullen and MacIntyre 1998). This has been the subject of debate (e.g., Siegel 1998) since the apparent paradox of coexistence of different species was proposed by Hutchinson (1961). The holist will likely balk at this point and will find little with which to agree in what follows.

## 3.1. Types of Growth

The prerequisite for effective experimental work with cultures is a clear understanding of what is being asked, an informed choice of the appropriate culturing system, and an appropriate analysis of results. Methods for culturing microalgae and for using cultures to investigate their physiological ecology can be organized in terms of whether the cells are acclimated to their environmental conditions and whether their growth is nutrient replete, nutrient starved, or nutrient limited.

### 3.1.1. Balanced vs. Unbalanced Growth

The specific growth rate is a rate of change of biomass that is set by the relative magnitude of anabolic processes (photosynthesis) and catabolic processes (respiration):

$$\mu = P - R, \tag{1}$$

where $\mu$ is the specific growth rate, $P$ is photosynthesis, and $R$ is respiration. Here, all rates are expressed in units of inverse time (e.g., g C/[g C]/d = $d^{-1}$). Cells are described as being acclimated (i.e., in balanced growth) when internal adjustments of metabolic pathways have ceased so the rate of change of all cellular components is the same (Shuter 1979, Eppley 1981):

$$\mu = \left(\frac{1}{X}\right)\left(\frac{dX}{dt}\right), \tag{2}$$

where $X$ is any index of biomass, such as cell number, Chl$a$, carbon, or nitrogen. (For a more detailed discussion of growth rates, see Chapter 18.) This condition can occur only when environmental conditions are held

constant for many generations. Such constant conditions are highly artificial; the diel cycle of irradiance imposes a daily imbalance of photosynthesis versus respiration ensuring that unbalanced growth is the hallmark of autotrophic existence. However, microalgae can readily acclimate to environmental conditions that are repeated day to day, in that they achieve acclimated balanced growth over a photocycle (Shuter 1979, Eppley 1981). Equation 2 would apply, then, to rates averaged over a photocycle. Growth is balanced but in dynamic, rather than simple, equilibrium. This can be illustrated by a comparison of within-day and between-day variability of Chl$a$, carbohydrate, and protein levels in cells grown on a light–dark cycle (Fig. 19.4). All indices increased during the light period, although the specific rates varied by more than threefold. In the following dark period, Chl$a$ increased by about 10%, protein and cell number increased by about 50%, and

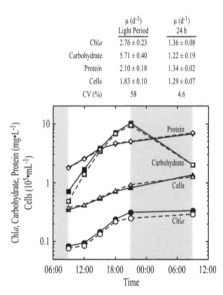

**FIGURE 19.4.** Variation in four indices of biomass in a semicontinuous culture of the diatom *Thalassiosira pseudonana* grown on a 12:12 L:D cycle (Xiaolong Yang, MacIntyre, and Cullen, unpublished, partially shown in Parkhill et al. 2001). The shaded bars indicate the dark period. Open and closed symbols are for replicate cultures. Exponential growth rates in the light period were determined by linear regression of ln-transformed data collected between 09:00 and 21:00 on the first day (Equation 3). Exponential growth rates over the 24-hour period were determined by linear regression of ln-transformed data collected at 09:00 on successive days. CV is the coefficient of variance (standard deviation/mean, expressed as a percentage). Note the good agreement between growth rates collected over the 24-hour period as opposed to those collected over the light period alone.

carbohydrate fell by 80%. The pattern is consistent with daytime synthesis of excess carbohydrate fueling nocturnal protein synthesis and cell division (Cuhel et al. 1985, Cullen 1985). However, when growth rates were calculated at the 24-hour interval, matching the length of the light–dark cycle, there was no significant difference between estimates based on the different indices. This is balanced growth in dynamic equilibrium.

### 3.1.2. Nutrient Stress: Limitation and Starvation

Because the process of acclimation is one of physiological and chemical change, cells that are acclimating to environmental conditions are fundamentally different from those that have already acclimated and are in balanced growth. It is important, therefore, to incorporate acclimation in the terminology for describing the effects of nutrients on microalgal growth. The various conditions under which nutrient supply imposes a level of stress on cellular physiology have been variously described as a limitation. We use the following definitions (Parkhill et al. 2001): *Limitation* refers to acclimated growth in which the growth rate is reduced as a result of nutrient availability; *starvation* refers to unbalanced growth in response to recent depletion of a nutrient; and *stress* is the inclusive generic term that covers both conditions.

Classically, *limitation* has been used to describe any restriction of growth by nutrient availability. It can refer to two very different circumstances; Liebig and Blackman limitation (Cullen et al. 1992b). Liebig's Law of the Minimum has its basis in agronomy and is properly applied to final yield, which is a measure of biomass, not the rate at which it accumulates (i.e., specific growth rate). In contrast, Blackman limitation has its basis in descriptions of kinetics and refers to the effect of a factor on a rate process, of which growth is an analog. This is illustrated in Figure 19.5, which shows the course of biomass changes in a closed system. Growth is initially exponential:

$$X_t = X_0 \exp(\mu t), \qquad (3)$$

where $X_0$ is the initial biomass, $X_t$ is the biomass at time $t$, and $\mu$ is the specific growth rate, with units of inverse time (e.g., $d^{-1}$). Note that this is *not* synonymous with the division rate, which is a growth rate expressed in divisions $\cdot d^{-1}$ and is calculated in Base 2 rather than Base $e$ (Guillard 1973b). The relationship between them is

$$Division\ Rate = \frac{\mu}{\ln(2)} = \frac{\mu}{0.693} \qquad (4)$$

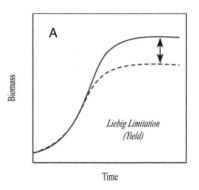

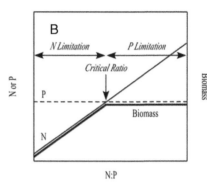

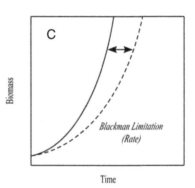

**FIGURE 19.5.** **(a)** Schematic representation of Liebig limitation of yield (biomass) in a culture. The final yield, which is the maximum biomass achieved, is determined by the abundance of a single factor, according to Liebig's Law of the Minimum. The divergence in final yields, caused by a difference in the abundance of the limiting nutrient, is indicated by the arrow. Note that the initial exponential growth rates are the same. **(b)** Schematic representation of Liebig limitation along a gradient of N : P, where the nutrient ratio was obtained by altering nitrogen (*light solid line*) against a fixed concentration of phosphorus (*dashed line*). The biomass at the Liebig yield (*heavy line*) is limited below the critical ratio by nitrogen and above it by phosphorus. **(c)** Schematic representation of Blackman limitation of growth in a culture. The growth rate is determined by the availability of a limiting nutrient. The divergence in growth rate, caused by a difference in the abundance of the limiting nutrient, is indicated by the arrow. Note that the exponential growth rates are different regardless of the final yield that is achieved.

As shown in Figure 19.5a, as the cells deplete the nutrients in the growth medium, the nutrient with the lowest concentration relative to the cells' requirements declines to the point at which the rate of uptake cannot match cellular demand. At this point, the cells enter *nutrient starvation*. The growth rate μ (see Equation 3) declines to zero as the limiting nutrient is exhausted and the cells enter stationary phase. The final yield is determined by the initial concentration of the nutrient that was present in the lowest availability relative to the cellular requirements. During the latter part of the growth phase, growth is neither exponential nor balanced: As the limiting nutrient is depleted, the rate of growth declines due to Blackman limitation (see following discussion), and once the limiting nutrient is gone, the yield is set by Liebig limitation. Cells do not acclimate to a particular nutrient concentration, because concentrations are always changing until the nutrient is exhausted, at which point neither growth nor acclimation can continue. The example in Figure 19.5a is appropriate for describing the transient process of nutrient starvation and is useful for exploring responses of microalgae to intermittent nutrient supply. It should be recognized, though, that the results of such nutrient starvation experiments may depend on preconditioning; for instance, the effects of nitrogen starvation on a marine diatom were delayed for nutrient-replete cultures as compared to acclimated nutrient-limited cultures (Smith et al. 1992, Parkhill et al. 2001).

The biomass yield in Liebig limitation is set by two factors: the intrinsic requirements of the cells (cell quota) and the external concentrations of nutrients. Microalgae regulate their internal nutrient quotas by reducing them in response to limited availability until they reach a minimum value (Droop 1968, 1974, Morel 1987). As they exhaust the concentration of a limiting nutrient, the cellular nutrient quota is reduced as cell division continues and the assimilation of limiting nutrient slows and then stops; that is, the quota reaches a minimum value. Ultimately, cell division stops and the population has reached the Liebig yield. The nutrient that sets the yield is the one whose quota declines to the cellular minimum first. This is illustrated for nitrogen versus phosphorus limitation in Figure 19.5b. When nitrogen is low relative to phosphorus, the cells reduce the nitrogen quota to the minimum value before the availability of phosphorus causes the phosphorus quota to decline to the minimum. That is, the yield is nitrogen limited. Conversely, when nitrogen is high relative to phosphorus, cellular phosphorus is reduced to the minimum before nitrogen is exhausted. That is, the yield is phosphorus limited. Under balanced nutrient-limited growth (e.g., in a chemostat; see following discussion), the crossover occurs at the critical ratio, a point at which the quotas of both nitrogen and phosphorus are at the minimum value (Terry et al. 1985).

The critical ratio has been measured for N : P in four species of microalgae and for vitamin B12 : P in one. The critical N : P ratio, expressed as an atomic ratio, was about 30 in the chlorophyte *Scenedesmus* spp. (Rhee 1978); 38 to 49 in the prymnesiophyte *Pavlova* (formerly *Monochrysis*) *lutheri* (Droop) Green (Terry et al. 1985); 25 to 33 in the diatom *Phaeodactylum tricornutum* (Terry et al. 1985); and 35 to 40 for the diatom *Chaetoceros muelleri* Lemmermann (Leonardos and Geider 2004). It is expected to and does vary with growth irradiance (Terry et al. 1985, Leonardos and Geider 2004). As discussed elsewhere (Geider and La Roche 2002), this is considerably higher than the Redfield ratio of 106 : 16 : 1 C : N : P (Redfield 1958, Goldman et al. 1979), arguably one of the most abused parameters in the field of aquatic ecology. The Redfield ratio is an approximation of composition averaged over very large scales of time and space and does not describe the conditions of individual cells or populations. Even so, it can be used as a guide to which nutrient will limit growth in culture; biasing N : P concentrations by fivefold to tenfold on either side of the ratio will ensure that a culture is limited by nitrogen or phosphorus. Note that in f/2 medium with 2 mM dissolved inorganic carbon (DIC), the C : N : P ratio is 55 : 24 : 1 (Guillard 1973a) and the cells will become carbon limited before exhausting nitrogen or phosphorus.

Unless the aim is to study carbon limitation, we recommend bubbling cultures with air. It should be passed through a cartridge filled with activated carbon (Guillard 1973a) to remove the volatile organics such as pump oils and solvents that are common in laboratory air and then through a sterile 0.2-μm capsule filter to maintain sterility in the culture. For N-limited cultures, the air should be bubbled through dilute phosphoric acid and then distilled water to trap atmospheric ammonium that is released is significant quantities from many cleaners and from people who smoke. We direct the reader to Chapters 2 through 5 for other considerations of nutrient composition in medium, including the relative merits of autoclaving and microwave sterilization on $CO_2$, ammonium, and silicate concentrations (Guillard 1973a, Keller et al. 1988); species-specific trace metal and buffering requirements (Keller et al. 1987); and the need to add selenium to cultures grown in polycarbonate or modern Pyrex culture vessels (Price et al. 1987) to avoid inadvertent trace-metal limitation.

In contrast to limitation of yield, Blackman limitation refers to the rate constant that describes a reaction (the analog of μ in Equation 3). This is illustrated in Figure 19.5c, which shows two cultures in exponential growth where one has a lower achieved growth rate than the other. If the slower growth is due to a reduction in the efficiency of a metabolic process because of restricted availability of a nutrient, the cells are *nutrient limited*. This can be an acclimated condition, provided that the rate of supply of the nutrient matches the growth rate, and growth can be both exponential and balanced. (See Kacser and Burns [1973] and Rees and Hill [1994] for biochemical limitation of rates.) We consider both nutrient starvation and nutrient limitation as they occur in different types of cultures.

### 3.2. Batch Culture

One of the simplest forms of culture is a batch culture, in which growth follows multiple phases (Fig. 19.6). Typically, there is an initial lag phase during which there is minimal growth. Experience shows that the lag is short if the inoculum is large and the parent culture has been acclimated to similar conditions; the lag phase is more an artifact of transfer than an inherent stage of growth. This is followed by an exponential growth phase (Equation 3) until a nutrient or $CO_2$ is depleted, after which growth slows and cells enter stationary phase (Fig. 19.6a). If the concentrations of nutrients are high and the incident irradiance relatively low, there is also the potential for the culture to shade itself and become light limited as cell density becomes very high.

It is very difficult to define a period of acclimated growth in batch cultures, because the progressive accumulation of biomass changes the availability of nutrients and light (Fig. 19.6c). Both changes drive acclimative responses in pigmentation and photosynthetic performance. The reduction in mean irradiance available to the cells as the culture starts to shade itself causes a response of increased pigmentation. As a result, even while growth can be described well by fitting to Equation 3, the apparent growth rate depends strongly on the index of biomass being used; the Chl*a*-based rate is twice that of the POC- or PON-specific rates (Fig. 19.6a). This is reflected in the variability in the biomass ratios (Fig. 19.6b) during the exponential phase; growth is clearly not balanced, as defined according to Equation 2. In this case, the nutrient setting the Liebig limit is nitrogen; after the culture has exhausted the supply

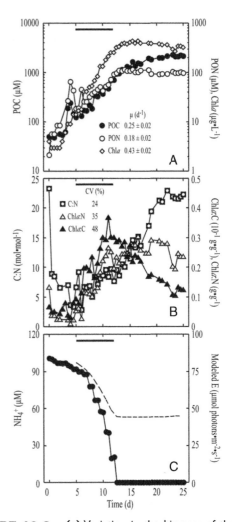

FIGURE 19.6. **(a)** Variation in the biomass of the prymnesiophyte *Isochrysis galbana* Parke during growth in batch culture. Note log scale. Exponential growth rates (Equation 3) were fitted to data collected between day 5 and day 12 (*black bar*), when ammonium concentrations were above the limit of detection **(c)**. Note that growth rates vary by 2.4 times, depending on the index of biomass used to calculate the rate. **(b)** Variation in ratios of the biomass indices over the same period. Coefficients of variance (standard deviation/mean, expressed as a percentage) are shown for these over the period over which the growth rates were calculated (*black bar*). Their variability indicates that growth was not balanced (Equation 2). **(c)** Variability in measured levels of ammonium, the nitrogen source in the medium, and in modeled light levels in the culture. (Light levels were modeled by Geider et al. 1998). (Data courtesy Kevin Flynn.)

of ammonium, on Day 12, accumulation of PON ceases. Accumulation of Chl*a* and POC does not. Batch cultures cover multiple growth phases, and growth is, at best, fleetingly acclimated, with growth rate and physiological state representative of nutrient-replete growth at the experimental irradiance. If cultures are to be maintained in balanced growth, an alternative method to batch cultures is needed.

### 3.3. Nutrient-Replete Growth in Semicontinuous and Turbidostat Cultures

To maintain cells in nutrient-replete exponential growth, the culture must be diluted regularly to keep cell concentrations low so that nutrient concentrations remain high and light levels are unaffected by self-shading (Beardall and Morris 1976). In semicontinuous cultures, a modification of the batch method, the culture is diluted repeatedly to restrict the range of biomass density (Fig. 19.7a). The growth rate is then calculated from Equation 3, using the change in biomass and interval between dilutions. The approach is labor intensive, because the biomass must be assayed before each dilution to maintain the culture within the desired range. (A description of means for monitoring biomass is given at right.) A more technologically complex way of maintaining the culture is to use a turbidostat, a continuous culture in which an optical sensor and automated pump are used to achieve the same effect (Myers and Clark 1944, Rhee et al. 1981). This is illustrated in Figure 19.7b, where the transparency of the culture was monitored continuously at 680 nm. This wavelength was chosen because it coincides with the red absorption peak of Chl*a* and so is a convenient means to monitor the abundance of cells containing the pigment.

A schematic representation of the turbidostat is shown in Figure 19.8a. A reservoir of fresh medium was connected to the culture via a pump, which was turned on when attenuation rose above a preset threshold and was turned off when it fell below a second. The supply of medium was, therefore, triggered on demand by growth of the culture. The preset limits that regulated pumping differed by about 3% of the dynamic range of the signal, so biomass and the apparent growth rate were maintained at a nearly constant level. In a turbidostat, the achieved growth rate, which is the nutrient-replete growth rate at the experimental irradiance and temperature $\mu_{E,T}$, is equivalent to the dilution rate, which is the inflow rate divided by the volume of the culture vessel

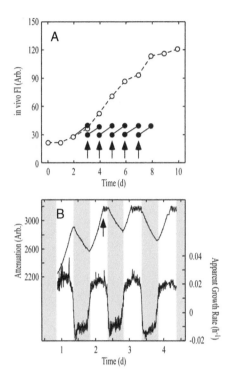

**FIGURE 19.7.** **(a)** Variation in the dark-adapted fluorescence of cultures of the pelagophyte *Aureococcus anophagefferens* Hargraves et Sieburth during growth in batch culture (*open symbols*) and in semicontinuous culture (*closed symbols*). Cultures were diluted daily (*arrows*) to maintain biomass (as indicated by fluorescence) within a restricted range. Note the limited variation in between-day slopes in semicontinuous culture as compared to the batch culture. (Data courtesy Fran Pustizzi.) **(b)** Variation in abundance and apparent growth rate of a culture of *A. anophagefferens* during growth in a turbidostat on a 12-12 L-D cycle (Todd Kana and MacIntyre, unpublished data.). The shaded bars indicate the dark period. Abundance was determined optically every 3 minutes, as transmission of light at 680 nm through the culture vessel. Apparent growth rate was calculated from the slope of ln-transformed data (Equation 3) over successive 9-minute intervals. Dilution with fresh medium was initiated when attenuation rose to more than 3200 units and was curtailed when it fell to 3100 units. The point at which dilution started is indicated by the arrow. Thereafter, data collected during the light period has the characteristic saw-tooth pattern of alternating intervals of growth and dilution. Declines in attenuation during the dark period are not due to dilution but to changes in the optical characteristics of the cells.

(Vol time$^{-1}$/Vol = time$^{-1}$). Because the sensor measures light attenuation across the sample vessel and initiates the supply of medium based on changes in it, it is very important that the attenuation reflects changes in the suspended culture alone. If the cells start to stick to the walls of the vessel, they will initiate dilution beyond

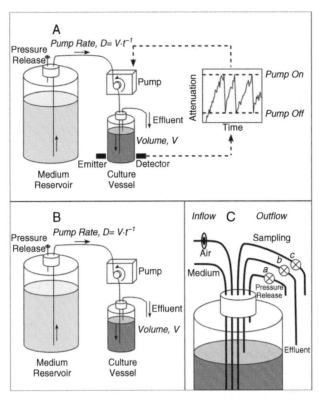

**FIGURE 19.8.** **(a)** Schematic representation of a turbidostat. For clarity, only the tubing through which medium and effluent from the culture flow are shown. In this configuration, an optical sensor interrogates the culture in the growth vessel. An alternative setup is to pump the culture through a separate detector loop, which has the advantage that the monitoring can be configured with alternating cleaning cycles and determination of background attenuation (Le Floc'h et al. 2002). The signal from the detector triggers an actuator that turns the pump on when attenuation exceeds a preset level. The influx of fresh sterile medium from the reservoir dilutes the culture until attenuation falls below the second preset level, at which point the pump is turned off. (The pressure relief on the medium reservoir should have an in-line sterile 0.2-μm filter to maintain sterility in the medium.) The culture is maintained at constant volume, $V$, by passive draining of effluent during the dilution. The growth rate is the intrinsic, nutrient-replete growth rate at the experimental irradiance and temperature, $\mu_{E,T}$. This is the rate of replacement of medium, $D$ (the volume pumped per unit of time per volume of culture; $D = V \cdot t^{-1}/V = t^{-1}$), which is equivalent to the mean rate of increase of attenuation. **(b)** Schematic representation of a chemostat. For clarity, only the tubing through which medium and effluent from the culture flow are shown. The pump that supplies fresh sterile medium runs continuously, at a rate, $D$, set by the experimenter. (The pressure relief on the medium reservoir should have an in-line sterile 0.2-μm filter to maintain sterility in the medium.) The culture is maintained at constant volume, $V$, by passive draining of effluent during the dilution. The growth rate is a predetermined nutrient-limited growth rate at the experimental irradiance and temperature, $\mu_{N,T}$. This is the rate of replacement of medium, $D$ (the volume pumped per unit of time per volume of culture; $D = V \cdot t^{-1}/V = t^{-1}$). **(c)** Schematic representation of the plumbing for a turbidostat or chemostat. The inflow tubes deliver fresh sterile medium and air that has been filtered through an in-line sterile 0.2-μm filter. There are three outflow tubes, each of which should have a gradient to prevent material in them flowing back into the culture and each of which has a pinch-valve ($a$, $b$, and $c$) that can completely occlude it. In normal operation, the pressure relief valve, $a$, and the effluent line, $b$, should have the valves open, and the sampling tube, $c$, should have the valve closed. Discrete samples of the culture can be taken by closing valves $a$ and $b$. A large-volume syringe is then attached to the sampling tube and valve $c$ is opened. When the sample has been drawn into the syringe, valve $c$ is closed and valves $a$ and $b$ are reopened. See Chapter 5 for more information on preserving sterility during culturing.

what is needed to maintain the growth of the cultures in suspension and cause it to be washed out. If wall growth starts in a turbidostat, the culture should be decanted into a clean vessel. (Also, the light measurement must be unaffected by ambient light, for example, by gating.) If the culture is to be grown on a light–dark cycle, the light that is used to interrogate the culture should be of a long enough wavelength that it is not photosynthetically active, ideally above 750 nm (see Section 5.0) and at as low an irradiance as the signal/noise ratio will permit. This will prevent the sensor from augmenting the growth irradiance during the day and inadvertently inducing photosynthesis and growth during the dark period.

The acclimated growth rate at the chosen temperature (and salinity) is then a function of irradiance and can be described in terms of three intrinsic characteristics of the cells. These are the maximum (i.e., light-saturated and nutrient-replete) growth rate, $\mu_m$ ($d^{-1}$), and two irradiances. The first, $K_C$, is one at which photosynthesis and respiration are equal so there is no net growth (Equation 1). The second, $K_E$, is the irradiance at which nutrient-replete growth becomes light saturated and is the analog of $E_k$, the saturating irradiance in a photosynthesis-irradiance curve (Talling 1957):

$$\mu_{E,T} = \mu_m \left( 1 - \exp\left[ -\frac{E - K_C}{K_E - K_C} \right] \right), \qquad (5)$$

where $\mu_{E,T}$ is the growth rate ($d^{-1}$) at irradiance $E$ and temperature $T$. In practice, most studies omit $K_C$ from the equation. Because balanced growth cannot be negative, Equation 5 only holds for the condition in which $E$ is greater than $K_C$. The same effect can be obtained by expressing growth in terms of ($\mu + R$) (Cullen et al. 1993), which is mathematically simpler. Both $K_E$ and $\mu_m$ are functions of temperature and salinity (Grzebyk and Berland 1996, Underwood and Provot 2000) and vary by about an order of magnitude between species (Langdon 1987, Geider et al. 1997, MacIntyre et al. 2002).

Irradiance is often expressed as photosynthetically active radiation (PAR, defined as 400 to 700 nm) in energetic or quantum-based units. Even if the possible influences of artificial versus natural light spectra on algal physiology are ignored, it is advisable to account for the spectral quality of the light source when quantifying experimental irradiance. This is done by calculating photosynthetically utilizable radiation (PUR) (Morel 1978), a measure of the light that can be absorbed by the microalgae (see following discussion).

One caveat in dealing with turbidostats is that the optical characteristics of the small sample used to monitor the culture must reflect the culture as a whole, and cells must be homogeneously distributed when culture volume is removed. In practical terms, this means that cells must be prevented from aggregating by sinking or swimming. The sinking rates of cells vary widely as a function of their size, shape, and physiological status (Smayda 1970, Bienfang and Szyper 1982, Bienfang et al. 1983, Richardson and Cullen 1995), as does their propensity for migratory swimming (Kamykowski 1995), but all will eventually become het-

erogeneously dispersed in the absence of any mixing. The mixing can be episodic or continuous and can be achieved by bubbling or using a mechanical stirrer. Though necessary to satisfy the assumption of homogeneity, the increased turbulence associated with stirring can impede the growth of some species, particularly dinoflagellates (Juhl et al. 2001, Sullivan and Swift 2003). Bubbling can also be problematic, because the bubbles interfere with the measurement of attenuation if they are in the light path. Conversely, unless concentrations of microalgae are kept low, a culture that is not bubbled with air is likely to enter carbon limitation, a form of nutrient limitation that is often overlooked. The choice between automated assessment of culture concentration and replacement of medium and periodic assessment and replacement is often dictated by the sensitivities of the microalgae.

### 3.4. Nutrient-Limited Growth in Chemostat and Cyclostat Cultures

In a turbidostat, medium is delivered when algal concentrations exceed a threshold, so algal growth is countered with dilution, and the dilution rate, which matches the growth rate, is determined by irradiance and temperature as they influence the intrinsic nutrient-replete growth rate, $\mu_{E,T}$ (Equation 5). In contrast, chemostat cultures are maintained in balanced growth under nutrient-limited conditions at a growth rate that is set by the experimenter (Rhee et al. 1981). A schematic representation of the chemostat is shown in Figure 19.8b. The medium, formulated to be deficient in one nutrient, is delivered continuously but at a rate, $D$, less than the nutrient-replete growth rate, $\mu_{E,T}$. The achieved growth rate, $\mu_N$, is equivalent to $D$, which is the inflow rate divided by the volume of the culture vessel (Vol time$^{-1}$/Vol = time$^{-1}$). Chemostat theory, the theoretical underpinning of the equivalence between $\mu_N$ and $D$, has been described by Rhee et al. (1981). A cyclostat is a chemostat grown under a repeatably varying regimen of light or (rarely) temperature. Cells in a cyclostat are also in balanced growth but are in dynamic equilibrium, rather than simple equilibrium. The growth rate is again equivalent to the dilution rate when averaged over the day, as for nutrient-replete growth on a light–dark cycle (see Fig. 19.4), although cell division can be entrained to the light–dark cycle in a diel rhythm (Chisholm 1981; see Chapter 21). Semicontinuous cultures can be run as quasi-chemostats so the culture is nutrient limited rather than nutrient replete (Sakshaug et al. 1989, MacIntyre et al. 1997b). Instead of diluting the culture on demand to maintain its density at a constant level, a regular (usually daily) dilution is applied at a rate below $\mu_{E,T}$. When the measured growth rate is equal to the dilution rate and acclimation is com-plete, the cultures are in balanced nutrient-limited growth. The approach differs from a chemostat in that the fresh medium is added episodically rather than continuously.

The continuous supply of nutrients in the chemostat introduces elements of both Liebig and Blackman limitation on growth. At all but the highest nutrient-limited growth rates, the residual concentration of the limiting nutrient in a chemostat is a very small fraction of the initial concentration in the medium (McCarthy 1981, Garside and Glover 1991). Biomass in the culture is determined by the initial concentration of limiting nutrient (Rhee 1978), which is consistent with Liebig limitation (see Fig. 19.5b). That is, if the input medium has 50 $\mu$g-atom $\cdot$ L$^{-1}$ nitrate and other nutrients are supplied in excess (and if excretion of dissolved organic nitrogen [DON] is low, which is typical), the concentration of particulate nitrogen in the culture will be close to 50 $\mu$g-atom $\cdot$ L$^{-1}$. However, the growth rate of microalgae in a chemostat is equal to the dilution rate, which is set by the experimenter, and the physiological adjustment by the cells to the supply of the limiting nutrient (see Fig. 19.2) reflects the degree of imposed Blackman limitation.

In a chemostat, the growth rate does not vary with irradiance, temperature, or the nutrient composition of the medium; rather, irradiance and temperature influence the degree of stress that is imposed at a particular nutrient-limited growth rate and the concomitant acclimative state that is achieved in response. Consequently, growth rates in chemostats are often expressed as relative to the maximal rate (i.e., $\mu_N/\mu_{E,T}$), thereby eliminating much of the variability in results that is a function of growth rate alone (Goldman 1980).

The concept of relative growth rate is critical to the interpretation of experiments in nutrient-limited continuous cultures because imposition of an added stress can induce responses that are difficult to interpret. By definition, the growth rate in the chemostat, $\mu_N$, will not vary even though an experimentally imposed stressor (e.g., a toxic compound or ultraviolet-B [UV-B] radiation) may have an adverse effect on the nutrient-limited cells. If it is sufficiently deleterious to cause a reduction in the nutrient-replete growth rate, $\mu_{E,T}$, the stressor would cause a highly counterintuitive *increase* in relative growth rate because $\mu_N$ remained constant at the experimental setting and $\mu_{E,T}$ declined. In our opinion, it is difficult to resolve the inhibitory effect of a stressor if

the growth rates in chemostat cultures grown with and without the stressor are the same so that the relative growth rate ($\mu_N/\mu_{E,T}$) is higher for the algae under stress. We do not recommend this experimental approach.

As in a turbidostat, when a culture is grown in a chemostat, there is no net change of biomass (Fig. 19.9a). If the medium is supplied at a rate higher than $\mu_{E,T}$, the cells are unable to divide rapidly enough to maintain a stable population and are washed out of the growth vessel. As with other cultures, the cells are in steady-state growth when there are no changes in the ratios of indices of biomass (Equation 1, compare Fig. 19.9b and Fig. 19.6b) or in indices of physiological

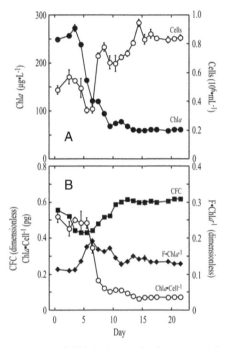

**FIGURE 19.9.** **(a)** Variation in the biomass of the diatom *Thalassiosira pseudonana* during growth in a chemostat culture (Xiaolong Yang, MacIntyre and Cullen, unpublished data). The chemostat was inoculated from a batch culture on day 0 and maintained at a dilution rate of 0.2 d$^{-1}$. The nutrient-replete growth rate, $\mu_{E,T}$ (Equation 5), at the same irradiance was 1.33 d$^{-1}$. **(b)** Variation in the cellular chlorophyll quota (the ratio of the two biomass indices) and in two physiological indices, cellular fluorescence capacity (*CFC*) and $F \cdot Chla^{-1}$. CFC is the ratio of variable fluorescence to maximum fluorescence of dark-adapted cells, as measured in the presence of the inhibitor DCMU, and is an analog of the quantum yield of PSII. The parameter $F \cdot Chla^{-1}$ is the ratio of dark-adapted Chla fluorescence of the culture *in vivo* to the fluorescence of its Chla content *in vitro* (Kiefer 1973a). Constancy in these indicators after day 15 indicates that the cells were fully acclimated to their environment and that growth was balanced (Equation 2).

response that are responsive to changes in growth rate and nutrient status (see following discussion). As a rule of thumb, plan on 10 generations of growth to be assured that the cells are fully acclimated. Much time can be saved by using an inoculum that has been grown under similar conditions. For nutrient-limited chemostats, this could include subjecting the parent culture to early phases of starvation before inoculation. This also serves to eliminate carryover of limiting nutrient from the inoculum to the chemostat.

The relative concentrations of nutrients that are needed to ensure that a given nutrient is limiting in the chemostat are usually chosen by assuming that the Redfield ratio (Redfield 1958) of 106 : 16 : 1 C : N : P (Goldman et al. 1979) or an extension of the Redfield ratio that includes trace metals (see Chapter 4) is an approximation of the critical ratio (see Fig. 19.5b). The concentration of the limiting nutrient is then set so that its relative abundance is well below the Redfield ratio. The absolute concentrations of nutrients should be set so that biomass is low enough to keep the sample optically thin, to avoid second-order effects of variability in irradiance within the growth vessel. In theory this depends on the optical characteristics of the cells (their numerical abundance and specific absorption and scattering coefficients), the diameter of the vessel, its optical geometry, and so on (Droop 1985). In practice, we try to keep biomass below $100\ \mu g$ Chl$a \cdot$ L$^{-1}$. Published relationships between nitrogen or phosphorus and Chl$a$ (see Fig. 19.2) can be used as a rough guide in determining conversion factors between nutrient concentrations and Chl$a$.

### 3.5. Cultures in Unbalanced Growth

The types of cultures we have discussed are most frequently used for determining the physiological status of microalgae under equilibrium conditions. An alternative approach is to impose varying environmental conditions on a culture and monitor its responses. The simplest case is one in which a single variable is changed when a culture is in balanced growth. The culture is then monitored during the acclimative process to determine the time course of establishing a new equilibrium. Such experiments have been done frequently with irradiance (Post et al. 1985, Cullen and Lewis 1988, Sukenik et al. 1990, Anning et al. 2000), and less often with nutrient concentrations (Sciandra and Amara 1994, Le Floc'h et al. 2002) or temperature (Anning et al. 2001). Although informative about the rate of acclimation, such conditions are still highly artificial.

Approaches that are more realistic and correspondingly more difficult to interpret impose environmental changes on time scales that are comparable to or shorter than the response time of physiological or behavioral acclimation. These include imposing changes in irradiance (Marra 1978a, b; Kroon et al. 1992a, b; Ibelings et al. 1994), nutrients (Sciandra 1991, Bernard et al. 1996), or species composition in competition experiments (Litchman 1998, 2003) and observing the responses. In some of these (Marra 1978a, b), the light source was natural sunlight. To the extent that the conditions in the culture vessel mimic those in nature, the outcomes can then be used to predict responses to variations in nature; such observations have been analyzed extensively, for example, to predict photosynthetic responses during mixing (Marra et al. 1985, Neale and Marra 1985, Denman and Marra 1986, Pahl-Wostle and Imboden 1990, Franks and Marra 1994). The approach has the attraction of relying less heavily on what a holist would regard as 'unnatural' conditions in the culture. Even so, there is a limit to its usefulness because it can be wildly inaccurate to extrapolate from one set of conditions to another; the physiological process that limits a microalga's photosynthetic rate is determined in large part by the rate of change in the environmental cue (Lewis et al. 1984, Cullen and Lewis 1988, MacIntyre et al. 2000). For instance, regulation of Chl$a$ content, accumulation and repair of photoinhibitory damage, and photosynthetic induction all have the potential to limit the rate of photosynthesis in a varying light field but have response times that vary over almost four orders of magnitude (Cullen and Lewis 1988, Baroli and Melis 1996, MacIntyre et al. 1997a, Neidhardt et al. 1998). Their relative importance as potential limiting rates for photosynthesis depends on whether the rate of change in the light field due to mixing is high (MacIntyre and Geider 1996) or low (Kamykowski et al. 1996) and applying the wrong rate constants in a simulation yields seriously inaccurate predictions.

### 4.0. MONITORING CULTURES

Sampling methods range from periodic analysis by hand to computer-driven continuous measurement via automated sensors. Manual analysis is much more common. A schematic representation of the plumbing required to remove small samples from a culture is shown in Figure 19.8c. Here, we recommend several techniques that we have found valuable; they are sensitive, reliable, and relatively fast. Sensitivity is important because if the culture is to be maintained, destructive analysis must not consume material more rapidly than the culture can produce it. The other major considerations are as follows:

1. If the monitoring is designed to ensure that the culture is acclimated to its growth conditions, it is important to measure more than one parameter to satisfy the requirement of Equation 2 that the rate of change of all parameters be equal, and in the case of continuous cultures, zero.
2. When microalgae are grown under light–dark cycles, physiological responses (Prézelin et al. 1977, Harding et al. 1981), cell division (Chisholm and Brand 1981), and such behaviors as migration (Palmer and Round 1965, Weiler and Karl 1979, Cullen 1985, Serôdio et al. 1997) are entrained to the photoperiod. Consequently, if the culture is being monitored to determine whether growth is balanced, it is critical that the culture be sampled at the same point in the cycle (see Fig. 19.4).

## 4.1. Cell Numbers

We recommend at least two estimates of biomass and/or physiological response when monitoring cultures. One that requires relatively little material is a cell count (see Chapter 16). Cells can be counted microscopically using bright-field microscopy with a hemocytometer or another counting chamber or by using epifluorescence after filtration onto a black membrane filter (MacIsaac and Stockner 1993). The latter is preferable where the cell density is low, so that the volume contained by the hemocytometer is too small to give a statistically acceptable estimate. Microscopic enumeration can be more tedious and less precise than some other methods, but experience has shown that it is an important part of experimentation with cultures; visual inspection provides important information on the morphology of cells and the presence of contaminants that could have a strong bearing on the interpretation of results.

Rapid counts can also be made using a Coulter counter or flow cytometer calibrated with fluorescent beads of a known concentration (Olson et al. 1993, see Chapter 17). The flow cytometer can measure the statistical distribution of several parameters (fluorescence of Chl$a$, phycobilin and/or a vital stain, and side and forward-angle scattering) simultaneously and could be particularly useful when used with a stain such as 4′,6-diamidino-2-phenylindole (DAPI) (Porter and Feig 1980) or fluorescein diacetate (FDA) (Jochem 1999). The forward-angle scattering also gives a first-order estimate of cell volume and, from that, cell carbon (DuRand et al. 2002). However, caution should be used in applying volume to carbon conversion factors, because diatoms, in particular, seem to have different carbon densities (carbon to volume relationships) from other protists (Menden-Deuer and Lessard 2000).

## 4.2. Chlorophyll $a$

Chl$a$ is ubiquitous in microalgae and the methods for measuring it are sensitive, precise, and (mostly) accurate (see Chapter 20). It is a highly useful index for monitoring growth in cultures, particularly in concert with *in vivo* fluorescence (see following discussion). For the purposes of monitoring, it is not necessary (and it would not be cost-effective) to determine Chl$a$ by the most accurate and discriminating method, high-performance liquid chromatography (HPLC). Fluorometric analysis after extraction in 90% (V/V) acetone is a sensitive, precise, and generally accurate means of assessing the concentration from small aliquots, although the technique is less reliable in species that have high levels of Chl$b$ and/or tough cell walls that impede pigment extraction (Holm-Hansen 1978, Welschmeyer 1994). Acceptably precise estimates (coefficients of variance of <5% between replicates) can usually be obtained from extracts with concentrations of as little as 1 to 5 µg Chl$a \cdot L^{-1}$. As most fluorometers require only 3 to 5 mL of the extract to make a measurement, this translates to a minimum of about 10 ng of Chl$a$ per aliquot (i.e., 0.1 mL of a culture with 100 µg Chl$a \cdot L^{-1}$). Measurement of Chl$a$ can be biased by the presence of Chl$b$ or the degradation product chlorophyllide $a$, which has similar optical qualities (Holt and Jacobs 1954, Jeffrey et al. 1997). A correction can be made for degradation products by measuring fluorescence before and after the sample has been acidified (Holm-Hansen 1978), but the method is still unreliable in the presence of Chl$b$. An alternative is to use filters that optimize for absorption and emission by Chl$a$ alone (Welschmeyer 1994). The latter is faster and acceptably accurate (Box 19.1, Fig. 19.10).

Our default technique for fluorometric assessment of Chl$a$ is to filter a small volume onto a glass-fiber filter and extract the pigments passively in the dark at −20° C for 24 hours, using 90% (V/V) acetone. Note that this is not recommended for HPLC protocols (see Chapter 20) because of the possibility of some degradation of the Chl$a$. However, the primary degradation

**BOX 19.1   A Comparison of Experimental Methods for Determining Chl*a*.**

A comparison of estimates of the Chl*a* content in samples of the prasinophyte *Pycnococcus provasolii* Guillard guides our discussion of techniques for measuring Chl*a* in cultures (Fig. 19.10). The alga has high levels of Chl*b* and Mg-2,4-divinyl pheoporphyrin $a_5$ monomethyl ester (Mg DVP; see Table 19.3), both of which can compromise assessment of Chl*a*, so it is a good experimental organism to test the methods. The benchmark estimate is obtained by high-performance liquid chromatography (HPLC) (see Chapter 20), as described by van Heukelem et al. (1992). The default rapid estimate is by passive extraction in 90% (v/v) acetone for 24 hours in darkness and at –4° C, followed by fluorometric determination against a pure Chl*a* standard (Welschmeyer 1994).

Alternative methods are as follows:

- *Spectrophotometry:* The extraction technique is the same as for fluorometry, so there is no gain in speed. It is less sensitive than the fluorometric method and tends to overestimate Chl*a* (MacIntyre et al. 1996, Van Heukelem et al. 2002), even when the estimate is corrected for interference by other chlorophylls (Jeffrey and Humphrey 1975, Jeffrey et al. 1997), as shown in Fig. 19.10a.
- *Grinding (not tested):* Grinding the sample in a mortar with a Teflon pestle or in a Ten-Broek tissue homogenizer obviates the need for the long passive extraction. Our frequent observation of the splatter-line of partially ground filter on the adjacent surfaces suggests that this is not an accurate technique in inexperienced hands.
- *Freeze/thaw:* Here, the filter is frozen in one part of distilled water and thawed with nine parts of acetone to give a final concentration of 90% (v/v) acetone (Glover et al. 1986). The method greatly accelerates extraction: As the sample freezes, ice crystals can break the cells' membranes and extraction is effectively complete in the time it takes to freeze a small aliquot of water. The method gives much better extraction in species with relatively tough cell walls such as the

cyanobacterium *Synechococcus* spp. (Glover et al. 1986). It may still underestimate Chl*a* (Fig. 19.10a), perhaps because of increased chlorophyllase activity in the aqueous phase during the freeze/thaw cycle: Chlorophyllase activity can increase with the aqueous content in acetone/water mixtures (Barrett and Jeffrey 1964).
- *Extraction in DMSO and acetone:* An alternative solvent is a 2 : 3 mixture of dimethyl sulfoxide (DMSO) and 90% (v/v) acetone (Shoaf and Lium 1976). Pigments are fully extracted in the mixture in 15 minutes of darkness and at room temperature, and the accuracy and precision compare favorably with HPLC (Fig. 19.10A and C). In a comparison of side-by-side fluorometer calibrations with 90% (v/v) acetone and the 90% acetone (v/v)/DMSO mixture, we found that there was no significant difference between their calibration slopes (Student's paired *t*-test: $P = 0.71$, $R^2 = 0.92$, n = 6). The same calibration can, therefore, be used for either method. Stramski and Morel (1990) report a similar result. DMSO must be handled with caution because it is readily absorbed through the skin and can transport other chemicals, such as the acetone, with it. Like acetone, it is a respiratory irritant and so should be used in a fume hood and handled with heavy gloves. Under no circumstances should either solvent be pipetted by mouth.
- *Direct injection:* An even more rapid extraction technique eliminates filtration; a small volume of culture (1% of the final extact volume) is pipetted directly into the solvent. The method compared favorably with HPLC when the solvent was 90% acetone (v/v)/DMSO and extraction was for 15 minutes in darkness at room temperature (Fig. 19.10A). When the same technique was used with 90% acetone and 24-hour extraction, a dense precipitate formed that gave an artifactually high estimate of Chl*a*. The estimate was lower and more accurate when the precipitate was removed by centrifugation (Fig. 19.10a,b).

product, chloropyllide *a*, has optical properties comparable to Chl*a*'s (Holt and Jacobs 1954), so the fluorometric technique is insensitive to minor degradation. Although it is important to keep the sample cold and in the dark during extraction to minimize degradation of Chl*a*, fluorescence yield in solvent is a function of tem-

perature and water condenses on cold cuvettes, so the extract should be warmed to room temperature before fluorescence is measured. This can be done quickly in a darkened water bath to prevent degradation of pigment. For cultures, the fluorometric technique compares favorably with HPLC estimates of Chl*a* (see Fig.

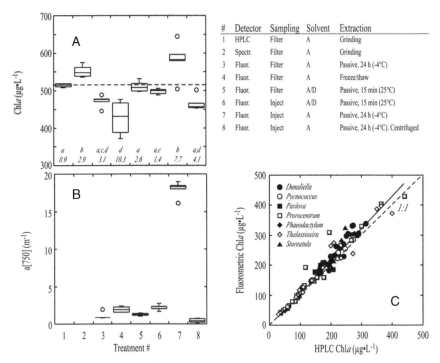

| # | Detector | Sampling | Solvent | Extraction |
|---|----------|----------|---------|------------|
| 1 | HPLC | Filter | A | Grinding |
| 2 | Spectr. | Filter | A | Grinding |
| 3 | Fluor. | Filter | A | Passive, 24 h (-4°C) |
| 4 | Fluor. | Filter | A | Freeze/thaw |
| 5 | Fluor. | Filter | A/D | Passive, 15 min (25°C) |
| 6 | Fluor. | Inject | A/D | Passive, 15 min (25°C) |
| 7 | Fluor. | Inject | A | Passive, 24 h (-4°C) |
| 8 | Fluor. | Inject | A | Passive, 24 h (-4°C). Centrifuged |

**FIGURE 19.10.** Comparison of chlorophyll *a* (Chl*a*) extraction by different analytical methods. **(a)** Box plot of estimated Chl*a* in a dense culture of the prymnesiophyte *Pycnococcus provasolii*, as determined by higher-performance liquid chromatography (HPLC) (Van Heukelem et al. 1992), spectrophotometry (Jeffrey and Humphrey 1975), and fluorometry (Welschmeyer 1994) (n = 6 for each treatment). Samples determined by fluorometry were either collected on Whatman GF/F filters or pipetted directly into the solvent (filter vs. injection; see key). All filtered samples had the same volume: Solvent volumes were varied over a 3x range to optimize for the detection method. Pigments were extracted in either 90% acetone or a 3 : 2 mixture of 90% acetone and dimethyl sulfoxide (DMSO) (Shoaf and Lium 1976). A, 90% acetone; A/D, 2 : 3 DMSO:90% acetone. The sample injected into 90% acetone (#7) was centrifuged at 3,100 *g* for 5 minutes to remove precipitates and re-read (#8). Numbers under each treatment are the coefficients of variance (%). Treatments identified with the same letter (*a–d*) were not significantly different (one-way ANOVA, Tukey's *a posteriori* test for significance, $P < 0.05$). **(b)** Scattering characteristics of the samples for fluorometric analysis in **(a)**, as determined by apparent absorption at 750 nm. Absorption was measured against 0.2 μm-filtered 90% acetone in a 1-cm cuvette at the port of an integrating sphere. Note the high value for culture injected directly into 90% acetone (#7). **(c)** Chl*a* estimated fluorometrically (Welschmeyer 1994) after 15 minutes of extraction in 3 : 2 90% acetone : DMSO and estimates made by HPLC after extraction by grinding in 90% acetone. The 100 samples were from cultures with diverse accessory pigments: Chl*b* and carotenoids (*Dunaliella tertiolecta* Butcher, chlorophyte; *P. provasolii*, prasinophyte); Chl*c* and carotenoids (*Pavlova lutheri*, prymnesiophyte; *Prorocentrum minimum* [Pavillard] Schiller, dinoflagellate; *Phaeodactylum tricornutum* and *Thalassiosira weissflogii*, diatoms); and Chl*c* and phycoerythrin (*Storeatula major* Hill, cryptophyte). The cultures were variously nutrient-replete and nutrient starved. On average, the fluorometric method overestimated Chl*a* by 5% ± 3% (*solid line*). The coefficient of determination ($R^2$) was 0.94.

19.10a) (MacIntyre et al. 1996), but passive extraction of pigments is too slow for decisions on whether to transfer or harvest a culture. We discuss the pros and cons of alternative techniques in Box 19.1.

### 4.3. Attenuation

Optical techniques in general lend themselves to non-invasive monitoring of microalgal cultures. Dedicated light detectors (e.g., Thermo Oriel, Ocean Optics) are relatively inexpensive and can be aligned with emitters

for continuous measurement of transmission through the culture vessel. The turbidostat data in Figure 19.7b were collected in this way. In practice, wall growth in front of the source and detector (or on the source and detectors themselves if they are inserted into a culture through ports) biases the data record. Formation of these biofilms is particularly problematic with benthic microalgae and cyanobacteria. Accumulation or consumption of colored nonalgal material in the culture (e.g., vitamin and trace-metal solutions) also biases the record unless the data are blank-corrected frequently. Continuous monitoring using optical techniques is best done by removal of discrete samples, as part of a cycle that includes determination of the blank and cleaning with a detergent solution followed by rinsing (Le Floc'h et al. 2002).

## 4.4. *In Vivo* Fluorescence

There are only a few means by which the abundance and/or physiological responses of microalgae can be measured without destructive sampling. Foremost among these is measurement of Chl*a* fluorescence. This pigment has the useful property that a proportion of absorbed light energy (excitation) is re-emitted with a longer wavelength (emission) as fluorescence. Chl*a* fluorescence can be measured from live cells (*in vivo*) or after extraction of the pigment into a solvent. The two measures, however, are not the same. In live cells, the excitation energy can be directly absorbed by the pigment itself or absorbed by accessory chlorophylls, carotenoids, or phycobilins and transferred to Chl*a*. When pigments are extracted in solvent, the disruption of the associated pigment–protein complexes ensures that only light absorbed directly by Chl*a* itself can be re-emitted.

Because fluorescence can be stimulated by relatively low light levels that induce no lasting physiological stress, *in vivo* fluorescence can be used as a nondestructive means of measuring growth rate (see Chapter 18). Monitoring is particularly easy if cultures are grown in 25-mm diameter (50-mL) tubes that can be fit into many conventional fluorometers (Brand et al. 1981). This method is particularly useful when cultures must be maintained under axenic conditions, where repeated removal of subsamples may compromise the culture. Assuming growth is balanced, changes in fluorescence correlate with changes in biomass. Some care has to be taken to ensure that this is so.

The amount of Chl*a* fluorescence depends on several factors:

$$Fl = \sum \left[ E_{Ex}(\lambda) \cdot a^{Cbl}(\lambda) \right] \phi_{Fl}, \qquad (6)$$

where *Fl* is fluorescence, $E_{Ex}(\lambda)$ is the scalar irradiance of the excitation beam at wavelength $\lambda$, and $a^{Cbl}(\lambda)$ is Chl*a*-specific absorption at wavelength $\lambda$, and $\phi_{Fl}$ is the quantum yield of fluorescence. None of these terms is constant, so fluorescence should be used as a measure of biomass only with caution.

1. The excitation characteristics of a fluorometer, $E_{Ex}(\lambda)$ (which determines absorption and can influence quenching; see later discussion), are set by its manufacturer and by the age of the lamp. Consequently, fluorescence is a relative property and cannot be compared directly either between two fluorometers of different types (Cullen et al. 1988) or even between different fluorometers that are the same model.

2. Chl*a*-specific absorption, $a^{Cbl}(\lambda)$, varies with the complement of accessory and photoprotective pigments, which in turn vary between species and within a species during acclimation to irradiance and under nutrient starvation and nutrient limitation (Sosik and Mitchell 1991, Wilhelm and Manns 1991, Latasa and Berdalet 1994, Goericke and Montoya 1998, MacIntyre et al. 2002). It also varies with the degree of pigment packaging imposed by variations in the cellular Chl*a* quota (Duysens 1956, Berner et al. 1989, Dubinsky 1992).

3. The quantum yield, $\phi_{Fl}$, can increase by as much as 50% when cells enter stationary phase (Kiefer 1973b, Beeler SooHoo et al. 1986, Beutler et al. 2002). The relationship between fluorescence and Chl*a* will, therefore, vary with growth conditions (see Fig. 19.9b). There is also variability in quantum yield due to fluorescence quenching, the reduction of the yield due to utilization of absorbed photons in photochemistry (photochemical quenching) or due to an enhanced probability of heat dissipation (nonphotochemical quenching, induced in bright light). With the exception of the relatively slow repair of photoinhibitory damage, quenching processes relax in darkness with a time constant of about 5 minutes or less (Kolber et al. 1990). The effects of short-term quenching on the fluorescence–Chl*a* relationship can, thus, be reduced to a minimum by placing the sample in darkness for 20 to 30 minutes before the measurement.

Fluorescence *in vivo* can be used only as a proxy for Chl*a* if the growth conditions are stable and the cells are acclimated to them. For instance, it is not a reliable measure of Chl*a* throughout growth in a batch culture, except in the early phase where the culture is optically 'thin' and nutrients are present in excess.

Because the relationship between Chl*a* and fluorescence is so sensitive to both short-term and long-term responses of the photosynthetic apparatus, the ratio of the two is a useful index of whether cells are in balanced growth. It can be parameterized in dimensionless terms as $F \cdot \text{Chl}a^{-1}$ (Kiefer 1973b), which is the ratio of dark-adapted fluorescence *in vivo* to the fluorescence of the same concentration of Chl*a* *in vitro* after extraction in solvent, measured with the same fluorometer. This is a sensitive indicator of acclimative state and reaches an equilibrium value when biomass-based indices also indicate that cells are in balanced growth (see Fig. 19.9). It also has the advantage that it can be compared between studies, provided that the excitation characteristics (spectral quality and intensity) of the fluorometers used to measure it are similar and that the same solvents are used in the *in vitro* measurements. Determination of Chl*a* directly to make the comparison requires destructive sampling of the culture, though.

Another dimensionless but highly sensitive fluorescence parameter is the ratio of variable to maximum fluorescence. This can be measured in the dark after destructive sampling, using the inhibitor 3-[3,4-dichlorophenyl]-1,1-dimethylurea (DCMU). We retain Vincent's (1980) description of this parameter as the cellular fluorescence capacity (CFC) (see Table 19.1, Fig. 19.9), to distinguish it from the analogous parameter, $F_v/F_m$, that is measured nondestructively using either a pulse-amplitude–modulated (PAM) (Schreiber et al. 1986, 1993) or a fast repetition rate (FRR) fluorometer (Falkowski et al. 1986, Kolber and Falkowski 1993). The PAM and FRR fluorometers differ in the intensity and duration of the exciting flash, so the former measures multiple PSII turnover events and the latter measures single turnover events. Both can also measure the ratio of variable to maximum fluorescence under actinic illumination, $F_v'/F_m'$, which can be equated with the quantum yield of PSII (Genty et al. 1989, 1990). Estimates of $F_v/F_m$ from PAM and CFC are correlated (Parkhill et al. 2001). Estimates of $F_v/F_m$ from PAM and FRR are also correlated, but the relationship is nonlinear and the estimates differ in magnitude (Koblízek et al. 2001, Suggett et al. 2003).

The nomenclature used in describing fluorescence is confusing and notoriously inconsistent between different research groups. Kromkamp and Forster (2003) provide a review of the nomenclature and a comparison of *CFC*, $F_v/F_m$, and $F_v'/F_m'$. Both *CFC* and $F_v/F_m$ can be measured rapidly on small amounts of biomass, making them ideal for assaying the physiological status of cultures. They are great indicators of a culture's status, being sensitive, quick, and easy to measure. $F_v/F_m$ in continuous cultures can be high, despite strong nutrient limitation, when cells are fully acclimated (Cullen et al. 1992b, Parkhill et al. 2001). This acclimation requires more time than it takes for other properties of the culture to reach a stable value, so reductions in $F_v/F_m$ are often cited as an indication of nutrient stress in both starved and limited cultures (Geider et al. 1993a, 1993b).

The PAM and FRR fluorometers could in theory interrogate a culture through the vessel walls and so be used to assess its physiological status noninvasively. It is easy to do with most PAMs (Lippemeier et al. 2001), which have fiberoptic emitter/detector assemblies. The configuration of commercially produced FRR fluorometers does not lend itself to this application, but modifications can be made to monitor cultures (Y. Huot, unpublished data). The idea should be treated with caution: *CFC* and $F_v/F_m$ are calculated as the difference between minimum and maximum fluorescence, normalized to the maximum ($[F_d - F]/F_d$ and $[F_m - F_0]/F_m$), so the numerator is insensitive to background fluorescence, but the denominator is not (Cullen and Davis 2003). Overestimation of the blank leads to underestimation of the ratio and vice versa, and the bias is amplified as biomass decreases. For the cultures shown in Figure 19.2a (40 to 140 µg Chl$a \cdot$ L$^{-1}$), the blank was only 1 to 3% of $F_d$; for the lower biomass natural samples in Figure 19.1 (0.1 to 4.6 µg Chl$a \cdot$ L$^{-1}$), the blank was 1 to 13% of $F_d$ (data not shown). This is high enough to warrant correction for background fluorescence. In practice, this is best done by gravity-filtering the sample and using the filtrate, whose fluorescence usually differs from both fresh medium and distilled water. Using vacuum filtration at anything above vanishingly low pressures can give an erroneously high blank, presumably through rupture of a small but significant fraction of the cells releasing fluorescing cell debris into the filtrate. Note that *CFC* can also be biased if the excitation intensity is too high when cultures are acclimated to low irradiance (Parkhill et al. 2001). This causes some photosynthetic induction, which causes $F$ to rise and so *CFC* is underestimated. The problem can be corrected by using appropriate neutral-density filters to attenuate the excitation beam.

## 5.0. PHOTOSYNTHESIS AND IRRADIANCE

Photosynthesis is a fundamental property of autotrophic microalgae that must be quantified to understand the role of aquatic systems in the global carbon cycle. The spatial and temporal scales of natural variability preclude measurement, so estimates are based on models of photosynthetic response to environmental factors (Behrenfeld and Falkowski 1997a, b). Photosynthesis varies with irradiance in a nonlinear manner, described experimentally as a photosynthesis–irradiance (PE) curve. Short-term changes in photosynthetic rate during exposure to different light intensities are due to regulation of the connectivity and activity of components of the photosynthetic apparatus. Long-term changes are due to acclimative regulation of their abundance (see Fig. 19.3). Describing both the short- and the long-term variability is central to our understanding of microalgal ecology. However, an unambiguous description of the causal factors and the magnitude and rate of the resulting changes in photosynthetic response is possible only under controlled conditions, so model development will depend on careful comparison between the variation in PE responses in cultures with descriptions of natural conditions. Variability in estimates of the PE response depends on the means used to measure it, the time course of the measurement, the biomass parameter used to normalize the data, and any spectral differences between the incubator and the growth vessel. We deal with these in turn.

## 5.1. Reconciling Light Fields: PAR vs. PUR

Photoautotrophic algae need light for photosynthesis, but not all photons are equal. For a given light source, the nutrient-replete growth rate can be described as a function of irradiance (Equation 5), or more accurately, spectral scalar irradiance. For measurement of irradiance itself, we refer the reader to Kirk (1994). A microalga's growth rate at a given intensity of PAR (measured as the average over 400 to 700 nm) depends both on its value relative to $K_E$ and on the spectral quality of light (Glover et al. 1987). This is largely because microalgae absorb light of different wavelengths with different efficiencies (Fig. 19.11).

Measurement of light absorption by microalgae is complicated by the fact that at all but high densities their absolute absorption is low and that, as particles, they both absorb and scatter light. Cultures can be grown to high enough densities to ensure adequate signal/noise ratios with a benchtop spectrophotometer. However, doing so inevitably produces gradients in irradiance within the culture vessel that complicate the description of growth irradiance; the least ambiguous estimates of growth irradiance are obtained when samples are optically thin. Scattering within a sample introduces two errors into estimates of absorption. The first is that the directional nature of scattering causes losses in the transmitted photon beam that are interpreted as absorption. The second is that multiple scattering at high particle densities increases the probability of subsequent absorption (absorption enhancement).

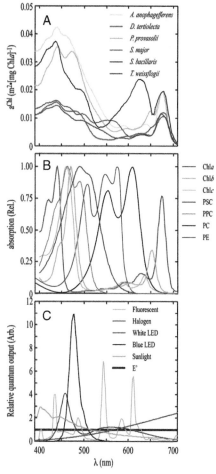

**FIGURE 19.11.** **(a)** Variation in pigment-specific absorption with wavelength for five species of microalgae grown at the same irradiance (80 μmol photons · m⁻² · s⁻¹ of fluorescent light). The species have diverse accessory pigments (see Table 19.3) characteristic of their taxa: *Aureococcus anophagefferens*, pelagophyte; *Dunaliella tertiolecta*, chlorophyte; *Pycnococcus provasolii*, prasinophyte; *Storeatula major*, cryptophyte: *Synechococcus bacillaris*, cyanobacterium; *Thalassiosira weissflogii*, diatom. **(b)** Variation in absorption *in vitro* for the pigments found in the species in **(a)**: chlorophylls *a*, *b*, and *c*; photosynthetic and photoprotective carotenoids (PSC and PPC); and R-phycoerythrin (PE) and R-phycocyanin (PC). Spectra are expressed relative to their maximum absorption. Spectra for chlorophylls and carotenoids are from Bidigare et al. (1990). Spectra for phycobilins are redrawn from Kirk (1994). **(c)** Spectral output of four light sources commonly used in growth and physiology experiments, sunlight, and a theoretical flat spectrum, *E′*, used to estimate the correction factor $A_e$ (see text for details). The artificial light sources were a cool white fluorescent tube (Osram Sylvania F40T12), a quartz halogen lamp (General Electric ENH250), a white LED (Emissive Energy Inova 5X), and a blue LED (Chelsea Instruments FAST^tracka excitation LED). Sunlight was measured under clear sky conditions at Dauphin Island, Alabama. All sources have been scaled to the same value of PAR (the integrated output between 400 and 700 nm).

1. If cultures are grown at low densities, the low signal/noise ratio can be improved by increasing the effective pathlength by collecting the culture on a filter and scanning the filter (Yentsch 1962). The absorption enhancement on the filter caused by scattering must be corrected, either by using an empirical comparison with a suspension in a cuvette (Mitchell and Kiefer 1984, Bricaud and Stramski 1990, Cleveland and Weidmann 1993) or by modeling (Roesler 1998). Published corrections for absorption enhancement vary by almost an order of magnitude (Roesler 1998).

2. The error caused by directional scattering losses in suspensions can be reduced by placing an opal-glass diffuser between the sample and the detector in a spectrophotometer (Shibata 1958), placing the sample immediately adjacent to the detector (Bricaud et al. 1983), or using either an integrating sphere (Nelson and Prézelin 1993) or a reflecting tube instrument (Kirk 1992, Zaneveld et al. 1994). Even so, scattering errors in suspensions may persist, as indicated by the non-zero values of apparent absorption at 750 nm, where pigments do not absorb (Roesler 1998).

Interspecific and intraspecific differences in the absorption characteristics of microalgae are a consequence of their size and pigment complement:

1. The absolute amount of light absorbed by a cell depends on its size and the amount of pigment in it (Agustí 1991, Finkel 2001).

2. The chlorophyll-specific absorption depends on how densely the photosynthetic pigments are packed within the cell. As the internal concentration of pigments rises, Chla-specific absorption decreases because of internal self-shading, particularly in large cells (Morel and Bricaud 1981, Berner et al. 1989, Dubinsky 1992). This can be seen as the absolute magnitude of Chla-specific absorption, $a^{Chl}$, in Figure 19.11. The effect of pigment packaging on absorption is most easily discerned at the Chla red peak (674 nm), where absorption by other pigments is negligible; specific absorption is much higher in the similarly sized small species, *Aureococcus anophagefferens*, *Pycnococcus provasolii*, and *Synechococcus bacillaris*, which have Chla quotas of less than 0.1 pg · Cell⁻¹, than in the larger species, which have Chla quotas of 2 to 3 pg · Cell⁻¹ (Table 19.2).

3. The primary regulation of the pigment complement is under genotypic control. Although all microalgae contain Chla, the distribution of accessory pigments (Chlb vs. one or more types of Chlc, specific carotenoids, xanthophylls, and/or phycobilins) is taxonomically determined. Jeffrey et

**TABLE 19.2** Pigment data for the cells shown in Fig. 19.11: *Aureococcus anophagefferens* (pelagophyte); *Dunaliella tertiolecta* (chlorophyte); *Pycnococcus provasolii* (prasinophyte); *Storeatula major* (cryptophyte); *Synechococcus bacillaris* (cyanobacterium) and *Thalassiosira weissflogii* (diatom). Size is approximate diameter (µm). Chla · cell⁻¹ in (pg). Pigment:Chla ratios in (g · g⁻¹).

| Species | Size | Chla | Chlb | Chlc | MG-DVP | [a]PSC | [b]PPC | [c]Xanth |
|---|---|---|---|---|---|---|---|---|
| | | Cell | Chla | Chla | Chla | Chla | Chla | Chla |
| A. anophagefferens | 2 | 0.09 | | 0.378 | | 0.693 | 0.016 | 0.100 |
| D. tertiolecta | 8 | 2.00 | 0.256 | | | 0.159 | 0.253 | 0.071 |
| P. provasolii | 2 | 0.07 | 0.793 | | 0.054 | 0.449 | 0.510 | 0.072 |
| S. major | 5 | 1.70 | | 0.118 | | | 0.338 | |
| S. bacillaris | 1 | 0.03 | | | | | 0.075 | 0.280 |
| T. weissflogii | 16 | 2.78 | | 0.093 | | 0.340 | 0.021 | 0.063 |

Chlc, the sum of all c chlorophylls; MG-DVP, Mg-2,4-divinyl pheoporphyrin $a_5$ monomethyl ester; PSC, photosynthetic carotenoids; PPC, photoprotective carotenoids; Xanth, xanthophylls.

[a]The distribution by taxon is: 19′-butanoyloxyfucoxanthin, fucoxanthin (A. anophagefferens); neoxanthin (D. tertiolecta); neoxanthin, prasinoxanthin (P. provasolii); fucoxanthin (T. weissflogii). S. major and S. bacillaris also contain phycobilins, which were not quantified.

[b]The distribution by taxon is β-carotene (A. anophagefferens); lutein, β-carotene (D. tertiolecta); lutein, β-carotene, astaxanthin (P. provasolii); β-carotene, alloxanthin, monadoxanthin, crocoxanthin (S. major); β-carotene (S. bacillaris); β-carotene (T. weissflogii).

[c]The distribution by taxon is diadinoxanthin, diatoxanthin (A. anophagefferens and T. weissflogii); violaxanthin, antheraxanthin, zeaxanthin (D. tertiolecta and P. provasolii); zeaxanthin (S. bacillaris).

al. (1997) provide a comprehensive review of the structures and distributions of the pigments. The increase in Chl*a*-specific absorption between 425 and 525 nm in *A. anophagefferens* relative to the similarly sized *P. provasolii* (see Fig. 19.11a) is due to its higher quota of photosynthetic carotenoids and the presence of Chl*c* rather than Chl*b* (see Fig. 19.11b, Table 19.2). The high absorption between 540 and 570 nm in *Storeatula major* and between 550 and 650 nm in *S. bacillaris* relative to the other species is a consequence of their being the only species shown that have phycobilins (see Fig. 19.11a,b).

4. Secondary regulation of the pigment complement is under phenotypic control, because microalgae respond to changes in their environment. Within a species, the relative abundance of both photosynthetic and photoprotective pigments depends on growth irradiance (reviewed by MacIntyre et al. 2002), on nutrient status (Sosik and Mitchell 1991, Wilhelm and Manns 1991, Latasa and Berdalet 1994, Goericke and Montoya 1998), and in the short-term, on recent exposure to bright light (Demmig-Adams 1990, Arsalane et al. 1994, Olaizola et al. 1994, Casper-Lindley and Björkman 1998).

The consequence of the size- and pigment-determined efficiency of absorption is variability in the efficiency with which the cells utilize light of different spectral quality. The relative quantum output of sunlight and several common light sources is shown in Figure 19.11c. The spectra have been weighted so that they have the same integrated value between 400 and 700 nm (i.e., they have the same value of PAR). The efficiency with which each light source is absorbed by the cells is not equivalent. It is intuitively obvious from comparison of the output spectra and absorption spectra that all of the species will absorb the light from the blue LED efficiently because its output is concentrated in the part of the spectrum where absorption by chlorophylls and carotenoids (which all the species have) is highest. Conversely, none will absorb light from the quartz halogen lamp as efficiently; its output increases with wavelength and only *S. bacillaris* has comparable efficiencies of absorption at the red and blue ends of the spectrum. The blue LED will, therefore, have a higher value of PUR than the quartz halogen lamps, although both have the same value of PAR.

Quantitative treatment of PUR relies on convolution of the output and absorption spectra. There is no meter that measures PUR, although the spectral weighting that is involved in calculating it is not complex. There are two ways in which this can be done. The first, which we designate as a "biological" description because it is expressed explicitly in terms of the cell's pigment content, is by simple convolution of the output and absorption spectra (Gallegos et al. 1990).

$$PUR^{Chl} = E_a = \sum_{\lambda=400}^{\lambda=700} E(\lambda)a^{Chl}(\lambda), \qquad (7)$$

where $E_a$ is the light absorbed by the cells per unit Chl*a*, $E(\lambda)$ is the quantum output at wavelength $\lambda$ and $a^{Chl}(\lambda)$ is the corresponding Chl*a*-specific absorption coefficient. As $E$ has units of $\mu$ mol photons $m^{-2} \cdot s^{-1}$, and $a^{Chl}$ has units of $m^2 \cdot (mg\ Chl)^{-1}$, $PUR^{Chl}$ has units of mol photons $\cdot (g\ Chl)^{-1} \cdot s^{-1}$ after reconciliation of units of quantity. Although the units may seem peculiar initially, they represent the efficiency with which photons are absorbed by the pigments present in the cells.

An alternative expression of PUR is in "physical" terms, without explicit reference to the absorption characteristics of the cell, as a flux of photons per unit area. We designate this simply PUR. It has the same units as PAR and is obtained by applying a dimensionless weighting function, $A_e$, to it (Markager and Vincent 2001):

$$PUR = A_e \cdot PAR \qquad (8)$$

The numerical value of the weighting function is the ratio of the light absorbed under the spectral regimen of interest ($E_a$, Equation 7) to $E_a'$, the light absorbed under a theoretical, spectrally flat regimen of equal photon flux over PAR, $E'$ (see Fig. 19.11c). Because the output of the theoretical regimen is wavelength independent, $E_a'$ is equal to the product of $E'$ and the mean value of $a^{Chl}(\lambda)$ between 400 and 700 nm, $a^{Chl}$:

$$A_e = \frac{E_a}{E_a'} = \frac{\sum_{\lambda=400}^{\lambda=700} E(\lambda) \cdot a^{Chl}(\lambda)}{E' \cdot a^{Chl}} \qquad (9)$$

Although PUR$^{Chl}$ and PUR differ in presentation and units, there is no significant difference in their computation; the same convolution is involved in both. PUR$^{Chl}$ preserves interspecific and intraspecific differences in the efficiency of Chl*a*-specific absorption: Note that PUR$^{Chl}$ (Table 19.3) is substantially higher under each light regimen for the three small cells that have high $a^{Chl}$ in Figure 19.11a than for their larger counterparts. In contrast, removal of pigment density and pigment package effects from PUR (Table 19.3) makes it easier to compare the relative absorption characteristics of cells with different pigment complements. For both measures, matching the light output of two different sources is a simple matter of taking the ratio of either PUR$^{Chl}$ or PUR. For example, the culture of *A. anophagefferens*

**TABLE 19.3**   Light absorption and PUR. Data in the upper panel are calculated values of $PUR^{Chl}$ (mol photons · [g Chl$a$]$^{-1}$ · s$^{-1}$) for the cells in Fig. 19.11a and the sources in Fig. 19.11c, all with an intensity of 2000 µmol photons · m$^{-2}$ · s$^{-1}$, measured as PAR (400–700 nm). Data in the lower panel are the dimensionless spectral weighting coefficients, $A_e$, used to convert PAR to PUR (Equation 8) by reference to the flat spectrum $E'$ in Fig. 19.11C.

$PUR^{chl}$ ($E_a$ and $E'_a$) by source

| Species | Sunlight | Fluorescent | Quartz Halogen | White LED | Blue LED | $E'_a$ |
|---|---|---|---|---|---|---|
| A. anophageffens | 58.3 | 37.7 | 23.9 | 46.1 | 66.4 | 38.9 |
| D. tertiolecta | 21.0 | 14.0 | 12.4 | 16.5 | 23.4 | 15.6 |
| P. provasolii | 47.0 | 31.4 | 22.6 | 39.2 | 60.7 | 33.2 |
| S. major | 24.6 | 18.5 | 14.2 | 20.6 | 24.6 | 18.8 |
| S. bacillaris | 50.2 | 39.8 | 33.0 | 41.0 | 44.1 | 39.6 |
| T. weissflogii | 20.1 | 14.0 | 11.7 | 15.9 | 19.5 | 15.0 |

$A_e(E_a/E'_a)$ by source

| Species | Sunlight | Fluorescent | Quartz Halogen | White LED | Blue LED | $E'$ |
|---|---|---|---|---|---|---|
| A. anophageffens | 1.50 | 0.97 | 0.61 | 1.18 | 1.71 | 1.00 |
| D. tertiolecta | 1.35 | 0.90 | 0.80 | 1.06 | 1.50 | 1.00 |
| P. provasolii | 1.42 | 0.95 | 0.68 | 1.18 | 1.83 | 1.00 |
| S. major | 1.31 | 0.98 | 0.76 | 1.10 | 1.31 | 1.00 |
| S. bacillaris | 1.27 | 1.01 | 0.83 | 1.04 | 1.12 | 1.00 |
| T. weissflogii | 1.34 | 0.93 | 0.78 | 1.06 | 1.30 | 1.00 |

shown in Figure 19.11a was grown under a fluorescent lamp at an irradiance of 80 µmol photons m$^{-2}$ · s$^{-1}$ PAR. If its photosynthetic rate were to be measured under a quartz halogen lamp, the equivalent flux of absorbed photons would be achieved at 127 µmol photons · m$^{-2}$ · s$^{-1}$ PAR (= 80 · [Fluorescent PUR$^{Chl}$/Quartz Halogen PUR$^{Chl}$] = 80 · [Fluorescent $A_e$/Quartz Halogen $A_e$]; see Table 19.3 for coefficients). For the culture of *S. bacillaris*, which was also grown at 80 µmol photons · m$^{-2}$ · s$^{-1}$ PAR under the fluorescent lamps, the equivalent flux of absorbed photons under the quartz halogen lights would be at 97 µmol photons · m$^{-2}$ · s$^{-1}$ PAR (Table 19.3).

Although this equivalence holds for total absorption, it does not hold strictly for absorption of light that is photosynthetically active. This is given by the photosynthetic cross-section, which describes the absorption by photosynthetic pigments alone, and which is equivalent to total absorption less the contributions of photoprotective pigments and pigmented non-photosynthetic material such as cell walls and other absorbing materials such as flavins and haemes (Sosik and Mitchell 1995, Suggett et al. 2003, 2004). In those

taxa that have an approximately equal distribution of pigments between PSII and PSI, the efficiency of utilization for incident irradiance follows the absorption spectrum, which differs little in shape from the excitation spectrum for Chl*a* fluorescence. Those photons that are absorbed contribute approximately equally to photosynthesis and growth so the quantum yield of photosynthesis (the amount of carbon fixed per photon absorbed) is either independent of or only weakly dependent on wavelength. This relationship has been reported for representatives of several chromophytic groups, the dinoflagellates, diatoms, and prymnesiophytes (Schofield et al. 1990, 1996, Nielsen and Sakshaug 1993). Consequently when these cells are acclimated to different spectral regimens, the growth rates show a strong dependence on both irradiance and spectral quality when irradiance is expressed as PAR. However, there is only a very weak dependence of the quantum yield of growth (the achieved growth rate per photon absorbed) on spectral quality.

In contrast, the photosynthetic action spectrum cannot be approximated from the absorption spectrum in those taxa that have phycobilins as accessory pigments, the cyanobacteria, cryptophytes, and rhodophytes (Haxo and Blinks 1960, Neori et al. 1986). In these taxa, the phycobilins are associated with PSII, whereas Chl$a$ is predominantly associated with PSI (Myers et al. 1980, Bidigare et al. 1989). Consequently, light of a spectral quality that is absorbed by phycobilins but not Chl$a$ or by Chl$a$ but not phycobilins cannot stimulate both photosystems, so the quantum yield is dependent on wavelength. Comparison of both absorption spectra, which include the contributions of both PSII and PSI and fluorescence excitation spectra, which are specific to PSII, may allow irradiances of different spectral quality to be reconciled for the phycobilin-containing taxa (Lutz et al. 2001).

## 5.2. Photosynthesis-Irradiance Curves

Photosynthesis expressed as a function of irradiance (PE) is highly responsive to environmental conditions (see Fig. 19.3). Understanding this variability and the kinetics of its change in response to environmental perturbation (Post et al. 1985, Prézelin et al. 1986b, Cullen and Lewis 1988, MacIntyre and Geider 1996, Anning et al. 2000, 2001) is central both to interpretation of comparable measurements made in the field and to building and testing models of PE response and growth. Research extends from descriptions of regulation under nutrient-replete (Sukenik et al. 1990) or nutrient-limited growth (Laws and Bannister 1980, Herzig and Falkowski 1989, Greene et al. 1991) to descriptions of regulation under conditions designed to simulate natural variability in irradiance and nutrient availability during mixing (Kroon et al. 1992a, Smith et al. 1992, Kromkamp and Limbeek 1993, Ibelings et al. 1994). In all cases, an important objective is to develop an insight into regulation of photosynthesis and productivity in nature, based on an improved understanding of microalgal physiology.

Photosynthesis can be measured as rates of carbon accumulation or oxygen evolution. The two can be reconciled by applying the photosynthetic quotient, which is the molar ratio of $O_2$ evolved per $CO_2$ assimilated. Falkowski and Raven (1997) review variability in the stoichiometry of photosynthetic and respiratory quotients. The high sensitivity of the $^{14}$C method compared to oxygen exchange has led to its widespread adoption in limnology and oceanography since its introduction by Steemann-Nielsen (1952) because microalgal biomass is typically low in most waters. The technique has been widely used in both the laboratory and the field

and interpreting the acreage of data collected to date has been a driving force in the development of models of photosynthesis and growth (Bannister and Laws 1980, Geider et al. 1996, 1998, Flynn and Hipkin 1999). Numerous factors should be considered when analyzing PE data, whether comparing within or across studies.

1. When comparing photosynthesis and growth, unless the PE incubator and the growth chamber use the same kind of lights, light data collected with a broad-band light sensor (PAR) must be reconciled to account for the proportion that can be absorbed by the algae (PUR, see earlier discussion).

2. The apparent dynamics in PE response across growth conditions are driven in part by the way in which they are normalized to biomass (Anning et al. 2000, MacIntyre et al. 2002). Cell-, Chl$a$-, and carbon-specific responses behave in different ways because of systematic variability in Chl$a$:C (see Fig. 19.2) and the cellular quotas of Chl$a$ and carbon (Thompson et al. 1991).

3. The carbon-uptake rate in a PE curve is time dependent (MacIntyre et al. 2000, 2002) so that the values of the fitted parameters depend on the duration of the incubation. The $^{14}$C method usually estimates production as the difference between labeling at the beginning and end of an incubation. Because it integrates changing rates over its duration, it is relatively insensitive to fluctuations in the rate when compared with a method that monitors changes continuously (MacIntyre et al. 2002). Cumulative uptake can also yield misleading results when the objective is to describe inhibition of photosynthesis as a function of exposure (Neale 2000); uptake during the initial part of an incubation will be measured even if the cells are dead by the end of it.

4. If the duration of the incubation is short enough that labeled carbon is not lost by respiration, $^{14}$C uptake should approximate gross photosynthesis (Williams 1993). The $^{14}$C method itself cannot measure the respiration rate. There have been relatively few comparisons of $^{14}$C uptake with net and gross photosynthesis, as measured by oxygen exchange (Williams et al. 1979, Grande et al. 1989b), but $^{14}$C uptake approximates gross photosynthesis when incubations are short. It is closer to net photosynthesis in longer (24-hour) incubations (Eppley and Sharp 1975), because the labeled carbon is cycled internally and some is lost

by respiration. Short-term $^{14}C$ uptake normalized to cellular carbon is, therefore, likely to overestimate growth rate unless corrected for an estimate of respiratory loss. In practice, this is not as easy as it might appear. The respiration rate increases with irradiance (Fig. 19.12) as other anabolic pathways such as nitrogen assimilation increase in activity to balance carbon fixation (Turpin et al. 1988) and, at high irradiance, as oxygen-consuming energy-dissipating pathways such as the Mehler cycle are induced (Grande et al. 1989a, Kana 1993, Lewitus and Kana 1995). Correcting short-term $^{14}C$ uptake by an estimate of respiration in the dark, measured as $O_2$

consumption by an oxygen electrode, will tend to overestimate net productivity, and it still requires an estimate of the photosynthetic quotient to reconcile the rates.

5. The $^{14}C$ method either does or does not measure loss of dissolved organic carbon (DOC), depending on the protocol. Where uptake is measured by acidifying a whole water sample and adding scintillation cocktail directly to it, the method measures the sum of POC synthesis and any DOC loss, subject to the labeling pattern (Smith and Horner 1981). Where uptake is measured by filtering the sample after the incubation and counting $^{14}C$ retention on the filter, the method

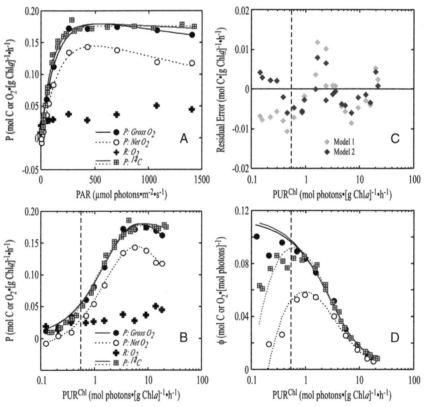

**FIGURE 19.12.** **(a)** Variation of gross and net photosynthesis and respiration as a function of irradiance (Todd Kana and Mac-Intyre, unpublished data). Rates were measured using an $^{18}O/^{16}O$ technique that measures both gross photosynthesis and respiration simultaneously with a membrane-inlet mass spectrometer (Kana 1990), subject to the limitations of the isotope dilution equations. Superimposed are data on $^{14}C$ uptake, measured in a photosynthetron (Lewis and Smith 1983). The fits are either a fit with two exponents (Platt et al. 1980) (Model 1, *solid lines*) or the same fit modified with an intercept term (Model 2, *dashed lines*). The light sources for both techniques differed. Irradiance is expressed as PAR. **(b)** The same data in which incident irradiance has been corrected for the spectral quality of the two light sources and recast as $PUR^{Chl}$ (see text for details). Note the log scale to emphasize differences between the data and fits at low irradiance. The vertical dashed line is growth irradiance. **(c)** Irradiance-dependence of residual errors (observed—expected values) for the $^{14}C$ data, as fit by Models 1 and 2. Note the magnitude and divergence at and below growth irradiance (*the dashed line*). **(d)** Variation in quantum yields of photosynthesis with irradiance. Note the divergence between the prediction of Model 1 and the $^{14}C$ data.

measures only POC synthesis, although there is some retention of DOC on the filter (Maske and Garcia-Mendoza 1994). This will cause an underestimate of productivity if there is uniform labeling of pools (Smith and Horner 1981). Although DOC production is likely low under conditions of nutrient-replete growth (Sharp 1977), it may rise under nitrogen starvation (Otero and Vincenzini 2004). Excretion of DOC also increases at high irradiance (Mague et al. 1980, Joint and Pomroy 1981), where it appears to be an energy-dissipating mechanism (Wood and Van Valen 1990, Gordillo et al. 2001), particularly under low nutrient conditions, as cells dump carbon fixed in excess of their ability to synthesis proteins. In benthic diatoms and in many cyanobacteria, almost 50% of carbon fixed may be excreted as extracellular polymeric substance (EPS) (Alcoverro et al. 2000, Smith and Underwood 2000, Staats et al. 2000). Failing to account for this will bias the estimated relative assimilation rates of carbon and other elements.

6. The influence of the model (hyperbolic tangent, exponential, quadratic, etc.) used to fit the PE curve on the value of the parameters has been reviewed extensively (Jassby and Platt 1976, Platt et al. 1980, Zimmerman et al. 1987, Eilers and Peeters 1988, Frenette et al. 1993, MacIntyre et al. 2002). The consensus is that all models give similar estimates of $P_m$ but differ in their estimates of $\alpha$ because of differences in their architecture. It is important, therefore, to be consistent in using a single model when comparing estimates of $\alpha$ or the saturating irradiance $E_k$ ($= P_m/\alpha$) in cultures grown under different conditions. However, most models give accurate estimates of the rates (as indicated by the distribution of residual errors) over most of the irradiance range.

7. Because the cellular organic carbon being respired during short-term incubations is not labeled with isotope, the $^{14}C$ method tends to overestimate net photosynthesis at low irradiance (Steemann Nielsen 1953). This is illustrated in Figure 19.12, which shows gross photosynthesis, determined by the $^{18}O/^{16}O$ method (Kana 1990) and $^{14}C$ incorporation. The fits given by Model 1 (Platt et al. 1980) are almost identical for the $O_2$ and $^{14}C$ data (Table 19.4), although there is a pronounced divergence between the fits and the $^{14}C$ data at low irradiance, as the fit is forced through zero. As a result, the residual errors in low light are high (see Fig. 19.12c), and the rate of carbon fixation at

**TABLE 19.4** Curve fit parameters for the PE data in Fig. 19.12.

Model 1: $= P_s\left[1 - \exp\left(\dfrac{-\alpha E}{P_s}\right)\right]\exp\left(\dfrac{-\beta E}{P_s}\right)$; $\quad P_m = P_s\left(\dfrac{\alpha}{\alpha+\beta}\right)\left(\dfrac{\beta}{\alpha+\beta}\right)^{\beta/\alpha}$

Model 2: $= P_s\left[1 - \exp\left(\dfrac{-\alpha E}{P_s}\right)\right]\exp\left(\dfrac{-\beta E}{P_s}\right) + P_0$; $\quad P_m = P_s\left(\dfrac{\alpha}{\alpha+\beta}\right)\left(\dfrac{\beta}{\alpha+\beta}\right)^{\beta/\alpha} - P_0$

| Data | $P_m$ | $P_s$ | $\alpha$ | $\beta$ | $P_0$ |
|---|---|---|---|---|---|
| **Model 1** | | | | | |
| Gross $O_2$ | 0.180 | 0.197 ± 0.008 | 0.113 ± 0.005 | 0.0021 ± 0.0007 | — |
| $^{14}C$ | 0.180 | 0.186 ± 0.007 | 0.115 ± 0.006 | 0.0006 ± 0.0005 | — |
| **Model 2** | | | | | |
| Net $O_2$ | 0.144 | 0.194 ± 0.008 | 0.104 ± 0.008 | 0.0039 ± 0.0009 | 0.021 ± 0.003 |
| $^{14}C$ | 0.177 | 0.194 ± 0.005 | 0.144 ± 0.010 | 0.0003 ± 0.0003 | 0.015 ± 0.004 |

Data were fit to Model 1 (Platt et al. 1980) or Model 2. $P_m$ is the light-saturated rated of photosynthesis; $P_s$ is the light-saturated rate of photosynthesis that would be achieved if $\beta$ were 0; $P_0$ is an intercept; $\alpha$ is the light-limited rate of photosynthesis; and $\beta$ is a term describing the decrease in photosynthesis. $P_m$, $P_s$, and $P_0$ are in (mol $O_2$ or C · [g Chla]$^{-1}$ · hr$^{-1}$). $\alpha$ and $\beta$ are in (mol $O_2$ or C · [g Chla]$^{-1}$ · hr$^{-1}$) (mol photons · [g Chla]$^{-1}$ · hr$^{-1}$)$^{-1}$ = (mol $O_2$ or C · [mol photons]$^{-1}$). Note that no errors are given for $P_m$, as the fitting protocol that was used did not return the covariation matrix for the output parameters, information that is needed to calculate the error correctly (Davis 1986, Zimmerman et al. 1987).

growth irradiance is overestimated relative to the observed data. This might reflect an increase in the respiration rate (which cannot be measured by the $^{14}C$ technique) relative to gross photosynthesis at low irradiance or might simply be an artifact of the $^{14}C$ method. When the $^{14}C$ data are fitted with a model modified by inclusion of a non-zero intercept (Table 19.4), the residual error is decreased at all irradiances but particularly at low ones (including in this case the growth irradiance) (see Fig. 19.12c). This provides a better description of the observed $^{14}C$ uptake at low irradiance. Model 2 (Table 19.4) is clearly the appropriate one for fitting net photosynthesis, as measured by oxygen exchange (see Fig. 19.12), in which case the intercept term is the dark respiration rate. A conservative approach for short-term $^{14}C$-uptake data is to force the fit through the origin (Sakshaug et al. 1997); it may be prudent to avoid making many measurements at low irradiance, where the $^{14}C$ method is prone to systematic error.

An alternative treatment of the PE relationship is to recast it in terms of quantum yield, $\phi$ (see Fig. 19.12d), which is the efficiency with which absorbed light is converted to carbon (mol C · [mol photons]$^{-1}$):

$$\phi = \frac{P_E}{E_a} = \frac{P_E}{PUR^{Cbl}} = \frac{P_E}{E \cdot A_c \cdot a^{Cbl}} \qquad (10)$$

Part of the impetus for doing so is the potential for using active fluorescence methods to predict productivity: $F_v/F_m$ and $F_v'/F_m'$ are the maximum and achieved quantum yields of Photosystem II activity (Genty et al. 1989, 1990). The relationship between $\phi$ and irradiance can be described in terms of the maximum value, $\phi_m$, a theoretical limit reached where irradiance tends to zero, and the saturation irradiance, $E_k$. Where the PE curve is expressed in terms of PAR, $\phi_m$ is equivalent to the quantum yield in the region of the initial slope, $\alpha$, when it is expressed in chlorophyll-specific units:

$$\phi_m = \frac{\alpha^{Cbl}}{A_c \cdot a^{Cbl}}, \qquad (11)$$

where a PE curve is expressed in terms of $PUR^{Cbl}$, the ratio of $P$ to $PUR^{Cbl}$ at any point is $\phi$ (compare Fig. 19.12b and 19.12d) and the initial slope of the curve fit is $\phi_m$. Note that if the curve is fitted with $P_0$, $\phi$ is calculated as $[P-P_0]/PUR^{Cbl}$. Because the photosynthetic rate at low irradiance is $[\alpha E-P_0]$, rather than $[\alpha E]$, $\phi_m$ cannot be calculated with Equation 11 and doing so can overestimate it by a large margin (see Table 19.4).

Attempting to model growth rate from PE data is also subject to uncertainties because of the temporal variability of the PE relationship. Both Chl$a$- and cell-specific $P_m$ and $\alpha$ vary through the light period when cells are grown on a light–dark cycle (McCaull and Platt 1977, Prézelin and Sweeney 1977, Harding et al. 1982, Prézelin et al. 1986a). Any attempt to model growth of cells on a light–dark cycle must account for the periodicity in PE response and attempt to integrate it over the same period that growth measurements are made.

### 5.3. Ultraviolet Radiation

Much research has been conducted to assess the effects of UV radiation (UVR) on microalgae. Although there have been descriptions of important mechanisms (Vincent and Neale 2000), interactions (Neale et al. 1998b), and adaptations to minimize damage (Roy 2000), and although many different sensitivities of microalgae to UV have been demonstrated (Vernet 2000), many experimental studies of microalgae cultures using UVR are difficult or impossible to relate to each other or to nature. Vernet (2000) concluded that "in spite of a large body of studies available, we are far from understanding how microalgae respond to environmental UVR." Reasons for this included the following:

1. The use of unrealistic irradiances in laboratory studies (sometimes vastly exceeding natural exposures).
2. A lack of relevant hypotheses about the influence of UVR in natural environments, where exposures vary with time and where the competing processes of damage and repair to the essential (but not immutably linked) processes of photosynthesis, nutrient assimilation, and cell division must be characterized very carefully to make any sense of what occurs in the real world.

Many of the problems with intercomparison between UV experiments can be avoided through careful selection and measurement of irradiance sources (lamps) (Cullen and Neale 1994). The objective should be to ensure that the experimental exposure can be related *quantitatively* to conditions in nature. Further, the spectrum should not unnaturally target one physiological process over another (e.g., DNA damage vs. photosynthesis) unless that is an intentional part of the experimental design.

The concept of spectral weighting functions (biological weighting functions and action spectra) is centrally important to UV research (Coohill 1989, Neale 2000). Simply put, radiant exposures cannot be compared in a biologically relevant way unless the relative effectiveness of different wavelengths is taken into account. This is also the case with PUR versus PAR (see previous discussion), but with UV the errors associated with using unweighted or incorrectly weighted irradiance can be much larger. The requirements for equipment are daunting, as is the complexity of quantitative analysis. Accurate measurement of UVR is extremely demanding and expensive; cheap spectrometers are incapable of excluding stray photons from their detectors when measuring rare but destructive UVB in the presence of natural visible radiation and so are inaccurate. (Manufacturers of optical instruments, such as scanning spectrophotometers, provide materials to explain the challenges of accurate radiometry.)

A comparison of different measures of UVR illustrates the complexity of matching experimental spectral irradiance with natural exposures (Table 19.5) (Cullen and Neale 1997). The table does not lend itself to rapid comprehension; neither does the quantitative comparison of experimental exposures to natural regimens of radiation. For any of the four measures in Table 19.5, values within a row represent relative exposure from the different sources: the UVB $(W \cdot m^{-2})$ from a bare sunlamp is nearly the same as surface conditions at 45° latitude; the solar simulator emits nearly twice the natural UVB. Comparison of the patterns between rows shows how important weightings are. When the spectra are weighted for DNA damage, the bare sunlamp is nearly threefold more damaging than surface irradiance at 45°, but when weighted for inhibition of photosynthesis, it is closer to the natural exposure rate. The implication is that a bare sunlamp would induce disproportionately high DNA damage as compared to nature, even though unweighted UVB $(W \cdot m^{-2})$ is about the same. In turn, the filtered sunlamp (a common treat-

**TABLE 19.5** Different measures of UV radiation for three sources of irradiance, compared to three natural situations.

| | Artificial Sources | | | Natural Spectra | | |
|---|---|---|---|---|---|---|
| | Bare lamp | Filtered lamp | Solar Simulator | 45° Surface | 60° S Ozone Hole | 60° S Ozone Hole 5 m |
| **Measure of Damaging Irradiance** | | | | | | |
| UVB $(W \cdot m^{-2})$ | 1.61 | 0.31 | 2.80 | 1.62 | 1.47 | 0.26 |
| DNA$_{300}$ | 1.28 | 0.15 | 0.76 | 0.45 | 0.42 | 0.08 |
| P. tricornutum | 0.96 | 0.12 | 1.18 | 0.89 | 0.67 | 0.18 |
| A. sanguinea (low light) | 0.34 | 0.04 | 0.61 | 0.50 | 0.35 | 0.10 |
| **Effectiveness Ratio** | | | | | | |
| (DNA/PHS)$_{diatom}$ | 1.33 | 1.28 | 0.65 | 0.50 | 0.63 | 0.47 |
| (DNA/PHS)$_{dinoflagellate(LL)}$ | 3.79 | 3.82 | 1.25 | 0.89 | 1.23 | 0.82 |

*Sources:* Spectra from two sunlamps (FS40 T12-UVB, Light Sources Inc., illuminated 200 hours before use), measured at 60 cm, are typical for laboratory studies on cultures. Data are from Cullen and Lesser (1991). Cellulose acetate film (0.13 mm thickness, aged 100 hours at 3 cm from the lamps) was used to attenuate shorter wavelengths that are not encountered in nature. The spectrum for a solar simulator (filtered through a Schott WG-295 long-pass filter) represents a more realistic simulation of solar radiation (Browman et al. 2003). The midday spectrum at 45° N is from a model; those at 60° S as influenced by the ozone hole (140 DU) are from measurements and a model for the Weddell Sea on a typically cloudy day (Cullen and Neale 1997).

*Measures of damaging irradiance:* Unweighted UV-B radiation (280–320 nm) is reported in $W \cdot m^{-2}$. DNA$_{300}$ is weighted for the action spectrum of DNA damage in alfalfa seedlings (Quaite et al. 1992) and normalized to 1.0 nm$^{-1}$ at 300 nm. Photosynthesis-inhibiting radiation, $E^*$ (Cullen et al. 1992a, Neale 2000) is calculated from biological weighting functions (BWFs) for the diatom, *Phaeodactylum tricornutum* (Cullen et al. 1992a) and the dinoflagellate, *Akashiwo sanguinea* (Hirasaka) Hansen et Moestrup (formerly *Gymnodinium sanguineum* Hirasaka), grown in low light (Neale et al. 1998a).

*Effectiveness ratio:* For each source of irradiance, the ratio of effectiveness for DNA damage (likely to affect growth rate), as compared to inhibition of photosynthesis, is calculated for both the diatom and the dinoflagellate. All effectiveness ratios are relative but comparable. For example, when compared to any of the natural spectra, each artificial source is more damaging to DNA then to photosynthesis, but the difference depends on the BWF for photosynthesis. Extrapolation of experimental results to nature requires explicit consideration of biologically weighted irradiance in both the laboratory and the field (after Cullen and Neale 1997).

ment) can induce only a fraction of near-surface inhibition of photosynthesis, closer to what would be encountered at a depth of several meters. If time is not invested in getting it right, results will not be comparable to nature. More fundamentally, if spectral irradiance is not measured, biologically weighted, and reported, results will not be comparable between studies or with nature.

Relative ratios of DNA damaging radiation to photosynthesis inhibiting radiation are reported for each spectrum in Table 19.5. The absolute values have no meaning because we have no basis for comparing DNA damage directly to inhibition of photosynthesis (but see Neale 2000), although a comparison of ratios within rows shows how unnatural some lamps can be. Generally, sunlamps are much more damaging to DNA than to photosynthesis, as compared to nature. The bias depends on the sensitivity of algal photosynthesis to UV, which depends in turn on species and preconditioning (Neale 2000). These calculations do not even consider the unnatural consequences of high ratios of UV : PAR common in many experiments (Vernet 2000) and the kinetics of photoinhibition versus DNA damage. Photoinhibition may rapidly reach a balance between damage and repair, so the reduction in photosynthesis is a function of irradiance (Cullen et al. 1992a), or it may not, so inhibition is a function of cumulative exposure (Neale et al. 1998c). In contrast, DNA damage is well described as a function of cumulative exposure on short time scales (Huot et al. 2000). Clearly, when embarking on studies of the physiological effects of UV, careful preparation is warranted.

---

## 6.0. Future Research

Much research in microalgal ecology has been driven by the development of more rapid and more sensitive techniques for detecting biomass or some characteristic of the physiological responses of the organisms. Whether there is a corresponding advance in our understanding of the organisms' ecological role in their natural habitat is debatable. We argue that advances are most likely to come from hypothesis-driven research (Platt 1964) and that framing a suitable question is more important than collecting huge quantities of data whose interpretation is largely conjectural. That said, the development of new technologies does permit observation on temporal and spatial scales that were unachievable two decades ago. For instance, it is now possible to monitor a culture continuously using automated sam-

plers and analyze nutrient concentrations (in filtered samples), cell number and particle size distributions, and spectral absorption in UV and visible by the cells (Bernard et al. 1996, Le Floc'h et al. 2002). Application of modeled relationships to these data then allows absorption at 678 nm to be corrected for pigment packaging, based on particle size distributions, to yield Chl*a* concentration (Le Floc'h et al. 2002), and for particle number and differential nutrient concentrations to be converted to cell nutrient quotas (Bernard et al. 2001), both with impressive accuracy. Techniques such as these are immensely powerful in determining physiological responses to the stochastic changes in nutrients or irradiance that are more representative of the natural environment than the stable conditions typically used in physiological studies. Such observations can and should be used to test and either refine or refute the predictions of first-principles models of physiological response to environmental change (Geider et al. 1998). The rapid development of autonomous moorings in oceanic, coastal, and estuarine waters provides a natural control for observations of natural populations themselves while describing the rate of change of the environmental characteristics to which the organisms respond (Sosik et al. 2003). A clearer understanding of the mechanisms that drive natural cycles of microalgal productivity will come largely from quantitative comparisons between natural assemblages and cultures. Although regulation of nutrient uptake, pigments, PE response, and growth appears to have common patterns across the taxa that have been studied, the coefficients that describe species-specific characteristics are few in number and span ranges of up to an order of magnitude (Raven and Johnston 1991, Smayda 1997, MacIntyre et al. 2002). Much basic descriptive work remains to be done.

---

## 7.0. Acknowledgments

We dedicate this chapter with affection and respect to Bob and Ruth Guillard. R. L. L. Guillard's contributions to microalgal research have been fundamental and Bob's inventiveness, curiosity, and enthusiasm for topics as diverse as trace metals and medieval weaponry have been an inspiration. As impressive is Ruth's grace and good humor, nowhere more evident than in the fact that she has yet to strangle Bob. We also thank other colleagues and friends whose discussions of culturing over the years have guided our efforts, in particular Wendy

Bellows, Dick Eppley, Richard Geider, Maureen Keller, Ed Renger, and Chase van Baalen. This work was supported by NSF (HLM) and by NSERC Research Partnerships and the Office of Naval Research (JJC).

## 8.0. References

Agustí, S. 1991. Allometric scaling of light absorption and scattering by phytoplankton cells. *Can. J. Fish. Aquat. Sci.* 48:763–7.

Alcoverro, T., Conde, E., and Mazzella, L. 2000. Production of mucilage by the Adriatic epipleic diatom *Cylindrotheca closterium* (Bacillariophyceae) under nutrient limitation. *J. Phycol.* 36:1087–95.

Anning, T., Harris, G., and Geider, R. J. 2001. Thermal acclimation in the marine diatom *Chaetoceros calcitrans* (Bacillariophyceae). *Eur. J. Phycol.* 36:233–41.

Anning, T., MacIntyre, H. L., Pratt, S. M., Sammes, P. J., and Geider, R. J. 2000. Photoacclimation in the marine diatom *Skeletonema costatum*. *Limnol. Oceanogr.* 45:1807–17.

Arsalane, W., Rousseau, B., and Duval, J. C. 1994. Influence of the pool size of the xanthophyll cycle on the effects of light stress in a diatom: competition between photoprotection and photoinhibition. *Biochem. Photobiol.* 60:237–43.

Bannister, T. T., and Laws, E. A. 1980. Modeling phytoplankton carbon metabolism. In: Falkowski, P. G., ed. *Primary Productivity in the Sea*. Plenum Press. New York, pp. 243–8.

Baroli, I., and Melis, A. 1996. Photoinhibition and repair in *Dunaliella salina* acclimated to different growth irradiances. *Planta* 198:640–46.

Barrett, J., and Jeffrey, S. W. 1964. Chlorophyllase and formation of an atypical chlorophyllide in marine algae. *Pl. Physiol.* 33:44–47.

Beardall, J., and Morris, I. 1976. The concept of light intensity adaptation in marine phytoplankton: some experiments with *Phaeodactylum tricornutum*. *Mar. Biol.* 37:377–87.

Beeler SooHoo, J., Kielfer, D. A., Collins, D. J., and McDermid, I. S. 1986. *In vivo* fluorescence excitation and absorption spectra of marine phytoplankton: I. Taxonomic characteristics and responses to photoadaptation. *J. Plankt. Res.* 8:197–214.

Behrenfeld, M. J., and Falkowski, P. G. 1997a. Photosynthetic rates derived from satellite-based chlorophyll concentration. *Limnol. Oceanogr.* 42:1–20.

Behrenfeld, M. J., and Falkowski, P. G. 1997b. A consumer's guide to phytoplankton primary productivity models. *Limnol. Oceanogr.* 7:1479–91.

Bernard, O., Malara, G., and Sciandra, A. 1996. The effects of controlled fluctuating nutrient environment on continuous culture of phytoplankton monitored by computers. *J. Exp. Mar. Biol. Ecol.* 197:263–78.

Bernard, O., Sciandra, A., and Sallet, G. 2001. A non-linear software sensor to monitor the internal nitrogen quota of phytoplanktonic cells. *Oceanol. Acta* 24:435–42.

Berner, T., Dubinsky, Z., Wyman, K., and Falkowski, P. G. 1989. Photoadaptation and the "package" effect in *Dunaliella tertiolecta* (Chlorophyceae). *J. Phycol.* 25:70–78.

Beutler, M., Wiltshire, K. H., Meyer, B., Moldaenke, C., Lüring, C., Meyerhöfer, M., Hansen, U. P., and Dau, H. 2002. A fluorometric method for the differentiation of algal populations *in vivo* and *in situ*. *Photosynth. Res.* 72:39–53.

Bidigare, R. B., Schofield, O., and Prézelin, B. B. 1989. Influence of zeaxanthin on quantum yield of photosynthesis of *Synechococcus* clone WH7803 (DC2). *Mar. Ecol. Prog. Ser.* 56:177–88.

Bidigare, R. R., Ondrusek, M. E., Morrow, J. H., and Kiefer, D. A. 1990. *In vivo* absorption properties of algal pigments. *SPIE* 1302:290–302.

Bienfang, P. K., and Szyper, J. P. 1982. Effects of temperature and salinity on sinking rates of the centric diatom *Ditylum brightwelli*. *Biol. Oceanogr.* 1:211–23.

Bienfang, P. K., Szyper, J., and Laws, E. 1983. Sinking rate and pigment responses to light limitation of a marine diatom: implications to dynamics of chlorophyll maximum layers. *Oceanol. Acta* 6:55–62.

Brand, L. E., Guillard, R. R. L., and Murphy, L. S. 1981. A method for the rapid and precise determination of acclimated phytoplankton reproduction rates. *J. Plankt. Res.* 3:193–201.

Bricaud, A., Morel, A., and Prieur, L. 1983. Optical efficiency factors of some phytoplankters. *Limnol. Oceanogr.* 28:816–32.

Bricaud, A., and Stramski, D. 1990. Spectral absorption coefficients of living phytoplankton and nonalgal biogenous matter: a comparison between the Peru upwelling area and the Sargasso Sea. *Limnol. Oceanogr.* 35:562–82.

Browman, H. I., Vetter, R. D., Rodriguez, C. A., Cullen, J. J., Davis, R. F., Lynn, E., and St-Pierre, J. F. 2003. Ultraviolet (280–400 nm) induced DNA damage in the eggs and larvae of *Calanus finmarchicus* G. (Copepoda) and Atlantic Cod (*Gadus morhua*). *Photochem. Photobiol.* 77:397–404.

Burger-Wiersma, T., and Post, A. F. 1989. Functional analysis of the photosynthetic apparatus of *Prochlorothrix hollandica* (Prochlorales), a chlorophyll *b* containing procaryote. *Pl. Physiol.* 91:770–74.

Casper-Lindley, C., and Björkman, O. 1998. Fluorescence quenching in four unicellular algae with different light-harvesting and xanthophyll-cycle pigments. *Photosynth. Res.* 56:277–89.

Chan, A. T. 1980. Comparative physiological study of marine diatoms and dinoflagellates in relation to irradiance and cell size. II. Relationship between photosynthesis, growth and carbon/chlorophyll *a* ratio. *J. Phycol.* 16:428–32.

Chisholm, S. W. 1981. Temporal patterns of cell division in unicellular algae. In: Platt, T. ed. *Physiological Bases of Phytoplankton Growth. Can. Bull. Fish. Aquat. Sci.* 210: 150–81

Chisholm, S. W., and Brand, L. E. 1981. Persistence of cell division phasing in marine phytoplankton in continuous light after entrainment to light/dark cycles. *J. Exp. Mar. Biol. Ecol.* 51:107–18.

Cleveland, J. S., and Weidmann, A. D. 1993. Quantifying absorption by aquatic particles: A multiple scattering correction for glass-fiber filters. *Limnol. Oceanogr.* 38:1321–7.

Coles, J. F., and Jones, R. C. 2000. Effect of temperature on photosynthesis-light responses and growth of four phytoplankton species isolated from a tidal freshwater river. *J. Phycol.* 36:7–16.

Coohill, T. P. 1989. Ultraviolet action spectra (280 to 380 nm) and solar effectiveness spectra for higher plants. *Photochem. Photobiol.* 50:451–7.

Cuhel, R. L., Ortner, P. B., and Lean, D. R. S. 1985. Night synthesis of protein by algae. *Limnol. Oceanogr.* 29:731–44.

Cullen, J. J. 1985. Diel vertical migration by dinoflagellates: roles of carbohydrate metabolism and behavioural flexibility. In: Rankin, M. A., ed. *Migration: Mechanisms and Adaptive Significance.* Contr. Mar. Sci. 27 Supplement, University of Texas Marine Science Institute, Port Aransas, pp. 135–142.

Cullen, J. J., and Davis, R. F. 2003. The blank can make a big difference in oceanographic measurements. *Limnol. Oceanogr. Bull.* 12:29–31.

Cullen, J. J., Geider, R. J., Ishizaka, J., Kiefer, D. A., Marra, J., Sakshaug, E., and Raven, J. A. 1993. Toward a general description of phytoplankton growth for biogeochemical models. In: Evans, G. T., and Fasham, M. J. R., eds. *Towards a Model of Ocean Biogeochemical Processes.* Springer-Verlag, Berlin, pp. 153–67.

Cullen, J. J., and Lesser, M. P. 1991. Inhibition of photosynthesis by ultraviolet radiation as a function of dose and dosage rate: Results for a marine diatom. *Mar. Biol.* 111:183–90.

Cullen, J. J., and Lewis, M. L. 1995. Biological processes and optical measurements near the sea surface: some issues relevant to remote sensing. *J. Geophys. Res.* 100:13255–66.

Cullen, J. J., and Lewis, M. R. 1988. The kinetics of algal photoadaptation in the context of vertical mixing. *J. Plankt. Res.* 10:1039–63.

Cullen, J. J., and MacIntyre, J. G. 1998. Behaviour, physiology and the niche of depth-regulating phytoplankton. In: Anderson, D. M., Cembella, A. D., and Hallegraef, G. M., eds. *Physiological Ecology of Harmful Algal Blooms.* Springer-Verlag, Berlin, pp. 559–79.

Cullen, J. J., Neale, P. J., and Lesser, M. P. 1992a. Biological weighting function for the inhibition of phytoplankton photosynthesis by ultraviolet radiation. *Science* 258:646–50.

Cullen, J. J., and Neale, P. J. 1994. Ultraviolet radiation, ozone depletion, and marine photosynthesis. *Photosynth. Res.* 39:303–20.

Cullen, J. J., and Neale, P. J. 1997. Biological weighting functions for describing the effects of ultraviolet radiation on aquatic systems. In: Häder, D. P., ed. *The Effects of Ozone Depletion on Aquatic Ecosystems.* R.G. Landes, Austin, Texas, pp. 97–118.

Cullen, J. J., Yang, X., and MacIntyre, H. L. 1992b. Nutrient limitation and marine photosynthesis. In: Falkowski, P. G., and Woodhead, A. D., eds. *Primary Productivity and Biogeochemical Cycles in the Sea.* Plenum Press, New York, pp. 69–88.

Cullen, J. J., Yentsch, C. M., Cucci, T. L., and MacIntyre, H. L. 1988. Autofluorescence and other optical properties as tools in biological oceanography. *SPIE* 925:149–56.

Davis, R. F. 1986. Measurement of primary production in turbid waters [M.A. Thesis], University of Texas at Austin, Austin, 122 pp.

Delwiche, C. F. 1999. Tracing the thread of plastic diversity through the tapestry of life. *Am. Nat.* 154 (suppl.):164–77.

Demmig-Adams, B. 1990. Carotenoids and photoprotection in plants: A role for xanthophyll zeaxanthin. *Biochim. Biophys. Act.* 1020:1–24.

Denman, K. L., and Marra, J. 1986. Modelling the time dependent photoadaptation of phytoplankton to fluctuating light. In: Nihoul, J. C. J., ed. *Marine Interfaces Ecohydrodynamics.* Elsevier, Amsterdam, pp. 341–59.

Droop, M. R. 1968. Vitamin $B_{12}$ and marine ecology. IV. The kinetics of uptake, growth and inhibition in *Monochrysis lutheri. J. Mar. Biol. Assoc. U.K.* 48:689–733.

Droop, M. R. 1974. The nutrient status of algal cells in continuous culture. *J. Mar. Biol. Assoc. U.K.* 54:825–55.

Droop, M. R. 1985. Fluorescence and the light/nutrient interaction in *Monochrysis. J. Mar. Biol. Assoc. U.K.* 65:221–37.

Dubinsky, Z. 1992. The functional and optical absorption cross-sections of phytoplankton photosynthesis. In: Falkowski, P. G., and Woodhead, A. D., ed. *Primary Productivity and Biogeochemical Cycles in the Sea.* Plenum Press, New York, pp. 31–45.

DuRand, M. D., Green, R. E., Sosik, H. M., and Olson, R. J. 2002. Diel variation in optical properties of *Micromonas pusilla* (Prasinophyceae). *J. Phycol.* 38:1132–42.

Duysens, L. N. M. 1956. The flattening of the absorption spectrum of suspensions, as compared to that of solutions. *Biochim. Biophys. Acta* 19:1–12.

Eilers, P. H. C., Peeters, J. C. H. 1988. A model for the relationship between light intensity and the rate of photosynthesis in phytoplankton. *Ecol. Monogr.* 42:199–215.

Eppley, R. W., and Sharp, J. H. 1975. Photosynthetic measurements in the central North Pacific: The dark loss of carbon in 24-h incubations. *Limnol. Oceanogr.* 20:981–7.

Eppley, R. W., Harrison, W. G., Chisholm, S. W., and Stewart, E. 1977. Particulate organic matter in surface waters off Southern California and its relationship to phytoplankton. *J. Mar. Res.* 35:671–96.

Eppley, R. W. 1981. Relations between nutrient assimilation and growth in phytoplankton with a brief review of estimates of growth rate in the ocean. *Can. Bull. Fish. Aquat. Sci.* 210:251–63.

Escoubas, J. M., Lomas, M., La Rovhe, J., and Falkowski, P. G. 1995. Light-intensity regulation of CAB gene transcription is signaled by the redox state of the plastiquinone pool. *Proc. Nat. Acad. Sci.* 92:10237–41.

Falkowski, P. G., Dubinsky, Z., and Wyman, K. 1985. Growth-irradiance relationships in phytoplankton. *Limnol. Oceanogr.* 30:311–21.

Falkowski, P. G., and Raven, J. A. 1997. *Aquatic Photosynthesis.* Blackwell Science, Malden, Massachusetts, 375 pp.

Falkowski, P. G., Wyman, K., Ley, A. C., and Mauzerall, D. C. 1986. Relationship of steady-state photosynthesis to fluorescence in eucaryotic algae. *Biochim. Biophys. Acta* 849:183–92.

Field, C. B., Behrenfeld, M. J., Randerson, J. T., and Falkowski, P. G. 1998. Primary production of the biosphere: Integrating terrestrial and oceanic components. *Science* 281:237–40.

Finkel, Z. V., and Irwin, A. J. 2000. Modelling size-dependent photosynthesis-light absorption and the allometric rule. *J. Theor. Biol.* 204:361–69.

Finkel, Z. V. 2001. Light absorption and size scaling of light-limited metabolism in marine diatoms. *Limnol. Oceanogr.* 46:86–94.

Flynn, K. J., and Fasham, M. J. R. 2003. Operation of light-dark cycles within simple ecosystem models of primary production and the consequences of using phytoplankton models with different abilities to assimilate N in darkness. *J. Plankt. Res.* 25:83–92.

Flynn, K. J., and Hipkin, C. R. 1999. Interactions between iron, light, ammonium, and nitrate: Insights from the construction of a dynamic model of algal physiology. *J. Phycol.* 35:1171–90.

Franks, P. J. S., and Marra, J. 1994. A simple new formulation for phytoplankton photoresponse and an application in a wind-driven mixed-layer model. *Mar. Ecol. Prog. Ser.* 111: 143–53.

Frenette, J. J., Demers, S., Legendre, L., and Dodson, J. 1993. Lack of agreement amongst models for estimating the photosynthetic parameters. *Limnol. Oceanogr.* 38:679–87.

Gallegos, C. L., Correll, D. L., and Pierce, J. W. 1990. Modeling spectral diffuse attenuation, absorption and scattering coefficients in a turbid estuary. *Limnol. Oceanogr.* 35: 1486–503.

Garside, C., and Glover, H. E. 1991. Chemiluminescent measurements of nitrate kinetics: I. *Thalassiosira pseudonana* (clone 3H) and neritic assemblages. *J. Plankt. Res.* 13:5–19.

Geider, R. J. 1992. Respiration: Taxation without representation? In: Falkowski, P. G. and Woodhead, A. D. eds. *Primary Productivity and Biogeochemical Cycles in the Sea.* Plenum Press, New York, pp. 333–60.

Geider, R. J., Greene, R., Kolber, Z., MacIntyre, H. L., and Falkowski, P. G. 1993a. Fluorescence assessment of the maximum quantum efficiency of photosynthesis in the Western North Atlantic. *Deep Sea Res.* 40:1205–24.

Geider, R. J., La Roche, J., Greene, R. M., and Olaizola, M. 1993b. Response of the photosynthetic apparatus of *Phaeodactylum tricornutum* (Bacillariophyceae) to nitrate, phosphate or iron starvation. *J. Phycol.* 29:755–66.

Geider, R. J., and La Roche, J. 2002. Redfield revisited: variability of C : N : P in marine microalgae and its biochemical basis. *Eur. J. Phycol.* 37:1–17.

Geider, R. J., MacIntyre, H. L., and Kana, T. M. 1996. A dynamic model of photoadaptation in phytoplankton. *Limnol. Oceanogr.* 41:1–15.

Geider, R. J., MacIntyre, H. L., and Kana, T. M. 1997. Dynamic model of phytoplankton growth and acclimation: responses of the balanced growth rate and the chlorophyll *a*:carbon ratio to light, nutrient-limitation and temperature. *Mar. Ecol. Prog. Ser.* 148:187–200.

Geider, R. J., MacIntyre, H. L., and Kana, T. M. 1998. A dynamic regulatory model of phytoplanktonic acclimation to light, nutrients and temperature. *Limnol. Oceanogr.* 43:679–94.

Geider, R. J., Osborne, B. A., and Raven, J. A. 1985. Light dependence of growth and photosynthesis in *Phaeodactylum tricornutum* (Bacillariophyceae). *J. Phycol.* 21:609–19.

Geider, R. J., Osborne, B. A., and Raven, J. A. 1986. Growth, photosynthesis and maintenance metabolic cost in the diatom *Phaeodactylum tricornutum* at very low light levels. *J. Phycol.* 22:39–48.

Geider, R. J., and Osborne, B. A. 1987. Light absorption by a marine diatom: experimental observations and theoretical calculations of the package effect in a small *Thalassiosira* species. *Mar. Biol.* 96:299–308.

Genty, B., Briantais, J. M., and Baker, N. R. 1989. The relationship between the quantum yield of photosynthetic electron transport and quenching of chlorophyll fluorescence. *Biochim. Biophys. Acta* 990:87–92.

Genty, B., Harbinson, J., Briatais, J. M., and Baker, N. R. 1990. The relationship between non-photochemical quenching of chlorophyll fluorescence and the rate of photosystem 2 photochemistry in leaves. *Photosynth. Res.* 25:249–57.

Glover, H., Keller, M. D., and Spinrad, R. W. 1987. The effects of light quality and intensity on photosynthesis and growth of marine eukaryotic and prokaryotic phytoplankton clones. *J. Exp. Mar. Biol. Ecol.* 105:137–59.

Glover, H. E., Campbell, L., and Prézelin, B. B. 1986. Contribution of *Synechococcus* spp. to size-fractionated primary productivity in three water masses in the Northwest Atlantic Ocean. *Mar. Biol.* 91:193–203.

Goericke, R., and Montoya, J. P. 1998. Estimating the contribution of microalgal taxa to chlorophyll *a* in the field: Variations of pigment ratios under nutrient- and light-limited growth. *Mar. Ecol. Prog. Ser.* 169:97–112.

Goldman, J. C., McCarthy, J. J., and Peavey, D. W. 1979. Growth rate influence on the chemical composition of phytoplankton in oceanic waters. *Nature* 279:210–15.

Goldman, J. C. 1980. Physiological processes, nutrient availability, and the concept of relative growth rate in marine phytoplankton ecology. In: Falkowski, P. G., ed. *Primary Productivity in the Sea.* Plenum Press, New York, pp. 179–94.

Gordillo, F. J. L., Jiménez, C., Chaverría, J., and Niell, F. X. 2001. Photosynthetic acclimation to photon irradiance and its relation to chlorophyll fluorescence and carbon assimilation in the halotolerant green alga *Dunaliella viridis. Proc. Nat. Acad. Sci.* 68:225–35.

Grande, K. D., Marra, J., Langdon, C., Heinemann, K., and Bender, M. L. 1989a. Rates of respiration in the light measured in marine phytoplankton using an $^{18}$O isotope-labeling technique. *J. Exp. Mar. Biol. Ecol.* 129: 95–120.

Grande, K. D., Williams, P. J. I., Marra, J., Purdie, D. A., Heinemann, K., Eppley, R. W., and Bender, M. L. 1989b. Primary production in the North Pacific gyre: a comparison of rates determined by the $^{14}$C, $O_2$ concentration and $^{18}$O methods. *Deep Sea Res.* 36:1621–34.

Greene, R. M., Geider, R. J., and Falkowski, P. G. 1991. Effect of iron limitation on photosynthesis in a marine diatom. *Limnol. Oceanogr.* 36:1772–82.

Grzebyk, D., and Berland, D. 1996. Influences of temperature, salinity and irradiance on growth of *Prorocentrum minimum* (Dinophyceae) from the Mediterranean Sea. *J. Plankt. Res.* 18:1837–49.

Guillard, R. R. L. 1973a. Methods for microflagellates and nannoplankton. In: Stein, J. R., ed. *Handbook of Phycological Methods—Culture Methods and Growth Measurements.* Cambridge University Press, Cambridge, pp. 69–85.

Guillard, R. R. L. 1973b. Division rates. In: Stein, J. R., ed. *Handbook of Phycological Methods: Culture Methods and Growth Measurements.* Cambridge University Press, Cambridge, pp. 289–311.

Harding, L. W. J., Meeson, B. W., Prézelin, B. B., and Sweeney, B. M. 1981. Diel periodicity of photosynthesis in marine phytoplankton. *Mar. Biol.* 61:95–105.

Harding, L. W. J., Prézelin, B. B., Sweeney, B. M., and Cox, J. L. 1982. Diel oscillations of the photosynthesis-irradiance (P-I) relationship in natural assemblages of phytoplankton. *Mar. Biol.* 67:167–78.

Haxo, F. T., and Blinks, L. R. 1960. Photosynthetic action spectra of marine algae. *J. Gen. Physiol.* 33:389–422.

Herzig, R., and Falkowski, P. G. 1989. Nitrogen limitation in *Isochrysis galbana* (Haptophyceae). I. Photosynthetic energy conversion and growth efficiencies. *J. Phycol.* 25:462–71.

Holm-Hansen, O. 1978. Chlorophyll *a* determination: improvements in methodology. *Oikos* 30:438–47.

Holt, A. S., and Jacobs, E. E. 1954. Spectroscopy of plant pigments. 1. Ethyl chlorophyllides *A* and *B* and their phaeophorbides. *Am. J. Bot.* 41:710–22.

Huot, Y., Jeffrey, W. H., Davis, R. F., and Cullen, J. J. 2000. Damage to DNA in bacterioplankton: A model of damage by ultraviolet radiation and its repair as influenced by vertical mixing. *Photochem. Photobiol.* 72:62–74.

Hutchinson, G. E. 1961. The paradox of the plankton. *Am. Naturalist* 95:137–45.

Ibelings, B. W., Kroon, B. M. A., and Mur, L. R. 1994. Acclimation of photosystem II in a cyanobacterium and a eukaryotic green alga to high and fluctuating photosynthetic photon flux densities, simulating light regimes induced by mixing in lakes. *New Phytol.* 128:407–24.

Jassby, A. D., and Platt, T. 1976. Mathematical formulation of the relationship between photosynthesis and light for phytoplankton. *Limnol. Oceanogr.* 21:540–7.

Jeffrey, S. W., and Humphrey, G. F. 1975. New spectrophotometric equations for determining chlorophylls *a*, *b*, $c_1$ and $c_2$ in higher plants, algae and natural phytoplankton. *Biochem. Physiol. Pflantz.* 167:191–94.

Jeffrey, S. W., Mantoura, R. F. C., and Wright, S. W. 1997. *Phytoplankton Pigments in Oceanography: Guidelines to Modern Methods.* UNESCO Publishing, Paris, 661 pp.

Jochem, F. 1999. Dark survival strategies in marine phytoplankton assessed by cytometric measurement of metabolic activity with fluorescein diacetate. *Mar. Biol.* 135:721–28.

Joint, I. R., and Pomroy, A. J. 1981. Primary production in a turbid estuary. *Est. Coast. Shelf Sci.* 13:303–16.

Juhl, A. R., Trainer, V. L., and Latz, M. I. 2001. Effect of fluid shear and irradiance on population growth and cellular toxin content of the dinoflagellate *Alexandrium fundyense. Limnol. Oceanogr.* 46:758–64.

Kacser, H., and Burns, J. A. 1973. The control of flux. *Symposia of the Society for Experimental Biology.* 28:65–104.

Kamykowski, D. 1995. Trajectories of autotrophic marine dinoflagellates. *J. Phycol.* 31:200–8.

Kamykowski, D., Janowitz, G. S., Kirkpatrick, G. J., and Reed, R. E. 1996. A study of time-dependent primary productivity in a natural upper-ocean mixed layer using a biophysical model. *J. Plankt. Res.* 18:1295–322.

Kana, T. M., and Glibert, P. M. 1987a. Effect of irradiances up to 2000 µE m$^{-2}$ s$^{-1}$ on marine *Synechococcus* WH7803: II. Photosynthetic responses and mechanisms. *Deep Sea Res.* 34:497–516.

Kana, T. M., and Glibert, P. M. 1987b. Effect of irradiances up to 2000 µE m$^{-2}$ s$^{-1}$ on marine *Synechococcus* WH7803: I. Growth, pigmentation, and cell composition. *Deep Sea Res.* 34:479–95.

Kana, T. M. 1990. Light-dependent oxygen cycling measured by an oxygen-18 isotope dilution technique. *Mar. Ecol. Prog. Ser.* 64:293–300.

Kana, T. M. 1993. Rapid oxygen cycling in *Trichodesmium thiebautii*. *Limnol. Oceanogr.* 38:18–24.

Keller, M. D., Selvin, R. C., Claus, W., and Guillard, R. R. L. 1987. Media for the culture of oceanic ultraphytoplankton. *J. Phycol.* 23:633–38.

Keller, M. D., Bellows, W. K., and Guillard, R. R. L. 1988. Microwave treatment for sterilization of phytoplankton culture media. *J. Exp. Mar. Biol. Ecol.* 117: 279–83.

Kiefer, D. A. 1973a. Fluorescence properties of natural phytoplankton populations. *Mar. Biol.* 22:263–9.

Kiefer, D. A. 1973b. Chlorophyll *a* fluorescence on marine centric diatoms: responses of chloroplasts to light and nutrient stress. *Mar. Biol.* 23:49–56.

Kirk, J. T. O. 1992. Monte Carlo modeling of the performance of a reflective tube absorption meter. *Appl. Opt.* 31:6463–8.

Kirk, J. T. O. 1994. *Light and Photosynthesis in Aquatic Ecosystems.* Cambridge University Press, Cambridge, 509 pp.

Koblízek, M., Kaften, D., and Nedbal, L. 2001. On the relationship between the non-photochemical quenching of the chlorophyll fluorescence and the Photosystem II light harvesting efficiency. A repetitive flash fluorescence induction study. *Photosynth. Res.* 68:141–52.

Kolber, Z., Wyman, K. D., and Falkowski, P. G. 1990. Natural variability in photosynthetic energy conversion efficiency: a field study in the Gulf of Maine. *Limnol. Oceanogr.* 35:72–9.

Kolber, Z., and Falkowski, P. G. 1993. Use of active fluorescence to estimate phytoplankton photosynthesis *in situ*. *Limnol. Oceanogr.* 38:1646–65.

Kromkamp, J. C., and Forster, R. M. 2003. The use of variable fluorescence measurements in aquatic ecosystems: Differences between multiple and single turnover measuring protocols and suggested terminology. *Eur. J. Phycol.* 38:103–12.

Kromkamp, J., and Limbeek, M. 1993. Effect of short-term variation in irradiance on light harvesting and photosynthesis of the marine diatom *Skeletonema costatum*: A laboratory study simulating vertical mixing. *J. Gen. Microbiol.* 139:2277–84.

Kroon, B. M. A., Burger-Wiersma, T., Visser, P., and Mur, L. R. 1992a. The effect of dynamic light regimes on *Chlorella*. I. Minimum quantum requirement and photosynthesis-irradiance parameters. *Hydrobiology* 238:79–88.

Kroon, B. M. A., Latasa, M., Ibelings, B. W., and Mur, L. R. 1992b. The effect of dynamic light regimes on *Chlorella*. II. Pigments and cross sections. *Hydrobiology* 238:71–78.

Langdon, C. 1987. On the causes of interspecific differences in the growth-irradiance relationship for phytoplankton. Part I. A comparative study of the growth-irradiance relationship of three marine phytoplankton species: *Skeletonema costatum*, *Olisthodiscus luteus* and *Gonyaulax tamarensis*. *J. Plankt. Res.* 9:459–82.

Latasa, M., and Berdalet, E. 1994. Effect of nitrogen or phosphorus starvation on pigment composition of cultured *Heterocapsa* sp. *J. Plankt. Res.* 16:83–94.

Laws, E. A., and Bannister, T. T. 1980. Nutrient- and light-limited growth of *Thalassiosira fluviatilis* in continuous culture, with implications for phytoplankton growth in the sea. *Limnol. Oceanogr.* 25:457–73.

Le Floc'h, E., Malara, G., and Sciandra, A. 2002. An automatic device for *in vivo* absorption spectra acquisition and chlorophyll estimation in phytoplankton cultures. *J. Appl. Phycol.* 14:435–44.

Lefèvre, N., Taylor, A. H., Gilbert, F. J., and Geider, R. J. 2003. Modeling carbon to nitrogen and carbon to chlorophyll *a* ratios in the ocean at low latitudes: evaluation of the role of physiological plasticity. *Limnol. Oceanogr.* 48: 1796–807.

Legendre, L., and Demers, S. 1984. Towards dynamic biological oceanography and limnology. *Can. J. Fish. Aquat. Sci.* 41:2–19.

Leonardos, N., and Geider, R. J. 2004. Responses of elemental and biochemical composition of *Chaetoceros muelleri* to growth under varying light and nitrate : phosphate supply ratios and their influence on critical N : P. *Limnol. Oceanogr. Methods.* 49 (in press).

Lewis, M. R., Cullen, J. J., and Platt, T. 1984. Relationships between vertical mixing and photoadaptation of phytoplankton: similarity criteria. *Mar. Ecol. Prog. Ser.* 15:141–49.

Lewis, M. R., and Smith, J. C. 1983. A small-volume, short-incubation-time method for measurement of photosynthesis as a function of incident irradiance. *Mar. Ecol. Prog. Ser.* 13:99–102.

Lewitus, A. J., and Kana, T. M. 1995. Light respiration in six estuarine phytoplankton species: contrasts under photoautotrophic and mixotrophic growth conditions. *J. Phycol.* 31:754–61.

Li, W. K. W., and Morris, I. 1982. Temperature adaptation in *Phaeodactylum tricornutum* Bohlin: Photosynthetic rate compensation and capacity. *J. Exp. Mar. Biol. Ecol.* 58:135–50.

Lippemeier, S., Hintze, R., Vanselow, K. H., Hartig, P., and Colijn, F. 2001. In-line recording of PAM fluorescence of phytoplankton cultures as a new tool for studying effects of fluctuating nutrient supply on photosynthesis. *Eur. J. Phycol.* 36:89–100.

Litchman, E. 1998. Population and community responses of phytoplankton to fluctuating light. *Oecologia* 117:247–57.

Litchman, E. 2003. Competition and coexistence of phytoplankton under fluctuating light: experiments with two cyanobacteria. *Aquat. Microb. Ecol.* 31:241–48.

Lutz, V. A., Sathyendranath, S., Head, E. J. H., and Li, W. K. W. 2001. Changes in the *in vivo* absorption and fluorescence excitation spectra with growth irradiance in three species of phytoplankton. *J. Plankt. Res.* 23:555–69.

MacIntyre, H. L., and Geider, R. J. 1996. Regulation of Rubisco activity and its potential effect on photosynthesis during mixing in a turbid estuary. *Mar. Ecol. Prog. Ser.* 144:247–64.

MacIntyre, H. L., Geider, R. J., and Miller, D. C. 1996. Microphytobenthos: the ecological role of the "secret garden" of unvegetated shallow-water marine habitats. I. Distribution, abundance and primary production. *Estuaries* 19:186–201.

MacIntyre, H. L., Sharkey, T. D., and Geider, R. J. 1997a. Activation and deactivation of ribulose-1,5-bisphosphate carboxylase/oxygenase (Rubisco) in three marine microalgae. *Photosynth. Res.* 51:93–106.

MacIntyre, H. L., Kana, T. M., and Geider, R. J. 2000. The effect of water motion on short-term rates of photosynthesis by marine phytoplankton. *Trends Plant Sci.* 5:12–7.

MacIntyre, H. L., Kana, T. M., Anning, T., and Geider, R. J. 2002. Photoacclimation of photosynthesis irradiance response curves and photosynthetic pigments in microalgae and cyanobacteria. *J. Phycol.* 38:17–38.

MacIntyre, J. G., Cullen, J. J., and Cembella, A. D. 1997b. Vertical migration, nutrition and toxicity of the dinoflagellate, *Alexandrium tamarense. Mar. Ecol. Prog. Ser.* 148:210–16.

MacIsaac, E. A., and Stockner, J. G. 1993. Enumeration of phototrophic picoplankton by autofluorescence microscopy. In: Kemp, P. F., Sherr, B. F., Sherr, E. B., and Cole, J. J., eds. *Handbook of Methods in Aquatic Microbial Ecology.* Lewis Publishers, Boca Raton, Florida, pp. 187–97.

Mague, T. H., Friberg, E., Hughes, D. J., and Morris, I. 1980. Extracellular release of carbon by marine phytoplankton: A physiological approach. *Limnol. Oceanogr.* 25:262–79.

Margalef, R. 1978. Life-forms as survival alternatives in an unstable environment. *Oceanol. Acta* 1:493–509.

Markager, S., and Vincent, W. F. 2001. Light absorption by phytoplankton: development of a matching parameter for algal photosynthesis under different spectral regimes. *J. Plankt. Res.* 23:1373–84.

Marra, J. 1978a. Effect of short-term variations in light intensity on photosynthesis of a marine phytoplankter: a laboratory simulation study. *Mar. Biol.* 46:191–202.

Marra, J. 1978b. Phytoplankton photosynthetic response to vertical movement in a mixed layer. *Mar. Biol.* 46:203–8.

Marra, J., Heinemann, K., and Landriau, G., Jr. 1985. Observed and predicted measurements of photosynthesis in a phytoplankton culture exposed to natural irradiance. *Mar. Ecol. Prog. Ser.* 24:43–50.

Maske, H., and Garcia-Mendoza, E. 1994. Adsorption of dissolved organic matter to the inorganic filter substrate and its implication for $^{14}$C uptake measurements. *Appl. Environ. Microbiol.* 60:3887–9.

Maxwell, D. P., Laudenbach, D. A., and Huner, H. P. A. 1995. Redox regulation of light-harvesting complex II and *cab* mRNA abundance in *Dunaliella salina. Pl. Physiol.* 109:787–95.

McCarthy, J. J. 1981. The kinetics of nutrient utilization. *Can. J. Fish. Aquat. Sci.* 210:83–102.

McCaull, W. A., and Platt, T. 1977. Diel variations in the photosynthetic parameters of coastal marine phytoplankton. *Limnol. Oceanogr.* 22:723–31.

Menden-Deuer, S., and Lessard, E. J. 2000. Carbon to volume relationships for dinoflagellates, diatoms and other protist plankton. *Limnol. Oceanogr.* 45:569–79.

Mitchell, B. G., and Kiefer, D. A. 1984. Determination of absorption and fluorescence excitation spectra for phytoplankton. In: Holm-Hansen, O., Bolis, L., and Giller, R., ed. *Marine Phytoplankton and Productivity.* Springer-Verlag, New York, pp. 157–69.

Morel, A. 1978. Available, usable, and stored radiant energy in relation to marine photosynthesis. *Deep Sea Res.* 25:673–88.

Morel, A., and Bricaud, A. 1981. Theoretical results concerning light absorption in a discrete medium, and application to specific absorption of phytoplankton. *Deep Sea Res. I* 28:1375–93.

Morel, F. M. M. 1987. Kinetics of nutrient uptake and growth in phytoplankton. *J. Phycol.* 23:137–50.

Myers, J., and Clark, L. B. 1944. Culture conditions and development of the photosynthetic mechanism. II. An apparatus for the continuous culture of *Chlorella. J. Gen. Physiol.* 28:103–12.

Myers, J., Graham, J. R., and Wang, R. T. 1980. Light harvesting in *Anacystis nidulans* studied in pigment mutants. *Pl. Physiol.* 66:1144–9.

Neale, P. J., and Marra, J. 1985. Short-term variation of $P_{max}$ under natural irradiance conditions: a model and its implications. *Mar. Ecol. Prog. Ser.* 26:113–24.

Neale, P. J., Banaszak, A. T., and Jarriel, C. R. 1998a. Ultraviolet sunscreens in *Gymnodinium sanguineum* (Dinophyceae): Mycosporine-like amino acids protect against inhibition of photosynthesis. *J. Phycol.* 34:928–38.

Neale, P. J., Davis, R. F., and Cullen, J. J. 1998b. Interactive effects of ozone depletion and vertical mixing on photosynthesis of Antarctic phytoplankton. *Nature* 392: 585–9.

Neale, P. J., Davis, R. F., and Cullen, J. J. 1998c. Inhibition of marine photosynthesis by ultraviolet radiation: Variable sensitivity of phytoplankton in the Weddell-Scotia Sea during austral spring. *Limnol. Oceanogr.* 43:433–48.

Neale, P. J. 2000. Spectral weighting functions for quantifying effects of UV radiation in marine systems. In: de Mora, S., Demers, S., and Vernet, M., ed. *The Effects of UV Radiation in the Marine Environment.* Cambridge University Press, Cambridge, pp. 72–100.

Neidhardt, J., Benemann, J. R., Zhang, L., and Melis, A. 1998. Photosystem-II repair and chloroplast recovery from irradiance stress: Relationship between chronic photoinhibition, light-harvesting chlorophyll antenna size and photosynthetic productivity in *Dunaliella salina* (green algae). *Photosynth. Res.* 56:175–84.

Nelson, N. B., and Prèzelin, B. B. 1993. Calibration of an integrating sphere for determining the absorption coefficient of scattering suspensions. *Appl. Opt.* 32: 6710–7.

Neori, A., Vernet, M., Holm-Hansen, O., and Haxo, F. T. 1986. Relationship between action spectra for chlorophyll *a* fluorescence and photosynthetic $O_2$ evolution in algae. *J. Plankt. Res.* 8:537–48.

Nielsen, M. V., and Sakshaug, E. 1993. Photobiological studies of *Skeletonema costatum* adapted to spectrally different light regimes. *Limnol. Oceanogr.* 38:1576–81.

Olaizola, M., La Roche, J., Kolber, Z., and Falkowski, P. G. 1994. Non-photochemical fluorescence quenching and the diadinoxanthin cycle in a marine diatom. *Photosynth. Res.* 41:357–70.

Olson, R. J., Zettler, E. R., and DuRand, M. D. 1993. Phytoplankton analysis using flow cytometry. In: Kemp, P. F., Sherr, B. F., Sherr, E. B., and Cole, J. J., eds. *Handbook of Methods in Aquatic Microbial Ecology.* Lewis Publishers, Boca Raton, Florida, pp. 187–97.

Otero, A., and Vincenzini, M. 2004. *Nostoc* (Cyanophyceae) goes nude: extracellular polysaccharides serve as a sink for reducing power under unbalanced C/N metabolism. *J. Phycol.* 40:74–81.

Pahl-Wostle, C., and Imboden, D. M. 1990. DYPHORA—a dynamic model for the rate of photosynthesis of algae. *J. Plankt. Res.* 12:1207–21.

Palmer, J. D., and Round, F. E. 1965. Persistent, verticalmigration rhythms in benthic microflora. I. The effect of light and temperature on the rhythmic behaviour of *Euglena obtusa. J. Mar. Biol. Assoc. U.K.* 45:567–82.

Parkhill, J. P., Maillet, G., and Cullen, J. J. 2001. Fluorescence-based maximal quantum yield for PSII as a diagnostic of nutrient stress. *J. Phycol.* 37:517–29.

Platt, J. R. 1964. Strong inference. *Science* 146:347–53.

Platt, T., Gallegos, C. L., and Harrison, W. G. 1980. Photoinhibition of photosynthesis in natural assemblages of marine phytoplankton. *J. Mar. Res.* 38:687–701.

Platt, T., and Sathyendranath, S. 1993. Estimators of primary production for interpretation of remotely sensed data on ocean color. *J. Geophys. Res.* 98 (C8):14561–76.

Porter, K. G., and Feig, Y. S. 1980. The use of DAPI for identifying and counting aquatic microflora. *Limnol. Oceanogr.* 25:943–48.

Post, A. F., Dubinsky, Z., Wyman, K., and Falkowski, P. G. 1985. Physiological responses of a marine planktonic diatom to transitions in growth irradiance. *Mar. Ecol. Prog. Ser.* 24:141–49.

Prèzelin, B. B., Meeson, B. W., and Sweeney, S. M. 1977. Characterization of photosynthetic rhythms in marine dinoflagellates. I. Pigmentation, photosynthetic capacity and respiration. *Pl. Physiol.* 60:384–87.

Prèzelin, B. B., Putt, M., and Glover, H. E. 1986a. Diurnal patterns in photosynthetic capacity and depth-dependent photosynthesis-irradiance relationships in *Synechococcus* spp. and larger phytoplankton in three water masses in the Northwest Atlantic. *Mar. Biol.* 91:205–17.

Prèzelin, B. B., Samuelsson, G., and Matlick, H. A. 1986b. Photosystem II photoinhibition and altered kinetics of photosynthesis during nutrient-dependent high-light photoadaptation in *Gonyaulax polyedra. Mar. Biol.* 93:1–12.

Prèzelin, B. B., and Sweeney, B. M. 1977. Characterization of photosynthetic rhythms in marine dinoflagellates. II. Photosynthesis-irradiance curves and *in vivo* chlorophyll a fluorescence. *Pl. Physiol.* 60:388–92.

Price, N. M., Thompson, P. A., and Harrison, P. J. 1987. Selenium: an essential element for growth of the coastal marine diatom *Thalassiosira pseudonana* (Bacillariophyceae). *J. Phycol.* 23:1–9.

Quaite, F. E., Sutherland, B. M., and Sutherland, J. C. 1992. Action spectrum for DNA damage in alfalfa lowers predicted impact of ozone depletion. *Nature* 358:576–8.

Raps, S., Wyman, K., Siegelman, H. W., and Falkowski, P. G. 1983. Adaptation of the cyanobacterium *Microcystis aeruginosa* to light intensity. *Pl. Physiol.* 72:829–32.

Raven, J. A., and Johnston, A. M. 1991. Mechanisms of inorganic-carbon acquisition in marine phytoplankton and their implications for the use of other resources. *Limnol. Oceanogr.* 36:1701–14.

Raven, J. A., and Kübler, J. E. 2002. New light on the scaling of metabolic rate with the size of algae. *J. Phycol.* 38:11–16.

Redfield, A. C. 1958. The biological control of chemical factors in the environment. *Am. Sci.* 46:205–21.

Rees, T., and Hill, S. A. 1994. Metabolic control analysis of plant metabolism. *Plant Cell Environ.* 17:587–99.

Reynolds, C. S., and Bellinger, E. G. 1992. Patterns of abundance and dominance of the phytoplankton of Rostherne Mere, England: evidence from an 18-year data set. *Aquatic Sci.* 54:10–36.

Rhee, G. Y. 1978. Effects of N : P atomic ratios and nitrate limitation on algal growth, cell composition, and nitrate uptake. *Limnol. Oceanogr.* 23:10–25.

Rhee, G. Y., Gotham, I. J., and Chisholm, S. W. 1981. Use of cyclostat cultures to study phytoplankton ecology. In: Calcott, P. H., ed. *Continuous Culture of Cells*, vol II. CRC Press, Boca Raton, Florida, pp. 159–186.

Richardson, T. L., and Cullen, J. J. 1995. Changes in buoyancy and chemical composition during growth of a coastal marine diatom: ecological and biogeochemical consequences. *Mar. Ecol. Prog. Ser.* 128:77–90.

Riley, G. A. 1947. Factors controlling phytoplankton populations on Georges Bank. *J. Mar. Res.* 6:54–73.

Roesler, C. S. 1998. Theoretical and experimental approaches to improve the accuracy of particulate absorption coefficients derived from the quantitative filter technique. *Limnol. Oceanogr.* 43:1649–60.

Roy, S. 2000. Strategies for the minimisation of UV-induced damage. In: de Mora, S., Demers, S., and Vernet, M., eds. *The Effects of UV Radiation in the Marine Environment.* Cambridge University Press, Cambridge, pp. 177–205.

Sakshaug, E., Andresen, K., and Kiefer, D. A. 1989. A steady state description of growth and light absorption in the marine planktonic diatom *Skeletonema costatum*. *Limnol. Oceanogr.* 34:198–205.

Sakshaug, E., Bricaud, A., Dandonneau, Y., Falkowski, P. G., Kiefer, D. A., Legendre, L., Morel, A., Parslow, J., and Takahashi, M. 1997. Parameters of photosynthesis: Definitions, theory and interpretation of results. *J. Plankt. Res.* 19:1637–70.

Schofield, O., Bidigare, R. R., and Prèzelin, B. B. 1990. Spectral photosynthesis, quantum yield and blue-green light enhancement of productivity rates in the diatom *Chaetoceros gracilis* and the prymensiophyte *Emiliania huxleyi*. *Mar. Ecol. Prog. Ser.* 64:175–86.

Schofield, O., Prèzelin, B., and Johnsen, G. 1996. Wavelength dependency of the maximum quantum yield of carbon fixation for two red tide dinoflagellates, *Heterocapsa pygmaea* and *Prorocentrum minimum* (Pyrrophyta): implications for measuring photosynthesis rates. *J. Phycol.* 32:574–83.

Schreiber, U., Neubauer, C., and Schliwa, U. 1993. PAM fluorometer based on medium-frequency pulsed Xe-flashed measuring light: a highly sensitive new tool in basic and applied photosynthesis research. *Photosynth. Res.* 36:65–72.

Schreiber, U., Schliwa, U., and Bilger, W. 1986. Continuous recording of photochemical and non-photochemical fluo-

rescence quenching with a new type of modulation fluorometer. *Photosynth. Res.* 10:51–62.

Sciandra, A. 1991. Coupling and uncoupling between nitrate uptake and growth rate in *Prorocentrum minimum* (Dinophyceae) under different frequencies of pulsed nitrate supply. *Mar. Ecol. Prog. Ser.* 72:261–69.

Sciandra, A., and Amara, R. 1994. Effects of nitrogen limitation on growth and nitrite excretion rates of the dinoflagellate *Prorocentrum minimum*. *Mar. Ecol. Prog. Ser.* 105: 301–9.

Serôdio, J., da Silva, J. M., and Catarino, F. 1997. Nondestructive tracing of migratory rhythms of intertidal benthic microalgae using *in vivo* chlorophyll *a* fluorescence. *J. Phycol.* 33:542–53.

Sharp, J. H. 1977. Excretion of organic matter by marine phytoplankton: Do healthy cells do it? *Limnol. Oceanogr.* 22:381–98.

Shibata, K. 1958. Spectrophotometry of intact biological materials. *J. Biochem.* 45:599–623.

Shoaf, W. T., and Lium, B. W. 1976. Improved extraction of chlorophyll *a* and *b* from algae using dimethyl sulfoxide. *Limnol. Oceanogr.* 21:926–28.

Shuter, B. 1979. A model of physiological adaptation in unicellular algae. *J. Theor. Biol.* 78:519–52.

Siegel, D. A. 1998. Resource competition in a discrete environment: Why are plankton distributions paradoxical? *Limnol. Oceanogr.* 43:1133–46.

Smayda, T. J. 1970. The suspension and sinking of phytoplankton in the sea. *Oceanogr. Mar. Biol. A. Rev.* 8:353–414.

Smayda, T. J. 1997. Harmful algal blooms: Their ecophysiology and general relevance to phytoplankton blooms in the sea. *Limnol. Oceanogr.* 42:1137–53.

Smith, D. F., and Horner, S. M. J. 1981. Tracer kinetic analysis applied to problems in marine biology. *Can. J. Fish. Aquat. Sci.* 210:113–29.

Smith, D. J., and Underwood, G. J. C. 2000. The production of extracellular carbohydrates by estuarine benthic diatoms: the effects of growth phase and light and dark treatment. *J. Phycol.* 36:321–33.

Smith, G. J., Zimmerman, R. C., and Alberte, R. S. 1992. Molecular and physiological responses of diatoms to variable levels of irradiance and nitrogen availability: growth of *Skeletonema costatum* in simulated upwelling conditions. *Limnol. Oceanogr.* 37:989–1007.

Sosik, H. M., and Mitchell, B. G. 1991. Absorption, fluorescence, and quantum yield for growth in nitrogen-limited *Dunaliella tertiolecta*. *Limnol. Oceanogr.* 36:910–21.

Sosik, H. M., and Mitchell, B. G. 1995. Light absorption by phytoplankton, photosynthetic pigments and detritus in the California Current System. *Deep Sea Res.* I 42:1717–48.

Sosik, H. M., Olson, R. J., Neubert, M. G., Shalapyonok, A., and Solow, A. R. 2003. Growth rates of coastal phytoplankton from time-series measurements with a submersible flow cytometer. *Limnol. Oceanogr.* 48:1756–65.

Staats, N., Stal, L. J., and Mur, L. R. 2000. Exopolysaccharide production by the epipelic diatom *Cylindrotheca closterium*: effects of nutrient conditions. *J. Exp. Mar. Biol. Ecol.* 249:13–27.

Steemann Nielsen, E. 1952. The use of radio-active carbon ($C^{14}$) for measuring organic production in the sea. *J. Conseil Int. l'Exploration Mer* 16:117–40.

Steemann Nielsen, E. 1953. Carbon dioxide concentration, respiration during photosynthesis, and maximum quantum yield of photosynthesis. *Physiol. Plant.* 6:316–32.

Stramski, D., and Morel, A. 1990. Optical properties of photosynthetic picoplankton in different physiological conditions as affected by growth irradiance. *Deep Sea Res.* 37:245–66.

Suggett, D. J., MacIntyre, H. L., and Geider, R. J. 2004. Evaluation of biophysical and optical determinations of light absorption by photosystem II in phytoplankton. *Limnol. Oceanogr. Methods.* 2:316–32.

Suggett, D. J., Oxborough, K., Baker, N. R., MacIntyre, H. L., Kana, T. M., and Geider, R. J. 2003. Fast repetition rate and pulse amplitude modulation chlorophyll *a* fluorescence measurements for assessment of photosynthetic electron transport in marine phytoplankton. *Eur. J. Phycol.* 38:371–84.

Sukenik, A., Bennett, J., Mortain-Bertrand, A., and Falkowski, P. G. 1990. Adaptation of the photosynthetic apparatus to irradiance in *Dunaliella tertiolecta*. *Pl. Physiol.* 92:891–98.

Sullivan, J. M., and Swift, E. 2003. Effects of small-scale turbulence on net growth and size of ten species of marine dinoflagellates. *J. Phycol.* 39:83–94.

Sverdrup, H. U. 1953. On conditions for the vernal blooming of phytoplankton. *J. Cons. Int. Explor. Mer.* 18:287–95.

Talling, J. F. 1957. The phytoplankton population as a compound photosynthetic unit. *New Phytol.* 56:133–49.

Taylor, A. H., Geider, R. J., and Gilbert, F. J. H. 1997. Seasonal and latitudinal dependencies of phytoplankton carbon-to-chlorophyll *a* ratios: results of a modelling study. *Mar. Ecol. Prog. Ser.* 152:51–66.

Terry, K. L., Laws, E. A., and Burns, D. J. 1985. Growth rate variation in the N : P requirement ratio of phytoplankton. *J. Phycol.* 21:323–29.

Thomas, W. H., and Dodson, A. N. 1972. On nitrogen deficiency in tropical Pacific oceanic phytoplankton. II. Photosynthetic and cellular characteristics of a chemostat-grown diatom. *Limnol. Oceanogr.* 17:515–23.

Thompson, P. A., Harrison, P. J., and Parslow, J. S. 1991. Influence of irradiance on cell volume and carbon quota

for ten species of marine phytoplankton. *J. Phycol.* 27:351–60.

Turpin, D. H., Elrifi, I. R., Birch, D. G., Weger, H. G., and Holmes, J. J. 1988. Interactions between photosynthesis, respiration, and nitrogen assimilation in microalgae. *Can. J. Bot.* 66:2083–97.

Underwood, G. J. C., and Provot, L. 2000. Determining the environmental preferences of four estuarine epipelic diatom taxa: growth across a range of salinity, nitrate and ammonium conditions. *Eur. J. Phycol.* 35:173–82.

Van Heukelem, L., Lewitus, A. J., Kana, T. M., and Craft, N. E. 1992. High performance liquid chromatography of phytoplankton pigments using a polymeric reversed phase $C_{18}$ column. *J. Phycol.* 28:867–72.

Van Heukelem, L., Thomas, C. S., and Glibert, P. M. (2002). Sources of variability in chlorophyll analysis by fluorometry and high-performance liquid chromatography in a SIMBIOS inter-calibration exercise. NASA, Greenbelt, Maryland, 50 pp.

Vernet, M. 2000. Effects of UV radiation on the physiology and ecology of marine phytoplankton. In: de Mora, S., Demers, S., and Vernet, M., eds. *The Effects of UV Radiation in the Marine Environment.* Cambridge University Press, Cambridge, pp. 237–78.

Vincent, W. F. 1980. Mechanisms of rapid photosynthetic adaptation in natural phytoplankton communities. II. Changes in photochemical capacity as measured by DCMU-induced chlorophyll fluorescence. *J. Phycol.* 16:568–77.

Vincent, W. F., and Neale, P. J. 2000. Mechanisms of UV damage to aquatic organisms. In: de Mora, S. J., Demers, S., and Vernet, M. eds. *The Effects of UV Radiation in the Marine Environment.* Cambridge University Press, Cambridge, pp. 149–76.

Weiler, C. S., and Karl, D. M. 1979. Diel changes in phased-dividing cultures of *Ceratium furca* (Dinophyceae): nucleotide triphosphates, adenylate energy charge, cell carbon, and patterns of vertical migration. *J. Phycol.* 15:384–91.

Welschmeyer, N. A. 1994. Fluorometric analysis of chlorophyll *a* in the presence of chlorophyll *b* and phaeopigments. *Limnol. Oceanogr.* 39:1985–92.

Wilhelm, C., and Manns, L. 1991. Changes in pigmentation of phytoplankton species during growth and stationary phase: Consequences for reliability of pigment-based methods of biomass determination. *J. Appl. Phycol.* 3:305–10.

Williams, P. J. L., Raine, R. C. T., and Bryan, J. R. 1979. Agreement between the $^{14}C$ and oxygen methods of measuring phytoplankton production: Reassessment of the photosynthetic quotient. *Oceanol. Acta.* 2:411–16.

Williams, P. J. l. 1993. Chemical and tracer methods of measuring plankton production. *ICES Mar. Sci. Symp.* 197:20–36.

Wood, A. M., and Van Valen, L. M. 1990. Paradox lost? On the release of energy-rich compounds by phytoplankton. *Mar. Microb. Food Webs* 4:103–16.

Yentsch, C. M. 1962. Measurement of visible light absorption by particulate matter in the ocean. *Limnol. Oceanogr.* 7: 207–17.

Zaneveld, J. R. V., Kitchen, J. C., and Moore, C. C. 1994. Scattering error correction of reflecting-tube absorption meters. *SPIE* 2258:44–58.

Zimmerman, R. C., SooHoo, J. B., Kremer, J. N., and D'Argenio, D. Z. 1987. Evaluation of variance approximation techniques for non-linear photosynthesis-irradiance models. *Mar. Biol.* 95:209–15.

CHAPTER 20

# ANALYSIS OF ALGAL PIGMENTS BY HIGH-PERFORMANCE LIQUID CHROMATOGRAPHY

ROBERT R. BIDIGARE
*Department of Oceanography, University of Hawaii*

LAURIE VAN HEUKELEM
*Horn Point Laboratory, University of Maryland Center for Environmental Science*

CHARLES C. TREES
*Center for Hydro-Optics and Remote Sensing, San Diego State University*

## CONTENTS

Key Index Words: Phytoplankton, Pigments, Chlorophylls, Carotenoids, HPLC, Quality Assurance, Quality Control

## 1.0. INTRODUCTION

Phytoplankton use chlorophyll (Chl) *a* as their major light harvesting pigment for photosynthesis. Accessory pigment compounds (e.g., Chls *b* and *c*, carotenoids, and phycobiliproteins) also play a significant role either in photosynthesis, by extending the organism's optical collection window, or in photoprotection, by preventing cellular damage at high growth irradiances. Important chlorophyll degradation products are also found in the aquatic environment, including the chlorophyllides, phaeophorbides, phaeophytins, and steryl chlorin esters.

The unique optical properties of Chl *a* have been used to develop spectrophotometric (Jeffrey and Humphrey 1975) and fluorometric (Holm-Hansen et al. 1965) measurement techniques. With the commercial availability of fluorometers for routine measurements of Chl *a*, this pigment has become a universal parameter for estimating phytoplankton biomass and productivity. These optical methods have the potential to significantly underestimate or overestimate Chl *a* concentrations, because of the overlap of the absorption and fluorescence bands of co-occurring Chls *b* and *c*, chlorophyll degradation products, and accessory pigments (Trees et al. 1985, Smith et al. 1987, Hoepffner and Sathyendranath 1992, Bianchi et al. 1995, Tester et al. 1995). Spectrophotometric and fluorometric methods are nevertheless commonly used for many applications because analyses are inexpensive, simple, and rapid.

High-performance liquid chromatography (HPLC) has made it possible to simultaneously determine the concentrations of a wide range of carotenoids and chlorophylls and their degradation products. Consequently, HPLC has provided researchers with a powerful tool for studying the processes that affect the phytoplankton pigment pool. HPLC pigment analysis can be used to aid in the determination of phytoplankton growth rates (see Chapter 18), zooplankton grazing activities, and phytoplankton physiological processes (see Chapter 19). The presence or absence of individual pigments helps differentiate the major algal groups in natural waters. Pigments that are unique to one algal class or are present in only two or three classes (Jeffrey and Vesk 1997) can be used for quantitative assessment of phytoplankton community composition.

Numerous HPLC techniques have been published to date, and deciding upon the method of choice can be overwhelming. No single HPLC method is suitable for all applications, and each method has its own advantages and limitations. Nineteen HPLC methods published between 1983 and 1998 are reviewed in Jeffrey et al. (1999), whereas other methods (Zapata et al. 2000, Van Heukelem and Thomas 2001) have since been published. Before attempting to select a method, the analyst should identify the target pigments to choose a technique that provides the best separations. Because many factors affect a method's sensitivity, assurances should be made that the methods for sample collection, extraction, and HPLC analysis in combination yield adequate detection for pigments present at low concentrations.

This chapter includes methodological information published in a NASA technical memorandum, *Ocean Optics Protocols for Satellite Ocean Color Sensor Validation* (Bidigare et al. 2003), that describes the Joint Global Ocean Flux Study (JGOFS)–based HPLC protocol (UNESCO 1994). This protocol is derived from the reverse-phase $C_{18}$ HPLC method of Wright et al. (1991) that was recommended by the Scientific Committee on Oceanographic Research (SCOR) for the analysis of phytoplankton pigments. Although this method is recognized for its ability to resolve many chemotaxonomically important pigments, it is limited by its inability to separate *normal* (monovinyl) chlorophyll *a* (MV Chl *a*) from divinyl chlorophyll *a* (DV Chl *a*), the primary constituents of total Chl *a* (TChl *a*, defined as MV Chl *a* + DV Chl *a* + chlorophyllide *a*).

Because TChl *a* represents the most important measurement in many HPLC pigment applications, it is important to understand the unique factors affecting its accurate quantification. Divinyl Chl *a*, the major photosynthetic pigment found in *Prochlorococcus*, accounts for 10 to 60% of the TChl *a* in subtropical and tropical oceanic waters (Goericke and Repeta 1993, Letelier et al. 1993, Andersen et al. 1996, Bidigare and Ondrusek 1996, Gibb et al. 2000). The use of HPLC methods that do not chromatographically separate DV Chl *a* and MV Chl *a* can result in a 15 to 25% overestimation of TChl *a* concentration (Latasa et al. 1996).

Total Chl *a* can be quantified by HPLC in several ways. MV Chl *a* and DV Chl *a* can be separated chromatographically and individually quantified with a $C_8$-based HPLC technique. The chromatographic separation of MV Chl *a* and DV Chl *a* is typically poor for C18-based HPLC methods. In the latter, the HPLC detector can be set to collect data at two different wavelengths (436 and 450 nm) and a dichromatic equation can be used to spectrally resolve MV Chl *a* and DV Chl *a* (Latasa et al. 1996). Alternatively, TChl *a* can be accurately quantified, even if DV Chl *a* and MV Chl *a* are both present and not chromatographically separated, by optimizing detector parameters such that both chlorophylls exhibit the same detector response (Van Heukelem et al. 2002). This simple approach is useful if the relative proportions of each chlorophyll type are unimportant to the analysis objectives.

HPLC analysis provides a plethora of pigment data that is unattainable by other analytical techniques (e.g., spectrophotometry and fluorometry). However, HPLC uncertainties can be high if the operator is unaware of the procedures required to ensure data quality. The focus of this chapter is to familiarize the reader with such procedures and suggest quality control (QC) measurements that are useful for assessing uncertainties associated with various steps in the analysis.

## 2.0. Laboratory Methods

Ancillary procedures that minimize analytical uncertainties are described below. In addition, several QC measurements are recommended to ensure data quality.

### 2.1. Sample Collection

Laboratory culture work requires culture harvesting and often requires some modification of the following volumes (for field collections), because the cultures are usually dense. Seawater samples should be collected in opaque collection bottles (e.g., Niskin and Go-Flo bottles). Pigment samples collected for the validation of ocean color products should be taken simultaneously with surface in-water upwelled radiance and reflectance measurements and at depth increments sufficient to resolve variability within at least the top optical depth. The diffuse attenuation coefficient ($K[z,\lambda]$) profiles over this layer are used to compute optically weighted, near-surface pigment concentration for bio-optical algorithm development (Gordon and Clark 1980). When possible, samples should be acquired at several depths distributed throughout the upper 200 m of the water column (or in turbid water, up to seven diffuse attenuation depths [i.e., $\ln(E[z,\lambda]\ E[z,\lambda]^{-1}) = 7$]) to provide a basis for relating optical measurements to pigment mass concentrations.

If filtration must be delayed for 1 hour or more after collection, sample bottles should be held on ice or in a refrigerator at 4°C and protected from light, because even brief exposure to light during sampling or storage can alter pigment concentrations. Sample collection and filtration during physiological investigations (e.g., xanthophyll cycling studies) must be immediate because of the rapid reaction rates and short time constants for pigment transformation (MacIntyre et al. 2000).

### 2.2. Filtration

Whatman GF/F glass fiber filters (nominal porosity of 0.7 μm) are preferred for concentrating phytoplankton from natural waters. The glass fibers assist in breaking the cells during grinding and accommodate larger sample volumes. Twenty-five-mm-diameter GF/F glass fiber filters should be used with vacuum (7–8 inches of mercury) or positive pressure (1–2 psi). Positive pressure filtration is recommended, because larger volumes of water can be filtered at reduced filtration times. The limitation with vacuum filtration is that unobservable air leaks may occur around the filtration holder, and as a result the pressure gradient across the filter is much less than what is indicated on the vacuum gauge. When positive filtration is used, any leakage around the filter holder results in observable dripping water.

Inert membrane filters, such as polyester filters, may be used when size fraction filtration is required. When this is done, it is recommended to also filter a replicate sample through a GF/F to determine the total concentration. Summing the various size-fractionated concentrations may not produce an accurate estimate of the total concentration because of the potential for cell disruption during filtration.

There has been an ongoing discussion of filter types and retention efficiencies for natural samples. Phinney and Yentsch (1985) showed the inadequacy of GF/F filters for retaining Chl *a* in oligotrophic waters, as did Dickson and Wheeler (1993) for samples from the North Pacific. In response to Dickson and Wheeler (1993), Chavez et al. (1995) compared samples collected in the Pacific Ocean using GF/F and 0.2-μm membrane filters with small filtered volumes (100–540 mL). Their results showed a very close agreement between the two filter types, with GF/F filters having only a slightly positive 5% bias.

Filtration volume can directly affect the retention efficiency for GF/F filters. Particles are retained by filters via various mechanisms, such as filter sieving, filter adsorption, and electrostatic and van der Waals attractions (Brock 1983). When water flows through the pores of a Nuclepore filter, streamlines are formed that can align small particles longitudinally, with the result that cell diameter becomes important with these filters. It is known, on the other hand, that GF/F filters can retain particles much smaller than their rated pore size. Generally, at small volumes (100–300 mL), filter adsorption and electrostatic and van der Waals attractions are important, whereas at larger volumes (>2000 mL) sieving dominates. This has been tested in oligotrophic waters off Hawaii in which small (<500 mL) and large volumes (>2–4 liters) retained similar amounts of TChl *a* on the two types of filters, whereas for intermediate sample volumes the GF/F filters yielded lower concentrations. During several cruises in the vicinity of the Hawaiian Islands, differences in GF/F filter retention efficiencies were found to be a function of sample volume; large sample volumes (2 and 4 liters) retained about 18% more TChl *a* than replicate 1-liter samples.

Filtration volumes are typically limited by the concentration of particles present in each sample. For HPLC analysis it is important to filter as large a volume as possible, so as to accurately quantify the minor pigment constituents. A qualitative check to determine whether a large enough volume has been filtered is to count the number of accessory pigments (Chls *b*, $c_1$, $c_2$, $c_3$, and carotenoids) quantified, excluding chlorophyll degradation products (Trees et al. 2000). Most algal groups (excluding phycobiliprotein-containing groups) contain at least *four* HPLC-measurable accessory pigments (Jeffrey and Vesk 1997). Therefore, pigment samples that do not meet this minimum accessory pigment criterion may have detection limit problems related to low signal-to-noise ratios for the HPLC detectors and/or inadequate concentration techniques. It is generally recommended that the following volumes be filtered for HPLC pigment analyses: 3 to 4 liters for oligotrophic waters, 1 to 2 liters for mesotrophic waters, and 0.5 to 1 liter for eutrophic waters.

There is no need to prefilter cultures to remove zooplankton. It is recommended that field samples *not* be prefiltered to remove mesozooplankton, because this practice may exclude pigment-containing colonial and chain-forming phytoplankton (e.g., diatoms and raft-forming cyanobacteria). Forceps may be used to remove mesozooplankton from the GF/F filters following filtration.

A select number of samples should be filtered in duplicate (or triplicate) to assess representativeness, filter homogeneity, and uncertainty in the method and instrumentation. In multi-ship/investigator studies, replicate samples should be collected and archived for future intercalibration checks.

### 2.3. Sample Handling and Storage

Filtered samples should be placed immediately in liquid nitrogen, even if fluorometric analyses are to be done soon after collection, because liquid nitrogen assists cell disruption and pigment extraction. For minimal pigment degradation, liquid nitrogen is the best method for storing samples for short- and long-term periods (e.g., 1 year). Van Heukelem et al. (2002) found that MV Chl *a* concentrations in laboratory-prepared filters stored for 1 year at –80°C varied by no more than ± 5% (95% confidence interval [CI]). Ultracold freezers (–80°C to –90°C) can be used for storage, although field samples have not been tested for more than 60 days (Mantoura et al. 1997b). Conventional deep freezers should not be used for storing samples for more than 20 hours.

Samples should be folded in half, with the filtered halves facing in to eliminate particle loss during handling. Folded filters can be individually stored in cryogenic tubes, HistoPrep tissue capsules, or folded packets of heavy-duty aluminum foil. The latter is the least expensive and occupies very little storage volume per sample. A short sample identifier should be written on each sample storage tube (or foil) with a fine-point permanent marker. When foil is used, writing should be done before the filter is folded inside to avoid puncturing the foil with the marker and to improve legibility. Information regarding sample identification should immediately be logged in a laboratory notebook with the analyst's initials.

### 2.4. Sample Extraction

Several factors affect pigment extraction efficiency and are discussed in depth in Wright et al. (1997). Extraction steps include (1) addition of water-miscible organic solvent (e.g., acetone or methanol) to the sample, (2) extraction of pigments from particulate samples, (3) clarification of the sample extract by centrifugation or filtration to remove cellular debris and other fine particles that can cause HPLC column blockage (HPLC syringe cartridge filters with polytetrafluoroethylene [PTFE] membranes are recommended, and nylon membrane filters should be avoided because they may bind certain hydrophobic pigments), and (4) determination of extraction volume.

The ease with which pigments are extracted from cells varies among different species (Wright et al. 1997, Latasa et al. 2001). In all cases, freezing the sample filters in liquid nitrogen improves extraction efficiency. To facilitate pigment extraction, cells are disrupted, as described in Bidigare et al. (2003), with either a sonic probe or a mechanical tissue grinder, after which the extract is allowed to soak at 0°C for 24 hours. Samples must be kept cold to minimize pigment degradation.

During extraction with acetone, the final water content in the sample extract should not exceed 10% because Chl *a* degredation can be enhanced (Wright et al. 1997, Latasa et al. 2001). It should also be noted that soaking of extracted pigments in an organic solvent in the presence of cell debris may promote the formation of chlorophyll derivatives, particularly chlorophyllides, because the enzyme chlorophyllase is activated upon cell disruption (Suzuki and Fujita 1986).

Uncertainties associated with HPLC extraction volume measurements can be large, as the small volumes of solvent added to the filter are markedly affected by other variables (e.g., retention of water by the filter and solvent evaporation). Although the water retained by filters is variable, it is estimated to be 0.2 mL and 0.7 mL, respectively, for 25- and 47-mm GF/F filters (Bidigare et al. 2003). Extraction volumes are more accurate if the water in filters is accounted for, the measuring device used to add solvent to filters is calibrated gravimetrically, and the solvent is at room temperature when dispensed. An internal standard (IS) can be used during extractions to account for these variables.

Details of how to use an IS can be found in Snyder and Kirkland (1979). In brief, a known amount of IS is added to the filter just prior to extraction, and the concentration ratio (or area ratio) of the IS before it is added to the filter and after extraction is complete is used to calculate the true extraction volume (see Section 4.3.5). The IS is quantitatively added to the filter in two ways. The extraction solvent can be amended with the IS, and a known volume of spiked solvent is added (with a calibrated measuring device) to each filter. Alternatively, the extraction solvent and IS can be added separately in two steps, in which the IS is quite concentrated and a small volume (50–250 μL) is added quantitatively to each filter. *Trans*-β-apo-8′-carotenal, D,L-α-tocopheryl acetate (vitamin E) (Thomas and Van Heukelem 2003), and canthaxanthin have been used successfully as internal standards with HPLC pigment analyses. Although canthaxanthin is a naturally occurring pigment, its distribution is limited in ocean waters and it has proven useful with such samples. Nevertheless, it is important, with any IS, to first verify that it does not occur naturally in samples to be analyzed.

For the advantages of using an IS to be fully realized, it is crucial that the HPLC injector performance is optimal, because two peak area measurements (before and after extraction) are required and precision error is increased when more than one such measurement is required (Snyder and Kirkland 1979). It is suggested that the IS concentration be measured, at least in trip-

licate, on the same day the samples to which it was added are analyzed.

Several QC measurements can be used to evaluate extraction procedures. A *system blank* consists of a filter, reagents, glassware, and hardware used in the analytical scheme and reflects the combined effects of injector carryover and cross-contamination of samples during extraction. A *spiked blank* is a *system blank* that has been spiked with an authentic external standard containing the pigments of interest and is used to assess pigment stability and recovery during extraction. A *spiked sample* is a sample filter that has been spiked with an authentic external standard containing the pigments of interest and is used to assess pigment stability and recovery in the presence of the sample matrix. Calculations for recoveries from *spiked samples* require that an unspiked duplicate sample filter be analyzed on the same day. These QC samples are quantified under extraction and instrumental conditions identical to those used for the sample extracts (Clesceri et al. 1998). It is highly recommended that these QC measurements be performed at regularly scheduled intervals.

## 2.5. HPLC Methods

At a minimum, the HPLC system consists of solvent pumps, sample injector, analytical column, detector(s), and data recording device, preferably a computer (Fig. 20.1). Temperature controllers for an injection autosampler and column compartments are optional. For a review of hardware and software requirements for measuring carotenoids, chlorophylls, and their degradation products, see Wright and Mantoura (1997).

### 2.5.1. Injection

HPLC injections are performed manually or with an automated injector. Typically, the water-miscible, filtered sample extract is combined with water or buffer before injection. If this mixing step is omitted, pigments that elute early in the chromatogram are broad and asymmetrical, causing their quantification to be less accurate. The type of injection buffer is specific to the particular HPLC method, and the proportion of sample extract to buffer is based on the volumes required to attain symmetrical peak shapes.

With manual injectors and automated HPLC injectors that are not programmable, the analyst combines the sample extract and buffer, waits 5 minutes for equilibration, and then either injects the sample mixture manually or places the sample mixture in the HPLC

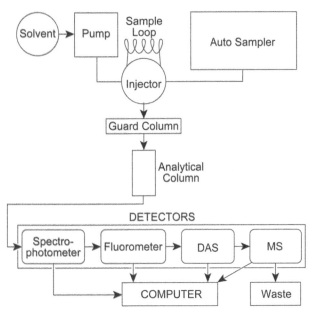

**FIGURE 20.1.** Major components of a high-performance liquid chromatograph (HPLC) system (DAS, diode array spectrophotometer; MS, mass spectrometer).

autosampler compartment (ASC) for immediate injection. Pigment losses and poor injector precision can occur if these mixtures are allowed to sit for long durations before injection (Mantoura et al. 1997a, Wright and Mantoura 1997, Latasa et al. 2001). The sample extract and buffer should be measured with pipetting devices that have been calibrated gravimetrically.

With programmable HPLC injectors, separate vials of buffer and sample extract are placed in a refrigerated ASC and the injector is programmed to combine buffer with sample immediately before injection. Programming features vary tremendously among different instrument manufacturers, and considerable effort is required when initially determining what factors produce the desired results. For example, early eluting peaks may be asymmetrical, split, retained poorly, or exhibit poor area reproducibility if injection conditions are not optimized. When using a programmable injector, it is important to equilibrate all vials to the temperature of the ASC before initiating injections.

Injection accuracy and precision are improved by minimizing or eliminating pigment carryover between injections and using HPLC vials that prevent evaporation. Injector carryover is evaluated by injecting a reagent blank immediately after a very concentrated sample extract. Peaks that appear when the reagent blank is analyzed reveal that pigments from the previous injection have contaminated results. If carryover occurs, calibration standards that are injected in ascend-ing order of concentration produce different linear regressions than when injected in descending order of concentration (Dolan 2001). Often carryover can be eliminated by using a "needle-wash" step with programmable automated injectors or, with manual injections, by flushing the sample loop thoroughly between injections.

Once injection conditions have been established, injection precision should be evaluated, after which no changes to injection procedures should be made without subsequent reevaluation. Such changes may alter retention times, peak shapes, and resolution between and response factors of early eluting pigments. A QC measurement for evaluating injection precision is described by the percent coefficient of variation (% CV) associated with replicate ($n \geq 3$), sequential injections of the same standard.

### 2.5.2. Detection

Pigments dissolved in organic solvent absorb strongly between 430 and 450 nm. Chlorophyll–related pigments also absorb between approximately 650 and 670 nm, and they also fluoresce. Chromatograms frequently contain many pigments and their degradation products; therefore, it can be useful to discriminate between chlorophylls and carotenoids by monitoring chlorophylls with an HPLC fluorescence detector and carotenoids with an HPLC absorption detector. Absorption detectors that monitor multiple wavelengths are also useful for a similar purpose; for example, 450 nm can be used for carotenoids and 665 nm for Chl *a*-related pigments. HPLC photo diode array detectors gather data from multiple wavelengths, and absorbance spectra can be obtained for each peak. Such spectra are used to corroborate pigment identities and to determine whether pigments co-elute, as absorbance spectra on the peak up slope and down slope would be dissimilar.

With HPLC detectors that monitor only one wavelength, 436 or 440 nm is commonly used because these wavelengths provide good detection for most pigments. If the HPLC method does not separate DV and MV Chl *a*, 436 nm (with a narrow bandwidth) is preferred because the detector responses of DV and MV Chl *a* are similar and errors in TChl *a* are small if DV Chl *a* is present. These two detector settings were evaluated by four laboratories that analyzed a DV Chl *a* standard with their own HPLC methods and quantified the results based on MV Chl *a* calibration standards, which were provided to them in this intercalibration exercise. DV Chl *a* concentration was accurately measured with a detector setting of 436 ± 2 nm but was overestimated by 20 and 30%, respectively, when 436 ± 4 nm and 436

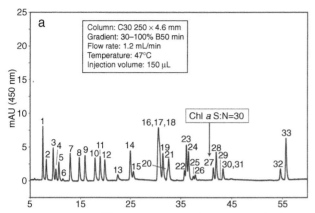

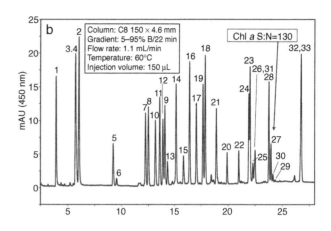

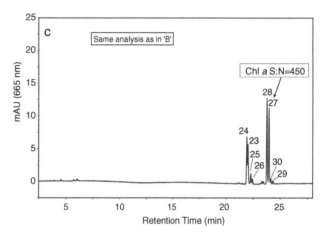

**FIGURE 20.2.** The same mixture of pigment standards was analyzed on the same HPLC with a $C_{30}$ column **(a)** and a $C_8$ column **(b, c)**. Solvent A was 70:30 methanol: 28 mM aqueous tetrabutyl ammonium acetate, pH 6.5 **(a, b, c)**, and solvent B was ethanol **(a)** and methanol **(b, c)**. Pigment code numbers correspond to the following: 1, chlorophyll $c_3$; 2, chlorophyll $c_1$; 3, chlorophyll $c_2$; 4, Mg 3,8-divinyl pheoporphyrin $a_5$ monomethyl ester; 5, peridinin; 6, peridinin isomer; 7, 19′-butanoyloxy fucoxanthin; 8, fucoxanthin; 9, 19′-hexanoyloxy fucoxanthin; 10, neoxanthin; 11, prasinoxanthin; 12, violaxanthin; 13, astaxanthin; 14, diadinoxanthin; 15, antheraxanthin; 16, alloxanthin; 17, diatoxanthin; 18, lutein; 19, zeaxanthin; 20, gyroxanthin diester–like; 21, canthaxanthin; 22, gyroxanthin diester–like; 23, monovinyl chlorophyll b; 24, divinyl chlorophyll b; 25, divinyl chlorophyll b′; 26, crocoxanthin; 27, monovinyl chlorophyll a; 28, divinyl chlorophyll a; 29, monovinyl chlorophyll a′; 30, divinyl chlorophyll a′; 31, monovinyl chlorophyll b′; 32, β, ε carotene; 33, β, β carotene.

± 5 nm were used. With 440 ± 4 nm, DV Chl *a* concentration was overestimated by 40% (Van Heukelem et al. 2002).

Once detector settings have been selected, the HPLC method should be tested to see if the detector response is linear over the range of concentrations expected in samples. Likewise, the linearity of the detector response of the IS should also be evaluated if one is to be used. If the concentration of a sample exceeds a detector's upper limit of linearity (LOL), the sample can simply be diluted and reanalyzed. However, to be suitable for low-concentration samples, the HPLC method must be optimized for sensitivity before it is used with real samples. Sensitivity can be improved by enhancing the signal (or peak height) and decreasing the noise. The instrument detection limit (IDL) is defined as the amount of injected analyte that produces a signal-to-noise ratio (S : N) of 5 (Clesceri et al. 1998). The IDL is affected by both pump and detector noise and is a useful parameter for comparing sensitivities among different HPLC methods. Noise can be reduced by optimizing the spectral bandwidth and data-sampling rate, replacing old lamps, and using mobile phases that cause little baseline disturbance.

Signal can be increased by creating a concentrated sample extract (attained with high filtration volumes and low extraction volumes), increasing the injection volume, using chromatographic conditions that yield sharp, narrow peaks, and using detector wavelengths near the pigments' absorbance maxima.

The effects of HPLC separation conditions and detector settings on sensitivity are illustrated in Figure 20.2, for which the same sample extract was analyzed on the same HPLC and detector settings but with two different separation methods. The method using a $C_{30}$ column (Fig. 20.2a) resulted in broad, short peaks and a Chl *a* S : N ratio about four times less than that observed with the method using a $C_8$ column (Fig. 2b). Data were acquired from two different wavelengths during the analysis on the $C_8$ column. The Chl *a* S : N ratio at 665 nm (Fig. 20.2c) was almost four times greater than the S : N observed at 450 nm (Fig. 20.2b). Comparing Chl *a* response at 450 and 665 nm was done here merely to show the effects of chromatographic conditions and wavelength on S : N ratios, and 450 nm is not commonly used or recommended for Chl *a* quantification.

### 2.5.3. Chromatographic Separations

Pigment separations are typically conducted with use of reverse-phase conditions and HPLC columns packed with stationary phases having an aliphatic chain length of $C_8$, $C_{18}$, or $C_{30}$ (Jeffrey et al. 1999, Van Heukelem and Thomas 2001). Column stationary phases also vary with regard to such specifications as particle size and shape, end capping, bonding phase chemistry, and pore size. Hence, columns of the same aliphatic chain length but from different manufacturing processes should not be expected to yield similar results. Columns also vary with respect to internal diameter and length. Nevertheless, some generalizations are possible regarding column performance. For example, MV and DV Chl $a$ are easily separated on $C_8$ and $C_{30}$ columns but not on $C_{18}$ columns, and carotene pigments are easily separated on $C_{18}$ and $C_{30}$ columns but not on $C_8$ columns. The exaggerated separations attainable with $C_{30}$ columns prove useful for unique applications, such as pigment isolations (Van Heukelem and Thomas 2001), but analysis times with $C_{30}$ columns are prohibitively long if routine separations of complex mixtures are required. Pigment separations and resolution can also be manipulated by column temperature, mobile phase composition, and gradient shape (e.g., whether the gradient is shallow, steep, or segmented).

The choice of HPLC method should be based on the analysis objectives and adequate separation between key pigments. Separation, or resolution ($R_s$), between pigments is calculated on the basis of peak widths at half-height or baseline. The following formula uses peak width at baseline:

$$R_s = \frac{2(t_{R2} - t_{R1})}{w_{B1} + w_{B2}} \tag{1}$$

where $t_{R1}$ and $t_{R2}$ are the retention times (minutes) of peaks 1 and 2, and $w_{B1}$ and $w_{B2}$ are the widths (minutes) of peaks 1 and 2 at their respective bases (Wright 1997).

Pigments unimportant to the objectives need not be separated. For example, a method that separates as many pigments as possible wastes resources if quantification of only Chl $a$–related pigments is required. Other factors to consider when choosing a method include costs associated with solvent purchase and disposal, solvent toxicity, method sensitivity, column availability and cost, reproducibility, and whether existing HPLC equipment is sufficient.

After selecting a HPLC method, test it for reproducibility by comparing values for key separation parameters (e.g., retention time and resolution between pigments). Method reproducibility is improved by keeping the following parameters constant: (1) column temperature, (2) column re-equilibration conditions between injections, (3) procedures for solvent preparation, and (4) injection conditions. Reproducibility declines with extended column use, as peak broadening naturally occurs and resolution worsens. When methods are transferred between laboratories, reproducibility can be adversely affected by differences in HPLC configurations pertaining to dwell volume (Snyder et al. 1988), detector flow cell volume, column temperature controller (Wolcott et al. 2000), injector type, and pumping capabilities. Reproducibility between laboratories is more difficult with methods involving complex mobile phases and segmented gradients. It is therefore necessary to validate the performance of an HPLC method, even if validation studies have been done previously in a different laboratory.

### 2.5.4. Pigment Identification

HPLC-separated pigment peaks are routinely identified by (1) comparison of retention time ($t_R$) values with those of standards (e.g., DHI Water & Environment) and extracts prepared from phytoplankton cultures of known pigment composition (Jeffrey et al. 1997) and (2) "on-line" diode array UV/VIS spectroscopy. In the latter, absorption spectra (350–750 nm), absorption maxima (nm), and carotenoid (% III : II) and cyclic tetrapyrrole (Soret : red) band ratios can be compared with published values (see Part IV in Jeffrey et al. 1997) and/or spectral libraries prepared "in-house" with use of pigment standards. Pigment identifications can be confirmed by performing liquid chromatography/mass spectrometry (LC/MS, Fig. 20.1) and comparing the resulting mass spectra with published fragment ion abundances and/or mass spectral libraries prepared in-house with pigment standards. In the case of unknown peaks, the use of on-line diode array UV/VIS spectroscopic and LC/MS/MS analyses can provide valuable optical data and structural information to aid in pigment identification (Goericke et al. 2000, Chen et al. 2001, Koblízek et al. 2003).

### 2.5.5. Pigment Quantification

Pigment quantification is based on the linear relationship between the weight of standard injected and the resulting peak height or area, the latter of which is less sensitive to small variations in chromatographic conditions and therefore more often used. With $R_s \geq 1.5$, adjacent pigments are baseline resolved and can be accurately quantified, regardless of their relative concentrations. However, if adjacent pigment concentrations are dissimilar and $R_s < 1.5$, large inaccuracies may occur when quantifying the more dilute pigment, especially if it elutes second in the pigment pair (Dolan 2002). Pigment chromatograms are often very complex, and it is unrealistic to expect that $R_s \geq 1.5$ will be possible with all pigments. At a minimum, however, $R_s$ should be $\geq 1.0$ for quantification by peak area, and quantification by peak height may be necessary for pigments that are not as well resolved (Snyder and Kirkland 1979). It is possible to quantify two co-eluting pigments if a multiwavelength absorption detector is used and if the two pigments respond differently at two dissimilar wavelengths, as described in Latasa et al. (1996) for the co-eluting pigments MV and DV Chl $a$ and in Hooker et al. (2000) for Chl $c_1$ and chlorophyllide $a$.

HPLC pigment calibrations begin with *primary pigment standards*, whose concentrations are very concentrated and determined spectrophotometrically using published extinction coefficients. The chromatographic purity of these standards should be assessed by analyzing them without dilution, for which findings often reveal the presence of pigment isomers at 10% or less of the total peak area. Primary pigment standards may exceed the detector's LOL, in which case they are diluted for use as HPLC *calibration standards*.

Calibration standards of various concentrations (usually $n = 5$) are analyzed to derive multipoint calibration curves, the correlation coefficients and $y$ intercepts of which should be near 0.999 and zero, respectively (Claustre et al. 2004). Once it has been demonstrated that the pigment standard exhibits the same response factor over all concentrations expected in samples, single-point calibration can be used. Knowledge of the limit of detection (LOD) and limit of quantification (LOQ) gives assurance to the analyst that low-level concentrations are useful, because injected amounts <LOD are not significantly different from zero, and those between the LOQ and LOL provide quantitatively meaningful results (Taylor 1987). LOD and LOQ can be defined in various ways, one of which requires measuring the mean ± standard deviation ($\bar{\chi} \pm s$ for the weight of standard injected) from replicate injections ($n = 7$) of the same low-concentration pigment solution. LOD ($3s$) and LOQ ($10s$) (Taylor 1987) values are influenced by method sensitivity and uncertainties associated with the formulation of calibration standards but are unaffected by other processes (e.g., sample storage, filtration, and extraction).

Short-term variability in HPLC performance is described by daily injections of QC standards, which are analyzed along with samples and whose concentrations are known. The percent difference between the concentration measured by the current calibration factors and the known concentration reflects stability of analyses on that day. The average percent difference $\pm 2s$ approximates the 95% confidence limits within which such daily measurements should fall. QC standard results that fall outside these "control limits" are considered "out of control," and corrective action should be taken to find the cause (Clesceri et al. 1998, Taylor 1987).

Long-term variability better describes reproducibility of the measurement process (Taylor 1987), and in HPLC pigment analysis it reflects uncertainties associated with spectrophotometric determination of primary pigment standard concentrations, formulation of calibration standards, and HPLC analyses, as affected by different batches of solvent, extended column use, and columns of varying serial numbers. If the "control limits" (average response factor $\pm 2s$) for various pigment standards are known, new batches of standards whose response factors fall outside these limits can be immediately identified and addressed before sample analysis begins.

## 2.6. Data Reporting

The ease with which data are transferred between analysts and data managers is facilitated by consistency in pigment nomenclature and measurement units and by reporting method detection limits (MDLs). Confusion regarding pigment nomenclature is easily avoided by using abbreviations given in Appendix A of Jeffrey et al. (1997). Pigment concentrations are usually reported in $\mu g \cdot L^{-1}$ or $mg \cdot m^{-3}$. In the JGOFS protocols, concentrations are reported as $ng \cdot Kg^{-1}$ (UNESCO 1994). In physiology studies, wherein pigment concentrations are related to cell density, $pg \cdot cell^{-1}$ is often used. How to report pigments of low concentration or those not found is a bit more complex. For certain, if the analyst has the ability to quantify a pigment and it is not found in a sample set, to omit that pigment from the report would cause the loss of valuable information.

Reporting data based on results less than the MDL has little purpose and introduces large uncertainties into a database. An important aspect of data reporting is to clarify what the MDL is for that laboratory, how it was calculated, and whether results less than the MDL are reported, reported and flagged, or merely reported as "not found." Inconsistencies exist in the literature regarding terminology that describes uncertainties for analytes of low concentration (Clesceri et al. 1998). MDL can describe uncertainties that arise strictly from instrumental conditions and calibrations and is analogous to the LOD (Section 2.5.5), or the MDL can describe uncertainties that arise from all aspects of the analytical process, which also includes sample collection, filtration, storage, and extraction. Thus, MDL for the analytes of interest can be determined from seven replicate injections of a standard mixture containing pigments in low concentrations or from seven replicate filters containing pigments in low concentrations (Taylor 1987, Glaser et al. 1981). For accuracy when calculating MDL, it is recommended that analyte concentrations not be greater than 5 times the MDL (Clesceri et al. 1998) or 20 times the standard deviation (Taylor 1987). The standard deviation of the seven replicate measurements ($S_c$) is calculated, and the MDL is computed as

$$MDL = t(6, 0.99)S_c,　(2)$$

where $t(6, 0.99)$ is the Student's $t$ value for a one-tailed test at the 99% confidence level, with $(N - 1) = 6$ degrees of freedom. For this particular sample size ($n = 7$) and the 99% confidence level, $t(6, 0.99) = 3.707$ (Abramowitz and Segun 1968, Table 26.10).

It is recommended that MDLs be calculated (Bidigare et al. 2003) and data less than the MDL be flagged, but in practice this is not done. Hence, unbeknownst to the data manager, values not significantly different from zero are given undue significance. With the complexities of MDL determinations and the many pigments to be quantified, it is not surprising that analysts have not applied MDL thresholds. The authors suggest a pragmatic alternative: to use the weight of injected pigment that results in an Signal : Noise of 10 as a proxy for MDL. This approach was tested by extracting seven replicate filters and comparing the MDL of four carotenoids that varied between 0.1 and 0.3 ng per injection, with the weight that resulted in an S : N of 10 that varied between 0.3 and 0.6 ng per injection (Van Heukelem, unpublished). The MDL and the proxy never differed by more than 0.3 ng. In all, the MDL proxy was measured for 14 carotenoids and varied between 0.3 and 0.6 ng per injection. We suggest that it may be practical to group carotenoids by peak height response factor and structural similarities, whereby the MDL for one pigment could be used as a proxy for related compounds. These groupings would be expected to vary with the HPLC method and detector settings employed. The MDL proxy describes only instrumental characteristics and does not reflect extraction precision error, which was 5% with the tests described above.

## 3.0. QUALITY ASSURANCE

It is beyond the scope of this chapter to describe all aspects of a quality assurance (QA) plan for pigment analysis. Two important aspects of a QA plan are (1) to ensure that QC measurements are conducted and properly recorded so that the degree of accuracy in sample analysis is quantified and (2) to delineate a means by which all aspects of sample handling and analysis are traceable. Pigment concentrations in natural samples cannot truly be known; therefore, it is not possible to ascertain if natural sample results are accurate. Instead, QC measurements demonstrate that the HPLC method is capable of producing accurate results and that uncertainties associated with various procedures are within expected limits.

Various QC measurements were described in Section 2.0 and are representative of what might be included in a QA plan. In Figure 20.3 these measurements are summarized as they apply to various steps involved in measuring pigment concentrations. Performance criteria are given merely as a guide for analysts tracking QC measurements for the first time or attempting to ascertain whether the performance of their HPLC method is typical of that attainable by others. These criteria were largely compiled from results of laboratories that participated in intercalibration exercises (Hooker et al. 2000, Van Heukelem et al. 2002) involving procedures similar to those described in this chapter. A common QC practice is to analyze all HPLC extracts fluorometrically (Holm-Hansen et al. 1965, Strickland and Parsons 1972) and compare results for TChl $a$ (Trees et al. 2003). These two analytical methods are not truly independent because they both rely on spectrophotometric measurements to determine the concentration of the MV Chl $a$ standard with which they are calibrated.

The use of appropriate reference materials (RMs) *should* be a key feature of the QA/QC structure in scientific investigations involving chemical measurements. Furthermore, the accuracy of chemical oceanographic measurements depends on calibration against RMs to ensure comparability over time and among laboratories. RMs can also be used to facilitate the development of new analytical methods and interpret results obtained from interlaboratory comparisons (Fig. 20.4). Unfortunately, several key parameters, including phytoplankton pigments, lack RMs for measurements in seawater, particles in the water column, and sediments. A recent report published by the National Research Council (2002) identified the most urgently required

| PROCESSING STEP | QUALITY CONTROL MEASUREMENT | PERFORMANCE CRITERIA |
|---|---|---|

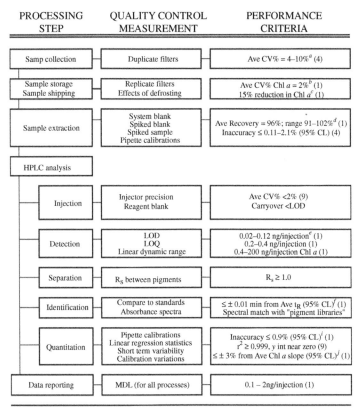

| | | |
|---|---|---|
| Samp collection | Duplicate filters | Ave CV% = 4–10%[a] (4) |
| Sample storage Sample shipping | Replicate filters Effects of defrosting | Ave CV% Chl a = 2%[b] (1) 15% reduction in Chl a[c] (1) |
| Sample extraction | System blank Spiked blank Spiked sample Pipette calibrations | Ave Recovery = 96%; range 91–102%[d] (1) Inaccuracy ≤ 0.11–2.1% (95% CL) (4) |
| HPLC analysis | | |
| Injection | Injector precision Reagent blank | Ave CV% <2% (9) Carryover <LOD |
| Detection | LOD LOQ Linear dynamic range | 0.02–0.12 ng/injection[e] (1) 0.2–0.4 ng/injection (1) 0.4–200 ng/injection Chl a (1) |
| Separation | $R_s$ between pigments | $R_s \geq 1.0$ |
| Identification | Compare to standards Absorbance spectra | ≤ ± 0.01 min from Ave $t_R$ (95% CL)[f] (1) Spectral match with "pigment libraries" |
| Quantitation | Pipette calibrations Linear regression statistics Short term variability Calibration variations | Inaccuracy ≤ 0.9% (95% CL)[i] (1) $r^2 \geq 0.999$, y int near zero (9) ≤ ± 3% from Ave Chl a slope (95% CL)[j] (1) |
| Data reporting | MDL (for all processes) | 0.1 – 2ng/injection (1) |

[a]The range describes average precision of key pigments in duplicate *in situ* filters from 11 sites.
[b]17 replicate filters were analyzed at intervals spanning 10 months. Filters were stored at –80°C.
[c]Chl a in duplicate sets of filters was compared (each duplicate set contained replicate filters from 12 sites). One entire set had been defrosted in transit (Hooker et al. submitted).
[d]Pigments evaluated were Chl $c_2$, Peri, Fuco, Diad, Zea, Lut, Chl b, β, β car, and Chl a.
[e]Calculated as in Section 2.5.5 from a standard mixture containing 17 pigments ranging from 1.5 to 5.6 ng/injection with S:N ratios from 30 to 85 (accessory pigments) and 140 (Chl a). LOD and LOQ based on standard deviation of peak heights, with approximate S:N ratios of 3 and 10, respectively.
[f]Retention time variability seen with 7 replicate injections on the same day.
[i]Class A glass volumetric pipettes (3, 5, 10 mL) accurate to within 0.1% and gas-tight glass syringes, 0.9%.
[j]Data acquired from 38 multipoint Chl a calibration curves analyzed over 2.5 years.

**FIGURE 20.3.** A summary of quality control measurements associated with various pigment analysis procedures. The performance criteria shown were primarily derived from results of intercalibration exercises (Hooker et al. 2000, Van Heukelem et al. 2002, Hooker et al. submitted) by laboratories whose procedures were consistent with those described in this chapter; the numbers of laboratories providing such data are in parentheses. A blank line indicates that no data were available (CV%, [standard deviation · mean⁻¹] × 100%; $R_s$, resolution; $t_R$, retention time; LOD, limits of detection; LOQ, limits of quantification; MDL, minimum detection limit).

chemical RMs on the basis of key themes for oceanographic research. For the biological matrices, it was recommended that diatom (*Thalassiosira pseudonana* [Hustedt] Hasle et Heimdal), dinoflagellate (*Scrippsiella trochoidea* [Stein] Loeblich III), and haptophyte (*Emiliania huxleyi* [Lohman] Hay et Mohler) mass cultures be developed as RMs to improve analytical performance. It was further recommended that these biological matrix materials should be used for the preparation of mixed pigment standards to aid in peak identification and to facilitate interlaboratory comparisons.

## 4.0. IMPLEMENTING AN HPLC METHOD

It is important to select an HPLC method that meets analysis objectives, yet different methods can yield similar results, as shown in an intercalibration exercise amongst four laboratories (Hooker et al. 2000), in which methods of Wright et al. (1991), Vidussi et al. (1996), Barlow et al. (1997), and Van Heukelem and Thomas (2001) were used with field samples. Chromatograms representing two of these methods depict the analysis of standard mixtures, in Figures 20.2b and c (Van

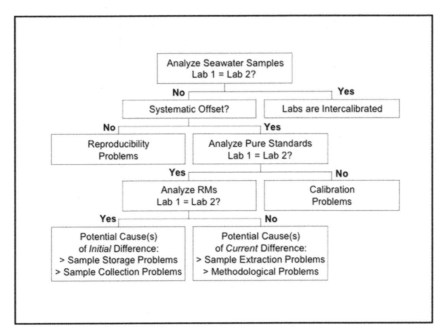

**FIGURE 20.4.** Flow chart for diagnosing interlaboratory differences. When comparing two sets of measurements, two systematic offsets are possible: constant offset and constant multiplier. Simple statistical comparisons (mean differences, mean ratios) are incapable of resolving which effect is present, as often both are operating to some degree. Model-2 linear regression techniques, such as Pearson's major axis (Pearson 1901) or the geometric mean regression (Ricker 1973), can resolve this issue. Note that reproducibility problems can be corrected with the use of an internal standard (see text for details).

Heukelem and Thomas 2001) and Figure 20.5 (Wright et al. 1991). Regardless of method, HPLC performance should be validated before and during use, and the method should be carefully implemented, details of which are discussed here with the method of Wright et al. (1991) as an example. Many such details are common to all methods. Notable exceptions are that three solvent systems are used with the Wright et al. (1991) method and MV and DV Chl *a* co-elute (a frequent occurrence with methods based on $C_{18}$ columns). This source of analytical uncertainty can be resolved by taking advantage of the different spectral signatures displayed by these pigments in the blue region of the visible spectrum. Response factors for these Chl *a*–related pigments are determined at two wavelengths (436 and 450 nm) by monitoring the absorption signal at each wavelength during separate injection of pure standards of these co-eluting pigments. These response factors are then used in conjunction with dichromatic equations to spectrally resolve mixtures of these co-eluting pigments:

$$A_1 = \varepsilon_{x1}C_x + \varepsilon_{y1}C_y \tag{3}$$

$$A_2 = \varepsilon_{x2}C_x + \varepsilon_{y2}C_y \tag{4}$$

where $\varepsilon_{x1}, \ldots, \varepsilon_{y2}$ are the reciprocal of the response factors calculated for pigments X and Y at $\lambda_1$ and $\lambda_2$, and

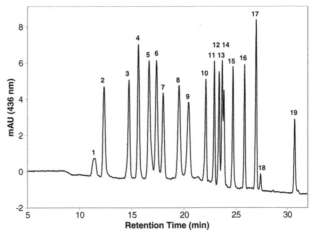

**FIGURE 20.5.** Representative reverse-phase chromatogram (436 nm) obtained from the analysis of a mixture of pigment standards by means of the HPLC method of Wright et al. (1991). Pigment code numbers correspond to the following: 1, chlorophyll $c_3$; 2, chlorophyll $c_1$ plus $c_2$; 3, peridinin; 4, 19′-butanoyloxy fucoxanthin; 5, fucoxanthin plus pheophorbide *a*; 6, neoxanthin; 7, 19′-hexanoyloxy fucoxanthin; 8, prasinoxanthin; 9, violaxanthin; 10, diadinoxanthin; 11, alloxanthin; 12, diatoxanthin; 13, lutein; 14, zeaxanthin; 15, canthaxanthin; 16, monovinyl chlorophyll *b* plus divinyl chlorophyll *b*; 17, monovinyl chlorophyll *a* plus divinyl chlorophyll *a*; 18, monovinyl plus divinyl chlorophyll *a*′; 19, β, β carotene.

$A_1$ and $A_2$ are the respective peak areas at those wavelengths. Solving these two simultaneous equations allows the calculation of individual concentrations of the co-eluting pigments, $C_x$ and $C_y$, from a single chromatographic analysis as follows:

$$C_x = (\varepsilon_{y2}A_1 - \varepsilon_{y1}A_2)(\varepsilon_{y2}\varepsilon_{x1} - \varepsilon_{y1}\varepsilon_{x2})^{-1} \quad (5)$$

$$C_y = (\varepsilon_{x1}A_2 - \varepsilon_{x2}A_1)(\varepsilon_{x1}\varepsilon_{y2} - \varepsilon_{x2}\varepsilon_{y1})^{-1} \quad (6)$$

This approach was evaluated by Latasa et al. (1996), who reported excellent agreement between this spectral approach and a physical separation of both pigments with use of a reverse-phase $C_8$ HPLC technique. However, the results from simultaneous equation calculations can be spurious when DV Chl $a$ is absent in natural samples. The SeaHARRE-2 intercalibration exercise, for example, revealed that laboratories utilizing simultaneous equations reported small quantities of DV Chl $a$ while laboratories using the $C_8$ HPLC technique reported none (Hooker et al. submitted).

## 4.1. Equipment and Supplies

The reader is referred to Figure 20.1 for review of the HPLC components. Components discussed here are required to implement the method of Wright et al. (1991). It should be noted that the Wright et al. (1991) technique was chosen as an example for HPLC implementation since it represents one of the most thoroughly evaluated pigment methods to date (Jeffrey et al. 1997).

### 4.1.1. HPLC Components

- Three pumps or one pump with three-way or four-way proportioning valve.
- High-pressure injector valve (with 200 µL sample loop) or an automated injector.
- Variable-wavelength, filter absorbance, or photodiode array detector that can monitor two wavelengths (436 and 450 nm) simultaneously. Bandwidth and slit width should be optimized to maximize response factor differences between MV and DV Chl $a$.
- Data recording device: a strip chart recorder or preferably an electronic integrator and computer equipped with hardware and software for chromatographic data analysis.
- Column temperature controller (optional) to facilitate reproducibility.

- Analytical HPLC column: reverse-phase with end capping (250 mm × 4.6 mm, 5 µm particle size, ODS-2 Spherisorb $C_{18}$ column).
- Guard column to extend life of analytical column: 50 mm × 4.6 mm with same stationary phase as analytical column.

### 4.1.2. Spectrophotometer

- Spectrophotometer with bandwidth, slit width, and optical resolution ≤2 nm (see discussion in Latasa et al. 1996, Clesceri et al. 1998, Latasa et al. 1999, Dunne 1999, and Reber 1997).
- Cuvettes: glass or quartz, 1 cm path length, preferably not less than 3 mL.
- NIST traceable neutral-density filters for validating spectrophotometer accuracy (optional, but recommended; see discussion in Latasa et al. 1999).

### 4.1.3. Dilution Devices

- Electronic balance that weighs to nearest µg for calibrating dilution devices.
- Class A glass volumetric pipettes (suggested volumes: 2, 3, 4, 5, and 10 mL).
- Class A glass volumetric flasks (suggested volumes: 25 and 50 mL).
- Gas-tight glass syringes (suggested volumes: 100, 250, 500, and 1,000 µL).
- Dilution devices are calibrated by weighing (at 25°C) the volume of 100% acetone delivered (repeat seven times). This weight is corrected for the specific gravity of acetone to calculate the precise volume delivered. One hundred percent acetone is used instead of water because some measuring devices are accurate with water but not with organic solvents. One hundred percent acetone is used instead of 90% acetone because the specific gravity of the former is known. (If methanol is the extraction solvent, it would be used to calibrate measuring devices.) Calculate the 95% confidence limits for expected accuracy.

### 4.1.4. Other Supplies

- Injector syringe for manual injector: 500 µL.
- Tabletop centrifuge (optional).
- HPLC syringe cartridge filters: PTFE membranes, may have glass fiber prefilters, diameter suitable for volumes filtered (3–25 mm), and acetone-compatible syringes for use with HPLC cartridge filters.
- HPLC vials and storage bottles for standards (glass or PTFE; nylon bottles cause pigment losses).

## 4.2. Reagents

- Reagents include HPLC-grade acetone (for pigment extraction) and HPLC-grade water, methanol, acetonitrile, and ethyl acetate; 0.5 M ammonium acetate aq. (pH = 7.2); and BHT (2,6-di-tert-butyl-p-cresol, Sigma Chemical Co.) for HPLC.

### 4.2.1. HPLC Solvents

- Solvent A (80 : 20, by volume, methanol : 0.5 M ammonium acetate aq.; pH = 7.2; 0.01% BHT, [w/v]), solvent B (87.5 : 12.5, by volume, acetonitrile : water; 0.01% BHT, [w/v]), and solvent C (ethyl acetate). Solvents A and B contain BHT to prevent the formation of Chl *a* allomers. Use HPLC-grade solvents. Measure volumes before mixing. Filter solvents through a solvent-resistant 0.4 μm filter before use.

### 4.2.2. Primary Pigment Standards

- Chlorophylls *a* and *b* and β, β-carotene can be purchased from Sigma Chemical Co. Other pigment standards can be purchased in solution with concentrations provided from the International Agency for [14]C Determination, DHI Water & Environment.
- Concentrations of standards prepared by the analyst are determined spectrophotometrically with use of published extinction coefficients found in Appendix E of Jeffrey et al. (1997). The absorbance of the pigment standards should be within 0.2 to 0.8 at $\lambda_{max}$ (Marker et al. 1980) for maximal accuracy.
- Absorbance is measured in a 1-cm cuvette at the standard's wavelength $\lambda_{max}$ (given with the published extinction coefficient) and at 750 nm to correct for light scattering. Concentrations are calculated as

$$C_{STD}^i = \frac{10^6[A^i(\lambda_{max}^i) - A^i(750)]}{bE_{1cm}^i}, \qquad (7)$$

where $C_{STD}^i$ is the concentration (μg L−1) of the standard for pigment I; $A^i(\lambda_{max}^i)$ and $A^i(750)$ are absorbances at $\lambda_{max}$ and 750 nm, respectively; $b$ is the path length of the cuvette (cm); and $E_{1cm}^i$ is the weight-specific absorption coefficient (L g$^{-1}$·cm$^{-1}$) of pigment $i$ (see Clesceri et al. 1998). Concentrations of standards are relatively stable for 1 month if stored under nitrogen in the dark at −20°C in containers that are proven to prevent evaporation (e.g., some glass or PTFE bottles or vials).

### 4.2.3. Calibration and Quality Control Standards

- Calibration standards are prepared by diluting the primary pigment standards with 90% acetone with use of calibrated dilution devices. Single-point response factors can be used as long as LOLs have previously been established, response factors are linear at all concentrations between the LOQ and LOL, and concentrations of samples fall within these ranges. Nevertheless, it is prudent to always do multipoint calibrations (*n* = 5) when calibrating for MV Chl *a*.
- QC standards are prepared like calibration standards but may include several pigments and are prepared independently from calibration standards.
- Retention-time mixtures contain pigments that will be found in samples so that retention times for those pigments can be updated daily. They need not be formulated quantitatively.

## 4.3. Procedures

It may require more than 1 day to complete all injections needed for calibration and method validation. It is typical to conduct these analyses on the day before sample analysis begins. In the following examples, it is assumed that calibration and method validation injections are already complete.

### 4.3.1. Initiate Start-Up

- Fill solvent reservoirs; turn on HPLC components (including autosampler chiller, solvent degasser, pumps, detector lamps).
- Prime pumps; bring flow to initial conditions slowly (keep pressure within column and HPLC tolerances). Turn on column oven compartment (if used).
- Check available disk space or paper in chart recorder.
- Check for leaks. Place buffer, needle-wash, and sample vials in chilled autosampler compartment of programmable injection system.
- Load method file: flow, 1 mL·min$^{-1}$; column temperature, ambient. To achieve desired separations, the published gradient times may need to be changed for HPLC systems with differing dwell volumes. Examples of gradient times and proportions of solvents A, B, and C used by R. R. B. are shown (Table 20.1).

**TABLE 20.1.** Solvent gradient used by R. R. B. as adapted from Wright et al. (1991).

| Time (min) | Solvent proportions | | | Activity (flow rate ) mL · min⁻¹ |
|---|---|---|---|---|
| | %A | %B | %C | |
| 0.0 | 90 | 10 | 0 | Start analysis (1.0) |
| 1.0 | 0 | 100 | 0 | Pigment separation (1.0) |
| 11.0 | 0 | 78 | 22 | Pigment separation (1.0) |
| 27.5 | 0 | 10 | 90 | Pigment separation (1.0) |
| 29.0 | 0 | 100 | 0 | Pigment separation (1.0) |
| 30.0 | 0 | 100 | 0 | Analysis complete (1.0) |
| 31.0 | 95 | 5 | 0 | Return to initial conditions (1.8) |
| 37.0 | 95 | 5 | 0 | Re-equilibration to initial conditions |
| 38.0 | 90 | 10 | 0 | (1.0) |

### 4.3.2. Plan Injection Sequence

- Plan sequence of injections, such as equilibration injection (data discarded), retention time mixture, QC standard(s), internal standard, 10–20 samples. Repeat: QC standard, internal standard, 10–20 samples, etc. Reinject retention time mixture if retention time shifts occur.

### 4.3.3. Start Analyses

- Manual injector (for 200-µL sample loop): mix 1,000 µL of sample or standard with 300 µL of distilled water (use calibrated measuring devices), shake, and equilibrate for 5 minutes before injecting. Rinse 500-µL injection syringe twice with 300 µL of injection mixture. Draw 500 µL into the syringe, place the syringe in the injector valve, and inject, overfilling the loop 2.5-fold.
- Auto injector (not programmable): mix sample (or standard) and water as for manual injections. Place HPLC vial with sample mixture in position and inject. When analysis is complete, repeat for second injection and so forth. Do not allow samples or standards premixed with water to sit for long durations because poor injector precision will result (see Section 2.5.1).
- Fully programmable automated injection systems: start sequence after all preparation steps are complete, instrument is "ready," and all vials have equilibrated to temperature of the refrigerated auto sample compartment.

- After the initial retention time mixture and QC standard analyses are complete, inspect for correct peak shapes, expected retention times, $R_s$ between peaks, and expected accuracy. Then finish remaining injections.

### 4.3.4. Reprocess Data

- Update retention times in calibration table. To determine pigment identities, overlay chromatograms of each sample with the retention time mixture. If in doubt, confirm identities by spectrophotometric analysis (350–750 nm) by collecting eluting peaks from the column outlet (or directly with an on-line diode array spectrophotometer). Absorption maxima for the various phytoplankton pigments can be found in Part IV of Jeffrey et al. (1997).
- Select integration functions that yield correct baselines for each peak (each analysis should be integrated individually, as frequent adjustments to integration functions may be required). Integrate samples and standards similarly.

### 4.3.5. Quantify Results

- For each pigment $i$, plot absorbance peak areas (arbitrary system units) against working standard pigment masses (concentrations multiplied by injection volume). The HPLC system response factor $F^i$ (area · µg⁻¹) for pigment $i$ is calculated as the slope of the regression of the peak areas of the parent pigment (plus areas of peaks for structurally related isomers if present) against the pigment masses of the injected working standards (µg). Structurally related isomers (e.g., Chl $a$ allomer) contribute to the absorption signal of the standards, and disregarding them will result in the overestimation of analytes in sample extracts (Bidigare 1991).
- If the extraction solvent was amended with internal standard (IS; e.g., canthaxanthin) before addition to the filter, calculate individual pigment concentration as follows:

$$C^i_{Sample} = \frac{A^i_{Sample} V_{Added} A^{Cantha}_{Extraction\ Solvent}}{F^i V_{Injected} V_{Sample} A^{Cantha}_{Sample}}, \qquad (8)$$

where $C^i_{Sample}$ is the individual pigment concentration (µg · L⁻¹), $A^i_{Sample}$ is the area of individual pigment peak for a sample injection, $V_{Added}$ is the volume of extraction solvent added (mL, to nearest 0.1 mL), $V_{Injected}$ is the volume injected (mL, measured to the nearest 0.001 mL), $V_{Sample}$ is the sample volume filtered (liters, measured to the nearest 0.001 liter), $A_{\text{Extraction Solvent}}^{\text{Cantha}}$ is the area of the canthaxanthin peak in the extraction solvent, $A_{\text{Sample}}^{\text{Cantha}}$ is the area of the canthaxanthin peak in the sample, and $F^i$ is the HPLC system response factor (area $\cdot \mu g^{-1}$).

- If the IS was added as a spike (e.g., 0.050 mL canthaxanthin IS), calculate individual pigment concentration as follows:

$$C_{\text{Sample}}^i = \frac{A_{\text{Sample}}^i V_{\text{IS}} A_{\text{IS}}^{\text{Cantha}}}{F^i V_{\text{Injected}} V_{\text{Sample}} A_{\text{Sample}}^{\text{Cantha}}}, \qquad (9)$$

where $C_{\text{Sample}}^i$ is the individual pigment concentration ($\mu g \cdot L^{-1}$), $A_{\text{Sample}}^i$ is the area of individual pigment peak for a sample injection, $V_{\text{IS}}$ is the volume of the IS spike added to the sample (mL, measured to the nearest 0.001 mL), $V_{\text{Injected}}$ is the volume injected (mL, measured to the nearest 0.001 mL), $V_{\text{Sample}}$ is the sample volume filtered (liter, measured to the nearest 0.001 liter), $A_{\text{IS}}^{\text{Cantha}}$ is the area of the canthaxanthin peak in the IS spike solution, $A_{\text{Sample}}^{\text{Cantha}}$ is the area of the canthaxanthin peak in the sample, and $F^i$ is the HPLC system response factor (area $\cdot \mu g^{-1}$).

### 4.3.6. Assess Uncertainties

- Estimate injector precision error by calculating coefficient of variation (% CV: [standard deviation $\cdot$ mean$^{-1}$] $\times$ 100%) from triplicate injections of a standard or sample extract. Typical % CV values for injector precision error with this HPLC method (assessed with manual injections of sample extracts) ranged from 0.6 to 6.0% (R. R. B.).
- Compare response factors to control limits for calibration variance.
- Calculate CV% for pigments in duplicate filters.
- Calculate accuracy of QC injections and compare to control limits.

## 5.0. FUTURE DIRECTIONS

Recent studies have identified the presence of novel bacterial phototrophs in coastal and oceanic waters. These include proteorhodopsin-containing bacteria (Béjà et al. 2000, 2001) and aerobic anoxygenic phototrophic bacteria (Kolber et al. 2000, 2001). Sequence analysis of BAC clone libraries prepared from Monterey Bay, Station ALOHA, and the Southern Ocean revealed that numerous uncultivated members of the γ-proteobacteria contain genes that code for prote-

orhodopsin. This membrane-bound pigment contains *trans*-retinal, absorbs at blue-green to green wavelengths, and functions as a light-driven proton pump. In an unrelated study, Kolber et al. (2000) used an infrared fast repetition rate (IRFRR) fluorometer to document the widespread occurrence of aerobic anoxygenic phototrophs (AAPs) in the world oceans. These microbes possess low amounts of bacteriochlorophyll *a* ($\lambda_{\text{max}}$ = 358, 581, and 771 nm) and unusually high levels of bacteriocarotenoids ($\lambda_{\text{max}}$ = 454, 465, 482, and 514 nm). They require molecular oxygen for growth. One of us (R. R. B.) has initiated HPLC pigment analysis of these latter clones and retinal-related compounds to determine if the Wright et al. (1991) method can be used for their separation and quantification.

## 6.0. ACKNOWLEDGMENTS

This work was supported by NASA grant NAG5-9757 (R. R. B.); NSF grants OCE-9617409 (R. R. B.) and EEC-9731725 (R. R. B.); the NASA SIMBIOS Project (L. V. H.) and the SeaWiFS field program (L. V. H.); and original efforts by NSF US JGOFS, with refinements/improvements by NASA's SIMBIOS Project. The authors acknowledge a series of NASA technical memorandums that revised and updated these pigment protocols for ocean color validation. They are grateful to Louise Schlüter (DHI Water & Environment) for Figure 20.5 and Edward T. Peltzer (Monterey Bay Aquarium Research Institute), who provided valuable input for Figure 20.4. This is US-JGOFS Contribution No. 1024, SOEST Contribution No. 6282, and University of Maryland Center for Environmental Science Contribution No. 3693.

## 7.0. REFERENCES

Abramowitz, A., and Segun, I. A. 1968. *Handbook of Mathematical Functions*. Dover, New York, 1046 pp.

Andersen, R. A., Bidigare, R. R., Keller, M. D., and Latasa, M. 1996. A comparison of HPLC pigment signatures and electron microscopic observations for oligotrophic waters of the North Atlantic and Pacific Oceans. *Deep-Sea Res. II* 43:517–37.

Barlow, R. G., Cummings, D. G., and Gibb, S. W. 1997. Improved resolution of mono- and divinyl chlorophylls *a* and *b* and zeaxanthin and lutein in phytoplankton extracts using reverse phase C-8 HPLC. *Mar. Ecol. Prog. Ser.* 161:303–7.

Béjà, O., Aravind, L., Koonin, E. V., Suzuki, M. T., Hadd, A., Nguyen, L. P., Jovanovich, S. B., Gates, C. M., Feldman, R. A., Spudich, J. L., Spudich, E. N., and DeLong, E. F. 2000. Bacterial rhodopsin: Evidence for a new type of phototrophy in the sea. *Science* 289:1902–6.

Béjà, O., Spudich, E. N., Spudich, J. L., LeClerc, M., and DeLong, E. F. 2001. Proteorhodopsin phototrophy in the ocean. *Nature* 411:786–9.

Bianchi, T. S., Lambert, C., and Biggs, D. C. 1995. Distribution of chlorophyll *a* and pheopigments in the northwestern Gulf of Mexico: A comparison between fluorometric and high-performance liquid chromatography measurements. *Bull. Mar. Science* 56:25–32.

Bidigare, R. R. 1991. Analysis of algal chlorophylls and carotenoids. In: Hurd, D. C., and Spencer, D. W., eds. *Marine Particles: Analysis and Characterization.* American Geophysical Union, Washington, DC, pp. 119–23.

Bidigare, R. R., and Ondrusek, M. E. 1996. Spatial and temporal variability of phytoplankton pigment distributions in the central equatorial Pacific Ocean. *Deep-Sea Res. II* 43:809–33.

Bidigare, R. R., Van Heukelem, L., and Trees, C. C. 2003. HPLC phytoplankton pigments: Sampling, laboratory methods, and quality assurance procedures. In: Mueller, J. L., Fargion, G. S., and McClain, C. R., eds. *Ocean Optics Protocols for Satellite Ocean Color Sensor, Revision 4, Volume V.* NASA Technical Memorandum 2003-211621. NASA Goddard Space Flight Center, Greenbelt, Maryland, pp. 5–14.

Brock, T. D. 1983. *Membrane Filtration: A User's Guide and Reference Manual.* Science Technology, Madison, Wisconsin, 381 pp.

Chavez, F., Buck, K. R., Bidigare, R. R., Karl, D. M., Hebel, D., Latasa, M., Campbell, L., and Newton, J. 1995. On the chlorophyll *a* retention properties of glass-fiber GF/F filters. *Limnol. Oceanogr.* 40:428–33.

Chen, N., Bianchi, T. S., McKee, B. A., and Bland, J. M. 2001. Historical trends of hypoxia on the Louisiana shelf: Application of pigments as biomarkers. *Org. Geochem.* 32:543–61.

Claustre, H., Hooker, S. B., Van Heukelem, L., Berthon, J.-F., Barlow, R., Ras, J., Sessions, H., Targa, C., Thomas, C. S., van der Linde, D., and Marty, J.-C. 2004. An intercomparison of HPLC phytoplankton methods using *in situ* samples: Application to remote sensing and database activities. *Mar. Chem.* 85:41–61.

Clesceri, L. S., Greenberg, A. E., and Eaton, A. D., eds. 1998. Part 10000, Biological Examination, Section 1020 B. In: *Standard Methods for the Examination of Water and Wastewater.* 20th ed. American Public Health Association, American Water Works Association, Water Environment Federation, Baltimore, Maryland.

Dickson, M.-L., and Wheeler, P. A. 1993. Chlorophyll *a* concentrations in the North Pacific: Does a latitudinal gradient exist? *Limnol. Oceanogr.* 38:1813–8.

Dolan, J. W. 2001. Autosampler carryover. *LC/GC.* 19:164–8.

Dolan, J. W. 2002. Resolving minor peaks. *LC/GC.* 20:594–8.

Dunne, R. P. 1999. Spectrophotometric measurement of chlorophyll pigments: A comparison of conventional monochromators and a reverse optic diode array design. *Mar. Chem.* 66:245–51.

Gibb, S. W., Barlow, R. G., Cummings, D. G., Rees, N. W., Trees, C. C., Holligan, P., and Suggett, D. 2000. Surface phytoplankton pigment distribution in the Atlantic: An assessment of basin scale variability between 50°N and 50°S. *Progr. Oceanogr.* 45:339–68.

Glaser, J. A., Foerst, D. L., McKee, G. D., Quave, S. A., and Budde, W. L. 1981. Trace analyses for wastewaters. *Environ. Sci. Technol.* 15:1426–35.

Goericke, R., and Repeta, D. J. 1993. Chlorophylls *a* and *b* and divinyl chlorophylls *a* and *b* in the open subtropical North Atlantic Ocean. *Mar. Ecol. Prog. Ser.* 101:307–13.

Georicke, R., Olson, R. J., and Shalapyonok, A. 2000. A novel niche for *Prochlorococcus* sp. in low-light suboxic environments in the Arabian Sea and the Eastern Tropical North Pacific. *Deep-Sea Res. I* 47:1183–205.

Gordon, H. R., and Clark, D. K. 1980. Remote sensing optical properties of a stratified ocean: An improved interpretation. *Appl. Optics* 19:3428–30.

Hoepffner, N., and Sathyendranath, S. 1992. Bio-optical characteristics of coastal waters: Absorption spectra of phytoplankton and pigment distribution in the western North Atlantic. *Limnol. Oceanogr.* 37:1660–79.

Holm-Hansen, O., Lorenzen, C. J., Holmes, R. W., and Strickland, J. D. H. 1965. Fluorometric determination of chlorophyll. *J. du Cons. Intl. Pour l'Expl. de la Mer.* 30:3–15.

Hooker, S. B., Claustre, H., Ras, J., Van Heukelem, L., Berthon, J.-F., Targa, C., van der Linde, D., Barlow, R., and Sessions, H. 2000. *The First SeaWiFS HPLC Analysis Round-Robin Experiment (SeaHARRE-1).* NASA Technical Memorandum 2000-206892. Vol. 14 (Hooker, S. B., and Firestone, E. R., eds.). NASA Goddard Space Flight Center, Greenbelt, Maryland, 42 pp.

Hooker, S. B., Van Heukelem, L., Thomas, C. S., Claustre, H., Ras, J., Barlow, R., Sessions, H., Schluter, L., et al. 2004. The Second SeaWiPS HPLC Analysis Round-Robin Experiment (SeaHARRE-2). NASA Goddard Space Flight Center, Greenbelt, Maryland, submitted.

Jeffrey, S. W., and Humphrey, G. F. 1975. New spectrophotometric equations for determining chlorophylls *a*, *b*, $c_1$ and $c_2$ in higher plants, algae and natural phytoplankton. *Biochem. Physiol. Pflanzen* 167:191–4.

Jeffrey, S. W., Mantoura, R. F. C., and Wright S. W., eds. 1997. *Phytoplankton Pigments in Oceanography*. Monographs on Oceanographic Methodology. UNESCO Publishing, Paris, 661 pp.

Jeffrey, S. W., and Vesk, M. 1997. Introduction to marine phytoplankton and their pigment signatures. In: Jeffrey, S. W., Mantoura, R. F. C., and Wright, S. W., eds. *Phytoplankton Pigments in Oceanography: Guidelines to Modern Methods*. Vol. 10, Monographs on Oceanographic Methodology. UNESCO Publishing, Paris, pp. 37–84.

Jeffrey, S. W., Wright, S. W., and Zapata, M. 1999. Recent advances in HPLC pigment analysis of phytoplankton. *Mar. Freshwater Res.* 50:879–96.

Koblízek, M., Béjà, O., Bidigare, R. R., Christensen, S., Benitez-Nelson, B., Vetriani, C., Kolber, M. K., Falkowski, P. G., and Kolber, Z. S. 2003. Isolation and characterization of *Erythrobacter* sp. strains from the upper ocean. *Arch. Microbiol.* 180:327–38.

Kolber, Z. S., Van Dover, C. L., Niederman, R. A., and Falkowski, P. G. 2000. Bacterial photosynthesis in surface waters of the open ocean. *Nature* 407:177–9.

Kolber, Z. S., Plumley, F. G., Lang, A. S., Beatty, J. T., Blankenship, R. E., VanDover, C. L., Vetriani, C., Koblízek, M., Rathgeber, C., and Falkowski, P. G. 2001. Contribution of aerobic photoheterotrophic bacteria to the carbon cycle in the ocean. *Science* 292:2492–5.

Latasa, M., Bidigare, R. R., Ondrusek, M. E., and Kennicutt II, M. C. 1996. HPLC analysis of algal pigments: A comparison exercise among laboratories and recommendations for improved analytical performance. *Mar. Chem.* 51:315–24.

Latasa, M., Bidigare, R. R., Ondrusek, M. E., and Kennicutt II, M. C. 1999. On the measurement of pigment concentrations by monochromator and diode-array spectrophotometers. *Mar. Chem.* 66:253–4.

Latasa, M., van Lenning, K., Garrido, J. L., Scharek, R., Estrada, M., Rodriguez, F., and Zapata, M. 2001. Losses of chlorophylls and carotenoids in aqueous acetone and methanol extracts prepared for RPHPLC analysis of pigments. *Chromatographia*. 53:385–91.

Letelier, R. M., Bidigare, R. R., Hebel, D. V., Ondrusek, M. E., Winn, C. D., and Karl, D. M. 1993. Temporal variability of phytoplankton community structure at the U.S.-JGOFS time-series Station ALOHA (22°45′N, 158°W) based on HPLC pigment analysis. *Limnol. Oceanogr.* 38:1420–37.

MacIntyre, H. L., Kana, T. M., and Geider, R. J. 2000. The effect of water motion on short-term rates of photosynthesis by marine phytoplankton. *Trends Plant Sci.* 5:12–17.

Mantoura, R. F. C., Barlow, R. G., and Head, E. J. H. 1997a. Simple isocratic HPLC methods for chlorophylls and their degradation products. In: Jeffrey, S. W., Mantoura, R. F. C.,

and Wright, S. W., eds. *Phytoplankton Pigments in Oceanography: Guidelines to Modern Methods*. Vol. 10, Monographs on Oceanographic Methodology. UNESCO Publishing, Paris, pp. 307–26.

Mantoura, R. F. C., Wright, S. W., Jefrrey, S. W., Barlow, R. G., and Cummings, D. E. 1997b. Filtration and storage of pigments from microalgae. In: Jeffrey, S. W., Mantoura, R. F. C., and Wright, S. W., eds. *Phytoplankton Pigments in Oceanography: Guidelines to Modern Methods*. Vol. 10, Monographs on Oceanographic Methodology. UNESCO Publishing, Paris, pp. 283–306.

Marker, A. F. H., Nusch, E. A., Rai, H., and Riemann, B. 1980. The measurement of photosynthetic pigments in freshwaters and standardization of methods: Conclusion and recommendations. *Arch. Hydrobiol. Beih. Ergebn. Limnol.* 14:91–106.

National Research Council. 2002. *Chemical Reference Materials: Setting the Standards for Ocean Science*. National Academies Press, Washington, D.C., 130 pp.

Pearson, K. 1901. On lines and planes of closest fit to systems of points in space. *Phil. Mag.* 2:559–72.

Phinney, D. A., and Yentsch, C. S. 1985. A novel phytoplankton chlorophyll technique: Toward automated analysis. *J. Plankton Res.* 7:633–42.

Reber, C., 1997. The effect of fluorescence on measured absorbance with diode-array and conventional spectrophotometers. Hewlett Packard Application Note #5965-9988E.

Ricker, W. E. 1973. Linear regressions in fishery research. *J. Fish. Res. Board Can.* 30:409–34.

Smith, R. C., Bidigare, R. R., Prezelin, B. B., Baker, K. S., and Brooks, J. M. 1987. Optical characterization of primary productivity across a coastal front. *Mar. Biol.* 96:575–91.

Snyder, L. R., Glajch, J. L., and Kirkland, J. J. 1988. Gradient elution. In: Kirkland, J. J. *Practical HPLC Method Development*. John Wiley & Sons, New York, pp. 153–78.

Snyder, L. R., and Kirkland, J. J., eds. 1979. *Introduction to Modern Liquid Chromatography*, John Wiley & Sons, New York, pp. 541–74.

Strickland, J. D. H., and Parsons, T. R. 1972. *A Practical Handbook of Sea Water Analysis*. 2nd Ed. Fisheries Research Board of Canada, Ottawa, Canada, 310 pp.

Suzuki, R., and Fujita, Y. 1986. Chlorophyll decomposition in *Skeletonema costatum*: A problem in chlorophyll determination of water samples. *Mar. Ecol. Prog. Ser.* 28:81–5.

Taylor, J. K. 1987. *Quality Assurance of Chemical Measurements*. Lewis Publishers, Boca Raton, Florida, 328 pp.

Tester, P. A., Geesey, M. E., Guo, C., Paerl, H. W., and Millie, D. F. 1995: Evaluating phytoplankton dynamics in the Newport River estuary (North Caroline, USA) by HPLC-derived pigment profiles. *Mar. Ecol. Prog. Ser.* 124:237–45.

Thomas, C. S., and Van Heukelem, L. 2003. Vitamin E acetate as an internal standard for use with pigment analysis by high performance liquid chromatography. *The Earth's Eyes: Aquatic Sciences Through Space and Time.* American Society of Limnology and Oceanography, Salt Lake City, Utah.

Trees, C. C., Kennicutt II, M. C., and Brooks, J. M. 1985. Errors associated with the standard fluorometric determination of chlorophylls and pheopigments. *Mar. Chem.* 17:1–12.

Trees, C. C., Bidigare, R. R., Karl, D. M., Van Heukelem L., and Dore J. 2003. Fluorometric chlorophyll *a*: sampling, laboratory methods, and data analysis protocols. In: Mueller, J. L., Fargion, G. S., and McClain, C. R., eds. *Ocean Optics Protocols for Satellite Ocean Color Sensor, Revision 4, Volume V.* NASA Technical Memorandum 2003-211621. NASA Goddard Space Flight Center, Greenbelt, Maryland, pp. 15–25.

Trees, C. C., Clark, D. C., Bidigare, R. R., Ondrusek, M. E., and Mueller, J. L. 2000. Accessory pigments versus chlorophyll *a* concentrations within the euphotic zone: A ubiquitous relationship. *Limnol. Oceanogr.* 45:1130–43.

UNESCO. 1994. *Protocols for the Joint Global Ocean Flux Study (JGOFS) core measurements.* Intergovernmental Oceanographic Commission, Scientific Committee on Oceanic Research, Manual and Guides, 29, UNESCO, Paris, 170 pp.

Van Heukelem, L., and Thomas, C. S. 2001. Computer-assisted high-performance liquid chromatography method development with applications to the isolation and analysis of phytoplankton pigments. *J. Chromatogr. A.* 910:31–49.

Van Heukelem, L., Thomas, C. S., and Glibert, P. M. 2002. Sources of variability in chlorophyll analysis by fluorometry and high-performance liquid chromatography in a SIMBIOS inter-calibration exercise. In: Fargion, G. S., and McClain, C. R., eds. *NASA Technical Memorandum 2002-211606.* NASA Goddard Space Flight Center, Greenbelt, Maryland, 50 pp.

Vidussi, F., Claustre, H., Bustillos-Guzman, J., Cailliau, D., and Marty, J. C. 1996. Determination of chlorophylls and carotenoids of marine phytoplankton: Separation of chlorophyll *a* from divinyl-chlorophyll *a* and zeaxanthin from lutein. *J. Plankton Res.* 18:2377–82.

Wright, S. W., Jeffrey, S. W., Mantoura, R. F. C., Llewellyn, C. A., Bjornland, T., Repeta, D., and Welschmeyer, N. 1991. Improved HPLC method for the analysis of chlorophylls and carotenoids from marine phytoplankton. *Mar. Ecol. Prog. Ser.* 77:183–96.

Wright, S. W. 1997. Summary of terms and equations used to evaluate HPLC chromatograms. Appendix H. In: Jeffrey, S. W., Mantoura, R. F. C., and Wright, S. W., eds. *Phytoplankton Pigments in Oceanography: Guidelines to Modern Methods.* Vol. 10, Monographs on Oceanographic Methodology. UNESCO Publishing, Paris, pp. 622–30.

Wright, S. W., Jeffrey, S. W., and Mantoura, R. F. C. 1997. Evaluation of methods and solvents for pigment extraction. In: Jeffrey, S. W., Mantoura, R. F. C., and Wright, S. W., eds. *Phytoplankton Pigments in Oceanography: Guidelines to Modern Methods.* Vol. 10, Monographs on Oceanographic Methodology. UNESCO Publishing, Paris, pp. 261–82.

Wright, S. W., and Mantoura, R. F. C. 1997. Guidelines for selecting and setting up an HPLC system and laboratory. In: Jeffrey, S. W., Mantoura, R. F. C., and Wright, S. W., eds. *Phytoplankton Pigments in Oceanography: Guidelines to Modern Methods.* Vol. 10, Monographs on Oceanographic Methodology. UNESCO Publishing, Paris, pp. 383–406.

Wolcott, R. G., Dolan, J. W., Snyder, L. R., Bakalyar, S. R., Arnold, M. A., and Nichols, J. A. 2000. Control of column temperature in reversed-phase liquid chromatography. *J. Chromatogr. A.* 869:211–30.

Zapata, M., Rodriquez, F., and Garrido, J. L. 2000. Separation of chlorophylls and carotenoids from marine phytoplankton: a new HPLC method using a reversed phase $C_8$ column and pyridine-containing mobile phases. *Mar. Ecol. Prog. Ser.* 195:29–45.

# CHAPTER 21

# ENDOGENOUS RHYTHMS AND DAYLENGTH EFFECTS IN MACROALGAL DEVELOPMENT

KLAUS LÜNING
*Alfred Wegener Institute for Polar and Marine Research*

## CONTENTS

Key Index Words: Algae, Cell Division, Circadian, Circannual, Semilunar, Gene Expression, Growth, Macroalgae, Photoperiodism, Photosynthesis

## 1.0. INTRODUCTION

This chapter describes methods for visualizing the action of endogenous clocks and environmental day length in cultures of macroalgae. The day-night cycle and the succession of the seasons are the dominant environmental oscillators shaping the conditions for life in our world. Endogenous cell clocks prepare, for example, the algal cell before dawn for photosynthesis and before dusk for various reactions that occur preferably during night such as cell division or growth. Mutant plants of

*Arabidopsis* that had lost the clock were less viable under very short-day conditions than their wild-type counterparts, probably because they could not anticipate the coming of the short-light phase, so as to capture the first light at dawn (Green et al. 2002). Period mutants of cyanobacteria outcompeted wild type when grown in a light-dark cycle corresponding to the mutant period in a light-dark cycle, but not when grown on a 24-hour cycle (Ouyang et al. 1998). The action of an endogenous clock becomes visible to the observer as an endogenous rhythm continuing under constant conditions in the laboratory, for example, with a period of approximately 24 hours (circadian rhythm), 2 weeks (biweekly or circa-semilunar rhythm), or almost a year (circannual rhythm). Circatidal rhythms, with a period of 12.4 hours, were found in several marine intertidal animals (Neumann 1981, Palmer 1995), but so far not in any algal species except for some tidal phenomena (Sweeney and Hastings 1962, Sweeney 1987). Cell cycles may be cited as examples for ultradian rhythms, with periods shorter than 1 day (Hastings et al. 1991).

Knowledge about endogenous rhythms is important for the algal cultivator. For example, temporal minima of algal activities during the day, month, or year may simply reflect the minima of endogenous rhythms, and in such cases, it might be wise to wait for or to be able to predict the next maximum.

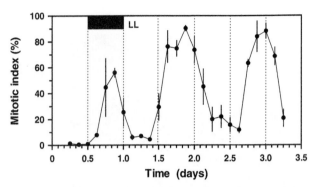

**FIGURE 21.1.** Rhythm of mitotic index in the top portion of *Kappaphycus* in continuous light (LL; 200 ± 50 μmol photons · m⁻² · s⁻¹). Mitotic index denotes frequency of all mitotic stages, from prophase to telophase. Vertical bars indicate standard deviation (n = 10 microscope eye fields). Day 0 represents the last day of the preceding light-dark cycle (LD 12 : 12), day 1 the start of LL. Dark phases are shaded. (From Schubert et al. 2004.)

rhythm (e.g., $\tau = 23$ hours or $\tau = 25$ hours), or T in case of the entrained rhythm (e.g., T = 24 hours). A *phase* is an arbitrarily chosen reference point or part in a cycle, for example, the day phase or the night phase, or the maximum (acrophase). *Subjective day* and *subjective night* designate the phases corresponding to the preceding light and dark phases in a LD cycle (Fig. 21.1).

## 1.1. Terminology

Chronobiologists have developed their own terminology, which can be found in introductory chapters in the textbooks or reviews by Aschoff (1981), Hastings et al. (1991), Sweeney (1987), Lumsden and Millar (1998), or Dunlap et al. (2003). A few major terms are explained here. A *free-running rhythm* is an endogenous rhythm recorded under constant conditions, for example, in constant light (LL). A rhythm recorded under a light-dark (LD) regimen, for example, LD 12 : 12, is called an *entrained rhythm*, because the endogenous rhythm is entrained or *synchronized* by the environmental LD rhythm. During entrainment, the free-running period (see following discussion) of the circadian rhythm is reset to exactly the 24-hour period of the LD regimen. The synchronizing environmental factor, in this example the light-dark cycle, is called a *zeitgeber*.

The *period* is the time to complete one cycle, for example, the time distance from one maximum to the next, and is designated $\tau$ in the case of a free-running

## 2.0. CIRCADIAN RHYTHMS

Circadian rhythms of various organismic activities have been reported to occur in a wide variety of eukaryotic taxa ranging from unicellular organisms to mammals and were discovered in cyanobacteria among prokaryotes (Aschoff 1981, Hastings et al. 1991, Lumsden and Millar 1998, McClung 2001, Suzuki and Johnson 2001, Johnson and Kondo 2001). The ubiquity of circadian rhythmicity among algae mainly has been demonstrated and reviewed for unicellular algal species, with rhythmic phenomena such as phototaxis, timing of cell division, photosynthetic capacity, bioluminescence, gene expression, or sensitivity to ultraviolet (UV) radiation (Sweeney 1987, Hastings et al. 1991, Johnson et al. 1998, Johnson 2001, Mittag 2001, Suzuki and Johnson 2001), although the present chapter deals with examples in macroalgae. Circadian rhythms are lost under certain conditions, such as in constant light, in bright light, or when *Euglena* is grown on organic medium (Sweeney 1987, Hastings et al. 1991).

Endogenous circadian rhythms are produced by a molecular feedback loop (Dunlap 1999, Somers 1999) and entrained to precisely 24 hours predominantly by environmental LD cycles by means of positive or negative phase shifts according to the phase-response curve (PRC) of a particular response (Sweeney 1987, Hastings et al. 1991). The clock-controlled anticipation of dawn and dusk has been demonstrated by high-density oligonucleotide microarrays in the higher plant *Arabidopsis* and revealed orchestrated transcription of key pathway genes (Harmer et al. 2000). For example, a large cluster of genes encoding proteins implicated in photosynthesis oscillated with messenger RNA (mRNA) levels rising well before dawn and peaking at midday, while genes involved in growth processes such as cell wall relaxation, expansion, and reinforcement by cellulose peaked in the evening or early night.

## 2.1. Circadian Cell Division

### 2.1.1. Test Organisms

Shifting nuclear and cell division into the night and control of this process by circadian rhythmicity is well known for many algal species including *Oedogonium* sp. (Bühnemann 1955), *Mougeotia* sp. (Neuscheler-Wirth 1970), dinoflagellates (Sweeney and Hastings 1958), juvenile sporophytes of the Laminariales (Makarov et al. 1995), the green alga *Ulva pseudocurvata* Koeman et Hoek (Titlyanov et al. 1996), or the red alga *Porphyra umbilicalis* (L.) J. Agardh (Lüning et al. 1997). According to the "escape from light" hypothesis by Pittendrigh (1993), an initial driving force for the early evolution of circadian clocks could have been the necessity to transfer sunlight-sensitive processes such as the UV-sensitive DNA replication into the night, especially in view of much higher UV doses on the surface of the earth during early evolution of life, with UV-C still present.

### 2.1.2. Equipment

*Cultivation equipment:* Aerated 2-liter glass beakers filled with Provasoli's enriched seawater (ES medium) (Starr and Zeikus, 1987; see Chapter 3, Appendix A).

*Lighting:* Cool white fluorescent light at 25 to 50 $\mu mol \cdot m^{-2} \cdot s^{-1}$, except for *Kappaphycus alvarezii*, which requires high light, for example, from a halogen lamp (Osram Power Star HQI-TS 150 W/NDL Neutral White) at 200 $\mu mol \cdot m^{-2} \cdot s^{-1}$.

*Hardware:* Zeiss standard photomicroscope equipped with an epi-illumination system (HBO 50-W mercury lamp, Zeiss filter combination 487702, 60× oil immersion objective).

### 2.1.3. Procedures

*Precultivation and cultivation:* Algal thalli cultured in the laboratory or collected in the field should be precultivated for 1 to 2 weeks for optimal synchronization (entrainment) in LD cycles, for example, with 12 hours of light per day, and then transferred to continuous white light (LL) or continuous darkness (DD). The medium should be changed weekly.

*DAPI staining:* Samples for determination of mitotic frequency should be harvested at regular intervals, for example, 3 hours, during the last LD cycle and subsequent LL or DD (see Fig. 21.1). For each determination of mitotic frequency with 4',6-diamidino-2-phenylindole (DAPI) (Hull et al. 1982, Goff and Coleman 1984), three algal pieces of a few millimeters in length should be harvested at a particular time. For juvenile sporophytes of Laminariales, the following procedure was successful (Makarov et al. 1995). The algal samples are fixed in 3 : 1 (95% ethanol : acetic acid) for 12 hours and then soaked in distilled water for 10 minutes, softened for 10 minutes in 5% (w/v) sodium EDTA solution, and heated on a microscope slide for a few seconds in a drop of 1% Triton-X 100 in phosphate-buffered saline (PBS). The samples are transferred to fresh PBS for 10 minutes and then stained for 30 minutes with 0.5 $\mu g \cdot mL^{-1}$ DAPI in McIlvaine's pH 4.0 buffer. Buffer stock solutions (Hull et al. 1982): (A) 21.01 g citric acid monohydrate $\cdot L^{-1}$ (0.1 M) and (B) 28.4 g anhydrous $Na_2HPO_4 \cdot L^{-1}$ (0.2 M). For pH levels of 4.0, add 61.4 mL of (A) to 38.6 mL of (B). Abbreviated procedures were successful for *Ulva* or *Porphyra* (Titlyanov et al. 1996, Lüning et al. 1997) by staining minidisks for 5 to 10 minutes directly in 0.5 $\mu g \cdot mL^{-1}$ DAPI in McIlvaine's pH 4.0 buffer after fixation for 1 minutes in 20 mL of distilled water in a 1,400-W microwave oven at 80% of its maximum power (*Ulva, Porphyra*) or fixation in boiling seawater for some seconds (*Kappaphycus*).

*Detecting mitotic stages:* The algal pieces stained with DAPI can be immediately scored for mitotic stages using a Zeiss standard microscope equipped with an epi-illumination system for observation of fluorescence excited by the beam of an HBO 50-W mercury lamp. A Zeiss filter combination 487702 giving a peak excitation in the range 400–500 nm should be used, with a ×100 oil immersion objective. For the counting procedure, minipieces with clearly recognizable nuclei should be selected, and 10 microscopic eye fields (e.g., each 300–500 cells) should be scored.

### 2.1.4. Results

The mitotic index designates the frequency of all mitotic stages at a given time and includes all mitotic stages, from prophase to telophase. Standard deviation can be based on percentages determined from the 10 eye fields. The resulting time curve of the mitotic index may look similar to Fig. 21.1, with the mitotic index peaking around midnight during the preceding LD cycle and a subsequent free run in LL with maxima during subjective night. A first glance at the free-running rhythm presented in Fig. 21.1 reveals that its period, τ, is longer than 24 hours, because the time dis-tance between the two maxima is greater than 1 day on the x-axis. There are programs for exact determination of τ (see Section 3.3), but more than two free-running cycles are required for that purpose.

### 2.2. Circadian Photosynthesis Rhythms

Circadian rhythms of photosynthesis in algae were mainly detected at higher irradiances. This is in contrast to the common experience that many circadian rhythms in animals and plants tend to disappear in bright light and turn into continuous arrhythmic activities (Aschoff 1981). Rhythms of photosynthetic capacity were found, for example, in white light in unicellular algae such as *Lingulodinium polyedrum* (Stein) Dodge (formerly called *Gonyaulax polyedra* Stein) (Hastings et al. 1961) or *Euglena* sp. (Walther and Edmunds 1973) or in red light in the filamentous brown alga *Ectocarpus* sp. (Schmid et al. 1992). A more recent example is the tropical carrageenophytic red macroalga *Kappaphycus alvarezii*, with a circadian rhythm of photosynthetic oxygen production in white continuous light (LL) throughout the range of 100 to 1,000 µmol photons · m$^{-2}$ · s$^{-1}$, but not at 40 µmol photons · m$^{-2}$ · s$^{-1}$ (Granbom et al. 2001).

### 2.2.1. Test Organisms

Among multicellular algae, circadian rhythms of photosynthesis have been detected in continuous white light. For example, such rhythms were found in the green algae *Acetabularia acetabulum* (L.) Silva (Driessche 1966), *Ulva* sp. (Britz and Briggs 1976), the brown alga *Spatoglossum pacificum* Yendo (Kageyama et al. 1979), and the red alga *Porphyra yezoensis* Ueda (Oohusa 1980). A very robust rhythm exists in *Kappaphycus alvarezii* (Granbom et al. 2001) for which the procedure is described in this chapter. *Kappaphycus* and the related *Eucheuma* are economically important carrageeno-phytes (Doty 1973, Pérez 1992). The test organisms should be collected from nature.

### 2.2.2. Equipment

*Cultivation equipment:* Aerated 10-liter glass bottles for batch culture or aerated 100-liter plastic tanks of various dimensions for flow-through culture.
*Lighting:* Halogen lamp (e.g., Osram Power Star HQI-TS 150 W/NDL Neutral White)
*Hardware:* Oxygen electrode and oxygen meter (e.g., WTW CellOx 325 meter; WTW Wissenschaftlich-Technische Werkstätten). PC with data acquisition card and data recording program. Projector

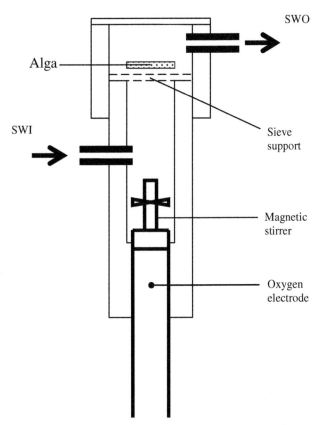

**FIGURE 21.2.** Oxygen-measuring chamber mounted on an oxygen electrode equipped with a stirrer (not to scale). SWI, seawater inlet; SWO, seawater outlet. (From Granbom et al. 2001.)

(24 V 250 W quartz-iodine lamp). Quantum sensor (Licor).

### 2.2.3. Procedures

*Precultivation and cultivation:* This should be performed for *Kappaphycus* according to Section 2.1.3, using a halogen lamp providing an irradiance of approximately 200 μmol photons · m$^{-2}$ · s$^{-1}$ at the water surface.

*Oxygen measurements:* Oxygen exchange rates are measured in a closed system by mounting a thallus branch of approximately 300 mg fresh weight by means of a thin nylon filament onto a sieve support within a cylindrical Plexiglas measuring chamber (see Fig. 21.2) with a volume of approximately 20 mL. The chamber can be fixed on top of a WTW CellOx 325 oxygen electrode tightened with an O-ring and a WTW magnetic stirrer attachment secures turbulence for the electrode (WTW Wissenschaftlich-Technische Werkstätten). A submersible magnetic pump (e.g., Eheim, type 1048; Eheim) pumps seawater from a 1-liter reservoir bottle through the measuring chamber for 5 minutes every hour. A security valve should prevent water flow through the measuring chamber during the hourly 55-minute period of oxygen measurement. The equipment described so far should be immersed in a circular Plexiglas container (e.g., 30-cm diameter, 40-cm high) filled with tap water to secure thermic stability, either in a constant temperature room or by use of a thermostat. The electrode is connected to a WTW OXI 597 oxygen meter, allowing automatic calibration of the electrode in moist air at the temperature of the experiment, with less than 1% deviation from the 100% saturation value in stirred seawater of the same temperature.

*Irradiation procedure:* The measuring chamber is illuminated from above, via a mirror, with light from a projector (24 V 250-W quartz-iodine lamp) run at a constant voltage. Underwater light, poor in red and rich in blue, may be imitated (Lüning 1980) by using the Schott filter 2-mm BG38 (Schott). Neutral-density filters can be used in addition for obtaining the required irradiances. Irradiance can be measured by means of a flat quantum sensor (Licor).

*Data recording and evaluation:* The oxygen measuring values can be transferred, for example, every 2 minutes to a computer via a 14-bit AD/DA card. Any data recording program may be used (e.g., MedeaLab by Dr. Kurt Vogel using linear regression for calculation of oxygen concentration at fixed time intervals, e.g., 30 minutes, and graphing oxygen exchange rates per hour from the successive differences of calculated oxygen concentrations; Medea/AV). For calculation of absolute rates of net photosynthesis or respiration, the oxygen content of air-equilibrated seawater can be obtained from tables in Green and Carritt (1967).

### 2.2.4. Results

A typical indicator for an oncoming free-running rhythm of photosynthesis can be seen already during the initial LD 12 : 12 cycle, with increases toward noon and decreases by about one third during the afternoon (Fig. 21.3a–d; days 0–1). In subsequent continuous light, the circadian rhythm should become obvious, in *Kappaphycus* at irradiances ranging from 200 to 1,000

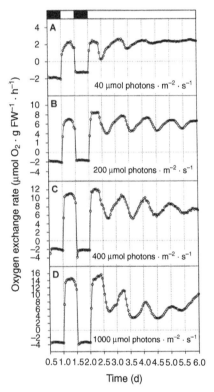

**FIGURE 21.3.** Oxygen exchange kinetics in the red alga *Kappaphycus alvarezii* in cool white fluorescent light at LD 12 : 12 (days 0–1) and in subsequent constant light (LL; days 2–5), at 40 **(a)**, 200 **(b)**, 400 **(c)**, or 1,000 μmol photons · m$^{-2}$ · s$^{-1}$ **(d)**. Filled blocks indicate dark phases. Position of numbers on the abscissa refer to 0600 hours and start of light phase during the preceding light-dark regimen of LD 12 : 12. The experiments were performed at 23°C. The light intensity during the light phase of LD was the same as in LL. (From Granbom et al. 2001.)

$\mu mol \cdot m^{-2} \cdot s^{-1}$ (Fig. 21.3b–d), whereas at $40 \, \mu mol \cdot m^{-2} \cdot s^{-1}$, the rhythm attained a very low amplitude after one cycle (Fig. 21.3a).

## 2.3. Circadian Growth Rhythms

When do algae grow preferably, during day or night? The answer to this question can be found by monitoring area expansion of flat macroalgal thalli in LD cycles using computer-assisted image analysis. With this approach, algae such as *Porphyra* sp. (Fig. 21.4) or *Ulva* sp. were found to behave as typical "night growers" and in subsequent LL, as controlled by circadian growth rhythmicity (Titlyanov et al. 1996, Lüning 2001). The circadian clock controls growth activity also in brown algae (Lüning 1994, Makarov et al. 1995), and the ubiquity of circadian growth rhythms in macroalgae may reflect a general pattern, particularly in view of the aforementioned findings of genes involved in growth processes peaking in early night in the higher plant *Arabidopsis* (Harmer et al. 2000).

### 2.3.1. Test Organisms

Suitable macroalgal species for observing circadian growth rhythms in LL are the red algae *Porphyra umbilicalis* (Lüning et al. 1997, Lüning 2001), and *Palmaria palmata* (L.) O. Kuntze (Lüning 1992), as well as juvenile laminarian sporophytes (Lüning 1994, Makarov et al. 1995), and in constant darkness (DD) for a few

cycles, the green alga *Ulva pseudocurvata* Koeman et van den Hoek (Titlyanov et al. 1996). The test organisms should be collected from nature.

### 2.3.2. Equipment

*Lighting:* Culture light: projector (24 V 250-W quartz-iodine lamp); measuring radiation: infrared diodes (>850 nm)

*Hardware:* Translucent algal growth chamber in dark cabinet (self-made according to Lüning 1992); black-and-white charged coupled device (CCD) camera (various types available) with macro objective (e.g., Ernitec MS18Z6); frame grabber card (e.g., Matrox Meteor-II) in a PC; image analysis software program (e.g., MedeaLab; Medea/AV GmbH)

### 2.3.3. Procedures

*Precultivation and cultivation:* The same procedures are recommended as scheduled in 2.1.3.

*Measuring and irradiation procedures:* The measuring cabinet (Fig. 21.5) should be installed in a dark constant temperature room. A thallus piece (approximately $20 \times 20$ mm) is cut from the growing part of a blade thallus precultivated in LD cycles and speared on a 1-mm thick Plexiglas rod in the middle of the lower plate of a translucent algal growth chamber (15-cm diameter; Fig. 21.5) consisting of two Plexiglas plates separated by an O-ring or a flat soft PVC sealing. The algal growth chamber is contained in a translucent water bath constantly supplied with flowing seawater from an immersible pump in a seawater reservoir on the floor beside the measuring cabinet (Fig. 21.5). Culture light (e.g., imitated underwater light using Schott filter 2-mm BG38; see Section 2.2.3) is supplied from the side by a projector and guided to the alga by a semitransparent mirror (Fig. 21.5). The radiation from nine infrared diodes (>850 nm) arranged within a grid ($15 \times 15$ cm) at a distance of 60 cm above the algal growth chamber (Fig. 21.5) passes through a semitransparent diffusing screen (e.g., Röhm and Haas Plexiglas GS White No. 017) at 30 cm and through the semitransparent mirror used for the projector light at 10 cm. The infrared image of the alga is produced by the macro objective of a black and white CCD camera. An infrared cutoff filter at 850 nm (e.g., Schott glass filter RG 850) on top of the macro objective secures that no visible light enters the CCD camera and the same radiation is available during light and dark

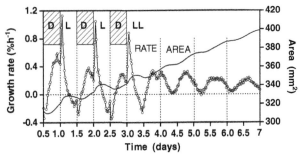

**FIGURE 21.4.** *Porphyra umbilicalis.* **(a)** Growth kinetics at LD 12 : 12 (days 0–3) in green light at 15 $\mu mol \cdot m^{-2} \cdot s^{-1}$ (Schott glass filter combination SFK 11) and in subsequent continuous green light (days 3–7). Hatched blocks indicate dark phases. Thallus area (———) measured every 2 minutes; relative growth rate (o———o) derived for successive 30-minute intervals from thallus area measurements. Day 0 represents the day of transfer from precultivation to measurement conditions. Position of numbers on the abscissa refers to 0600 hours and start of light phase during preceding light-dark regimen. (From Lüning 2001.)

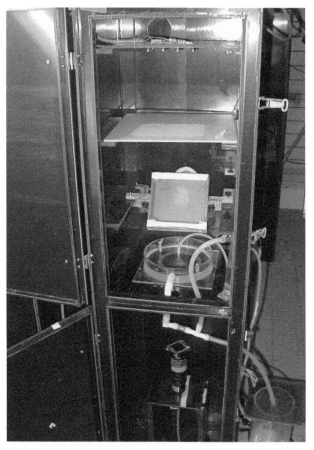

**FIGURE 21.5.** Measuring cabinet for growth measurements by means of image analysis. A circular translucent algal growth chamber within a translucent water bath contains a piece of a flat algal thallus illuminated by culture light from a projector via a semitransparent mirror (*middle*). The radiation from infrared diodes (*top*) passes a diffusing plate, the semitransparent mirror, and produces an infrared image of the algal piece recorded by a CCD camera (*bottom*) coupled to an image analysis system. See text for further details.

phases. For image digitizing, the CCD camera is coupled to a frame grabber card in a PC with an image analysis software program (e.g., MedeaLab). If the thallus area is recorded at intervals of, say, 30 minutes (or averaged for 30 minutes from 1-minute measurements), it should be possible to obtain safe relative growth rates (R), without too much background noise. R is calculated according to $R = (\ln A_{30}) - (\ln A_0)\, 2$, where $A_{30}$ is the area at the measuring time, and $A_0$ is the area 30 minutes earlier. Percentage increase per hour is regarded as equal to 100 R (Lüning 1992).

*Chronobiological analysis:* Chronobiological parameters such as the period of free-running rhythms,

circadian acrophase (maximum), or phase shifts for constructing a PRC can be determined by means of computer programs such as Bioclock (Refinetti 1999) (including software for chronobiological analysis and simulation), Clocklab (available from ferster@actimetrics.com), or Chrono II (written by Dr. Till Roenneberg, Institut für Medizinische Psychologie, University of Munich).

### 2.3.4. Results

In the example given for *Poryphyra umbilicalis*, growth rate decreased during the light and increased during the dark phase (see Fig. 21.4, days 0–2). The positive or negative transient rapid rate changes upon the onset of light or darkness reflected reversible osmotic changes (Lüning 1992, Lüning et al. 1997, Aguilera et al. 1997). In subsequent continuous light (LL), persistent free-running growth rhythms became evident (see Fig. 21.4, days 3–7). The descending and ascending parts of the free-running growth rhythm corresponded to the growth kinetics occurring during the preceding light and dark phases and thus indicated subjective day and night, respectively (see Fig. 21.4). Using different irradiances in different spectral ranges, one may change the period of the clock and obtain hints leading to the possible photoreceptor(s) used by the circadian clock(s) of a specific alga such as *Porphyra* (Lüning 2001) or the intensively investigated *Lingulodinium polyedra* (Roenneberg and Hastings 1988, Roenneberg and Morse 1993, Roenneberg 1996).

## 2.4. Circadian Gene Expression

Circadian rhythms of gene expression were detected in cyanobacteria, *Chlamydomonas*, the dinoflagellate *Pyrocystis lunula* Hulburt (Suzuki and Johnson 2001, Mittag 2001, Okamoto and Hastings 2003), and higher plants (Piechulla 1999, McClung 2001), with mRNA abundances of most clock-controlled genes in photosynthetic organisms oscillating over the circadian cycle. For marine macroalgae, the red alga *Kappaphycus alvarezii* provides a first example for circadian regulation of two photosynthetic transcripts (Jacobsen et al. 2003).

### 2.4.1. Test Organisms

Potential test organisms are all macroalgal species for which circadian rhythmicity of cell division, growth, photosynthesis, or other cellular processes has been ascertained (see preceding sections). The test organisms should be collected from nature.

### 2.4.2. Equipment

Equipment for cultivation has been described in Section 2.2.2. Standard molecular biological equipment is required for analysis of circadian gene expression.

### 2.4.3. Procedures

The procedure for analysis of gene expression in macroalgae was described initially in the red alga *Kappaphycus* (Jacobsen et al. 2003).

*Cultivation and sampling procedures:* The algae were cultivated in circular 25-liter Plexiglas containers under strictly defined conditions at an irradiance sufficient to obtain circadian rhythmicity, for example, in the case of *K. alvarezii* 180 µmol photons · m$^{-2}$ · s$^{-1}$, as measured at the water surface level. The seawater was exchanged continuously by timer-controlled supply and heated to 23°C to maintain constant conditions. After acclimation in LD 12 : 12 for a few weeks, light conditions were changed to constant illumination (LL) for sampling. At defined times during the last LD and the subsequent LL cycles, small pieces of algal tissue (young parts) were harvested, rinsed very briefly with freshwater, and frozen immediately in liquid nitrogen. The samples were stored at −80°C until used for RNA extraction.

*Extraction of total RNA:* Total RNA was successfully isolated from algal samples using the RNeasy Plant Mini Kit (Qiagen) with slight modifications. Algal tissue was ground in liquid nitrogen, and about 200 mg of tissue powder was resuspended in the guanidine thiocyanate containing buffer RLT for the extraction procedure. No clear correlation between the amount of starting material and RNA yield was observed within a range of 50 to 450 mg of tissue powder. The yield of total RNA varied between 6 and 38 µg per sample.

*RNA gel electrophoresis and Northern blotting:* RNA gel electrophoresis was carried out using denaturing 1% agarose gels containing 0.8% formaldehyde; 2 µg of total RNA was loaded per sample, but depending on the expression level of the gene analyzed, it may be necessary to increase the amount of RNA. To assess RNA quality and equal loading, the rRNA bands were visualized under UV light after electrophoresis. Therefore, the sample loading buffer contained 0.05 µg · µL$^{-1}$ ethidium bromide. RNA was blotted to a positive charged nylon membrane (Schleicher and Schuell) by neutral upward capillary transfer over night.

*Northern hybridization:* Northern hybridization was performed using the digoxigenin detection system (Roche Molecular Biochemicals) according to the instruction manual. The immunological detection was carried out in sealed plastic bags using 4 mL of NBT/BCIP color substrate solution per approximately 150 cm$^2$ of membrane. When other detection methods like chemiluminescent visualization are used, it is important to develop several x-ray films to ensure accurate analysis of both weak and strong signals on the blot.

*Probe preparations:* Gene-specific probes were obtained by labeling of polymerase chain reaction (PCR) products amplified using specific or degenerated primers. For primer design, sequences available in databases were used. Template DNA for the PCR experiments was isolated using the DNeasy Plant Mini Kit (Qiagen). PCR amplifications were carried out using standard Taq DNA polymerase (Roche Molecular Biochemicals) and common PCR programs and conditions. Successful amplification was confirmed by gel electrophoresis and sequence analysis of the PCR products. Probes were prepared by labeling 0.3 to 1 µg of purified PCR product (QIAquick PCR purification kit; Qiagen) with digoxigenin-dUTP (digoxigenin high prime labeling; Roche Molecular Biochemicals).

### 2.4.4. Results

Transcript abundance of the analyzed genes in *Kappaphycus alvarezii* coding for phycoerythrin α- and β-subunit and for ribulose-1,5-biphosphate carboxylase exhibited oscillations under LD and under LL conditions over four free-running cycles (Jacobsen et al. 2003). One may expect more results of this type in view of the ubiquity of circadian rhythms in marine macroalgae.

---

## 3.0. BIWEEKLY (CIRCA-SEMILUNAR) RHYTHMS

Wave action in the sea may reduce gamete concentrations of marine organisms with external fertilization to the point where reproductive success becomes too low for sustaining a population (Levitan and Petersen 1995,

Brawley and Johnson 1992). Synchronous reproductive development evolved in several intertidal animals, and gamete release occurs twice each lunar month, or every 14.8 days (e.g., at spring tides, at extremely low water levels) (Neumann 1981). This improved the chances of mating. One example is the lower intertidal chironomid midge *Clunio marinus* Haliday, a rare case of a "terrestrial" insect species in the sea. In this species, the short-lived male imagoes are able to detect the narrow temporal window of 1 to 2 hours of spring tide every 2 weeks, when the habitat of *C. marinus* falls dry, so the males can unfold their wings after eclosion and mate with the wingless females (Neumann 1976). This is possible because of an endogenous biweekly (circa-semilunar) clock that runs free in the laboratory with a period of approximately 15 days (Neumann 1976). The clock is synchronized every second cycle by moonlight, that is, the continuous-light signal during the few full-moon nights occurring every 29.5 days, and the various successive larval stages of *C. marinus* develop, thus, in synchrony under water. This pattern also was successfully repeated in the laboratory, with four successive illuminated nights simulating the full-moon phase (Neumann 1976). The light intensity of simulated moonlight is not important; it may be as low as in moonlight but also attain normal intensity used for culturing during the day phase (H. D. Franke, personal communication).

Similarly, the brown alga *Dictyota dichotoma* (Hudson) Lamouroux releases eggs twice a month in the field, and Müller (1962) synchronized the alga to do this in the laboratory. Female gametophytes were cultivated in LD cycles but exposed every 28 days to a synchronizing artificial moonlight, that is, one LL cycle of normal culture light intensity of 24 µmol photons $\cdot$ m$^{-2}$ $\cdot$ s$^{-1}$, or even only of moonlight intensity (0.06 µmol photons $\cdot$ m$^{-2}$ $\cdot$ s$^{-1}$; values calculated from original light intensities given in lux). After a synchronization phase of 2 months, the synchronizing moonlight signal was omitted, and the alga exhibited a free-running rhythm of biweekly egg release at LD 12 : 12 for at least five cycles with intervals of 16 to 17 days (Müller 1962).

### 3.1. Test Organisms

The egg release rhythm of *Dictyota dichotoma* is the only algal example for which a free-running biweekly (circa-semilunar) rhythm was documented (Fig. 21.6). There is, however, quite a number of field observations on several species of intertidal macroalgae discharging gametes at biweekly (semilunar) intervals during spring tides. This happens preferably at dawn, so in addition to the

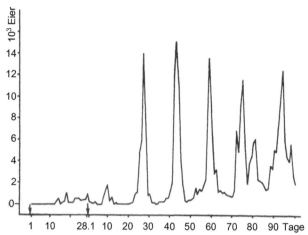

**FIGURE 21.6.** Biweekly (circa-semilunar) rhythm in egg release in the brown alga *Dictyoa dichotoma*. Experimental gametophytes were cultivated at LD 14 : 10. Number of eggs released was counted daily. Continuous light (simulating moonlight) was given for 24 hours at days 1 and 28. Thereafter, a free-running biweekly rhythm of egg release became apparent. (From Müller 1962).

gametangia ripening at biweekly intervals, gamete expulsion is narrowed to a few minutes when the first traces of light occur in the morning at spring tides, probably because of additional circadian rhythmicity. Examples for such potential test organisms for discovery of more biweekly free-running rhythms of gamete release among macroalgae are the green algae *Ulva* sp. (Smith 1947) and *Enteromorpha* sp. (Christie and Evans 1962) and the brown algae *Sargassum muticum* (Yendo) Fensholt (Fletcher 1980, Hales and Fletcher 1990) and *Fucus ceranoides* L. (Brawley and Johnson 1992). The test organisms should be collected from nature.

### 3.2. Equipment

No special equipment is required, but it is important that the experimental algae be cultivated in a constant manner over several months. Renewal of the culture medium should be in a regular way (e.g., every 5 days or, preferably, in a flow-through culture with plain seawater or with automatic mixing with additional nutrients).

### 3.3. Procedures and Possible Results

One may use the same approach as described previously for *Clunio* or *Dictyota*, if one wants to test an alga for the presence of endogenous biweekly rhythmicity. As a first step, during the synchronizing phase, the algal thalli (e.g., gametophytes of *Ulva* sp.) should be synchronized in the laboratory in LD 12 : 12 or LD 16 : 8 by 2 to 4 LL cycles every 28 days, at least for 2 to 3 months. The simulated moonlight may have the same light intensity as used during the day phase, as stated previously. Reproductive development (i.e., presence of fertile margins in the case of *Ulva* sp.) should be monitored daily and should peak every 2 weeks. Note that in *Dictyota*, no rhythm developed after only one artificial moonlight signal, only after two signals (see Fig. 21.6). As a second step, after the synchronizing phase, the moonlight signal should be omitted, and a free-running biweekly rhythm of gamete discharge should evolve.

## 4.0. DAY-LENGTH EFFECTS (PHOTOPERIODISM)

The importance of day length (photoperiod) for plant development was first realized in 1920, when Garner and Allard (1920) discovered that day length controls seasonal flowering in several higher plant species. Numerous photoperiodic effects were subsequently found in higher plants and animals on reproduction, growth activity, and many other seasonal activities (Thomas and Vince-Prue 1997, Lumsden and Millar 1998). Surprisingly, very few genuine photoperiodic effects were discovered among unicellular algae (Suzuki and Johnson 2001).

The mechanism of photoperiodic responses is thought to consist of a circadian (daily) timer that somehow measures the length of the night and triggers the photoperiodic response (Thomas and Vince-Prue 1997, Lumsden and Millar 1998, Suzuki and Johnson 2001, Bastow and Dean 2002). There is genetic evidence from *Arabidopsis* mutants with aberrant time of flowering and parallel altered properties of circadian rhythms that the photoperiodic time measurement (PTM) timer shares molecular components with circadian clocks controlling daily rhythms (Carré 1998, Johnson 2001). Furthermore, the molecular basis of the photoperiodic flowering response in *Arabidopsis* may consist of an external coincidence mechanism based on the endogenous circadian control of a transcription factor expression (CO) and the modulation of CO function by light (Bastow and Dean 2002). Bünning (1936) had proposed the external coincidence model for the

circadian oscillation as a basis for the photoperiodic response, with a light-sensitive fraction of the circadian oscillation being illuminated in long days in summer and not in short days in winter. As an additional possibility, recent work using mutant mice suggests a set of two molecular oscillators tracing dusk and dawn, respectively, and translating in this way day length to the organism (Oster et al. 2002).

Macroalgal phycologists, who started to search for photoperiodic effects in seaweed species in the 1970s, put together an impressively long list of genuine photoperiodic effects, mostly on induction of upright thalli or reproductive organs by short-day (SD) treatment in various red, brown, and green algae, but a few long-day (LD) effects also were discovered (Dring 1984, 1988). (Note that *LD* means "long day," whereas *LD* followed by numbers [e.g., 12 : 12] means "light–dark regimen"). SD plants require a long uninterrupted night, and the SD effect (e.g., the formation of tetrasporangia in a red alga at 8 hours of light per day) is prevented in individuals cultivated in a nightbreak (NB) regimen (i.e., a daily light regimen of 8 hours of light, 7.5 hours of darkness, 1 hour of light, and 7.5 hours of darkness), just as in algae cultivated in the LD regimen with 16 hours of light per day. In other words, the interruption of a long dark period by a short nightbreak converts SD into LD conditions, an effect discovered by Hamner and Bonner (1938). Obtaining the same result in NB as in an LD regimen crucially confirms a genuine photoperiodic effect, because this result shows that the effect is caused by time measurement (hours of darkness) and not by amount of photosynthate during a 24-hour cycle (Dring 1984, 1988).

As an example of a photoperiodic response enabling an organism to avoid unfavorable seasons, the formation of upright macroscopic thalli from a prostrate system is illustrated in Fig. 21.7. In LD conditions in the laboratory, the zoospores of the brown alga *Scytosiphon lomentaria* (Lyngbye) Link form a small, regular, disclike crust (Fig. 21.7a). This is the oversummering life form in the field, whereas in SD conditions the upright thallus is formed (Fig. 21.7b), ascending from an irregular holdfast system and growing into the new macroscopic thallus as observed in the field in spring (Dring and Lüning 1975). Similarly, the induction of upright thalli from prostrate systems is induced by SD and prevented by LD or NB conditions in approximately 10 smaller algae species of the known 50 or so genuine photoperiodic responses in macroalgae tabulated by Dring (1984). As with flowering plants, insects, and birds, there are photoperiodic ecotypes in

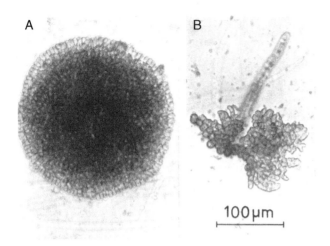

**FIGURE 21.7.** Brown alga *Scytosiphon lomentaria*, cultivated from zoospores for 25 days in 16 hours **(a)** or 8 hours **(b)** of white light per day at 16°C. (From Dring and Lüning 1975.)

**FIGURE 21.8.** A 300-liter tank with light-tight PVC hood and slide door in a constant-temperature room. Two halogen lamps on top (not visible) provide culture light. The seawater is aerated by a 20-mm wide PVC pipe on the bottom of the tank with a 2-mm opening at every 10 cm. Note white plastic shields on top of the water carrying marked algal individuals and circulating with the whole biomass within the water column.

*Scytosiphon lomentaria* with 9 to 12 hours of light per day, allowing upright formation in Mediterranean isolates already in autumn, or 13 to 16 hours of light per day in an isolate from Iceland, securing that upright thalli are produced from the prostrate discs only after March into Arctic summer (Lüning 1980). Among larger algae, *Sargassum muticum* (Yendo) Fensholt produces erect thalli from germlings only in SD and not in LD or NB conditions (Hwang and Dring 2002).

The bulk of known genuine photoperiodic responses in marine macroalgae refers to induction of reproductive organs in SD conditions, mainly of tetrasporangia in red algae (Dring 1984), securing that the new generation of gametophytes is available in the field in early spring with optimal environmental conditions for growth of many algal species such as high nutrients, ample light supply, and low temperature. Among larger algae, in laminarian species, formation of sporangia is restricted to SD conditions in *Laminaria saccharina* (L.) Lamouroux (Lüning 1988) and *L. setchellii* Silva (Dieck 1991). As to fucalean species, gametangia are formed only in SD conditions in *Fucus distichus* (Bird and McLachlan 1976) and *Ascophyllum nodosum* (L.) Le Jolis (Terry and Moss 1980).

### 4.1. Test Organisms

Potential test organisms are macroalgal species exhibiting seasonal responses in the field. The test organisms should be collected from nature.

### 4.2. Equipment

If only one constant-temperature room is available, one may use black boxes with automatically opening or shutting covers for smaller algae in petri or crystallizing dishes. In self-made boxes, one may integrate a commercial pneumatically operated lever normally used for opening cellar windows controlled by an automatic time switch (see construction in Lüning 1980). The fluorescent lamps for culture light are mounted above the black boxes.

For larger algae such as laminarian sporophytes, 300-liter tanks with circulating or flow-through seawater may be used, and different day-length regimens in the same constant-temperature room are possible due to black, light-tight PVC hoods mounted on the tanks containing inside fluorescent lamps or on top of halogen lamps shining into the tanks (Fig. 21.8). Individual experimental algae may be fastened to small numbered PVC plates mounted on Styrofoam shields of appropriate size (Fig. 21.8) so that the algal individuals may circulate with the whole biomass within the water column but can be identified for measurement and photographing (e.g., at intervals of 1 month). Again for larger algae, one may alternatively fix an automatic blind on a 2000-liter tank, as illustrated in Fig. 21.9. The blind consists of a flexible, light-tight

**FIGURE 21.9.** A 2,000-liter tank with automatic blind consisting of a light-tight plastic cover on a motor-driven roll.

plastic cover on a motor-driven roll (manufactured by Mariscope).

## 4.3. Procedures and Possible Results

For detection of a photoperiodic response, one may use unialgal cultures starting from gametes, zygotes, or spores and follow the development under different daylength conditions. Alternatively, one may start with uniform adult thalli and transfer, say, adult vegetative sporophytes of *Laminaria saccharina* in June into SD tanks to obtain sori on the thallus surface a few weeks later, instead of October as in the field in the North Sea (Lüning 1988). As emphasized earlier, it is important to use not only SD and LD, but also NB conditions (usually 1 hour of light in the middle of the long night) to ascertain a genuine photoperiodic response. In the above example, no sorus will be formed in experimental sporophytes of *L. saccharina* cultivated in tanks with an LD or NB regimen.

It may be important to run the experiments at more than one temperature, because an SD response may be confined to the upper temperature range, for example, as a measure of the alga to discriminate autumn (SD and high temperature) from winter and early spring (SD and low temperature). For example, the *Trailliella* phase of the red alga *Bonnemaisonia hamifera* Hariot was found to form tetrasporangia in SD conditions only at 15°C, not at 5 or 10°C (Lüning 1980), and a similar case was reported for the *Falkenbergia* phase of the red alga *Asparagopsis armata* Harvey (Guiry and Dawes 1992).

## 5.0. CIRCANNUAL RHYTHMICITY

The seasonal development of a perennial photoperiodic organism is dictated by the sequence of short and long days over the years; for example, using some above examples, SD conditions will induce gametangia in *Ascophyllum nodosum* and sporangia in *Laminaria saccharina*, LD conditions in summer will prevent that, and the algae will have to wait for the next SD signal to become reproductive. In contrast, a perennial circannual organism will perform its seasonal changes autonomously because of an endogenous circannual rhythm. As an example, experimental sporophytes of the kelp *Pterygophora californica* Ruprecht were cultivated in the laboratory at LD 16 : 8 for more than a year in constant conditions of temperature and nutrient supply (Lüning 1991), although researchers should plan to do such an experiment for 2 years or more. The algae were able to produce autonomously the natural sequence of active growth during the spring and growth dormancy during summer resulting in one terminal blade per year separated from the previous year's blade by a constriction (Fig. 21.10a: free-running growth curves of three individuals; Fig. 21.11a: photographs of the same three individuals at the end of the experiment).

Interestingly, it was a macroalgal phycologist who intuitively anticipated circannual rhythmicity or thought that it should exist. It was Henrik Printz who wrote the following sentences in his investigation of the algal flora of the Trondhjemsfjord, in the chapter on algal periodicity (Printz 1926, pp. 15–22; translated from original German):

> *The external factors are hardly sufficient to explain the annual periodicity of the algal vegetation. . . . In areas with extensive climatic variations, the vegetation exhibits a yearly rhythm that matches the environment extremely well—this trait is valid for the whole plant world—and one might easily succumb to the idea that the periodicity would be caused by the environment. . . . One may say that a periodic climate cannot cause the ability of organisms to react adequately to periodic changes, but that this reactive ability must be present beforehand; it is the primary system, and the environment turns it into certain directions.*

Circannual rhythms are known mostly for animal species, with numerous examples among migrating birds, hibernating mammals, and many other animal groups (Gwinner 1986), but nothing is known about the mechanisms of such long-term "clocks." It is assumed that circannual rhythmicity may result from a temporal

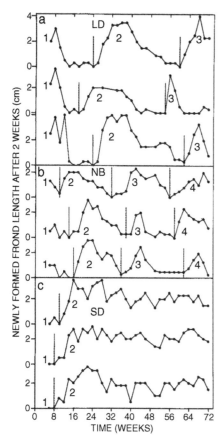

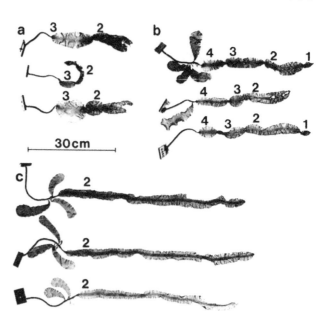

**FIGURE 21.11.** Two-year-old experimental sporophytes of *Pterygophora californica*, whose growth curves are represented in Fig. 21.10. **(a)** long day (LD); **(b)** nightbreak (NB); **(c)** short day (SD). See Fig. 21.10 legend for further details. (From Lüning 1991.)

**FIGURE 21.10.** **(a)**, **(b)** Free-running circannual growth rhythms in a long day (LD 16 : 8) and nightbreak (NB) regimen (8 hours of light per day plus 1 hour of light in the middle of the night) in the terminal blade of the kelp *Pterygophora californica* cultivated in 300-liter tanks at 10°C. **(c)** continuous growth in short days (SDs), at LD 8 : 16. Vertical bars: start of new terminal blade. Numbers along the blades: successive blade generations, with number 1 referring to blade formed during pretreatment phase in 2-liter beakers. Week 0 equates with start of treatment in tanks. (From Lüning 1991.)

sequence of different physiological states rather than a molecular mechanism as in circadian rhythmicity (Dawson et al. 2001). Interestingly, circannual rhythmicity may reside in a single cell, because it was found to govern cyst formation and germination capacity in certain dinoflagellate species, with an in-built resting phase of several months in the resting stages, and growth rates of dinoflagellates exhibit circannual rhythmicity (Anderson and Keafer 1987, Costas and Varela 1988, 1989, Costas et al. 1990).

The free-running period of a circannual rhythm is often shorter than a year (Gwinner 1986) (e.g., only 7 to 8 months in *Pterygophora californica*) (see Fig. 21.10a).

Synchronization to the 12-month period of the natural year is achieved in nature or in the laboratory by the environmental cycle of day length as a zeitgeber. Thus, when *P. californica* was cultivated in a light-tight tank with the lamp timers switched weekly to sunrise and sunset, imitating the local annual cycle of day length (Fig. 21.12a–c), the sporophytes started new blade formation when day length was at a minimum, and thus, like in nature in winter, reduced growth in late spring when day length still increased and stopped growth at maximum day length in summer. Shifting the zeitgeber curve in parallel tanks resulted in shifting of the annual growth curves (Figs. 21.12a–c, 21.13a–c). Reducing in other tanks the period of the zeitgeber curve to 6 months by omitting every second week (Figs. 21.12d, 21.13d) or to 3 months by using only the sunrise/sunset data of the first week of each month (Figs. 21.12e, 21.13e) resulted in a new "year's" blade every 6 or 3 months, correspondingly, which was shorter of course than "12-month blades" (Fig. 21.13).

In certain animal species, it was observed that the circannual "clock" functions properly only in a restricted range of constant day length. For example, the European starling exhibits the seasonal sequence of molt, and the sequence of testicular increases in spring and decrease in summer autonomously over several years only if reared at LD 12 : 12 (Gwinner 1986). At 8

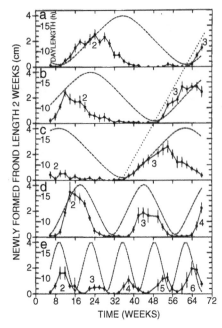

**FIGURE 21.12.** Synchronized growth rhythms in five groups (n = 10) of *Pterygophora californica* (solid lines; vertical bars are confidence limits, *P* = 0.05) exposed to sinusoidal changes in day length (broken lines) at 5°C. Period of day-length cycles: 12 **(a-c)**, 6 **(d)**, or 3 months **(e)**. Day-length cycle **(a)** represents a simulated natural annual cycle of day length at 54° N, adjusted once a week. Day-length cycles **(b)** and **(c)** correspond to the same cycle but advance shifted by 3 or 6 months, respectively, as indicated by the dotted line. See Fig. 21.10 legend for further details. (From Lüning 1991.)

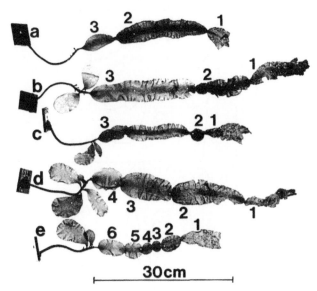

**FIGURE 21.13.** Two-year-old experimental sporophytes of *Pterygophora californica*, each representing an example of the five experimental groups, whose growth curves are represented in Fig. 21.12. See Fig. 21.10 legend for further details. (From Lüning 1991.)

hours of light per day, testicular size never decreases and the gonads remain continually active, and at 16 hours of light per day, the gonad size goes through one cycle and remains small with no subsequent size increase. These effects may mean that certain day-length ranges stop the circannual "clock" from ticking or, at least, prevent it from controlling the system (Gwinner 1986). Similarly in kelps, *P. californica* grows continuously in SD conditions (Figs. 21.10c, 21.11c), which means that its circannual "clock" is no longer able to provide the growth-off signal, because it can do it in LD conditions (Figs. 21.10a, 21.11a). The critical day-length range separating continuous from periodic growth in *P. californica* is situated at 9 to 10 hours (Lüning and Kadel 1993). Interestingly, like in a genuine photoperiodic response, cultivation in NB conditions (Figs. 21.10b, 21.11b) gave the same result of periodic growth as in LD conditions, which may indicate that photoperiodic time measurement is deeply entrenched in the mechanism of circannual rhythmicity. The kelp *Laminaria hyperborea* (Gunnerus) Foslie exhibited circannual peri-

odic growth only at LD 12 : 12, continuous growth in SD at LD 8 : 16, and only one blade cycle in long days at LD 16 : 8 (Schaffelke and Lüning 1994), and thus showed a similarly narrow range for expression of circannual rhythmicity as the European starling (Gwinner 1986).

### 5.1. Test Organisms

Whereas circannual rhythmicity has been searched for and detected in many animal species (Gwinner 1986), as stated previously, the list of plant species exhibiting circannual rhythmicity is extremely short. For higher plants, there are a few examples such as *Lemna minor* L. (Bornkamm 1966) and the seagrass *Posidonia oceanica* (L.) Delile (Ott 1979), for seaweeds, the three kelp species *Pterygophora californica* (Lüning 1991), *Laminaria setchellii* (Dieck 1991), and *L. hyperborea* (Schaffelke and Lüning 1994; the case of *L. digitata* does not appear quite clear at present), and for the dinoflagellates examples are cited above. The reason for the scarcity of circannual plant species may partially reflect the lag of botany behind zoology in this case, because all animal species with circannual reactions were at first regarded as photoperiodic species, and only long-term studies revealed their circannual character, which may happen in the future with plant species as well. In addition, it may have appeared more attractive and promising to follow in constant conditions for a few years the development of a deer, sheep, or bird than the development of a plant or alga. Nevertheless, among marine macroalgae potential test organisms for detecting circannual rhythmicity are all perennial species exhibiting seasonal traits. The test organisms should be collected from nature.

## 5.2. Equipment

The equipment is the same as described previously for photoperiod experiments.

## 5.3. Procedures and Possible Results

If one plans to follow the long-term development of a macroalgal species for detection of a potential circannual response, one should secure at least n = 10 and hold conditions as constant possible. If only one constant LD regimen in a constant-temperature room is available, one might preferably select LD 12 : 12, because LD conditions might inhibit the expression of the circannual rhythm, as in *Laminaria hyperborea* (Schaffelke and Lüning 1994) and SD conditions might lead to arrhythmic continuous growth activity. The latter aspect is, of course, interesting for aquaculture and might be worth testing, especially for this purpose (Gomez and Lüning 2001, Lüning and Pang 2003).

After an initial active growth phase for several months, one may expect a phase of growth dormancy (see Fig. 21.10a,b), and it will be exciting for the algal cultivator to see whether the experimental algae will enter the next phase of active growth "by themselves" (see Fig. 21.10a,b), indicating the presence of circannual rhythmicity, or never again grow actively, indicating cultural disaster. It may be exactly this point, before the circannual "comeback," at which earlier algal cultivators threw away their algae and thus potentially missed the discovery of a circannual rhythm.

## 6.0. REFERENCES

Aguilera, J., Figueroa, F. L., and Niell, F. X. 1997. Photocontrol of short-term growth in *Porphyra leucosticta*. *Eur. J. Phycol.* 32:417–24.

Anderson, D. M., and Keafer, B. A. 1987. An endogenous annual clock in the toxic marine dinoflagellate *Gonyaulax tamarensis*. *Nature* 325:616–7.

Aschoff, J. 1981. Free-running and entrained circadian rhythms. In: Aschoff, J., ed. *Handbook of Behavioral Neurobiology*, vol 4, *Biological Rhythms*. Plenum Press, New York, pp. 81–93.

Bastow, R., and Dean, C. 2002. The molecular basis of photoperiodism. *Dev. Cell* 3:461–2.

Bird, N. L., and McLachlan, J. 1976. Control of formation of receptacles in *Fucus distichus* L. subsp. *distichus*. *Phycologia* 15:79–84.

Bornkamm, R. 1966. Ein Jahresrhythmus des Wachstums bei *Lemna minor* L. *Planta (Berl.)* 69:178–86.

Brawley, S. H., and Johnson, L. E. 1992. Gametogenesis, gametes and zygotes: An ecological perspective on sexual reproduction in the algae. *Br. Phycol. J.* 27:233–52.

Britz, S. J., and Briggs, W. R. 1976. Circadian rhythms of chloroplast orientation and photosynthetic capacity in *Ulva*. *Plant Physiol.* 58:22–7.

Bühnemann, F. 1955. Das endodiurnale System der Oedogoniumzelle IV. Die Wirkung verschiedener Spektralbereiche auf die Sporulations-und Mitoserhythmik. *Planta (Berl.)* 46:227–55.

Bünning, E. 1936. Die endogene Tagesrhythmik als Grundlage der photoperiodischen Reaktion. *Ber. Dtsch. Bot. Ges.* 54:590–607.

Carré, I. A. 1998. Genetic dissection of the photoperiod-sensing mechanism in the long-day plant *Arabidopsis thaliana*. In: Lumsden, P. J., and Millar, A. J., eds. *Biological Rhythms and Photoperiodism in Plants*. Bios, Oxford, pp. 257–69.

Christie, A. O., and Evans, L. V. 1962. Periodicity in the liberation of gametes and zoospores of *Enteromorpha intestinalis* Linn. *Nature (London)* 193:193–4.

Costas, E., Navarro, M., and López-Rodas, V. 1990. An environment-synchronized internal clock controlling the annual cycle of dinoflagellates. In: Graneli, E., Sundstrom, B., Edler, L., and Anderson, D. M., eds. *Toxic Marine Phytoplankton*, Elsevier, Amsterdam, pp. 280–3.

Costas, E., and Varela, M. 1988. Evidence of an endogenous circannual rhythm in growth rates on dinoflagellates. *Chronobiologia* 15:223–6.

Costas, E., and Varela, M. 1989. Circannual rhythm in cyst formation and growth rates in the dinoflagellate *Scripsiella trochoidea* Stein. *Chronobiologia* 16:265–70.

Dawson, A., King, V. M., Bentley, G. E., and Ball, G. F. 2001. Photoperiodic control of seasonality in birds. *J. Biol. Rhythms* 16:365–80.

Dieck (Bartsch), I. tom. 1991. Circannual growth rhythm and photoperiodic sorus induction in the kelp *Laminaria setchellii* (Phaeophyta). *J. Phycol.* 27:341–50.

Doty, M. S. 1973. Farming the red seaweed, *Eucheuma*, for carrageenans. *Micronesica* 9:59–73.

Driessche, T. V. 1966. Circadian rhythms in *Acetabularia*: Photosynthetic capacity and chloroplast shape. *Exp. Cell Res.* 42:18–30.

Dring M. J. 1984. Photoperiodism and phycology. *Prog. Phycol. Res.* 3:159–92.

Dring M. J. 1988. Photocontrol of development in algae. *Annu. Rev. Plant Physiol. Plant Mol. Biol.* 39:157–74.

Dring, M. J., and Lüning, K. 1975. A photoperiodic response mediated by blue light in the brown alga *Scytosiphon lomentaria*. *Planta (Berl.)* 125:25–32.

Dunlap, J. C. 1999. Molecular bases for circadian clocks. *Cell* 96:271–90.

Dunlap, J. C., Ioros, J. J., and DeCoursey, P. J. 2003. *Chronobiology: Biological Timekeeping*. Sinauer, Sunderland, Massachusetts, 382 pp.

Fletcher, R. L. 1980. Studies of the recently introduced brown alga *Sargassum muticum* (Yendo) Fensholt. III. Periodicity in gamete release and "incubation" of early germling stages. *Bot. Mar.* 31:425–32.

Garner, W. W., and Allard, H. A. 1920. Effect of the relative length of day and night and other factors of the environment of growth and reproduction in plants. *J. Agric. Res.* 18:553–606.

Goff, L. J., and Coleman, A. W. 1984. Elucidation of fertilization and development in a red alga by quantitative DNA microspectrofluorometry. *Dev. Biol.* 102:173–94.

Gomez, I., and Lüning, K. 2001. Constant short-day treatment of outdoor-cultivated *Laminaria digitata* prevents summer drop in growth rate. *Eur. J. Phycol.* 36:391–5.

Granbom, M., Pedersén, M., Kadel, P., and Lüning, K. 2001. Circadian rhythm of photosynthetic oxygen evolution in *Kappaphycus alvarezii* (Rhodophyta): Dependence on light quantity and quality. *J. Phycol.* 37:1020–5.

Green, E. J., and Carritt, D. E. 1967. New tables for oxygen saturation of seawater. *J. Mar. Res.* 25:140–7.

Green, R. M., Tingay, S., Wang, Z., and Tobin, E. M. 2002. Circadian rhythms confer a higher level of fitness to *Arabidopsis* plants. *Plant Physiol.* 129:576–84.

Guiry, M. D., and Dawes, C. J. 1992. Daylength, temperature and nutrient control of tetrasporogenesis in *Asparagopsis armata* (Rhodophyta). *J. Exp. Mar. Biol. Ecol.* 158:197–217.

Gwinner, E. 1986. *Circannual Rhythms. Endogenous Annual Clocks in the Organization of Seasonal Processes*. Springer, Berlin, 154 pp.

Hales, J. M., and Fletcher, R. L. 1990. Studies of the recently introduced brown alga *Sargassum muticum* (Yendo) Fensholt. V. Receptacle initiation and growth, and gamete release in laboratory culture. *Bot. Mar.* 33:241–9.

Hamner, K. C., and Bonner, J. 1938. Photoperiodism in relation to hormones as factors on floral initiation and development. *Bot. Gaz.* 100:388–431.

Harmer, S. L., Hogenesch, J. B., Straume, M., Chang, H., Han, B., Zhu, T., Wang, X., Kreps, J. A., and Kay, S. A. 2000. Orchestrated transcription of key pathways in *Arabidopsis* by the circadian clock. *Science* 290:2110–3.

Hastings, J. W., Astrachan, L., and Sweeney, B. M. 1961. A persistent daily rhythm in photosynthesis. *J. Gen. Physiol.* 45:69–76.

Hastings, J. W., Rusak, B., and Boulos, Z. 1991. Circadian rhythms: The physiology of biological timing. In: Prosser, C. L., ed. *Neural and Integrative Animal Physiology. Comparative Animal Physiology*, 4th ed. Wiley-Liss, New York, pp. 435–546.

Hull, H. H., Hoshaw, R. W., and Wang, J.-C. 1982. Cytofluorometric determination of nuclear DNA in living and preserved algae. *Stain Technol.* 57:273–82.

Hwang, E. K., and Dring, M. J. 2002. Quantitative photoperiodic control of erect thallus production in *Sargassum muticum*. *Bot. Mar.* 45:471–5.

Jacobsen, S., Lüning, K., and Goulard, F. 2003. Circadian changes in relative abundance of two photosynthetic transcripts in the marine macroalga *Kappaphycus alvarezii*. *J. Phycol.* 39:888–96.

Johnson, C. H. 2001. Endogenous timekeepers in photosynthetic organisms. *Annu. Rev. Physiol.* 63:695–728.

Johnson, C. H., Knight, M., Trewavas, A., and Kondo, T. 1998. A clockwork green: Circadian programs in photosynthetic organisms. In: Lumsden, P. J., and Millar, A. J., eds. *Biological Rhythms and Photoperiodism in Plants*. Bios, Oxford, pp. 1–34.

Johnson, C. H., and Kondo, T. 2001. Circadian rhythms in unicellular organisms. In: *Handbook of Behavioral Neurobiology*. Plenum Press, New York, pp. 61–77.

Kageyama, A., Yokohama, Y., and Nisizawa, K. 1979. Diurnal rhythm of apparent photosynthesis of a brown alga, *Spatoglossum pacificum*. *Bot. Mar.* 22:199–201.

Levitan, D. R., and Petersen, C. 1995. Sperm limitation in the sea. *Trend Ecol. & Evol.* 10:228–31.

Lumsden, P. J., and Millar, A. J., eds. 1998. *Biological Rhythms and Photoperiodism in Plants*. Bios, Oxford, 284 pp.

Lüning, K. 1980. Control of algal life-history by daylength and temperature. In: Price, J. H., and Farnham, W. F., eds. *The Shore Environment*, Vol. 2, *Ecosystems*. Academic Press, London, pp. 915–45.

Lüning, K. 1988. Photoperiodic control of sorus formation in the brown alga *Laminaria saccharina*. *Mar. Ecol. Progr. Ser.* 45:137–44.

Lüning, K. 1991. Circannual growth rhythm in a brown alga, *Pterygophora californica*. *Bot. Acta* 104:157–62.

Lüning, K. 1992. Day and night kinetics of growth rate in green, brown, and red seaweeds. *J. Phycol.* 28:794–803.

Lüning, K., 1994. Circadian growth rhythm in juvenile sporophytes of Laminariales (Phaeophyta). *J. Phycol.* 30:193–9.

Lüning, K. 2001. Circadian growth rhythm in the red macroalga *Porphyra umbilicalis*: Spectral sensitivity of the circadian system. *J. Phycol.* 37:52–8.

Lüning, K., and Kadel, P. 1993. Daylength range for circannual rhythmicity in *Pterygophora californica* (Alariaceae, Phaeophyta) and synchronization of seasonal growth by daylength cycles in several other brown algae. *Phycologia* 32:379–87.

Lüning, K., and Pang, S. 2003. Mass cultivation of seaweeds: Current aspects and approaches. *J. Appl. Phycol.* 15:115–9.

Lüning, K., Titlyanov, E. A., and Titlyanova, T. 1997. Diurnal and circadian periodicity of mitosis and growth in marine macroalgae. III. The red alga *Porphyra umbilicalis*. *Eur. J. Phycol.* 32:167–73.

Makarov, V. N., Schoschina, E. V., and Lüning, K. 1995. Diurnal and circadian periodicity of mitosis and growth in marine macroalgae. I. Juvenile sporophytes of Laminariales (Phaeophyta). *Eur. J. Phycol.* 30:261–6.

McClung, C. R. 2001. Circadian rhythms in plants. *Annu. Rev. Plant Physiol. Plant Mol. Biol.* 52:139–62.

Mittag, M. 2001. Circadian rhythms in microalgae. *Int. Rev. Cytol.* 206:213–47.

Müller, D. G. 1962. Über jahres- und lunarperiodische Erscheinungen bei einigen Braunalgen. *Bot. Mar.* 4:140–55.

Neumann D. 1976. Entrainment of a semilunar rhythm. In: DeCoursey, P. J., ed. *Biological Rhythms in the Marine Environment*. University of South Carolina Press, Columbia, South Carolina, pp. 115–127.

Neumann, D. 1981. Tidal and lunar rhythms. In: Aschoff, J., ed. *Handbook of Behavioral Neurobiology*, Vol. 4. Plenum Press, New York, pp. 351–80.

Neuscheler-Wirth, H. 1970. Wachstumsgeschwindigkeit und Wachstumsrhythmik bei *Mougeotia*. *Z. Pflanzenphysiol.* 63:352–69.

Okamoto, O. K., and Hastings, J. W. 2003. Novel dinoflagellate clock-related genes identified through microarray analysis. *J. Phycol.* 39:519–26.

Oohusa, T. 1980. Diurnal rhythm in rates of cell division, growth and photosynthesis of *Porphyra yezoensis* (Rhodophyceae) cultured in the laboratory. *Bot. Mar.* 23:1–5.

Oster, H., Marone, E., and Albrecht, U. 2002. The circadian clock as a molecular calendar. *Chronobiol. Int.* 19:507–16.

Ott, J. A. 1979. Persistence of a seasonal growth rhythm in *Posidonia oceanica* (L.) Delile under constant conditions of temperature and illumination. *Mar. Biol. Lett.* 1:99–104.

Ouyang, Y., Andersson, C. R., Kondo, T., Golden, S. S., and Johnson, C. H. 1998. Resonating circadian clocks enhance fitness in cyanobacteria. *Proc. Natl. Acad. Sci. U.S.A.* 95:8660–4.

Palmer, J. D. 1995. *The Biological Rhythms and Clocks of Intertidal Animals*. Oxford University Press, New York, 375 pp.

Pérez, R. 1992. *La Culture des Algues Marines dans le Monde*. Ifremer, Plouzaine, 614 pp.

Piechulla, B. 1999. Circadian expression of the light-harvesting complex protein genes in plants. *Chronobiology Int.* 16:115–28.

Pittendrigh, C. S. 1993. Temporal organization: Reflections of a Darwinian watcher. *Annu. Rev. Physiol.* 1993:17–54.

Printz, H. 1926. Die Algenvegetation des Trondhjemsfjordes. *Skr. Norske Vidensk. Akad. I. Mat-Naturv. Kl.* 5: 1–274.

Refinetti, R. 1999. *Circadian Physiology*. CRC Press, Boca Raton, Florida, 200 pp.

Roenneberg, T. 1996. The complex circadian system of *Gonyaulax polyedra*. *Physiol. Plantarum* 96:733–7.

Roenneberg, T., and Hastings, J. W. 1988. Two photoreceptors control the circadian clock of a unicellular alga. *Naturwissenschaften* 75:206–7.

Roenneberg, T., and Morse, D. 1993. Two circadian oscillators in one cell. *Nature* 362:362–4.

Schaffelke, B., and Lüning, K. 1994. A circannual rhythm controls seasonal growth in the kelps *Laminaria hyperborea* and *L. digitata* from Helgoland (North Sea). *Eur. J. Phycol.* 29:49–56.

Schmid, R., Forster, R., and Dring, M. J. 1992. Circadian rhythm and fast responses to blue light of photosynthesis in *Ectocarpus* (Phaeophyta, Ectocarpales). II. Light and $CO_2$ dependence of photosynthesis. *Planta* 187:60–6.

Schubert, H., Gerbersdorf, S., Titlyanov, E., Titlyanov, T., Granbom, M., Pape, C., and Lüning, K. 2004. Circadian rhythm of photosynthesis in *Kappaphycus alvarezii* (Rhodophyta): Independence of the cell cycle and possible photosynthetic clock targets. *Eur. J. Phycol.* (in press).

Smith, G. M. 1947. On the reproduction of some Pacific coast species of *Ulva*. *Am. J. Bot.* 30:80–7.

Somers, D. E. 1999. The physiology and molecular bases of the plant circadian clock. *Plant Physiol.* 121:9–19.

Starr R., and Zeikus J. A. 1987. UTEX—The culture collection of algae at the University of Texas at Austin. *J. Phycol.* 23(Suppl):1–47.

Suzuki, L., and Johnson, C. H. 2001. Algae know the time of the day: circadian and photoperiodic programs. *J. Phycol.* 37:933–42.

Sweeney, B. M. 1987. *Rhythmic Phenomena in Plants*. Academic Press, London, 172 pp.

Sweeney, B. M., and Hastings, J. W. 1958. Rhythmic cell division in populations of *Gonyaulax polyedra*. *J. Protozool.* 5:217–24.

Sweeney, B. M., and Hastings, J. W. 1962. Rhythms. In: Lewin, R. A., ed. *Physiology and Biochemistry of Algae.* Academic Press, New York, pp. 687–700.

Terry, L. A., and Moss, B. L. 1980. The effect of photoperiod on receptacle formation in *Ascophyllum nodosum* (L.) Le Jol. *Br. Phycol. J.* 15:291–301.

Thomas, B., and Vince-Prue, D. 1997. *Photoperiodism in Plants*, 2nd ed. Academic Press, San Diego, 428 pp.

Titlyanov, E. A., Titlyanova, T., and Lüning, K. 1996. Diurnal and circadian periodicity of mitosis and growth in marine macroalgae. II. The green alga *Ulva pseudocurvata. Eur. J. Phycol.* 31:181–8.

Walther, W. G., and Edmunds, L. N. 1973. Studies on the control of the rhythm of photosynthetic capacity in synchronized cultures of *Euglena gracilis* (Z). *Plant Physiol.* 51:250–8.

# CHAPTER 22

# VIRAL CONTAMINATION
# OF ALGAL CULTURES

JANICE LAWRENCE

*Department of Biology, University of New Brunswick*

## CONTENTS

Key Index Words: Microalgae, Macroalgae, Phytoplankton, Cyanobacteria, Virus, Infection, Phycodnaviridae, Cyanophage, Viral Contamination

## 1.0. INTRODUCTION

For many years, investigators have been aware of the existence of viruses in algae. In most instances this stemmed from observations of viruslike particles (VLPs) when cells or tissue were examined by electron microscopy (e.g., Lee 1971, Chapman and Lang 1973). In a comprehensive review of viruses of algae, Van Etten et al. (1991) reported that viruses or VLPs from at least 44 eukaryotic algal taxa had been reported. Since this report, the number of taxa has grown considerably. As awareness of the occurrence of viruses in prokaryotic and eukaryotic algae has increased and their significance has been recognized (e.g., Suttle 2000a, 2000b), investigators have become interested in assessing for the presence of viruses and VLPs in their cultures and samples. It is recognized that viruses can lead to difficulties in culturing, particularly if a lytic viral pathogen has been inadvertently introduced and the algal culture is lost. Even when viruses do not cause culture lysis, their presence can be problematic because of contamination of nucleic acid preparations in molecular and genomic studies.

This chapter provides phycologists with some tools for assessing algal cultures for viral infection; it outlines the strengths and weaknesses of a number of approaches and supplies simplified methods adapted from those developed and used by algal virologists. There are other reviews of methods for enumerating and detecting viruses in cultures and environmental samples (e.g., Suttle 1993), but none have specifically focused on viruses infecting algae. The methods described in this chapter do not replace those already available for studying viruses but are intended to introduce phycologists to the tools needed to perform a simple diagnosis.

This chapter first provides an overview of virus biology and then a brief history of algal virus research. More thorough examinations of algal viruses are available elsewhere, such as extensive reviews by Van Etten et al. (1991, 2002) and Suttle (2000a, 2000b). As the field of algal virology is in its infancy, there will undoubtedly be further developments and discoveries leading to a better understanding of the relationship between algae and the viruses that infect them. I hope this chapter will serve to illuminate those curious about this fascinating relationship and provide them with the tools to partake in these discoveries.

## 1.1. General Biology of Viruses

Viruses are ultramicroscopic entities measuring only 20 to 400 nm. They comprise a nucleic-acid core surrounded by a protein coat (capsid) and sometimes an outer lipid envelope. Individual viral particles can be icosahedral (polygonal), helical, or complex in structure, and contain only one type of nucleic acid, either RNA or DNA, in single-stranded (ss) or double-stranded (ds) form. As viruses cannot reproduce independently of a host cell, they are considered to be obligate intracellular parasites and, as such, are relegated to their own taxonomic group. The classification system of the International Committee on the Taxonomy of Viruses

(ICTV) is used to define families of viruses, and it is based largely on the nucleic-acid type and the presence or absence of an envelope. Further classification is based on capsid symmetry, host, pathology of the disease, site of virus replication, and other properties.

The term *algal virus* is often used to group together the viruses that infect all types of algae. There are two major types of algal viruses: the cyanophage and the eukaryotic algal viruses. Cyanophage are viruses that infect the prokaryotic algae, or cyanobacteria. They are divided among three families of tailed phage: Myoviridae, which have contractile tails that are separated from the capsid by a neck; Podoviridae, which have short, noncontractile tails; and Siphoviridae, which have long, noncontractile tails. All of the phage in these families contain dsDNA and infect bacteria and archaea (Ackermann and DuBow 1987, Murphy et al. 1995). Unlike the cyanophage, there is currently only one family of eukaryotic algal viruses, the Phycodnaviridae, which is currently subdivided into four genera—*Chlorovirus*, *Prasinovirus*, *Prymnesiovirus*, and *Phaeovirus*—according to the algae they infect. The members of this family are tailless and contain dsDNA. There are, however, a number of viruses that infect eukaryotic algae that do not fit into these genera or the family Phycodnaviridae, such as an ssRNA virus that infects *Heterosigma akashiwo* (Tai et al. 2003), a dsRNA virus that infects *Micromona pusilla* (Brussaard et al. 2004), a ssRNA virus that infects *Rhizosolenia setigera* Brightwell (Nagasaki et al. 2004) and a novel virus system that produces two distinct viral particles in *Heterosigma akashiwo* (Lawrence and Suttle, unpublished).

The basic life-strategy of a virus is as an obligate intracellular parasite. The three main components of this strategy are to find a suitable host (adsorption, penetration, and uncoating), to utilize the host's reproductive machinery for proliferate reproduction (transcription, translation, and replication), and to lyse the host to release the progeny back into the environment (assembly and release). Aquatic algal viruses are likely transmitted between hosts by passive diffusion, although other vectors, such as invertebrates, have not yet been examined. The probability of contact between a virus and host, and therefore the potential infection rate, are directly proportional to the product of the viral and host abundances (Murray and Jackson 1992). Adsorption of the virus to the host will only then occur if there is contact between molecular structures on both. Individual proteins or a group of proteins may act as a recognition structure on a virus. The adsorption structure may be protein, carbohydrate, or glycolipid, and it often has other known functions in host cellular metabolism. For example, T5 uses a siderophore to gain entry into its host, vaccinia virus uses an epidermal growth factor receptor, and rabies virus uses an acetylcholine receptor. While these host structures are often referred to as viral receptors, this is perhaps misleading, as the primary function of these components is not for viral attachment. There are no energy requirements for attachment, but adsorption may be pH-dependent or require additional cofactors. For example, $Ca^{2+}$ and $Mg^{2+}$ are frequently required in millimolar concentrations for cyanophage adsorption (Ackerman and DuBow 1987). This adsorption process occurs rapidly relative to the collision rate.

Once a virus has attached to the host, viral DNA enters the host either by surface fusion and injection or endocytosis. The infection can follow one of two basic cycles: lytic or latent. When considering these viral replication cycles it is important to bear in mind that viruses are parasites and their success is entirely dependent on the host. The most familiar cycle is lytic infection, which is well documented in microalgae. Once viral nucleic acid has entered a host cell, transcription, translation, and replication of progeny viruses begin. Viral pathogenesis is highly variable and may result in apoptotic or necrotic cell death and lysis. The cytopathic effects observed during necrosis include disrupted mitochondria, cell swelling, organizational disruption, and lysis. During apoptosis the effects include DNA condensation and fragmentation, cell shrinkage, and membrane blebbing. Progeny viruses are released during lysis of the host (Fig. 22.1a) and are then available to other host cells. The lytic process contributes significant organic matter to the surrounding environment through cell debris and is an important source of organic matter for the microbial community. The recycling of this material through viral lysis is called the microbial shunt (Wilhelm and Suttle 1999).

During latent infections the viral genome may remain dormant in the host for many generations. This has been demonstrated in macroalgae, bacteria, and mammalian cells. In bacteria, which include cyanobacteria, this phage cycle is called lysogeny, and the bacterial cells whose chromosomes contain phage DNA are lysogens. Latency provides the virus with protection from environmental conditions and could, as in the case of prokaryotic systems, confer the host with protection from infection by similar viruses. Latent viral DNA is replicated indefinitely along with the DNA of host cells until an environmental or biochemical trigger induces lytic reproduction (see Fig. 22.1b). Induction of a temperate virus or lysogenic phage is usually associated with

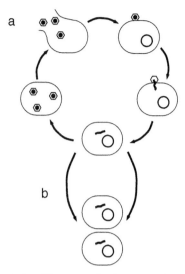

**FIGURE 22.1.** The reproductive strategies of viruses. In lytic infection **(a)**, a free virus infects a host, replicates, and lyses the host. During latent infection **(b)**, the viral genome is incorporated into the host and is passed on to daughter host cells until the virus is triggered to undergo lytic reproduction.

a threat to the survival of the host. Integration of viral DNA with genomic, chloroplastid, or mitochondrial DNA also means these infections could be important as mechanisms of lateral gene transfer, as has been demonstrated in prokaryotic systems. Lysogenic associations between cyanophage and filamentous cyanobacteria have been reviewed by Sherman and Brown (1978) and Martin and Benson (1988), and new lysogens have been isolated since those reviews (Bisen et al. 1986, Franche 1987, Ohki and Fujita 1996). Known lysogenic associations with unicellular cyanobacteria are fewer, with the only well-documented species to date infecting the marine *Synechococcus* strain NKBG 042902 (Sode et al. 1994). Latent infections among algae have been found only in macroalgal species thus far (Table 22.1).

The existence of chronic or persistent infections in algae is a matter of debate at present. Chronic infections, during which viral progeny are continually produced and shed by an otherwise healthy host, have been identified in multicellular hosts. During chronic infections, either infected cells survive infection and continue to produce infectious viruses, or else a minority of cells in an otherwise healthy (i.e., resistant) population is infected and lyses, and therefore the spread of the virus is limited. In this situation cell death is counterbalanced by new cell production, and there may be only subtle effects on the health of the host or population, rendering this type of infection difficult to detect. Within unicellular host populations, identifying true

**TABLE 22.1** Characteristics of the eukaryotic algal viruses isolated and characterized to data.

| Virus | Host alga | Nucleic acid | Particle size | Morphology | Replication | Burst size | Latent/lytic | References |
|---|---|---|---|---|---|---|---|---|
| CCV | Chara corallina Willdenow Charophyceae | ssRNA | 18 nm× 532 nm | Rod | Cytoplasm | | Lytic, 10–12 d | Gibbs et al. 1975, Skotnicki et al. 1976 |
| | Chlorococcum minutum Starr Chlorophyceae | dsDNA | 220 nm | Icosahedral | Cytoplasm | | | Gromov and Mamkaeva 1981 |
| CVLP | Cylindrocapsa geminella Wolle Chlorophyceae | dsDNA | 200 nm | Icosahedral | Nucleus/ cytoplasm | | Latent | Hoffman and Stanker 1976, Stanker et al. 1981 |
| | Uronema gigas Vischer Chlorophyceae | dsDNA | 390 nm | Icosahedral | Nucleus | | Lytic | Cole et al. 1980, Dodds and Cole 1980 |
| PBCV | Chlorella-like green alga symbiotic with Paramecium bursaria Ehrenberg | dsDNA | 185 nm | Icosahedral | Nucleus | 200–350 | Lytic, 12–20 h | Meints et al. 1981, Van Etten et al. 1982 |
| HVCV | Chlorella-like green alga symbiotic with Hydra viridis Linnaeus | dsDNA | 185 nm | Icosahedral | | | Lytic, 24 h | Meints et al. 1981, Van Etten et al. 1981 |
| ChlV | Chlorella sp. Chlorophyceae | dsDNA | 190 nm 330 kb | Icosahedral | Nucleus | 200–350 | Lytic, 4–6 h | Van Etten et al. 1991, Van Etten and Meints 1999 |
| BtV | Aureococcus anophagefferens Hargraves and Seiburth Pelagophyceae | dsDNA | 150 nm | Icosahedral | Nucleus | >500 | Lytic, 48 h | Milligan and Cosper 1994, Gastrich et al. 1998 |
| AaV | Aureococcus anophagefferens[a] Pelagophyceae | | | Cytoplasm | | | Lytic | Garry et al. 1998 |
| HcV | Heterocapsa circularisquama Horiguchi Dinophyceae | dsDNA | 200 nm | Icosahedral | Cytoplasm | >1,300 | Lytic, 48–72 h | Taruntani et al. 2001 |
| EhV | Emiliania huxleyi Lohman (Hay and Mohler) Prymnesiophyceae | dsDNA | 170 nm 415 kb | Icosahedral | Cytoplasm | 400–1,000 | Lytic, 12–15 h | Castberg et al. 2002 |
| PpV | Phaeocystis pouchetii (Hariot) Langerheim Prymnesiophyceae | dsDNA | 120 nm 485 kb | Icosahedral | Cytoplasm | 350–600 | Lytic, 12–18 h | Jacobsen et al. 1996, Sandaa et al. unpub. |
| MclaV | Myriotrichia clavaeformis Harvey Phaeophyceae | dsDNA | 195 nm 320 kb | Icosahedral | Nucleus/cytoplasm | | Latent | Kapp et al. 1997, Wolf et al. 2000 |
| HincV | Hincksia hinksiae (Harvey) Silva Phaeophyceae | dsDNA | 140–170 nm 240 kb | Icosahedral | Nucleus/cytoplasm | $1-2 \times 10^6$ | Latent | Parodi and Müller 1994, Kapp et al. 1997, Wolf et al. 1998 |

| Abbreviation | Host | Nucleic acid | Size / Genome | Morphology | Location | Burst size | Infection | Reference |
|---|---|---|---|---|---|---|---|---|
| PlitV | *Pilayella littoralis* (Linnaeus) Kjellman Phaeophyceae | dsDNA | 160 nm 280 kb | Icosahedral | Cytoplasm | $2.5 \times 10^5$ | Latent | Maier et al. 1998 |
| EsV | *Ectocarpus siliculosus* (Dillwyn) Lyngbye Phaeophyceae | dsDNA | 130 nm 340 kb | Icosahedral | Cytoplasm | $2 \times 10^6$ | Latent | Müller et al. 1990 |
| EfasV | *Ectocarpus fasciculatus* Harvey Phaeophyceae | dsDNA | 140 nm 320 kb | Icosahedral | Nucleus | | Latent | Müller et al. 1996, Kapp et al. 1997 |
| FsV | *Feldmannia* sp. Phaeophyceae | dsDNA | 150 nm | Icosahedral | Nucleus | $1–5 \times 10^6$ | Latent | Henry and Meints 1992 |
| FlexV | *Feldmannia simplex* (Crouan and Crouan) Hamel Phaeophyceae | dsDNA | 120 nm 220 kb | Icosahedral | | | Latent | Müller and Stache 1992, Friess-Klebl et al. 1994 |
| FirrV | *Feldmannia irregularis* (Kützing) Hamel Phaeophyceae | dsDNA | 150 nm 180 kb | Icosahedral | | | Latent | Kapp et al. 1997 |
| PoV | *Pyramimonas orientalis* Butcher ex McFadden, Hill and Wetherbee Prasinophyceae | dsDNA 560 kb | 200 nm | Icosahedral | Cytoplasm | 800–1,000 | Lytic, 14–19 h | Sandaa et al. 2001 |
| MpV | *Micromonas pusilla* (Butcher) Manton and Parke Prasinophyceae | dsDNA | 115 nm 200 kb | Icosahedral | Cytoplasm | | Lytic, 7–14 h | Cottrell and Suttle 1991 |
| MpV | *Micromonas pusilla* Prasinophyceae | dsRNA | 50 nm | Icosahedral | Cytoplasm | 400 | Lytic | Brussaard et al. 2004 |
| CbV | *Chrysochromulina brevifilum* Parke and Manton Prymnesiophyceae | dsDNA | 145–170 nm | Icosahedral | Cytoplasm | >320 | Lytic, ~7 d | Suttle and Chan 1995 |
| CeV | *Chrysochromulina ericina* Parke and Manton Prymnesiophyceae | dsDNA | 160 nm 510 kb | Icosahedral | Cytoplasm | 1,800–4,100 | Lytic, 14–19 h | Sandaa et al. 2001 |
| HaV | *Heterosigma akashiwo* Hada Raphidophyceae | dsDNA | 200 nm | Icosahedral | Cytoplasm | 770 | Lytic, 30–33 h | Nagasaki and Yamaguchi 1997 |
| HaNIV | *Heterosigma akashiwo* Raphidophyceae | | 30 nm | Icosahedral | Nucleus | $1 \times 10^5$ | Lytic, 42 h | Lawrence et al. 2001 |
| HaRNAV | *Heterosigma akashiwo* Raphidophyceae | ssRNA | 25 nm 9100 bp | Icosahedral | Cytoplasm | | Lytic, 96 h | Tai et al. 2003 |

[a]AaV and BtV are reported to be the same virus, although they differ in size and morphology.

chronic infection is especially difficult. Chronic infections will appear very similar to latent infections of populations in which certain cells have been induced to undergo a lytic cycle, as well as lytic infections in which only a subset of the population is susceptible to infection. The only microalga in which chronic infection has been suspected is *Brachiomonas submarina* Bohlin (L. R. Hoffman, University of Illinois, personal communication).

Viruses are restricted with respect to the hosts they can infect and are generally believed to infect hosts only within a single species. Confusion surrounding the taxonomy of cyanobacteria has clouded attempts to accurately determine the host ranges of the cyanophage. In the future, advancements in the molecular taxonomy of cyanobacteria will help to elucidate the host ranges and resolve the level of host specificity among the cyanophage. The few studies that have examined viruses of eukaryotic unicellular algae suggest that the host ranges of these viruses are limited to within a species but are complex at the strain level (Jacobsen et al. 1996; Nagasaki and Yamaguchi 1997, 1998; Tai et al. 2003; Lawrence et al. 2001). In contrast, there is evidence that viruses infecting multicellular eukaryotic algae have an intergenic host range (Müller and Schmid 1996).

Through their host specificity, viruses can exert considerable selective pressure within a population of algae. Furthermore, through natural succession and species shifts, these changes in population dynamics have the potential to influence community structure and have cascading effects in aquatic ecosystems (Larsen et al. 2001). The impact of viruses is density-dependent, and as such, the most drastic impacts of viruses on algal populations are evident during algal blooms. Indeed, there are accumulating observations of VLPs in coincidence with the collapse of algal blooms (Sieburth et al. 1988; Bratbak et al. 1993; Nagasaki et al. 1994a, 1994b; Bratbak et al. 1996; Brussaard et al. 1996). Alternatively, when host densities are low the spread of viral infection will be limited and possibly even eliminated.

## 1.2. History of Algal Virus Research

In the 1800s the term *virus*, which is Latin for *poison* or *venom*, was used to describe any agent harmful to an organism. Development of the porcelain bacterial filter in 1884 by Charles Chamberland opened the doors to virology, permitting the first investigations of the filterable agent causing tobacco mosaic disease, now known as the tobacco mosaic virus (TMV). By the late 1930s

it was evident that viruses are complexes of nucleic acid and proteins, and that they are able to replicate only inside living host cells. Given that living organisms, by definition, are capable of self-reproduction, this sparked an ongoing debate over whether or not viruses are living.

The first reports of eukaryotic algal viruses appeared in the literature in the early 1970s. These citations identified VLPs by electron microscopy of cultured and wild-collected algae. Very rapidly, VLPs were observed in almost every eukaryotic algal phyla, including the rhodophytes (Lee 1971, Chapman and Lang 1973), chromophytes (Toth and Wilce 1972, Baker and Evans 1973, Clitheroe and Evans 1974, Markey 1974, LaClaire and West 1977, Hoffman 1978, Sicko-Goad and Walker 1979), chlorophytes (Tikhonenko and Zavarsina 1966, Mattox et al. 1972, Pickett-Heaps 1972, Swale and Belcher 1973, Pearson and Norris 1974, Pienaar 1976), cryptophytes (Pienaar 1976), dinophytes (Soyer 1978, Sicko-Goad and Walker 1979), and prymnesiophytes (Manton and Leadbeater 1974, Pienaar 1976). Most of these VLPs were large polygonal or icosahedral particles found in either the cytoplasm or the nucleus of infected algal cells and, based on their general characteristics, were most similar to iridoviruses. In addition, long, flexuous structures have been found in some of these algae (Mattox et al. 1972, Clitheroe and Evans 1974, Markey 1974, Tripodi and Beth 1976). These VLPs were not confirmed as viruses often because of the technical difficulties associated with isolating, purifying, and confirming their infectivity. To prove the causal relationship between VLPs and diseases observed in algae containing VLPs, a number of criteria must be fulfilled (Koch 1884). The virus must be present in all diseased cases but not in any healthy cases, the virus must be isolated and purified from the diseased host, the same disease must result from inoculating a healthy host with the purified virus, and the same virus must be isolated again from the diseased host.

Safferman and Morris isolated the first cyanophage in 1963. This virus infected a filamentous freshwater cyanobacterium, and it wasn't until 1981 that the first marine cyanophage was isolated (Moisa et al. 1981). During the 25 years after the isolation of the first cyanophage, most research concentrated on understanding basic cyanophage biology (reviewed by Brown 1972, Sherman and Brown 1978, Martin and Benson 1988). In the 1990s the ecological significance of cyanophage became a focus of interest when abundances of viruses in natural marine waters were documented to be in excess of $10^7 \cdot mL^{-1}$ (Bergh et al. 1989,

Børsheim et al. 1990). It was also determined that a large proportion of natural marine cyanobacterial populations were visibly infected (Proctor and Fuhrman 1990). Shortly thereafter, a number of research groups began to report on the high abundances of cyanophages in seawater, frequently finding over $10^4$ to $10^5$ infectious viruses per milliliter of seawater (Suttle and Chan 1993, Suttle et al. 1993, Waterbury and Valois 1993; see also Suttle 2000a, 2000b).

In 1975 the first eukaryotic algal virus was isolated, purified, and characterized (Gibbs et al. 1975, Skotnicki et al. 1976). This ssRNA virus infects the green alga *Chara corallina*, but it is unlike other algal viruses described to date. *Chara corallina* virus (CCV) most closely resembles the tobamoviruses or furoviruses (cited in Van Etten et al. 1991), two families of rod-shaped RNA viruses that infect higher plants. Interestingly, the Charales are believed to be an ancestral link between the algae and higher plants.

In 1978, Oliveira and Bisalputra isolated and purified the first representative of the large icosahedral VLPs from the brown alga *Sorocarpus uvaeformis* (Lyngbye) Pringsheim. Much of the eukaryotic algal virus research conducted in the 1980s and 1990s focused on viruses that infect *Chlorella*-like algae that are symbionts of *Hydra viridis* and *Paramecium bursaria* (Meints et al. 1981; Van Etten et al. 1981, 1982, 1991), leading to the establishment of the Phycodnaviridae (Van Etten and Ghabrial 1991). Significant progress was also made in the 1990s toward understanding a group of viruses that infect ectocarpalean brown algae (Oliveira and Bisalputra 1978, Müller 1991a, Henry and Meints 1992, Müller and Frenzer 1993, Van Etten and Meints 1999).

The Phycodnaviridae include the most intensively studied algal viruses (Van Etten and Ghabrial 1991). These viruses have several properties in common, including large size (>100 nm), icosahedral shape, dsDNA, and no obvious membrane (Van Etten et al. 1991). They also share several conserved regions within their DNA polymerase genes (Chen and Suttle 1996). Phylogenetic analysis has shown that the phycodnaviruses, which include both microalgal and macroalgal viruses, are more closely related to each other than to other dsDNA viruses (Lee et al. 1998), and they form a distinct monophyletic group, suggesting they originated from a common ancestor.

Until recently the only known non-Phycodnaviridae algal virus was CCV (Gibbs et al. 1975, Skotnicki et al. 1976). However, a number of novel viruses have recently been isolated and characterized. These include a dsRNA virus that infects *Micromonas pusilla* (Brussaard et al. 2004), an ssRNA that infects *Heterosigma*

*akashiwo* (Tai et al. 2003), a small (30-nm) virus that infects *Heterosigma akashiwo* (Lawrence et al. 2001), and a novel virus system that produces two distinct viral particles in *Heterosigma akashiwo* (Lawrence and Suttle, unpublished observations). There are also a number of tailed icosahedral viruses (Swale and Belcher 1973, Pienaar 1976, Dodds and Cole 1980, Gromov and Mamkaeva 1981), including the *Brachiomonas* virus, which causes a persistent infection in *Brachiomonas submarina* (Hoffman, personal communication; see Table 22.1). These viruses are not available from general culture collections, and their distribution is managed by the researchers who have isolated them and retain them in private collections.

The ecology of viruses in aquatic systems employs a number of new approaches. Flow-cytometric methods are used for detecting and enumerating viruses (Marie et al. 1999, Brussaard et al. 2000, Chen et al. 2001). Pulsed-field-gel electrophoresis is used to examine the diversity of viral genome sizes (Wommack et al. 1999, Steward et al. 2000). Restriction fragment length polymorphism analysis is used to study genetic variation in cyanophage (Muradov et al. 1990, Wilson et al. 1993, Lu et al. 2001) and eukaryotic algal virus isolates (Cottrell and Suttle 1991). Perhaps the most significant advances are being made as genetic sequence data become available for algal viruses, thus enabling researchers to develop methods for identifying algal viruses within mixed virioplankton samples and establishing diversity and abundance. Sequence data aided PCR methods for amplifying algal-specific gene fragments (Chen and Suttle 1995a, Fuller et al. 1998, Zhong et al. 2002), which in turn furthered our understanding of phylogenetic relationships (Chen et al. 1996, Chen and Suttle 1996, Wilson et al. 2000, Short and Suttle 2002, Frederickson et al. 2003). It has also allowed investigators to detect viral DNA in latently infected host algae (Bräutigam et al. 1995, Sengco et al. 1996).

## 2.0. METHODS FOR DETECTING VIRAL INFECTIONS

### 2.1. Visual Detection

Perhaps the most obvious way is to observe the virus particles. This, however, is not a trivial task when the particles range in size from 20 to 400 nm. The following section outlines three methods for visualizing virus particles using epifluorescence microscopy (EFM) or transmission electron microscopy (TEM). Visual detection, however, has a number of drawbacks. Observation of VLPs intracellularly or extracellularly does not confirm a causal relationship without fulfilling the rest of Koch's postulates. Relying on visual detection is especially problematic when working with bacterized algal cultures. The presence of bacteriophage in algal cultures is highly likely, and outside of their host the phage may be mistaken for algal viruses. Conversely, false-negative results can occur, especially if algal viral abundance is low and/or cell debris is abundant. False-positives can be reduced by examining uninfected cultures in the same manner and looking for differences that are consistent with infection.

## 2.2. Fluorescent Dyes and Epifluorescence Microscopy

Relatively simple, rapid techniques for detecting free viruses have been developed using fluorescent dyes that bind to nucleic acids. The fluorescent dyes have a high specificity for nucleic acids; therefore, only viruses and other particles containing nucleic acids will be illuminated. EFM detection does not work well with cultures that contain contaminating bacteria and associated bacteriophage as these particles are difficult to discriminate when fluorescently stained (Fig. 22.2). However, these dyes will permit detection of nonspherical viruses, such as rods, which are difficult to identify with TEM. Since the host cells contain genomic, chloroplastid, and mitochondrial DNA, all of these will be stained with the nucleic acid stains. However, the size, shape, and staining intensity of the host cells make them easy to distinguish from free viral particles. EFM examination provides no information on the type of virus, so its application is usually restricted to monitoring virus cultures in algal virology laboratories. Results are not generally accepted as proof of viral infection and therefore are not often included in reports on the isolation and characterization of new viruses.

The first EFM methods for virus detection used DAPI (4'6-diamidino-2-phenylindole) (Sigma-Aldrich), an ultraviolet-excited dye that fluoresces blue (Coleman et al. 1981, Suttle et al. 1990, Hara et al. 1991). The method is rapid, but DAPI fluorescence is relatively weak and requires image enhancement or high-quality optics to detect some viruses, especially small ones (Weinbauer and Suttle 1997). Chlorophyll autofluorescence of host cells in EFM preparations can be omitted

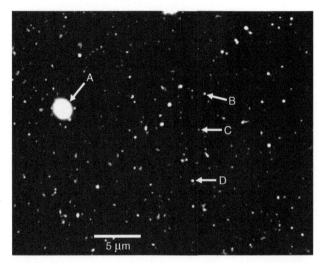

**FIGURE 22.2.** Epifluorescence micrograph of *Heterosigma akashiwo* virus system Ols20 lysate. This novel system produces two different-sized algal virus particles (~35 and ~200 nm). Non-axenic lysate was stained with SYBR Green I and photographed digitally (1-second exposure) under blue excitation. Arrows indicate as follows: **(a)** bacterial cell (near saturation); **(b)** large algal virus particle; **(c)** small algal virus particle; and **(d)** contaminating bacteriophage. Photo courtesy of A. M. Comeau and K. C. Wen.

with filters, allowing detection of virus particles in association with algal material (Maier and Müller 1998).

Yo-Pro (4-[3-methyl-2,3-dihydro-(benzo-1,3-oxazole)-2-methylmethyledene]-1-1[3'-trimethylammoniumpropyl]-quinolinium diiodide) (Molecular Probes) is a cyanine-based dye that fluoresces green when excited with blue light (Hirons et al. 1994) and was first described for staining marine viruses by Hennes and Suttle (1995). While Yo-Pro is a high-fluorescence-yield dye making visual detection simple, it is incompatible with many aldehydes, so samples must be prepared from unfixed material, rinsed, and then incubated with stain for 48 hours for complete staining. Further developments of this method have reduced the incubation time by microwave irradiation (Xenopolous and Bird 1997) and allowed staining of fixed samples. As this dye provides similar results to the following dyes and provides no obvious improvements for viral detection, it will not be discussed further.

A new generation of high-affinity nucleic acid stains has become available and has been applied to virus detection in the last 5 years (Noble and Fuhrman 1998). The SYBR generation dyes (Molecular Probes) include SYBR Green I and II and SYBR Gold (Lebaron et al. 1998, Noble and Fuhrman 1998, Marie et al. 1999,

Chen et al 2001). These dyes have a high fluorescence yield, provide the most rapid staining, and may be used to stain fixed samples. Also, a number of these dyes, such as SYBR Green II, have a high binding affinity for both RNA and DNA, permitting detection of all types of viruses (see Fig. 22.2).

Flow cytometry may also be employed (see Chapters 7, 17). Flow cytometry permits rapid, accurate detection of viruses in solution and can distinguish up to three virus populations (Marie et al. 1999, Brussaard et al. 2000). Therefore, the investigator may be able to examine for contaminating bacteriophage in addition to algal viruses.

### 2.2.1. DAPI Staining Technique

For a DAPI stock solution (5 $\mu g \cdot mL^{-1}$), make up DAPI (Sigma-Aldrich) in McIlvaine's buffer (pH 4.4). To make buffer, dissolve 3.561 g of $Na_2HPO_4 \cdot 2H_2O$ in 100 mL of distilled water (solution A) and 2.101 g of citric acid in 100 mL distilled water (solution B); combine 8.82 mL of solution A with 11.18 mL of solution B. Store at 4°C in the dark.

1. Filter (≥0.45 $\mu m$ low-protein-binding filter; e.g., Durapore polyvinylidene difluoride filters [Millipore]), or low-speed-centrifuge a 2-mL aliquot of infected culture medium to remove host algae and debris.
2. Add DAPI to a final concentration of 1 $\mu g \cdot mL^{-1}$ and incubate in dark for 30 minutes.
3. Filter 0.8 mL (maximum volume retained within outer support ring) through a 0.02-$\mu m$ $Al_2O_3$ filter (Anodisc, Whatman Int. Ltd.) at <20 kPa. Use a 0.45-$\mu m$ nitrocellulose filter as a backing filter for even filtration. Turn vacuum off as soon as filter dries.
4. Lay filter over a drop of low-fluorescence immersion oil on microscope slide. Place a drop of oil on filter and cover with coverslip.
5. View at 1,000× magnification with 334 to 365 nm excitation and >420–nm emission filters.

### 2.2.2. SYBR Staining Technique

1. Filter as for DAPI staining.
2. Place 100 $\mu L$ of 2.5% SYBR in deionized water in the bottom of a petri dish. Place Anodisc sample side up on stain solution and incubate 15 minutes in dark.
3. Remove Anodisc from stain and blot excess fluid from back and sides of filter, being careful not to touch the sample.

4. Mount Anodisc on microscope slide with a drop of 50% glycerol, 50% phosphate buffered saline (0.05 M $Na_2HPO_4 \cdot 2H_2O$, 0.85% NaCl; pH, 7.5) with 0.1% p-phenylenediamine (Sigma-Aldrich) made fresh daily from a frozen 10% aqueous stock. Cover with coverslip.
5. View at 1,000× magnification with blue excitation (peak excitation is 498 nm). Red and yellow autofluorescence can be limited by using an emission filter.
6. If the sample is too concentrated, dilute the sample 1 : 10 or 1 : 100 with sterile 0.02 $\mu m$ filtered culture medium.

### 2.3. Transmission Electron Microscopy

An alternative is to stain free virus particles with heavy metals (negative staining) and observe them under TEM (Fig. 22.3). This method provides more information than EFM, e.g., morphological characteristics and the potential to discriminate between contaminating phage and algal viruses. It requires a trained eye to identify VLPs among cellular debris. Furthermore, if the

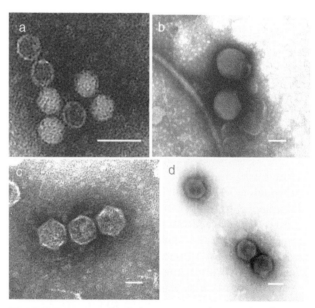

FIGURE 22.3. Algal viruses negatively stained with uranyl acetate and viewed under transmission electron microscopy. Algal viruses occur in various shapes and sizes. These viruses were isolated against *Heterosigma akashiwo* from environmental samples. Extracellularly viruses may be small, nearly spherical particles (a); large, membrane enveloped particles with short tails (b); large icosahedrals lacking tails (c); and medium-sized icosahedrals with short tails (d). Scale bars are 50 nm.

algal culture produces significant cell debris or the virus yield is low, detection may be nearly impossible and lead to false-negative findings. This being said, TEM examination of lysates produced by pathogens suspected to be viruses has become routine and is generally required to confirm Koch's postulates. Gibbs et al. (1975) examined purified preparations of infected *Chara corallina* to verify that the pathogenic agent was a rod-shaped virus, closely resembling the tobacco mosaic virus. Mayer and Taylor (1979) prepared negatively stained samples of medium from lysed *Micromonas pusilla* cultures and identified hexagonal or pentagonal VLPs. New isolates infecting cyanobacteria have also been found to be similar to bacteriophage (Safferman and Morris 1963, Suttle and Chan 1993). Negative staining has also been used in viral ecology studies. Suttle et al. (1990) used TEM to confirm that VLPs were present in seawater samples that, when added to phytoplankton cultures, resulted in reduced primary productivity.

There are two general methods for preparing negatively stained viruses from cultures with low viral titers. Either a viral concentrate is prepared by ultrafiltration and applied to a TEM specimen grid (Proctor and Fuhrman 1990) or viruses are sedimented onto the specimen grid by ultracentrifugation (Bergh et al. 1989, Børsheim et al. 1990). If the algal culture produces a high viral titer, the concentration steps will not be needed. The method outlined below is a simplified adaptation of the method outlined in Børsheim et al. (1990) that permits the detection of viruses but does not allow the investigator to calculate viral abundance. The following procedure should be carried out under dim light.

### 2.3.1. Concentration and Adsorption

1. Filter a 10-mL aliquot of infected algal culture through a 0.45-μm low-protein-binding filter (e.g., Durapore) to remove host algae, and concentrate filtrate by ultracentrifugation (see Step 8). For macroalgal cultures, collect 1 to 5 g of algal culture on a nylon mesh. After removing capillary water, add 1 mL of culture medium (same as used for culturing host) at 37°C and manually squeeze sample, repeating 5 to 10 times to produce a suspension of virus particles. Filter and centrifuge as above.
2. Holding Formvar-coated 300-mesh copper grid (Electron Microscopy Supplies) in electron microscopy (EM) forceps, drop 10 μL of concentrate on grid and allow to adsorb for 10 minutes.

3. Wick off with hardened Whatman No.1 filter paper (Whatman Int. Ltd.).
4. Drop 10 μL of 1% uranyl acetate or 2% uranyl sulfate or 2% phosphotungstic acid (Structure Probe) on grid and incubate for 10 seconds.
5. Wick off excess stain with filter paper and allow to air-dry for 15 minutes.
6. View in TEM at approximately 10,000× magnification.
7. If the sample is too concentrated, dilute the viral concentrate 1 : 10 or 1 : 100 with sterile 0.02-μm-filtered culture medium. If there are too few particles on the grid, allow the concentrate to adsorb for up to 1 hour in a humid chamber. A chamber can be constructed by placing a wetted piece of filter paper inside a large covered petri dish. The sample should not be allowed to dry during the adsorption period.
8. To pellet small viruses, centrifuge samples until particles of 80S are sedimented with 100% efficiency. For samples in distilled water at 20°C, this can be calculated using the following formula:

$$T = (1/s)(ln[r_{max}/r_{min}]/[\omega^2 \times 60])$$

where T = time in minutes; s = sedimentation coefficient in seconds (an 80S particle has a sedimentation coefficient of $80 \times 10^{-13}$); $r_{min}$ = distance (cm) from the center of the rotor to the top of the sample in the centrifuge tube; $r_{max}$ = distance (cm) from the center of the rotor to the bottom of the centrifuge tube; ω (angular velocity in radians) = 0.10472 × rpm. Note that sedimentation times will vary as a function of temperature and salinity. Increase sedimentation times by ~12% if sample is 3.5% salt, and 25% for every 5°C below 20°C.

### 2.3.2. Thin Sectioning

The most convincing visual evidence of viral infection is the observation of virus particles inside cells of putatively infected cultures but not in the cells of uninfected cultures. The earliest TEM observations of algal infections were made with thin sectioning, and it is still regularly used to confirm infection in both macroalgae and microalgae. Unfortunately, thin sectioning of cells and viewing with EM are also the most labor-intensive methods and require the most expertise and infrastructure of all the methods described in this chapter.

Packaged virus particles are visible during only the last part of the lytic cycle and will not likely be present in all cells at the same time. This, combined with the

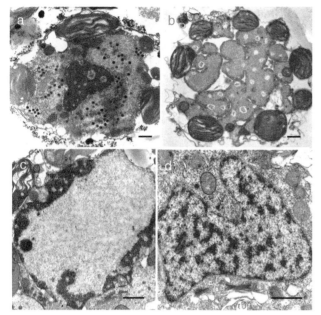

**FIGURE 22.4.** Sectioning cells and viewing with TEM provides tangible evidence of algal infections. Pathologies of infected *Heterosigma akashiwo* cells include large icosahedral viruses in the cytoplasm of infected cells **(a)**; swelling of the endoplasmic reticulum resulting in blebbing and the disintegration of cytoplasm **(b)**; and margination of heterochromatin and swelling of the nucleus **(c)**, as compared with an uninfected nucleus **(d)**.

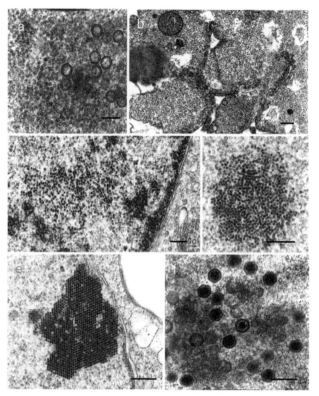

**FIGURE 22.5.** Intracellularly, virus particles exist in a variety of forms. In sectioned *Heterosigma akashiwo* cells, occasionally two distinct particles may be seen in an individual host cell **(a)**. Otherwise particles may be randomly dispersed in the cytoplasm **(b)** or nucleus **(c)** and may be found in unordered aggregations **(d)**, inclusions, or paracrystalline arrays **(e)** and appear hexagonal or pentagonal when sectioned **(f)**, suggesting 3-D icosahedral symmetry.

fact that only a small proportion of a cell is viewed at a time (i.e., 10 µm diameter cell cut in 100 nm sections = 100 sections per cell), requires that many cells on numerous sections must be viewed. Furthermore, it can be difficult to identify unknown viruses in cells, and this is compounded if the investigator is not familiar with the ultrastructure of the algal species in question (Figs. 22.4 and 22.5). At least as many uninfected cells as infected cells should be examined to ensure the proper identification of viruses and pathogenic effects.

Two general methods are provided below. The optimal preparation methods are best obtained from ultrastructural investigations of the algal species in question, while considering the general guidelines for preparing infected algal specimens provided below. Supplies for EM can be purchased through a number of companies, including Structure Probe Inc., Electron Microscopy Supplies, and Ted Pella Inc.

### 2.3.3. Macroalgal Method (from Müller et al. 1990)

1. Fix specimens for 2 hours on ice in an aqueous solution with a final concentration of 3% glutaraldehyde, 25% medium (that used to culture

the algae), and 0.25% caffeine in 0.05 M sodium cacodylate (pH, 7.7).
2. Wash in 50% medium with 0.5% caffeine and 0.1 M sodium cacodylate.
3. Post-fix for 1 hour in 1% osmium tetroxide.
4. Wash in distilled water, dehydrate in acetone, and embed in Spurr's resin (Spurr 1969).
5. Cut thin sections and stain with 2% aqueous uranyl acetate, followed by 0.5% lead citrate in 0.1 N sodium hydroxide (Reynolds 1963).
6. View in TEM at 10,000 to 20,000× magnification.

### 2.3.4. Microalgal Method (from Lawrence et al. 2001)

1. Fix cells for 2 hours on ice with 1% glutaraldehyde in 0.2 M sodium cacodylate and 0.25 M sucrose (pH, 7.2). A lower-osmolarity buffer may be required for some specimens.

2. Harvest cells by gentle centrifugation at 1,200 g for 5 minutes.
3. Without resuspending pellet rinse with 0.2 M sodium cacodylate, then resuspend cells in 1% osmium tetroxide in 0.1 M sodium cacodylate and post-fix for 1 hour.
4. Pellet cells and dehydrate in ethanol; embed in EPON 812 without resuspending.
5. Cut thin sections and stain with 2% aqueous uranyl acetate, followed by 0.5% lead citrate (Reynolds 1963).
6. View in TEM at 10,000 to 20,000× magnification.

When preparing infected algal specimens for thin sectioning and TEM examination, the following should be considered.

### 2.3.5. Preparing Infected Cells for Thin Sectioning

Because infected cells nearing lysis are extremely fragile, proper osmolarity and fixatives are important to ensure cells are not ruptured during processing. Cultures must be harvested, fixed, and dehydrated gently, by minimizing handling. In the case of unicellular algal cultures, this can be achieved by enrobing cells in agar (Proctor and Fuhrman 1990). Gently pelleted cells are covered in 50 μL of 1% 0.22-μm-filtered 38°C purified agar, allowed to solidify, and handled as an agar plug thereafter. Viruses are visible at only the end of the lytic cycle. To increase the probability of processing and observing infected cells, harvest microalgal cultures just prior to the projected decline in biomass or selectively harvest moribund cells by collecting nonmotile cells at the bottom of the culture flask, as some motile algae lose their motility when infected (Lawrence and Suttle 2004). When investigating macroalgal cultures, examine specimens for signs of infection, especially in sporangia, and prepare for EM. Small viruses are smaller than average thin sections (90–100 nm). If sections are more than one virus thick, viruses may not appear as clear and distinct particles. Cutting ultrathin sections (<90 nm) will help with this problem, but requires increased staining time to account for the decreased contrast of ultrathin sections. Small RNA-containing viruses may look like ribosomes if distributed in the cytoplasm. The viral nature of the particles can be confirmed by applying RNase to sectioned material. The RNase will digest ribosomal nucleic acids, leaving viral nucleic acids protected by their capsids (Hatta and Francki 1981).

## 2.4. Bioassays

Bioassays are a simple and intuitive approach to confirm viral infection. They do not require special equipment or genetic information and are relatively inexpensive. In order to confirm Koch's third postulate—that infection will result from the inoculation of uninfected culture with purified pathogen—a bioassay must be conducted. The only drawback with bioassays, especially with macroalgal cultures that have long life cycles, is that in order to obtain conclusive results, the host and virus must go through a number of life cycles. Also, to conduct bioassays the investigator must have access to uninfected cultures of the same strain as those suspected of being infected. The premise of a bioassay lies in propagating the effect seen in putatively infected cultures to uninfected cultures (Fig. 22.6). It is also important to eliminate the possibilities that the agents being propagated are abiotic or nonviral pathogens, as studies suggest that there are dissolved substances that adversely affect algal productivity (Suttle 1992) and bacterial pathogens that cause algal lysis (Chan et al. 1997, Kim et al. 1998). Assaying autoclave-sterilized and filtered material from infected cultures allows these possibilities to be tested.

### 2.4.1. Virus Propagation

Assaying for the presence of viruses essentially requires the same facilities and equipment as culturing algae. If

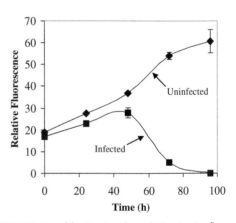

**FIGURE 22.6.** Monitoring the relative *in vivo* fluorescence of cultures can easily identify lytic viral activity in microalgal cultures. A rapid decline in fluorescence indicates a concomitant decrease in biomass. *Heterosigma akashiwo* cultures with (■) and without (◆) the addition of the algal virus HaNIV. Error bars are 1 SD; n = 3.

the microalga can be cultured on a solid medium, it permits the enumeration of infectious viruses and rapid isolation of virus clones. Each infective unit will form a clearing (plaque) on a lawn of host cells. A number of algal viruses have been isolated and purified on solid media, including cyanophages and eukaryotic algal viruses. The plaque assay for cyanophage isolation was first described by Safferman and Morris (1963) and is still the preferred method of isolation and purification of cyanobacterial species that can be grown on solid media (Suttle and Chan 1993). The plaque assay has been adapted for isolating a number of different eukaryotic microalgae, including viruses that infect *Chlorella*-like algae (Van Etten et al. 1983) and *Emiliania huxleyi* (Bratbak et al. 1996, Schroeder et al. 2002).

If the microalgae cannot be cultured on a solid medium, liquid cultures are used. Many viruses have been isolated in liquid culture, including those that infect cyanobacteria (Suttle and Chan 1993), dinoflagellates (Taruntani et al. 2001), prymnesiophytes (Suttle and Chan 1995, Jacobsen et al. 1996, Sanda et al. 2001), prasinophytes (Sandaa et al. 2001), and raphidophytes (Nagasaki and Yamaguchi 1997, Lawrence et al. 2001, Tai et al. 2003). The enumeration of infectious units in liquid culture may still be achieved by adding serial dilutions of a suspect culture to replicate cultures and examining for lysis at each dilution level. In this assay it is assumed that cultures that lyse must contain at least one infectious virus particle. Statistical analysis permits the investigator to then estimate the number of viruses present in the dilution series (see Suttle 1993 for complete method).

### 2.4.2. Microalgal Liquid Culture Method

Grow uninfected host under the same conditions as potentially infected cultures, monitoring biomass by cell counts or *in vivo* fluorescence. When the uncontaminated population has reached early exponential growth, inoculate with 5% (v/v) contaminated culture, and retain control cultures with no addition. Monitor growth of cultures. A decrease in biomass relative to the control indicates the presence of an algal pathogen.

### 2.4.3. Microalgal Solid Culture Method

This method is adapted from Van Etten et al. (1983), Suttle (1993), and Schroeder et al. (2002). Grow uninfected host cells under appropriate culture conditions and harvest cells during early exponential growth.

Concentrate, if needed, to obtain approximately $10^9$ cells $\cdot$ mL$^{-1}$. Combine 500 μL of host culture with 500 μL of sample to be assayed for the presence of viruses, and incubate 30 to 45 minutes to allow viruses to absorb to hosts. Mix sample with 2 to 2.5 mL 0.6% agar (in same medium used to culture algae) at ~47°C and immediately pour over plate containing 15 mL of 1% agar in culture medium. Incubate under conditions used for liquid cultures, monitoring for plaques (clearings in algal lawn) after an even host lawn appears.

### 2.4.4. Macroalgal Method

This method is adapted from Kapp et al. (1997). Harvest macroalgae (5–10 g) with symptoms of infection on nylon mesh screen. Remove capillary water and add 1 mL culture medium at 37°C and manually squeeze sample. Repeat Step 2 five to 10 times to produce a turbid suspension of virus particles. Inoculate uninfected algal culture containing swimming spores with virus suspension and allow to adsorb for 1 hour in culture dish. When spores have settled, remove algal filaments and add culture medium. Allow host spores to develop and examine for symptoms of infection with low-magnification microscopy.

### 2.4.5. Biotic Nature and Viral Size of Pathogens

Once the presence of a pathogen has been confirmed by bioassay, the biotic nature and size range of the pathogen may be determined. To determine if the pathogen is biotic, the lysate is heat-sterilized and assayed to determine if the pathogenic effect is retained post heat treatment. Heat treatment is achieved by autoclaving an aliquot of infected algal culture for 15 minutes and cooling prior to conducting a bioassay as outlined previously. It is imperative that cultures with no addition and untreated additions be assayed as controls.

To determine if the pathogen is viral-sized, filter aliquots of infected algal culture through 0.45-μm and 0.22-μm low-protein-binding filters (e.g., Durapore) and conduct bioassays with filtered, unfiltered, and no addition, as described previously. It is important to note that very large viruses may be retained on 0.45-μm filters and, conversely, very small bacteria may pass through 0.22-μm filters. Thus, filterability is not a definitive viral assay, but in most cases it will provide additional information on the size and/or nature of the pathogen (e.g., Milligan and Cosper 1994, Taruntani et al. 2001).

## 2.5. Molecular Techniques

The availability of genetic sequence data from a number of algal viruses led to the development of PCR-based methods to detect viruses. Chen and Suttle (1995a, 1995b) designed PCR primers against the DNA polymerase gene, which is highly conserved among eukaryotic algal viruses, and they amplified an algal virus-specific gene fragment (Fig. 22.7). This method has been used to detect algal viruses in the Gulf of Mexico (Chen et al. 1996) and has been adapted for use with denaturing gradient gel electrophoresis (DGGE) to examine algal virus diversity (Short and Suttle 2000, 2002). Brautigam et al. (1995) developed primers specific to the coat protein gene 30 of the *Ectocarpus siliculosus* virus (EsV), permitting the detection of EsV from around the world (Sengco et al. 1996). A number of primers have also been developed for detecting the prokaryotic cyanophage. Fuller et al. (1998) initially developed cyanophage-specific primers (CPSs) to amplify a fragment from gene 20, which is a capsid assembly protein. These primers were also used to develop a competitive PCR assay for quantifying viruses. Other primers for detecting cyanophage have since been developed against gene 20 (Zhong et al. 2002), and the basic PCR methods have been developed for use with DGGE to permit examination of diversity (Wilson et al. 2000, Frederickson et al. 2003).

While PCR provides a rapid diagnostic tool, it has a number of limitations and drawbacks. First, the method is only as exclusive and inclusive as the PCR primers themselves. Since primers are designed from known nucleic acid sequences, only viruses that have genetic sequences similar to the primers will be detected; novel viruses may not be detected. For example, the primers used by Chen and Suttle (1995a) target a highly conserved region in the DNA polymerase gene. While this gene is found in most DNA-containing viruses sequenced to date, it is not found in all and will clearly not be present in RNA-containing viruses.

The following method is adapted from Chen and Suttle (1995a, 1995b). The use of a nested primer (POL) verifies that the amplified DNA product is polymerase-specific. The development of universal primers and other molecular detection methods await the availability of sequence data from other viruses.

### 2.5.1. Eukaryotic PCR-Based Detection

This method is adapted from Chen et al. (1996). The primer sequences are:

AVS-1: (5′-GA[A/G]GGIGCIACIGTI[T/C]TIGA[T/C]GC-3′)
AVS-2: (5′-GCIGC[A/G]TAIC[G/T][T/C]TT[T/C]TTI[G/C][A/T][A/G]TA-3′)
POL: (5′-[G/C][A/T][A/G]TCIGT[A/G]TCICC[A/G]TA-3′)

1. Process culture material as described in 2.3.1 Step 1 and concentrated as in 2.3.1 Step 8.
2. Resuspend pellet in 250 µL of distilled water, chill in ice bath for 2 minutes, heat in boiling water for 2 minutes, and rechill in ice bath for 2 minutes.
3. Add a 2-µL aliquot of sample to a PCR mixture containing *Taq* DNA polymerase assay buffer (50 mM KCl, 10 mM Tris-HCl [pH, 9.0], 0.1% Triton X-100, 1.5 mM MgCl$_2$), a 0.16 mM concentration of each deoxyribonucleoside triphosphate, 100 pmol

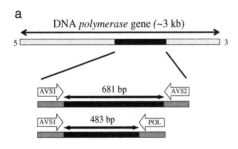

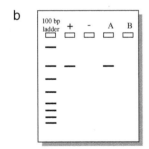

**FIGURE 22.7.** PCR-based methods for detecting phycodnaviruses have been developed with use of degenerate primers designed against a fragment of the DNA *polymerase* gene. Upstream (AVS1) and downstream (AVS2) primers **(a)** are used to amplify a 681-bp product, which can be detected by running the PCR products on standard gel electrophoresis **(b)**. Use of a nested primer set (AVS1 and POL) in a second round of PCR allows the investigator to confirm the 681-bp product is a *polymerase* gene fragments, as only *pol* gene fragments will contain the *pol* binding site.

of each primer (AVS1 and AVS2), and 0.625 U of Platinum *Taq* DNA polymerase (Invitrogen Life Technologies) in a PCR reaction tube.

Prepare negative (no sample) and positive (an algal virus with a DNA-pol gene) controls at the same time.

4. Perform PCR in a thermocycler as follows: denaturation at 95°C for 30 seconds, annealing at 50°C for 45 seconds, and extension at 72°C for 1 minute, for 30 cycles.

5. Run 10 μL of PCR product on 2% agarose in 0.5× TBE buffer (0.045 M Tris-borate, 1 mM EDTA; pH, 8.0) at 90 V for 1 hour. Visualize with ethidium bromide staining and ultraviolet illumination.

6. For second-step amplification, excise plugs from gel with glass Pasteur pipet and transfer into new reaction tube. Set up reaction as in Step 3, using primers AVS-1 and POL in place of AVS-1 and AVS-2.

7. Perform and visualize PCR as in Steps 4 and 5.

### 2.5.2. *Prokaryotic PCR-Based Detection*

The following method was developed by Fuller et al. (1998) for detecting cyanophage. The primers were designed against gene 20, which codes for the portal vertex protein.

The primer sequences are

CPS1: (5'-GTAG[T/A]ATTTTCTACATTGA[C/T] GTTGG -3')

CPS2: (5'-GGTA[G/A]CCAGAAATC[C/T]TC[C/A] AGCAT -3')

1. Process samples as described earlier.
2. Add a 2-μL aliquot of sample to a PCR mixture, containing *Taq* DNA polymerase assay buffer (50 mM KCl, 20 mM Tris-HCl [pH, 8.4], 10 mM MgCl$_2$), a 0.2-mM concentration of each deoxyribonucleoside triphosphate, and 2 pmol of each primer (CPS1 and CPS2), in a PCR reaction tube. Prepare negative (no sample) and positive controls at the same time.
3. Perform PCR in a thermocycler as follows: hot start at 94°C for 5 minutes; hold at 80°C while 1.25 U of Platinum *Taq* DNA polymerase (Invitrogen Life Technologies) is added. Follow with 35 cycles of denaturation at 94°C for 1 minute, annealing at 55°C for 1 minute, extension at 72°C for 1 minute, and a final extension.

4. Run 10 μL of PCR product on 3% agarose in 0.5× TBE buffer (0.045 M Tris-borate, 1 mM EDTA; pH, 8.0) at 90 V for 1 hour. Visualize with ethidium bromide staining and ultraviolet illumination.

## 3.0. AVOIDING VIRAL INFECTION

As viral infections of algae are extremely difficult to cure, it is essential they be avoided at the outset. Lytic infections of microalgal cultures usually become apparent quite rapidly. Host cells often lose motility and sink to the bottom of culture flasks, change color as pigmentation is lost, and aggregate as dying and dead cells adhere to one another. However, if the viral genome is integrated into the host or infection is restricted to a susceptible subset of the cultured population, the effects of infection may not be apparent under general culturing conditions. For macroalgae, visual inspection of infected cultures may reveal signs of infection such as swollen reproductive organs and pigment loss, and growth rate is often reduced (Kapp 1998).

Viruses are most likely introduced to laboratory cultures through contaminated culture medium. To kill viruses, the only reliable method for sterilizing medium is autoclaving at 121°C (see Chapter 5). Pasteurization, microwaving, and other heating techniques do not provide a high enough sterilization temperature to completely eliminate infectious viruses. Filter sterilization is also ineffective against viruses. At 20 nm in diameter, a virus particle will pass through any filter used to remove/reduce other contaminants.

The sterile culturing techniques outlined next will reduce or eliminate the chance of introducing a viral pathogen to a culturing facility. However, they will not guard against the transmission of infection between uninfected and infected cultures. If infection is suspected, care should be taken to isolate the contaminated culture from healthy cultures immediately. This includes using separate incubators, pipetting devices, and any other laboratory equipment that may come in contact with aerosolized virus particles, as well as ensuring complete decontamination by heat-sterilizing equipment and isolating work spaces for contaminated and uncontaminated cultures. Decontamination of surfaces not autoclavable can be achieved by spraying with 70% ethanol and allowing the alcohol to evaporate, applying a dilute solution of bleach, or washing with a solution of laboratory detergent.

## 4.0. SEARCHING FOR NEW ALGAL VIRUSES

Much of the research being currently conducted by algal virologists entails isolating, purifying, and characterizing new viruses. Relatively few algal-virus systems have been described, although viruses are believed to infect most if not all species of algae, and unrelated viruses have been found that infect the same species (Nagasaki and Yamaguchi 1997, Lawrence et al. 2001, Tai et al. 2003). The number of novel systems that potentially remain to be discovered is staggering, and until we have an understanding of the number and diversity of these viruses it will be difficult to fully assess their ecological consequence.

### 4.1. Lytic Viruses

#### 4.1.1. Clonal Isolation

A number of different tactics should be employed when hunting for viruses, including the use of raw and concentrated water samples (Nagasaki and Yamaguchi 1997, Tai et al. 2003) and sediments (Lawrence and Suttle 2002, Lawrence and Suttle 2004). Whole-water isolations have led to the discovery of a number of large viruses (Nagasaki and Yamaguchi 1997, Taruntani et al. 2001, Castberg et al. 2002). Screening whole-water samples leaves no chance that large viruses will be lost by the processes of filtration and concentration. This also means that pathogens other than viruses may be isolated, such as bacteria. Whole-water screening is the simplest and least-time-consuming approach, and a few considerations will enhance the probability of successful isolation. Since no means of viral concentration are used, large volumes of water should be screened to ensure the detection of rare viruses. For example, by adding 250 mL of sample water to 250 mL of culture, the investigator should theoretically be able to detect viruses as rare as $4 \cdot L^{-1}$. Also, screening with relatively dense, rapidly growing cultures reduces the chance of the host being outcompeted by contaminating microorganisms added with the sample.

Using concentrated water samples to screen for viruses greatly increases the detection limits of screening, although the processes of prefiltration and concentration may both remove substantial numbers of viruses. Vortex flow filtration, tangential flow filtration, and ultracentrifugation have all been used to concentrate

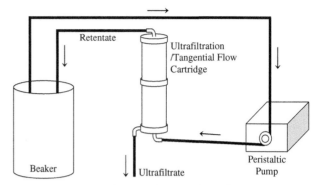

**FIGURE 22.8.** Large volumes of viral lysate can be concentrated with ultrafiltration of tangential flow filtration. In this process the sample is recirculated past a ~30-kDa membrane. Virus-free water passes through the cartridge and flows out as ultrafiltrate, and viruses are retained by the membrane and recirculated as retentate. As more virus-free ultrafiltrate is produced over time, the volume of the virus-rich retentate is reduced, thus concentrating the viruses.

viruses from aquatic systems (Suttle et al. 1991, Paul et al. 1991, Wommack et al. 1995). Figure 22.8 outlines methods for concentration by ultrafiltration. Screening concentrates will enhance the possibility of detecting rare viruses and reduce the possibility that pathogens other than viruses are isolated, but will selectively remove large viruses. For example, a 200-liter sample can be concentrated by ultrafiltration to 250 mL and screened for the presence of viruses. This will permit detection of viruses as rare as $0.005 \cdot L^{-1}$, an 800% increase over the detection limits for screening unconcentrated samples of the same volume.

Sediments have recently been shown to be important reservoirs of algal viruses (Lawrence and Suttle 2002). Sediment extracts can be prepared relatively simply and provide investigators with the ability to screen historical material for the presence of pathogens. Sediment extracts can be prepared by adding an extraction buffer such as PBS (0.05 M $Na_2HPO_4$, 0.85% NaCl; pH, 7.5) to the sample (1 : 1 [v/v]) and shaking. The buffer and agitation will help remove viruses from sediment particles, which are then separated from the extract by centrifugation (4,000 g for 5 minutes) and serial filtration (glass-fiber filter, nominal pore size of 1.2 μm; Advantec MFS; polyvinylidene difluoride, nominal pore size of 0.45 μm, e.g., Millipore). The sediment extract is then screened against a rapidly growing culture for the presence of lytic agents.

Clonal isolates of viruses are then obtained by screening serial dilutions of lysate for the propagation of viral infection. Theoretically only one virus particle is required to cause the ultimate lysis of an entire population, so the highest dilution that causes lysis will contain only one virus and therefore provide a clonal isolate. Because endpoint dilutions are not very accurate, this process should be carried out a minimum of three consecutive times to ensure a clonal isolate has been obtained. The process is labor-intensive and time-consuming, but may be reduced if cultures can be grown in microtiter plates and culture growth can be monitored by microplate readers. Obtaining isolates of viruses using plaque assays, as described above in the section on propagating viruses, is the most efficient method for obtaining virus clones since individual plaques represent the propagation of single virus. Unfortunately, growth in microtiter plates and on solid media is not possible with many algae, especially macroalgae. Plaque assays and microtiter plates have most commonly been used for the isolation of viruses that infect cyanobacteria (Safferman and Morris 1963, Suttle and Chan 1993, Fuller et al. 1998) or other robust algae, such as *Chlorella*-like algae (Van Etten et al. 1983) and *Micromonas pusilla* (Cottrell and Suttle 1991).

### 4.1.2. Concentration

Algal viruses are present at a relatively low abundance in liquid suspension, and large volumes of lysate must often be produced and concentrated for characterization. Viral concentrates can be produced by many methods. For example, for concentrating small volumes of viral lysate, ultracentrifugation (see Section 2.3.1 for calculating ultracentrifugation times) (Gibbs et al. 1975, Cole et al. 1980, Stanker et al. 1981, Henry and Meints 1992, Taruntani et al. 2001) or precipitation with polyethylene glycol (PEG) (Müller et al. 1990, Friess-Klebl et al. 1994, Jacobsen et al. 1996, Kapp et al. 1997, Maier et al. 1998) or ammonium sulphate (Skotnicki et al. 1976) can be used. Larger volumes require other approaches, such as ultrafiltration (Suttle and Chan 1995, Garry et al. 1998, Sandaa et al. 2001, Castberg et al. 2002, Paul et al. 2002). A diagram of ultrafiltration filtration is shown in Figure 22.8.

### 4.1.3. Purification

Viral concentrates may be purified by a number of methods, such as rate-zonal centrifugation (Skotnicki et al. 1976, Cole et al. 1980, Stanker et al. 1981, Suttle and Chan 1995, Sandaa et al. 2001, Castberg et al. 2002), density gradient centrifugation (Müller et al. 1990, Friess-Klebl et al. 1994, Jacobsen et al. 1996, Garry et al. 1998), affinity chromatography (Gibbs et al. 1975), or ion exchange chromatography. Diagrams of rate-

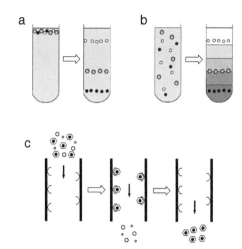

**FIGURE 22.9.** Concentrated virus samples can be purified by a number of means. In rate-zonal centrifugation **(a)**, viruses are sedimented through a dense fluid medium and are separated on a basis of sedimentation velocity and density. Density gradient centrifugation **(b)** utilizes solutions that form gradients when subjected to high g-forces. Viruses are separated into bands along isopycnal gradients and are therefore separated by density alone. Affinity chromatography **(c)** allows viruses to be purified from contaminating materials by binding viruses to a solid substrate by antibodies or charge and washing away impurities.

zonal and density gradient centrifugation and affinity chromatography are shown in Figure 22.9.

### 4.1.4. Characterization

Virus taxonomy is based on general characteristics. These include, but are not limited to, the nature of the host (type of alga), nucleic acid characteristics (DNA/RNA, double/single-stranded, molecular weight, linear/circular), capsid morphology (icosahedral, helical, etc.), number and size of capsid proteins, presence of envelope and ether/chloroform sensitivity, diameter of capsid, intracellular site of replication, latent period, and burst size. Most-probable-number assays are used for enumerating the number of viruses and thereby for inferring the burst size, or number of viruses produced by an infected host (outlined in Suttle 1993). One-step growth curves can be used to determine the length of the lytic cycle (Proctor et al. 1993).

Complete sequences are available only for a virus that infects the *Chlorella*-like alga (PBCV, reviewed in Van Etten et al. 2002) and a virus that infects *Ectocarpus siliculosus* (EsV, Delaroque et al. 2001). DNA polymerase gene fragment sequences are available for many viruses, including *Micromonas pusilla* virus (Chen and Suttle 1996), *Chlorella* virus (Grabherr et al. 1992),

*Chyrsochromulina brevifilum* virus (Chen and Suttle 1996), *Feldmannia simplex* virus (Lee et al. 1998), and *Emiliania huxleyi* virus (Castberg et al. 2002). This information can be used to examine phylogenetic and evolutionary relationships between these viruses.

### 4.2.1. Latent Viruses

Latent viruses, called temperate viruses, are believed to be the most abundant type of virus in the aquatic environment (Heldal and Bratbak 1991). This is based on their overwhelming abundance in other environments (Freifelder 1987). Until better methods for detecting latent viruses are developed, they can be detected only by inducing lytic infection. The biggest obstacle remains finding methods for reliable induction. In other systems, lytic infections are inducible from latent or lysogenic infections by DNA-damaging agents or other environmental stimuli. For example, bacteriophage can be induced by mitomycin C or UV-C (Jiang and Paul 1994, Weinbauer and Suttle 1996, Tapper and Hicks 1998), which are known to damage DNA or interfere with DNA replication. Cyanobacterial prophages may be induced by mitomycin C (Ohki and Fujita 1996, McDaniel et al. 2002, Ortmann et al. 2002), phosphate addition (Wilson et al. 1996), temperature (Rimon and Oppenheim 1975), and high copper levels (Sode et al. 1997). Latent infections of symbiotic dinoflagellates of sea anemone (Wilson et al. 2001) and some green algae (Stanker et al. 1981) have been induced by heat shock. Ultraviolet treatment is also used to enhance virus production in some eukaryotic systems (Bratbak et al. 1996, Sandaa et al. 2001). Fortunately, the latent infections found in many species of brown algae are automatically induced when the host plant reaches maturity. While the somatic cells of lysogenized macroalgae do not exhibit signs of infection, viral DNA is integrated in their genome and has been transmitted to all cells (Müller 1991a, Delaroque et al. 1999). When the macroalgal reproductive organs are induced, viruses are produced in the sporangia or gametangia and released during conditions that are optimal for the release of spores and gametes, such as changes in light, temperature, and seawater composition (Müller 1991b). Thus, the release of infectious virus particles is synchronized with the release of zooids, which—surely by no accident—are the only macroalgal cell types susceptible to viral infection. In unicellular algae, individual hosts cannot be selected and monitored for signs of infection, making latency difficult to detect. Also, it may be more difficult to detect the successful induction of unicellular algae if the temperate viruses released by induction

immediately adsorb to new hosts, integrate, and resume dormancy.

### 4.2.2. Chronic Viruses

Persistent or chronic infections are very difficult to detect, as they will not necessarily cause a net loss to a host population; cell death is counterbalanced by the production of new cells. Chronic infections have not been confirmed to exist in algal systems, yet chronic infection of *Brachiomonas submarina* is suspected (Hoffman, personal communication). Cultures of this alga can be maintained by normal culturing methods, and infection can be detected only by examining individual cells for the presence of VLPs, which occur in fewer than one in 1,000 cells (Hoffman, personal communication).

## 5.0. SUMMARY *OR* WHERE THIS FIELD IS HEADING AND WHY IT'S TAKING SO LONG . . .

Over the last 20 years the field of algal virology has experienced a great number of advances due to the discovery of novel virus-host systems and a heightened awareness of the importance of viruses in aquatic ecology. Accurate assessments of algal virus abundance, diversity, and quantification of virally induced algal mortality await improved methods for detection and characterization. The study of algal virology requires healthy doses of patience, innovation, and optimism and perhaps more important, a little green magic. The early works of R. Meints, D. Müller, J. Van Etten, and numerous others opened the doors to a fascinating field of discovery. The doors are still open.

## 6.0. ACKNOWLEDGMENTS

The author thanks C. A. Suttle for the support and opportunity to write this chapter while completing a fellowship in his laboratory. The manuscript benefited greatly from discussions with and input from many members of the Suttle Lab, including Azeem Ahmad, Jessie Clasen, André Comeau, Cindy Frederickson, Emma Hambly, Alice Ortmann, and Kevin Wen.

## 7.0. REFERENCES

Ackerman, H-W., and DuBow, M. S. 1987. *Viruses of Prokaryotes (Volume 1): General Properties of Bacteriophages.* CRC Press, Boca Raton, Florida.

Baker, J. R. J., and Evans, L. V. 1973. The ship fouling alga *Ectocarpus.* I. Ultrastructure and cytochemistry of plurilocular reproductive stages. *Protoplasma* 77:1–13.

Bergh, Ø., Børsheim, K. Y., Bratbak, G., and Heldal, M. 1989. High abundance of viruses found in aquatic environments. *Nature* 340:467–8.

Bisen, P. S., Bagchi, S. N., and Audholia, S. 1986. Nitrate reductase activity of a cyanobacterium *Phormidium uncinatum* after cyanophage LPP-1 infection. *FEMS Microbiol. Lett.* 33:69–72.

Børsheim, K. Y., Bratbak, G., and Heldal, M. 1990. Enumeration and biomass estimation of planktonic bacteria and viruses by transmission electron microscopy. *Appl. Environ. Microbiol.* 56:352–6.

Bratbak, G., Egge, J. K., and Heldal, M. 1993. Viral mortality of the marine alga *Emiliania huxleyi* (Haptophyceae) and termination of algal blooms. *Mar. Ecol. Prog. Ser.* 93:39–48.

Bratbak, G., Wilson, W., and Heldal, M. 1996. Viral control of *Emiliania huxleyi* blooms? *J. Mar. Systems.* 9:75–81.

Bräutigam, M., Klein, M., Knippers, R., and Müller, D. G. 1995. Inheritance and meiotic elimination of a virus genome in the host *Ectocarpus siliculosus* (Phaeophyceae). *J. Phycol.* 31:823–7.

Brown, R. M., Jr. 1972. Algal viruses. *Adv. Virus Res.* 17:243–77.

Brussaard, C. P. D., Kempers, R. S., Kop, A. J., Riegman, R., and Heldal, M. 1996. Virus-like particles in a summer bloom of *Emiliania huxleyi* in the North Sea. *Aquat. Microb. Ecol.* 10:105–13.

Brussaard, C. P. D., Marie, D., and Bratbak, G. 2000. Flow cytometric detection of viruses. *J. Virol. Methods* 85:175–82.

Brussaard, C. P. D., Noodeloos, A. A. M., Sandaa, R-A., Heldal, M., and Bratbak, G. 2004. Discovery of a dsRNA virus infecting the marine unicellular phytoplankter *Micromonas pusilla. Virology* 319:280–91.

Castberg, T., Thyrhaug, R., Larsen, A., Sandaa, R-A., Heldal, M., Van Etten, J. L., and Bratbak, G. 2002. Isolation and characterization of a virus that infects *Emiliania huxleyi* (Haptophyta). *J. Phycol.* 38:767–74.

Chan, A. M., Kacsmarska, I., and Suttle, C. A. 1997. Isolation and characterization of a species-specific bacterial pathogen which lyses the marine diatom *Navicula pulchripora. Abstracts of the American Society of Limnology and Oceanography,* Santa Fe, New Mexico, 121 pp.

Chapman, R. L., and Lang, N. J. 1973. Virus-like particles and nuclear inclusions in the red alga *Porphridium purpureum* (Bory) Drew et Ross. *J. Phycol.* 9:117–22.

Chen, F., Lu, J.-R., Binder, B. J., Liu, Y.-C., and Hodson, R. E. 2001. Application of digital image analysis and flow cytometry to enumerate marine viruses stained with SYBR Gold. *Appl. Environ. Microbiol.* 67:539–45.

Chen, F., and Suttle, C. A. 1995a. Amplification of DNA polymerase gene fragments from viruses infecting microalgae. *Appl. Environ. Microbiol.* 61:1274–8.

Chen, F., and Suttle, C. A. 1995b. Nested PCR with three highly degenerate primers for amplification and identification of DNA from related organisms. *Biotechniques* 18:609–12.

Chen, F., and Suttle, C. A. 1996. Evolutionary relationships among large double-stranded DNA viruses that infect microalgae and other organisms as inferred from DNA polymerase genes. *Virology* 219:170–8.

Chen, F., Suttle, C. A., and Short, S. M. 1996. Genetic diversity in marine algal virus communities as revealed by sequence analysis of DNA polymerase genes. *Appl. Environ. Microbiol.* 62:2869–74.

Clitheroe, S. B., and Evans, L. V. 1974. Viruslike particles in the brown alga. *Ectocarpus. J. Ultrastruct. Res.* 49:211–7.

Cole, A., Dodds, J. A., and Hamilton, R. I. 1980. Purification and some properties of a double-stranded DNA containing virus-like particles from *Uronema gigas,* a filamentous eucaryotic green alga. *Virology* 100:166–74.

Coleman, A. W., Maguire, M. J., and Coleman, J. R. 1981. Mithramycin-and 4′-6-diamidino-2-phenylindole (DAPI): DNA staining for fluorescence microspectrophotometric measurement of DNA in nuclei, plastids, and virus particles. *J. Histochem. Cytochem.* 29:959–68.

Cottrell, M. T., and Suttle, C. A. 1991. Wide-spread occurrence and clonal variation in viruses which cause lysis of a cosmopolitan, eukaryotic marine phytoplankter, *Micromonas pusilla. Mar. Ecol. Prog. Ser.* 78:1–9.

Delaroque, N., Maier, I., Knippers, R., and Müller, D. G. 1999. Persistent virus integration into the genome of its algal host, *Ectocarpus siliculosus* (Phaeophyceae). *J. Gen. Virol.* 80:1367–70.

Delaroque, N., Wolf, S., Müller, D. G., Bothe, G., Pohl, T., Knippers, R., and Boland, W. 2001. The complete DNA sequence of the *Ectocarpus siliculosus* virus genome. *Virology* 287:112–32.

Dodds, J. A., and Cole, A. 1980. Microscopy and biology of *Uronema gigas,* a filamentous eucaryotic green alga, and its associated tailed virus-like particle. *Virology* 100:156–65.

Franche, C. 1987. Isolation and characterization of a temperate cyanophage for a tropical *Anabaena* strain. *Arch. Microbiol.* 148:172–7.

Frederickson, C. M., Short, S. M., and Suttle, C. A. 2003. The physical environment affects cyanophage communities in British Columbia inlets. *Microb. Ecol.* 46:348–57.

Freifelder, D., 1987. *Molecular Biology: A Comprehensive Introduction to Prokaryotes and Eukaryotes.* Jones and Bartlett Publishers, Boston, 979 pp.

Friess-Klebl, A-K., Knippers, R., and Müller, D. G. 1994. Isolation and characterization of a DNA virus infecting *Feldmannia simplex* (Phaeophyceae). *J. Phycol.* 30:653–8.

Fuller, N. J., Wilson, W. H., Joint, I. R., and Mann, N. H. 1998. Occurrence of a sequence in marine cyanophages similar to that of T4 g20 and its application to PCR-based detection and quantification techniques. *Appl. Environ. Microb.* 64:2051–60.

Garry, R. T., Hearing, P., and Cosper, E. M. 1998. Characterization of a lytic virus infectious to the bloom-forming microalga *Aureococcus anophagefferens* (Pelagophyceae). *J. Phycol.* 34:616–21.

Gastrich, M. D., Anderson, O. R., Benmayor, S. S., and Cosper, E. M. 1998. Ultrastructural analysis of viral infection in the brown-tide alga, *Aureococcus anophagefferens* (Pelagophyceae). *Phycologia* 37:300–6.

Gibbs, A., Skotnicki, A. H., Gardiner, J. E., and Walker, E. S. 1975. A tobamovirus of a green alga. *Virology* 64:571–4.

Grabherr, R., Strasser, P., and Van Etten, J. L. 1992. The DNA polymerase gene from *Chlorella* viruses PBCV-1 and NY-2A contains an intron with nuclear splicing sequences. *Virology* 188:721–31.

Gromov, B. V., and Mamkaeva, K. A. 1981. A virus infection in the synchronized population of the *Chlorococcum minutum* zoospores. *Arch. Hydrobiol. Suppl.* 60:252–9.

Hara, S., Terauchi, K., and Koike, I. 1991. Abundance of viruses in marine waters: Assessment by epifluorescence and transmission electron microscopy. *Appl. Environ. Microb.* 57:2731–4.

Hatta, T., and Franki, R. I. B. 1981. Identification of small polyhedral virus particles in thin sections of plant cells by an enzyme cytochemical technique. *J. Ultrastructure Res.* 74:116–29.

Heldal, M., and Bratbak, G. 1991. Production and decay of viruses in aquatic environments. *Mar. Ecol. Prog. Ser.* 72:205–12.

Hennes, K. P., and Suttle, C. A. 1995. Direct counts of viruses in natural waters and laboratory cultures by epifluorescence microscopy. *Limnol. Oceanogr.* 40:1050–5.

Henry, E. C., and Meints, R. H. 1992. A persistent virus infection in *Feldmannia* (Phaeophyceae). *J. Phycol.* 28:517–26.

Hirons, G. T., Fawcett, J. J., and Crissman, H. A. 1994. TOTO and YOYO: New very bright fluorochromes for DNA content analyses by flow cytometry. *Cytometry* 15:129–40.

Hoffman, L. R. 1978. Virus-like particles in *Hydrurus* (*Chrysophyceae*). *J. Phycol.* 14:110–4.

Hoffman, L. R., and Stanker, L. H. 1976. Virus-like particles in the green alga *Cylindrocapsa*. *Can. J. Bot.* 54:2827–41.

Jacobsen, A., Bratbak, G., and Heldal, M. 1996. Isolation and characterization of a virus infecting *Phaeocystis pouchetii* (Prymnesiophyceae). *J. Phycol.* 32:923–7.

Jiang, S. C., and Paul, J. H. 1994. Seasonal and diel abundance of viruses and occurrence of lysogeny/bacteriocinogy in the marine environment. *Mar. Ecol. Prog. Ser.* 104:163–72.

Kapp, M. 1998. Viruses infecting marine brown algae. *Virus Genes* 16:111–7.

Kapp, M., Knippers, R., and Müller, D. G. 1997. New members of a group of DNA viruses infecting brown algae. *Phycol. Res.* 45:85–90.

Kim, M-C., Yoshinaga, I., Imai, I., Nagasaki, K., Itakura, S., and Ishida, Y. 1998. A close relationship between algicidal bacteria and termination of *Heterosigma akashiwo* (Raphidophyceae) blooms in Hiroshima Bay, Japan. *Mar. Ecol. Prog. Ser.* 170:25–32.

Koch, R. 1884. Die aetiologie der tuberkulose. *Mittheilungen aus dem Kaiserlichen Gesundheitsamte* 2:1–88.

LaClaire, J. W., and West, J. A. 1977. Virus-like particles in the brown alga *Streblonema*. *Protoplasma* 93:127–30.

Larsen, A., Castberg, T., Sandaa, R. A., Brussaard, C. P. D., Egge, J., Heldal, M., Paulino, A., Thyrhaug, R., van Hannen, E. J., and Bratbak, G. 2001. Population dynamics and diversity of phytoplankton, bacteria and viruses in a seawater enclosure. *Mar. Ecol. Prog. Ser.* 221:47–57.

Lawrence, J. E., and Suttle, C. A. 2002. Viruses causing lysis of the toxic bloom-forming alga *Heterosigma akashiwo* (Raphidophyceae) are widespread in coastal sediments of British Columbia, Canada. *Limnol. Oceanogr.* 47:545–50.

Lawrence, J. E., and Suttle, C. A. 2004. The effect of viral infection on sinking rates of *Heterosigma akashiwo* and its implications for bloom termination. *Aquat. Microbial. Ecol.* In press.

Lawrence, J. E., Chan, A. M., and Suttle, C. A. 2001. A novel virus (HaNIV) causes lysis of the toxic bloom-forming alga *Heterosigma akashiwo* (Raphidophyceae). *J. Phycol.* 37:216–22.

Lebaron, P., Parthuisot, N., and Catala, P. 1998. Comparison of blue nucleic acid dyes for flow cytometric enumeration of bacteria in aquatic systems. *Appl. Environ. Microbiol.* 64:1725–30.

Lee, R. E. 1971. Systemic viral material in the cells of the freshwater red alga *Sirodotia tenuissima* (Holden) Skuja. *J. Cell. Sci.* 8:623–31.

Lee, A. M., Ivey, R. G., Meints, R. H. 1998. The DNA polymerase gene of a brown algal virus: Structure and phylogeny. *J. Phycol.* 34:608–15.

Lu, F., Chen, F., and Hodson, R. E. 2001. Distribution, isolation, host specificity, and diversity of cyanophages infecting marine *Synechococcus* spp. in river estuaries. *Appl. Environ. Microbiol.* 67:3285–90.

Maier, I., and Müller, D. G. 1998. Virus binding to brown algal spores and gametes visualized by DAPI fluorescence microscopy. *Phycologia* 37:60–5.

Maier, I., Wolf, S., Delaroque, N., Müller, D. G., and Kawai, H. 1998. A DNA virus infecting the marine brown alga *Pilayella littoralis* (Ectocarpales, Phaeophyceae) in culture. *Eur. J. Phycol.* 33:213–20.

Manton, I., and Leadbeater, B. S. C. 1974. Fine-structural observations on six species of *Chyrsochromulina* from wild Danish marine nanoplankton, including a description of *C. campanulifera* sp. nov. and a preliminary summary of the nanoplankton as a whole. *Dan. Vidensk. Sel. Biol. Skr.* 20:1–26.

Marie, D., Brussaard, C. P. D., Thyrhaug, R., Bratbak, G., and Vaulot, D. 1999. Enumeration of marine viruses in culture and natural samples by flow cytometry. *Appl. Environ. Microb.* 65:45–52.

Markey, D. R. 1974. A possible virus infection in the brown alga *Pylaiella littoralis*. *Protoplasma* 80:223–32.

Martin, E. L., and Benson, R. 1988. Phages of cyanobacteria. In: Calendar, R., ed. *The Bacteriophages*. Plenum Press, New York, pp. 607–45.

Mattox, K. R., Stewart, K. D., and Floyd, G. L. 1972. Probable virus infections in four genera of green algae. *Can. J. Microbiol.* 18:1620–1.

Mayer, J. A., and Taylor, F. J. R. 1979. A virus which lyses the marine nanoflagellate *Micromonas pusilla*. *Nature* 281:299–301.

McDaniel, L. Houchin, L. A., Williamson, S. J., and Paul, J. H. 2002. Lysogeny in marine *Synechococcus*. *Nature* 415:496–7.

Meints, R. H., Van Etten, J. L., Kuczmarski, D., Lee, K., and Ang, B. 1981. Viral infection of the symbiotic *Chlorella*-like alga present in *Hydra viridis*. *Virology* 113:698–703.

Milligan, K. L. D., and Cosper, E. M. 1994. Isolation of virus capable of lysing the brown tide microalga, *Aureococcus anophagefferens*. *Science* 266:805–7.

Moisa, I., Sotropa, E., and Velehorschi, V. 1981. Investigations on the presence of cyanophages in fresh and sea waters of Romania. *Rev. Roum. Med.-Virol.* 32:127–32.

Müller, D. G. 1991a. Marine virioplankton produced by infected *Ectocarpus siliculosus* (Phaeophyceae). *Mar. Ecol. Prog. Ser.* 76:101–2.

Müller, D. G. 1991b. Mendelian segregation of a virus genome during host meiosis in the marine brown alga *Ectocarpus siliculosus*. *J. Plant. Physiol.* 137:739–43.

Müller, D. G., and Stache, B. 1992. Worldwide occurrence of virus-infections in filamentous marine brown algae. *Helgolander Meeresunters.* 46:1–8.

Müller, D. G., and Frenzer, K. 1993. Virus infections in three marine brown algae: *Feldmannia irregularis*, *F. simplex*, and *Ectocarpus siliculosus*. *Hydrobiologia* 260/261:37–44.

Müller, D. G., and Schmid, C. E. 1996. Interfeneric infection and persistence of *Ectocarpus* virus DNA in *Kuckuckia* (Phaeophyceae, Ectocarpales). *Bot. Mar.* 39:401–4.

Müller, D. G., Kawai, H., Stache, B., and Lanka, S. 1990. A virus infection in the marine brown alga *Ectocarpus siliculosus* (Phaeophyceae). *Bot. Acta.* 103:72–82.

Müller, D. G., Sengco, M., Wolf, S., Bräutigam, M., Schmid, C. E., Kapp, M., and Knippers, R. 1996. Comparison of two DNA viruses infecting the marine brown algae *Ectocarpus siliculosus* and *E. fasciculatus*. *J. Gen. Virol.* 77:2329–33.

Muradov, M. M., Cherkasova, G. V., Akhmedova, D. U., Kmilova, F. D., Mukhameov, R. S., Abdukarimov, A. A., and Khalmuradov, A. G. 1990. Comparative study of NP-IT cyanophages, which lysogenize nitrogen-fixing bacteria of the genera *Nostoc* and *Plectonema*. *Microbiology* (Translation of *Mikrobiologiya*) 59:558–63.

Murphy, F. A., Fauquet, C. M., Bishop, D. H. L., Ghabrial, S. A., Jarvis, A. W., Martelli, G. P., Mayo, M. A., and Summers, M. D. 1995. The classification and nomenclature of viruses: Sixth report of the International Committee on the Taxonomy of Viruses. *Archiv. Virol. (Suppl. 10)*. Springer-Verlag, Vienna.

Murray, A. G., and Jackson, G. A. 1992. Viral dynamics: A model of the effects of size, shape, motion and abundance of single-celled planktonic organisms and other particles. *Mar. Ecol. Prog. Ser.* 89:103–16.

Nagasaki, K., Ando, M., Itakura, S., Imai, I., and Ishida, Y. 1994a. Viral mortality in the final stage of *Heterosigma akashiwo* (Raphidophyceae) red tide. *J. Plankton Res.* 16:1595–9.

Nagasaki, K., Ando, M., Imai, I., Itakura, S., and Ishida, Y. 1994b. Virus-like particles in *Heterosigma akashiwo* (Raphidophyceae): A possible red tide disintegration mechanism. *Mar. Biol.* 119:307–12.

Nagasaki, K., Tomaru, Y. Katanosaka, N., Shirai, Y., Nishida, K., Itajura, S., and Yamaguchi, M. 2004. Isolation and characterization of a novel single-stranded RNA virus infecting the bloom-forming diatom *Rhizosolenia setigera*. *Appl. Environ. Micobiol.* 70:704–11.

Nagasaki, K., and Yamaguchi, M. 1997. Isolation of a virus infectious to the harmful bloom causing microalga *Heterosigma akashiwo* (Raphidophyceae). *Aquat. Microb. Ecol.* 13:135–40.

Nagasaki, K., and Yamaguchi, M. 1998. Intra-species host specificity of HaV (*Heterosigma akashiwo* virus) clones. *Aquat. Microb. Ecol.* 14:109–12.

Noble, R. T., and Fuhrman, J. A. 1998. Use of SYBR Green I for rapid epifluorescence counts of marine viruses and bacteria. *Aquat. Microb. Ecol* 14:113–18.

Ohki, K., and Fujita, Y. 1996. Occurrence of a temperate cyanophage lysogenizing the marine cyanophyte *Phormidium persicinum*. *J. Phycol.* 32:365–70.

Oliveira, L., and Bisalputra, T. 1978. A virus infection in the brown alga *Sorocarpus uvaeformis* (Lyngbye) Pringsheim (Phaeophyta, Ectocarpales). *Ann. Bot.* 42:439–45.

Ortmann, A. C., Lawrence, J. E., and Suttle, C. A. 2002. Lysogeny and lytic viral production during a bloom of the cyanobacterium *Synechococcus* spp. *Microb. Ecol.* 43:225–31.

Parodi, E. R., and Müller D. G. 1994. Field and culture studies on virus infections in *Hincksia hincksiae* and *Ectocarpus fasciculatus* (Ectocarpales, Phaeophyceae). *Eur. J. Phycol.* 29:113–17.

Paul, J. H., Houchin, L., Griffin, D., Slifko, T., Guo, M., Richardson, B., and Steidinger, K. 2002. A filterable lytic agent obtained from a red tide bloom that caused lysis of *Karenia brevis* (*Gymnodinium breve*) cultures. *Aquat. Microb. Ecol.* 27:21–7.

Paul, J. H., Jiang, S. C., and Rose, J. B. 1991. Concentration of viruses and dissolved DNA from aquatic environments by Vortex Flow Filtration. *Appl. Environ. Microbiol.* 57:2197–204.

Pearson, B. R., and Norris, R. E. 1974. Intranuclear virus-like particles in the marine alga *Platymonas* sp. (*Chlorophyta, Prasinophyceae*). *Phycologia* 13:5–9.

Pienaar, R. N. 1976. Virus-like particles in three species of phytoplankton from San Juan Island, Washington. *Phycologica* 15:185–90.

Pickett-Heaps, J. D. 1972. A possible virus infection in the green alga *Oedogonium*. *J. Phycol.* 8:44–7.

Proctor, L. M., and Fuhrman, J. A. 1990. Viral mortality of marine bacteria and cyanobacteria. *Nature* 343:60–2.

Proctor, L. M., Ocubo, A., and Fuhrman, J. A. 1993. Calibrating estimates of phage-induced mortality in marine bacteria: Ultrastructural studies of marine bacteriophage development from one-step growth experiments. *Microb. Ecol.* 25:161–82.

Reynolds, E. S. 1963. The use of lead citrate at high pH as an electron-opaque stain in electron microscopy. *J. Cell Biol.* 17:208–12.

Rimon, A., and Oppenheim, A. B. 1974. Isolation and genetic mapping of temperature-sensitive mutants of cyanophage LPP2-SP1. *Virology* 62:567–9.

Safferman R. S., and Morris, M. E. 1963. Algal virus: Isolation. *Science* 140:679–80.

Sandaa, R-A., Heldal, M., Castberg, T., Thyrhaug, R., and Bratbak, G. 2001. Isolation and characterization of two viruses with large genome size infecting *Chrysochromulina ericina* (Prymnesiophyceae) and *Pyramimonas orietalis* (Prasinophyceae). *Virology* 290:272–80.

Schroeder, D. C., Oke, J., Malin, G., and Wilson, W. H. 2002. Coccolithovirus (*Phycodnaviridae*): Characterization of a new large dsDNA algal virus that infects *Emiliania huxleyi*. *Arch. Virol.* 147:1685–98.

Sengco, M. R., Bräutigam, M., Kapp, M., and Müller, D. G. 1996. Detection of virus DNA in *Ectocarpus siliculosus* and *E. fasciculatus* (Phaeophyceae) from various geographic areas. *Eur. J. Phycol.* 31:73–8.

Sherman, L. A., and Brown, R. M., Jr. 1978. Cyanophages and viruses of eukaryotic algae. In: Fraenkel-Conrat, H., and Wagner, R. R., eds. *Comprehensive Virology*, Vol. 23. Plenum Press, New York, pp. 145–234.

Short, S. M., and Suttle, C. A. 2000. Denaturing gradient gel electrophoresis resolves virus sequences amplified with degenerate primers. *BioTechniques* 28:20–6.

Short, S. M., and Suttle, C. A. 2002. Sequence analysis of marine virus communities reveals that groups of related algal viruses are widely distributed in nature. *Appl. Environ. Microb.* 68:1290–6.

Sicko-Goad, L., and Walker, G. 1979. Viroplasm and large virus-like particles in the dinoflagellate *Gymnodinium uberrimum*. *Protoplasma* 99:203–10.

Sieburth, J. McN., Johnson, P. W., and Hargraves, P. E. 1988. Ultrastructure and ecology of *Aureococcus anophagefferens* gen. et sp. nov. (Chrysophyceae): The dominant picoplankter during a bloom in Narragansett Bay, Rhode Island, summer 1985. *J. Phycol.* 24:416–25.

Skotnicki, A., Gibbs, A., and Wrigley, N. G. 1976. Further studies on *Chara corallina* virus. *Virology* 75:457–68.

Sode, K., Oonari, R., and Oozeki, M. 1997. Induction of a temperate marine cyanophage by heavy metal. *J. Mar. Biotechnol.* 5:178–80.

Sode, K., Oozeki, M., Asakawa, K., Burgess, J. G., and Matsunaga, T. 1994. Isolation of a marine cyanophage infecting the marine unicellular cyanobacterium, *Synechococcus* sp. NKBG 042902. *J. Mar. Biotech.* 1:189–92.

Soyer, M-O. 1978. Particules de type viral et filaments trichocystoides chez les dinoflagellés. *Protistologica* 14:53–8.

Spurr, A. R. 1969. A low-viscosity epoxy resin-embedding medium for electron microscopy. *J. Ultrastr. Res.* 26:31–43.

Stanker, L. H., Hoffman, L. R., and MacLeod, R. 1981. Isolation and partial chemical characterization of a virus-like particle from a eukaryotic alga. *Virology* 114:357–69.

Steward, G. F., Montiel, J. L., and Azam, F. 2000. Genome size distributions indicate variability and similarities among marine viral assemblages from diverse environments. *Limnol. Oceanogr.* 45:1697–1706.

Suttle, C. A. 1992. Inhibition of photosynthesis in phytoplankton by the submicron size fraction concentrated from seawater. *Mar. Ecol. Prog. Ser.* 87:105–12.

Suttle, C. A. 1993. Enumeration and isolation of viruses. In: Kemp, P. R., Sherr, B. R., Sherr, E. B., and Cole, J. J., eds.

*Current Methods in Aquatic Microbial Ecology.* Lewis Publishers, Boca Raton, Florida, pp. 121–34.

Suttle, C. A. 2000a. Ecological, evolutionary, and geochemical consequences of viral infection of cyanobacteria and eukaryotic algae. In: Hurst, C., ed. *Viral Ecology.* Academic Press, New York, pp. 247–96.

Suttle, C. A. 2000b. Cyanophages and their role in the ecology of cyanobacteria. In: Whitton B. A., and Potts M., eds. *The Ecology of Cyanobacteria: Their Diversity in Time and Space.* Kluwer Academic Publishers, Boston, pp. 563–89.

Suttle, C. A., and Chan, A. M. 1993. Marine cyanophages infecting oceanic and coastal strains of *Synechococcus*: Abundance, morphology, cross-infectivity and growth characteristics. *Mar. Ecol. Prog. Ser.* 92:99–109.

Suttle, C. A., and Chan, A. M. 1995. Viruses infecting the marine Prymnesiosphyte *Chrysochromulina* spp.: isolation, preliminary characterization and natural abundance. *Mar. Ecol. Prog. Ser.* 118:275–82.

Suttle, C. A., Chan, A. M., and Cottrell, M. T. 1990. Infection of phytoplankton by viruses and reduction of primary production. *Nature.* 347:467–9.

Suttle, C. A., Chan, A. M., and Cottrell, M. T. 1991. Use of ultrafiltration to isolate viruses from seawater which are pathogens of marine phytoplankton. *Appl. Environ. Microbiol.* 57:2197–204.

Suttle, C. A., Chan, A. M., Chen, F., and Garza, D. R. 1993. Cyanophages and sunlight: A paradox. In: Guerrero, R., and Pedros-Alio, C., eds. *Trends in Microbial Ecology.* Spanish Society Microbiology, Barcelona, pp. 303–7.

Swale, E. M. F., and Belcher, J. H. 1973. A light and electron microscope study of the colourless flagellate *Aulacomonas* Skuja. *Arch. Mikrobiol.* 92:91–103.

Tai, V., Lawrence, J. E., Lang, A. S., Chan, A. M., Culley, A. I., and Suttle, C. A. 2003. Characterization of HaRNAV, a ssRNA virus causing lysis of *Heterosigma akashiwo* (Raphidophyceae). *J. Phycol.* 39:343–52.

Tapper, M. A., and Hicks, R. E. 1998. Temperate viruses and lysogeny in Lake Superior bacterioplankton. *Limnol. Oceanogr.* 43:95–103.

Taruntani, K., Nagasaki, K., Itakura, S., and Yamaguchi, M. 2001. Isolation of a virus infecting the novel shellfish-killing dinoflagellate *Heterocapsa circularisquama. Aquat. Microbial. Ecol.* 23:103–11.

Tikhonenko, A. S., and Zavarsina, N. B. 1966. Morphology of a lytic agent of *Chlorella purenoidosa. Mickrobiologia* 35: 850–2.

Toth, R., and Wilce, R. T. 1972. Viruslike particles in the marine alga *Chorda tomentosa* Lyngbye (*Phaeophyceae*). *J. Phycol.* 8:126–30.

Tripodi, G., and Beth, K. 1976. Unusual cell structures in tumor-like formations of *Gracilaria* (Rhodophyta). *Arch. Microbiol.* 108:167–74.

Van Etten, J. L., and Ghabrial, S. A. 1991. Phycodnaviridae. In: Francki, R. I. B., Fauget, C. M., Knudson, D. L., and Brown, F., eds. *Classification and Nomenclature of Viruses. Arch. Virol. Suppl. 2.* Springer-Verlag, Vienna, pp. 137–9.

Van Etten, J. L., Burbank, D. E., Kuczmarski, D., and Meints, R. H. 1983. Virus infection of culturable *Chlorella*-like algae and development of a plaque assay. *Science* 219:994–6.

Van Etten, J. L., Graves, M. V., Müller, D. G., Boland, W., and Delaroque, N. 2002. Phycodnaviridae—Large DNA algal viruses. *Arch. Virol.* 147:1479–516.

Van Etten, J. L., Lane, C. L., and Meints, R. H. 1991. Viruses and viruslike particles of eukaryotic algae. *Microbiol. Rev.* 55:586–620.

Van Etten, J. L., and Meints, R. H. 1999. Giant algal viruses. *Annu. Rev. Microbiol.* 53:447–94.

Van Etten, J. L., Meints, R. H., Burbank, D. E., Kuczmarski, D., Cuppels, D. A., and Lane, L. C. 1981. Isolation and characterization of a virus from the intracellular green alga symbiotic with *Hydra viridis. Virology* 113:704–11.

Van Etten, J. L., Meints, R. H., Kuczmarski, D., Burbank, D. E., and Lee, K. 1982. Viruses of symbiotic *Chlorella*-like algae isolated from *Paramecium bursaria* and *Hydra viridis. Proc. Natl. Acad. Sci.* 79:3867–71.

Waterbury, J. B., and Valois, F. W. 1993. Resistance to co-occurring phages enables marine *Synechococcus* communities to coexist with cyanophages abundant in seawater. *Appl. Environ. Microbiol.* 59:3393–9.

Weinbauer, M. G., and Suttle, C. A. 1996. Potential significance of lysogeny to bacteriophage production and bacterial mortality in coastal waters of the Gulf of Mexico. *Appl. Environ. Microbiol.* 62:4374–80.

Weinbauer, M. G., and Suttle, C. A. 1997. Comparison of epifluorescence and transmission electron microscopy for counting viruses in natural marine waters. *Aquat. Microb. Ecol.* 13:225–32.

Wilhem, S. W., and Suttle, C. A. 1999. Viruses and nutrient cycles in the sea. *Bioscience* 49:781–8.

Wilson, W. H., Fuller, N. J., Joint, I. R., and Mann, N. H. 2000. Analysis of cyanophage diversity in the marine environment using denaturing gradient gel electrophoresis. In: Bell, C. R., Brylinsky, M., and Johnson-Green, P., eds. *Microbial Biosystems: New Frontiers. Proceedings of the 8th International Symposium on Microbial Ecology, Halifax, Nova Scotia, Canada.* Atlantic Canada Society for Microbial Ecology, Halifax, Canada, pp. 565–71.

Wilson, W. H., Francis, I., Ryan, K., and Davy, S. K. 2001. Temperature induction of viruses in symbiotic dinoflagellates. *Aquat. Microb. Ecol.* 25:99–102.

Wilson, W. H., Joint, I. R., Carr, N. G., and Mann, N. H. 1993. Isolation and molecular characteristics of five marine cyanophages propagated on *Synechococcus* sp. strain WH7803. *Appl. Environ. Microbiol.* 59:3736–43.

Wilson, W. H., Carr, N. G., and Mann, N. H. 1996. The effect of phosphate status on the kinetics of cyanophage infection of the oceanic cyanobacterium Synechococcus sp. WH7803. *Journal of Phycology*. 32:506–16.

Wolf, S., Maier, I., Katsaros, C., and Müller, D. G. 1998. Virus assembly in *Hincksia hincksiae* (Ectocarpales, Phaeophyceae): An electron and fluorescence microscopic study. *Protoplasma*. 203:153–67.

Wolf, S., Müller, D. G., and Maier, I. 2000. Assembly of a large icosahedral DNA virus, MclaV-1, in the marine alga *Myriotrichia clavaeformis* (Dictyosiphonales, Phaeophyceae). *Eur. J. Phycol.* 35:163–71.

Wommack, K. E., Hill, R. T., and Colwell, R. R. 1995. A simple method for the concentration of viruses from natural water samples. *J. Microbiol. Methods* 22:57–67.

Wommack, K. E., Ravel, J., Hill, R. T., Chun, J., and Colwell, R. R. 1999. Population dynamics of Chesapeake Bay virioplankton: Total community analysis by pulsed-field gel electrophoresis. *Appl. Environ. Microb.* 65:231–40.

Xenopoulos, M. A., and Bird, D. F. 1997. Virus à la sauce la sauce Yo-Pro: Microwave-enhanced staining for counting viruses by epifluorescence microscopy. *Limnol. Oceanogr.* 42:1648–50.

Zhong, Y., Chen, F., Wilhelm, S. W., Poorvin, L., and Hodson, R. E. 2002. Phylogenetic diversity of marine cyanophage isolates and natural virus communities as revealed by sequences of viral capsids assembly protein gene g20. *Appl. Environ. Microbiol.* 68:1576–84.

CHAPTER 23

# CONTROL OF
# SEXUAL REPRODUCTION
# IN ALGAE IN CULTURE

ANNETTE W. COLEMAN

*Department of Molecular Biology, Cell Biology, and Biochemistry,*
*Brown University*

THOMAS PRÖSCHOLD

*Department of Molecular Biology, Cell Biology, and Biochemistry,*
*Brown University*

## CONTENTS

Key Index Words: Agglutination, *Chlamydomonas*, Gametogenesis, Pheromone, Sexual Reproduction, Zygote

## 1.0. INTRODUCTION

### 1.1. Purpose

The ability to manipulate sexual reproduction in algal cultures is valuable not only to illuminate the life cycle, but also to check relatedness of isolates, to manipulate their genomes, and to evaluate the physiology and genetics of the process itself. Yet, for only a few types of algae have mating conditions been defined sufficiently to produce zygotes with confidence at every attempt.

Sexual reproduction obviously requires opening a new set of developmental pathways in the life cycle and often requires the simultaneous influence of a compatible mating strain from the early stages, but in addition, many "environmental" factors influence the success. These include physical factors such as temperature, chemical factors such as medium composition, and even such unusual biological factors as a requirement for the presence of a symbiont. In general, conditions for successful gametogenesis are narrower than those permitting vegetative growth. We have tried to include here all the pertinent gleanings from the literature, including observations that might appear to be quite organism specific but are nevertheless suggestive of phenomena that do occur in nature. The only algal group for which we have deliberately omitted observations is the dinoflagellates, whose sexuality is treated in Chapter 24. A sexual cycle in euglenophytes and cryptophytes has not been shown clearly.

### 1.2. Basic Sexuality in Eukaryotes

Here, *sexual reproduction* refers to the unique invention of eukaryotes in which two specially differentiated haploid cells (gametes) fuse, their nuclei fuse, and either sooner or later, the complex events of meiosis produce haploid progeny that contain reassorted chromosome sets. The algae are remarkable for their ability to avoid this process by using mitotic cell division for prolonged periods, and this can be a characteristic of either the haploid generation, the diploid generation, or both. Thus, manipulation of sexual reproduction requires methods both for obtaining active gametes and for inducing a diploid generation to undergo meiosis. Many excellent general reviews of sexuality in algae (Dring 1974, Gantt 1980) are mentioned in the course of this chapter. They often include descriptions of observations from nature, rather than culture manipulation in the laboratory, information of potential value.

## 1.3. Definition of Terms: Isogamy vs. Oogamy

A great many technical terms have been engendered by studies of sexual reproduction, some influenced by parallels with terrestrial plant biology and others by parallels with protozoans. Most of these are ignored here but can be checked in Coleman (1979). For the specialty terms in diatoms, see Geitler (1932) and von Stosch (1950), as well as the review of diatom sexuality by Drebes (1977). Of the terms we use, *isogamy* refers to the morphological similarity of two gametes that fuse, ignoring the physiological differences that are presumably present but unobserved. *Oogamy*, by contrast, traditionally describes the situation in which a large immotile egg cell is fertilized by a motile and much smaller sperm cell. In between lie all degrees of *anisogamy*, in which both gametes are flagellated but differ in size (at least on the average).

## 2.0. CONTROLLING STAGES OF SEXUAL REPRODUCTION

The successive stages of sexual reproduction in algae can be grouped into three subsets: induction of gametogenesis, gamete interaction to fusion, and zygote maturation and germination (Fig. 23.1). The first subset is concerned with how to induce the switch from vegetative cell division to the production of gametes. For diplonts such as diatoms, this involves induction of meiosis that leads directly to gamete formation, whereas in haplonts or haploid generations of haplodiplonts, it requires a switch from haploid vegetative reproduction to the pathways that generate gametes. The known trig-

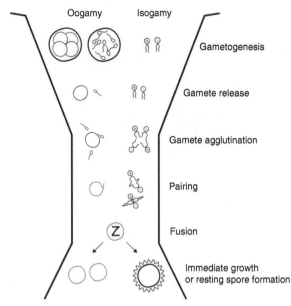

**FIGURE 23.1.** Diagrammatic representation of the series of steps in the developmental pathway leading to zygote formation, both in isogamous and in oogamous forms. The ever-narrowing funnel represents the increasing level of molecular specificity (reflecting close genetic relatedness) necessary for successful passage. Where the induction of gametogenesis includes a biological component, as in *Volvox* where a female colony forms only in response to a chemical signal from the male colonies, this molecular specificity is continuous upward.

gers for these changes are manifold, but many could be categorized as "stress signals," stress under conditions acceptable for mating.

Despite 100 years of experiments with algae in culture, for only a few representative species is there anything like a protocol guaranteeing production of zygotes. These are sufficiently reproducible in some cases that, for example, biological supply houses sell kits with mating strains and instructions for class demonstration of *Chlamydomonas reinhardtii* Dangeard mating and protocols can be found in Harris (1989). However, even among unicellular green algae, such can be said for only 12 to 13 genera, with *Chlamydomonas* most prominent, but still represented only by seven species (Trainor 1959, Pröschold 1997). For field-collected *Fucus*, standard protocols are also available (Quatrano 1980).

### 2.1. Induction of Gametogenesis in Culture

#### 2.1.1. Physical Conditions of the Environment

In general, conditions to trigger the formation of gametes and then permit zygote formation are a nar-

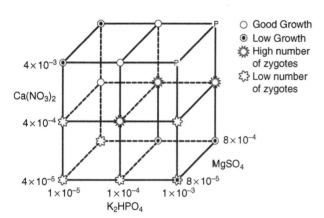

**FIGURE 23.2.** Media supporting growth and mating. Triplicate tubes of 5 mL of medium containing the given molarity of three inorganic components, calcium nitrate, potassium phosphate, and magnesium sulfate, and 1 mL of trace-element solution per liter were inoculated with a mixture of two compatible mating types of *Pandorina morum* (UTEX 853 and UTEX 854) and allowed to grow in an LD 16:8 light regimen. Growth and zygote production were monitored after 3 weeks. P, initial precipitate in the medium, invalidating the test. For initial and final pH levels, see Wilbois (1958).

rower subset of those permitting vegetative growth (Fig. 23.2). In Figure 23.2, representing a simple exploration of media preliminary to designing one in which mating should be expected consistently, note that abundant zygotes can be formed even in conditions in which growth is suboptimal, and that zygotes can fail to form even when growth is optimal. Overall, a vigorously growing culture, in ample light and in a medium containing adequate trace elements and calcium, best serves generally as a source of gametes. Because, so far as is known, gametes are in $G_1$ of the cell cycle when they fuse, some consideration of synchronization of growth (Necas and Tetik 1984) is worth pondering, particularly for organisms that do not divide every day. Procedures for obtaining gametes from field-collected material of such algae as *Fucus* and red algae have been described (Quatrano 1980, Polanshek and West 1980).

Temperature tolerance for sexual reproduction is generally fairly broad and lies within the limits for vegetative growth. However, a temperature shift may evoke sexuality in some cases, and/or the temperature range permitting sexual reproduction may be significantly narrower than that for growth. The requirement for growth at a lower temperature range to induce gametogenesis has been reported for various brown algae (Peters 1987). Avoidance of temperatures higher than

28°C was useful for inducing gametogenesis in *Scenedesmus* (Trainor 1996). Likewise, higher statospore yield was obtained by Sandgren and Flanagin (1986) in *Synura* at temperatures less than 20°C, and *Dinobryon* forms higher numbers of zygotes (statospores, with two nuclei) at temperatures less than 20°C (Sandgren 1986). Similarly, growth at 4°C was used to induce gametogenesis in *Urospora* (Kornmann 1961), and for *Chaetoceros*, use of small cells and transfer to 5°C from 15°C induced spermatogenesis (Jensen et al. 2003). By contrast, a period at more than 30°C was successful in inducing gametogenesis in *Haematococcus pluvialis* Flotow (Schulze 1927). A striking effect of a brief elevated temperature treatment was reported by Kirk and Kirk (1986) for induction of pheromone release in the male colonies of *Volvox*.

Light presents multiple aspects for variation, including day length, intensity, and wavelength. Gamete formation in many algae requires or is promoted by light, and likewise maintenance of gametic activity requires the continued presence of light. The exceptions include such mixotrophic organisms as the colonial alga *Astrephomene* and certain substrains of *Chlamydomonas reinhardtii* growing on acetate that will continue to pair and fuse if placed in the dark, a character controlled by a single genetic locus (Saito et al. 1998).

A requirement for short days to induce gametangia in some brown and red algae and a few cases of a requirement for long days are described by Dring (1984) and further covered in Chapter 21. Gamete formation and release may follow a semilunar cycle in attached shore algae such as *Derbesia*, *Enteromorpha*, *Ulva*, and *Dictyota* (see Chapter 21). The diatom *Stephanopyxis palmeriana* (Greville) Grunow begins sexual reproduction only under long-day conditions, not neutral or short day, in ASP-6 medium (Steele 1965). A different approach induces spermatogenesis in *Thalassiosira*. A dramatic downshift in light intensity evokes this response, and flow cytometry data indicate that it is cells in $G_1$ that respond (Armbrust et al. 1990).

Gametogenesis may require light for some processes in addition to photosynthesis. Such is known in *Chlamydomonas reinhardtii*, where there is a specific requirement for blue wavelengths to activate gametes to the agglutination stage (Beck and Haring 1996), perhaps acting through a flavin type of sensory receptor (Hegemann et al. 2001). A specific requirement during gametogenesis for blue light has been described for various brown algae (Dring 1987).

Culture medium components may be important. The classic trigger for induction of gamete formation in such freshwater green algae as *Chlamydomonas* and its relatives is to reduce the level of the available nitrogen source (Sager and Granick 1954). For example, transfer to a low-nitrogen medium promotes gametogenesis in *Oedogonium* (Hill 1980) and *Coelastrum* (Trainor 1996), and dilution with two parts of distilled water induces sexual reproduction in *Dunaliella* (Lerche 1937). The effect on metabolic pathways is still not understood after all these years, but the method works on many freshwater green algae, and gametic activity is rapidly repressed upon addition of an assimilatable nitrogen source.

There are two major approaches to preparing gametes and producing zygotes in culture, using the low nitrogen approach. The first involves a nutritional step-down procedure and is the standard protocol used by *Chlamydomonas* workers (Harris 1989). One grows the two mating types separately on agar to approximately late exponential stage and then resuspends the cells in a low-nitrogen version of the same medium, or even in distilled water with added calcium, for some hours in the light, at which point cells behave as gametes when the mating types are mixed.

In the second and less laborious method for both *Chlamydomonas* and most of the freshwater green algae, two compatible mating types can be inoculated into a single tube of suitable culture medium (soil water or medium with relatively low nitrogen) and permitted to grow in the light. After growth, when the nitrogen level falls, mating ensues, and copious zygotes, with their distinctive orange walls, accumulate on the walls of the tube or fall to the bottom. Co-inoculation of compatible clones into a suitable low-nitrogen medium works also for chrysophytes (Sandgren 1986). A variant particularly helpful for desmids is to mix the mating types in a shallow glass depression dish and enclose in a transparent damp chamber, perhaps with $NaHCO_3$ added to the liquid reservoir to ensure adequate atmospheric $CO_2$ (Starr 1955, Ichimura 1983). Mating and zygote formation will ensue in a situation optimal for microscopic monitoring.

Lowered nitrogen seems not to be required to trigger sexual reproduction in *Synura*, although the standard medium is not very concentrated (Sandgren and Flanagin 1986). Likewise, Mann and Stickle (1995) obtained sexual reproduction in their diatom species under conditions of continuous replenishment of medium in constant light. Perhaps related to this is the report by Schultz and Trainor (1968) that spermatogenesis and auxospore formation in *Cyclotella* spp. were correlated with an increase in sodium concentration in the medium and that a 5% increase in salinity induced spermatogenesis in *Coscinodiscus* (Jensen et al. 2003).

Reported effects of trace elements on sexuality, both their requirement and their necessary absence, are scattered through the literature. In the diatom *Ditylum brightwellii* (T. West) Grunow ex Van Heurck grown in Provasoli's ASP-6 medium, sexuality occurred only when manganese was absent from the medium (Steele 1965). After *Pandorina morum* Bory cells had mated in soil water medium, the recovered medium supported little growth when re-inoculated. The major component limiting growth, as revealed by supplementations, was the nitrogen source. However, even better growth ensued when both nitrogen and trace elements were added, and the factor in trace elements that satisfied the depletion was iron (Wilbois 1958), suggesting the particular importance of that element.

### 2.1.2. Biological Requirements for Gametogenesis and Mating

The primary biological factor controlling whether sexuality will ensue is the genetic control of sexuality in the species of choice. Fundamentally, eukaryotes are either *homothallic* ("selfers"), where mating can occur within a culture derived from a single cell, or *heterothallic*, where two different clonal cultures must first be mixed. This is not always predictable from general knowledge, because within a genus or even a species, there may exist both heterothallic and homothallic clones. What does appear true of all systems analyzed so far in algae, with the possible exception of dinoflagellates (see Chapter 24), is that heterothallic clones come in just two types. Where the two types are not morphologically recognizable as female and male, they are typically designated "plus" and "minus" mating type. The genetic potential for gamete formation is limited to only one type in a heterothallic clone, and the genetic control behaves as a single mendelian locus with only two alleles (see Ferris et al. 2002 for a review of the known genetics of mating type loci). Obviously, for zygote formation to occur using heterothallic clones, two sexually compatible clones must be mixed. Using single inocula as controls will reveal any homothallism.

In addition, keep in mind that aspects of gamete interactions may set a requirement for a certain minimal gamete concentration (Necas 1981), although positive phototaxis of gametes may largely overcome this. Also, a reasonably similar proportion of the two gamete types may be necessary for successful zygote formation (Richards and Sommerfeld 1974).

Sexuality plays a special crucial role in many diatom species where repeated vegetative cell divisions result in ever-smaller cells in the population. Restoration of cell size results from generation of a zygote, the auxospore. Among such species, some are known to become susceptible to gametogenesis only when cell size approaches its lower limit.

The third major biological factor is whether there is a requirement in heterothallic algae for the presence of the opposite mating type to induce gametogenesis. In some cases, the requirement for the presence of a partner early in the process is absolute, whereas in other cases it may only enhance the outcome by precipitating and synchronizing a mating reaction between two strains. The best studied example of the requirement for a soluble substance produced by one mating type that affects the developmental pathway of the mate is the inducer substances of the *Volvox carteri* Iyengar strains (glycoproteins active at $10^{-16}$ M; Starr and Jaenicke 1974). The pheromone appears to be absolutely required for inducing the female gametogenesis pathway and is species and strain specific.

A final major step in gametogenesis is the release of gametes from their mother wall. This process may be triggered by gamete interaction or may occur independently of the opposite mating type. Thus, the release of gametes from a gametangium may occur in the total absence of any biological clue, as in *Ulva*. At the other extreme, release may not be completed until after the initial membrane fusion to make a zygote has created a bridge between the two gametes, as in *Chlamydomonas moewusii* Gerloff (= *C. eugametos* Moewus).

Environmental factors can affect gamete release. In the marine alga *Bryopsis plumosa* (Hudson) C. Agardh, gamete release is a response to blue light with spectral optima that might suggest a flavin receptor (Mine et al. 1996). By contrast, gamete release in *Ulva mutabilis* Foyn is inhibited by a substance synthesized and released in red light during gametogenesis (Wichard 2001), which is then presumably diluted by the incoming tide.

Many algae show diel periodicity in gamete release/gametic potential, that is, selectivity as to the time of day or night. On a standard light–dark cycle, Volvocales typically are at their highest gametic potential about an hour after the lights come on (Coleman 1977), and this is the timing of gamete release for certain brown algae (Peters 1987) and green marine algae (see Chapter 21).

The biochemistry of the enzymes responsible for degrading the gamete cell wall has been studied most extensively in the chlamydomonads and proves more complex than might have been anticipated. The gametolysins or gamete autolysins are different from the wall degradation enzymes active during vegetative growth. In many green algae, gametolysins are activated by gamete–gamete contact and act only on the same or a closely related subgroup of organisms (Claes 1971, Schlösser 1976, Matsuda 1998). At least some have unusual metal cofactor requirements (Heitzer and Hallmann 2002) that highlight the possible need to examine trace-element nutrition.

### 2.1.3. Gamete Interaction to Fusion

Once gametes have been formed/released, they must (1) find and (2) agglutinate with a suitable partner and (3) activate the membrane region responsible for the final gamete cell fusion.

In algae in which both gamete types are motile, phototaxis may suffice to bring gametes together. In oogamous forms, pheromones may play a role in attracting sperm to the egg. Where one gamete type (e.g., the egg) releases a substance that can be detected and that affects directional swimming in the other gamete type (e.g., the sperm), the pheromone substance is called an *attractant*. Early work indicates that several chemoattractants are involved in the sexual reproduction of *Oedogonium* (Rawitscher-Kunkel and Machlis 1962). More recently studied examples include the pheromones exuded by the female strain of *Chlamydomonas allensworthii* Starr, Marner, et Jaenike (Starr et al. 1995) and the pheromones produced by eggs of brown algae, serving as attractants to sperm. These latter pheromones tend to serve families or orders, rather than being species or genus specific (Maier and Müller 1986).

In some algae, gametes are never released free in the medium (desmids, *Spirogyra*) or at least form only in immediate proximity to a partner (pennate diatoms), so that additional requirements for bringing together two gametes are not obvious. Yet, even in conjugating green algae such as *Closterium*, there is evidence for a diffusible glycoprotein that both attracts a partner and induces it to begin producing a wall release substance (Sekimoto 2000). The wall release substance, in turn, diffuses to the mate, leading to release of the gamete from its wall. Pairing between diatom cells, suggesting the action of a pheromone, was reported by Mann and Stickle (1995). For the nonmotile sperm of red algae, cultivation of male and female gametophytes in a moving container leads to fertilization (van der Meer 1990).

The two final and most specific stages of zygote formation, agglutination and gamete fusion, have been studied in far greater detail in *Chlamydomonas* than any other algae, but comparable steps are present in essentially all eukaryote–gamete interactions (Pan and Snell

2000). For further descriptions of these processes, see Beck and Haring (1996). Both of these steps, where analyzed in detail, have proven to involve molecules with high specificity, not just in terms of the mating type but also being effective only with extremely closely related organisms.

Gamete agglutination is a recognition process via substances on the surface of flagella. The interaction of sperm invading an egg sheath or colonial matrix is a recognition process of the same nature. This agglutination process appears to require a minimal level ($>10^{-6}$ M) of calcium. In the case of biflagellate gametes, mixing the two gamete types leads initially to formation of massive clumps of gametes, interlinked with each other by their temporary agglutination with more than one partner. Where this "clumping" of multiple gametes occurs, as in the biflagellate *Chlamydomonas*, it is followed by a sorting process such that both flagella of one gamete finally become agglutinated exclusively with both flagella of a potential partner. From studies of *Chlamydomonas*, it is clear that there are multiple sites of agglutination on flagella, and site interaction and turnover are necessary to sort partners into pairs (Demets et al. 1988).

Once achieved, successful pairing triggers a series of physiological changes, involving cyclic adenosine monophosphate (cAMP) increase (Pasquale and Goodenough 1987) and kinase and phosphatase activities (Pan and Snell 2000) in *C. reinhardtii*. The cell wall digestion is now activated. Finally, one further and critical morphological change occurs. This is the appearance of a mating papillum at the base of the flagella that serves as the initial site of cell fusion, forming a bridge with the partner gamete that rapidly expands to amalgamate the two gametes into one cell, the zygote.

### 2.1.4. Zygote Maturation and Germination

For organisms in which the zygote immediately resumes growth, no special changes in nutrition or conditions have been reported, except for the necessity of light, presumably to support photosynthesis. For many freshwater unicells, the formation of the zygote is tightly associated with the onset of a new developmental pathway, that of heavy wall formation to produce a resting spore. This proceeds without requiring new nutrients, but light is necessary. Mature resting zygotes of freshwater green algae can remain viable for years. Chrysophyte statospores also remain viable at refrigeration temperatures and in darkness for more than 1 year.

VanWinkle-Swift and Thuerauf (1991) have used this characteristic of the *Chlamydomonas monoica* Strehlow life cycle—the requirement for zygote growth and maturation—to study the genetics of different isolates of the homothallic species *Chlamydomonas monoica*. The ability of different isolates to cross was detected by deriving from each isolate a mutant unable to carry out zygote maturation and then pairing the mutants to obtain zygotes, an indication of complementation of the nonallelic mutants in the diploid cell.

Those organelles with their own genomes, the plastid and the mitochondrion, display predominantly uniparental inheritance in all algae investigated. In oogamous forms, inheritance is through the egg, with the sperm variously lacking or not propagating any organelle genomes. For isogamous forms, there seems to be no hard and fast rule. *C. reinhardtii* inherits the plastid genome from the plus mating type and the mitochondrial genome from the minus mating type parent, whereas *C. eugametos* inherits both plastid and mitochondrial genomes from the same parental type (Harris 1989).

For phototrophs, zygote germination generally requires light. *Dinobryon* statospores (Sandgren 1986) have a several-month dormancy and then can germinate when put out in the light. Typically, for heavy-walled spores of freshwater green algae there is a period of several weeks or months of dormancy before germination is successful, and the germination process requires light and is enhanced by transfer to fresh medium. Suzuki and Johnson (2002) have reported that long days enhanced zygote germination in *Chlamydomonas reinhardtii*. The zygote type unique to diatoms, the auxospore, does not seem to require a prolonged resting stage.

The first method of mating described in Section 2.1.1 for *C. reinhardtii* is useful particularly for avoiding a dormancy period. By removing newly formed zygotes from the light after only the few hours necessary for the nuclei to fuse, no heavy wall forms. Such zygotes germinate readily when put out into the light with fresh medium (Harris 1989).

---

## 3.0. CONCLUDING ISSUES

### 3.1. Finding Clones That Can Mate

Total absence of mating and zygote formation can be discouraging when dealing with a set of algal isolates.

The problem could be the wrong conditions for sexuality, or it could be that the isolates are basically heterothallic and happen all to be of the same mating type, or they might not be sufficiently closely related to interact sexually. One shortcut approach to resolve the problem is to mix as many as 10 vigorously growing isolates of the alga in question and then subject aliquots of the mixture to different media, temperatures, and light conditions. The object is to obtain some sign of mating, after which one can go back and sort out exactly which strains contribute. Experience with Volvocaceae suggests that testing multiple isolates simultaneously does not interfere with their potential interaction (Coleman and Maguire 1983).

The ability to interact sexually has so far proven to be the most reliable predictor of genetic relatedness as confirmed by DNA sequencing (Fabry et al. 1999). The converse of this is that those clones sharing the greatest DNA sequence similarity are also those most likely to mate with each other. The second Internal Transcribed Spacer (ITS2) region of the nuclear ribosomal repeats shows a rate of evolutionary divergence that is particularly appropriate for this determination. Where a broad span of collected strains is available, ITS2 sequencing can be used to assess which ones are so similar genetically that they are likely to mate, because only rarely do ITS2 sequences from sexually compatible strains differ by as much as a single nucleotide (Coleman 2001).

## 3.2. Sterility of Clones after Prolonged Culture

For decades, there have accumulated reports that algal and fungal clones under prolonged vegetative reproduction no longer express any sign of sexual reproduction. Some authors (DaSilva and Bell 1992) have assumed that a mutation has finally arisen that interferes with a sexual pathway, and that this mutation has spread through the clonal population because there is no selection for sexual reproduction during asexual cultivation. This is a possible explanation, but innately untestable, because one cannot do the test cross. Another possible explanation is that under prolonged vegetative growth conditions, some gradual epigenetic change, perhaps a chromatin remodeling, occurs such that the cells no longer respond to the environmental cues inducing the sexual pathway. If this is so, then it should be possible to find treatments that could restore the normal sensitivity.

Kirk and Kirk (1986) reported that a short period at elevated temperature induces sexuality in *Volvox* clones where it was otherwise unexpected; this procedure has not yet been reported to succeed with other algae. In at least some clones of colonial green algae, where infertility arises, a series of rapid transfers in bright-light conditions can restore sexuality to an otherwise asexual clone. Furthermore, within the single morphological species *Pandorina morum*, where there are multiple sexually isolated pairs of mating types, some of these tend to lose sexual capability after only 2 to 3 years in culture (Coleman 1975), whereas others are still sexual after nearly 50 years in culture. The infertility occurs in both mating types and is not associated with any detectable failure at a particular stage of sexual interaction, as with known mutations, but appears to reflect a total failure of gametogenesis *per se*. Fertile clones can always be reisolated from germinating zygotes. It has become clear that the same mating type pairs that show the tendency to become asexual in prolonged culture do so with each reisolation from zygotes, arguing for an epigenetic rather than a mutational basis for the phenomenon. Perhaps when we learn more of the genetics of gametogenesis, we will also discover the remedy for clonal sterility.

In fact, the major mystery to be solved for controlling sexual reproduction in culture concerns the question of which metabolic pathways pertain particularly to gametogenesis. It is in this area that the greatest diversity of phenomena has been reported for algae. Two bizarre examples are that sexual reproduction of the green unicell *Chlorocystis cohnii* (Wright) Reinhard was achieved only in the presence of a particular associated diatom (Kornmann and Sahling 1983), and similarly, associated bacteria seemed to promote sperm formation in *Coscinodiscus* (Nagai and Imai 1998). For the whole of the green algae, little more has been learned about gametogenesis since the paper of Sager and Granick in 1954. Perhaps with the advent of genomic sequencing, we can now dissect this stage and begin to make the kind of progress already achieved for the subsequent steps of gamete interaction.

## 4.0. REFERENCES

Armbrust, E., Chisholm, S., and Olson, R. 1990. Role of light and the cell cycle on the induction of spermatogenesis in a centric diatom. *J. Phycol.* 26:470–8.

Beck, C. F., and Haring, M. A. 1996. Gametic differentiation in *Chlamydomonas*. *Int. Rev. Cytol.* 168:259–99.

Claes, H. 1971. Autolyse der Zellwand bei den Gameten von *Chlamydomonas reinhardtii. Arch. Mikrobiol.* 78:180–8.

Coleman, A. W. 1975. The long-term maintenance of fertile algal clones: Experience with the genus *Pandorina* (Chlorophyceae). *J. Phycol.* 11:282–6.

Coleman, A. W. 1977. Sexual and genetic isolation in the cosmopolitan algal species *Pandorina morum*. *Am. J. Bot.* 64:361–8.

Coleman, A. W. 1979. Sexuality in colonial Volvocales. In: Hutner, S. H., and Levandowsky, M., eds. *The Protozoa*, Vol. 1, 2nd ed. Academic Press, New York, pp. 307–40.

Coleman, A. W. 2001. Biogeography and speciation in the *Pandorina/Volvulina* (Chlorophyta) superclade. *J. Phycol.* 37:836–51.

Coleman, A. W., and Maguire, M. J. 1983. A simplified, rapid method for identifying mating type in algae: Results with the *Pandorina morum* (Chlorophyceae) species complex. *J. Phycol.* 19:536–9.

Da Silva, J., and Bell, G. 1992. The ecology and genetics of fitness in *Chlamydomonas*. VI. Antagonism between natural selection and sexual selection. *Proc. Roy. Soc. Lond. B* 249:227–33.

Demets, R., Tomson, A. M., Homan, W. L., Stegwee, D., and van den Ende, H. 1988. Cell–cell adhesion in conjugating *Chlamydomonas* gametes: A self-enhancing process. *Protoplasma* 145:27–36.

Drebes, G. 1977. Sexuality. In: Werner, D., ed. *The Biology of Diatoms*. Blackwell, Oxford, pp. 250–83.

Dring, M. J. 1974. Reproduction. In Stewart, W. D. P., ed. *Algal Physiology and Biochemistry*. University of California Press, Los Angeles, California, pp. 814–37.

Dring, M. J. 1984. Photoperiodism and phycology. *Prog. Phycol. Res.* 3:159–92.

Dring, M. J. 1987. Marine plants and blue light. In: Senger, H., ed. *Blue Light Responses: Phenomena and Occurrence in Plants and Microorganisms*. CRC Press, Boca Raton, Florida, pp. 121–40.

Fabry, S., Köhler, A., and Coleman, A. W. 1999. Intraspecies analysis: Comparison of ITS sequence data and gene intron sequence data with breeding data for a world-wide collection of *Gonium pectorale*. *J. Mol. Evol.* 48:94–101.

Ferris, P. J., Armbrust, E. V., and Goodenough, U. W. 2002. Genetic structure of the mating-type locus of *Chlamydomonas reinhardtii*. *Genetics* 160:181–200.

Gantt, E. ed. 1980. *Handbook of Phycological Methods. Developmental and Cytological Methods*. Cambridge University Press, Cambridge, 425 pp.

Geitler, L. 1932. Der Formwechsel der pennaten Diatomeen (Kieselalgen). *Arch. Protistenk.* 78:1–226.

Harris, E. H. 1989. *The* Chlamydomonas *Sourcebook*. Academic Press, San Diego, 780 pp.

Hegemann, P., Fuhrmann, M., and Kateriya, S. 2001. Algal sensory photoreceptors. *J. Phycol.* 37:668–76.

Heitzer, M., and Hallmann, A. 2002. An extracellular matrix-localized metalloproteinase with an exceptional QEXXH metal binding site prefers copper for catalytic activity. *J. Biol. Chem.* 277:28280–6.

Hill, G. J. C. 1980. Mating induction in *Oedogonium*. In: Gantt, E., ed. *Handbook of Phycological Methods. Developmental and Cytological Methods*. Cambridge University Press, Cambridge, pp. 25–36.

Ichimura, T. 1983. Hybrid inviability and predominant survival of mating type minus progeny in laboratory crosses between two closely related mating groups of *Closterium ehrenbergii*. *Evolution* 37:252–60.

Jensen, K. G., Moestrup, Ø., and Schmid, A.-M. M. 2003. Ultrastructure of the male gametes from two centric diatoms, *Chaetoceros laciniosus* and *Coscinodiscus wailesii* (Bacillariophyceae). *Phycologia* 42:98–105.

Kirk, D. L., and Kirk, M. M. 1986. Heat shock elicits production of sexual inducer in *Volvox*. *Science* 231:51–4.

Kornmann, P. 1961. Über *Codiolum* und *Urospora*. *Helgol. wiss. Meeresunters.* 8:42–57.

Kornmann, P., and Sahling, P.-H. 1983. Die Meeresalgen von Helgoland: Ergänzung. *Helgol. Meeresunters.* 36:1–65.

Lerche, W. 1937. Untersuchungen über Entwicklung und Fortpflanzung in der Gattung *Dunaliella*. *Arch. Protistenk.* 88:236–68.

Maier, I., and Müller, D. G. 1986. Sexual pheromones in algae. *Biol. Bull.* 170:145–75.

Mann, D. G., and Stickle, A. J. 1995. Sexual reproduction and systematics of *Placoneis* (Bacillariophyta). *Phycologia* 34:74–86.

Matsuda, Y. 1998. Gametolysin. In: Barrett, A., Woessner, F., and Rawlings, N., eds. *Handbook of Proteolytic Enzymes*. Academic Press, London, pp. 1140–3.

Mine, I., Okudoa, K., and Tatewaki, M. 1996. Gamete discharge by *Bryopsis plumosa* (Codiales, Chlorophyta) induced by blue and UV-A light. *Phycol. Res.* 44:185–91.

Nagai, S., and Imai, I. 1998. Enumeration of bacteria in seawater and sediment from the Seto Inland Sea of Japan that promote sperm formation in *Coscinodiscus wailesii* (Bacillariophyceae). *Phycologia* 37:363–8.

Necas, J. 1981. Dependence of the gametogenesis induction, zygote formation and their germination on the culture density of the homothallic alga *Chlamydomonas geitleri* Ettl. *Biol. Plant.* 23:278–84.

Necas, J., and Tetik, K. 1984. The use of synchronization for higher yields of zygotes in the culture of the homothallic alga *Chlamydomonas geitleri* Ettl. *Arch. Protistenk.* 128:55–68.

Pan, J., and Snell, W. J. 2000. Signal transduction during fertilization in the unicellular green alga, *Chlamydomonas*. *Curr. Opin. Microbiol.* 3:596–602.

Pasquale, S. M., and Goodenough, U. W. 1987. Cyclic AMP functions as a primary sexual signal in gametes of *Chlamydomonas reinhardtii*. *J. Cell Biol.* 105:2279–92.

Peters, A. F. 1987. Reproduction and sexuality in the Chordariales (Phaeophyceae). A review of culture studies. In: Round, F. E., and Chapman, D. J., eds. *Prog. Phycol. Res.* 5:223–63.

Polanshek, A. R., and West, J. A. 1980. Hybridization of marine red algae. In: Gantt, E., ed. *Handbook of Phycological Methods. Developmental and Cytological Methods.* Cambridge University Press, Cambridge, pp. 77–93.

Pröschold, T. 1997. *Gametogenese bei einzelligen Grünalgen* [PhD Thesis]. Universität Göttingen, Cuvillier Publishers, Göttingen, Germany, 122 pp.

Quatrano, R. S. 1980. Gamete release, fertilization, and embryogenesis in the Fucales. In: Gantt, E., ed. *Handbook of Phycological Methods. Developmental and Cytological Methods.* Cambridge University Press, Cambridge, pp. 59–68.

Rawitscher-Kunkel, E., and Machlis, L. 1962. The hormonal integration of sexual reproduction in *Oedogonium*. *Am. J. Bot.* 49:177–83.

Richards, J. S., and Sommerfeld, M. R. 1974. Gamete activity in mating strains of *Chlamydomonas eugametos*. *Arch. Mikrobiol.* 98:69–75.

Sager, R., and Granick, S. 1954. Nutritional control of sexuality in *Chlamydomonas reinhardtii*. *J. Gen. Physiol.* 37:729–42.

Saito, T., Inoue, M., Yamada, M., and Matsuda, Y. 1998. Control of gametic differentiation and activity by light in *Chlamydomonas reinhardtii*. *Plant Cell Physiol.* 39:8–15.

Sandgren, C. 1986. Effects of environmental temperature on the vegetative growth and sexual life history of *Dinobryon cylindricum* Imhof. In: Kristiansen, J., and Andersen, R. A., eds. *Chrysophytes: Aspects and Problems.* Cambridge University Press, Cambridge, pp. 207–25.

Sandgren, C., and Flanagin, J. 1986. Heterothallic sexuality and density dependent encystment in the chrysophycean alga *Synura petersenii* Korsh. *J. Phycol.* 22:206–16.

Schlösser, U. G. 1976. Entwicklungsstadien-und sippenspezifische Zellwand-Autolysine bei der Freisetzung von Fortpflanzungszellen in der Gattung *Chlamydomonas. Ber. Dtsch. Bot. Ges.* 89:1–56.

Schultz, M. E., and Trainor, F. R. 1968. Production of male gametes and auxospores in the centric diatoms *Cyclotella meneghiniana* and *C. cryptica*. *J. Phycol.* 4:85–8.

Schulze, B. 1927. Zur Kenntnis einiger Volvocales (*Chlorogonium, Haematococcus, Stephanosphaera*, Spondylomoraceae und *Chlorobrachis*). *Arch. Protistenkd.* 58:508–76.

Sekimoto, H. 2000. Intercellular communication during sexual reproduction of *Closterium* (Conjugatophyceae). *J. Plant Res.* 113:343–52.

Starr, R. C. 1955. Isolation of sexual strains of placoderm desmids. *Bull. Torrey Bot. Club* 82:261–5.

Starr, R. C., and Jaenicke, L. 1974. Purification and characterization of the hormone initiating sexual morphogenesis in *Volvox carteri f. nagariensis* Iyengar. *Proc. Natl. Acad. Sci. USA* 71:1050–4.

Starr, R. C., Marner, F. J., and Jaenicke, L. 1995. Chemoattraction of male gametes by a pheromone produced by female gametes of *Chlamydomonas. Proc. Natl. Acad. Sci. USA* 92:641–5.

Steele, R. L. 1965. Induction of sexuality in two centric diatoms. *BioScience* 15:298.

Suzuki, L., and Johnson, C. 2002. Photoperiodic control of germination in the unicell *Chlamydomonas. Naturwissenschaften* 89:214–20.

Trainor, F. R. 1959. A comparative study of sexual reproduction in four species of *Chlamydomonas. Am. J. Bot.* 46:65–70.

Trainor, F. R. 1996. Reproduction in *Scenedesmus. Algae Korean J. Phycol.* 11:183–201.

Van der Meer, J. P. 1990. Genetics. In: Cole, K. M., and Sheath, R. G., eds. *Biology of the Red Algae.* Cambridge University Press, Cambridge, pp. 103–22.

VanWinkle-Swift, K. P., and Thuerauf, D. J. 1991. The unusual sexual preferences of a *Chlamydomonas* mutant may provide insight into mating-type evolution. *Genetics* 127:103–15.

von Stosch, H. A. 1950. Oogamy in a centric diatom. *Nature* 165:531–2.

Wichard, T. 2001. Biochemische und physiologische Charakterisierung des Schwärminhibitors von *Ulva mutabilis* Føyn [Diploma Thesis]. Universität Regensburg, Regensburg, Germany (in Hegemann et al. 2001).

Wilbois, A. D. 1958. *Sexual isolation in* Pandorina morum *Bory* [PhD Thesis], Indiana University, Bloomington, Indiana, 99 pp.

# MICROALGAL LIFE CYCLES: ENCYSTMENT AND EXCYSTMENT

*CSIRO Marine Research*

*CSIRO Marine Research and Australian Government Department of
Agriculture, Fisheries and Forestry*

---

CONTENTS

1.0. Introduction
2.0. Dinoflagellates: A Case Study
2.1. Establishing Cultures
2.2. Temporary Cysts
2.3. Investigating Sexual Life Cycles
    2.3.1. Encystment
    2.3.2. Crosses
    2.3.3. Observations of Life Stages
    2.3.4. Excystment
    2.3.5. Nuclear Events and the Sexual Life Cycle
2.4. Dinoflagellate Mating Systems
    2.4.1. Mating Systems and Mating Types
    2.4.2. Investigating Mating Systems in Culture
    2.4.3. Assessment of Reproductive Compatibility
    2.4.4. Analysis of Mating Systems
3.0. Complex Interactions
3.1. Cell-to-Cell Communication
3.2. Bacteria and Microalgal Life Cycles
4.0. From Culture to Nature
5.0. References

Key Index Words: Microalgae, Dinoflagellate, Resting Cyst, Temporary Cyst, Encystment, Excystment, Dormancy, Homothallic, Heterothallic, Life Cycle, Mating System

## 1.0. INTRODUCTION

Microalgae most commonly reproduce vegetatively or asexually (vegetative cell division or vegetative reproduction). One cell forms two identical daughter vegetative cells, two cells form four cells, and so on, leading to typical growth and proliferation of a microalgal population. Many microalgae are also able to reproduce sexually, enabling genetic exchange and maintaining genetic variation within a population. Sexual reproduction involves a number of different life stages that form a sexual life cycle or life history. This can enhance survival and therefore evolutionary success of a species (Maynard Smith 1976).

Resting stages may be formed by vegetative cells or as a part of the sexual life cycle. They come in a variety of forms, with differing degrees of environmental resistance, metabolic activity, and genetic characteristics, depending on the microalgal class and whether they are produced sexually or vegetatively. The term *cyst* has been used to describe some microalgal resting stages with an environmentally resistant outer coating. Resting stages, including cysts, are important for persistence and survival, in some cases over long time periods (years) and in conditions of environmental extremes (Fryxell 1983), and they can be significant as transport vectors within and between geographic

*Algal Culturing Techniques*

399

Copyright © 2005 by Academic Press
All rights of reproduction in any form reserved.

locations on local to intercontinental scales (Hallegraeff 1993, 2003). Resting stages link the benthos and plankton—a key element in the ecology of a system (Boero et al. 1996). They form a "seed bank" in the sediments, and germination may contribute to algal blooms, including toxic and other harmful blooms (Anderson and Wall 1978). Conversely, the formation of resting stages can contribute significantly to decline of algal blooms (Heiskanen 1993, Ishikawa and Taniguchi 1996). Considering the important roles of resting stages, there have been few overall syntheses, notable exceptions being Fryxell (1983) and Steidinger and Walker (1984). General information on resting stages for all algal classes are given in Bold and Wynne (1985).

Cysts are produced by the Dinophyceae (dinoflagellates). Those formed as part of the sexual life cycle are known as resting cysts, while vegetative cysts are known as temporary cysts (Pfiester and Anderson 1987). Encystment is the suite of processes involved in cyst formation from a vegetative cell or cells, and excystment is the germination of the cyst and the other processes that result in reestablishment of the vegetative cell in the life cycle. Details of these processes in dinoflagellates are a focus of this chapter (see following sections).

Chrysophytes form resistant siliceous statospores or statocysts. Sandgren (1983) unraveled some of the confusion about the development and biology of statospores, which may be produced vegetatively or sexually. He also described the factors influencing encystment. Like dinoflagellate resting cysts, chrysophycean statospores are maintained in sediments and can be useful paleoecological and biogeographic tools (Cronberg 1986, Sandgren 1991, Duff et al. 1997). For the diatoms (Bacillariophyceae), resting stages provide survival mechanisms for many species. The morphology and physiology of the resting stages show a range of deviations from the vegetative parent cells, leading to the distinction of spores (hypnospores), winter stages, and resting cells (McQuoid and Hobson 1996). Spores are heavily silicified (Hargraves and French 1983, Hargraves 1976). The ecological roles of all these resting stages are thought to be similar (Garrison 1984). Only a few diatom species are known to form resting stages as part of sexual reproduction (Hargraves and French 1983, Hargraves 1990). For a review of diatom resting stages, see McQuoid and Hobson (1996).

Other microalgae that form resting stages are the chlorophytes (see Coleman 1983). They may be vegetative or formed by sexual reproduction (e.g., thick walled hypnozygotes or hypnospores). One type of vegetative resting stage is also thick-walled, an akinete, where the wall thickens far more than in a metabolically active cell and may incorporate additional kinds of wall materials. Hypnozygote and sexual reproduction in the chlorophytes is covered in detail in Chapter 23.

Cyanobacteria (blue-green algae), in the filamentous heterocyst-forming orders Nostocales and Stigonematales, form akinetes that are resistant to cold and desiccation. Although akinetes can form when there are no heterocysts, when present heterocysts can influence the position of akinetes. Stulp and Stam (1982), Herdman (1988), and Adams and Duggan (1999) give comprehensive information on the structure and roles of akinetes in the Cyanophyceae.

The elucidation of microalgal life cycles is difficult when based on sampling and observations in the field alone. Indeed, to follow a single cell or small group of cells through morphological, physiological, and genetic transformations in nature is usually impossible. It is therefore necessary to study the processes in controlled culture conditions with careful observations of individual living cells. This chapter uses dinoflagellates as a case study to describe the methods that can be used to investigate microalgal life cycles, with a focus on the processes of encystment and excystment. Many of the culturing techniques are equally appropriate to life-cycle studies and investigation of resting stages in other microalgal classes. The methods described for investigating reproductive compatibility and mating systems have applicability to other microalgae as well as other protists. We also address general poorly understood phenomena, such as microalgal cell-to-cell communication and microalgal-bacterial interactions.

## 2.0. DINOFLAGELLATES: A CASE STUDY

Reproduction in dinoflagellates is usually vegetative (see Pfiester and Anderson 1987). In some species mass proliferation by vegetative reproduction leads to the formation of dense "red tides" or algal blooms, including toxic harmful blooms (Hallegraeff 1993, 2003). The ecological success of dinoflagellates is enhanced by cyst formation. Temporary cysts are common; however, sexually produced resting cysts have a profound influence on species autecology and system ecology. A number of reviews have considered the ecological bases for the development and morphology of dinoflagellate resting cysts (Dale 1983, Pfiester 1984, Walker 1984, Sarjeant et al. 1987, Pfiester and Anderson 1987).

Micropaleontologists carried out the first observations on dinoflagellate resting cysts, known from fossil sediments that are hundreds of millions of years old (Dale 2001). It was not until the 1960s, through observations of modern equivalents, that these fossil dinoflagellates were identified as resting cysts rather than fossils of vegetative cells (Dale 1983). An understanding of the role that resting cysts played in the sexual life cycle of living dinoflagellates came with careful observation and recording of dinoflagellate behavior. Early research provided the first elucidation of life cycle phases and cellular details, and remains an impressive example of careful observation (von Stosch 1964, 1965, 1972, 1973; Pfiester 1975, 1976, 1977; Beam and Himes 1974; Walker and Steidinger 1979). Most were based on cultured strains, with confirmatory results from natural populations in some cases (Walker 1984). By 1989 the sexual life cycles of 22 dinoflagellate species had been detailed (Pfiester and Anderson 1987), and now more than 50 have been described (Table 24.1). The fossil record suggests that an even greater number of dinoflagellates may produce resting cysts—several hundred species (Head 1996). An example of one "typical" dinoflagellate life cycle is shown by *Gymnodinium catenatum* (Fig. 24.1; Blackburn et al. 1989). It is a relatively simple cycle that has been observed (with some variations) in a number of different species. Gametes are isogamous (morphologically identical to one another, as compared with anisogamous, i.e., morphologically different). Fusion of gametes forms a young planozygote, a large cell with a double longitudinal flagellum (posteriorly biflagellate or "skiing track" flagellum, *sensu* von Stosch 1973) and a single large nucleus. In the older planozygote there is loss of one of the longitudinal flagella and rounding of the cell. This is followed by formation of a nonmotile resting cyst, or hypnozygote, with a resistant wall (Anderson et al. 1988). After a short period of mandatory dormancy, and if environmental conditions are suitable, the resting cyst germinates to produce a germling cell or planomeiocyte. The planomeiocyte divides to form two and then four vegetative cells, with an accompanying reduction nuclear division (meiosis). This reestablishes vegetative cells, and typical asexual reproduction can proceed. A number of variations have been reported. For example, some, such as *Noctiluca scintillans*, do not include a resting cyst (Zingmark 1970, Schnepf and Drebes 1993). The diversity of dinoflagellate life cycles has been reviewed several times (Walker 1984, Pfiester 1989, Pfiester and Anderson 1987).

## 2.1. Establishing Cultures

The first step for investigating encystment and excystment, both sexual and vegetative, is the establishment of culture strains. These strains can be obtained by isolating vegetative cells from field samples and estab-

lishing single-cell isolates (see Chapter 6). Micromanipulation (Throndsen 1973) of motile cells is a good way to isolate healthy vegetative cells: if cells are moving, they are vital. Cultures can also be established by germinating wild resting cysts (Nehring 1997). One approach is to add various culture media to sediment samples to promote excystment followed by isolation of the germling cells (Lewis et al. 1999). Alternatively, if the history and identity are known, individual resting cysts are isolated and incubated in culture medium. Bolch (1997) developed a process for separating and concentrating living dinoflagellate cysts from marine sediments, using aqueous solutions of the nontoxic chemical sodium polytungstate. Strains that are established from resting cysts may be nonclonal if they contain all the germination products (see Section 2.3.5).

Culture strains are a valuable resource, and depositing strains with a culture collection is a good option. See Andersen (2003) for a list of culture collections. Culture studies are a powerful scientific approach because they not only allow investigations of known strains under controlled conditions but also permit strains from different locations to be studied together.

## 2.2. Temporary Cysts

Temporary cysts are vegetative resting cells. They are not usually the product of reproduction, vegetative or sexual; instead, they are formed by ecdysis (i.e., shed-

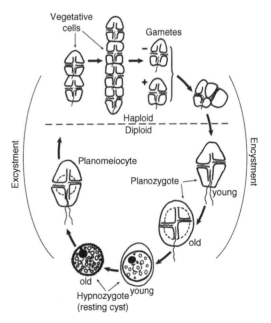

**FIGURE 24.1.** The sexual life cycle of *Gymnodinium catenatum* (after Blackburn et al. 1989).

**TABLE 24.1** Dinoflagelates for which sexual life cycles are known, their thallism, and conditions for induction of sexuality species. Species names in brackets indicate synonyms/basionyms under which life cycle literature has been published for these species.

| Species | Thallism | Induction of sexuality | References |
|---|---|---|---|
| *Alexandrium catenella* (Whedon et Kofoid) Balech | Heterothallic | Nutrient depletion | Yoshimatsu 1981, Yoshimatsu 1984 |
| *Alexandrium hiranoi* Kita et Fukuyo | Homothallic | No cue required | Kita et al. 1985, Kita et al. 1993 |
| *Alexandrium lusitanicum* Balech | Homothallic | N and Fe limitation | Silva and Faust 1995, Blanco 1995 |
| *Alexamdrium minutum* Halim | Heterothallic | | Bolch et al. 1991 |
| *Alexandrium monilatum (monilata)* (Howell) Balech nom. prov. | — | N limitation | Walker and Steidinger 1979 |
| *Alexandrium ostenfeldii* (Paulsen) Balech et Tangen | Heterothallic | | Jensen and Moestrup 1997 |
| *Alexandrium pseudogonyaulax* (Biecheler) Horiguchi ex Yuki et Fukuyo | — | Nutrient depletion | Montresor 1995, Montresor and Marino 1996 |
| *Alexandrium tamarense* (Labour) Balech | Heterothallic and homothallic | Increased temperature, nutrient depletion | Dale 1977, Anderson and Wall 1978, Dale et al. 1978, Turpin et al. 1978, Anderson 1980, Destombe and Cembella 1990 |
| *Alexandrium taylori* Balech | Homothallic | N limitation | Giacobbe and Yang 1999, Garcés et al. 1998 |
| *Amphidinium carterae* Hulburt | Homothallic | Aging cultures | Cao Vien 1967, 1968 |
| *Amphidinium klebsii* Kofoid et Swezy | — | Inc. salinity | Barlow and Triemer 1988 |
| *Ceratium comutum* (Ehr.) Claparède et Lachmann | Heterothallic | Decreased temperature, day-length and light intensity | von Stosch 1965 |
| *Ceratium horridum* (Cleve) Gran | — | | von Stosch 1964 |
| *Coolia monotis* Meunier | Homothallic | Addition of mangrove extract | Faust 1992 |
| *Crypthecodinium cohnii* (Seligo) Chatton | Homothallic | N and P limitation | Beam and Himes 1974, Tuttle and Loeblich 1974 |
| *Cystodinium bataviense* Klebs | — | | Pfiester and Lynch 1980 |
| *Dinophysis fortii* Pavillard | — | | Uchida et al. 1999 |
| *Dinophysis norvegica* Claparède et Lachmann | — | | Subba Rao 1995 |
| *Dinophysis pavillardi* Schröder | Homothallic | Transfer into nutrient replete media | Giaccobe and Gangemi 1997 |
| *Gambierdiscus toxicus* Adachi et Fukuyo | — | | Taylor 1979 |
| *Glenodinium lubiniensiforme* Diwald | Heterothallic | Nutrient depletion | Diwald 1937 |
| *Gloeodinium montanum* Klebs | Homothallic | | Kelley 1989, Kelley and Pfiester 1990 |
| *Gonyaulax apiculata* (Penard) Entz | — | | Hickel and Pollingher 1986 |
| *Gymnodinium catenatum* Graham | Heterothallic | | Blackburn et al. 1989, Blackburn et al. 2001 |
| *Gymnodinium fungiforme* Anisomova | — | Decreased food | Spero and Morée 1981 |
| *Gymnodinium mikimotoi* Miyake et Kominami | — | | Ouchi et al. 1994 |
| *Gymnodinium nolleri* Ellegaard et Moestrup | Heterothallic | Temperature window, P limitation | Ellegaard et al. 1998 |
| *Gymnodinium paradoxum* Schilling | Heterothallic | | von Stosch 1972 |
| *Gymnodinium pseudopalustre* Schiller | — | Low temperature, short day; N, P limitation | von Stosch 1973 |
| *Gyrodinium uncatenum* Hulburt | — | | Tyler et al. 1982, Coats et al. 1984 |
| *Helgolandidinium subglobosum* von Stosch | Homothallic | Aging cultures | von Stosch 1972 |
| *Karenia brevis* (Davis) Hansen et Moestrup | Heterothallic and homothallic | N limitation, cold temperature, blue light | Walker 1982, Steidinger et al. 1998 |

**TABLE 24.1** *Continued*

| Species | Thallism | Induction of sexuality | References |
| --- | --- | --- | --- |
| *Noctiluca scintillans* (Macartney) Kofoid et Swezy | Homothallic, Heterothallic | | Zingmark 1970, Hoefker 1930, Schnepf and Drebes 1993 |
| *Oxyrrhis marina* Dujardin | Homothallic | Change in food source | von Stosch 1972 |
| *Peridinium balticum* (Levander) Lemmermann | — | N deficiency | Chesnick 1987, Chesnick and Cox 1989, Zhang and Li 1990 |
| *Peridinium bipes* Stein | — | | |
| *Peridinium cinctum f. ovoplanum* Lindemann | Homothallic | N deficiency | Pfiester 1975 |
| *Peridinium cunningtonii* (Lemmemann) Lemmermann | Homothallic | N and P deficient medium | Sako et al. 1984 |
| *Peridinium gatunense* Nygaard | Homothallic | N deficient medium | Pfiester 1977 |
| *Peridinium inconspicuum* Lemmermann | Homothallic | N deficiency | Pfiester et al. 1984 |
| *Peridinium limbatum* (Stokes) Lemmermann | Homothallic | N deficiency | Pfiester and Skvarla 1980 |
| *Peridinium penardii* (Lemmermann) Lemmermann | Homothallic | N and P deficient medium | Sako et al. 1987 |
| *Peridinium volzii* Lemmermann | Heterothallic | N deficiency | Pfiester and Skvarla 1979 |
| *Peridinium willei* Huitfeldt-Kaas | Homothallic | N deficiency | Pfiester 1976 |
| *Pfiesteria piscicida* Steidinger et Burkholder | — | | Steidinger et al. 1996 |
| *Polykrikos kofoidii* Chatton | — | | Morey-Gaines and Ruse 1980, Nagai et al. 2002 |
| *Polykrikos schwartzii* Bütschli | — | Starvation | Nagai et al. 2002 |
| *Pyrocystis noctiluca* Murray et Haeckel | Homothallic | | Seo and Fritz 2001 |
| *Pyrocystis lunula* Hulburt | Homothallic | | Seo and Fritz 2001 |
| *Prorocentrum lima* (Ehr.) Dodge | — | | Faust 1993 |
| *Pyrodinium bahamense* Plate | Heterothallic | | Corrales et al. 1995 |
| *Pyrophacus steinii* (Schiller) Wall et Dale | Heterothallic | | Pholpunthin et al. 1999 |
| *Scrippsiella cf. lachrymosa* Lewis | — | Nutrient depletion | Olli and Anderson 2002 |
| *Scrippsiella hangoei* (Schiller) Larsen | — | | Kremp and Heiskanen 1999 |
| *Scrippsiella minima* Gao et Dodge | — | | Gao and Dodge 1991 |
| *Scrippsiella trochoidea* (Stein) Loeblich III | — | N, P limitation | Watanare et al. 1982, Uchida 1991 |
| *Symbiodinium microadriaticum* Freudenthal | — | | Taylor 1973 |
| *Woloszynskia apiculata* von Stosch | Heterothallic | N, P limitation, lower light | von Stosch 1973 |

ding of the vegetative cell thecal plates) and formation of a thin cyst wall (thinner than that of sexually produced resting cysts) (Fig. 24.2; Anderson and Wall 1978). However, in *Alexandrium hiranoi* and *Peridinium quinquecorne* Abé, temporary cysts are formed as part of vegetative reproduction, i.e., they can complete cell division only through the formation of temporary cysts (Anderson et al. 2003). Temporary cysts form in response to exposure to unfavorable conditions such as decreased temperature, nutrient depletion, or exposure to copper (Anderson and Wall 1978, Parker 2002). Normal, motile vegetative cells are usually reestablished once the environment becomes favorable for growth again. Temporary cysts are often produced in stationary-phase cultures or otherwise stressed cultures,

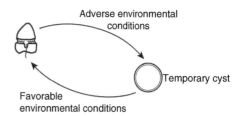

**FIGURE 24.2.** Encystment and excystment of temporary (vegetative) cysts.

e.g., those grown in conditions of very high or low light intensity, suboptimal temperatures, or nutrient depletion. In nature some species, such as Alexandrium taylori (Garcés et al. 1989), form temporary cysts in the evening, giving rise to motile cells the next morning, such as *Alexandrium taylori* (Garces et al. 1989).

## 2.3. Investigating Sexual Life Cycles

### 2.3.1. Encystment

Encystment encompasses the processes involved in cyst formation. Investigating encystment in culture requires self-fertile or compatible strains (see below) and appropriate environmental conditions (see following sections). Mixtures of strains or nonclonal strains can be useful for making observations of the behavior and interactions of heterogeneous groups of cells, including the type of reproductive compatibility. While not the focus of this chapter, there are also a number of methods that have been developed to study encystment in the field, including the use of sediment traps and discrete water column and net sampling (Godhe et al. 2001).

*Reproductive Compatibility: Homothallism/Heterothallism.* Homothallic or self-fertilizing dinoflagellates can reproduce sexually within a clonal culture derived from a single-cell isolate (self-crosses); heterothallism requires compatible strains of opposite mating type (Fig. 24.3). Mating type is the genetically determined physiological and biochemical difference between individuals or strains that determines their reproductive compatibility with other individuals or strains (Sonneborn 1957). Mating types are usually designated plus and minus, i.e., if one strain is labeled "+" then a

strain with which it is compatible is therefore of the "–" mating type. Whether a strain is homothallic or heterothallic is determined by making self-crosses and intercrosses (Section 2.3.2). Note that a number of species have been found to have both homothallic and heterothallic strains (Destombe and Cembella 1990, Blackburn et al. 2001, Parker 2002).

In our laboratory we have occasionally observed a few cysts in crosses of *Gymnodinium catenatum* that have not previously produced cysts. These cysts resemble sexually produced resting cysts but are thinner-walled and more fragile. Although unconfirmed, these could be the products of parthenogenesis (reproduction by development of an unfertilized gamete). Parthenogenesis was discussed by Pfiester and Anderson (1987) from reports by von Stosch (personal communication) and Happach-Kasam (1980).

Over the last 30 years, a range of potential environmental cues for encystment have been proposed and investigated. Many of these cues have been based on the premise that encystment is stimulated by an environment that is suboptimal for vegetative reproduction. This has resulted in an emphasis on nutrient depletion, particularly nitrogen, in experimental studies (Pfiester 1975, Walker and Steidinger 1979, Yoshimatsu 1981, Walker 1982). However, other laboratory and field observations indicate that resting cysts can form under apparently optimal nutrient conditions (Anderson et al. 1983, Pfiester et al. 1984, Parker 2002). Other factors that have been found to influence encystment include temperature, with defined temperature "windows" for encystment identified for several species, e.g., 12° to 25°C for *Alexandrium tamarense* (Anderson et al. 1984) and 13° to 33°C for *Gymnodinium nolleri* (Ellegaard et al. 1998). Effects on encystment from micronutrients have also been observed. For example, Blanco (1995) found that a deficiency in iron increased encystment in *Alexandrium lusitanicum*. Light intensity can also have a significant impact on encystment in some dinoflagellates; von Stosch (1964) and Blackburn et al. (1989) found that using higher-than-optimal light intensities enhanced encystment in *Ceratium* and *Gymnodinium catenatum*, respectively. Encystment has also been found to be inversely related to day length in some species, with shorter photoperiods (e.g., 8 : 16 light-dark cycles) increasing encystment in four calcareous cyst–forming dinoflagellates (Sgrosso et al. 2001). Natural bacterial assemblages associated with an *Alexandrium tamarense* bloom have also been found to enhance encystment of *A. catenella* in culture (Adachi et al. 1999).

Cell-to-cell communication has been hypothesized to influence encystment. The presence of pheromone-like

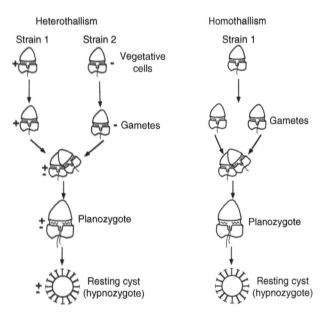

**FIGURE 24.3.** Homothallism and heterothallism in dinoflagellates.

substances or chemical cues that facilitate sexual reproduction have been suggested for dinoflagellates (Destombe and Cembella 1990, Wyatt and Jenkinson 1997; see Section 3.1.). Endogenous factors such as declining internal nutrient pools also potentially trigger encystment (e.g., declining internal nitrogen has been correlated with gamete fusion in *Alexandrium minutum*) (Probert et al. 1998). Next we outline some simple experimental methods for investigating the influence of different factors on encystment.

1. *Media/nutrients/micronutrients.* A number of different media have been used for encystment studies. These include GPM (Loeblich 1975), GSe (a variation of GPM; Blackburn et al. 2001), f/2-Si (Guillard 1975), and K media (Keller et al. 1987), all of which are good media for dinoflagellate growth (see Appendix A). Modifications of these media have also been used to investigate effects on encystment. A common method is to exclude a particular nutrient or nutrients from the medium, e.g., without phosphate and/or nitrate or trace metals. Anderson et al. (1985) used f/2-Si with f/30 concentrations of nitrate, ammonia, and phosphate in studies of *Gyrodinium uncatenum*, while Blanco (1995) used a number of variations of f/2-Si, including a preparation without iron. Another method is the reduction of the overall concentration of nutrients in the medium (e.g., to a $\frac{1}{10}$, $\frac{1}{20}$, or $\frac{1}{1,000}$ concentration of nutrients). Parker (2002) modified GSe in this way for experimental observations of encystment and excystment in *Gymnodinium catenatum* and *Alexandrium minutum*.

2. *Temperature.* A temperature gradient table is simple and useful (Watras et al. 1982). This is commonly a solid cast aluminum block with wells to hold culture flasks, held in place by the rims of the culture caps and illuminated from either above or below; the temperature gradient is established with circulating water baths at each end (one warm and one cold). Crosses can be set up in these flasks and differences in encystment under the different temperatures ascertained. Cultures need to be gradually acclimated to the more extreme temperatures of the gradient; for example, Anderson et al. (1985) acclimated cultures for at least 10 generations in their study of *Gyrodinium uncatenum*. Temperatures should be checked regularly throughout experiments.

3. *Light.* Light quality, quantity, and duration can influence encystment. Light quantity can be easily varied using glassine paper or neutral density filters to decrease light intensity. Changing the photoperiod alters the total irradiance received over 24 hours. Filters of different spectral output can be used to alter the light quality. While we are not aware of life history studies of dinoflagellates with different spectra, they have been used in studies of cyanobacteria akinetes (Yamamoto 1976).

4. *Cell-to-cell communication.* Options for studying the effects that other cells have on encystment through pheromones (cell-to-cell communication) include using filtrates obtained by gently filtering either field samples or cultures of single and/or mixed strains and using these as a medium for crossing strains (Parker 2002). There is also a range of filtration and dialysis membranes and agar preparations used to study ciliate pheromones that could be adapted for dinoflagellate studies (Kuhlmann et al. 1997).

### 2.3.2. Crosses

A *self-cross* is the incubation of a particular strain under experimental conditions, while an *intercross* is the incubation of two strains together under experimental conditions (Fig. 24.4). Tissue culture multiwell plates, petri dishes, and tissue culture flasks are the most useful cultureware because they allow microscopic observations without having to subsample. An example of an intercross is 1 mL of each strain at a cell density of approximately $10^6$ cells $\cdot$ L$^{-1}$ in a 90-mm-diameter sterile Petri dish containing 10 mL culture medium (Blackburn et al. 2001). A matrix of crosses is set up to determine the sexual compatibility of a number of strains (Fig. 24.5). Cultures are observed over time for evidence of sexual reproduction (e.g., gamete movement and behavior patterns such as fusion, planozygotes, or resting cysts). Crosses can also be made in large-scale cultures (several liters) if large numbers of cysts are required for study

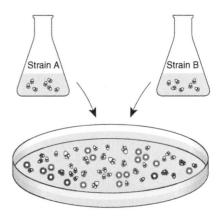

**FIGURE 24.4.** Intercross between strains A and B, showing formation of sexual stages in the cross.

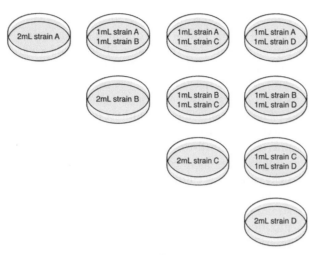

**FIGURE 24.5.** A matrix of crosses between strains A, B, C, and D. Strain inoculum to be added to 10 mL of crossing medium.

(Parker et al. 2002). However, large-scale crosses are difficult to monitor, as *in situ* microscopic observation is not possible. They are therefore not recommended as an initial step in studying encystment.

### 2.3.3. Observations of Life Stages

Careful, frequent, and sometimes prolonged microscopic observations are the key to identifying different life stages and the process of transformation from one stage to another. This may involve daily, hourly, or other short time-lapse observations. However, if the aim is only to achieve encystment, then observations every 3 to 4 days are sufficient, and in some cases, such as when crosses are run for several months, weekly observations may suffice. It is useful to observe crosses at a range of magnifications from very low (e.g., 20×) to 400× or more. In our laboratory we use a range of microscopes, including inverted and compound microscopes and a stereomicroscope equipped with dark-field optics. Observations at low magnification with a stereomicroscope may facilitate identifying changes in behavior and size differences in cells, e.g., the early detection of gametes. An inverted microscope with long-working-distance objectives is very useful for observing cell behavior *in situ* in petri dishes or small flasks placed directly onto the microscope stage. Individual or small groups of cells can be micropipetted onto either a cavity or a normal glass microscope slide for observations on a compound microscope at higher magnification with optical superiority.

Photography is the key method of data acquisition in life cycle studies. The records obtained with simple

photomicrographic technology by such workers as von Stosch (1964, 1972, 1973) and Pfiester (1975, 1976, 1977) are testimony to the powers of careful and astute observation and recording. Examples of high-caliber scientific data in the form of stunning photographic records, obtained with both still photography and video imagery, of microalgal behavior and processes are those of Pickett-Heaps (see www.cytographics.com). Video records are an excellent way of documenting transitional stages in the life cycle. By recording the process—not just the different stages—unambiguous evidence is acquired. Given the value of this powerful technique, it has been relatively underutilized.

Low-magnification (20–100×) observations can give an early indication of sexuality and encystment. Von Stosch (1973) described "dancing gametes," referring to the clustering of gametes that precedes gamete fusion in the species studied. In our laboratory we have often observed unusual movements such as spinning or erratically moving cells and concentric layering of cells. These are indicative of the presence of gametes. As well, potentially fusing gametes may be distinguished from dividing vegetative cells by the orientation of the cells (e.g., *Gymnodinium catenatum* cells fuse laterally but divide obliquely) (Blackburn et al. 1989).

To facilitate photomicrography, there are a number of methods to slow down or "freeze-frame" the cells. These include short-term refrigeration, gradual drying of the slide, adding very dilute glutaraldehyde (1 : 1,000) drop by drop to the edge of the slide (Parker 2002), and addition of small amounts of viscous substances such as Polyox (polyethylene oxide) (Spoon et al. 1977) and sodium polytungstate (SPT) (Bolch 1997) to the side of the coverslip. Viscous substances have the disadvantage of lowering optical clarity for photomicrography.

### 2.3.4. Excystment

Excystment is the process of resting cyst germination, and it results in new vegetative cells (see Fig. 24.1). Cysts can be individually isolated in multiwell plates or other small culture vessels, and their germination and subsequent development can be tracked. The individual products of germination can be isolated and cultured. Valuable observations can also be obtained by maintaining and observing intercrosses or self-crosses after resting cyst formation. These may contain hundreds to thousands of resting cysts that may germinate either synchronously or at different times. Important regulating parameters for excystment are the dormancy period and the environmental conditions (see the following).

Dormancy is the period between resting cyst formation and germination. It is controlled by both endogenous and exogenous factors. The mandatory or minimum dormancy period is that period within which germination cannot occur because of internal processes or inhibition (Pfiester and Anderson 1987). Once this mandatory dormancy period has elapsed, germination can take place if environmental conditions are suitable. If conditions are not suitable, the resting cyst enters a period of quiescence and does not germinate until conditions become more favorable. The mandatory dormancy period varies greatly from species to species. For example, *Gymnodinium catenatum* and *Alexandrium minutum* have short mandatory dormancy periods (2 and 4 weeks, respectively) (Blackburn et al. 1989, Parker 2002). In contrast, the dormancy period for *Ceratium hirundinella* (Müller) Dujardin (Rengefors and Anderson 1998) is 4.5 months and that for *Scrippsiella hangoei* is 6 months (Kremp and Anderson 2000). The mandatory dormancy period may also differ between different populations of the one species, e.g., Australian versus Japanese *Alexandrium catenella* (Hallegraeff et al. 1998).

Dormancy has been described and measured in a number of ways, but it is important to carefully define what has been measured as dormancy. The mandatory dormancy period is quite difficult to measure. It requires tracking both encystment and excystment or having a cross where both encystment and excystment are synchronized. We have rarely observed the latter situation. Other, more easily measured values are the median dormancy, i.e., the time from cross-inoculation to when 50% of cysts have excysted (Binder and Anderson 1987), and the standard dormancy, i.e., the number of days from first cyst production to first cyst germination (Blackburn et al. 2001). Blackburn et al. (2001) found median dormancies of between 36 and 81 days for *G. catenatum* resting cysts formed from Tasmania/Tasmania, Japan/Spain, Japan/Tasmania, and Spain/Tasmania population intercrosses. This suggests that dormancy may have a genetic basis and may differ among populations. Standard dormancies for the same study were 15 to 37.5 days. These differences in the two dormancy measures (standard and median) illustrate the importance of being very specific about how dormancy is assessed.

## 2.3.4.1. Environment for Excystment

Many dinoflagellates have a "temperature window" outside of which they are unable to germinate (Bravo and Anderson 1994, Kremp and Anderson 2000, Parker 2002). Low temperature usually inhibits excystment

(Bravo and Anderson 1994, Hallegraeff et al. 1998); however, *G. catenatum* resting cysts germinate at 4°C— a temperature quite unsuitable for vegetative growth. Most resting cysts either are unable to germinate or have very low excystment in the dark (Blackburn et al. 1989, Bravo and Anderson 1994, Nuzzo and Montresor 1999). Anoxic conditions inhibit excystment in several dinoflagellates (Bravo and Anderson 1994, Kremp and Anderson 2000), and salinity may influence excystment success, such as with *A. minutum* (Cannon 1993). Growth factors (platelet-derived growth factor and fetal bovine serum) also affect excystment (Costas et al. 1993). The macronutrient concentration in the culture medium does not appear to be a regulating factor for excystment (Bravo and Anderson 1994, Rengefors and Anderson 1998).

If the primary aim of an experiment is simply to achieve excystment, it is best to initially try the conditions that sustain vegetative growth. A temperature gradient table (see Section 2.3.1.) is useful; resting cysts can be isolated into individual wells along the temperature gradient and monitored for germination. Light intensity can be varied by adding layers of glassine paper or neutral-density filters, and by using different-colored filters or a light source with a different spectral output, one can also test the effect of light quality. To test the effects of anaerobic conditions, anaerobic chambers can be used, (e.g., BBL GasPak System anaerobic chamber with GasPak Plus anaerobic system envelopes and Palladium catalyst [$H_2$ and $CO_2$] [available from scientific supply companies]). Petri dishes containing resting cysts are placed in the chambers for varying periods of time. Subsampling can be performed under anaerobic conditions using an $N_2$-filled glove bag (a flexible, inflatable polyethylene chamber with built-in gloves that lets you work in an isolated and controlled environment).

Salinity can be modified by diluting full-strength natural or artificial seawater with sterilized distilled or Milli-Q water in calculated proportions. The culture medium is the same that supports vegetative growth; it is best to start with a nutrient-replete medium.

## 2.3.4.2. Observations of Excystment

Resting cysts can be isolated with standard micromanipulation techniques (Throndsen 1973; Chapter 6) from the cross. If the cysts adhere to the crossing dish, they can be aseptically rinsed and new media added. This removes vegetative cells while the majority of the cysts remain adhered to the petri dish. Alternatively, "sticky" cysts can be carefully lifted from the surface of the petri dish with a small Teflon scraper (Parker 2002). Because crosses with resting cysts are often old cultures, bacterial numbers may be high and individual cysts should be well rinsed in sterile seawater or medium prior to incubation for excystment.

Physiological changes can be monitored preceding excystment. In *Scrippsiella trochoidea*, metabolic activity increases (Binder and Anderson 1990), and differences in fluorescence appear to relate to germination ability (Blanco 1990). Another characteristic is how the germling cell emerges from the resting cyst, i.e., the nature of the archeopyle (cyst wall opening where the germling cell exits). The archeopyle may be a small slit or an opening in the wall reflecting a plate boundary of the vegetative cell, or the wall may split in half (Bolch 1997, Matsuoka and Fukuyo 2003). Indisputable evidence of excystment is the presence of vegetative cells coupled with an empty cyst wall.

### 2.3.5. Nuclear Events and the Sexual Life Cycle

The dinoflagellate nucleus is unusual, large and obvious (Spector 1984, Treimer and Fritz 1984, Taylor 1989). Nuclear fusion coincides with gamete fusion to form a zygote (planozygote; see Fig. 24.1), and germination is accompanied by meiosis. A rotational nuclear phenomenon called "cyclosis" described by Biecheler (1952) is associated with meiosis (von Stosch 1972, 1973; Pfiester and Anderson 1989). The timing and nature of nuclear fusion and meiosis can vary between different species. In particular, meiosis can take place prior to, during, or after excystment; in the latter case the germling cell is a diploid planomeiocyte with the vegetative cells forming with subsequent mitotic divisions (see Fig. 24.1).

Meiosis in dinoflagellates is usually a two-step process in which genetic recombination can happen at both the first and second meiotic divisions. Indeed, Beam and Himes (1980) consider the two steps uncoordinated in many dinoflagellates. More unusual is one-step meiosis, in which the genetic recombination occurs only during the first division of meiosis, leading to a 1 : 1 segregation of genotypic characters. Pfiester and Anderson (1989) describe these differences based on work of Beam and Himes (Himes and Beam 1975, Beam et al. 1977, Beam and Himes 1980) and personal communications from von Stosch, Theil, and Happach-Kasam. The recombination and segregation of genetic characteristics during meiosis can be studied by isolating the vegetative cells formed after meiosis (e.g., the heritability of toxin production was investigated in this way by Oshima et al. [1993]).

To facilitate observations of nuclear events, including chromosome discrimination during encystment and excystment, there are a number of nuclear stains such as aceto-carmine and fluorochromes such as 4–6-diamidino-2-phenylindole (DAPI) (Loper et al. 1980, Coleman et al. 1981). Recently, Litaker et al. (2002) used the nuclear-specific dyes DAPI and Hoechst 33258 to elegantly discriminate the different nuclear stages of the life cycle of *Pfiesteria piscicida* in concert with the different morphological stages of the life cycle.

### 2.4. Dinoflagellate Mating Systems

### 2.4.1. Mating Systems and Mating Types

The term "mating system" is used here to describe the mating interactions in a dinoflagellate species. The simplest is self-fertility (homothallic systems), followed by heterothallic binary systems in which the mating type of all individuals is either positive (+) or negative (−) (e.g., *Alexandrium catenella*, Yoshimatsu 1981, 1984). Multiple or complex mating systems may have varying levels of compatibility between individuals and groups of individuals, including tendencies toward (+) and (−) mating types rather than discrete (+) and (−) mating types, as well as having different sets of mating types (e.g., *Alexandrium excavatum* [Braarud] Balech et Tangen—now considered a synonym of *A. tamarense* [Balech 1995]), Destombe and Cembella 1990; *Gymnodinium catenatum*, Blackburn et al. 2001). Complex mating systems may also include partial compatibility between strains (i.e., only some life cycle processes are completed successfully). Examples include planozygote formation that is not followed by resting cyst formation or the formation of resting cysts that are not able to germinate. The mating system not only is important in its control over whether sexual reproduction can be induced in cultured strains but also influences the population dynamics in nature (see Section 4.0).

### 2.4.2. Investigating Mating Systems in Culture

Investigating mating systems in culture requires a series of intercrosses and self-crosses. A pairwise matrix (see Fig. 24.5) is prepared by combining each strain with all other strains; self-crosses should also be made as a control and to test for homothallism. Evidence of sexual reproduction and quantitative estimates of sexual reproduction (e.g., cysts per volume) should be determined (see Section 2.4.3.). It is important to have similar cell numbers of each strain in an intercross to optimize the opportunity for compatible strains to meet. Replication of pairwise crosses is important for statistical reliability; it is not uncommon for variation in a single cross (i.e., the cross may form resting cysts in some but not all replicates; Yoshimatsu 1984, Blackburn et al. 1989, Destombe and Cembella 1990). Negative results (no cysts) do not mean that the strains are necessarily incompatible, but positive results demonstrate reproductive compatibility to the level of successful encystment. If one is testing a large number of strains (more than 20), the number of crosses required may be formidable (e.g., 210 crosses for a single replicate of a 20-strain crossing matrix). In such cases, making a large crossing matrix without replication followed by replication of those which initially had negative results decreases the total number of crosses required (Blackburn et al. 2001).

### 2.4.3. Assessment of Reproductive Compatibility

Usually the formation of a resting cyst is taken to demonstrate compatibility (Yoshimatsu 1981; Blackburn et al. 1989, 2001); however, other degrees of compatibility are relevant in examining the genetic relatedness of strains. Destombe and Cembella (1990) assessed the frequency of gamete recognition (i.e., pairs of fusing cells). Successful germination, meiosis, and subsequent cell divisions indicate viability of the products of germination and meiosis. Blackburn (2001) used successful division to the eight-cell stage as an indication of viability for *G. catenatum*. Backcrosses to parental strains and $F_1$ viability are important when assessing complete gene flow and/or the inheritance of certain traits. Oshima et al. (1993) made backcrosses to investigate heritability of toxin production in *Gymnodinium catenatum*. Destombe and Cembella (1990) considered gamete recognition, meiosis, and zygote germination to be distinct phenomena; success in one does not necessarily imply success in another. Deciding the level of compatibility depends on the questions being asked. For example, resting cyst formation is evidence of genetic closeness of the parental strains, in that they are able to form a successful zygote stage; resting cyst germination, viability of the progeny, and successful backcrosses of the progeny to the parent strains are a test of genetic similarity at the biological species level (Mayr 1940), a much more stringent test of interbreeding success than resting cyst formation alone.

Reproductive compatibility can be assessed quantitatively. For example, if mating is observed in four replicate wells of a total of eight wells in the same crossing experiment, the frequency of mating for these crossing

pairs is calculated to be 50%. Percentage encystment (Sgrosso et al. 2001) is calculated as

$$\% \; Encystment = \left( \frac{2 \times a}{b + 2 \times a} \right) \times 100 \qquad (1)$$

where $a$ (resting cysts) is the resting cyst concentration per mL and $b$ (vegetative cells) is the vegetative cell concentration per mL. This formula is based on the assumption that resting cysts are the result of the fusion of two cells (functional gametes). Cyst production can be determined either by microscopically scanning the entire culture vessel or by counting intact and germinated cysts along transects and estimating the total (Blackburn et al. 2001). Blackburn et al. (2001) used a scoring of 0 to 4 for increasing cyst concentrations up to $>10^5 \cdot L^{-1}$. A score of 0 was a cyst concentration up to $3 \times 10^2 \cdot L^{-1}$; 1 was 300 to $2 \times 10^3 \cdot L^{x1}$; 2 was $2 \times 10^3 \cdot L^{-1}$ to $1 \times 10^4 \cdot L^{-1}$; 3 was $1 \times 10^4 \cdot L^{-1}$ to $1 \times 10^5 \cdot L^{-1}$; and 4 was $>10^5 \cdot L^{-1}$. These scores were incorporated in analyses of reproductive compatibility (see next paragraph). Measuring cyst concentrations can be difficult (or sometimes impossible) when the cysts are mucilaginous and sticky and/or form dense aggregates on surfaces (e.g., *Alexandrium minutum*, Parker 2002). See Chapter 16 for further information about counting clumping cells. Finally, frequency of germination was defined by Destombe and Cembella (1990) as the number of successful germinations relative to the total number of crossing attempts.

### 2.4.4. Analysis of Mating Systems

The simplest form of mating system analysis is presence/absence of resting cysts in self- and intercrosses, involving sorting the results by eye into groups of strains with similar mating characteristics (e.g., Fig. 24.6). With more than a handful of strains, using presence/absence rapidly becomes confusing and unwieldy. The Jaccard similarity coefficient ($S_J$) was used for *Alexandrium excavatum* to represent similarity in reproductive behavior between isolates and therefore allocate a "mating type tendency" to each strain (Destombe and Cembella 1990). The tendency was used because they did not find a straightforward system of homothallic strains or heterothallic (+) and (−) mating types but instead found gradations of mating between (+) and (−) for the strains. The coefficient was calculated as $S_J$ a/a + u, where "a" is the number of occasions on which similar reproductive behavior (germination or nongermination of zygote) was observed between two isolates, and "u" is the number of mismatched pairs in crossing

| Strain | | 1 | 2 | 3 | 4 | 5 | 6 | 7 | 8 |
|---|---|---|---|---|---|---|---|---|---|
| | | + | + | − | + | − | − | − | − |
| 1 | + | | − | Z | − | Z | Z | Z | Z |
| 2 | + | − | | Z | − | Z | Z | Z | Z |
| 3 | − | Z | Z | | Z | − | − | − | − |
| 4 | + | − | − | Z | | Z | Z | Z | Z |
| 5 | − | Z | Z | − | Z | | − | − | − |
| 6 | − | Z | Z | − | Z | − | | − | − |
| 7 | − | Z | Z | − | Z | − | − | | − |
| 8 | − | Z | Z | − | Z | − | − | − | |

**FIGURE 24.6.** The summary result of four intercrossing experiments. Strain numbers 1–8: +, plus mating type; –, minus mating type. Z, resting cysts; -, no resting cysts (after Yoshimatu 1981.)

| Group | Strain | 1 | | | | | | | | 2 | | | | | | 3 | | | 4 | 5 | 6 | 7 |
|---|---|---|---|---|---|---|---|---|---|---|---|---|---|---|---|---|---|---|---|---|---|---|
| 1 | HU15 | 0 | 0 | 0 | 0 | 0 | 0 | 0 | 0 | 1 | 1 | 1 | 0 | 1 | 1 | 1 | 0 | 1 | 0 | 0 | 1 | 1 |
| | DE06 | 0 | 0 | 0 | 0 | 0 | 0 | 0 | 0 | 1 | 1 | 0 | 0 | 1 | 1 | 1 | 0 | 0 | 1 | 1 | 0 | 1 |
| | DE07 | 0 | 0 | 0 | 0 | 0 | 0 | 0 | 0 | 0 | 1 | 0 | 0 | 1 | 1 | 1 | 0 | 0 | 1 | 0 | 0 | 1 |
| | DE01 | 0 | 0 | 0 | 0 | 0 | 0 | 0 | 0 | 0 | 1 | 0 | 0 | 1 | 1 | 0 | 0 | 0 | 1 | 0 | 0 | 1 |
| | HU02 | 0 | 0 | 0 | 0 | 0 | 0 | 0 | 0 | 0 | 1 | 0 | 1 | 1 | 1 | 1 | 0 | 0 | 1 | 0 | 0 | 1 |
| | DE05 | 0 | 0 | 0 | 0 | 0 | 0 | 0 | 0 | 0 | 0 | 0 | 0 | 1 | 0 | 0 | 1 | 0 | 1 | 0 | 1 | 1 |
| | PT02 | 0 | 0 | 0 | 0 | 0 | 0 | 0 | 0 | 0 | 0 | 0 | 1 | 1 | 0 | 1 | 0 | 1 | 0 | 0 | 0 | 1 |
| | DE09 | 0 | 0 | 0 | 0 | 0 | 0 | 0 | 0 | 0 | 0 | 0 | 0 | 0 | 0 | 0 | 0 | 0 | 0 | 0 | 0 | 1 |
| 2 | HU07 | 1 | 1 | 0 | 0 | 0 | 0 | 0 | 0 | 0 | 0 | 0 | 0 | 0 | 0 | 1 | 1 | 0 | 0 | 1 | 0 | 1 |
| | HU11 | 1 | 1 | 1 | 1 | 1 | 0 | 0 | 0 | 0 | 0 | 0 | 0 | 0 | 0 | 1 | 1 | 0 | 0 | 1 | 1 | 1 |
| | HU08 | 1 | 0 | 0 | 0 | 0 | 0 | 0 | 0 | 0 | 0 | 0 | 0 | 0 | 0 | 0 | 0 | 0 | 0 | 1 | 0 | 0 |
| | PT01 | 0 | 0 | 0 | 0 | 1 | 0 | 0 | 0 | 0 | 0 | 0 | 0 | 0 | 0 | 0 | 0 | 1 | 1 | 0 | 1 | 0 |
| | SP08 | 1 | 1 | 1 | 1 | 1 | 1 | 1 | 0 | 0 | 0 | 0 | 0 | 0 | 0 | 0 | 1 | 1 | 1 | 1 | 1 | 1 |
| | SP09 | 1 | 1 | 1 | 1 | 1 | 0 | 1 | 0 | 0 | 0 | 0 | 0 | 0 | 0 | 0 | 1 | 1 | 1 | 0 | 1 | 1 |
| 3 | HU06 | 1 | 1 | 1 | 0 | 1 | 0 | 0 | 0 | 1 | 1 | 0 | 0 | 0 | 0 | 0 | 0 | 0 | 1 | 1 | 1 | 0 |
| | JP01 | 0 | 0 | 0 | 0 | 0 | 1 | 1 | 0 | 1 | 1 | 0 | 0 | 1 | 1 | 0 | 0 | 0 | 1 | 1 | 1 | 1 |
| | JP10 | 1 | 0 | 0 | 0 | 0 | 0 | 0 | 0 | 0 | 0 | 0 | 1 | 1 | 1 | 0 | 0 | 0 | 1 | 1 | 1 | 1 |
| 4 | HU10 | 0 | 1 | 1 | 1 | 1 | 1 | 1 | 0 | 0 | 0 | 1 | 1 | 1 | 1 | 1 | 1 | 1 | 0 | 1 | 1 | 1 |
| 5 | DE02 | 0 | 1 | 0 | 0 | 0 | 0 | 0 | 0 | 1 | 1 | 0 | 0 | 1 | 0 | 1 | 1 | 1 | 1 | 0 | 1 | 1 |
| 6 | DE08 | 1 | 0 | 0 | 0 | 0 | 1 | 0 | 0 | 0 | 1 | 0 | 1 | 1 | 1 | 1 | 1 | 1 | 1 | 1 | 0 | 1 |
| 7 | HU09 | 1 | 1 | 1 | 1 | 1 | 1 | 1 | 1 | 1 | 1 | 1 | 0 | 1 | 1 | 0 | 1 | 1 | 1 | 1 | 1 | 0 |

**FIGURE 24.7.** *Gymnodinium catenatum* multiple mating groups, based on the criterion of largest possible groups. Seven groups are shown (shaded), the largest containing 8, 6, and 3 strains, the other four comprising single strains; 1, resting cysts; 0, unsuccessful crosses (after Blackburn et al. 2001).

experiments with all other isolates. Blackburn et al. (2001) used several computer-aided approaches to sort strains of *G. catenatum* into multiple incompatibility groups (Fig. 24.7; see also Dini and Nyberg 1993). The compatibility index ($CI_s$) was the number of compatible pairings (score of $\geq 1$) divided by the total number of possible crosses other than self-crosses. This provides a measure of "willingness to cross" with other strains. Average vigor ($AV_s$) was the average of the scores (0–4) for maximum cyst production for all successful crosses involving a particular strain (i.e., an average measure of the number of cysts produced per cross in successful

crosses). Note that the cyst production scores approximate a logarithmic scale, and so the average score corresponds to a geometric mean of cyst numbers. This is more appropriate, because the cyst score distribution was skewed toward larger scores. Reproductive compatibility ($RC_s$) was calculated as the product of the strain's CI and AV in all crosses (i.e., $RC_s = CI_s \times AV_s$). To assess between-population compatibility, the same measures were calculated from the corresponding subset of pairwise interpopulation crosses. The between-population measures were designated as the between-population compatibility index ($CI_P$), between-population average vigor ($AV_P$) and between-population reproductive compatibility ($RC_P$).

Parker (2002) analyzed crossing matrices for several dinoflagellates, using hierarchical cluster analysis with average linkage (JMP statistical package, Version 4.0; SAS Institute, Inc.). This analysis clusters strains whose reproductive compatibilities are close to each other relative to strains of other clusters. The subsequent arrangement of strains (represented in dendrograms; Fig. 24.8a,b) and the numerical distances between groups can be used to determine the potential numbers of mating types and to describe the mating system. Strains separated by a distance of less than 1 have identical reproductive compatibility and are considered to be of the same mating type. In complex mating systems there is the potential for a mating type to consist of multiple strains whose mating compatibilities are not identical (i.e., with distances of greater than 1) but are similar. Different mating types containing strains of "variable closeness" in reproductive compatibility can be identified depending on the distance chosen to define mating type.

Cluster analysis also provides a maximum distance between strains, which is indicative of the amount of variability in reproductive compatibility within a crossing matrix. Homothallic strains should not be included in cluster analysis because some or all resting cysts produced may be the result of self-crosses. Differences in means of compatibility measures for each crossing matrix or species can be tested for significance with Mann-Whitney rank sum tests (SigmaStat version 2.0; SPSS, Inc.).

## 3.0. Complex Interactions

### 3.1. Cell-to-Cell Communication

Since von Stosch (1973) first described the "dancing gametes" of *Gymnodinium pseudopalustre* and *Woloszyn-*

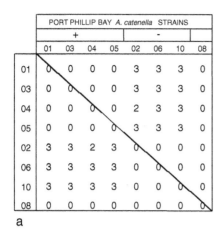

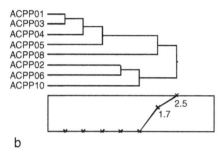

**FIGURE 24.8.** **(a)** Matrix of reproductive compatibility for six strains of *Alexandrium catenella* isolated from Port Phillip Bay, Victoria, Australia. Scores of 1 to 4 indicate increasing numbers of resting cysts per cross. **(b)** Upper dendogram is a cluster analysis of mating compatibility between strains in **a**. Lower plot indicates the distance between different clusters, based on average linking (after Parker 2002).

*skia apiculata*, there has been speculation that there may be cell-to-cell communication in dinoflagellates through pheromone-like substances that attract compatible cells to one another (Pfiester 1989, Wyatt and Jenkinson 1997). An alternative hypothesis is that contact between compatible cells is a random event. For either scenario, the physical environment (e.g., water column stability and currents) and dinoflagellate swimming speed influence the likelihood of cell-to-cell contact. If cell-to-cell contact is random, then the other primary factor controlling encystment would be cell density. If there is cell-to-cell communication, the likelihood and efficiency of compatible gamete contact should be significantly enhanced.

While dinoflagellate cell-to-cell communication and pheromones are currently restricted to speculation, they are relatively well studied in other microalgae, macroalgae, and in other protozoa (Maier and Mueller 1986, Maier 1993, Fukumoto et al. 2002, Kuhlmann et al. 1997). Dinoflagellates are alveolates, a diverse group

that includes ciliates (Taylor 1980, Pfiester and Anderson 1989, Fast et al. 2002). In the ciliates, *Euplotes raikovi* Agamaliev releases pheromones in complementary mating types (nonself pheromones) that induce sexual behavior, while self pheromones either have no effect (Stock et al. 1999) or have even been found to promote vegetative growth (Vallesi et al. 1995).

Pheromone-like substances in dinoflagellates could act at various different life stages and their efficacy and function might be linked to cell density (Wyatt and Jenkinson 1997). Indeed, it is possible that cell density may act as a trigger for the release of pheromone-like substances, which then trigger gametogenesis or attract compatible gametes. Chemical signals may also influence planozygote development into a resting cyst or impact the location of cyst deposition (Holt et al. 1994, Wyatt and Jenkinson 1997). Nonrandom deposition of *A. minutum* resting cysts suggests that this may indeed occur in some species (Parker 2002).

We have studied the effect of filtrates from natural blooms of *G. catenatum* on encystment in cultured crosses (using no filtrate as the control) and observed a significant enhancement of encystment (Parker 2002). Dialysis or filtration membranes could also be used for studying the effect of one strain on the behavior of another without direct physical contact.

### 3.2. Bacteria and Microalgal Life Cycles

The influence of bacteria on algal cells and vice versa is little studied, although interest in this area has increased because of the potential role of bacteria in toxicity (Doucette et al. 1998). Interactions between dinoflagellates and bacteria may influence life cycle dynamics and have a significant effect on the algal ecology overall. Lewis et al. (2001) found bacteria associated with all stages of the life cycle of *Alexandrium tamarense* and *A. fundyense*, Balech both on the surface of the cells and within. Adachi et al. (1999) showed that natural planktonic bacterial assemblages (which were not identified) enhanced encystment of *A. catenella* in culture.

Algicidal bacteria may play a role in structuring phytoplankton communities, and many bacteria are able to lyse harmful algal species, e.g., lysis of *Heterocapsa circularisquama* Horiguchi by *Cytophaga* sp. (Imai et al. 1993), *Gymnodinium catenatum* by *Pseudoalteromonas* sp. (Lovejoy et al. 1998), and *Karenia brevis* by unidentified bacterial species (Doucette et al. 1999). *Gymnodinium catenatum* and *A. minutum* resting cysts were resistant to algicidal bacteria isolated from the Huon Estuary, in southeast Tasmania, while vegetative cells of both species rapidly lysed (Parker and Skerratt, personal communication). Encystment is a potentially powerful ecological strategy for avoiding algicidal bacteria, but currently there is no evidence that these bacteria cause encystment.

## 4.0. FROM CULTURE TO NATURE

Culture studies enable description and analyses of processes that cannot be obtained easily, if at all, from field studies. They elucidate phases of the life cycle that may be cryptic, transitory, or rare in nature, and strains from different locations can be cultured side by side or intercrossed. Controlled comparison of environmental variables and their effects on encystment and excystment can be made. Culture studies can then be used to interpret field studies, giving insights into both autecology and system ecology. For example, encystment or excystment temperature ranges offer evidence of seasonal "windows" for bloom initiation and decline (Bravo and Anderson 1994, Kremp and Anderson 2000). Culture studies may also aid in predicting harmful algal blooms or managing natural environments and aquaculture sites. Culture-based mating studies give insights on genetic relationships between different populations globally (e.g., Australian, Japanese, and Spanish *G. catenatum* populations; Blackburn et al. 2001) and provide information to evaluate both the origins and introduction of microalgae into new geographic areas.

## 5.0. REFERENCES

Adachi, M., Kanno, T., Matsubara, T., Nishijima, T., Itaura, S., and Yamaguchi, M. 1999. Promotion of cyst formation in the toxic dinoflagellate *Alexandrium* (Dinophyceae) by natural bacterial assemblages from Hiroshima Bay, Japan. *Marine Ecol. Progr. Series* 191:175–85.

Adams, D. G., and Duggan, P. S. 1999. Heterocyst and akinete differentiation in cyanobacteria. *New Phytol.* 144:3–33.

Andersen, R. A. 2003. A world list of culture collections. In: Hallegraeff G. M., Anderson, D. M., and Cembella, A. D., eds. *Manual on Harmful Marine Microalgae.* UNESCO Publishing, Paris, pp. 753–66.

Anderson, D. M. 1980. Effects of temperature conditioning on development and germination of *Gonyaulax tamarensis* (Dinophyceae) hypnozygotes. *J. Phycol.* 16:166–72.

Anderson, D. M. 1984. Shellfish toxicity and dormant cysts in toxic dinoflagellate blooms. In: Ragelis E. P., ed. *Seafood Toxins.* American Chemical Society, Washington, D.C., pp. 125–38.

Anderson, D. M., Chisholm, S. W., and Watras, C. J. 1983. Importance of life cycle events in the population dynamics of *Gonyaulax tamarensis. Marine Biol.* 76:179–89.

Anderson, D. M., Coats, D. W., and Tyler M. A. 1985. Encystment of the dinoflagellate *Gyrodinium uncatenum*: temperature and nutrient effects. *J. Phycol.* 21:200–6.

Anderson, D. M., Fukuyo, Y., and Matsuoka, K. 1995. Cyst Methodologies. In: Hallegraeff, G. M., Anderson, D. M., Cembella, A. D., eds. *Manual on Harmful Marine Microalgae.* UNESCO, Paris, pp. 229–49.

Anderson, D. M., Fukuyo, Y., and Matsuoka, K. 2003. Cyst methodologies. In: Hallegraeff G. M., Anderson, D. M., and Cembella, A. D., eds. *Manual on Harmful Marine Microalgae.* UNESCO, Paris, pp. 165–91.

Anderson, D. M., Jacobson, D., Bravo, I., and Wrenn, J. H. 1988. The unique, microreticulate cyst of the naked dinoflagellate *Gymnodinium catenatum. J. Phycol.* 24:255–62.

Anderson, D. M., Kulis, D. M., and Binder, B. J. 1984. Sexuality and cyst formation in the dinoflagellate *Gonyaulax tamarensis*: Cyst yield in batch cultures. *J. Phycol.* 20:418–25.

Anderson, D. M., and Wall, D. 1978. Potential importance of benthic cysts of *Gonyaulax tamarensis* and *G. excavata* in initiating toxic dinoflagellate blooms. *J. Phycol.* 13:224–34.

Balech, E. 1995. The genus *Alexandrium* Halim (Dinoflagellata), Sherkin Island Press, Sherkin Island, Ireland, 151 pp.

Barlow, S., and Triemer, R. E. 1988. Alternate life history stages in *Amphidinium klebsii* (Dinophyceae, Pyrrophyta). *J. Phycol.* 27:413–20.

Beam, C. A., and Himes, M. 1974. Evidence for sexual fusion and recombination in the dinoflagellate *Crypthecodinium cohnii. Nature* 250:435.

Beam, C. A., and Himes, M. 1980. Sexuality and meiosis in dinoflagellates. In: Levandowsky, and Hunter, S. H., eds. *Biochemistry and Physiology of Protozoa.* Vol. 3, Academic Press, New York, pp. 171–206.

Beam, C. A., Himes, M., Himelfarb, J., Link, C., Shaw, K. L. 1977. Genetic evidence of meiosis in the dinoflagellate *Crypthecodinium cohnii. Genetics* 87:19–32.

Biecheler, B. 1952. Recherches sur les Péridiniens. *Bull. Biol. France Belg. Suppl.* 36:1–149.

Binder, B. J., and Anderson, D. M. 1987. Physiological and environmental control of germination in *Scrippsiella trochoidea* (Dinophyceae) resting cysts. *J. Phycol.* 23:99–107.

Binder, B. J., and Anderson, D. M. 1990. Biochemical composition and metabolic activity of *Scrippsiella trochoidea* (Dinophyceae) resting cysts. *J. Phycol.* 26:289–98.

Blackburn, S. I., Bolch, C. J., Haskard, K. A., and Hallegraeff, G. M. 2001. Reproductive compatibility among four global populations of the toxic dinoflagellate *Gymnodinium catenatum* (Dinophyceae). *Phycologia* 40:78–87.

Blackburn, S. I., Hallegraeff, G. M., and Bolch, C. J. 1989. Vegetative reproduction and sexual life cycle of the toxic dinoflagellate *Gymnodinium catenatum* from Tasmania, Australia. *J. Phycol.* 25:577–90.

Blanco, J. 1990. Cyst germination of two dinoflagellate species from Galicia (NW Spain). *Sci. Mar.* 54:287–91.

Blanco, J. 1995. Cyst production in four species of neritic dinoflagellates. *J. Plankton Res.* 17:165–82.

Boero, F., Belmonte, G., Fanelli, G., Piraino, S., and Rubino, F. 1996. The continuity of living matter and the discontinuities of its constituents: Do plankton and benthos really exist? *Trends Ecol. Evolution* 11:177–80.

Bolch, C. J. S. 1997. The use of sodium polytungstate for the separation and concentration of living dinoflagellate cysts from marine sediments. *Phycologia* 36:472–8.

Bolch, C. J., Blackburn, S. I., Cannon, J. A., and Hallegraeff, G. M. 1991. The resting cyst of the red-tide dinoflagellate *Alexandrium minutum* (Dinophyceae). *Phycologia* 30: 215–9.

Bold, C. H., and Wynne, M. J. 1985. *Introduction to the Algae.* Prentice-Hall, Princeton, New Jersey, 720 pp.

Bravo, I., and Anderson, D. M. 1994. The effects of temperature, growth medium and darkness on excystment and growth of the toxic dinoflagellate *Gymnodinium catenatum* from northwest Spain. *J. Plankton Res.* 16:513–25.

Cannon, J. A. 1993. Germination of the toxic dinoflagellate, *Alexandrium minutum*, from sediments in the Port River, South Australia. In: Smayda, T. J., and Shimizu, Y., eds. *Toxic Phytoplankton Blooms in the Sea.* Elsevier, Amsterdam, pp. 109–14.

Cao Vien, M. 1967. Sur l'existence de phénomenes sexuels chez un Péridinien libre, *l'Amphidinium carteri. Compte rendu hebdomadaire des seances de l'Academie des Sciences. Paris, Series D,* 264:1006–8.

Cao Vien, M. 1968. Sur la germination du zygote et sur un mode particular de multiplication vegetative chez le peridinien libre, *l'Amphidinium carteri. Compte rendu hebdomadaire des seances de l'Academie des Sciences. Paris, Series D,* 267:701–3.

Chesnick, J. M. 1987. *Ultrastructural investigations of three stages of the sexual life history of Peridinium balticum (Pyrrhophyta).* PhD Dissertation, Texas A & M University. Dissertation Abstracts International Part B: Science and Engineering 47, 179 pp.

Chesnick, J. M., and Cox, E. R. 1989. Fertilization and zygote development in the binucleate dinoflagellate *Peridinium balticum* (Pyrrhophyta). *Am. J. Bot.* 76:1060–72.

Coats, D. W., Tyler, M. A., and Anderson, D. M. 1984. Sexual processes in the life cycle of *Gyrodinium uncatenum* (Dinophyceae): A morphogenetic overview. *J. Phycol.* 20:351–61.

Coleman, A. W. 1983. The roles of resting spores and akinetes in chlorophyte survival. In: Fryxell, G. A., ed. *Survival Strategies of the Algae.* Cambridge University Press, Cambridge, pp. 1–22.

Coleman, A. W., Maguire, M. J., and Coleman, J. R. 1981. Mithramycin- and 4-6-diamidino-2-phenylindole (DAPI)-DNA staining for fluorescence microspectrophotometric measurement of DNA in nuclei, plastids, and virus particles. *J Histochem. Cytochem.* 29:959–68.

Corrales, R. A., Reyes, M., and Martin, M. 1995. Notes on the encystment and excystment of *Pyrodinium bahamense* var. *compressum in vitro*. In: Lassus, P., Arzul, G., Erard-Le Denn, E., Gentien, P., and Marcaillou-Le Baut, C., eds. *Harmful Marine Algal Blooms.* Lavoisier Publishing, Paris, pp. 573–8.

Costas, E., Gonzalez, G. S., Aguilera, A., Lopez, R. V. 1993. An apparent growth factor modulation of marine dinoflagellate excystment. *J. Exper. Marine Biol. Ecol.* 166:241–9.

Cronberg, G. 1986. Chrysophycean cysts and scales in lake sediments: A review. In: Kristiansen, J., and Andersen, R. A., eds. *Chrysophytes: Aspects and Problems,* Cambridge University Press, Cambridge, pp. 281–315.

Dale, B. 1977. Cysts of the toxic red-tide dinoflagellate *Gonyaulax excavata* (Braarud) Balech from Oslofjorden, Norway. *Sarsia* 63:29–34.

Dale, B. 1983. Dinoflagellate resting cysts: "Benthic plankton." In: Fryxell, G. A., ed. *Survival Strategies of the Algae.* Cambridge University Press, Cambridge, pp. 69–136.

Dale, B. 2001. The sedimentary record of dinoflagellate cysts: Looking back into the future of phytoplankton blooms. *Sci. Mar.* 65(Suppl 2):257–72.

Dale, B., Yentch, C. M., and Hurst, J. 1978. Toxicity in resting cysts of the red tide dinoflagellate *Gonyaulax excavata* from deeper water coastal sediments. *Science* 201:1223–5.

Destombe, C., and Cembella, A. 1990. Mating-type determination, gametic recognition and reproductive success in *Alexandrium excavatum* (Gonyaulacales, Dinophyta), a toxic red-tide dinoflagellate. *Phycologia* 29:316–25.

Dini, F., and Nyberg, D. 1993. Sex in ciliates. *Adv. Microb. Ecol.* 13:85–153.

Diwald, K. 1937. Die ungeshlechtiliche und geschlechtiche Fortpflanzung von *Glenodinium lubiniensiforme* spec. nov. *Flora* 132:174–92.

Doucette, G. J., Kodama, M., Franca, S., Gallacher, S. 1998. Bacterial interactions with harmful algal bloom species: Bloom ecology, toxigenesis and cytology. In: Anderson, D. M., Cembella, A. D., Hallegraeff, G. M., eds. *Physiological Ecology of Harmful Algal Blooms.* NATO ASI series. Vol. G41. Springer-Verlag, Berlin, pp. 619–47.

Doucette, G. J., McGovern, E. R., and Babinchak, J. A. 1999. Algicidal bacteria active against *Gymnodinium breve* (Dinophyceae). I. Bacterial isolation and characterization of killing activity. *J. Phycol.* 35:1447–54.

Duff, K. E., Zeeb, B. A., and Smol, J. P. 1997. Chrysophyte cyst biogeographical and ecological distributions: A synthesis. *J. Biogeogr.* 24:791–812.

Ellegaard, M., Kulis, D. M., and Anderson, D. M. 1998. Cysts of Danish *Gymnodinium nolleri* Ellegaard et Moestrup sp. ined. (Dinophyceae): studies on encystment, excystment and toxicity. *J. Plankton Res.* 20:1743–55.

Fast, N. M., Xue, L., Bingham, S., and Keeling, P. K. 2002. Re-examining alveolate evolution using multiple protein molecular phyloogenies. *J. Eukaryot. Microbiol.* 49:30–7.

Faust, M. A. 1992. Observations on the morphology and sexual reproduction of *Coolia monotis* (Dinophyceae). *J. Phycol.* 28:94–104.

Faust, M. A. 1993. Sexuality in a toxic dinoflagellate, *Prorocentrum lima*. In: Smayda T. J. and Shimizu Y., eds. *Toxic Phytoplankton Blooms in the Sea*. Elsevier, Amsterdam, pp. 121–6.

Fryxell, G. A., ed. 1983. *Survival Strategies of the Algae*. Cambridge University Press. Cambridge, 144 pp.

Fukumoto, R.-H., Dohmae, N., Takio, K., Satoh, S., Fujii, T., and Sekimoto, H. 2002. Purification and characterization of a pheromone that induces sexual cell division in the unicellular green alga *Closterium ehrenbergii*. *Plant Physiol. Biochem.* 40:183–8.

Gao, X., and Dodge, J. D. 1991. The taxonomy and ultrastructure of a marine dinoflagellate, *Scrippsiella minima* sp. nov. *Br. Phycol. J.* 26:21–31.

Garcés, E., Delgado, M., Maso, M., and Camp, J. 1998. Life history and *in situ* growth rates of *Alexandrium taylori* (Dinophyceae, Pyrrophyta). *J. Phycol.* 34:880–7.

Garrison, D. L. 1984. Planktonic diatoms. In: Steidinger, K. A., and Walker, L. M., eds. *Marine Plankton Life Cycle Strategies*. CRC Press, Boca Raton, Florida, pp. 1–18.

Giacobbe, M. G., and Gangemi, E. S. O. 1997. Vegetative and sexual aspects of *Dinophysis pavillardi* (Dinophyceae). *J. Phycol.* 33:73–80.

Giacobbe, M. G., and Yang, X. 1999. The life history of *Alexandrium taylori* (Dinophyceae). *J. Phycol.* 35:331–8.

Godhe, A., Noren, F., Kuylenstierna, M., Ekberg, C., and Karlson, B. 2001. Relationship between planktonic dinoflagellate abundance, cysts recovered in sediment traps and environmental factors in the Gullmar Fjord, Sweden. *J. Plankton Res.* 23:923–38.

Guillard, R. R. L. 1975. Culture of phytoplankton for feeding marine invertebrates. In: Smith, W. L., and Chanley, M. H., eds. *Culture of Marine Invertebrate Animals*. Plenum, New York, pp. 29–60.

Hallegraeff, G. M. 1993. A review of harmful algal blooms and their apparent global increase. *Phycologia* 32:79–99.

Hallegraeff, G. M. 2003. Harmful algal blooms: A global overview. In: Hallegraeff G. M., Anderson, D. M., and Cembella, A. D., eds. *Manual on Harmful Marine Microalgae*. UNESCO Publishing, Paris, pp. 1–50.

Hallegraeff, G. M., Marshall, J. A., Valentine, J., and Hardiman, S. 1998. Short cyst-dormancy period of an Australian isolate of the toxic dinoflagellate *Alexandrium catenella*. *Mar. and Freshwat. Res.* 49:415–20.

Happach-Kasam, C. 1980. *Beobachtungen zur Entwicklungsgeschichte der Dinophycee* Ceratium cornutum *Sexualitat*. Dissertation, Phillips University, Marburg, Germany.

Hargraves, P. E. 1976. Studies on marine planktonic diatoms. II. Resting spore morphology. *J. Phycol.* 12:118–28.

Hargraves, P. E., and French, F. W. 1983. Diatom resting spores: Significance and strategies. In: Fryxell, G. A., ed. *Survival Strategies of the Algae*. Cambridge University Press, Cambridge, pp. 49–68.

Hargraves, P. E. 1990. Studies on marine planktonic diatoms. V. Morphology and distribution of *Leptocylindrus minimus* Gran. *Beih. Nova Hedwigia* 100:47–60.

Heiskanen, A.-S. 1993. Mass encystment and sinking of dinoflagellates during a spring bloom. *Marine Biology* 116:161–7.

Head, M. J. 1996. Late Cenozoic dinoflagellates from the Royal Society borehole at Ludham, Norfolk, eastern England. *J. Paleontol.* 70:543–70.

Herdman, M. 1988. Cellular differentiation: Akinetes. *Methods Enzymol.* 167:222–32.

Hickel, B., and Pollingher, U. 1986. On the morphology and ecology of *Gonyaulax apiculata* (Penard) ENTZ from the Selenter See (West Germany). *Arch. Hydrobiol.* 73:227–32.

Himes, M., and Beam, C. 1975. Genetic analysis in the dinoflagellate *Crypthecodinium* (*Gyrodinium*) *cohnii*: Evidence for unusual meiosis. *Proc. Natl. Acad. Sci. USA* 72:4546–9.

Hoefker, I. 1930. Uber *Noctiluca scintillans* (Macartney). *Arch. Protist.* 71:57–78.

Holt, J. R., Merrell, J. R., Seaborn, D. W., and Hartranft, J. L. 1994. Population dynamics and substrate selection by three *Peridinium* species. *J. Freshwater Ecol.* 9:17–128.

Imai, I., Ishida, Y., and Hata, Y. 1993. Killing of marine phytoplankton by a gliding bacterium *Cytophaga* sp. isolated from the coastal sea of Japan. *Marine Biol.* 116:527–32.

Ishikawa, A., and Taniguchi, A. 1996. Contribution of benthic cysts to the population dynamics of *Scrippsiella* spp. (Dinophyceae) in Onagawa Bay, northeast Japan. *Mar. Ecol. Prog. Ser.* 140:169–78.

Jensen, M. O., and Moestrup, Ø. 1997. Autecology of the toxic dinoflagellate *Alexandrium ostenfeldii*: Life history and growth at different temperatures and salinities. *Eur. J. Phycol.* 32:9–18.

Keller, M. D., Selvin, R. C., Claus, W., and Guillard, R. R. L. 1987. Media for the culture of oceanic ultraplankton. *J. Phycol.* 23:633–8.

Kelley, I. 1989. *Observations on the life cycle and ultrastructure of the freshwater dinoflagellate* Gloeodinium montanum K *(Dinophyceae)*. PhD Dissertation, University of Oklahoma, Norman. Dissertation Abstracts International Part B: Science and Engineering 49, 149 pp.

Kelley, I., and Pfiester, L. A. 1990. Sexual reproduction in the freshwater dinoflagellate *Gloeodinium montanum. J. Phycol.* 26:167–73.

Kita, T., Fukuyo, Y., Tokuda, H., and Hirano, R. 1985. Life history of *Goniodoma pseudogoniaulax* (Pyrrophyta) in a rockpool. *Bull. Mar. Sci.* 37:643–51.

Kita, T., Fukuyo, Y., Tokuda, H., and Hirano, R. 1993. Sexual reproduction of *Alexandrium hiranoi* (Dinophyceae). *Bull. Plankton Soc. Japan* 39:79–85.

Kremp, A., and Anderson, D. M. 2000. Factors regulating germination of resting cysts of the spring bloom dinoflagellate *Scrippsiella hangoei* from the northern Baltic Sea. *J. Plankton Res.* 22:1311–27.

Kremp, A., and Heiskanen, A.-S. 1999. Sexuality and cyst formation of the spring-bloom dinoflagellate *Scrippsiella hangoei* in the coastal northern Baltic Sea. *Marine Biol.* 134:771–7.

Kuhlmann, H.-W., Bruenen-Nieweler, C., and Heckmann, K. 1997. Pheromones of the ciliate *Euplotes octocarinatus* not only induce conjugation but also function as chemoattractants. *J. Exper. Zool.* 277:38–48.

Lewis, J., Harris, A. S. D., Jones, K. J., and Edmonds, R. L. 1999. Long-term survival of marine planktonic diatoms and dinoflagellates in stored sediment samples. *J. Plankton Res.* 21:343–54.

Lewis, J., Kennaway, G., Franca, S., and Alverca, E. 2001. Bacterium-dinoflagellate interactions: Investigative microscopy of *Alexandrium* spp. (Gonyaulacales, Dinophyceae). *Phycologia* 40:280–5.

Litaker, R. W., Vandersea, M. W., Kibler, S. R., Madden, V. J., Noga, E. J., and Tester, P. A. 2002. Life cycle of the heterotrophic dinoflagellate *Pfiesteria piscicida* (Dinophyceae). *J. Phycol.* 38:442–63.

Loeblich, A. R. 1975. A seawater medium for dinoflagellates and the nutrition of *Cachonina niei. J. Phycol.* 11:80–6.

Loper, C. L., Steidinger, K. A., and Walker, L. M. 1980. A simple chromosome spread technique for unarmored dinoflagellates and implications of polyploidy in algal cultures. *Trans. Am. Micros. Soc.* 99:343–6.

Lovejoy, C., Bowman, J. P., and Hallegraeff, G. M. 1998. Algicidal effects of a novel marine *Pseudoalteromonas* isolate (Class Proteobacteria, Gamma subdivision) on harmful algal bloom species of the genera *Chattonella, Gymnodinium*, and *Heterosigma. App. Environ. Microbiol.* 64:2806–13.

Maier, I. 1993. Gamete orientation and induction of gametogenesis by pheromones in algae and plants. *Plant, Cell and Environment* 16:891–907.

Maier, I., and Mueller, D. G. 1986. Sexual pheromones in algae. *Bio. Bull.* 170:145–72.

Matsuoka, K., and Fukuyo, Y. 2003. Taxonomy of cysts. In: Hallegraeff, G. M., Anderson, D. M., and Cembella, A. D., eds. *Manual on Harmful Marine Microalgae*. UNESCO Publishing, Paris, pp. 562–92.

Maynard Smith, J. 1976. *The Evolution of Sex*. Cambridge University Press, Cambridge, 222 pp.

Mayr, E. 1940. Speciation phenomena in birds. *Am. Nat.* 74:249–78.

McQuoid, M. R., and Hobson, L. A. 1996. Diatom resting stages. *J. Phycol.* 32:889–902.

Montresor, M., and Marino, D. 1996. Modulating effect of cold-dark storage on excystment in *Alexandrium pseudogonyaulax* (Dinophyceae). *Mar. Biol.* 127:55–60.

Montresor, M. 1995. The life history of *Alexandrium pseudogonyaulax* (Gonyaulacales, Dinophyceae). *Phycologia* 34:444–8.

Morey-Gaines, G., and Ruse, R. H. 1980. Encystment and reproduction of the predatory dinoflagellate *Polykrikos kofoidii* Chatton (Gymnodiniales). *Phycologia* 19:230–6.

Nagai, S., Matsuyama, Y., Haruyoshi, T., Kotani, Y. 2002. Morphology of *Polykrikos kofoidii* and *P. schwartzii* (Dinophyceae, Polykrikaceae) cysts obtained in culture. *Phycologia* 41:319–27.

Nehring, S. 1997. Dinoflagellate resting cysts from recent German coastal sediments. *Bot. Mar.* 40:307–24.

Nuzzo, L., and Montresor, M. 1999. Different excystment patterns in two calcareous cyst-producing species of the dinoflagellate genus *Scrippsiella. J. Plankton Res.* 21:2009–18.

Olli, K., and Anderson, D. M. 2002. High encystment success of the dinoflagellate *Scrippsiella* cf. *lachrymosa* in culture experiments. *J. Phycol.* 38:145–56.

Oshima, Y., Itakura, H., Lee, K.-C., Yasumoto, T., Blackburn, S. I., and Hallegraeff, G. 1993. Toxin production by the dinoflagellate *Gymnodinium catenatum*. In: Smayda, T. J. and Shimizu, Y., eds. *Toxic Phytoplankton Blooms in the Sea*. Elsevier Science, Amsterdam, pp. 907–12.

Ouchi, A., Aida, S., Uchida, T., and, Honjo, T. 1994. Sexual reproduction of a red tide dinoflagellate *Gymnodinium mikimotoi. Fisheries Sci.* 60:125–6.

Parker, N. S. 2002. *Sexual reproduction and bloom dynamics of toxic dinoflagellates from Australian estuarine waters*. PhD thesis, University of Tasmania, Hobart, Australia, 276 pp.

Parker, N. S., Negri, A. P., Frampton, D. M. F., Rodolfi, L., Tredici, M. R., Blackburn, S. I. 2002. Growth of the toxic dinoflagellate *Alexandrium minutum* (Dinophyceae) using high biomass culture systems. *J. Appl. Phycol.* 14:313–24.

Pfiester, L. A. 1975. Sexual reproduction of *Peridinium cinctum* f. *ovoplanum* (Dinophyceae) *J. Phycol.* 11:259–65.

Pfiester, L. A. 1976. Sexual reproduction of *Peridinium willei* (Dinophyceae). *J. Phycol.* 12:234–8.

Pfiester, L. A. 1977. Sexual reproduction of *Peridinium gatunense* (Dinophyceae). *J. Phycol.* 13:92–5.

Pfiester, L. A. 1984. Sexual reproduction. In: Spector, D. L., ed. *Dinoflagellates*. Academic Press, Orlando, Florida, pp. 181–99.

Pfiester, L. A. 1989. Dinoflagellate Sexuality. *Intl. Rev. Cytol.* 114:249–72.

Pfiester, L. A., and Anderson, D. M. 1987. Dinoflagellate reproduction. In: Taylor, F. J. R., ed. *The Biology of Dinoflagellates*. Blackwell Scientific Publications, Oxford, pp. 611–48.

Pfiester, L. A., and Lynch, R. A. 1980. Amoeboid stages and sexual reproduction of *Cystodinium bataviense* and its similarity to *Dinococcus* (Dinophyceae). *Phycologia* 19:178–83.

Pfiester, L. A., and Skvarla, J. J. 1979. Heterothallism and thecal development in the sexual life history of *Peridinium volzii* (Dinophyceae). *Phycologia* 18:13–8.

Pfiester, L. A., and Skvarla, J. J. 1980. Comparative ultrastructure of vegetative and sexual thecae of *Peridinium limbatum* and *Peridinium cinctum* (Dinophyceae). *Am. J. Bot.* 67:955–8.

Pfiester, L. A., Timpano, P., Skvarla, J. J., and Holt, J. R. 1984. Sexual reproduction and meiosis in *Peridinium inconspicuum* Lemmermann (Dinophyceae). *Am. J. Bot.* 71:1121–7.

Pholpunthin, P., Fukuyo, Y., Matsuoka, K., and Nimura, Y. 1999. Life history of a marine dinoflagellate *Pyrophacus steinii* (Schiller) Wall et Dale. *Bot. Mar.* 42:189–97.

Probert, I. P., Erard-le Denn, E., and Lewis, J. 1998. Intracellular nutrient status as a factor in the induction of sexual reproduction in a marine dinoflagellate. In: Reguera, B., Blanco, J., Fernandez, M. L., Wyatt, T., eds. *Harmful Algae*. Xunta de Galicia and Intergovernmental Oceanographic Commission of UNESCO, Paris, pp. 343–4.

Rengefors, K., and Anderson, D. M. 1998. Environmental and endogenous regulation of cyst germination in two freshwater dinoflagellates. *J. Phycol.* 34:568–77.

Sako, Y., Ishida, Y., Kadota, H., and Hata, Y. 1984. Sexual reproduction and cyst formation in the freshwater dinoflagellate *Peridinium cunningtonii*. *Bull. Jap. Soci. Fish.* 50:743–50.

Sako, Y., Ishida, Y., Nishijima, T., and Hata, Y. 1987. Sexual reproduction and cyst formation in the freshwater dinoflagellate *Peridinium penardii*. *Nippon Suisan Gakkaishi* 53:473–8.

Sandgren, C. D. 1983. Survival strategies of chrysophycean flagellates: Reproduction and the formation of resistant resting cysts. In: Fryxell, G. A., ed. *Survival Strategies of the Algae*. Cambridge University Press, Cambridge, pp. 23–48.

Sandgren, C. D. 1991. Chrysophyte reproduction and resting cysts: A paleolimnologist's primer. *J. Paleolimnol.* 5:1–9.

Sarjeant, W. A. S., Lacalli, T., and Gaines, G. 1987. The cysts and skeletal elements of dinoflagellates: Speculations on the ecological causes for their morphology and development. *Micropaleontology* 33:1–36.

Schnepf, E., and Drebes, G. 1993. Anisogamy in the dinoflagellate *Noctiluca*? *Helgolander Meeresuntersuchungen* 47:265–73.

Seo, K. S., and Fritz, L. 2001. Evidence of sexual reproduction in the marine dinoflagellates, *Pyrocystis noctiluca* and *Pyrocystis lunula* (Dinophyta). *J. Phycol.* 37:530–5.

Sgrosso, S., Esposito, F., and Montresor, M. 2001. Temperature and daylength regulate encystment in calcareous cyst-forming dinoflagellates. *Mar. Ecol. Prog. Ser.* 211: 77–87.

Silva, E. S., and Faust, M. A. 1995. Small cells in the life history of dinoflagellates (Dinophyceae): A review. *Phycologia* 34:396–408.

Sonneborn, T. M. 1957. Breeding systems, reproductive methods, and species problems in protozoa. In: Mayr, E., ed. *The Species Problem*. American Association for the Advancement of Science, Washington, D.C., pp. 155–324.

Spector, D. L. 1984. Dinoflagellate nuclei. In: Spector, D. L., ed. *Dinoflagellates*. Academic Press, Orlando, Florida, pp. 107–41.

Spero, H. J., and Morée, M. D. 1981. Phagotrophic feeding and its importance to the life cycle of the holozoic dinoflagellate, *Gymnodinium fungiforme*. *J. Phycol.* 17:43–51.

Spoon, D. M., Feise, C. O. II., and Youn, R. S. 1977. Poly(Ethylene Oxide), a new slowing agent for protozoa. *J. Protozool.* 24:471–4.

Steidinger, K. A., and Walker, L. M. eds. 1984. *Marine Plankton Life Cycle Strategies*. CRC Press, Boca Raton, Florida, 158 pp.

Steidinger, K. A., Burkholder, J. M., Glasgow, H. B. Jr., Hobbs, C. W., Garrett, J. K., Truby, E. W., Noga, E. J., and Smith, S. A. 1996. *Pfiesteria piscicida* gen. et sp. nov. (Pfiesteriaceae fam. nov.), a new toxic dinoflagellate with a complex life cycle and behavior. *J. Phycol.* 32:157–64.

Steidinger, K. A., Landsberg, J. H., Truby, E. W., and Roberts, B. S. 1998. First report of *Gymnodinium pulchellum* (Dinophyceae) in North America and associated fish kills in the Indian River, Florida. *J. Phycol.* 34:431–7.

Stock, C., Kruppel, T., Key, G., Lueken, W. 1999. Sexual behaviour in *Euplotes raikovi* is accompanied by pheromone-induced modifications of ionic currents. *J. Exper. Biol.* 202:475–83.

Stulp, B. K., and Stam, W. T. 1982. General morphology and akinete germination of a number of *Anabaena* strains (Cyanophyceae) in culture. *Arch. Hydrobiol., Suppl.* 63:35–52.

Subba Rao, D. V. 1995. Life cycle and reproduction of the dinoflagellate *Dinophysis norvegica*. *Aquat. Microb. Ecol.* 9:199–201.

Taylor, D. L. 1973. The cellular interactions of algae-invertebrate symbiosis. *Adv. Marine Biol.* 11:1–56.

Taylor, F. J. R. 1980. On dinoflagellate evolution. *BioSystems* 13:65–108.

Taylor, F. J. R. ed. 1987. *The Biology of Dinoflagellates. Botanic Monographs* 21. Blackwell Scientific Publications, Oxford, 785 pp.

Taylor, F. J. R. 1979. A description of the benthic dinoflagellate associated with maitotoxin and ciguatoxin, including observations of Hawaiian material. In: Taylor, D. L., and Seliger, H. H., eds. *Toxic Dinoflagellate Blooms.* Elsevier/North Holland, New York, pp. 71–6.

Throndsen, J. 1973. Special methods—micromanipulators. In: Stein, J. R., ed. *Handbook of Phycological Methods, Culture Methods and Growth Measurements.* Cambridge University Press, Cambridge, pp. 139–44.

Triemer, R. E., and Fritz, L. 1984. Cell cycle and mitosis. In: Spector, D. L., ed. *Dinoflagellates.* Academic Press, Orlando, Florida, pp. 149–76.

Turpin, D. H., Dobell, P. E. R., and Taylor, F. J. R. 1978. Sexuality and cyst formation in Pacific strains of the toxic dinoflagellate *Gonyaulax tamarensis. J. Phycol.* 14:235–8.

Tuttle, R. C., and Loeblich, A. R. III. 1974. Genetic recombination in the dinoflagellate *Crypthecodinium cohnii. Science* 185:1061–2.

Tyler, M. A., Coats, D. W., and Anderson, D. M. 1982. Encystment including estuarine environments: selective deposition of dinoflagellate cysts by a frontal convergence. *Mar. Ecol. Prog. Ser.* 7:163–78.

Uchida, T. 1991. Sexual reproduction of *Scrippsiella trochoidea* isolated from Muroran Harbor, Hokkaido. *Nippon Suisan Gakkaishi* 57:1215.

Uchida, T., Matsuyama, Y., and Kamiyama, T. 1999. Cell fusion in *Dinophysis fortii. Bulletin of Fisheries and Environment of the Inland Sea* 1:163–5.

Vallesi, A., Giuli, G., Bradshaw, R. A., Luporini, P. 1995. Autocrine mitogenic activity of pheromones produced by the protozoan ciliate *Euplotes raikovi. Nature* 376:522–4.

von Stosch, H. A. 1964. Zum problem der sexuellen Fortpflanzung in der Peridineengattung *Ceratium. Helgol. Wiss. Meeresunters.* 10:140–52.

von Stosch, H. A. 1965. Sexualitat bei *Ceratium cornutum* (Dinophyta). *Naturwissenschaftern* 52:112–3.

von Stosch, H. A. 1972. La signification cytologique de la 'cyclose nucléaire' dans le cycle de vie des Dinoflagalles. *Mem. Soc. Bot. Fr.* 201–12.

von Stosch, H. A. 1973. Observation on vegetative reproduction and sexual life cycles of two freshwater dinoflagellates, *Gymnodinium pseudopalustre* Schiller and *Woloszynskia apiculata* sp. nov. *Br. Phycol. J.* 8:105–34.

Walker, L. M., and Steidinger, K. A. 1979. Sexual reproduction in the toxic dinoflagellate *Gonyaulax monilata. J. Phycol.* 15:312–5.

Walker, L. M. 1982. Evidence for a sexual cycle in the Florida red tide dinoflagellate, *Ptychodiscus brevis* (= *Gymnodinium breve*). *Trans. Am. Microsc. Soc.* 101:809–10.

Walker, L. M. 1984. Life histories, dispersal and survival in marine, planktonic dinoflagellates. In: Steidinger K. A., and Walker, L. M., eds. *Marine Plankton Life Cycle Strategies.* CRC Press, Boca Raton, Florida, pp. 19–34.

Watanabe, M. M., Watanabe, M., and Fukuyo, Y. 1982. *Scrippsiella trochoidea*: encystment and excystment of red tide flagellates. I. Induction of encystment of *Scrippsiella trochoidea. Res. Rep. Natl. Inst. Environ. Stud. Jpn.* 30:27–41.

Watras, C. J., Chisholm, S. W., and Anderson, D. M. 1982. Regulation of growth in an estuarine clone of *Gonyaulax tamarensis*: salinity-dependent temperature responses. *J. Exper. Marine Biol. Ecol.* 62:25–37.

Wyatt, T., and Jenkinson, I. R. 1997. Notes on *Alexandrium* dynamics. *J. Plankton Res.* 19:551–75.

Yamamoto, Y. 1976. Effect of some physical and chemical factors on the germination of akinetes of *Anabaena cylindrica. J. Gen. Appl. Microbiol.* 22:311–23.

Yoshimatsu, S. 1981. Sexual reproduction of *Protogonyaulax catenella* in culture. I. Heterothallism. *Bull. Plankton Soc. Japan* 28:131–9.

Yoshimatsu, S. 1984. Sexual reproduction of *Protogonyaulax catenella* in culture. II. Determination of mating type. *Bull. Plankton Soc. Japan* 31:107–11.

Zhang, S., and Li, S. 1990. Sexual reproduction and hypnozygote formation in the freshwater dinoflagellate *Peridinium bipes* Stein. *Acta Hydrobiologica Sinica* 14:1–9.

Zingmark, R. G. 1970. Sexual reproduction in the dinoflagellate *Noctiluca miliaris* Suriray. *J. Phycol.* 6:122–6.

# CHAPTER 25

# CULTURES AS A MEANS OF PROTECTING BIOLOGICAL RESOURCES: *EX SITU* CONSERVATION OF THREATENED ALGAL SPECIES

MAKOTO M. WATANABE

*Environmental Biology Division, National Institute for Environmental Studies*

## CONTENTS

Key Index Words: Threatened Species, IUCN Red Data List Categories, *Ex Situ* Conservation, Charales, *Nemalionopsis tortuosa, Thorea okadae, Compsogonopsis japonica*

## 1.0. INTRODUCTION

### 1.1. Structure of the IUCN Categories of Threat

During the long history of life earth has experienced several periods of mass extinction, and the extinction of a species is an inevitable destiny. However, the current extinction crisis differs in that it is the direct result of human activities. Since the year 1600, half of 1298 known species extinctions occurred during the twentieth century. The International Union for Conservation of Nature and Natural Resources (IUCN) has been assessing the existing status of species on a global scale. The IUCN Red List of Threatened Species highlights taxa threatened with extinction and promotes their conservation. For almost 30 years before 1994, more subjective threatened species categories in the IUCN Red Data Book and Red Lists had been in place. A new system, IUCN Red Data List Categories (Fig. 25.1),

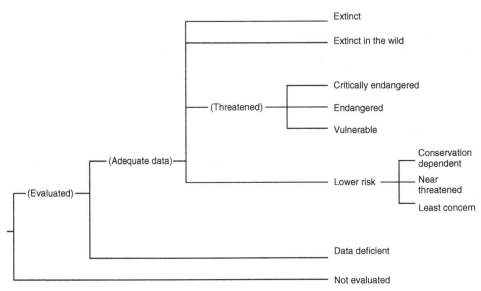

**FIGURE 25.1.** Structure of the IUCN categories of threat. (From IUCN 1994.)

was developed in 1994. All taxa listed as Critically Endangered qualify for Vulnerable and Endangered status, and all listed as Endangered qualify for Vulnerable status. Together these categories are described as "threatened." The threatened categories form a part of the overall scheme, and all taxa can be placed into one of the categories.

## 1.2. Criteria for Each Category

A taxon is *Extinct* (EX) when there is no reasonable doubt that the last individual has died. A taxon is presumed Extinct when exhaustive surveys in known and/or expected habitat, at appropriate times (diel, seasonal, annual), throughout its historic range have failed to record an individual. Surveys should be over a time frame appropriate to the taxon's life cycle and life form. A taxon is *Extinct in the Wild* (EW) when it is known only to survive in cultivation, in captivity, or as a naturalized population (or populations) well outside the past range. A taxon is *Critically Endangered* (CR), *Endangered* (EN), or *Vulnerable* (VU) when the best available evidence indicates that it meets any of the criteria, respectively, shown in Table 25.1. A taxon is *Near Threatened* (NT) when it has been evaluated against the criteria but does not qualify for CR, EN, or VU but is close to qualifying or is likely to qualify for a threatened category in the near future. A taxon is *Least Concern* (LC) when it has been evaluated against the criteria and does not qualify for CR, EN, VU, or NT. Widespread and abun-

dant taxa are included in this category. A taxon is *Data Deficient* (DD) when there is inadequate information to make a direct or indirect assessment of its risk of extinction based on its distribution and/or population status. A taxon in this category may be well studied and its biology well known, but appropriate data on abundance and/or distribution are lacking. DD is, therefore, not a category of threat. Listing of taxa in this category indicates that more information is required and acknowledges the possibility that future research will show that threatened classification is appropriate. A taxon is *Not Evaluated* (NE) when it has not yet been evaluated against the criteria.

## 1.3. Algal Species on the Red List of Threatened Species: IUCN 2003

The marine red alga, *Vanvoorstia bennettiana* (Harvey) Papenfuss (Delesseriaceae, Ceramiales) was assessed as Extinct by IUCN. This species was only collected from two sites in Australia. The initial discovery, the type locality, was from the eastern end of Spectacle Island in the Parramatta River at 1855. The other site was on the seabed between Point Piper and Shark Island in Port Jackson, Sydney Harbor, from which this species was collected at 1886. In spite of the extensive surveys from 1913 to the present, no one has found this species. It seems most unlikely that the species would remain in existence but undetected for more than 100 years. Therefore, as no specimens have been seen or collected

**TABLE 25.1**   Threatened category thresholds.

| Criteria | Critical | Endangered | Vulnerable |
|---|---|---|---|
| A. Rapid decline | >80% over 10 years or three generations | >50% over 10 years or 3 generations | >50% over 20 years or five generations |
| B. Small range (fragmented, declining, or fluctuating) | Extent of occurrence <100 km² or area of occupancy <10 km² | Extent of occurrence <5,000 km² or area of occupancy <500 km² | Extent of occurrence <20,000 km² or area of occupancy <2,000 km² |
| C. Small population (declining) | <250 mature individuals | <2,500 mature individuals | <10,000 mature individuals |
| D1. Very small population | <50 mature individuals | <250 mature individuals | <1,000 mature individuals |
| D2. Very small range | — | — | <100 km² or less than five locations |
| E. Unfavorable population viability analysis | Probability of extinction >50% within 5 years | Probability of extinction >20% within 20 years | Probability of extinction >10% within 100 years |

*Source*: Department of Environment, U.K. 1996.

in the intervening 116 years, despite numerous collections made by algologists in that period, this species can be considered truly classified Extinct. Besides this species, no other algae have been included in the Red List of IUCN of extinctions because the amount and quality of the data are regarded as being not yet adequate to give a reasonable global overview.

## 1.4.  Threatened Algal Species

Different from "lovely or charismatic wildlife," such as vertebrates, beetles, butterflies, and flowering plants, algae have not received extensive attention. Almost all of the algae have insufficient information regarding their status in the IUCN Red List Categories and Criteria, and, at present, the assessment must be rather subjective. Geographical studies, ecological studies, and population genetics must be further promoted in the field of phycology to improve this situation. In particular, population genetics of algae are far behind those of higher plants and animals. We should become more aware of the coastal zones of lakes, reservoirs, and oceans, habitats that have been heavily exploited and altered. Many phycologists fear that a number of macroalgal species, especially freshwater ones, have already become extinct or exist in critically endangered states. Therefore, the most urgent matter for conservation of algal diversity is the assessment of "threatened" species. Even if this is a subjective process, it must be

made for each algal species and population based on present knowledge. Next, the appropriate *in situ* and/or *ex situ* conservation methods should be developed for each threatened algal species. This should help control rapid loss of algal diversity accelerated by human activities.

In Australia, *Lynchnothamnus barbatus* (Meyen) Leonhardi (Charales, Chlorophyta) has been declared a rare and endangered species under relevant legislation for the Commonwealth of Australia and for the state of Queensland (Casanova et al. 2003; Casanova, personal communication). This species was particularly diverse and widespread in the Pliocene. The extant species was still widespread until the last decades of the twentieth century with localities in Europe, India, China, Papua New Guinea, and Australia. Despite thorough searches for *L. barbatus* in the 1990s, it was absent from previously recorded localities until it was rediscovered in Australia in 1996. Habitat modifications of lakes and reservoirs have contributed to the disappearance of *L. barbatus*.

As shown in the following, many charophyte species, as well as several kinds of freshwater algae, have declined in Japan in response to habitat modification. They have been included in the Japan Red Data Book published in 2000. Also, Germany placed many freshwater algae in the Red Lists, Germany (www.rote-listen.de/rlonline/index.html), including all of the existing genera of Charales (*Chara, Nitella, Nitellopsis, Tolypella, Lamprothamnium,* and *Lychnothamnus*), red algae (*Batrachospermum, Thorea, Bangia, Audouinella,*

*Chantransia, Porphyridium, Hildenbrandia*, etc.), and brown algae (*Heribaudiella*). However, many countries still exclude algae from lists of endangered and threatened organisms. For instance, no alga is listed in the United States as threatened or endangered (see http://endangered.fws.gov/wildlife.html). Each country's experiences should be accumulated to develop a national/regional list of threatened species.

## 2.0. JAPAN'S EXPERIENCE

The Environmental Agency of Japan published *Threatened Wildlife of Japan-Red Data Book*, second edition, volume 9, in 2000. This volume is the first Red List of nonvascular plants (mosses, fungi, algae, and lichens) prepared according to the IUCN *Red List Categories and Criteria* (1994), but it is often based on insufficient information and data. The Red List of algae in Japan includes 5 extinct species, 1 species extinct in the wild, 35 critically endangered or endangered species, 6 vulnerable species, and 24 near-threatened species (Table 25.2). Several kinds of Japanese endangered and threatened algae also occur on Red Lists for Germany.

### 2.1. Determining Categories of Threat

Four genera and seventy-four species and varieties of Charales have been reported from lakes, reservoirs, and ponds in Japan (Hirose and Yamagishi 1977). Of these, 35 taxa are endemic to Japan. However, recent surveys are lacking, and little information is available in most of the previously occupied lakes, reservoirs, and ponds. For other freshwater algae, targeted algal taxa were selected in the same way. It is usually difficult to survey all habitats of Charales in a short time because they live in various habitats in lakes, reservoirs, and ponds. Therefore, it is first necessary to select an appropriate habitat for surveys for species with a high risk of extinction. Kasaki (1964) surveyed the distribution of Charales in 46 lakes in Japan and found 31 taxa. Since 1995, field explorations have been implemented for Charales in the 46 lakes (Watanabe et al. 2005).

Higher plants have been investigated, so the occurrence and abundance on a geographic grid are often available. However, there are no quantitative studies on the distribution of Charales and other freshwater algae; only the names of lakes and rivers from which the algae were collected are known. Therefore, the threatened categories were determined by calculating a ratio using the number of lakes or rivers where the alga survives and those where they previously survived. Among the 46 lakes surveyed by Kasaki in 1964, 38 lakes were re-surveyed for Charales from 1995 to 1997 (Fig. 25.2). No charophycean plants were found in 26 lakes; in 7 lakes, at least some Charales were extinct; all Charales still existed in only 5 lakes. Only 6 of 30 taxa originally reported in the 38 lakes were found during this survey. Even these existing taxa were extinct in more than 50% of the lakes where they were previously found in 1964. Based on the results, 5 taxa of Charales were classified extinct, 1 taxon was classified extinct in the wild, and 24 were classified as critically endangered or endangered species (see Table 25.2). The details were shown in the book *Threatened Wildlife of Japan—Red Data Book* (2000).

## 3.0. NEED FOR *EX SITU* CONSERVATION OF ALGAL DIVERSITY

It had previously been estimated that 15 to 20% of all species would become extinct and 25 to 35% of the genetic diversity would possibly be lost over the 12-year period leading up to the year 2000 (Lugo 1988, Maxted et al. 1997). Although these losses have not been verified, the threat of extinction for many species increases as whole ecosystems vanish, the human population proliferates, and human-mediated interference increases. Current views of recovery time from anthropogenic extinctions suggest that many millions of years will be required (Kirchner and Weill 2000).

Major efforts to preserve biological diversity are underway through two strategies of *ex situ* and *in situ* conservations. Maxted et al. (1997) clarified and enhanced the methodologies by proposing a model for plant and animal genetic diversity conservation. The flow model is composed of (1) selection of target taxa, (2) project commission, (3) ecogeographic survey/preliminary survey mission, (4) conservation objectives, (5) field exploitation, (6) conservation strategies, (7) *ex situ* techniques, (8) *in situ* techniques, (9) conservation products, (10) conserved product deposition and dissemination, (11) characterization/evaluation, (12) utilization, and (13) utilization products. This model can be applied to algae. For algae, *in situ* conservation involves the maintenance of genetic variation at the location where it is encountered, either in the wild or in traditional farming systems. The establishment and scientific management of nature reserves and national parks in many parts of the world is imperative.

**TABLE 25.2**   The red list of threatened species of algae in Japan.

**Extinct (EX)**

Charophyceae — *Chara fibrosa* Agardh var. *brevibracteata* Kasaki
*Chara globularis* Thuillier var. *hakionensis* Kasaki
*Nitella flexilis* (Linnaeus) Agardh var. *bifurcata* Kasaki
*Nitella furcata* Agardh var. *fallosa* (Morioka) Imahori
*Nitella minispora* Imahori

**Extinct in Wild (EW)**

Charophyceae — *Nitellopsis obtusa* Groves[a]

**Critically Endangered (CR) or Endangered (EN)**

Cyanobacteria — *Aphanothece sacrum* (Suringar) Okada

Rhodophyceae — *Porphyra kuniedae* Kurogi
*Porphyra tenera* Kjellman
*Porphyra tenuuipedalis* Miura
*Nemalionopsis tortuosa* Yoneda et Yagi
*Thorea gaudichaudii* C. Agardh
*Compsopogonopsis japonica* Chihara
*Pseudodichotomosiphon constrictus* (Yamada) Yamada (= *Vaucheria constricta* Yamada)

Chlorophyceae — *Cladophora aegagropila* (Linnaeus) Rabenhorst
*Dasycladus vermicularis* (Scopoli) Krasser (= *Spomgia vermicularis* Scopoli; *Dasycladus clavaeformis* [Roth] C. Agardh)
*Acetabularia caliculus* Lamouroux

Charophyceae — *Chara benthamii* Zaneveld var. *benthamii*
*Chara braunii* Gmelin[a]
*Chara corallina* Willdenow var. *corallina*
*Chara globularis* Thuillier var. *globularis*[a]
*Chara sejuncta* Braun
*Chara zeylanica* Willdenow
*Lamprothamnium succinctum* Wood

*Nitella acuminata* Braun var. *capitulifera* Imahori
*Nitella acuminata* Braun var. *subglomerata* Braun
*Nitella batrachosperma* (Reichenbach) A. Braun[a]
*Nitella flexilis* (Linnaeus) Agardh var. *flexilis*[a]
*Nitella flexilis* (Linnaeus) Agardh var. *longifolia* Braun
*Nitella furcata* (Roxbburgh ex Bruzelius) Agardh var. *furcata*[a]
*Nitella gracilens* Moroika
*Nitella hyalina* Agardh[a]
*Nitella imahorii* Wood
*Nitella microcarpa* A. Braun
*Nitella mirabilis* Nordstedt var. *inokasiraensis* Kasaki ex Wood
*Nitella morongii* T. F. Allen var. *spiciformis* (Morioka) Imahori
*Nitella orientalis* T. F. Allen
*Nitella pseudoflabellata* A. Braun var. *mucosa* (Nordst.) Bailey
*Nitella pseudoflabellata* A. Braun var. *pseudoflabellata*
*Nitella pseudoflabellata* A. Braun var. *pseudoflabellata* f. *macrophylla* Kasaki
*Nitella pulchella* T. F. Allen

**Vulnerable (VU)**

Rhodophyceae — *Thorea okadae* Yamada
*Compsopogon aeruginosus* (J. Agardh) Kützing
*Compsopogon coeruleus* Montagne (= *C. oishii* Okamura)
*Compsopogon corticrassus* Chihara et Nakamura
*Compsopogon hookeri* Montagne
*Compsopogon prolificus* Yadava et Kumano

*Source:* Environmental Agency, Japan 2000.
[a] Species also on Red Lists, Germany.

For *ex situ* conservation, algal samples conserved either as living collections (e.g., culture collections) or as spores, cysts, DNA, and so on are maintained under special artificial conditions. The *ex situ* material can be used for reproduction of threatened and endangered species and for biotechnology. These two strategies should not be seen as alternatives or in opposition to each another, but as being complementary. One conservation strategy or technique can act as a backup to another, with emphasis placed on each depending on the conservation aims, the type of species being con-

served, the resources available, and whether the species has utilization potential.

There are several situations in which *ex situ* conservation may be of great importance: (1) Unpredictable events may threaten rare genotypes or species in nature, (2) some threatened species may require cultivation or breeding in captivity to build their numbers for reintroduction to the wild, and (3) some genotypes and species can survive only *ex situ* because of total loss or alteration of their habitats. Because ecogeographical study results and population genetics on algae are insuf-

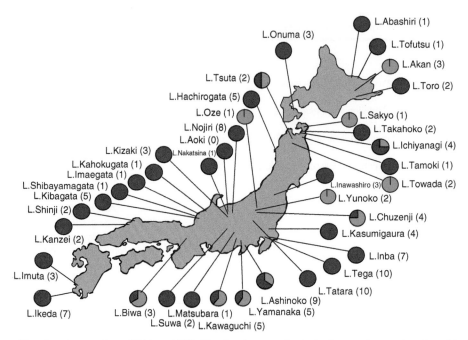

**FIGURE 25.2.** Charales changes from 1964 to 1997. Number in parenthesis is the species number found in 1964 in each lake. Extinction status during the last 30 years is shown as the ratio of existing species to extinct species. Green, existing species; red, extinct species. For example, 10 taxa existed in Lake Inba at 1964, but now all are extinct (100% red circle); all species found in 1964 still exist in Lake Akan (100% green circle); 10 were found in Lake Ashinoko in 1964, but only 3 still exist today (30% green, 70% red) (Watanabe et al. 2005).

ficient to assess and evaluate their existing status, it is important to accumulate the quantitative data on the existing status of these algal species. Also, knowledge of the genetic diversity of each targeted algal species is important. One needs to define and understand the processes involved with the threatened situation and to attempt to develop practical techniques to conserve the algae. Loss or alteration of lake/river shores has drastically reduced the diversity of freshwater macroalgae in those habitats. It is highly possible that some genotypes and species will survive only through *ex situ* conservation.

The advantages of culture collections as *ex situ* methods are that they are efficient and reproducible and feasible for short-, medium-, and long-term secure storage. To realize long-term secure storage, it is particularly important to preserve cultures of threatened algae in more than two culture collections and to establish or strengthen networks between culture collections, including those of medicinal, agricultural, aquacultural, and industrial importance. Space limitations make it difficult to maintain individuals of each macroalgal species that are necessary to protect all the genetic diversity surviving in the wild. However, genetic diversity can be easily preserved in the form of purified high-molecular-

weight DNA. Sometime soon, DNA sequencing will be fully automated, and our descendants will be able to rapidly derive the sequence of any organism whose DNA has been appropriately collected and stockpiled. The environmental and economic values of DNA banks should not be overlooked and will undoubtedly provide materials that will contribute to advances in sustainable development and human welfare. The ultimate aim of biodiversity conservation in our generation is to pass on to later generations the essentials of our immense biodiversity. Culture collections including DNA banks will surely contribute to our objective. Existing culture collections should expand their funds and facilities to play an important role for *ex situ* conservation of threatened algal species.

## 4.0. CULTURES OF THREATENED AND ENDANGERED SPECIES OF ALGAE

Generally speaking, the species concept for algae mostly depends on morphological characteristics, but biosystematic studies clearly demonstrate that even morphologically defined species are composed of several reproductively isolated groups, namely biological species (Proctor 1971, 1975, Grant and Proctor 1972, Watanabe and Ichimura 1978a, 1978b, 1982, Ichimura 1983). In addition, ecosystem-dependent adaptive radiation of picocyanobacteria has been postulated (Ernst et al. 2003). It is likely that traditional species of algae are composed of biologically and genetically different species or populations. The loss of a local population may mean the loss of a cryptic species or the genetic diversity of the species. Therefore, the culture of threatened algal species and populations is critically needed for conservation purposes.

A considerable number of species of algae are threatened and endangered (see Table 25.1). The freshwater rhodophycean and charalean algae seem to be especially endangered because of environmental deterioration of inland waters of Japan. It is true also in Germany (www.rote-listen.de/rlonline/index.html). Currently, 22 taxa of Charales and 3 species of freshwater Rhodophyceae (*Nemalionopsis tortuosa*, *Thorea okadae*, and *Compsogonopsis japonica*) are maintained as unialgal culture strains in the Microbial Culture Collection, National Institute for Environmental Studies, Japan (NIES Collection). Culture methods of these endangered or threatened algae are given in Section 4.1.1.

**FIGURE 25.3.** **(a)** *Chara braunii* in culture; **(b)** *C. braunii* with oogonium and archegonium.

## 4.1. Charales

Charales have macroscopic thalli that are differentiated into nodes and internodes; the nodes bear whorls of branches (Fig. 25.3). All are oogamous, either dioecious or monoecious. Recent 18S rDNA sequence analyses demonstrate that Charales are a sister group to land plants. Ecologically, Charales form in the charaphyte zone below the aquatic macrophyte zone in lakes and offer habitat for life and spawning of aquatic animals. Charales are sensitive to eutrophication (Blindow 1988, 1992, Simons et al. 1994, Simons and Nat 1996). In addition, reproductively isolated groups, namely biological species, have been recognized within morphologically defined species in Charales (Proctor 1971, 1975, Grant and Proctor 1972). The extinction of a local population may therefore imply the extinction of one biological species.

### 4.1.1. Culturing Information

Soil water medium is satisfactory for culture and maintenance of almost all of the species of Charales (Proctor 1970, Nozaki et al. 1998). If possible, use sandy soil from lake beaches where the targeted charalean species grew. If not possible, commercial sandy loam soil is also available (Proctor 1970). Place 400 to 600 mL of sandy soil in a 2-liter glass vessel and add 1,500 to 1,800 mL distilled water. Settle in the dark at room temperature for 1 to 2 days (until the liquid phase is clear). This should never be autoclaved or steamed. Plant a single thallus of the charalean alga to a soil water medium, and incubate at 20°C under about 10 to 20 μmol photons · $m^{-2}$ · $sec^{-1}$ with an LD 12 : 12 cycle to establish a unicharalean culture. Transfer a part of the cultured thalli to a new soil water medium every 2 to 3 months to maintain the cultures. The addition of snails is recommended because they eat attached algae on the *Chara* plant. No other attempt should be made to prevent contamination from other algae.

Medium II of Forsberg (1965), a modification of Chu-10 nutrient solution (Gerloff et al. 1950), includes a buffer, chelater, and micronutrients (see Chapter 2, Appendix A). It is suitable for the growth of *Chara globularis*, *C. aspera*, *C. zeylanica*, and probably other charalean species. *Nitella flexilis* was also cultured with a 1 : 1 mixture of 1.8% soil extract and Medium II (Foissner 1989). The medium can be autoclaved, and it is suitable for axenic culture.

Establish a unicharalean culture in soil water (see previous discussion). To establish an axenic culture, collect mature oospores and sterilize them by micropipette washing method (see Chapters 6 and 8). Inoculate each of sterilized oospores into a culture flask; they usually require about 4 weeks for germination. After germination, inoculate the obtained sporelings into a sterile medium containing beef extract and peptone to check for bacterial contamination. In this solution, nonsterile sporelings produce growth of common bacterial contaminants after a few days. If the sporeling is axenic, it can be stored for months with the test solution. Maintain axenic cultures at 20° to 25°C under about 10 to 20 μmol photons · m$^{-2}$ · sec$^{-1}$ with an LD 12 : 12 cycle. Aseptically transfer a piece of the plant, including two or three nodes, into a new medium every month.

## 4.2. Nemalionopsis tortuosa *and* Thorea okadae

*Nemalionopsis tortuosa* is a freshwater red alga endemic to southern Japan (Fig. 25.4). This species was known from 10 localities, but it still grows in only three localities. It is categorized as CR/EN (see Table 25.1). *Thorea okadae*, endemic to Japan, grows in several rivers of Kyushu district and is rarely recorded from Honshu district. The population of this species decreased because of habitat deterioration caused by water pollution and urban development. It is categorized as VU (see Table 25.1).

### 4.2.1. Culturing Information

Use Bold's Basal Medium for culture (see Chapter 2, Appendix A). Plant a single thallus to 30 mL of medium in a 50-mL plastic culture vessel. Incubate it at 20°C under about 30 μmol photons · m$^{-2}$ · sec$^{-1}$ with an LD 12 : 12 cycle to establish a unialgal culture. Transfer a piece of thallus to a new culture medium every month to maintain the cultures.

## 4.3. Compsogonopsis japonica

*Compsogonopsis japonica* was first reported from several rivers in the Kanto district and later from Ishigaki Island (Fig. 25.5). Because of urban development, it has now disappeared from many localities of the Kanto district. It is categorized as CR/EN.

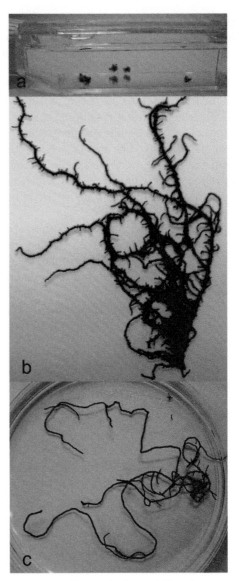

**FIGURE 25.4.** *Thorea okadae and Nemalionopsis tortuosa.* **(a)** *T.* okadae in culture; **(b)** plant of *T.* okadae; **(c)** plant of *N. tortuosa.*

### 4.3.1. Culturing Information

Isolate and culture as described for *Nemalionopsis tortuosa* and *Thorea okadae*. To maintain established cultures, recognize the formation of monospores and transfer monospores to a new medium every month.

## 5.0. Acknowledgments

I want to thank Dr. Masanobu Kawachi for his useful information about the cultures of freshwater red algae

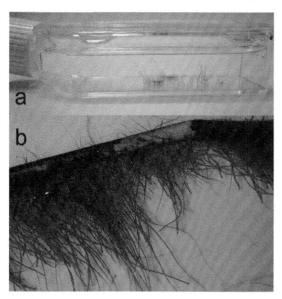

**FIGURE 25.5. (a)** *Compsogonopsis japonica* in culture; **(b)** plant of *C. japonica*.

and for providing photographs of threatened species maintained in the NIES Collection.

## 6.0. REFERENCES

Blindow, I. 1988. Phosphorus toxicity in *Chara*. *Aquatic Bot.* 32:393–5.

Blindow, I. 1992. Decline of charophytes during eutrophication: Comparison with angiosperms. *Freshwater Biol.* 28: 9–14.

Casanova, M. T., Garcia, A., and Feist, M. 2003. The ecology and conservation of *Lymchnothamnus barbatus* (Characeae). *Acta Micropalaeo. Sinica* 20:119–29.

Environmental Agency of Japan, ed. 2000. *Threatened Wildlife of Japan—Red Data Book*, 2nd ed, Vol. 9, *Bryophytes, Algae, Lichens, Fungi*. Japan Wildlife Research Center, Tokyo, 429 pp.

Ernst, A., Becker, S., Wollenzein, U. I., and Postius, C. 2003. Ecosystem-dependent adaptive radiations of picocyanobacteria inferred from 16S rRNA and ITS-1 sequence analysis. *Microbiology* 149:217–28.

Foissner, I. 1989. pH-dependence of chlortetracycline (CTC)-induced plug formation in *Nitella flexilis* (Characeae). *J. Phycol.* 25:313–8.

Forsberg, C. 1965. Nutritional studies of *Chara* in axenic cultures. *Physiol. Plantarum* 18:275–90.

Gerloff, G. C., Fitzgerald, G. P., and Skoog, F. K. 1950. The isolation, purification, and culture of blue-green algae. *Am. J. Bot.* 37:216–8.

Grant, M. C., and Proctor, V. W. 1972. *Chara vulgaris* and *C. contraria*: Patterns of reproductive isolation for two cosmopolitan species complexes. *Evolution* 26:267–81.

Hirose, H., and Yamagishi, T. eds. 1977. *Illustrations of the Japanese Fresh-Water Algae*. Uchidarokakuho Publishing Co., Ltd., Tokyo, 933 pp.

Ichimura, T. 1983. Hybrid inviability and predominant survival of mating type minus progeny in laboratory crosses between two closely related mating groups of *Closterium ehrenbergii*. *Evolution* 32:252–60.

IUCN. 1994. *IUCN Red List Categories and Criteria*. IUCN Species Survival Commission, Gland, Switzerland.

Kasaki, H. 1964. The Charophyta from the lakes of Japan. *J. Hattori Bot. Lab.* 27:217–314.

Kirchiner, J. W., and Weil, A. 2000. Delayed biological recovery from extinctions throughout the fossil record. *Nature* 404:177.

Lugo, A. E. 1988. Estimating reductions in the diversity of tropical forest species. In: Wilson, E. O., ed. *Biodiversity*. National Academy Press, Washington, D.C., pp. 58–70.

Maxted, N., Ford-Lloyed, B. V., and Hawks, J. G. 1997. *Plant Genetic Conservation: The* In Situ *Approach*. Chapman & Hall, London.

Nichols, H. W., and Bold, H. C. 1965. *Trichosarcina polymorpha* gen. et. sp. nov. *J. Phycol.* 1:34–8

Nozaki, H., Kodo, M., Miyaji, K., Kato, M., Watanabe, M. M., and Kasaki, H. 1998. Observations on the morphology and oospore wall ornamentation in culture of the rediscovered Japanese endemic *Nitella gracilens* (Charales, Chlorophyta). *Eur. J. Phycol.* 33:357–9.

Proctor, V. W. 1970. Taxonomy of *Chara braunii*: An experimental approach. *J. Phycol.* 6:317–21.

Proctor, V. W. 1971. *Chara globularis* Thuillier (C. fragilis Desvaux): Breeding patterns within a cosmopolitan complex. *Limnol. Oceanogr.* 16:422–36.

Proctor, V. W. 1975. The nature of charophyte species. *Phycologia* 14:97–113.

Simons, J., Ohm, M., Daalder, R., Boers, P., and Rip, W. 1994. Restoration of Botshol (The Netherlands) by reduction of external nutrient load: Recovery of a characean community, dominated by *Chara connivens*. *Hydorobiologia* 275/276: 243–53.

Simons, J., and Nat, E. 1996. Past and present distribution of stoneworts (Characeae) in The Netherlands. *Hydrobiologia* 340:127–35.

Watanabe, M. M., and Ichimura, T. 1978a. Biosystematic studies of the *Closterium peracerosum-strigosum-littorale* complex. II. Reproductive isolation and morphological variation among several populations from the Northern Kanto area in Japan. *Bot. Mag. Tokyo* 91:1–10.

Watanabe, M. M., and Ichimura, T. 1978b. Biosystematic studies of the *Closterium peracerosum-strigosum-littorale* complex. III. Degrees of sexual isolation among the three population groups from the Northern Kanto area. *Bot. Mag. Tokyo* 91:11–24.

Watanabe, M. M., and Ichimura, T. 1982. Biosystematic studies of the *Closterium peracerosum-strigosum-littorale* complex. IV. Hybrid breakdown between two closely related groups, Group II-A and Group II-B. *Bot. Mag. Tokyo* 95:241–247.

Watanabe, M. M., Nozaki, H., Kasaki, H., Sano, S., Kato, N., Omori, Y., and Nohara, S. 2005. Threatened states of the Charales in the lakes of Japan. In Kasai, F., ed. *Algal Culture Collections and the Environment*, Tokai University Press, Tokyo, (in press).

# Appendix A—Recipes for Freshwater and Seawater Media

Prepared by
R. A. Andersen
J. A. Berges
P. J. Harrison
M. M. Watanabe

## Contents

The following culture medium recipes and culturing protocols are an incomplete compilation from the literature and the contributors to this volume. Although we attempted to edit this information into a convenient and consistent format, we did not attempt to either test or verify individual recipes. Even so, the compilation that follows should prove to be a useful guide to a variety of widely used algal culturing media. The culture medium recipes provide the components and their concentrations; for more preparation details, see Chapters 2 to 4. The medium recipes are designed for growing established algal strains; for isolation or purification of single organisms, the nutrients may require a reduction by 10- to 10,000-fold, depending upon the sensitivity of the alga (see Chapters 6–9). For subculturing of established strains, see Chapters 10 and 11. Additional recipes can be found on Web sites (see Table A.1).

Water quality, including natural seawater, is very important. The abbreviation $dH_2O$ refers generally to distilled, deionized, distilled/deionized water, Milli-Q water (Millipore Corp.), etc. As the culture application becomes more critical, the quality of water also becomes more critical. Natural seawater (and natural freshwater) should be obtained from a nonpolluted source. For oligotrophic (open ocean) phytoplankton, natural water should be obtained from an open ocean site. When dissolving chemicals, wait for the first component to dissolve before adding the second. Stirring (and sometimes heat) is often necessary to dissolve chemicals efficiently. In many cases, preparation of stock solutions is not only convenient but also necessary to avoid errors from weighing very tiny quantities. Attention should be given to the pH of the final medium. In most cases, if a pH adjustment is required, this occurs before steriliza-

---

**TABLE I.** Major service culture collections with culture medium recipes on their Internet Web sites. For additional culture collections, see: http://wdcm.nig.ac.jp/hpcc.html.

| Culture Collection | Internet Website URL |
| --- | --- |
| CCAP | www.ife.ac.uk/ccap |
| CCMP | ccmp.bigelow.org |
| NIES | www.nies.go.jp/biology/mcc/home.htm |
| PCC | www.pasteur.fr/recherche/banques/PCC |
| SAG | www.epsag.uni-goettingen.de/html/sag.html |
| UTCC | www.botany.utoronto.ca/utcc |
| UTEX | www.bio.utexas.edu/research/utex |

---

tion. Note that autoclaving drives carbon dioxide out of media lacking carbonate stabilizing buffers, and this makes the medium very alkaline immediately after removal from the autoclave. In these cases, wait approximately 24 hours for gaseous equilibrium before inoculating cells into the medium. Natural seawater may produce precipitates following autoclave sterilization (see Chapter 3). In general, inorganic stocks should be autoclaved and handled aseptically. A silicate stock is best stored in a dark Teflon-lined bottle. If silicate is not required for growth of the alga, potential precipitation problems can be avoided by leaving it out of the medium. Most trace metal solutions are chelated with EDTA (see Chapter 4). When preparing an EDTA-buffered trace metal solution, dissolve the EDTA before dissolving the metals. Stock solutions of common vitamins (thiamine · HCl, $B_1$, $B_{12}$) should be dispensed into small volumes in sterile, freeze-safe containers and then frozen (e.g., 1-mL Eppendorf tubes or 10-mL polycarbonate screw-top tubes). Vitamins retain more potency if filter-sterilized and added aseptically after autoclaving (see Chapters 2, 3, and 5). The stock solutions can be auto-claved if first acidified. Aliquots from vitamin stock solutions may be added to the final medium and autoclaved, usually with little damaging effect on algal growth. However, the vitamins should be the last component added when preparing a culture medium, and the other components should be well mixed.

Generally, the recipes have been normalized to factors of 1,000 (e.g., g, mg, μg; 1 mL, 1 L), and stock solutions are described in 1-liter quantities. However, by reducing the quantity of the component by 10, a 100-mL stock solution is easily prepared. We have calculated the molar concentration of each component in the final medium for easy comparisons among various media. Furthermore, the 1-mL quantity used for most stock solutions allows the use of traditional sterile glass (and disposable plastic) pipettes as well as 1,000 μL automatic pipetters with sterile tips.

The media are organized according to the following outline, and within each heading individual culture media are arranged alphabetically. Algae that are known to grow in the media are often listed; occasionally simple modifications are listed after the recipe.

## 1.0. FRESHWATER CULTURE MEDIA

### 1.1. Freshwater Synthetic Media

#### AF6 Medium, Modified

(Kato 1982, Watanabe et al. 2000)

AF-6 medium was developed by Kato (1982) for culturing *Colacium* (Euglenophyceae) but has widespread application for algae requiring slightly acidic medium. It has been modified (Watanabe et al. 2000) by removing the calcium carbonate ($1.00 \times 10^{-4}$ M), adding MES buffer, and using a different trace metals mixture. It has been used to culture volvocalean algae (e.g., *Carteria, Gonium, Chlorogonium, Pandorina, Paulschulzia, Platydorina, Pleodorina, Pteromonas, Pseudocarteria*, and some species of *Volvox*), the xanthophytes *Botrydiopsis arrhiza* Borzi and *Botrydium granulatum* (L.) Greville, many cryptophytes, the dinoflagellate *Peridnium bipes* Stein f. *globosum* Lindermann, the synurophyte *Synura sphagnicola* (Korsh.) Korsh., the euglenoids *Phacus agilis* Skuja, *Euglena clara* Skuja and *E. mutabilis* Schmitz, and the green ciliate *Paramecium bursaria* Ehr. A half-strength AF-6 medium is used for maintenance of the freshwater dinoflagellates *Hemidinium nasutum* Stein and *Peridinium plonicum* Woloszynska (Watanabe et al. 2000). **Prepare** the stock solutions. **Into 950 mL** of dH$_2$O, first dissolve the MES, Fe citrate, and citrate; then add the stock solutions and bring the final volume up to 1 liter. The pH is adjusted to 6.6. Autoclave.

| Component | Stock Solution ($g \cdot L^{-1}$ dH$_2$O) | Quantity Used | Concentration in Final Medium (M) |
|---|---|---|---|
| MES buffer | — | 400 mg | $2.05 \times 10^{-3}$ |
| Fe-citrate | 2 | 1 mL | $8.17 \times 10^{-6}$ |
| Citric acid | 2 | 1 mL | $1.04 \times 10^{-5}$ |
| NaNO$_3$ | 140 | 1 mL | $1.65 \times 10^{-3}$ |
| NH$_4$NO$_3$ | 22 | 1 mL | $2.75 \times 10^{-4}$ |
| MgSO$_4 \cdot$ 7H$_2$O | 30 | 1 mL | $1.22 \times 10^{-4}$ |
| KH$_2$PO$_4$ | 10 | 1 mL | $7.35 \times 10^{-5}$ |
| K$_2$HPO$_4$ | 5 | 1 mL | $2.87 \times 10^{-5}$ |
| CaCl$_2 \cdot$ 2H$_2$O | 10 | 1 mL | $6.80 \times 10^{-5}$ |
| Trace metals solution | (See following recipe) | 1 mL | — |
| Vitamin solution | (See following recipe) | 1 mL | — |

*Trace Metals Solution*

Prepare the primary stock solutions. **Into 950 mL** of $dH_2O$, dissolve the EDTA, then individually dissolve the metals, and finally add 1 mL of the two primary stocks.

| Component | 1° Stock Solution $(g \cdot L^{-1}\ dH_2O)$ | Quantity Used | Concentration in Final Medium (M) |
|---|---|---|---|
| $Na_2EDTA \cdot 2H_2O$ | — | 5.00 g | $1.34 \times 10^{-5}$ |
| $FeCl_3 \cdot 6H_2O$ | — | 0.98 g | $3.63 \times 10^{-6}$ |
| $MnCl_2 \cdot 4H_2O$ | — | 0.18 g | $9.10 \times 10^{-7}$ |
| $ZnSO_4 \cdot 7H_2O$ | — | 0.11 g | $3.83 \times 10^{-7}$ |
| $CoCl_2 \cdot 6H_2O$ | 20.0 | 1 mL | $8.41 \times 10^{-8}$ |
| $Na_2MoO_4 \cdot 2H_2O$ | 12.5 | 1 mL | $5.17 \times 10^{-8}$ |

*AF6 Vitamin Solution*

Prepare the primary stocks. **Into 950 mL** of $dH_2O$, add and dissolve the thiamine · HCl, and then add 1 mL from each of the two primary stocks. Filter-sterilize. Refrigerate or freeze.

| Component | 1° Stock Solution $(g \cdot L^{-1}\ dH_2O)$ | Quantity Used | Concentration in Final Medium (M) |
|---|---|---|---|
| Thiamine · HCl (vitamin $B_1$) | — | 10 mg | $2.96 \times 10^{-8}$ |
| Biotin (vitamin H) | 2.0 | 1 mL | $8.19 \times 10^{-9}$ |
| Pyridoxine · HCl (vitamin $B_6$) | 1.0 | 1 mL | $5.91 \times 10^{-9}$ |
| Cyanocobalamin (vitamin $B_{12}$) | 1.0 | 1 mL | $7.38 \times 10^{-10}$ |

## Allen's Blue-Green Algal Medium

(M. M. Allen 1968)

*See* BG-11 Medium.

## Allen's *Cyanidium* Medium, Modified

(M. B. Allen 1959, Watanabe et al. 2000)

Mary Belle Allen (1959) developed this medium for culturing *Cyanidium caldarium* (Tilden) Geitler (Rhodophyceae), and it can be used to grow the related genera *Cyanidioshyzon* and *Galdieria* (Watanabe et al. 2000). Our recipe uses the original macronutrient concentrations; however, the trace metals solution is from Watanabe et al. (2000) because it includes a chelator. Allen added $10^{-3}$ M $H_2SO_4$ to acidify the medium. We recommend adjusting the pH after all components are added because of the presence of EDTA. For the growth in the dark, 1% glucose is added to the medium. **Into 900 mL** of $dH_2O$, individually dissolve the following components and add 100 µL of the trace metals solution. Adjust to pH 2.5, using 1 N sulfuric acid. Autoclave.

| Component | Stock Solution $(g \cdot L^{-1}\ dH_2O)$ | Quantity Used | Concentration in Final Medium (M) |
|---|---|---|---|
| $(NH_4)_2SO_4$ | — | 1.320 g | $1.00 \times 10^{-2}$ |
| $KH_2PO_4$ | — | 0.272 g | $2.00 \times 10^{-3}$ |
| $MgSO_4 \cdot 7H_2O$ | — | 0.247 g | $1.00 \times 10^{-3}$ |
| $CaCl_2$ | — | 0.055 g | $5.00 \times 10^{-4}$ |
| Trace metals solution | (See following recipe) | 1 mL | — |

*Trace Metals Solution*

(Watanabe et al. 2000)

Prepare the primary stock solutions. **Into 900 mL** of $dH_2O$, dissolve the Fe-EDTA, then individually add the metal compounds. Bring final volume to 1 liter.

| Component | 1° Stock Solution $(g \cdot L^{-1} \ dH_2O)$ | Quantity Used | Concentration in Final Medium (M) |
|---|---|---|---|
| Fe-Na-EDTA $3H_2O$ | — | 30.16 g | $7.16 \times 10^{-5}$ |
| $H_3BO_3$ | — | 2.86 g | $4.63 \times 10^{-5}$ |
| $MnCl_2 \cdot 4H_2O$ | — | 1.79 g | $9.04 \times 10^{-6}$ |
| $(NH_4)_6Mo_7O_{24} \cdot 4H_2O$ | — | 0.13 g | $1.05 \times 10^{-7}$ |
| $ZnSO_4 \cdot 7H_2O$ | — | 0.22 g | $7.65 \times 10^{-7}$ |
| $CuSO_4 \cdot 5H_2O$ | — | 0.079 g | $3.16 \times 10^{-7}$ |
| $NH_4VO_3$ | — | 0.023 g | $1.97 \times 10^{-7}$ |

## BG-11 Medium, Modified

### (Allen 1968, Allen and Stanier 1968, Rippka et al. 1979)

BG-11 medium (Allen and Stanier 1968) was derived from medium no. 11 (Hughes et al. 1958) for the culture of freshwater, soil, thermal, and marine cyanobacteria (see Stanier and Cohen-Bazire 1977, Allen 1968). The major changes from medium no. 11 are a threefold increase of $NaNO_3$ (nitrate and phosphate levels are exceptionally high in this medium) and the replacement of Gaffron's minor trace elements with a modified A5 trace elements solution. However, the nitrate value can be lowered considerably, and $N_2$-fixing cyanobacteria have been cultured routinely in a modified $BG-11_0$ medium that omits nitrate entirely. Note that this recipe separates the EDTA from the trace metals solution. It would be more sensible to add the EDTA to the trace metals solution and, conversely, add the boron with other macronutrients. The $MgNa_2EDTA \cdot H_2O$ may be replaced by the molar equivalent of $Na_2EDTA \cdot 2H_2O$ (1.04 g) (Rippka et al.). The ferric citrate and the citric acid may be prepared as a combined stock solution. $Na_2SiO_3 \cdot 9H_2O$ can be added ($2.04 \times 10^{-4}$ M final concentration) (Allen 1968). Allen (1968) found that when the medium and agar were autoclaved together, the algae didn't grow. She prepared a 2× concentration of the medium and a 2× concentration of the agar (1%–2.5% final concentration), autoclaved them separately, and then aseptically mixed them when the agar was approximately 58°C. She also prepared the ferric citrate and manganese stocks separately, adding them aseptically to the sterilized medium. **Prepare** the ferric ammonium citrate stock solution (dissolve citrate and ferric citrate in 1 liter $dH_2O$) and other stock solutions. **Into 900 mL** of $dH_2O$, add 1 mL of the Fe citrate solution, and then add the remaining components. Autoclave. Final pH should be 7.4 after cooling and $CO_2$ equilibration.

| Component | Stock Solution $(g \cdot L^{-1} dH_2O)$ | Quantity Used | Concentration in Final Medium (M) |
|---|---|---|---|
| *Fe Citrate solution* | | 1 mL | |
| Citric acid | 6 | 1 mL | $3.12 \times 10^{-5}$ |
| Ferric ammonium citrate | 6 | 1 mL | $\sim 3 \times 10^{-5}$ |
| $NaNO_3$ | — | 1.5 g | $1.76 \times 10^{-2}$ |
| $K_2HPO_4 \cdot 3H_2O$ | 40 | 1 mL | $1.75 \times 10^{-4}$ |
| $MgSO_4 \cdot 7H_2O$ | 75 | 1 mL | $3.04 \times 10^{-4}$ |
| $CaCl_2 \cdot 2H_2O$ | 36 | 1 mL | $2.45 \times 10^{-4}$ |
| $Na_2CO_3$ | 20 | 1 mL | $1.89 \times 10^{-4}$ |
| $MgNa_2EDTA \cdot H_2O$ | 1.0 | 1 mL | $2.79 \times 10^{-6}$ |
| Trace metals solution | (See following recipe) | 1 mL | — |

*Trace Metals Solution*

This is also known as A5 + Co trace metals solution. **To 950 mL** of dH$_2$O, add the EDTA and other components; bring final volume to 1 liter.

| Component | 1° Stock Solution (g · L$^{-1}$ dH$_2$O) | Quantity Used | Concentration in Final Medium (M) |
|---|---|---|---|
| H$_3$BO$_3$ | — | 2.860 g | $4.63 \times 10^{-5}$ |
| MnCl$_2$ · 4H$_2$O | — | 1.810 g | $9.15 \times 10^{-6}$ |
| ZnSO$_4$ · 7H$_2$O | — | 0.220 g | $7.65 \times 10^{-7}$ |
| CuSO$_4$ · 5H$_2$O | 79.0 | 1 mL | $3.16 \times 10^{-7}$ |
| Na$_2$MoO$_4$ · 2H$_2$O | — | 0.391 g | $1.61 \times 10^{-6}$ |
| Co(NO$_3$)$_2$ · 6H$_2$O | 49.4 | 1 mL | $1.70 \times 10^{-7}$ |

## Bold's Basal Medium (BBM)

### (Bold 1949, Bischoff and Bold 1963)

This was derived from a modified version of Bristol's solution (see Bold 1949). The medium lacks vitamins, and some trace metal concentrations are high. This is a useful medium for many algae, especially chlorococcalean algae, volvo-calean algae (e.g., *Chlamydomonas*), the filamentous green alga *Klebsormidium flaccidum* (Kützing) S. Mattox et Black-well (Floyd et al. 1972), the xanthophycean alga *Heterococcus endolithicus* Darling et Friedmann (Darling et al. 1987), euglenoid *Colacium vesiculosum* Ehr. (Rosowski and Kugrens 1973), and the cyanobacterium *Microcystis aeruginosa* Kützing (Cole and Wynne, 1974). However, the formulation is unsuitable for algae with vitamin requirements. Several modified media have been developed (see descriptions beneath table). **Into 936 mL** of dH$_2$O, add 10 mL of the first six stock solutions. Add 1 mL each of the alkaline EDTA, acidified iron, boron, and trace metals solutions. Autoclave. The final pH should be 6.6.

| Component | Stock Solution ($g \cdot L^{-1}$ dH$_2$O) | Quantity Used | Concentration in Final Medium (M) |
|---|---|---|---|
| *Macronutrients* | | | |
| NaNO$_3$ | 25.00 | 10 mL | $2.94 \times 10^{-3}$ |
| CaCl$_2 \cdot$ 2H$_2$O | 2.50 | 10 mL | $1.70 \times 10^{-4}$ |
| MgSO$_4 \cdot$ 7H$_2$O | 7.50 | 10 mL | $3.04 \times 10^{-4}$ |
| K$_2$HPO$_4$ | 7.50 | 10 mL | $4.31 \times 10^{-4}$ |
| KH$_2$PO$_4$ | 17.50 | 10 mL | $1.29 \times 10^{-3}$ |
| NaCl | 2.50 | 10 mL | $4.28 \times 10^{-4}$ |
| *Alkaline EDTA Solution* | | 1 mL | |
| EDTA | 50.00 | | $1.71 \times 10^{-4}$ |
| KOH | 31.00 | | $5.53 \times 10^{-4}$ |
| *Acidified Iron Solution* | | 1 mL | |
| FeSO$_4 \cdot$ 7H$_2$O | 4.98 | | $1.79 \times 10^{-5}$ |
| H$_2$SO$_4$ | | 1 mL | |
| *Boron Solution* | | 1 mL | |
| H$_3$BO$_3$ | 11.42 | | $1.85 \times 10^{-4}$ |
| *Trace Metals Solution* | | 1 mL | |
| ZnSO$_4 \cdot$ 7H$_2$O | 8.82 | | $3.07 \times 10^{-5}$ |
| MnCl$_2 \cdot$ 4H$_2$O | 1.44 | | $7.28 \times 10^{-6}$ |
| MoO$_3$ | 0.71 | | $4.93 \times 10^{-6}$ |
| CuSO$_4 \cdot$ 5H$_2$O | 1.57 | | $6.29 \times 10^{-6}$ |
| Co(NO$_3$)$_2 \cdot$ 6H$_2$O | 0.49 | | $1.68 \times 10^{-6}$ |

**KBBM medium** (BBM + 0.25% sucrose + 1.0% proteose peptone) was developed for a *Chlorella*-like strain that lives exosymbiotically with *Paramecium bursaria* Focke (Schuster et al. 1990).

**BBM + GA medium** (BBM + 1% glucose + 0.01 M amino acid hydrolysates [e.g., acid hydrolysates of casein] or 0.01 M amino acids [e.g., proline, glutamine, or arginine]) was developed to grow the lichenaceous algae *Trebouxia* (Fox 1967).

**3NBBM medium** (BBM enriched with three times the nitrate) was developed to grow *Anabaena flos-aquae* Brébisson ex Bornet et Flahault and *Anacystis nidulans* Gardner (Thomas and Montes 1978).

**3NBBM + vitamins medium** (BBM with three times the nitrate plus the addition of three vitamins) was developed to culture for *Batrachospermum* (Aghajanian 1979). Vitamins: add 0.1525 μg · L$^{-1}$ thiamine · HCl (4.52 × 10$^{-10}$ M, final concentration); 0.125 μg · L$^{-1}$ biotin (5.12 × 10$^{-10}$ M, final concentration); 0.125 μg · L$^{-1}$ cyanocobalamin (9.22 × 10$^{-11}$ M, final concentration).

**Double-strength BBM + a sterilized wheat seed**. This medium was used to culture the cryptomonads *Chilomonas paramaecium* Ehr. and *Cyathomonas truncata* (Fres.) Fish. (Kugrens et al. 1987).

## C Medium, Modified

(Ichimura 1971, Watanabe et al. 2000)

C (*Closterium*) medium, a derivative of *Volvox* medium (Provasoli and Pintner 1960), was developed by Ichimura (1971) for the culture of desmids, especially *Closterium* species, from eutrophicated waters. The recipe uses a slightly modified version of the PIV trace metals solution described by Provasoli and Pintner (1960), and vitamins are added (Watanabe et al. 2000). C medium also sustains growth for most chlorococcalean algae, some volvocalean algae (e.g., *Chlamydomonas, Carteria, Chloromonas, Hafniomonas,* and *Haematococcus*), some other desmids (e.g., *Cosmarium, Cylindrocystis, Gonatozygon,* and *Mesotaenium*), some filamentous green algae (e.g., *Ulothrix, Uronema, Draparnaldia, Hyalotheca, Hydrodictyon,* and *Stigoeclonium*), a glaucophycean alga *Cyanophora pradoxa*, and the synurophycean algae *Synura petersenii* Korsh. and *S. spinosa* Korsh. (Watanabe et al. 2000). Several modified versions are listed below. **Into 900 mL** dH$_2$O, dissolve the TRIS buffer, then add the remaining components. Bring to 1 liter with dH$_2$O, and autoclave. Final pH should be 7.5.

| Component | Stock Solution (g · L$^{-1}$ dH$_2$O) | Quantity Used | Concentration in Final Medium (M) |
|---|---|---|---|
| Tris base | — | 0.50 g | $4.13 \times 10^{-3}$ |
| KNO$_3$ | — | 0.10 g | $9.89 \times 10^{-4}$ |
| Ca(NO$_3$)$_2$ · 4H$_2$O | — | 0.15 g | $6.35 \times 10^{-4}$ |
| Na$_2$ β-glycerophosphate · 5H$_2$O | 50 | 1 mL | $1.63 \times 10^{-4}$ |
| MgSO$_4$ · 7H$_2$O | 40 | 1 mL | $1.62 \times 10^{-4}$ |
| Trace metals solution | (See following recipe) | 3 mL | — |
| Vitamins solution | (See following recipe) | 1 mL | — |

*Modified PIV Trace Metals Solution*

(Provasoli and Pintner 1960, Watanabe et al. 2000)

PIV used HOEDTA (hydroxyethyl ethylenediamine triacetic acid), but we have substituted Na$_2$EDTA (= 13.98 g HOEDTA). **Into 950 mL** of dH$_2$O, dissolve the EDTA and then individually dissolve the metals. Bring to 1 liter with dH$_2$O.

| Component | 1° Stock Solution (g · L$^{-1}$ dH$_2$O) | Quantity Used | Concentration in Final Medium (M) |
|---|---|---|---|
| Na$_2$EDTA | — | 1.000 g | $8.06 \times 10^{-6}$ |
| FeCl$_3$ · 6H$_2$O | — | 0.194 g | $2.15 \times 10^{-6}$ |
| MnCl$_2$ · 4H$_2$O | 36.00 | 1 mL | $5.46 \times 10^{-7}$ |
| ZnCl$_2$ | 10.44 | 1 mL | $2.30 \times 10^{-7}$ |
| Na$_2$MoO$_4$ · 2H$_2$O | 12.62 | 1 mL | $1.56 \times 10^{-7}$ |
| CoCl$_2$ · 6H$_2$O | 4.04 | 1 mL | $5.09 \times 10^{-8}$ |

*Vitamins Solution*

(Watanabe et al. 2000)

**Into 950 mL** of $dH_2O$, dissolve the thiamine · HCl and then add 1 mL of each primary stock. Bring to 1 liter with $dH_2O$. Filter-sterilize and store frozen.

| Component | 1° Stock Solution $(g \cdot L^{-1} dH_2O)$ | Quantity Used | Concentration in Final Medium (M) |
|---|---|---|---|
| Thiamine · HCl (vitamin $B_1$) | — | 10 mg | $2.97 \times 10^{-8}$ |
| Biotin (vitamin H) | 0.1 | 1 mL | $4.09 \times 10^{-10}$ |
| Cyanocobalamin (vitamin $B_{12}$) | 0.1 | 1 mL | $7.38 \times 10^{-11}$ |

**CB medium** (C medium with pH adjusted to 9.0, buffered with bicine substituted for Tris). This has been used to culture strains of the water-bloom-forming cyanobacterium species *Microcystis* (Watanabe et al. 2000).

**Csi medium** (C medium with pH adjusted to 7.0, buffered with 500 mg HEPES in place of Tris + 100 mg · $L^{-1}$ $Na_2SiO_3 \cdot 9H_2O$; $3.52 \times 10^{-4}$ M final concentration). This has been used successfully to culture diatoms (Otsuki et al. 1987, Watanabe et al. 1987, Soma 1993), various kinds of freshwater benthic algae (Takamura et al. 1988, 1989) and two glaucophytes, *Cyanophora paradoxa* Korshikov and *C. tetracyanea* Korshikov (Watanabe 2000).

**CT medium** (C medium with pH adjusted to 8.2, buffered with 400 mg TAPS substituted for Tris). The medium has been used to grow water-bloom-forming *Anabaena* strains (Li et al. 1998) and oscillatorioid cyanobacteria (Suda et al. 2002).

**CYT medium** (1 liter C medium + 1 g yeast + 2 g tryptone). This medium provides good growth for the colorless cryptomonad *Chilomonas paramecium* Ehr. (Erata and Chihara 1987).

## CA Medium, Modified

(Ichimura and Watanabe 1974, Watanabe et al. 2000)

CA medium was developed for culturing oligotrophic desmids, and it contains both $KNO_3$ and $NH_4NO_3$ as nitrogen sources. It is modified here by adding vitamins and using a slightly different trace metals solution. *Closterium aciculare* T. West var. *subpronum* W. et G. S. West, which forms blooms in mesotrophic lakes and reservoirs (and sometimes is abundant in eutrophic waters), can be grown in a CA medium because *C. aciculare* can use only ammonium as the nitrogen source (Coesel 1991). One modification is described below. C medium (see previous recipe) was also developed for eutrophic desmids (e.g., *Closterium* species), and the choice among CA medium, C medium, and their modifications should be made after considering the habitat and environmental conditions of the collection site. **Into 900 mL** $dH_2O$, dissolve the HEPES buffer, then add the remaining components. Bring to 1 liter with $dH_2O$, and autoclave. Final pH should be 7.2.

| Component | Stock Solution ($g \cdot L^{-1}$ $dH_2O$) | Quantity Used | Concentration in Final Medium (M) |
|---|---|---|---|
| HEPES | — | 0.40 g | $1.68 \times 10^{-3}$ |
| $KNO_3$ | — | 0.10 g | $9.89 \times 10^{-4}$ |
| $Ca(NO_3)_2 \cdot 4H_2O$ | 20.0 | 1 mL | $8.47 \times 10^{-5}$ |
| $NH_4NO_3$ | 50.0 | 1 mL | $6.25 \times 10^{-4}$ |
| $Na_2$ β-glycerophosphate $\cdot 5H_2O$ | 30.0 | 1 mL | $9.80 \times 10^{-5}$ |
| $MgSO_4 \cdot 7H_2O$ | 20.0 | 1 mL | $8.11 \times 10^{-5}$ |
| Iron–EDTA solution | (See following recipe) | 1 mL | — |
| PIV trace metals solution | (See following recipe) | 1 mL | — |
| Vitamins solution | (See following recipe) | 1 mL | — |

*Iron–EDTA Solution*

| Component | 1° Stock Solution ($g \cdot L^{-1}$ $dH_2O$) | Quantity Used | Concentration in Final Medium (M) |
|---|---|---|---|
| $Na_2EDTA \cdot 2H_2O$ | — | 0.372 g | $1.00 \times 10^{-6}$ |
| $FeCl_3 \cdot 6H_2O$ | — | 0.270 g | $1.00 \times 10^{-6}$ |

*Modified PIV Trace Metals Solution*

(Provasoli and Pintner 1960, Watanabe et al. 2000)

PIV used HOEDTA (hydroxyethyl ethylenediamine triacetic acid), but we have substituted $Na_2EDTA$ (= 13.98 g HOEDTA). **Into 950 mL** of $dH_2O$, dissolve the EDTA and then individually dissolve the metals. Bring to 1 liter with $dH_2O$.

| Component | 1° Stock Solution ($g \cdot L^{-1}$ $dH_2O$) | Quantity Used | Concentration in Final Medium (M) |
|---|---|---|---|
| $Na_2EDTA$ | — | 1.000 g | $2.69 \times 10^{-6}$ |
| $FeCl_3 \cdot 6H_2O$ | — | 0.194 g | $7.18 \times 10^{-7}$ |
| $MnCl_2 \cdot 4H_2O$ | 36.00 | 1 mL | $1.82 \times 10^{-7}$ |
| $ZnCl_2$ | 10.44 | 1 mL | $7.66 \times 10^{-8}$ |
| $Na_2MoO_4 \cdot 2H_2O$ | 12.62 | 1 mL | $5.22 \times 10^{-8}$ |
| $CoCl_2 \cdot 6H_2O$ | 4.04 | 1 mL | $1.70 \times 10^{-8}$ |

*Vitamins Solution*

(Watanabe et al. 2000)

**Into 950 mL** of $dH_2O$, dissolve the thiamine · HCl and then add 1 mL of each primary stock. Bring to 1 liter with $dH_2O$. Filter-sterilize and store frozen.

| Component | 1° Stock Solution ($g \cdot L^{-1}$ $dH_2O$) | Quantity Used | Concentration in Final Medium (M) |
|---|---|---|---|
| Thiamine · HCl ($B_1$) | — | 10 mg | $2.97 \times 10^{-8}$ |
| Biotin (vitamin H) | 0.1 | 1 mL | $4.09 \times 10^{-10}$ |
| Cyanocobalamin (vitamin $B_{12}$) | 0.1 | 1 mL | $7.38 \times 10^{-11}$ |

**CAM medium** (CA medium, pH 6.5, buffered with MES instead of HEPES); designed for desmids living in oligotrophic and acidic waters (Watanabe et al. 2000).

## Carefoot's Medium

(Carefoot 1968)

This was developed to grow the freshwater dinoflagellate *Peridinium cincutum* f. *ovoplanum* Lindeman. It was derived from a modified Bold's Basal Medium (Starr 1964) by adding Provasoli's PIV trace metal solution (Provasoli and Pintner 1960). Carefoot (1968) tested many organic compounds and varied the medium pH. He found that malonate, pyruvate, and certain sugars enhanced growth. Watanabe et al. (2000) provide a modified recipe that includes vitamins, which are necessary for many freshwater dinoflagellates. However, even with vitamins, it is difficult to maintain freshwater dinoflagellates in a stable condition for the long term. Carefoot provides both weight addition and final molarity in his recipe, but because of rounding errors they vary slightly; we have used his weight addition and provided more precise molarities. **Into 950 ml** dH$_2$O, dissolve the NaNO$_3$ and add the appropriate volume of the stock solutions. Bring to 1 liter with dH$_2$O. Adjust pH to 7.5 and autoclave.

| Component | Stock Solution (g · L$^{-1}$ dH$_2$O) | Quantity Used | Concentration in Final Medium (M) |
|---|---|---|---|
| NaNO$_3$ | — | 0.25 g | $2.94 \times 10^{-3}$ |
| K$_2$HPO$_4$ | 9.7 | 1 mL | $5.57 \times 10^{-5}$ |
| KH2PO4 | 22.7 | 1 mL | $1.67 \times 10^{-4}$ |
| MgSO$_4$ · 7H$_2$O | 4.9 | 1 mL | $1.99 \times 10^{-5}$ |
| CaCl$_2$ · 2H$_2$O | 16.5 | 1 mL | $1.49 \times 10^{-4}$ |
| NaCl | 16.5 | 1 mL | $2.82 \times 10^{-4}$ |
| PIV trace metals soln. | (See following recipe) | 5 mL | — |

*PIV Trace Metals Solution*

(Provasoli and Pintner 1960)

PIV used HOEDTA (hydroxyethyl ethylenediamine triacetic acid), but we have substituted $Na_2EDTA \cdot 2H_2O$ (= 1.398 g HOEDTA). **Into 950 mL** of $dH_2O$, dissolve the EDTA and then individually dissolve the metals. Bring to 1 liter with $dH_2O$.

| Component | 1° Stock Solution $(g \cdot L^{-1} \, dH_2O)$ | Quantity Used | Concentration in Final Medium (M) |
|---|---|---|---|
| $Na_2EDTA \cdot 2H_2O$ | — | 1.512 g | $2.03 \times 10^{-5}$ |
| $FeCl_3 \cdot 6H_2O$ | — | 0.194 g | $3.58 \times 10^{-6}$ |
| $MnCl_2 \cdot 4H_2O$ | — | 0.036 g | $9.10 \times 10^{-7}$ |
| $ZnCl_2$ | — | 0.010 g | $3.82 \times 10^{-7}$ |
| $Na_2MoO_4 \cdot 2H_2O$ | — | 0.013 g | $2.61 \times 10^{-7}$ |
| $CoCl_2 \cdot 6H_2O$ | 3.64 | 1 mL | $7.65 \times 10^{-8}$ |

## Chu #10 Medium

(Chu 1942)

Chu #10 medium was the most popular of 17 media that Chu (1942) described. It is a synthetic medium designed to mimic lake water, but it lacks a chelator, vitamins, and trace metals (except for iron). It has been extensively used for a variety of algae, including green algae, diatoms, cyanobacteria (Chu 1942, Nalewajiko et al. 1995), and the glaucophycean alga *Glaucocystis nostochilearum* Itzigsohn (Hall and Claus 1967). Many synthetic freshwater media are derived from Chu #10, and several modified Chu #10 media have been developed. We list two modifications below. Note: Chu used anhydrous silicate, which is very difficult to dissolve. You can substitute 0.0583 g $Na_2SiO_3 \cdot 9H_2O$ for the anhydrous silicate. Chu gave two possible phosphate concentrations (0.005, 0.01); we used the smaller amount. **Into 950 mL** of $dH_2O$, individually dissolve each component. Bring the final volume to 1 liter and autoclave.

| Component | Stock Solution $(g \cdot L^{-1}\ dH_2O)$ | Quantity Used | Concentration in Final Medium (M) |
|---|---|---|---|
| $Ca(NO_3)_2$ | 40.0 | 1 mL | $2.44 \times 10^{-4}$ |
| $K_2HPO_4$ | 5.0 | 1 mL | $2.87 \times 10^{-5}$ |
| $MgSO_4 \cdot 7H_2O$ | 25.0 | 1 mL | $1.01 \times 10^{-4}$ |
| $Na_2CO_3$ | 20.0 | 1 mL | $1.89 \times 10^{-4}$ |
| $Na_2SiO_3$ | 25.0 | 1 mL | $2.05 \times 10^{-4}$ |
| $FeCl_3$ | 0.8 | 1 mL | $4.93 \times 10^{-6}$ |

**Chu-10 + 0.1 $\mu g \cdot L^{-1}$ vitamin $B_{12}$ medium** was used to culture *Cyclotella meneghiniana* Kützing and *C. cryptica* Reimann (Schultz and Trainor 1968).

## Half-Strength Chu #10 Medium

### (Nalewajko and O'Mahony 1989)

The original Chu #10 medium lacked both trace metals and vitamins, and the nutrient concentrations were somewhat high for sensitive oligotrophic organisms. Nalewajko and O'Mahony (1989) corrected all three problems with this half-strength medium. The medium was used for growing *Chlamydomonas vernalis* Skuja, *Nitzschia* sp., and *Oscillatoria utermoehlii* (Utermöhl) J. de Toni isolated from Plastic Lake, south-central Ontario, Canada (Nalewajko and O'Mahony 1989). See Chu #10 Medium for hydrated silicate substitution. **Into 950 mL** of dH$_2$O, individually dissolve each component. Bring the final volume to 1 liter and autoclave.

| Component | Stock Solution ($g \cdot L^{-1}$ dH$_2$O) | Quantity Used | Concentration in Final Medium (M) |
|---|---|---|---|
| Ca(NO$_3$)$_2$ | 20.0 | 1 mL | $1.22 \times 10^{-4}$ |
| K$_2$HPO$_4$ | 2.5 | 1 mL | $1.44 \times 10^{-5}$ |
| MgSO$_4 \cdot$ 7H$_2$O | 12.5 | 1 mL | $5.07 \times 10^{-5}$ |
| Na$_2$CO$_3$ | 10.0 | 1 mL | $9.43 \times 10^{-5}$ |
| Na$_2$SiO$_3$ | 12.5 | 1 mL | $1.02 \times 10^{-4}$ |
| FeCl$_3$ | 0.4 | 1 mL | $2.47 \times 10^{-6}$ |
| Trace metals solution | (See following recipe) | 1 mL | — |
| Vitamins solution | (See following recipe) | 1 mL | — |

### *Trace Metals Solution*

Note: the solution includes boron as a metal and lacks a chelator. **Into 950 mL** of dH$_2$O, individually dissolve the components. Bring to 1 liter with dH$_2$O.

| Component | 1° Stock Solution ($g \cdot L^{-1}$ dH$_2$O) | Quantity Used | Concentration in Final Medium (M) |
|---|---|---|---|
| H$_3$BO$_3$ | 2.48 | 1 mL | $4.01 \times 10^{-8}$ |
| MnSO$_4 \cdot$ H$_2$O | 1.47 | 1 mL | $8.70 \times 10^{-9}$ |
| ZnSO$_4 \cdot$ 7H$_2$O | 0.23 | 1 mL | $8.00 \times 10^{-10}$ |
| CuSO$_4 \cdot$ 5H$_2$O | 0.10 | 1 mL | $4.01 \times 10^{-10}$ |
| (NH$_4$)$_6$Mo$_7$O$_{24} \cdot$ 4H$_2$O | 0.07 | 1 mL | $5.66 \times 10^{-11}$ |
| Co(NO$_3$)$_2 \cdot$ 6H$_2$O | 0.14 | 1 mL | $4.81 \times 10^{-10}$ |

*Vitamins Solution*

**Into 950 mL** of dH$_2$O, dissolve the thiamine · HCl and then add 1 mL of each primary stock. Bring to 1 liter with dH$_2$O. Filter-sterilize and store frozen.

| Component | 1° Stock Solution (g · L$^{-1}$ dH$_2$O) | Quantity Used | Concentration in Final Medium (M) |
|---|---|---|---|
| Thiamine · HCl (vitamin B$_1$) | — | 50 mg | 1.48 × 10$^{-7}$ |
| Biotin (vitamin H) | 2.5 | 1 mL | 1.02 × 10$^{-8}$ |
| Cyanocobalamin (vitamin B$_{12}$) | 2.5 | 1 mL | 1.84 × 10$^{-9}$ |

## COMBO Medium

### (Kilham et al. 1998)

COMBO medium was derived from Guillard and Lorenzen's (1972) WC medium (e.g., removing the glycylglycine or Tris buffer, modifying the trace elements) and adding 1 mL of animal trace metals. COMBO medium was used to support robust growth of cyanobacteria, cryptophytes, green algae, and diatoms that served as food for many clado-cerans, planaria, and rotifers. COMBO medium supports rates of algal growth statistically comparable to those in WC medium because of its similarity to WC medium in major salt and nutrient compositions and concentrations. It also supports fecundity of zooplankton in a range similar to that in natural lake water. COMBO medium is completely defined, readily manipulated for nutrition research, and superior to other artificial zooplankton media. **Into 900 mL** of dH$_2$O, individually add 1 mL of each stock solution. Bring the final volume to 1 liter with dH$_2$O. Adjust the pH to 7.8 and filter-sterilize.

| Component | Stock Solution ($g \cdot L^{-1}$ dH$_2$O) | Quantity Used | Concentration in Final Medium (M) |
|---|---|---|---|
| NaNO$_3$ | 85.01 | 1 mL | $1.00 \times 10^{-3}$ |
| CaCl$_2 \cdot$ 2H$_2$O | 36.76 | 1 mL | $2.50 \times 10^{-4}$ |
| MgSO$_4 \cdot$ 7H$_2$O | 36.97 | 1 mL | $1.66 \times 10^{-4}$ |
| NaHCO$_3$ | 12.60 | 1 mL | $1.50 \times 10^{-4}$ |
| Na$_2$SiO$_3 \cdot$ 9H$_2$O | 28.42 | 1 mL | $1.00 \times 10^{-4}$ |
| K$_2$HPO$_4$ | 8.71 | 1 mL | $5.00 \times 10^{-5}$ |
| H$_3$BO$_3$ | 24.00 | 1 mL | $3.88 \times 10^{-4}$ |
| KCl | 7.45 | 1 mL | $1.00 \times 10^{-4}$ |
| Algal trace elements solution | (See following recipe) | 1 mL | — |
| Animal trace elements solution | (See following recipe) | 1 mL | — |
| Vitamins solution | (See following recipe) | 1 mL | — |

*Algal Trace Elements Solution*

**Into 950 mL** of dH$_2$O, first dissolve the EDTA and then add the components. Bring final volume to 1 liter with dH$_2$O. Autoclave.

| Component | 1° Stock Solution (g · L$^{-1}$ dH$_2$O) | Quantity Used | Concentration in Final Medium (M) |
|---|---|---|---|
| Na$_2$EDTA · 2H$_2$O | — | 4.36 g | $1.17 \times 10^{-5}$ |
| FeCl$_3$ · 6H$_2$O | — | 1.00 g | $3.70 \times 10^{-6}$ |
| CuSO$_4$ · 5H$_2$O | 1.0 | 1 mL | $4.01 \times 10^{-9}$ |
| ZnSO$_4$ · 7H$_2$O | 22.0 | 1 mL | $7.65 \times 10^{-8}$ |
| CoCl$_2$ · 6H$_2$O | 12.0 | 1 mL | $5.04 \times 10^{-8}$ |
| MnCl$_2$ · 4H$_2$O | 180.0 | 1 mL | $9.10 \times 10^{-7}$ |
| Na$_2$MoO$_4$ · 2H$_2$O | 22.0 | 1 mL | $9.09 \times 10^{-8}$ |
| H$_2$SeO$_3$ | 1.6 | 1 mL | $1.24 \times 10^{-8}$ |
| Na$_3$VO$_4$ | 1.8 | 1 mL | $9.79 \times 10^{-9}$ |

*Animal Trace Elements*

(Elendt and Bias 1990)

Note: these trace elements are necessary for animals (e.g., Cladocera) but not for algae. **Into 950 mL** of dH$_2$O, first dissolve the EDTA and then add the components. Bring final volume to 1 liter with dH$_2$O. Autoclave.

| Component | 1° Stock Solution (g · L$^{-1}$ dH$_2$O) | Quantity Used | Concentration in Final Medium (M) |
|---|---|---|---|
| LiCl | — | 0.31 g | $7.31 \times 10^{-6}$ |
| RbCl | — | 0.07 g | $5.79 \times 10^{-7}$ |
| SrCl$_2$ · 6H$_2$O | — | 0.15 g | $5.63 \times 10^{-7}$ |
| NaBr | 16.0 | 1 mL | $1.55 \times 10^{-7}$ |
| KI | 3.3 | 1 mL | $1.99 \times 10^{-8}$ |

*Vitamins Solution*

**Into 950 mL** of $dH_2O$, dissolve the thiamine · HCl, add 1 mL of the primary stocks, and bring final volume to 1 liter with $dH_2O$. Filter-sterilize and store frozen.

| Component | 1° Stock Solution ($g \cdot L^{-1}$ $dH_2O$) | Quantity Used | Concentration in Final Medium (M) |
|---|---|---|---|
| Thiamine · HCl ($B_1$) | — | 100 mg | $2.96 \times 10^{-7}$ |
| Biotin (vitamin H) | 0.50 | 1 mL | $2.05 \times 10^{-9}$ |
| Cyanocobalamin (vitamin $B_{12}$) | 0.55 | 1 mL | $4.06 \times 10^{-10}$ |

## D Medium

### (Sheridan 1966)

This was designed by Sheridan (1966) for growing thermophilic cyanobacteria from hot springs. It has ~1.2 g · L$^{-1}$ total dissolved solids, a typical amount for many hot springs. Although it contains rather high levels of nitrate and phosphate, the molar ratio is ~12. Many modifications of this medium have been made and applied to a variety of cyanobacteria, including thermal and nonthermal (Halfen 1973, Garcia-Pichel and Castenholz 1991) and even *Chroomonas* sp. (Cryptophyceae; Schmek et al. 1994). The natures of the variants are shown in detail in Castenholz (1969, 1988). **Into 950 mL** of dH$_2$O, dissolve the nitrilotriacetic acid (NTA), add the other components, and bring the final volume to 1 liter using dH$_2$O. Adjust pH to 6.8. Autoclave.

| Component | Stock Solution (g · L$^{-1}$ dH$_2$O) | Quantity Used | Concentration in Final Medium (M) |
|---|---|---|---|
| Nitrilotriacetic acid | — | 0.100 g | $5.23 \times 10^{-4}$ |
| KNO$_3$ | — | 0.103 g | $1.02 \times 10^{-3}$ |
| NaNO$_3$ | — | 0.689 g | $8.11 \times 10^{-3}$ |
| Na$_2$HPO$_4$ | — | 0.111 g | $7.82 \times 10^{-4}$ |
| MgSO$_4$ · 7H$_2$O | — | 0.100 g | $4.06 \times 10^{-4}$ |
| CaSO$_4$ · 2H$_2$O | 60.00 | 1 mL | $3.48 \times 10^{-4}$ |
| NaCl | 8.00 | 1 mL | $1.37 \times 10^{-4}$ |
| FeCl$_3$ | 0.29 | 1 mL | $1.79 \times 10^{-6}$ |
| Trace metals solution | — | 0.5 mL | — |

*Trace Metals Solution*

First, add 0.5 mL concentrated H$_2$SO$_4$ to 950 mL dH$_2$O, then dissolve or add the other components and bring the final volume to 1 liter with dH$_2$O.

| Component | 1° Stock Solution (g · L$^{-1}$ dH$_2$O) | Quantity Used | Concentration in Final Medium (M) |
|---|---|---|---|
| H$_2$SO$_4$ (concentrated) | — | 0.5 mL | — |
| MnSO$_4$ · H$_2$O | — | 2.28 g | $5.11 \times 10^{-6}$ |
| ZnSO$_4$ · 7H$_2$O | — | 0.50 g | $8.69 \times 10^{-7}$ |
| H$_3$BO$_3$ | — | 0.50 g | $4.04 \times 10^{-6}$ |
| CuSO$_4$ · 5H$_2$O | 25 | 1 mL | $5.00 \times 10^{-8}$ |
| Na$_2$MoO$_4$ · 6H$_2$O | 25 | 1 mL | $5.17 \times 10^{-8}$ |
| CoCl$_2$ · 6H$_2$O | 45 | 1 mL | $9.46 \times 10^{-8}$ |

## D11 Medium

### (Graham et al. 1982)

This medium was developed for culturing the freshwater *Cladophora* (see also Hoffman and Graham 1984). The medium has an inordinate number of trace elements, many of which are probably unnecessary. Furthermore, the concentration of some metals (e.g., copper) is very high. Rather than prepare all three trace element solutions, it may be worth trying an addition of the most likely elements (Se, Va, Co, Ni) to the first trace elements solution. The recipe calls for selenic acid (selenate, $Na_2SeO_4$), but selenious acid (selenite, $Na_2SeO_3$) is often the only form used by algae; therefore, selenite should be used ($1.09 \text{ g} \cdot \text{L}^{-1}$ $dH_2O$ $Na_2SeO_3$ gives the same molar concentration of Se). The sodium bicarbonate stock should be filter-sterilized, not autoclaved. D11 medium has also been used, with modifications, for *Spirogyra* and *Coleochaete* (Graham et al. 1986, 1995). For nonaxenic cultures of *Spirogyra*, three changes were made: glass-distilled water was used, B7 and C13 trace elements were omitted because they were detrimental to *Spirogyra* growth, and the pH was adjusted to 7 with 1N HCl after addition of stock vitamins and bicarbonate (Graham et al. 1995). **Into 900 mL** $dH_2O$, dissolve the magnesium sulfate and calcium nitrate, then add 1 mL of each stock solution except the sodium bicarbonate and vitamins. Bring the final volume to 1 liter with $dH_2O$. Adjust to pH 6.0. Autoclave and cool. Aseptically add 1 mL of sodium bicarbonate solution and 1 mL of vitamins solution. The final pH should be approximately 7.5.

| Component | Stock Solution $(g \cdot L^{-1} \, dH_2O)$ | Quantity Used | Concentration in Final Medium (M) |
|---|---|---|---|
| $MgSO_4 \cdot 7H_2O$ | — | 100 mg | $4.06 \times 10^{-4}$ |
| $Ca(NO_3)_2 \cdot 4H_2O$ | — | 150 mg | $6.35 \times 10^{-4}$ |
| $NaHCO_3$ | 100 | 1 mL | $1.19 \times 10^{-3}$ |
| KCl | 30 | 1 mL | $4.02 \times 10^{-4}$ |
| $Na_2SiO_3 \cdot 9H_2O$ | 60 | 1 mL | $2.11 \times 10^{-4}$ |
| $K_2HPO_4$ | 15 | 1 mL | $8.61 \times 10^{-5}$ |
| D11 Trace elements solution | (See following recipe) | 1 mL | — |
| B7 Trace elements solution | (See following recipe) | 1 mL | — |
| C13 Trace elements solution | (See following recipe) | 1 mL | — |
| Vitamins stock solution | (See following recipe) | 1 mL | — |

*D11 Trace Metals Solution*

**Into 900 mL** $dH_2O$, dissolve the EDTA, and then individually add the remaining components. Bring to 1 liter with $dH_2O$.

| Component | 1° Stock Solution ($g \cdot L^{-1} dH_2O$) | Quantity Used | Concentration in Final Medium (M) |
|---|---|---|---|
| $Na_2EDTA \cdot 2H_2O$ | — | 12.5 g | $3.36 \times 10^{-5}$ |
| $FeSO_4 \cdot 7H_2O$ | — | 5.0 g | $1.80 \times 10^{-5}$ |
| $H_3BO_3$ | — | 1.0 g | $1.62 \times 10^{-5}$ |
| $ZnSO_4 \cdot 7H_2O$ | — | 0.1 g | $3.48 \times 10^{-7}$ |
| $MnSO_4 \cdot H_2O$ | — | 0.2 g | $1.18 \times 10^{-6}$ |
| $Na_2MoO_4 \cdot 2H_2O$ | 25.0 | 1 mL | $1.03 \times 10^{-7}$ |
| $CuSO_4 \cdot 5H_2O$ | 23.0 | 1 mL | $9.21 \times 10^{-8}$ |

*C13 Trace Elements Solution*

(Arnon 1938)

Note: we recommend substituting selenite for selenate (see previous discussion). First, dissolve $As_2O_3$ in 2 or 3 drops of concentrated HCl and dilute with $dH_2O$ to 1 liter. Next, heat $PbCl_2$ in $dH_2O$ to dissolve; bring final volume to 1 liter. To prepare the C13 trace elements solution, begin with 900 mL $dH_2O$ and add 1 mL of each stock solution. Bring the final volume to 1 liter with $dH_2O$. Adjust the pH to 4.0 to 5.0, using 1 N NaOH.

| Component | 1° Stock Solution ($g \cdot L^{-1} dH_2O$) | Quantity Used | Concentration in Final Medium (M) |
|---|---|---|---|
| $Al_2(SO_4) \cdot 18H_2O$ | 6.175 | 1 mL | $9.27 \times 10^{-9}$ |
| $As_2O_3$ | 0.660 | 1 mL | $3.34 \times 10^{-9}$ |
| $CdCl_2$ | 0.820 | 1 mL | $4.47 \times 10^{-9}$ |
| $SrCl_2$ | 1.520 | 1 mL | $9.59 \times 10^{-9}$ |
| $HgCl_2$ | 0.680 | 1 mL | $2.50 \times 10^{-9}$ |
| $PbCl_2$ | 0.670 | 1 mL | $2.41 \times 10^{-9}$ |
| $LiCl$ | 3.060 | 1 mL | $7.22 \times 10^{-8}$ |
| $RbCl$ | 0.710 | 1 mL | $5.87 \times 10^{-9}$ |
| $KBr$ | 0.740 | 1 mL | $6.22 \times 10^{-9}$ |
| $KI$ | 0.650 | 1 mL | $3.91 \times 10^{-9}$ |
| $NaF$ | 1.100 | 1 mL | $2.62 \times 10^{-8}$ |
| $Na_2SeO_4$ | 1.190 | 1 mL | $6.30 \times 10^{-9}$ |
| $BeSO_4 \cdot 4H_2O$ | 9.820 | 1 mL | $5.54 \times 10^{-8}$ |

*B7 Trace Metals Solution*

(Arnon 1938)

First prepare the primary stock solutions. Dissolve the $NH_4VO_3$ in 1 mL of 1 : 1 concentrated nitric acid : $dH_2O$, and then bring final volume to 1 liter. Dissolve the $SnCl_2 \cdot 2H_2O$ in a couple drops of HCl and then dilute to 1 liter with $dH_2O$. To 800 mL of $dH_2O$, add 1 mL of each stock solution. Adjust to a pH of 3 to 4, using 1 N NaOH.

| Component | 1° Stock Solution (g · L⁻¹ dH₂O) | Quantity Used | Concentration in Final Medium (M) |
|---|---|---|---|
| $NH_4VO_3$ | 2.30 | 1 mL | $1.97 \times 10^{-8}$ |
| $K[Cr(SO_4)_2] \cdot 12H_2O$ | 9.60 | 1 mL | $1.92 \times 10^{-8}$ |
| $NiSO_4 \cdot 6H_2O$ | 4.50 | 1 mL | $1.71 \times 10^{-8}$ |
| $CoCl_2 \cdot 6H_2O$ | 4.00 | 1 mL | $1.68 \times 10^{-8}$ |
| $Na_2WO_4 \cdot 2H_2O$ | 1.80 | 1 mL | $5.46 \times 10^{-9}$ |
| $K_2TiO(C_2O_4)_2 \cdot 2H_2O$ | 7.39 | 1 mL | $2.20 \times 10^{-8}$ |
| $SnCl_2 \cdot 2H_2O$ | 1.90 | 1 mL | $8.42 \times 10^{-9}$ |

*Vitamins Stock Solution*

**Into 900 mL,** dissolve the thiamine · HCl and then add 1 mL of each primary stock solution. Bring final volume to 1 liter. Filter-sterilize and store frozen.

| Component | 1° Stock Solution (g · L⁻¹ dH₂O) | Quantity Used | Concentration in Final Medium (M) |
|---|---|---|---|
| Thiamine · HCl (vitamin $B_1$) | 10.0 | 1 mL | $2.96 \times 10^{-8}$ |
| Biotin (vitamin H) | 10.0 | 1 mL | $4.09 \times 10^{-8}$ |
| Cyanocobalamin (vitamin $B_{12}$) | 0.1 | 1 mL | $7.38 \times 10^{-11}$ |

## DyIII Medium

(Lehman 1976)

This was based on concentrations of anions and cations present in a pond during exponential or luxurious growth of *Dinobryon* (Lehman 1976). The medium contains both nitrate and ammonium as nitrogen sources, and it has a low level of potassium because of sensitivity of *Dinobryon* to potassium. In the original recipe, iron wire was dissolved in a small amount of concentrated HCl. Alternatively, a ferric chloride ($FeCl_3 \cdot 6H_2O$) separate stock solution can be prepared by adding 3.389 g to 1 liter $dH_2O$, and this will provide the same amount of iron; the additional chloride ($5 \times 10^{-6}$ M in final medium) is probably insignificant. The medium provides acceptable growth for some chrysophytes, but the growth of other species is not supported, apparently because of the lack of certain trace elements (see DY-V medium). **Into 950 mL** of $dH_2O$, dissolve the MES, add the other components, and bring the final volume to 1 liter, using $dH_2O$. Adjust pH to 6.8. Autoclave.

| Component | Stock Solution ($g \cdot L^{-1}$ $dH_2O$) | Quantity Used | Concentration in Final Medium (M) |
|---|---|---|---|
| MES buffer | — | 200 mg | $1.02 \times 10^{-3}$ |
| $CaCl_2 \cdot 2H_2O$ | 75.00 | 1 mL | $5.10 \times 10^{-4}$ |
| $Na_2SiO_3 \cdot 9H_2O$ | 15.00 | 1 mL | $5.28 \times 10^{-5}$ |
| $MgSO_4 \cdot 7H_2O$ | 50.00 | 1 mL | $2.03 \times 10^{-4}$ |
| $NH_4NO_3$ | 5.00 | 1 mL | $6.25 \times 10^{-5}$ |
| $NaNO_3$ | 20.00 | 1 mL | $2.35 \times 10^{-4}$ |
| KCl | 3.00 | 1 mL | $4.02 \times 10^{-5}$ |
| $Na_2$ β-glycerophosphate | 10.00 | 1 mL | $4.63 \times 10^{-5}$ |
| $H_3BO_3$ | 4.58 | 1 mL | $7.40 \times 10^{-5}$ |
| Trace metals solution | (See following recipe) | 1 mL | — |
| Vitamins solution | (See following recipe) | 1 mL | — |

*Trace Metals Solution*

The iron wire is dissolved in a small amount of concentrated HCl and added to $dH_2O$. (A ferric chloride [$FeCl_3 \cdot 6H_2O$] primary stock solution can be prepared by adding 3.389 g to 1 liter $dH_2O$.) **Into 950 mL** $dH_2O$, add 1 mL of each stock solution and bring the final volume to 1 liter with $dH_2O$.

| Component | 1° Stock Solution ($g \cdot L^{-1}$ $dH_2O$) | Quantity Used | Concentration in Final Medium (M) |
|---|---|---|---|
| $Na_2EDTA$ (anhydrous) | — | 8.000 g | $2.74 \times 10^{-5}$ |
| Fe (iron wire) | | 0.700 g | $1.25 \times 10^{-5}$ |
| $MnCl_2 \cdot 4H_2O$ | | 0.720 g | $3.64 \times 10^{-6}$ |
| $ZnSO_4 \cdot 7H_2O$ | | 0.176 g | $6.12 \times 10^{-7}$ |
| $Na_2MoO_4 \cdot 6H_2O$ | 50 | 1 mL | $2.08 \times 10^{-7}$ |
| $CoCl_2 \cdot 6H_2O$ | 29 | 1 mL | $1.22 \times 10^{-7}$ |

*Vitamins Solution*

**Into 950 mL** of $dH_2O$, dissolve the thiamine $\cdot$ HCl, then add 1 mL of the primary stocks and bring the final volume to 1 liter with $dH_2O$. Filter-sterilize; store in refrigerator or freezer.

| Component | 1° Stock Solution ($g \cdot L^{-1}$ $dH_2O$) | Quantity Used | Concentration in Final Medium (M) |
|---|---|---|---|
| Thiamine $\cdot$ HCl (vitamin $B_1$) | — | 200 mg | $5.93 \times 10^{-7}$ |
| Biotin (vitamin H) | 0.5 | 1 mL | $2.05 \times 10^{-9}$ |
| Cyanocobalamin (vitamin $B_{12}$) | 0.5 | 1 mL | $3.69 \times 10^{-10}$ |

## DY-V Medium

(Andersen, unpublished)

DY-V Medium was derived, via DY-IV (see Andersen et al. 1997), from Dy III Medium (Lehman 1976). DY IV medium contained lower concentrations of nitrate and ammonium than those of DY III, but it was supplemented with vanadium and selenium (other trace metals lower than Dy III). The nitrogen and phosphorus concentrations were increased for DY-V. It supports the growth of a wide range of heterokont algae, cryptophytes, and other algae that require slightly acidic to circumneutral pH conditions (see Andersen et al. 1997). **Into 950 mL** of dH$_2$O, dissolve the MES and add the other components. Bring the final volume to 1 liter, using dH$_2$O. Adjust pH to 6.8 with NaOH. Autoclave.

| Component | Stock Solution (g · L$^{-1}$ dH$_2$O) | Quantity Used | Concentration in Final Medium (M) |
|---|---|---|---|
| MES | — | 200 mg | $1.02 \times 10^{-3}$ |
| MgSO$_4$ · 7H$_2$O | 50.99 | 1 mL | $2.03 \times 10^{-4}$ |
| KCl | 3.00 | 1 mL | $4.02 \times 10^{-5}$ |
| NH$_4$Cl | 2.68 | 1 mL | $5.01 \times 10^{-5}$ |
| NaNO$_3$ | 20.00 | 1 mL | $2.35 \times 10^{-4}$ |
| Na$_2$ β-glycerophosphate | 2.16 | 1 mL | $1.00 \times 10^{-5}$ |
| H$_3$BO$_3$ | 0.80 | 1 mL | $1.29 \times 10^{-5}$ |
| Na$_2$SiO$_3$ · 9H$_2$O | 14.00 | 1 mL | $4.93 \times 10^{-5}$ |
| CaCl$_2$ | 75.00 | 1 mL | $6.76 \times 10^{-4}$ |
| Trace elements solution | (See following recipe) | 1 mL | — |
| Vitamins solution | (See following recipe) | 1 mL | — |

*Trace Element Solution*

**Into 900 mL** $dH_2O$, dissolve the EDTA, and then add the remaining components. Bring the final volume to 1 liter.

| Component | 1° Stock Solution $(g \cdot L^{-1} \, dH_2O)$ | Quantity Used | Concentration in Final Medium (M) |
|---|---|---|---|
| $Na_2EDTA \cdot 2H_2O$ | — | 8.0 g | $2.15 \times 10^{-5}$ |
| $FeCl_3 \cdot 6H_2O$ | — | 1.0 g | $3.70 \times 10^{-6}$ |
| $MnCl_2 \cdot 4H_2O$ | — | 200 mg | $1.01 \times 10^{-6}$ |
| $ZnSO_4 \cdot 7H_2O$ | — | 40 mg | $1.39 \times 10^{-7}$ |
| $CoCl_2 \cdot 6H_2O$ | 8.0 | 1 mL | $3.36 \times 10^{-8}$ |
| $Na_2MoO_4 \cdot 6H_2O$ | 20.0 | 1 mL | $8.27 \times 10^{-8}$ |
| $Na_3VO_4 \cdot 10H_2O$ | 2.0 | 1 mL | $5.49 \times 10^{-9}$ |
| $H_2SeO_3$ | 4.0 | 1 mL | $2.31 \times 10^{-8}$ |

*Vitamins Solution*

**Into 950 mL** of $dH_2O$, dissolve the thiamine $\cdot$ HCl, add 1 mL of the primary stocks, and bring final volume to 1 liter with $dH_2O$. Filter-sterilize; store in refrigerator or freezer.

| Component | 1° Stock Solution $(g \cdot L^{-1} \, dH_2O)$ | Quantity Used | Concentration in Final Medium (M) |
|---|---|---|---|
| Thiamine $\cdot$ HCl (vitamin $B_1$) | — | 100 mg | $2.96 \times 10^{-7}$ |
| Biotin (vitamin H) | 0.5 | 1 mL | $2.05 \times 10^{-9}$ |
| Cyanocobalamin (vitamin $B_{12}$) | 0.5 | 1 mL | $3.69 \times 10^{-10}$ |

## Forsberg's Medium II

### (Forsberg 1965)

This was derived from a modified Chu #10 medium (Gerloff et al. 1950). It is suitable for the growth of *Chara globularis* Thuillier, *C. aspera* Willdenow, and *C. zeylanica* Willdenow and probably other charalean species. *Nitella flexilis* (L.) C. Agardh was also cultured with a 1 : 1 mixture of 1.8% soil extract and this medium (Foissner 1989). The anhydrous silicate should be dissolved at high pH; alternatively, substitute 23.28 g of $Na_2SiO_3 \cdot 9H_2O$. **Into 950 mL** of $dH_2O$, first dissolve the Tris and NTA, then add other components or stock solutions and bring the final volume up to 1 liter. Adjust the pH to 7.0; autoclave.

| Component | Stock Solution $(g \cdot L^{-1} \, dH_2O)$ | Quantity Used | Concentration in Final Medium (M) |
|---|---|---|---|
| Tris base | — | 0.5 g | $4.13 \times 10^{-3}$ |
| Nitrilotriacetic acid | — | 20 mg | $1.05 \times 10^{-4}$ |
| $MgSO_4 \cdot 7H_2O$ | — | 0.1 g | $4.06 \times 10^{-4}$ |
| $Na_2CO_3$ | 20.00 | 1 mL | $1.89 \times 10^{-4}$ |
| $Ca(NO_3)_2 \cdot 4H_2O$ | 80.00 | 1 mL | $3.39 \times 10^{-3}$ |
| KCl | 30.00 | 1 mL | $4.02 \times 10^{-4}$ |
| $Na_2SiO_3$ (anhydrous) | 10.00 | 1 mL | $8.19 \times 10^{-5}$ |
| $K_2HPO_4$ | 0.56 | 1 mL | $3.22 \times 10^{-6}$ |
| $H_3BO_3$ | 2.29 | 1 mL | $3.70 \times 10^{-5}$ |
| $FeCl_3 \cdot 6H_2O$ | 1.94 | 1 mL | $7.18 \times 10^{-6}$ |
| $ZnCl_2$ | 0.21 | 1 mL | $1.54 \times 10^{-6}$ |
| $Na_2MoO_4 \cdot 2H_2O$ | 0.25 | 1 mL | $1.03 \times 10^{-6}$ |
| $CuCl_2$ | 0.00846 | 1 mL | $6.29 \times 10^{-8}$ |
| $MnCl_2 \cdot 4H_2O$ | 0.0072 | 1 mL | $3.64 \times 10^{-8}$ |
| $CoCl_2 \cdot 6H_2O$ | 0.00807 | 1 mL | $3.39 \times 10^{-8}$ |

## Fraquil Medium

(Morel et al. 1975)

This was designed for study of trace metal interactions with freshwater phytoplankton (Morel et al. 1975); for a modification named Fraquil* medium, see Chapter 4. The major salts are those of WC medium, but nitrate, phosphate, and silicate concentrations were lowered. The mixed trace metals solution was described by Morel et al. (1975), and the vitamins are identical to those of f/2 medium (Guillard and Ryther 1962). The six stock solutions ($CaCl_2 \cdot 2H_2O$, $MgSO_4$ $7H_2O$, $NaHCO_3$, $Na_2SiO_3 \cdot 9H_2O$, $NaNO_3$, and $KH_2PO_4$) are each passed through a resin Chelex 100 column (Bio-Rad Laboratories) to remove trace metal impurities (see Chapter 4 for details). The free ion activities and the speciation of the components can be calculated on the basis of thermodynamic equilibrium in the system with use of the computer program MINEQL (Westall et al. 1976). Fraquil medium was used for examining the effects of metals on growth, nutrient uptakes, photosynthetic activity, or morphology of *Scenedesmus quadricauda* (Turpin) Brébisson (Rueter and Ades 1987), *Anabaena* (Rueter 1988), *Asterionella ralfsii* var. *americana* Körner (Genesemer 1990, Genesmer et al. 1993), and *Selenastrum capricornutum* Printz (Errécalde and Campbell 2000). **Into 900 mL** of salt solution, add 1 mL each of nutrient solution, 0.5 mL of mixed vitamin solution, and 1 mL of mixed trace metal solution. Bring final volume to 1 liter.

| Component | Stock Solution ($g \cdot L^{-1}$ $dH_2O$) | Quantity Used | Concentration in Final Medium (M) |
|---|---|---|---|
| $CaCl_2 \cdot 2H_2O$ | 36.80 | 1 mL | $2.50 \times 10^{-4}$ |
| $MgSO_4 \cdot 7H_2O$ | 37.00 | 1 mL | $1.50 \times 10^{-4}$ |
| $NaHCO_3$ | 12.60 | 1 mL | $1.50 \times 10^{-4}$ |
| $Na_2SiO_3 \cdot 9H_2O$ | 3.55 | 1 mL | $1.25 \times 10^{-5}$ |
| $NaNO_3$ | 8.50 | 1 mL | $1.00 \times 10^{-4}$ |
| $K_2HPO_4$ | 1.74 | 1 mL | $1.00 \times 10^{-5}$ |
| Vitamin solution | (See following recipe) | 0.5 mL | — |
| Trace metals solution | (See following recipe) | 1 mL | — |

*Vitamin Solution*

**Into 950 mL** of dH$_2$O, dissolve the thiamine · HCl and then add 1 mL of each primary stock solution. Bring final volume to 1 liter. Filter-sterilize. Store in refrigerator or freezer.

| Component | 1° Stock Solution (g · L$^{-1}$ dH$_2$O) | Quantity Used | Concentration in Final Medium (M) |
|---|---|---|---|
| Thiamine · HCl (vitamin B$_1$) | — | 200 mg | $2.96 \times 10^{-7}$ |
| Biotin (vitamin H) | 1.0 | 1 mL | $2.05 \times 10^{-9}$ |
| Cyanocobalamin (vitamin B$_{12}$) | 1.1 | 1 mL | $4.06 \times 10^{-10}$ |

*Trace Metals Solution*

**Into 950 mL** of dH$_2$O, add 1 mL each of primary stock solution, followed by EDTA and FeCl$_3$ · 6H$_2$O. It is important to add Na$_2$EDTA before FeCl$_3$ in order to prevent precipitation of ferric hydroxide. Bring final volume to 1 liter.

| Component | 1° Stock Solution (g · L$^{-1}$ dH$_2$O) | Quantity Used | Concentration in Final Medium (M) |
|---|---|---|---|
| CuSO$_4$ · 5H$_2$O | 0.249 | 1 mL | $9.97 \times 10^{-10}$ |
| (NH$_4$)$_6$Mo$_7$O$_{24}$ · 4H$_2$O | 0.265 | 1 mL | $2.14 \times 10^{-10}$ |
| CoCl$_2$ · 6H$_2$O | 0.595 | 1 mL | $2.50 \times 10^{-9}$ |
| MnCl$_2$ · 4H$_2$O | 4.550 | 1 mL | $2.30 \times 10^{-8}$ |
| ZnSO$_4$ · 7H$_2$O | 1.150 | 1 mL | $4.00 \times 10^{-9}$ |
| Na$_2$EDTA · 2H$_2$O | — | 1.860 g | $5.00 \times 10^{-6}$ |
| FeCl$_3$ · 6H$_2$O | — | 0.122 g | $4.51 \times 10^{-7}$ |

## MA Medium

### (Ichimura 1979)

This was developed for culturing bloom-forming *Microcystis* strains. It is characterized by a high pH, a bicine buffer, and high amounts of trace elements except for Fe. Compared with the trace elements of BG-11 medium, there is approximately a sevenfold increase of boron, threefold increase of Mn, twofold increase of Zn, 20-fold increase of Mo, and 100-fold increase of Co. Conversely, the medium omits Cu and has a 10-fold lower concentration of Fe than does BG-11 medium. The molar ratio of the cationic metals (e.g., Fe, Mn, Zn, and Co) to EDTA is approximately 4 : 1, and therefore it is underchelated by EDTA, but note that the medium contains a significant amount of bicine. Orthophosphates, such as $K_2HPO_4$ and $Na_2HPO_4$, cause precipitation by complexing with divalent cations (e.g., $Mg^{++}$ and $Ca^{++}$). Therefore, $Na_2$ DL-β-glycerophosphate $\cdot$ $5H_2O$ is substituted, and it is used by algae with alkaline phosphatase activity (e.g., most cyanobacteria, including *Microcystis*). MA medium has been widely used by Japanese researchers for culturing *Microcystis* (Kusumi et al. 1987; Ooi et al. 1989; Kaya and Watanabe 1990; Otsuka et al. 1998a,b, 1999a,b, 2000, 2001; Kato et al. 1991). The medium yields a high cell density for *Microcystis*, up to $10^9$ cells $\cdot$ $mL^{-1}$. **Into 900 mL** $dH_2O$, dissolve the bicine and then add 1 mL of the stock solutions. Adjust to pH 8.6 (1 N NaOH) and then autoclave.

| Component | Stock Solution ($g \cdot L^{-1}$ $dH_2O$) | Quantity Used | Concentration in Final Medium (M) |
|---|---|---|---|
| bicine | | 0.5 g | $3.06 \times 10^{-3}$ |
| $NaNO_3$ | 50.0 | 1 mL | $5.88 \times 10^{-4}$ |
| $KNO_3$ | 100.0 | 1 mL | $9.89 \times 10^{-4}$ |
| $Ca(NO_3)_2 \cdot 4H_2O$ | 50.0 | 1 mL | $2.12 \times 10^{-4}$ |
| $Na_2$ β-glycerophosphate $\cdot$ $5H_2O$ | 50.0 | 1 mL | $1.63 \times 10^{-4}$ |
| $Na_2SO_4$ | 40.0 | 1 mL | $2.82 \times 10^{-4}$ |
| $MgCl_2 \cdot 6H_2O$ | 50.0 | 1 mL | $2.46 \times 10^{-4}$ |
| $H_3BO_3$ | 20.0 | 1 mL | $3.23 \times 10^{-4}$ |
| $Na_2EDTA \cdot 2H_2O$ | 5.0 | 1 mL | $1.34 \times 10^{-5}$ |
| $FeCl_3 \cdot 6H_2O$ | 0.5 | 1 mL | $1.85 \times 10^{-6}$ |
| $MnCl_2 \cdot 4H_2O$ | 5.0 | 1 mL | $2.53 \times 10^{-5}$ |
| $CoCl_2 \cdot 6H_2O$ | 5.0 | 1 mL | $2.10 \times 10^{-5}$ |
| $ZnCl_2$ | 0.5 | 1 mL | $3.67 \times 10^{-6}$ |
| $Na_2MoO_4 \cdot 2H_2O$ | 0.8 | 1 mL | $3.31 \times 10^{-6}$ |

## MW Medium

(Sako et al. 1984)

This was derived from W Medium (Watanabe 1983) by adding $NH_4Cl$, $CaCO_3$, $CaCl_2 \cdot 2H_2O$, $KHCO_3$, and a 10-fold increase in vitamins; W medium was derived from *Volvox* Medium by adding urea. W Medium contained a low level of potassium because of the potassium sensitivity of dinoflagellate *Peridinium*. MW Medium is organically buffered (glycylglycine), and it is also rich in inorganic and other organic nitrogen sources. MW medium supports excellent growth of freshwater bloom-forming dinoflagellates such as *Peridinium* (Sako et al. 1984, 1985, 1987). However, it is difficult to maintain stable long-term growth of *Peridinium* in this medium. MW Medium provides much better short-term growth with a high yield when compared with Carefoot's Medium. **Into 900 mL** $dH_2O$, dissolve the glycylglycine and calcium nitrate, then add the remaining ingredients. Bring final volume to 1 liter with $dH_2O$ and adjust the pH to 7.2 before autoclaving.

| Component | Stock Solution $(g \cdot L^{-1}\ dH_2O)$ | Quantity Used | Concentration in Final Medium (M) |
|---|---|---|---|
| Glycylglycine | — | 0.1 g | $7.57 \times 10^{-4}$ |
| $Ca(NO_3)_2 \cdot 4H_2O$ | — | 0.1 g | $4.23 \times 10^{-4}$ |
| Urea | 8.50 | 1 mL | $1.42 \times 10^{-4}$ |
| $CaCl_2$ | 14.00 | 1 mL | $1.26 \times 10^{-4}$ |
| $CaCO_3$ | 10.00 | 1 mL | $1.00 \times 10^{-4}$ |
| $KHCO_3$ | 9.00 | 1 mL | $8.99 \times 10^{-5}$ |
| $MgSO_4 \cdot 7H_2O$ | 15.00 | 1 mL | $6.09 \times 10^{-5}$ |
| $KNO_3$ | 10.00 | 1 mL | $9.89 \times 10^{-5}$ |
| $NaNO_3$ | 1.70 | 1 mL | $2.00 \times 10^{-5}$ |
| $Na_2$ β-glycerophosphate $\cdot 5H_2O$ | 20.00 | 1 mL | $6.53 \times 10^{-5}$ |
| $NH_4NO_3$ | 0.42 | 1 mL | $5.25 \times 10^{-6}$ |
| Trace metals solution | (See following recipe) | 500 μL | — |
| Vitamins solution | (See following recipe) | 1 mL | — |

*Modified PIV Trace Metals Solution*

(Provasoli and Pintner 1960, Watanabe et al. 2000)

PIV used HOEDTA (hydroxyethyl ethylenediamine triacetic acid), but we have substituted $Na_2EDTA$ (= 13.98 g HOEDTA). **Into 950 mL** of $dH_2O$, dissolve the EDTA and then individually dissolve the metals. Bring to 1 liter with $dH_2O$.

| Component | 1° Stock Solution ($g \cdot L^{-1}$ $dH_2O$) | Quantity Used | Concentration in Final Medium (M) |
|---|---|---|---|
| $Na_2EDTA$ | — | 15.120 g | $1.34 \times 10^{-6}$ |
| $FeCl_3 \cdot 6H_2O$ | — | 1.936 g | $3.59 \times 10^{-7}$ |
| $MnCl_2 \cdot 4H_2O$ | — | 0.370 g | $9.10 \times 10^{-8}$ |
| $ZnCl_2$ | — | 0.104 g | $3.83 \times 10^{-8}$ |
| $Na_2MoO_4 \cdot 2H_2O$ | — | 0.126 g | $2.61 \times 10^{-8}$ |
| $CoCl_2 \cdot 6H_2O$ | 36.39 | 1 mL | $8.49 \times 10^{-9}$ |

*Vitamins Solution*

**Into 950 mL** of $dH_2O$, first dissolve the thiamine $\cdot$ HCl and then add 1 mL of each primary stock. Bring to 1 liter with $dH_2O$, filter-sterilize, and store frozen.

| Component | 1° Stock Solution ($g \cdot L^{-1}$ $dH_2O$) | Quantity Used | Concentration in Final Medium (M) |
|---|---|---|---|
| Thiamine $\cdot$ HCl (vitamin $B_1$) | — | 20 mg | $5.93 \times 10^{-8}$ |
| Biotin (vitamin H) | 0.2 | 1 mL | $8.19 \times 10^{-10}$ |
| Cyanocobalamin (vitamin $B_{12}$) | 0.2 | 1 mL | $1.48 \times 10^{-10}$ |

## N-HS-Ca Medium

(Schlegel et al. 2000)

This was modified from N-HS Medium (Hepperle and Krienitz 1997) and was used to culture and investigate *Phacotus lenticularis* (Ehr.) Diesing, particularly the production of its calcified lorica. The medium contains extremely high calcium and magnesium concentrations compared with other common culture media. The original N-HS medium was based on the natural conditions of Lake Haussee and Lake Stechlin, Germany, and that first formulation was also used to study calcification in *P. lenticularis* (Hepperle and Krienitz 1997). The N-HS-Ca medium provides several advantages over N-HS medium. In contrast to N-HS medium, N-HS-Ca medium is easily prepared and can be stored in the refrigerator for at least 9 months without detectable alterations. Compared with the N-HS medium, N-HS-Ca medium contains a 100-fold higher concentration of free carbonic acid, which contributes to the stabilization of pH (9.0) for the medium. Nutrient limitation is minimized with N-HS-Ca medium because it has relatively high amounts of nitrogen and phosphorus. N-HS-Ca medium may be useful for investigating the behavior of other organisms tolerating highly alkaline conditions. Note: sodium metasilicate pentahydrate is used, but this is a high-pH medium that enhances dissolution. **Into 950 mL** $dH_2O$, add the calcium sulfate and magnesium carbonate and stir overnight. Dissolve the sodium bicarbonate and then add 1 mL of each stock solution. Filter-sterilize.

| Component | Stock Solution $(g \cdot L^{-1}\ dH_2O)$ | Quantity Used | Concentration in Final Medium (M) |
|---|---|---|---|
| $CaSO_4 \cdot 2H_2O$ | — | 137.80 mg | $8.00 \times 10^{-4}$ |
| $(MgCO_3)_4Mg(OH)_2 \cdot 5H_2O$ | — | 38.86 mg | $8.00 \times 10^{-5}$ |
| $NaHCO_3$ | — | 360.40 mg | $4.29 \times 10^{-3}$ |
| $Ca(NO_3)_2 \cdot 4H_2O$ | 177.10 | 1 mL | $7.50 \times 10^{-4}$ |
| $K_2HPO_4$ | 8.70 | 1 mL | $5.00 \times 10^{-5}$ |
| $CaCl_2$ | 47.80 | 1 mL | $4.3 \times 10^{-4}$ |
| $Na_2SiO_3 \cdot 5H_2O$ | 21.21 | 1 mL | $1.00 \times 10^{-4}$ |
| Trace metals solution | (See following recipe) | 1 mL | — |
| Vitamins stock solution | (See following recipe) | 1 mL | — |

*Trace Metals Solution*

This is a very dilute trace metals solution. The copper sulfate primary stock is 1,000 times more concentrated than the other stocks; use 1 μL. **Into 950 mL** of $dH_2O$, dissolve the EDTA and then individually dissolve the metals. Bring to 1 liter with $dH_2O$.

| Component | 1° Stock Solution ($g \cdot L^{-1}\ dH_2O$) | Quantity Used | Concentration in Final Medium (M) |
|---|---|---|---|
| EDTA | 4 | 1 mL | $1.37 \times 10^{-8}$ |
| $FeSO_4 \cdot 7H_2O$ | 3.5 | 1 mL | $1.26 \times 10^{-8}$ |
| $H_3BO_3$ | 0.05 | 1 mL | $8.09 \times 10^{-10}$ |
| $MgSO_4 \cdot 7H_2O$ | 0.01 | 1 mL | $4.06 \times 10^{-11}$ |
| $ZnSO_4 \cdot 7H_2O$ | 0.05 | 1 mL | $1.74 \times 10^{-11}$ |
| $Co(NO_3)_2 \cdot 6H_2O$ | 0.05 | 1 mL | $1.72 \times 10^{-11}$ |
| $Na_2MoO_4 \cdot 2H_2O$ | 0.05 | 1 mL | $2.07 \times 10^{-11}$ |
| $CuSO_4 \cdot 5H_2O$ | 0.25 | 1 μL | $1.00 \times 10^{-13}$ |

*Vitamin Stock Solution*

**Into 950 mL** of $dH_2O$, first dissolve the thiamine · HCl and then add 1 mL of each primary stock. Bring to 1 liter with $dH_2O$. Filter-sterilize and store frozen.

| Component | 1° Stock Solution ($g \cdot L^{-1}\ dH_2O$) | Quantity Used | Concentration in Final Medium (M) |
|---|---|---|---|
| Thiamine · HCl (vitamin $B_1$) | — | 100 mg | $2.96 \times 10^{-7}$ |
| Biotin (vitamin H) | 1.0 | 1 mL | $4.09 \times 10^{-9}$ |
| Cyanocobalamin (vitamin $B_{12}$) | 0.2 | 1 mL | $1.48 \times 10^{-10}$ |
| Nicotinic acid (niacin) | 0.1 | 1 mL | $8.19 \times 10^{-10}$ |

## *Spirulina* Medium, Modified

### (Aiba and Ogawa 1977, Schlösser 1994)

This medium was slightly modified from *Spirulina platensis* Medium of Ogawa and Terui (1970) to prevent the formation of precipitation (Aiba and Ogawa 1977, Schlösser 1994). Schlösser (1994) recommends making two trace metals solutions: (1) 0.4 g EDTA and 0.7 g $FeSO_4 \cdot 7H_2O$ in 100 mL $dH_2O$, and (2) 0.4 g EDTA and remaining elements in 900 mL $dH_2O$. The two solutions should be autoclaved separately and combined aseptically when cool. **Prepare Solutions I and II**; Solution II includes 1 mL of the trace metals stock solution but not the vitamin stock solution. Autoclave Solutions I and II separately and cool; aseptically combine the two solutions. Aseptically add 1 mL of the cyanocobalamin ($B_{12}$) solution.

| Component ($g \cdot L^{-1}$ $dH_2O$) | Stock Solution | Quantity Used | Concentration in Final Medium (M) |
|---|---|---|---|
| Solution I | 500 mL | — | — |
| $NaHCO_3$ | — | 13.61 g | $1.62 \times 10^{-1}$ |
| $Na_2CO_3$ | — | 4.03 g | $3.80 \times 10^{-2}$ |
| $K_2HPO_4$ | — | 0.50 g | $2.87 \times 10^{-3}$ |
| Solution II | 500 mL | — | — |
| $NaNO_3$ | — | 2.5 g | $2.94 \times 10^{-2}$ |
| $K_2SO_4$ | — | 1.0 g | $5.74 \times 10^{-3}$ |
| NaCl | — | 1.0 g | $1.71 \times 10^{-2}$ |
| $MgSO_4 \cdot 7H_2O$ | — | 0.2 g | $8.11 \times 10^{-4}$ |
| $CaCl_2 \cdot 2H_2O$ | — | 0.04 g | $2.72 \times 10^{-4}$ |
| $FeSO_4 \cdot 7H_2O$ | — | 0.01 g | $3.60 \times 10^{-5}$ |
| $Na_2EDTA \cdot 2H_2O$ | — | 0.08 g | $2.15 \times 10^{-4}$ |
| Trace metals solution | (See following recipe) | 1 mL | — |
| Vitamins solution | (See following recipe) | 1 mL | — |

*Trace Metals Solution*

**Into 900 mL** of $dH_2O$, dissolve the EDTA and then dissolve each remaining component; bring the final volume to 1 liter.

| Component | $1°$ Stock Solution $(g \cdot L^{-1}\ dH_2O)$ | Quantity Used Medium (M) | Concentration in Final |
|---|---|---|---|
| $Na_2EDTA \cdot 2H_2O$ | — | 0.8 g | $2.15 \times 10^{-6}$ |
| $FeSO_4 \cdot 7H_2O$ | — | 0.7 g | $2.52 \times 10^{-6}$ |
| $ZnSO_4 \cdot 7H_2O$ | 1.0 | 1 mL | $3.48 \times 10^{-9}$ |
| $MnSO_4 \cdot 7H_2O$ | 2.0 | 1 mL | $8.97 \times 10^{-9}$ |
| $H_3BO_3$ | 10.0 | 1 mL | $1.62 \times 10^{-7}$ |
| $Co(NO_3)_2 \cdot 6H_2O$ | 1.0 | 1 mL | $3.44 \times 10^{-9}$ |
| $Na_2MoO_4 \cdot 2H_2O$ | 1.0 | 1 mL | $4.13 \times 10^{-9}$ |
| $CuSO_4 \cdot 5H_2O$ | 0.005 | 1 mL | $2.00 \times 10^{-11}$ |

*Cyanocobalamin Stock Solution*

Dissolve the cyanocobalamin in 1 liter of $dH_2O$. Filter-sterilize and store frozen.

| Component | $1°$ Stock Solution $(g \cdot L^{-1}\ dH_2O)$ | Quantity Used | Concentration in Final Medium (M) |
|---|---|---|---|
| Cyanocobalamin (vitamin $B_{12}$) | — | 5 mg | $3.69 \times 10^{-9}$ |

## URO Medium

### (Nakahara 1981)

URO (*Uroglena*) medium was developed by Nakahara (1981) on the basis of concentrations of nutrients in Lake Biwa, Japan, during the seasonal blooms of *Uroglena* (= *Uroglenopsis*). The recipe presented here has been modified slightly by Watanabe et al. (2000). Since this medium contains very low concentrations of nutrients, it is suitable for investigating the morphology and life cycle of *Uroglena*. It is also available for maintenance culturing of red-tide-bloom-forming *Peridinium* species, although it provides only a low yield. Kimura and Ishida (1986) studied phagotrophy in *Uroglena*, and Kurata (1986) investigated the bacterial production of vitamins that are necessary for growth, correlating vitamin availability and *Uroglena* blooms. **Into 950 mL** $dH_2O$, add the stock solution quantities, and bring final volume to 1 liter. Adjust pH to 7.5 and then autoclave.

| Component | Stock Solution $(g \cdot L^{-1} \, dH_2O)$ | Quantity Used | Concentration in Final Medium (M) |
|---|---|---|---|
| $NH_4NO_3$ | 5.0 | I mL | $6.25 \times 10^{-5}$ |
| $Na_2$ β-glycerophosphate $\cdot$ $5H_2O$ | 4.0 | I mL | $1.31 \times 10^{-5}$ |
| $MgSO_4 \cdot 7H_2O$ | 10.0 | I mL | $4.06 \times 10^{-5}$ |
| KCl | 1.0 | I mL | $1.34 \times 10^{-5}$ |
| $CaCl_2$ | 10.0 | I mL | $9.01 \times 10^{-5}$ |
| Fe-Na-EDTA $3H_2O$ | 0.5 | I mL | $1.19 \times 10^{-6}$ |
| Trace metals solution | (See following recipe) | I mL | — |
| Vitamins solution | (See following recipe) | I mL | — |

*Modified PIV Trace Metals Solution*

(Provasoli and Pintner 1960, Watanabe et al. 2000)

PIV used HOEDTA (hydroxyethyl ethylenediamine triacetic acid), but we have substituted $Na_2EDTA$ (= 13.98 g HOEDTA). **Into 950 mL** of $dH_2O$, dissolve the EDTA and then individually dissolve the metals. Bring to 1 liter with $dH_2O$.

| Component | 1° Stock Solution $(g \cdot L^{-1}\ dH_2O)$ | Quantity Used | Concentration in Final Medium (M) |
|---|---|---|---|
| $Na_2EDTA$ | — | 1 g | $2.69 \times 10^{-6}$ |
| $FeCl_3 \cdot 6H_2O$ | — | 0.194 g | $7.18 \times 10^{-7}$ |
| $MnCl_2 \cdot 4H_2O$ | 36 | 1 mL | $1.82 \times 10^{-7}$ |
| $ZnCl_2$ | 10.44 | 1 mL | $7.66 \times 10^{-8}$ |
| $Na_2MoO_4 \cdot 2H_2O$ | 12.62 | 1 mL | $5.22 \times 10^{-8}$ |
| $CoCl_2 \cdot 6H_2O$ | 4.04 | 1 mL | $1.70 \times 10^{-8}$ |

*Vitamins Solution*

**Into 950 mL** of $dH_2O$, dissolve the thiamine $\cdot$ HCl and then add 1 mL of each primary stock. Bring to 1 liter with $dH_2O$. Filter-sterilize and store frozen.

| Component | 1° Stock Solution $(g \cdot L^{-1}\ dH_2O)$ | Quantity Used | Concentration in Final Medium (M) |
|---|---|---|---|
| Thiamine $\cdot$ HCl (vitamin $B_1$) | — | 10 mg | $2.96 \times 10^{-8}$ |
| Biotin (vitamin H) | 0.1 | 1 mL | $4.09 \times 10^{-10}$ |
| Cyanocobalamin (vitamin $B_{12}$) | 0.1 | 1 mL | $7.38 \times 10^{-11}$ |

### *Volvox* Medium

(Provasoli and Pintner 1960)

*Volvox* medium was designed for culturing *Volvox globator* L. emend. Ehr. and *V. tertius* Meyer. When formulated in 1960, *Volvox* medium was a major advance because glycylglycine was first used as a pH buffer. According to Provasoli and Pintner (1960), when Tris or TEA buffer was used in the alkaline range, they created toxic effects for the two *Volvox* species. The minor variations in pH from 7.0 to 7.6 with buffering by TEA were immediately reflected in a change in availability of trace metals, particularly Fe. However, when glycylglycine was used as a pH buffer in place of TEA, minor pH changes no longer affected metal availability. Thus, *Volvox* showed good growth with the same metal mixture (PIV metals) when buffered with glycylglycine. *Volvox* medium is suitable for most axenic strains of *Volvox* and some strains of *Eudorina*, *Pandorina*, and *Gonium*. However, it is not suitable for xenic cultures because many bacteria metabolize glycylglycine. *Volvox* medium is also suitable for culturing axenic freshwater *Cryptomonas* species (Watanabe et al. 2000). There are many modifications of *Volvox* medium, and we list some below. Provasoli and Pintner (1960) gave a range of trace metals additions (3 to 40 mL per liter of medium), and we provide a recipe that uses 3 mL. Several publications have varied the trace metals solution (e.g., Starr and Zeikus 1993, Watanabe et al. 2000); see C Medium for an alternative trace metals solution. **Into 900 mL** dH$_2$O, dissolve the calcium nitrate and glycylglycine, add the other stock solution quantities, and after thorough mixing, add the vitamins. Autoclave.

| Component | Stock Solution (g · L$^{-1}$ dH$_2$O) | Quantity Used | Concentration in Final Medium (M) |
|---|---|---|---|
| Glycylglycine | — | 500.0 mg | $2.46 \times 10^{-3}$ |
| Ca(NO$_3$)$_2$ · 4H$_2$O | — | 117.8 mg | $5.00 \times 10^{-4}$ |
| Na$_2$ β-glycerophosphate · 5H$_2$O | 50.0 | 1 mL | $1.63 \times 10^{-4}$ |
| MgSO$_4$ · 7H$_2$O | 40.0 | 1 mL | $1.62 \times 10^{-4}$ |
| KCl | 50.0 | 1 mL | $6.71 \times 10^{-4}$ |
| PIV trace metals solution | (See following recipe) | 3 mL | — |
| Biotin (vitamin H) | 0.0001 | 1 mL | $4.09 \times 10^{-10}$ |
| Cyanocobalamin (vitamin B$_{12}$) | 0.0001 | 1 mL | $7.38 \times 10^{-11}$ |

*PIV Trace Metals Solution*

(Provasoli and Pintner 1960)

The original recipe used HOEDTA (hydroxyethyl ethylenediamine triacetic acid), which is difficult to obtain; to use $Na_2EDTA \cdot 2H_2O$, substitute 15.12 g. Prepare the primary stock solution. **Into 950 mL** of $dH_2O$, dissolve the EDTA and then individually add and dissolve the metals. Bring to 1 liter with $dH_2O$.

| Component | 1° Stock Solution ($g \cdot L^{-1}$ $dH_2O$) | Quantity Used | Concentration in Final Medium (M) |
|---|---|---|---|
| HOEDTA | — | 1.398 g | $1.22 \times 10^{-5}$ |
| $FeCl_3 \cdot 6H_2O$ | — | 0.194 g | $2.15 \times 10^{-6}$ |
| $MnCl_2 \cdot 4H_2O$ | — | 0.036 g | $5.46 \times 10^{-7}$ |
| $ZnCl_2$ | — | 0.104 g | $2.29 \times 10^{-7}$ |
| $Na_2MoO_4 \cdot 2H_2O$ | — | 0.013 g | $1.56 \times 10^{-7}$ |
| $CoCl_2 \cdot 6H_2O$ | 4.04 | 1 mL | $5.09 \times 10^{-8}$ |

**HEPES-*Volvox* medium:** 0.94 g HEPES replaces glycylglycine; the pH is adjusted to 8.2 with 1 N NaOH. It is suitable for both axenic and xenic cultures, especially *Euglena*, *Pithophora*, and *Glaucosphaera vacuolata* Korsh. (McCracken et al. 1980).

**S *Volvox* medium (SVM):** *Volvox* medium + 20 mg $\cdot L^{-1}$ $Na_2CO_3$ + 30 mg $\cdot L^{-1}$ urea; either glycylglycine or 600 mg $\cdot L^{-1}$ HEPES can be used as buffer. It is especially suitable for growing *Volvox carteri* Stein (Kirk and Kirk 1983, Kobl et al. 1998).

***Volvox* TAC medium:** *Volvox* medium + 200 mg $\cdot L^{-1}$ sodium acetate. It has been a successful medium for culturing *Astrephomene gubernaculifera* Pocock, *A. perforata* Nozaki, *Gonium pectorale* Müller, *Yamagishiella unicocca* (Raybarn et Starr) Nozaki, and *Volvulina steinii* Playfair (Watanabe et al. 2000).

## MES *Volvox* Medium

### (Starr and Zeikus 1993)

MES *Volvox* Medium, described by Starr and Zeikus (1993), has perhaps the broadest applicability of the many modifications of the original *Volvox* Medium (Provasoli and Pintner 1960). MES buffer is used in place of glycylglycine, and ammonium chloride is added as a second nitrogen source. Starr and Zeikus (1993) refer to their trace metal solution as PIV trace metals solution, but it differs slightly from the original PIV trace metals (see *Volvox* Medium) (Provasoli and Pintner 1960). MES *Volvox* Medium grows a wide range of freshwater algae (Starr and Zeikus 1993). **Into 900 mL** dH$_2$O, dissolve the MES and calcium nitrate, and then add the other components. Adjust pH to 6.7 with NaOH. Autoclave.

| Component | Stock Solution ($g \cdot L^{-1}$ dH$_2$O) | Quantity Used | Concentration in Final Medium (M) |
|---|---|---|---|
| MES | — | 1.95 g | $1.00 \times 10^{-2}$ |
| Ca(NO$_3$)$_2$ · 4H$_2$O | — | 117.8 mg | $5.00 \times 10^{-4}$ |
| Na$_2$ β-glycerophosphate · 5H$_2$O | 60.0 | 1 mL | $1.96 \times 10^{-4}$ |
| MgSO$_4$ · 7H$_2$O | 40.0 | 1 mL | $1.62 \times 10^{-4}$ |
| KCl | 50.0 | 1 mL | $6.71 \times 10^{-4}$ |
| NH$_4$Cl | 26.7 | 1 mL | $5.00 \times 10^{-4}$ |
| Trace metals solution | (See following recipe) | 6 mL | — |
| Biotin (vitamin H) | 0.0025 | 1 mL | $1.02 \times 10^{-8}$ |
| Cyanocobalamin (vitamin B$_{12}$) | 0.0015 | 1 mL | $1.11 \times 10^{-9}$ |

*Trace Metals Solution*

**Into 950 mL** of dH$_2$O, dissolve the EDTA and then individually add and dissolve the metals. Bring to 1 liter with dH$_2$O.

| Component | 1° Stock Solution ($g \cdot L^{-1}$ dH$_2$O) | Quantity Used | Concentration in Final Medium (M) |
|---|---|---|---|
| Na$_2$EDTA | — | 750 mg | $1.54 \times 10^{-5}$ |
| FeCl$_3$ · 6H$_2$O | — | 97 mg | $2.15 \times 10^{-6}$ |
| MnCl$_2$ · 4H$_2$O | — | 41 mg | $1.24 \times 10^{-6}$ |
| ZnCl$_2$ | 5.0 | 1 mL | $2.20 \times 10^{-7}$ |
| Na$_2$MoO$_4$ · 2H$_2$O | 4.0 | 1 mL | $9.92 \times 10^{-8}$ |
| CoCl$_2$ · 6H$_2$O | 2.0 | 1 mL | $5.04 \times 10^{-8}$ |

## WC Medium

### (Guillard and Lorenzen 1972)

This was derived from Chu #10 and Wright's (1964) Medium, which Wright used to grow cryptophytes (WC = Wright's cryptophyte) (R. R. L. Guillard, personal communication). Wright used ferric citrate, citrate, and ammonium acetate as well as additional trace elements (Al, Br, I, Li, Ni, Sn). WC Medium is slightly alkaline (pH, 7.8). Glycylglycine is used as a buffer for axenic cultures, and Tris is used as a buffer for xenic cultures. Note the following: (1) nitrate and phosphate should be reduced twofold to 10-fold for sensitive organisms; (2) 2.65 to 26.5 mg · $L^{-1}$ $NH_4Cl$ should be used for organisms that cannot utilize nitrate; (3) the more complete trace metal solution using ferric citrate–citric acid described by Wright (1964) can also be used; and (4) ferric citrate–citric acid (3–9 mg · $L^{-1}$ each) can be used in place of ferric EDTA. WC medium with no added buffer was used to culture the xanthophytes *Tribonema aequale* Pascher, *Botrydium becherianum* Vischer, *Vaucheria sessilis* (Vaucher) De Candolle, and *Ophiocytium maius* Nägeli, the eustigmatophyte *Eustigmatos magna* Hibberd, the raphidophytes *Gonyostmum semen* Diesing and *Vacuolaria virescens* Cienkowski, and synurophytes *Synura* species (Guillard and Lorenzen 1972). It was noted that *Gonyostomum* grew best when ca. 2/3 distilled water and 1/3 filtered bog water were used and that *Synura* and *Vaucheria* grew better with 1% soil extract added. The medium also grows chlorococcalean and volvocalean algae (*Ankistrodesmus, Chlorella, Chlamydomonas*; Ahn et al. 2002), desmids (*Spondylosium, Cosmarium, Staurastrum*; Freire-Nordi et al. 1998, Healey and Hendzel 1988), diatoms (*Asterionella, Cyclotella, Sellaphora, Stephanodiscus*; Tilman & Kilham 1976, Donk and Kilham 1990, Mann et al. 1999), the cyanobacterium *Microcystis* (Olsen 1989, Olsen et al. 1989), and the cryptophyte *Cryptomonas* (Klaveness 1982). In order to culture cryptomonads such as *Cryptomonas rostratiformis* Skuja, *C. phaseolus* Skuja, *C. undulata* Gervais, *C. ovata* Ehr., and *Chroomonas* sp., WC medium was modified by Gervais (1997) as follows: no boron was added, and the nitrogen source was ammonium chloride instead of sodium nitrate. Also, because the growth of *C. rostratiformis* was suppressed by Tris, a phosphate buffer (0.008 mM $KH_2PO_4$ + 0.042 mM $Na_2HPO_4$) was substituted. **Into 900 mL** of $dH_2O$, dissolve one of the two possible buffers. Individually, add the remaining components and bring the final volume to 1 liter with $dH_2O$. Adjust the pH to 7.6–8.0 and autoclave.

| Component | Stock Solution ($g · L^{-1}$ $dH_2O$) | Quantity Used | Concentration in Final Medium (M) |
|---|---|---|---|
| *Buffer* (use one, not both) | | | |
| Glycylglycine | — | 500 mg | $3.78 \times 10^{-3}$ |
| Tris | — | 500 mg | $4.13 \times 10^{-3}$ |
| $NaNO_3$ | 85.01 | 1 mL | $1.00 \times 10^{-3}$ |
| $CaCl_2 · 2H_2O$ | 36.76 | 1 mL | $2.50 \times 10^{-4}$ |
| $MgSO_4 · 7H_2O$ | 36.97 | 1 mL | $1.50 \times 10^{-4}$ |
| $NaHCO_3$ | 12.60 | 1 mL | $1.50 \times 10^{-4}$ |
| $Na_2SiO_3 · 9H_2O$ | 28.42 | 1 mL | $1.00 \times 10^{-4}$ |
| $K_2HPO_4$ | 8.71 | 1 mL | $5.00 \times 10^{-5}$ |
| Trace metals solution | (See following recipe) | 1 mL | — |
| Vitamins solution | (See following recipe) | 1 mL | — |

*Trace Metals Solution*

**Into 950 mL** of dH$_2$O, add components and bring final volume to 1 liter with dH$_2$O. Autoclave.

| Component | 1° Stock Solution (g · L$^{-1}$ dH$_2$O) | Quantity Used | Concentration in Final Medium (M) |
|---|---|---|---|
| Na$_2$EDTA · 2H$_2$O | — | 4.36 g | 1.17 × 10$^{-5}$ |
| FeCl$_3$ · 6H$_2$O | — | 3.15 g | 1.17 × 10$^{-5}$ |
| CuSO$_4$ · 5H$_2$O | 10.0 | 1 mL | 4.01 × 10$^{-8}$ |
| ZnSO$_4$ · 7H$_2$O | 22.0 | 1 mL | 7.65 × 10$^{-8}$ |
| CoCl$_2$ · 6H$_2$O | 10.0 | 1 mL | 4.20 × 10$^{-8}$ |
| MnCl$_2$ · 4H$_2$O | 180.0 | 1 mL | 9.10 × 10$^{-7}$ |
| Na$_2$MoO$_4$ · 2H$_2$O | 6.0 | 1 mL | 2.48 × 10$^{-8}$ |
| H$_3$BO$_3$ | — | 1.00 g | 1.62 × 10$^{-5}$ |

*Vitamins Solution*

**Into 950 mL** of dH$_2$O, dissolve the thiamine · HCl, add 1 mL of the primary stocks, and bring final volume to 1 liter with dH$_2$O. Filter-sterilize and store in refrigerator or freezer.

| Component | 1° Stock Solution (g · L$^{-1}$ dH$_2$O) | Quantity Used | Concentration in Final Medium (M) |
|---|---|---|---|
| Thiamine · HCl (vitamin B$_1$) | — | 100 mg | 2.96 × 10$^{-7}$ |
| Biotin (vitamin H) | 0.5 | 1 mL | 2.05 × 10$^{-9}$ |
| Cyanocobalamin (vitamin B$_{12}$) | 0.5 | 1 mL | 3.69 × 10$^{-10}$ |

## 1.2. Enriched Natural Freshwater Media

### Alga-Gro Lake-Water Medium

(Carolina Biological Supply Co.)

This is prepared by adding 40 mL of "Alga-Gro" medium (Carolina Biology Supply Co.) to 960 mL lake water. A double-strength Bold's Basal Medium (see previous description) can be substituted. The medium has been successfully used to culture the tetrasporalian green alga *Octosporiella coloradoensis* Kugrens (Kugrens 1984), freshwater cryptomonads *Storeatula rhinosa* Kugrens et al., *Pyrenomonas ovalis* Kugrens et al. (Kugrens et al. 1999), and *Cryptomonas* and *Chroomonas* species (Kugrens et al. 1987)], and it may prove successful for a variety of other freshwater algae.

## VS Medium

(Gargiulo et al. 2001)

This medium was developed for culturing and investigating the life cycle of the freshwater red alga *Bangia atropur-purea* (Roth) C. Agardh. The medium uses natural river water as its base. Four different concentrations were used by Gargiulo et al. (2001), and these are referred to as VS, VS 2.5, VS 5, and VS 10 (1, 2.5, 5, and 10× concentration, respectively). We have provided a recipe for the basic VS medium, but with simple changes in stock solution additions (e.g., 2.5 mL for VS 2.5 medium), the other media can easily be prepared. **Into 993 mL** of sterilized freshwater (from a clean lake or stream), add the following components. The two vitamins could be prepared as primary stocks. Autoclave.

| Component | Stock Solution $(g \cdot L^{-1} \, dH_2O)$ | Quantity Used | Concentration in Final Medium (M) |
|---|---|---|---|
| $NaNO_3$ | 42.52 | 1 mL | $5.00 \times 10^{-4}$ |
| $Na_2$ β-glycerophosphate $\cdot$ $5H_2O$ | 5.36 | 1 mL | $1.75 \times 10^{-5}$ |
| $Na_2EDTA \cdot 2H_2O$ | 3.72 | 1 mL | $1.00 \times 10^{-5}$ |
| $FeSO_4 \cdot 7H_2O$ | 0.278 | 1 mL | $1.00 \times 10^{-6}$ |
| $MnCl_2 \cdot 4H_2O$ | 1.98 | 1 mL | $1.00 \times 10^{-5}$ |
| Biotin (vitamin H) | 0.10 | 1 mL | $4.09 \times 10^{-10}$ |
| Cyanocobalamin (vitamin $B_{12}$) | 0.0004 | 1 mL | $2.95 \times 10^{-10}$ |

## 1.3. Soil-Water Enriched Freshwater Media

### *Audouinella* Medium

(Glazer et al. 1997)

*Audouinella* medium was developed for culture of the freshwater red alga *Audouinella*. It uses peat extract, natural seawater, and an enrichment solution. The authors, referring to Starr and Zeikus (1993), indicate that the enrichment solution is Provasoli's (1968) ES enrichment solution. However, the history of ES medium is complicated (see ES medium below), and Starr and Zeikus reference the wrong original source (e.g., Provasoli 1963). Furthermore, the trace metals solution is claimed to be Provasoli's PII metals solution (see Starr and Zeikus 1993), but sulfated manganese, zinc, and cobalt are substituted for Provasoli's chlorinated metals, and the iron and boron concentrations are different (see Provasoli 1958, 1963). Glazer et al. (1997) delete the Tris buffer that was used by both Provasoli (1968) and Starr and Zeikus (1993), which seems sensible to us. **Into 950 mL** $dH_2O$, add the peat extract, the natural seawater, and the enrichment solution. Pasteurize and store at 4°C. Final pH should be 5.5, with conductance of 500–600 $\mu S \cdot cm^{-1}$.

| Component | Stock Solution ($g \cdot L^{-1}$ $dH_2O$) | Quantity Used | Concentration in Final Medium (M) |
|---|---|---|---|
| Dry peat moss extract | — | 30 mL | — |
| Natural seawater (30 psu) | — | 10 mL | — |
| Enrichment solution | — | 10 mL | — |

### *Enrichment Solution*

(Starr and Zeikus 1993)

Into 400 mL of $dH_2O$, dissolve the nitrate and phosphate quantities; add 250 mL each of the Fe solution and trace metal solution; and then finally dissolve the vitamins. Bring final volume to 1 liter with $dH_2O$. Autoclave and cool before using.

| Component | Stock Solution ($g \cdot L^{-1}$ $dH_2O$) | Quantity Used | Concentration in Final Medium (M) |
|---|---|---|---|
| $NaNO_3$ | — | 3.50 g | $4.12 \times 10^{-4}$ |
| $Na_2$ β-glycerophosphate $\cdot$ $5H_2O$ | — | 0.50 g | $1.63 \times 10^{-5}$ |
| Fe solution | (see below) | 250 mL | — |
| Trace metals solution | (see below) | 250 mL | — |
| Thiamine $\cdot$ HCl (vitamin $B_1$) | — | 5 mg | $1.48 \times 10^{-7}$ |
| Biotin (vitamin H) | 0.05 | 1 mL | $2.05 \times 10^{-9}$ |
| Cyanocobalamin (vitamin $B_{12}$) | 0.1 | 1 mL | $7.38 \times 10^{-10}$ |

*Fe Solution*

(Starr and Zeikus 1993)

Dissolve EDTA in 900 mL of dH$_2$O, dissolve iron sulfate and then bring to 1 liter with dH$_2$O.

| Component | 1° Stock Solution (g · L$^{-1}$ dH$_2$O) | Quantity Used | Concentration in Final Medium (M) |
|---|---|---|---|
| Na$_2$EDTA anhydrous | — | 0.6 g | $5.13 \times 10^{-6}$ |
| Fe(NH$_4$)$_2$(SO$_4$)$_2$ · 6H$_2$O | — | 0.7 g | $4.46 \times 10^{-6}$ |

*Trace Metals Solution*

(Starr and Zeikus 1993)

**Into 950 mL** dH$_2$O, dissolve the EDTA and then add the quantities for each of the other components. Bring the final volume to 1 liter.

| Component | 1° Stock Solution (g · L$^{-1}$ dH$_2$O) | Quantity Used | Concentration in Final Medium (M) |
|---|---|---|---|
| Na$_2$EDTA · 2H$_2$O | — | 1.000 g | $6.72 \times 10^{-6}$ |
| H$_3$BO$_3$ | — | 1.140 g | $4.61 \times 10^{-5}$ |
| FeCl$_3$ · 6H$_2$O | — | 0.049 g | $4.53 \times 10^{-7}$ |
| MnSO$_4$ · 4H$_2$O | — | 0.164 g | $1.84 \times 10^{-6}$ |
| ZnSO$_4$ · 7H$_2$O | 22.0 | 1 mL | $1.91 \times 10^{-7}$ |
| CoSO$_4$ · 7H$_2$O | 4.8 | 1 mL | $4.27 \times 10^{-8}$ |

**Biphasic Soil Water Medium**

This old, but reliable, medium provides good growth and normal morphology for many freshwater algae (e.g., Pringsheim 1946). The medium can be slightly acidic or alkaline, depending upon the soil source. For details regarding the soils and culture medium preparation methods, see Chapter 2.

## Diatom Medium, Modified

### (Cohn and Pickett-Heaps 1988, Cohn et al. 2003)

This was developed for culturing the freshwater diatom *Surirella* and has proved useful for other diatom species as well. It should not be confused with the "Diatom Medium" of Beakes et al. (1988) (see also Tompkins et al. 1995). The medium described here contains soil extract (see Chapter 2 for details) and a high concentration of phosphate. A standard $Na_2SiO_3 \cdot 9H_2O$ solution (e.g., 1 mL of 28.42 g $\cdot$ L$^{-1}$ dH$_2$O stock solution = $1.00 \times 10^{-4}$ final concentration) could be substituted for the commercial silica stock solution (Sigma #S1773, Sigma-Aldrich). Cohn (personal communication) recommends adding $NaHCO_3$ (1 mL of 8.4 g $\cdot$ L$^{-1}$ dH$_2$O stock solution = $1.00 \times 10^{-4}$ final concentration) after tyndallization. This medium has also been used for growing *Craticula cuspidata* (Kütz.) Mann (= *Navicula cuspidata* Kütz. in Cohn et al. 1989), *Nitzschia* species, *Pinnularia viridis* (Nitzsch) Ehr., and *Stauroneis* species (Cohn and Disparti 1994, Cohn and Weitzel 1996, Cohn et al. 2003). **Into 900 mL** of dH$_2$O, dissolve all components except the vitamins (and $NaHCO_3$ if added). Bring the final volume to 1 liter, using dH$_2$O. Sterilize the solution by tyndallization (see Chapter 5). Cool and then add the vitamins (and $NaHCO_3$). Adjust pH to 6.75. Filter-sterilize.

| Component | Stock Solution (g $\cdot$ L$^{-1}$ dH$_2$O) | Quantity Used | Concentration in Final Medium (M) |
|---|---|---|---|
| $Ca(NO_3)_2 \cdot 4H_2O$ | 70.85 | 1 mL | $3.00 \times 10^{-4}$ |
| $KH_2PO_4$ | 54.44 | 1 mL | $4.00 \times 10^{-4}$ |
| $MgSO_4 \cdot 7H_2O$ | 24.65 | 1 mL | $1.00 \times 10^{-4}$ |
| $Na_2SiO_3$ (27% aq. sat. soln.) | 20 mL, pH 8.5 | 5 mL | ~$3.00 \times 10^{-4}$ |
| $FeSO_4 \cdot 7H_2O$ | 0.278 | 1 mL | $1.00 \times 10^{-6}$ |
| $MnCl_2 \cdot 4H_2O$ | 0.02 | 1 mL | $1.00 \times 10^{-7}$ |
| Soil extract | — | 50 mL | — |
| Vitamins solution | (See following recipe) | 1 mL | — |

*Vitamin Solution*

**Into 950 mL**, dissolve the thiamine $\cdot$ HCl and then add 1 mL of each stock solution. Bring final volume to 1 liter. Filter-sterilize and store frozen.

| Component | 1° Stock Solution (g $\cdot$ L$^{-1}$ dH$_2$O) | Quantity Used | Concentration in Final Medium (M) |
|---|---|---|---|
| Thiamine $\cdot$ HCl (vitamin B$_1$) | — | 1 g | $2.97 \times 10^{-6}$ |
| Biotin (vitamin H) | — | 1 g | $4.09 \times 10^{-6}$ |
| Nicotinic acid (niacin) | — | 1 g | $8.12 \times 10^{-6}$ |
| Cyanocobalamin (vitamin B$_{12}$) | 1 | 1 mL | $7.38 \times 10^{-10}$ |

## 1.4. Freshwater Media Enriched with Organics

### *Polytoma* Medium

(Starr 1964)

*Polytoma* medium was reported by Starr (1964) on the basis of a personal communication from Pringsheim. Pringsheim (1955, 1963) demonstrated that this colorless *Chlamydomonas*-like alga will utilize acetate but not glucose. It is suitable for all axenic cultures of *Polytoma and Astasia* and colorless strains of *Euglena* (Starr and Zeikus 1993). **To prepare**, dissolve 2 g sodium acetate, 1 g yeast extract, and 1 g tryptone in 1 liter of dH₂O. Schlösser (1994) substitutes 2 g of glucose for 2 g of sodium acetate and 30 mL of soil water extract for 30 mL of dH₂O. Dispense into tubes or flasks. Autoclave; store at 4°C until ready for use.

## *Porphyridium* Medium (Modified)

(Sommerfeld and Nichols 1970)

This medium, used to study the morphology and growth characteristics of *Porphyridium aerugineum* Geitler, was modified by Sommerfeld and Nichols (1970). The medium apparently originated with Pringsheim. **To prepare,** dissolve 1 g yeast extract and 1 g tryptone in 900 mL of dH$_2$O. Add 100 mL of soil water extract (see Chapter 2). Autoclave.

## *Prototheca* Medium

### (Wolff and Kück 1990)

This is a solid (agar) medium containing malt that was developed to study the structural analysis of mitochondrial SSU rRNA of *Prototheca wickerhamii* Soneda et Tubaki. It has proved suitable for all *Prototheca* cultures. Schlösser (1994) substitutes 30 mL of soil water extract for 30 mL of dH₂O. See also Anderson (1945) and Patni and Aaronson (1974). **Add 45 g** of malt extract agar to 1 liter of dH₂O (see Chapter 2 for preparation of agar). Alternatively, 30 g malt extract and 15 g agar can be added to 1 liter of dH₂O.

## 1.5. Bacterial-Fungal Test Media for Freshwater Strains

### Bacterial Test Media

(Watanabe et al. 2000)

The following are prepared by adding the organic compounds to 1 liter of algal culture medium (used to grow the alga that is being tested). For oligotrophic bacteria, reduce the organic concentration substantially. If agar plates are required, prepare broth, add agar (10–15 g agar $\cdot$ L$^{-1}$) and heat to "dissolve" (see Chapter 2). Once agar is dissolved, autoclave in a flask and then dispense into petri dishes.

**B-1 Test Medium**: 1 g proteose peptone.
**B-II Test Medium**: 5 g yeast extract.
**B-III Test Medium**: 5 g peptone + 3 g beef extract.
**B-IV Test Medium**: 1 g glucose + 1 g peptone.
**B-V Test Medium**: 0.5 g sodium acetate + 0.5 g glucose + 0.5 g tryptone + 0.3 g yeast extract.
**YT Test Medium**: 1 g yeast extract + 2 g tryptone.

## Freshwater Test Medium

(Andersen et al. 1997)

This is a general purpose test medium to detect the presence of common bacteria and fungi in freshwater cultures. **To prepare** a broth medium, dissolve 5 g peptone and 10 g malt extract in 1 liter of $dH_2O$, dispense into test tubes, and autoclave. If agar plates are required, prepare broth, add agar (e.g., 10 g agar $\cdot L^{-1}$), and heat to "dissolve" (see Chapter 2). Once the agar is dissolved, autoclave in a flask and then dispense into petri plates.

## 2.0. MARINE CULTURE MEDIA

### 2.1. Artificial Seawater Media

#### Aquil* Medium

(See Chapter 4)

Aquil* is a new version of the original Aquil medium (Morel et al. 1979, Price et al. 1989) that is presented in Chapter 4. The medium, in all three versions, is designed for critical trace metal experimental work. To prepare properly, solutions are passed through a Chelex column to remove impurities. Details for column purification are thoroughly described in Chapter 4, and it is imperative that these details are followed when using the medium for critical trace metal experimental work. **To prepare** the synthetic ocean water, use 600 mL of high-quality $dH_2O$ (e.g., Milli-Q water) and individually dissolve each of the anhydrous salts. Next, use 300 mL of high-quality $dH_2O$, and individually dissolve each of the hydrous salts. Combine the two solutions. Add 1 mL of each of the major nutrient stock solutions, add 1 mL of the trace metals solution, and add 1 mL of the vitamins solution. Sterilize in a microwave oven or filter-sterilize to avoid metal contamination from an autoclave (see Chapter 4). Final salinity is 35 psu.

*Synthetic Ocean Water (SOW)*

| Component | Stock Solution $(g \cdot L^{-1} dH_2O)$ | Quantity Used | Concentration in Final Medium (M) |
|---|---|---|---|
| Anhydrous Salts | | | |
| NaCl | — | 24.540 g | $4.20 \times 10^{-1}$ |
| $Na_2SO_4$ | — | 4.090 g | $2.88 \times 10^{-2}$ |
| KCl | — | 0.700 g | $9.39 \times 10^{-3}$ |
| $NaHCO_3$ | — | 0.200 g | $2.38 \times 10^{-3}$ |
| KBr | — | 0.100 g | $8.40 \times 10^{-4}$ |
| $H_3BO_3$ | — | 0.003 g | $4.85 \times 10^{-5}$ |
| NaF | — | 0.003 g | $7.15 \times 10^{-5}$ |
| Hydrous Salts | | | |
| $MgCl_2 \cdot 6H_2O$ | — | 11.100 g | $5.46 \times 10^{-2}$ |
| $CaCl_2 \cdot 2H_2O$ | — | 1.540 g | $1.05 \times 10^{-2}$ |
| $SrCl_2 \cdot 6H_2O$ | — | 0.017 g | $6.38 \times 10^{-5}$ |

*Major Nutrients*

**Into 900 mL** of highest-quality deionized water, add 1 mL of each stock solution and bring final volume to 1 liter. Filter-sterilize or sterilize in microwave oven.

| Component | 1° Stock Solution ($g \cdot L^{-1}$ $dH_2O$) | Quantity Used | Concentration in Final Medium (M) |
|---|---|---|---|
| $NaH_2PO_4 \cdot H_2O$ | 1.38 | 1 mL | $1.00 \times 10^{-5}$ |
| $NaNO_3$ | 85.00 | 1 mL | $1.00 \times 10^{-4}$ |
| $Na_2SiO_3 \cdot 9H_2O$ | 28.40 | 1 mL | $1.00 \times 10^{-4}$ |

*Metal/Metalloid Stock Solution*

**Into 900 mL** of highest-quality deionized water, dissolve the EDTA (not $Na_2$ EDTA) and add 1 mL of each stock solution. Bring the volume up to 1 liter.

| Component | 1° Stock Solution ($g \cdot L^{-1}$ $dH_2O$) | Quantity Used | Concentration in Final Medium (M) |
|---|---|---|---|
| EDTA (anhydrous) | — | 29.200 g | $1.00 \times 10^{-5}$ |
| $FeCl_3 \cdot 6H_2O$ | — | 0.270 g | $1.00 \times 10^{-6}$ |
| $ZnSO_4 \cdot 7H_2O$ | — | 0.230 g | $7.97 \times 10^{-8}$ |
| $MnCl_2 \cdot 4H_2O$ | — | 0.0240 g | $1.21 \times 10^{-7}$ |
| $CoCl_2 \cdot 6H_2O$ | — | 0.0120 g | $5.03 \times 10^{-8}$ |
| $Na_2MoO_4 \cdot 2H_2O$ | — | 0.0242 g | $1.00 \times 10^{-7}$ |
| $CuSO_4 \cdot 5H_2O$ | 4.9 | 1 mL | $1.96 \times 10^{-8}$ |
| $Na_2SeO_3$ | 1.9 | 1 mL | $1.00 \times 10^{-8}$ |

*Vitamin Stock Solution*

**Into 950 mL** of highest-quality deionized water, dissolve the thiamine · HCL and add 1 mL of the stock solutions. Bring the final volume to 1 liter. Filter-sterilize.

| Component | 1° Stock Solution ($g \cdot L^{-1}$ $dH_2O$) | Quantity Used | Concentration in Final Medium (M) |
|---|---|---|---|
| Thiamine · HCl (vitamin $B_1$) | — | 100 mg | $2.97 \times 10^{-7}$ |
| Biotin (vitamin H) | 5.0 | 1 mL | $2.25 \times 10^{-9}$ |
| Cyanocobalamin (vitamin $B_{12}$) | 5.5 | 1 mL | $3.70 \times 10^{-10}$ |

## ASP-2 Medium + NTA

(Provasoli et al. 1957)

This is one of several artificial seawater media described by Provasoli et al. (1957); see original publication for other recipes. This medium is buffered by Tris and nitrilotriacetic acid. Exclude NTA for basic ASP-2 medium. **Into 900 mL** dH$_2$O, dissolve the components and stock solutions; bring final volume to 1 liter with dH$_2$O. Final pH should be 7.6 to 7.8.

| Component | Stock Solution ($g \cdot L^{-1}$ dH$_2$O) | Quantity Used | Concentration in Final Medium (M) |
|---|---|---|---|
| NaCl | — | 18.00 g | $3.08 \times 10^{-1}$ |
| MgSO$_4 \cdot$ 7H$_2$O | — | 5.00 g | $2.03 \times 10^{-2}$ |
| KCl | — | 0.60 g | $8.05 \times 10^{-3}$ |
| CaCl$_2$ | — | 0.10 g | $9.01 \times 10^{-4}$ |
| Tris base | — | 1.00 g | $8.25 \times 10^{-3}$ |
| Nitrilotriacetic acid | — | 0.10 g | $5.23 \times 10^{-4}$ |
| Na$_2$EDTA | — | 0.03 g | $1.03 \times 10^{-4}$ |
| NaNO$_3$ | — | 0.05 g | $5.88 \times 10^{-4}$ |
| H$_3$BO$_3$ | 6.0 | 1 mL | $5.55 \times 10^{-4}$ |
| Na$_2$SiO$_3 \cdot$ 9H$_2$O | 15.0 | 1 mL | $5.28 \times 10^{-5}$ |
| K$_2$HPO$_4$ | 5.0 | 1 mL | $2.87 \times 10^{-5}$ |
| FeCl$_3$ | 0.8 | 1 mL | $1.43 \times 10^{-5}$ |
| ZnCl$_2$ | 0.15 | 1 mL | $2.29 \times 10^{-6}$ |
| MnCl$_2 \cdot$ 4H$_2$O | 1.2 | 1 mL | $2.18 \times 10^{-5}$ |
| CoCl$_2 \cdot$ 6H$_2$O | 0.003 | 1 mL | $5.09 \times 10^{-8}$ |
| CuCl$_2$ | 0.0012 | 1 mL | $1.89 \times 10^{-8}$ |
| Cyanocobalamin (vitamin B$_{12}$) | 0.002 | 1 mL | $1.48 \times 10^{-9}$ |
| S3 vitamins solution | (See following recipe) | 1 mL | — |

*S3 Vitamins Solution*

**Into 950 mL** of highest-quality deionized water, dissolve the components or add 1 mL of the primary stock solutions. Bring the final volume to 1 liter. Filter-sterilize.

| Component | 1° Stock Solution ($g \cdot L^{-1}$ dH$_2$O) | Quantity Used | Concentration in Final Medium (M) |
|---|---|---|---|
| Inositol | — | 500 mg | $2.78 \times 10^{-6}$ |
| Thymine | — | 300 mg | $2.38 \times 10^{-6}$ |
| Thiamine · HCl (vitamin B$_1$) | — | 50 mg | $1.48 \times 10^{-7}$ |
| Nicotinic acid (niacin) | — | 10 mg | $8.12 \times 10^{-8}$ |
| Ca pantothenate | — | 10 mg | $4.20 \times 10^{-8}$ |
| p-Aminobenzoic acid | 1.0 | 1 mL | $7.29 \times 10^{-9}$ |
| Biotin (vitamin H) | 0.1 | 1 mL | $4.09 \times 10^{-10}$ |
| Folic acid | 0.2 | 1 mL | $4.53 \times 10^{-10}$ |

## ASP-M Medium

### (McLachlan 1964, Goldman and McCarthy 1978)

This is an artificial enriched seawater medium that was designed as a general medium for marine macroalgae and microalgae. It is derived from the earlier ASP Medium series (see Provasoli et al. 1957). The TMS-II trace metals solutions are derived from the S1 metals solution of Provasoli and Pintner (1953). The vitamins solution is complex, and some of the vitamins are unnecessary for common algal species. **To prepare**, dissolve the anhydrous salts in 500 mL $dH_2O$ and the hydrous salts in 300 mL $dH_2O$, and then combine the solutions. Dissolve the Tris base and the glycylglycine. Add the indicated quantity of stock solutions and bring the final volume to 1 liter. Autoclave or sterile-filter. The pH should be 7.5 at room temperature.

| Component | Stock Solution $(g \cdot L^{-1} \ dH_2O)$ | Quantity Used | Concentration in Final Medium (M) |
|---|---|---|---|
| **Anhydrous salts** | | | |
| NaCl | — | 23.38 g | $4.0 \times 10^{-1}$ |
| KCl | — | 0.75 g | $1.0 \times 10^{-2}$ |
| $CaCl_2$ | — | 1.120 g | $1.0 \times 10^{-2}$ |
| $NaHCO_3$ | — | 0.168 g | $2.0 \times 10^{-3}$ |
| **Hydrous salts** | | | |
| $MgSO_4 \cdot 7H_2O$ | — | 4.930 g | $2.0 \times 10^{-2}$ |
| $MgCl_2 \cdot 4H_2O$ | — | 4.060 g | $2.0 \times 10^{-2}$ |
| **Macronutrients** | | | |
| $NaNO_3$ | 85.0 | 1 mL | $1.0 \times 10^{-3}$ |
| $NaH_2PO_4 \cdot H_2O$ | 13.8 | 1 mL | $1.0 \times 10^{-4}$ |
| $Na_2SiO_3 \cdot 9H_2O$ | 56.8 | 1 mL | $1.0 \times 10^{-4}$ |
| **Other components** | | | |
| Fe-EDTA | 84.2 | 100 µL | $2.0 \times 10^{-6}$ |
| Tris base | — | 0.606 g | $5.0 \times 10^{-3}$ |
| Glycylglycine | — | 0.660 g | $5.0 \times 10^{-3}$ |
| TSM-I solution | (See following recipe) | 1 mL | — |
| TSM-II solution | (See following recipe) | 1 mL | — |
| S3 vitamin solution | (See following recipe) | 1 mL | — |

*Trace Metal Solution: TMS-I*

(McLachlin 1964)

**Into 900 mL** of $dH_2O$, first dissolve the EDTA and then individually dissolve the metals. Bring the final volume to 1 liter.

| Component | 1° Stock Solution ($g \cdot L^{-1} dH_2O$) | Quantity Used | Concentration in Final Medium (M) |
|---|---|---|---|
| EDTA | — | 14.026 g | $4.8 \times 10^{-5}$ |
| $FeCl_3$ | — | 0.324 g | $2.0 \times 10^{-6}$ |
| $H_3BO_3$ | — | 24.732 g | $4.0 \times 10^{-4}$ |
| $MnCl_2 \cdot 4H_2O$ | — | 1.979 g | $1.0 \times 10^{-5}$ |
| $ZnSO_4 \cdot 7H_2O$ | — | 10.064 g | $3.5 \times 10^{-5}$ |
| $NaMoO_4 \cdot 2H_2O$ | — | 1.210 g | $5.0 \times 10^{-6}$ |
| $CuSO_4 \cdot 5H_2O$ | — | 0.075 g | $3.0 \times 10^{-7}$ |
| $CoCl_2 \cdot 6H_2O$ | — | 0.071 g | $3.0 \times 10^{-7}$ |

*Trace Metal Solution: TMS-II*

(McLachlin 1964)

These are necessary only for certain marine macrophytes. **Into 900 mL** of $dH_2O$, dissolve individually components and bring the final volume to 1 liter.

| Component | 1° Stock Solution ($g \cdot L^{-1} dH_2O$) | Quantity Used | Concentration in Final Medium (M) |
|---|---|---|---|
| KBr | — | 51.450 g | $5.0 \times 10^{-4}$ |
| SrCl2 | — | 26.662 g | $1.0 \times 10^{-4}$ |
| Ru | — | 0.242 g | $2.0 \times 10^{-6}$ |
| Li | — | 0.424 g | $1.0 \times 10^{-5}$ |
| I | — | 0.030 g | $2.0 \times 10^{-7}$ |

*S3 Vitamins Solution*

(Provasoli 1963)

**Into 900 mL** of dH$_2$O, dissolve the first four components and then add 1 mL of each primary stock solution. Bring the final volume to 1 liter. Filter-sterilize and freeze.

| Component | 1° Stock Solution (g · L$^{-1}$ dH$_2$O) | Quantity Used | Concentration in Final Medium (M) |
|---|---|---|---|
| Inositol | — | 900.000 mg | $5.0 \times 10^{-6}$ |
| Thiamine · HCl (vitamin B$_1$) | — | 168.635 mg | $5.0 \times 10^{-7}$ |
| Ca pantethenate | | 23.830 mg | $1.0 \times 10^{-7}$ |
| Nicotinic acid (niacin) | — | 12.310 mg | $1.0 \times 10^{-7}$ |
| $p$-Aminobenzoic acid | 1.371 | 1 mL | $1.0 \times 10^{-8}$ |
| Biotin (vitamin H) | 0.244 | 1 mL | $1.0 \times 10^{-9}$ |
| Folic acid | 0.883 | 1 mL | $2.0 \times 10^{-9}$ |
| Cyanocobalamin (vitamin B$_{12}$) | 1.355 | 1 mL | $1.0 \times 10^{-9}$ |
| Thymine | 0.378 | 1 mL | $3.0 \times 10^{-6}$ |

## ESAW Medium

(Harrison et al. 1980, Berges et al. 2001)

This medium is designed for coastal and open ocean phytoplankton. The artificial seawater base is enriched with a modification of the enrichment solutions of Provasoli's (1968) ES medium. The recipe is from Berges et al. (2001) and has been modified from the earlier version (Harrison et al. 1980) by adding borate only in the salt solution (not in trace metals), replacing glycerophosphate with inorganic phosphate, and preparing the silicate stock solution at half-strength without acidification to facilitate dissolution. Three additional trace elements have been added: $Na_2MoO_4 \cdot 2H_2O$, $Na_2SeO_3$, and $NiCl_2 \cdot 6H_2O$. The anhydrous and hydrated salts must be dissolved separately; masses assume specific gravity = 1.021 at 20°C. The iron is now added solely as chloride (to remove ammonium) from a separate stock with equimolar EDTA. Filter sterilization is recommended (e.g., with a 147-mm Millipore GS filter [pore size, 0.22 µm] with a Gelman A/E prefilter). Autoclaving the final medium often causes precipitates to form. If autoclaving is necessary, autoclave the two salt solutions separately, and when they are completely cooled, aseptically combine them. The 1- or 2-mL nutrient additions should be added with a 0.2-µm-pore-size sterile filter and a syringe. The medium should be bubbled with filtered air for 12 hours before use. **Dissolve** the anhydrous salts in 600 mL of $dH_2O$ and dissolve the hydrated salts in 300 mL $dH_2O$. Combine salt solutions I and II, and add the indicated amounts of each stock solution. Bring the final volume to 1 liter with $dH_2O$. The final pH should be 8.2.

| Component | Stock Solution ($g \cdot L^{-1}$ $dH_2O$) | Quantity Used | Concentration in Final Medium (M) |
|---|---|---|---|
| Salt solution I: Anhydrous salts | | | |
| NaCl | — | 21.194 g | $3.63 \times 10^{-1}$ |
| $Na_2SO_4$ | — | 3.550 g | $2.50 \times 10^{-2}$ |
| KCl | — | 0.599 g | $8.03 \times 10^{-3}$ |
| $NaHCO_3$ | — | 0.174 g | $2.07 \times 10^{-3}$ |
| KBr | — | 0.0863 g | $7.25 \times 10^{-4}$ |
| $H_3BO_3$ | — | 0.0230 g | $3.72 \times 10^{-4}$ |
| NaF | — | 0.0028 g | $6.67 \times 10^{-5}$ |
| Salt solution II: Hydrated salts | — | | |
| $MgCl_2 \cdot 6H_2O$ | — | 9.592 g | $4.71 \times 10^{-2}$ |
| $CaCl_2 \cdot 2H_2O$ | — | 1.344 g | $9.14 \times 10^{-3}$ |
| $SrCl_2 \cdot 6H_2O$ | — | 0.0218 g | $8.18 \times 10^{-5}$ |
| Major Nutrients | — | | |
| $NaNO_3$ | 46.670 | 1 mL | $5.49 \times 10^{-4}$ |
| $NaH_2PO_4 \cdot H_2O$ | 3.094 | 1 mL | $2.24 \times 10^{-5}$ |
| $Na_2SiO_3 \cdot 9H_2O$ | 15.000 | 2 mL | $1.06 \times 10^{-4}$ |
| Iron-EDTA Stock Solution | | | |
| $Na_2EDTA \cdot 2H_2O$ | — | 2.44 g | $6.56 \times 10^{-6}$ |
| $FeCl_3 \cdot 6H_2O$ | 1.77 | 1 mL | $6.55 \times 10^{-6}$ |
| Trace metals stock solution II | (See following recipe) | 1 mL | — |
| Vitamins solution | (See following recipe) | 1 mL | — |

*Trace Metals Solution II*

**Into 900 mL** of $dH_2O$, add the indicated quantities of trace elements; bring the final volume to 1 liter with $dH_2O$.

| Component | 1° Stock Solution ($g \cdot L^{-1} dH_2O$) | Quantity Used | Concentration in Final Medium (M) |
|---|---|---|---|
| $Na_2EDTA \cdot 2H_2O$ | — | 3.09 g | $8.30 \times 10^{-6}$ |
| $ZnSO_4 \cdot 7H_2O$ | — | 0.073 g | $2.54 \times 10^{-7}$ |
| $CoSO_4 \cdot 7H_2O$ | — | 0.016 g | $5.69 \times 10^{-8}$ |
| $MnSO_4 \cdot 4H_2O$ | — | 0.54 g | $2.42 \times 10^{-6}$ |
| $Na_2MoO_4 \cdot 2H_2O$ | 1.48 | 1 mL | $6.12 \times 10^{-9}$ |
| $Na_2SeO_3$ | 0.173 | 1 mL | $1.00 \times 10^{-9}$ |
| $NiCl_2 \cdot 6H_2O$ | 1.49 | 1 mL | $6.27 \times 10^{-9}$ |

*Vitamins Stock Solution*

**Into 900 mL** $dH_2O$, dissolve the thiamine $\cdot$ HCl, add 1 mL of the other two vitamin solutions, and bring the final volume to 1 liter with $dH_2O$. Filter-sterilize.

| Component | 1° Stock Solution ($g \cdot L^{-1} dH_2O$) | Quantity Used | Concentration in Final Medium (M) |
|---|---|---|---|
| Thiamine $\cdot$ HCl (vitamin $B_1$) | — | 0.1 g | $2.96 \times 10^{-7}$ |
| Biotin (vitamin H) | 1.0 | 1 mL | $4.09 \times 10^{-9}$ |
| Cyanocobalamin (vitamin $B_{12}$) | 2.0 | 1 mL | $1.48 \times 10^{-9}$ |

## CCAP Artificial Seawater

(Tompkins et al. 1995)

This medium is prepared with commercially available Ultramarine Synthetica sea salts (Waterlife Research Industries). **Into 900 mL** dH$_2$O, dissolve the Ultramarine Synthetica sea salts, then dissolve the tricine and add the stock solution quantities. Bring final volume to 1 liter with dH$_2$O. Adjust to pH of 7.6 to 7.8 and then autoclave.

| Component | Stock Solution (g · L$^{-1}$ dH$_2$O) | Quantity Used | Concentration in Final Medium (M) |
|---|---|---|---|
| Ultramarine sea salts | — | 33.6 g | — |
| Tricine | — | 0.50 g | $2.79 \times 10^{-3}$ |
| Macronutrients | (See following recipe) | 1 mL | — |
| Soil water extract | (See Chapter 2) | 25 mL | — |
| Vitamins stock solution | (See following recipe) | 1 mL | — |

*Macronutrients Stock Solution*

| Component | 1° Stock Solution (g · L$^{-1}$ dH$_2$O) | Quantity Used | Concentration in Final Medium (M) |
|---|---|---|---|
| NaNO$_3$ | — | 112.50 g | $1.32 \times 10^{-3}$ |
| Na$_2$HPO$_4$ | — | 4.50 g | $2.58 \times 10^{-5}$ |
| K$_2$HPO$_4$ | — | 3.75 g | $2.76 \times 10^{-5}$ |

*Vitamins Stock Solution*

**Into 900 mL** of dH$_2$O, individually dissolve each component or stock solution and bring final volume to 1 liter. Filter-sterilize and store frozen.

| Component | 1° Stock Solution (g · L$^{-1}$ dH$_2$O) | Quantity Used | Concentration in Final Medium (M) |
|---|---|---|---|
| Inositol | — | 2.5 g | $1.39 \times 10^{-5}$ |
| Thymine | — | 1.5 g | $1.19 \times 10^{-5}$ |
| Thiamine · HCl (vitamin B$_1$) | — | 0.25 g | $7.41 \times 10^{-7}$ |
| Nicotinic acid (niacin) | — | 0.05 g | $4.06 \times 10^{-7}$ |
| Ca pantothenate | — | 0.05 g | $2.10 \times 10^{-7}$ |
| Cyanocobalamin (vitamin B$_{12}$) | — | 0.01 g | $7.38 \times 10^{-9}$ |
| Folic acid | 1.0 | 1 mL | $2.27 \times 10^{-9}$ |
| Biotin (vitamin H) | 0.5 | 1 mL | $2.05 \times 10^{-9}$ |

## ASN-III Medium

### (Rippka 1988)

**To prepare,** begin with 900 mL of dH$_2$O and dissolve the designated quantities for the following components. Bring final volume to 1 liter. The final medium should have a pH of 7.3 ± 0.2 at room temperature (before autoclaving).

| Component | Stock Solution (g · L$^{-1}$ dH$_2$O) | Quantity Used | Concentration in Final Medium (M) |
|---|---|---|---|
| NaCl | — | 25.0 g | $4.28 \times 10^{-1}$ |
| MgSO$_4$ · 7H$_2$O | — | 3.5 g | $1.42 \times 10^{-2}$ |
| MgCl$_2$ · 6H$_2$O | — | 2.0 g | $9.84 \times 10^{-3}$ |
| NaNO$_3$ | — | 0.75 g | $8.82 \times 10^{-3}$ |
| K$_2$HPO$_4$ · 3H$_2$O | — | 0.75 g | $3.29 \times 10^{-3}$ |
| CaCl$_2$ · 2H$_2$O | — | 0.5 g | $3.40 \times 10^{-3}$ |
| KCl | — | 0.5 g | $6.71 \times 10^{-3}$ |
| NaCO$_3$ | — | 0.02 g | $2.41 \times 10^{-4}$ |
| Citric acid | — | 3.0 mg | $1.56 \times 10^{-5}$ |
| Ferric ammonium citrate | — | 3.0 mg | (approx.) $9 \times 10^{-6}$ |
| Mg EDTA | — | 0.5 mg | $1.59 \times 10^{-6}$ |
| Cyanocobalamin (vitamin B$_{12}$) | — | 10.0 µg | $7.39 \times 10^{-9}$ |
| A-5 + Co trace metals sol. | (See following recipe) | 1.0 ml | — |

*A5 + Co Trace Metals Solution*

**Into 950 mL** of dH$_2$O, dissolve each of the following components, and then bring the final volume up to 1 liter.

| Component | 1° Stock Solution (g · L$^{-1}$ dH$_2$O) | Quantity Used | Concentration in Final Medium (M) |
|---|---|---|---|
| H$_3$BO$_3$ | — | 2.860 g | $4.60 \times 10^{-5}$ |
| MnCl$_2$ · 4H$_2$O | — | 1.810 g | $9.14 \times 10^{-6}$ |
| ZnSO$_4$ · 7H$_2$O | — | 0.222 g | $7.72 \times 10^{-7}$ |
| NaMoO$_4$ · 2H$_2$O | — | 0.390 g | $1.61 \times 10^{-3}$ |
| CuSO$_4$ · 5H$_2$O | — | 0.079 g | $3.16 \times 10^{-7}$ |
| Co(NO$_3$)$_2$ · 6H$_2$O | — | 49.40 mg | $1.70 \times 10^{-7}$ |

## YBC-II Medium

### (Chen et al. 1996)

This medium was developed to culture nitrogen-fixing *Trichodesmium* (no nitrogen source in medium) under critical culture conditions for physiological studies (Chen et al. 1996). It was based upon Ohki's medium (Ohki et al. 1992), but it is significantly different. Another more complex medium, YBC-III medium, is presented in Chen et al. (1996), and it more closely resembles oligotrophic seawater. **To prepare**, make the necessary stock solutions, using dH$_2$O. **Into 900 mL** of dH$_2$O, add the following components. Bring the final volume to 1 liter with high-quality dH$_2$O. Adjust to a pH of 8.15 to 8.2 with NaOH and filter-sterilize; do not autoclave.

| Component | Stock Solution (g · L$^{-1}$ dH$_2$O) | Quantity Used | Concentration in Final Medium (M) |
|---|---|---|---|
| *Anhydrous salts* | | | |
| NaCl | — | 24.5500 g | $4.20 \times 10^{-1}$ |
| KCl | — | 0.7500 g | $1.00 \times 10^{-2}$ |
| NaHCO$_3$ | — | 0.2100 g | $2.50 \times 10^{-3}$ |
| H$_3$BO$_3$ | — | 0.0360 g | $5.80 \times 10^{-4}$ |
| KBr | — | 0.1157 g | $9.72 \times 10^{-4}$ |
| NaF | 2.94 | 1 mL | $7.00 \times 10^{-5}$ |
| *Hydrous salts* | | | |
| MgCl$_2$ · 6H$_2$O | — | 4.067 g | $2.00 \times 10^{-2}$ |
| CaCl$_2$ · 2H$_2$O | — | 1.47 g | $1.00 \times 10^{-2}$ |
| MgSO$_4$ · 7H$_2$O | — | 6.16 g | $2.50 \times 10^{-2}$ |
| SrCl$_2$ · 6H$_2$O | 17.33 | 1 mL | $6.50 \times 10^{-8}$ |
| *Macronutrients* | | | |
| NaH$_2$PO$_4$ · H$_2$O | 6.9 | 1 mL | $5.00 \times 10^{-5}$ |
| Trace metals solution | (See following recipe) | 1 mL | — |
| Vitamins solution | (See following recipe) | 1 ml | — |

*Trace Metals Solution*

**Into 950 mL** of high-quality $dH_2O$, add the EDTA and 1 mL of each primary stock solution. Filter-sterilize.

| Component | 1° Stock Solution $(g \cdot L^{-1} \, dH_2O)$ | Quantity Used | Concentration in Final Medium (M) |
|---|---|---|---|
| $Na_2EDTA \cdot 2H_2O$ | — | 0.745 g | $2.00 \times 10^{-6}$ |
| $FeCl_3 \cdot 6H_2O$ | 0.11 | 1 mL | $4.07 \times 10^{-7}$ |
| $MnCl_2 \cdot 4H_2O$ | 3.96 | 1 mL | $2.00 \times 10^{-8}$ |
| $ZnSO_4 \cdot 7H_2O$ | 1.15 | 1 mL | $4.00 \times 10^{-9}$ |
| $CoCl_2 \cdot 6H_2O$ | 5.95 | 1 mL | $2.50 \times 10^{-9}$ |
| $Na_2MoO_2 \cdot 2H_2O$ | 2.66 | 1 mL | $1.10 \times 10^{-8}$ |
| $CuSO_4$ | 2.23 | 1 mL | $1.00 \times 10^{-9}$ |

*Vitamins Solution*

**Into 950 mL** of $dH_2O$, add the thiamine $\cdot$ HCl and primary stock solutions, and bring the final volume to 1 liter with $dH_2O$. Filter-sterilize and store frozen.

| Component | 1° Stock Solution $(g \cdot L^{-1} \, dH_2O)$ | Quantity Used | Concentration in Final Medium (M) |
|---|---|---|---|
| Thiamine $\cdot$ HCl (vitamin $B_1$) | — | 100 mg | $2.96 \times 10^{-7}$ |
| Biotin (vitamin H) | 0.5 | 1 mL | $2.05 \times 10^{-9}$ |
| Cyanocobalamin (vitamin $B_{12}$) | 0.5 | 1 mL | $3.69 \times 10^{-10}$ |

## 2.2. Enriched Natural Seawater Media

### ES Medium

(Provasoli 1968)

ES Medium (enriched natural seawater), in addition to nitrate and phosphate, includes Tris base buffer, trace metals, and vitamins in place of soil-water extract. The origin and composition of ES medium is confusing; recipes appeared in two publications, and they are different (D'Angostino and Provasoli 1968, Provasoli 1968). Provasoli himself cited the Provasoli (1968) publication (e.g., D'Agostino and Provasoli 1970, Provasoli et al. 1970, Provasoli and Pintner 1980). Unfortunately, the Provasoli (1968) publication contains several errors. Further confusion arose when some authors (e.g., Starr and Zeikus 1993) erroneously attributed ES Medium to Provasoli (1963). Provasoli (1963) describes a new enriched natural seawater medium (SWII Medium), derived from Iwasaki's SWI Medium (Iwasaki 1961); however, SWII Medium is not ES Medium. Finally, the PII trace metals solution in ES Medium is not the original formulation (see Provasoli 1958). The molar concentrations differ and he substitutes sulfated Mn, Z, and Co for the original chlorinated forms. The PII trace metals recipe is sometimes traced back to Provasoli et al. (1957) (e.g., by Provasoli himself in D'Angostino and Provasoli 1968), but it first appeared in Provasoli (1958) and subsequently in Provasoli (1963). See Provasoli (1968) for the version of trace metals used in ES medium described here. In addition to this recipe and the ES medium by D'Angostino and Provasoli (1968), McLachlin (1973) provides another (different) ES Medium recipe. McLachlin's version, based on personal communication from John West, has $6.6 \times 10^{-5}$ M Tris, $6.6 \times 10^{-5}$ M nitrate, $2.5 \times 10^{-6}$ M glycerophosphate, $7.2 \times 10^{-3}$ M iron–EDTA, and a different formulation of vitamins and trace metals; i.e., everything is different. West and McBride (1999) provide yet another version of ES Medium (see following recipe), but their version may be the best formulation for general use. **To prepare** the enrichment stock solution, begin with 900 mL of dH$_2$O, add the following components (vitamins should be added last, after mixing other ingredients), bring the final volume to 1 liter with dH$_2$O, and pasteurize (do not autoclave). To prepare the ES Medium, add 20 mL of the enrichment stock solution to 980 mL of filtered natural seawater. Pasteurize.

*Enrichment Stock Solution*

| Component | Stock Solution ($g \cdot L^{-1}$ dH$_2$O) | Quantity Used | Concentration in Final Medium (M) |
|---|---|---|---|
| Tris base | — | 5.0 g | $8.26 \times 10^{-4}$ |
| NaNO$_3$ | — | 3.5 g | $8.24 \times 10^{-4}$ |
| Na$_2$ β-glycerophosphate · H$_2$O | — | 0.5 g | $4.63 \times 10^{-5}$ |
| Iron–EDTA solution | (See following recipe) | 250 mL | — |
| Trace metals solution | (See following recipe) | 25 mL | — |
| Thiamine · HCl (vitamin B$_1$) | — | 0.500 mg | $2.96 \times 10^{-8}$ |
| Biotin (vitamin H) | 0.005 | 1 mL | $4.09 \times 10^{-10}$ |
| Cyanocobalamin (vitamin B$_{12}$) | 0.010 | 1 mL | $1.48 \times 10^{-10}$ |

*Iron–EDTA Solution*

**Into 900 mL** of dH$_2$O, dissolve the EDTA and then the iron sulfate. Bring the final volume to 1 liter. Pasteurize and store refrigerated.

| Component | 1° Stock Solution (g · L$^{-1}$ dH$_2$O) | Quantity Used | Concentration in Final Medium (M) |
|---|---|---|---|
| Na$_2$EDTA · 2H$_2$O | — | 0.841 g | $1.13 \times 10^{-5}$ |
| Fe(NH$_4$)$_2$(SO$_4$)$_2$ · 6H$_2$O | — | 0.702 g | $8.95 \times 10^{-6}$ |

*Trace Metals Solution*

(Provasoli 1968)

**Into 900 mL** of dH$_2$O, dissolve the EDTA and then individually dissolve the following components. (The boron is not necessary for enriching natural seawater and should be left out.) Bring the final volume to 1 liter and store refrigerated.

| Component | 1° Stock Solution (g · L$^{-1}$ dH$_2$O) | Quantity Used | Concentration in Final Medium (M) |
|---|---|---|---|
| Na$_2$EDTA · 2H$_2$O | — | 12.74 g | $1.71 \times 10^{-4}$ |
| FeCl$_3$ · 6H$_2$O | — | 0.484 g | $8.95 \times 10^{-6}$ |
| H$_3$BO$_3$ | — | 11.439 g | $9.25 \times 10^{-5}$ |
| MnSO$_4$ · 4H$_2$O | — | 1.624 g | $3.64 \times 10^{-5}$ |
| ZnSO$_4$ · 7H$_2$O | — | 0.220 g | $3.82 \times 10^{-6}$ |
| CoSO$_4$ · 7H$_2$O | — | 0.048 g | $8.48 \times 10^{-7}$ |

### West and McBride's Modified ES Medium

(West and McBride 1999)

This medium is derived from ES Medium (Provasoli 1968) and Modified ES Medium (McLachlan 1973) (see ES Medium). Compared with earlier ES media, this recipe removes the Tris buffer and has lower nitrate, phosphate, and trace metal levels; it has higher iron–EDTA and vitamin levels. The trace metals solution is referred to as PII trace metals in West and McBride (1999), but it varies substantially from Provasoli's original PII trace metal solution (see Provasoli 1958, 1963) as well as his formulation for ES Medium (Provasoli 1968). West and McBride (1999) do not specify the amount of enrichment, but the correct addition is 10 mL of enrichment stock for each liter of natural seawater (John West, personal communication). The stock solution quantities (but not final molar concentrations) have been adjusted in our recipe. **To prepare** the enrichment stock solution, begin with 800 mL of dH$_2$O, and then add the components and bring the final volume to 1 liter with dH$_2$O. Pasteurize (do not autoclave). To prepare the final Modified ES Medium, aseptically add 10 mL of the enrichment stock solution to 990 mL of filtered natural seawater that has been sterilized by pasteurization.

*Enrichment Stock Solution*

| Component | Stock Solution $(g \cdot L^{-1} dH_2O)$ | Quantity Used | Concentration in Final Medium (M) |
|---|---|---|---|
| NaNO$_3$ | — | 3.85 g | $4.53 \times 10^{-4}$ |
| Na$_2$ β-glycerophosphate · H$_2$O | — | 0.4 g | $1.31 \times 10^{-5}$ |
| Fe-EDTA solution | (See following recipe) | 100 mL | — |
| Trace metal solution | (See following recipe) | 20 mL | — |
| Thiamine · HCl (vitamin B$_1$) | 0.5 | 8.0 mL | $1.19 \times 10^{-7}$ |
| Biotin (vitamin H) | 0.05 | 8.0 mL | $1.78 \times 10^{-8}$ |
| Cyanocobalamin (vitamin B$_{12}$) | 0.025 | 3.5 mL | $6.46 \times 10^{-10}$ |

*Iron–EDTA Solution*

**Into 950 mL** of dH$_2$O, first dissolve the EDTA and then the iron compound. Bring the final volume to 1 liter with dH$_2$O and pasteurize; store in a refrigerator.

| Component | 1° Stock Solution (g · L$^{-1}$ dH$_2$O) | Quantity Used | Concentration in Final Medium (M) |
|---|---|---|---|
| Na$_2$EDTA · 2H$_2$O | — | 6.00 g | $1.61 \times 10^{-4}$ |
| Fe(NH$_4$)$_2$(SO$_4$)$_2$ · 6H$_2$O | — | 7.00 g | $1.78 \times 10^{-4}$ |

*Trace Metals Solution*

**Into 900 mL** of dH$_2$O, dissolve the EDTA and then individually add and dissolve the remaining components. (Boron is not necessary when enriching natural seawater, and it should be left out.) Bring the final volume to 1 liter with dH$_2$O.

| Component | 1° Stock Solution (g · L$^{-1}$ dH$_2$O) | Quantity Used | Concentration in Final Medium (M) |
|---|---|---|---|
| Na$_2$EDTA · 2H$_2$O | — | 2.548 g | $1.37 \times 10^{-5}$ |
| H$_3$BO$_3$ | — | 2.240 g | $7.25 \times 10^{-5}$ |
| MnSO$_4$ · 4H$_2$O | — | 0.240 g | $2.15 \times 10^{-6}$ |
| ZnSO$_4$ · 4H$_2$O | — | 0.044 g | $3.06 \times 10^{-7}$ |
| CoSO$_4$ · 7H$_2$O | — | 0.010 g | $7.11 \times 10^{-8}$ |

## ESNW Medium

(Harrison et al. 1980, Berges et al. 2001)

This medium is a modification of Provasoli's (1968) ES medium. For an artificial seawater version, see ESAW Medium. **Into 900 mL**, add the indicated amounts of each stock solution. Bring the final volume to 1 liter with dH$_2$O. Final pH should be 8.2. Filter-sterilize.

| Component | Stock Solution (g · L$^{-1}$ dH$_2$O) | Quantity Used | Concentration in Final Medium (M) |
|---|---|---|---|
| NaNO$_3$ | 46.67 | 1 mL | $5.49 \times 10^{-4}$ |
| NaH$_2$PO$_4$ · H$_2$O | 3.094 | 1 mL | $2.24 \times 10^{-5}$ |
| Na$_2$SiO$_3$ · 9H$_2$O | 15 | 2 mL | $1.06 \times 10^{-4}$ |
| Iron–EDTA Stock Solution | | | |
| Na$_2$EDTA · 2H$_2$O | — | 2.44 g | $6.56 \times 10^{-6}$ |
| FeCl$_3$ · 6H$_2$O | 1.77 | 1 mL | $6.55 \times 10^{-6}$ |
| Trace metals stock solution II | (See following recipe) | 1 mL | — |
| Vitamins solution | (See following recipe) | 1 mL | — |

*Trace Metals Solution II*

**Into 900 mL** of dH$_2$O, add the indicated quantities of trace elements; bring the final volume to 1 liter with dH$_2$O.

| Component | 1° Stock Solution (g · L$^{-1}$ dH$_2$O) | Quantity Used | Concentration in Final Medium (M) |
|---|---|---|---|
| Na$_2$EDTA · 2H$_2$O | — | 3.09 g | $8.30 \times 10^{-6}$ |
| ZnSO$_4$ · 7H$_2$O | — | 0.073 g | $2.54 \times 10^{-7}$ |
| CoSO$_4$ · 7H$_2$O | — | 0.016 g | $5.69 \times 10^{-8}$ |
| MnSO$_4$ · 4H$_2$O | — | 0.54 g | $2.42 \times 10^{-6}$ |
| Na$_2$MoO$_4$ · 2H$_2$O | 1.48 | 1 mL | $6.12 \times 10^{-9}$ |
| Na$_2$SeO$_3$ | 0.173 | 1 mL | $1.00 \times 10^{-9}$ |
| NiCl$_2$ · 6H$_2$O | 1.49 | 1 mL | $6.27 \times 10^{-9}$ |

*Vitamins Stock Solution*

**Into 900 mL** $dH_2O$, dissolve the thiamine $\cdot$ HCl, add 1 mL of the other two vitamin solutions, and bring the final volume to 1 liter with $dH_2O$. Filter-sterilize.

| Component | 1° Stock Solution ($g \cdot L^{-1}$ $dH_2O$) | Quantity Used | Concentration in Final Medium (M) |
|---|---|---|---|
| Thiamine $\cdot$ HCl (vitamin $B_1$) | — | 0.1 g | $2.96 \times 10^{-7}$ |
| Biotin (vitamin H) | 1.0 | 1 mL | $4.09 \times 10^{-9}$ |
| Cyanocobalamin (vitamin $B_{12}$) | 2.0 | 1 mL | $1.48 \times 10^{-9}$ |

## f/2 Medium

### (Guillard and Ryther 1962, Guillard 1975)

This is a common and widely used general enriched seawater medium designed for growing coastal marine algae, especially diatoms. The concentration of the original formulation, termed "f Medium" (Guillard and Ryther 1962), has been reduced by half (Guillard 1975). The original medium (Guillard and Ryther 1962) used ferric sequestrene; we have substituted $Na_2EDTA \cdot 2H_2O$ and $FeCl_3 \cdot 6H_2O$. **Into 950 mL** of filtered natural seawater, add the following components. Bring the final volume to 1 liter with filtered natural seawater. Autoclave. If silicate is not required, omit to reduce precipitation.

| Component | Stock Solution $(g \cdot L^{-1} dH_2O)$ | Quantity Used | Concentration in Final Medium (M) |
|---|---|---|---|
| $NaNO_3$ | 75 | 1 mL | $8.82 \times 10^{-4}$ |
| $NaH_2PO_4 \cdot H_2O$ | 5 | 1 mL | $3.62 \times 10^{-5}$ |
| $Na_2SiO_3 \cdot 9H_2O$ | 30 | 1 mL | $1.06 \times 10^{-4}$ |
| Trace metals solution | (See following recipe) | 1 mL | — |
| Vitamins solution | (See following recipe) | 0.5 mL | — |

*f/2 Trace Metals Solution*

**Into 950 mL** of $dH_2O$, dissolve the EDTA and other components. Bring the final volume to 1 liter with $dH_2O$.

| Component | 1° Stock Solution $(g \cdot L^{-1} dH_2O)$ | Quantity Used | Concentration in Final Medium (M) |
|---|---|---|---|
| $FeCl_3 \cdot 6H_2O$ | — | 3.15 g | $1.17 \times 10^{-5}$ |
| $Na_2EDTA \cdot 2H_2O$ | — | 4.36 g | $1.17 \times 10^{-5}$ |
| $MnCl_2 \cdot 4H_2O$ | 180.0 | 1 mL | $9.10 \times 10^{-7}$ |
| $ZnSO_4 \cdot 7H_2O$ | 22.0 | 1 mL | $7.65 \times 10^{-8}$ |
| $CoCl_2 \cdot 6H_2O$ | 10.0 | 1 mL | $4.20 \times 10^{-8}$ |
| $CuSO_4 \cdot 5H_2O$ | 9.8 | 1 mL | $3.93 \times 10^{-8}$ |
| $Na_2MoO_4 \cdot 2H_2O$ | 6.3 | 1 mL | $2.60 \times 10^{-8}$ |

*f/2 Vitamins Solution*

**Into 950 mL** of $dH_2O$, dissolve the thiamine $\cdot$ HCl, and add 1 mL of the primary stocks. Bring final volume to 1 liter with $dH_2O$. Filter-sterilize and store frozen.

| Component | 1° Stock Solution $(g \cdot L^{-1} dH_2O)$ | Quantity Used | Concentration in Final Medium (M) |
|---|---|---|---|
| Thiamine $\cdot$ HCl (vitamin $B_1$) | — | 200 mg | $2.96 \times 10^{-7}$ |
| Biotin (vitamin H) | 1.0 | 1 mL | $2.05 \times 10^{-9}$ |
| Cyanocobalamin (vitamin $B_{12}$) | 1.0 | 1 mL | $3.69 \times 10^{-10}$ |

**h/2 Medium** is f/2 medium plus 1 mL of $NH_4Cl$ solution (26.75 g $\cdot$ $L^{-1}$ $dH_2O$) to give a final medium concentration of $5.0 \times 10^{-4}$ M.

# K Medium

(Keller et al. 1987)

K medium was designed for oligotrophic (oceanic) marine phytoplankters. It uses a 10-fold higher EDTA chelation than most common marine media, and this reduces the availability of trace metals, thereby reducing the possibility of metal toxicity. The macronutrients are perhaps too high for some open ocean organisms. The necessity of Tris is questionable, and it may be omitted. If organisms do not require silica, the silicate solution should be omitted because it enhances precipitation. For best results, use natural oligotrophic ocean water rather than coastal seawater for the base. The original publications also described an artificial seawater version of K medium. **Into 950 mL** of filtered natural seawater, add the following components and bring the final volume up to 1 liter with filtered natural seawater. Autoclave.

| Component | Stock Solution $(g \cdot L^{-1} \, dH_2O)$ | Quantity Used | Concentration in Final Medium (M) |
|---|---|---|---|
| $NaNO_3$ | 75.00 | 1 mL | $8.82 \times 10^{-4}$ |
| $NH_4Cl$ | 2.67 | 1 mL | $5.00 \times 10^{-5}$ |
| $Na_2$ β-glycerophosphate | 2.16 | 1 mL | $1.00 \times 10^{-5}$ |
| $Na_2SiO_3 \cdot 9H_2O$ | 15.35 | 1 mL | $5.04 \times 10^{-4}$ |
| $H_2SeO_3$ | 0.00129 | 1 mL | $1.00 \times 10^{-8}$ |
| Tris-base (pH 7.2) | 121.10 | 1 mL | $1.00 \times 10^{-3}$ |
| Trace metals solution | (See following recipe) | 1 mL | — |
| Vitamins solution | (See following recipe) | 0.5 mL | — |

*Trace Metals Solution*

**Into 950 mL** of $dH_2O$, dissolve the EDTA and then add the components. Bring the final volume to 1 liter, using $dH_2O$.

| Component | 1° Stock Solution $(g \cdot L^{-1} \, dH_2O)$ | Quantity Used | Concentration in Final Medium (M) |
|---|---|---|---|
| $Na_2EDTA \cdot 2H_2O$ | — | 37.220 g | $1.00 \times 10^{-4}$ |
| $Fe\text{-}Na\text{-}EDTA \cdot 3H_2O$ | — | 4.930 g | $1.17 \times 10^{-5}$ |
| $FeCl_3 \cdot 6H_2O$ | — | 3.150 g | $1.17 \times 10^{-5}$ |
| $MnCl_2 \cdot 4H_2O$ | — | 0.178 g | $9.00 \times 10^{-7}$ |
| $ZnSO_4 \cdot 7H_2O$ | 23.00 | 1 mL | $8.00 \times 10^{-8}$ |
| $CoSO_4 \cdot 7H_2O$ | 14.05 | 1 mL | $5.00 \times 10^{-8}$ |
| $Na_2MoO_4 \cdot 2H_2O$ | 7.26 | 1 mL | $3.00 \times 10^{-8}$ |
| $CuSO_4 \cdot 5H_2O$ | 2.50 | 1 mL | $1.00 \times 10^{-8}$ |

*f/2 Vitamins Solution*

(Guillard and Ryther 1962, Guillard 1975)

**Into 950 mL** of $dH_2O$, dissolve the thiamine · HCl. Add 1 mL of the primary stocks and then bring the final volume to 1 liter with $dH_2O$. Filter-sterilize and store frozen.

| Component | 1° Stock Solution $(g \cdot L^{-1} \, dH_2O)$ | Quantity Used | Concentration in Final Medium (M) |
|---|---|---|---|
| Thiamine · HCl (vitamin $B_1$) | — | 200 mg | $2.96 \times 10^{-7}$ |
| Biotin (vitamin H) | 0.1 | 1 mL | $2.05 \times 10^{-9}$ |
| Cyanocobalamin (vitamin $B_{12}$) | 1.0 | 1 mL | $3.69 \times 10^{-10}$ |

## L1 Medium

### (Guillard and Hargraves 1993)

L1 medium is another in the alphabet medium series by Guillard (e.g., f, h, and k). It is derived from f/2 medium (Guillard and Ryther 1962, Guillard 1975) by adding more trace elements. It is a good general-purpose marine medium for growing coastal algae. **Into 950 mL** of filtered natural seawater, add the components and bring the final volume to 1 liter, using filtered natural seawater. Autoclave. Final pH should be 8.0 to 8.2.

| Component | Stock Solution ($g \cdot L^{-1}$ $dH_2O$) | Quantity Used | Concentration in Final Medium (M) |
|---|---|---|---|
| $NaNO_3$ | 75.00 | 1 mL | $8.82 \times 10^{-4}$ |
| $NaH_2PO_4 \cdot H_2O$ | 5.00 | 1 mL | $3.62 \times 10^{-5}$ |
| $Na_2SiO_3 \cdot 9H_2O$ | 30.00 | 1 mL | $1.06 \times 10^{-4}$ |
| Trace elements solution | (See following recipe) | 1 mL | — |
| Vitamins solution | (See following recipe) | 0.5 mL | — |

*L1 Trace Elements Solution*

**Into 950 mL $dH_2O$,** add the following components and bring the final volume to 1 liter with $dH_2O$. Autoclave.

| Component | 1° Stock Solution ($g \cdot L^{-1}$ $dH_2O$) | Quantity Used | Concentration in Final Medium (M) |
|---|---|---|---|
| $Na_2EDTA \cdot 2H_2O$ | — | 4.36 g | $1.17 \times 10^{-5}$ |
| $FeCl_3 \cdot 6H_2O$ | — | 3.15 g | $1.17 \times 10^{-5}$ |
| $MnCl_2 \cdot H_2O$ | 178.10 | 1 mL | $9.09 \times 10^{-7}$ |
| $ZnSO_4 \cdot 7H_2O$ | 23.00 | 1 mL | $8.00 \times 10^{-8}$ |
| $CoCl_2 \cdot 6H_2O$ | 11.90 | 1 mL | $5.00 \times 10^{-8}$ |
| $CuSO_4 \cdot 5H_2O$ | 2.50 | 1 mL | $1.00 \times 10^{-8}$ |
| $Na_2MoO_4 \cdot 2H_2O$ | 19.90 | 1 mL | $8.22 \times 10^{-8}$ |
| $H_2SeO_3$ | 1.29 | 1 mL | $1.00 \times 10^{-8}$ |
| $NiSO_4 \cdot 6H_2O$ | 2.63 | 1 mL | $1.00 \times 10^{-8}$ |
| $Na_3VO_4$ | 1.84 | 1 mL | $1.00 \times 10^{-8}$ |
| $K_2CrO_4$ | 1.94 | 1 mL | $1.00 \times 10^{-8}$ |

## f/2 Vitamins Solution

### (Guillard and Ryther 1962, Guillard 1975)

**Into 950 mL** of dH$_2$O, dissolve the thiamine · HCl, add 1 mL of the primary stocks, and bring final volume to 1 liter with dH$_2$O. Filter-sterilize.

| Component | 1° Stock Solution (g · L$^{-1}$ dH$_2$O) | Quantity Used | Concentration in Final Medium (M) |
|---|---|---|---|
| Thiamine · HCl (vitamin B$_1$) | — | 100 mg | $2.96 \times 10^{-7}$ |
| Biotin (vitamin H) | 0.5 | 1 mL | $2.05 \times 10^{-9}$ |
| Cyanocobalamin (vitamin B$_{12}$) | 0.5 | 1 mL | $3.69 \times 10^{-10}$ |

## MNK Medium

(Noël et al. 2004)

This enriched seawater medium was developed to grow oceanic coccolithophores (Noël et al. 2004), but it is also a good general medium for marine phytoplankton. Noël (personal communication) recommends storing filtered natural seawater in a cold (4°C) dark environment for at least 3 months before using it. Because open ocean phytoplankters are often sensitive, all glassware should be acid-cleaned and rinsed thoroughly with high-quality water (e.g., Milli-Q water) before use. The medium should be filter-sterilized rather than autoclaved; however, see Chapter 22 regarding viruses. Use natural oligotrophic ocean water. **Into 997 mL** of filtered natural seawater, add 1 mL of nutrient stock, trace metals stock, and vitamin stock solutions. Filter-sterilize. The final pH is 8.7.

*Nutrient Stock Solution*

**Into 950 mL** dH$_2$O, dissolve the components and bring the final volume to 1 liter with dH$_2$O. Filter-sterilize (0.2 µm pore size) and store refrigerated or frozen. For convenience, the solution can be distributed as 1-mL quantities into Eppendorf or cryovial tubes.

| Component | Stock Solution (g · L$^{-1}$ dH$_2$O) | Quantity Used | Concentration in Final Medium (M) |
|---|---|---|---|
| NaNO$_3$ | — | 20.00 g | $2.35 \times 10^{-4}$ |
| Na$_2$HPO$_4$, · 12H$_2$O | — | 0.28 g | $7.82 \times 10^{-7}$ |
| K$_2$HPO$_4$ (anhydrous) | — | 1.00 g | $5.74 \times 10^{-6}$ |

*Trace Metals Solution*

**Into 950 mL** $dH_2O$, add 1 mL each of the primary stock solution. Bring the final volume to 1 liter with $dH_2O$. Store refrigerated or frozen.

| Component | 1° Stock Solution ($g \cdot L^{-1} \, dH_2O$) | Quantity Used | Concentration in Final Medium (M) |
|---|---|---|---|
| Fe-Na-EDTA · $3H_2O$ | 25.900 | 1 mL | $6.15 \times 10^{-8}$ |
| Mn-EDTA · $3H_2O$ | 33.200 | 1 mL | $7.49 \times 10^{-8}$ |
| $Na_2EDTA \cdot 2H_2O$ | 3.723 | 1 mL | $1.00 \times 10^{-8}$ |
| $MnCl_2 \cdot 4H_2O$ | 9.000 | 1 mL | $4.55 \times 10^{-8}$ |
| $ZnSO_4 \cdot 7H_2O$ | 2.400 | 1 mL | $8.35 \times 10^{-9}$ |
| $CoSO_4 \cdot 7H_2O$ | 1.200 | 1 mL | $4.27 \times 10^{-9}$ |
| $Na_2MoO_4 \cdot 2H_2O$ | 0.720 | 1 mL | $2.98 \times 10^{-9}$ |
| $CuSO_4 \cdot 5H_2O$ | 0.060 | 1 mL | $2.40 \times 10^{-10}$ |
| $Na_2SeO_3$ | 0.030 | 1 mL | $1.73 \times 10^{-10}$ |

*Vitamins Solution*

**Into 950 mL** of $dH_2O$, add the thiamine · HCl and 1 mL of each primary stock solution. Bring the final volume to 1 liter with $dH_2O$. Filter-sterilize and freeze.

| Component | 1° Stock Solution ($g \cdot L^{-1} \, dH_2O$) | Quantity Used | Concentration in Final Medium (M) |
|---|---|---|---|
| Thiamine · HCl (vitamin $B_1$) | — | 20 mg | $5.93 \times 10^{-8}$ |
| Biotin (vitamin H) | 0.15 | 1 mL | $6.14 \times 10^{-10}$ |
| Cyanocobalamin (vitamin $B_{12}$) | 0.15 | 1 mL | $1.11 \times 10^{-10}$ |

## PC Medium

(Keller in Andersen et al. 1997)

PC medium is an enriched natural seawater medium developed by Maureen Keller specifically for growing *Prochloro-coccus*, but it can be used for other open ocean phytoplankters. All containers should be acid-cleaned and rinsed with high-quality $dH_2O$ (e.g., Milli-Q water). Seawater should be collected from the oligotrophic open ocean (e.g., Sargasso seawater), taking the usual precautions to avoid contamination. Only ultrapure chemicals and high-quality $dH_2O$ should be used in preparing nutrient stock solutions. A microwave oven is used for sterilization (see Chapter 5). **To prepare**, filter 1 liter of seawater and sterilize in a microwave oven. After cooling, aseptically add the following components.

| Component | Stock Solution $(g \cdot L^{-1} \, dH_2O)$ | Quantity Used | Concentration in Final Medium (M) |
|---|---|---|---|
| Na2 β-glycerophosphate | 2.16 | I mL | $1.00 \times 10^{-5}$ |
| $NH_4Cl$ | 2.68 | I mL | $5.01 \times 10^{-5}$ |
| Urea | 3.00 | I mL | $5.00 \times 10^{-5}$ |
| Trace metals solution | (See following recipe) | 100 μL | — |
| f/2 vitamins solution | (See following recipe) | 50 μL | — |

*PC Trace Metals Solution*

**Into 900 mL** of high-quality $dH_2O$, dissolve the EDTA and then individually dissolve the metals. Store at 4°C.

| Component | 1° Stock Solution $(g \cdot L^{-1} \, dH_2O)$ | Quantity Used | Concentration in Final Medium (M) |
|---|---|---|---|
| $Na_2EDTA \cdot 2H_2O$ | — | 41.60 g | $1.12 \times 10^{-4}$ |
| $FeCl_3 \cdot 6H_2O$ | — | 3.15 g | $1.17 \times 10^{-5}$ |
| $MnCl_2 \cdot 4H_2O$ | — | 0.18 g | $9.09 \times 10^{-8}$ |
| $ZnSO_4 \cdot 7H_2O$ | 22.0 | 1 mL | $7.65 \times 10^{-9}$ |
| $CoCl_2 \cdot 6H_2O$ | 10.0 | 1 mL | $4.20 \times 10^{-9}$ |
| $Na_2MoO_4 \cdot 2H_2O$ | 6.3 | 1 mL | $2.60 \times 10^{-9}$ |
| $H_2SeO_3$ | 1.3 | 1 mL | $1.01 \times 10^{-9}$ |
| $NiSO_4 \cdot 6H_2O$ | 2.7 | 1 mL | $1.03 \times 10^{-9}$ |

*f/2 Vitamins Solution*

(Guillard and Ryther 1962, Guillard 1975)

**Into 950 mL** of $dH_2O$, dissolve the thiamine $\cdot$ HCl. Add 1 mL of the primary stocks and bring final volume to 1 liter with $dH_2O$. Filter-sterilize and store frozen.

| Component | 1° Stock Solution $(g \cdot L^{-1} \, dH_2O)$ | Quantity Used | Concentration in Final Medium (M) |
|---|---|---|---|
| Thiamine $\cdot$ HCl (vitamin $B_1$) | — | 200 mg | $2.96 \times 10^{-8}$ |
| Biotin (vitamin H) | 1.0 | 1 mL | $2.05 \times 10^{-10}$ |
| Cyanocobalamin (vitamin $B_{12}$) | 1.0 | 1 mL | $3.69 \times 10^{-11}$ |

### Pro99 Medium

(Moore et al. 2002)

This medium was developed specifically for *Prochlorococcus* (Sallie Chisholm, personal communication) and is currently the best medium available for *Prochlorococcus*. No name was used in the publication (Moore et al. 2002), but Pro99 was used on a recipe informally distributed by Chisholm's group, and we have applied that name for our Appendix. It can also be used for other oceanic species tolerating high ammonia concentrations (e.g., *Bolidomonas*). All containers should be acid-cleaned and rinsed with high-quality $H_2O$ (e.g., Milli-Q). Seawater should be collected from the oligotrophic open ocean (e.g., Sargasso Sea) and only ultrapure-grade reagents should be used. Good sterile technique is required and a laminar flow hood is recommended. The oligotrophic seawater is autoclaved in a Teflon-lined container and cooled before aseptically adding nutrients. **To prepare**, filter 1 liter of natural oligotrophic seawater into a Teflon-lined container. Autoclave and cool. Aseptically, add 1 mL each of the $NaH_2PO_4$, $NH_4Cl$, and trace element solutions. Note: ammonium chloride and sodium phosphate solutions should be filter-sterilized and stored at 4°C.

| Component | Stock Solution ($g \cdot L^{-1}$ $dH_2O$) | Quantity Used | Concentration in Final Medium (M) |
|---|---|---|---|
| $NaH_2PO_4 \cdot H_2O$ | 6.90 | 1.0 mL | $5.0 \times 10^{-5}$ |
| $NH_4Cl$ | 42.80 | 1.0 mL | $8.0 \times 10^{-4}$ |
| Trace elements solution | (See following recipe) | 1.0 mL | — |

*Pro99 Trace Elements Solution*

**Into 900 mL** of high-quality $dH_2O$, dissolve EDTA and add other components. Bring final volume to 1 liter with high-quality $dH_2O$, filter-sterilize, and store at 4°C.

| Component | 1° Stock Solution ($g \cdot L^{-1}$ $dH_2O$) | Quantity Used | Concentration in Final Medium (M) |
|---|---|---|---|
| $Na_2EDTA \cdot 2H_2O$ | — | 0.436 g | $1.17 \times 10^{-6}$ |
| $FeCl_3 \cdot 6H_2O$ | — | 0.316 g | $1.17 \times 10^{-6}$ |
| $ZnSO_4 \cdot 7H_2O$ | 2.30 | 1 mL | $8.00 \times 10^{-9}$ |
| $CoCl_2 \cdot 6H_2O$ | 1.19 | 1 mL | $5.00 \times 10^{-9}$ |
| $MnCl_2 \cdot 4H_2O$ | 17.80 | 1 mL | $9.00 \times 10^{-8}$ |
| $Na_2MoO_4 \cdot 2H_2O$ | 0.73 | 1 mL | $3.00 \times 10^{-9}$ |
| $Na_2SeO_3$ | 1.73 | 1 mL | $1.00 \times 10^{-8}$ |
| $NiSO_4 \cdot 6H_2O$ | 2.63 | 1 mL | $1.00 \times 10^{-8}$ |

## SN Medium

### (Waterbury et al. 1986)

SN Medium is the most popular of three media developed by Waterbury et al. (1986) for culturing marine *Synchococcus sensu lato* (Cyanophyceae). The medium has very high nitrogen, and the trace metals are buffered with citrate rather than EDTA. If a solid medium (agar) is prepared, standard agar should be washed to remove impurities (see Chapter 2 and Waterbury et al. 1986). The salinity of the natural seawater is reduced by adding dH$_2$O. **To prepare**, autoclave 750 mL of filtered natural seawater in a Teflon-lined bottle and separately autoclave 236 mL of double-distilled H$_2$O; cool and aseptically combine the two solutions. Aseptically, add 10 mL of sodium nitrate solution and 1 mL of the other five stock solutions. Final salinity should be ~26.

| Component | Stock Solution (g · L$^{-1}$ dH$_2$O) | Quantity Used | Concentration in Final Medium (M) |
|---|---|---|---|
| NaNO$_3$ | 76.50 | 10 mL | $9.00 \times 10^{-3}$ |
| K$_2$HPO$_4$ | 15.68 | 1 mL | $9.90 \times 10^{-5}$ |
| Na$_2$EDTA · 2H$_2$O | 5.58 | 1 mL | $1.50 \times 10^{-5}$ |
| Na$_2$CO$_3$ | 10.70 | 1 mL | $1.00 \times 10^{-4}$ |
| Cyanocobalamin (vitamin B$_{12}$) | 0.001 | 1 mL | $7.38 \times 10^{-10}$ |
| Trace metals solution | (See following recipe) | 1 mL | — |

*Cyano Trace Metals Solution*

**Into 900 mL** of dH$_2$O, dissolve the citrates and then add the remaining components. Bring final volume to 1 liter with dH$_2$O.

| Component | 1° Stock Solution (g · L$^{-1}$ dH$_2$O) | Quantity Used | Concentration in Final Medium (M) |
|---|---|---|---|
| Citric Acid · H$_2$O | — | 6.250 g | $3.25 \times 10^{-5}$ |
| Ferric ammonium citrate | — | 6.000 g | — |
| MnCl$_2$ · 4H$_2$O | — | 1.400 g | $7.08 \times 10^{-6}$ |
| Na$_2$MoO$_4$ · 2H$_2$O | — | 0.390 g | $1.61 \times 10^{-6}$ |
| ZnSO$_4$ · 7H$_2$O | — | 0.222 g | $7.72 \times 10^{-7}$ |
| Co(NO$_3$)$_2$ · 6H$_2$O | — | 0.025 g | $8.59 \times 10^{-8}$ |

## von Stosch (Grund) Medium

### (Guiry and Cunningham 1984)

This enriched seawater medium was modified from Grund Medium (von Stosch 1963); the Tris buffer was removed in an effort to reduce bacterial growth. Guiry and Cunningham (1984) used it to grow the red seaweed *Gigartina*, and it is suitable for growing many different red seaweeds. **To prepare**, pasteurize 940 mL of filtered natural seawater and aseptically add 10 mL each of the following stock solutions. Autoclave.

| Component | Stock Solution ($g \cdot L^{-1}$ d$H_2O$) | Quantity Used | Concentration in Final Medium (M) |
|---|---|---|---|
| $Na_2$ β-glycerophosphate | 5.36 | 10 mL | $2.48 \times 10^{-4}$ |
| $NaNO_3$ | 42.52 | 10 mL | $5.00 \times 10^{-3}$ |
| $FeSO_4 \cdot 7H_2O$ | 0.28 | 10 mL | $1.00 \times 10^{-5}$ |
| $MnCl_2 \cdot 4H_2O$ | 1.96 | 10 mL | $1.00 \times 10^{-4}$ |
| $Na_2EDTA \cdot 2H_2O$ | 3.72 | 10 mL | $1.00 \times 10^{-4}$ |
| Vitamins stock solution | (See following recipe) | 10 mL | — |

*Vitamins Stock Solution*

**Into 950 mL** of d$H_2O$, dissolve the thiamine $\cdot$ HCl, and add 1 mL of each of the two primary stock solutions. Filter-sterilize and freeze.

| Component | 1° Stock Solution ($g \cdot L^{-1}$ d$H_2O$) | Quantity Used | Concentration in Final Medium (M) |
|---|---|---|---|
| Thiamine $\cdot$ HCl (vitamin $B_1$) | — | 200 mg | $5.93 \times 10^{-6}$ |
| Biotin (vitamin H) | 0.1 | 1 mL | $4.09 \times 10^{-9}$ |
| Cyanocobalamin (vitamin $B_{12}$) | 0.2 | 1 mL | $1.48 \times 10^{-9}$ |

## Walne's Medium

(Walne 1970)

This enriched seawater medium was designed for mass culture of marine phytoplankton used as feed for shellfish. The medium has a boron addition, which is completely unnecessary when using natural seawater as a base. Sodium metasilicate ($40 \, mg \cdot L^{-1}$) should be added for the growth of diatoms. The vitamins solution is extremely dilute and lacks biotin. It would be more sensible to move the EDTA, iron, and manganese to the trace metals solution (with $1,000\times$ increase in concentrations to compensate for the 1 mL addition). Note that the EDTA concentration is approximately $10\times$ higher than usual for coastal culture media; it is comparable to the oceanic K Medium (Keller et al. 1987). **To prepare**, pasteurize 1 liter of filtered natural seawater; after cooling, aseptically add 1 mL of the nutrient solution and 100 µL of the vitamins solution.

*Nutrient Solution*

**Into 900 mL** of high quality $dH_2O$, dissolve the components. Bring final volume to 1 liter with high-quality $dH_2O$, filter sterilize, and store at 4°C.

| Component | Stock Solution ($g \cdot L^{-1}$ $dH_2O$) | Quantity Used | Concentration in Final Medium (M) |
|---|---|---|---|
| $NaNO_3$ | — | 100.0 g | $1.18 \times 10^{-3}$ |
| $H_3BO_3$ | — | 33.6 g | $5.43 \times 10^{-4}$ |
| $Na_2$ EDTA (anhydrous) | — | 45.0 g | $1.54 \times 10^{-4}$ |
| $NaH_2PO_4 \cdot H_2O$ | — | 20.0 g | $1.28 \times 10^{-4}$ |
| $FeCl_3 \cdot 6H_2O$ | — | 1.3 g | $4.81 \times 10^{-6}$ |
| $MnCl_2 \cdot 4H_2O$ | — | 0.36 g | $1.82 \times 10^{-6}$ |
| Trace metals solution | (See following recipe) | 1 mL | — |

*Trace Metals Solution*

**Into 900 mL** of high quality $dH_2O$, dissolve the components. This solution is normally cloudy. Acidify with a few drops of concentrated HCl to give a clear solution. Bring final volume to 1 liter with high-quality $dH_2O$, filter sterilize, and store at 4°C.

| Component | 1° Stock Solution $(g \cdot L^{-1}\ dH_2O)$ | Quantity Used | Concentration in Final Medium (M) |
|---|---|---|---|
| $ZnCl_2$ | — | 21.0 g | $1.54 \times 10^{-7}$ |
| $CoCl_2 \cdot 6H_2O$ | — | 20.0 g | $8.41 \times 10^{-8}$ |
| $(NH_4)_6Mo_7O_{24} \cdot 4H_2O$ | — | 9.0 g | $7.28 \times 10^{-9}$ |
| $CuSO_4 \cdot 5H_2O$ | — | 20.0 g | $8.01 \times 10^{-8}$ |

*Vitamins Solution*

**Into 950 mL** of $dH_2O$, dissolve the thiamine $\cdot$ HCl and cyanocobalamin. Bring final volume to 1 liter, filter sterilize, and freeze.

| Component | 1° Stock Solution $(g \cdot L^{-1}\ dH_2O)$ | Quantity Used | Concentration in Final Medium (M) |
|---|---|---|---|
| Thiamine $\cdot$ HCl (vitamin $B_1$) | — | 1.0 g | $2.96 \times 10^{-10}$ |
| Cyanocobalamin (vitamin $B_{12}$) | — | 50 mg | $3.69 \times 10^{-12}$ |

## 2.3. Enriched Natural Seawater + Organics

### Antia's Medium

(Antia et al. 1969)

This medium was designed for heterotrophic growth of a cryptophyte by enriching with glycerol (Antia et al. 1969); it was the first demonstration of heterotrophic growth in cryptophytes. This is a reduced-salinity medium (~28). The original recipe gives a component weight of 1.125 mg $MnSO_4 \cdot 4H_2O$ to equal $5.00 \times 10^{-6}$ M; a weight of 1.25 mg gives a more precise $5.00 \times 10^{-6}$ M concentration, and perhaps 1.125 mg is a typographical error. The medium is not recommended for routine maintenance of cryptophytes. **Into 800 mL** of filtered natural seawater, add each component as indicated below, and bring the final volume to 1 liter, using $dH_2O$. Adjust the Tris · HCl stock solution to a pH of 6.8 to 6.9 to give a pH of 7.6 to 7.8 after autoclaving. Final salinity should be ~28.

| Component | Stock Solution $(g \cdot L^{-1}\ dH_2O)$ | Quantity Used | Concentration in Final Medium (M) |
|---|---|---|---|
| $KNO_3$ | 25.00 | 1 mL | $2.47 \times 10^{-3}$ |
| $NaH_2PO_4 \cdot 2H_2O$ | 34.50 | 1 mL | $2.50 \times 10^{-4}$ |
| $Na_2SiO_3 \cdot 9H_2O$ | 84.10 | 1 mL | $2.96 \times 10^{-4}$ |
| Tris · HCl (pH 6.8–6.9) | 1.31 | 1 mL | $8.31 \times 10^{-6}$ |
| Glycerol | — | 46.1 g | $5.00 \times 10^{-1}$ |
| Trace metals solution | (See following recipe) | 2.5 mL | — |
| Vitamins solution | (See following recipe) | 1 mL | — |

*Trace Metals Stock Solution*

**Into 900 mL** of $dH_2O$, dissolve the EDTA and add remaining components. Bring final volume to 1 liter with $dH_2O$ and adjust the pH to 7.6–7.8. Store frozen.

| Component | 1° Stock Solution $(g \cdot L^{-1}\ dH_2O)$ | Quantity Used | Concentration in Final Medium (M) |
|---|---|---|---|
| $Na_2$ EDTA · $2H_2O$ | — | 8.110 g | $2.18 \times 10^{-5}$ |
| $FeCl_3 \cdot 6H_2O$ | — | 2.700 g | $1.00 \times 10^{-5}$ |
| $MnSO_4 \cdot 4H_2O$ | — | 1.125 g | $4.56 \times 10^{-6}$ |
| $ZnSO_4 \cdot 7H_2O$ | — | 0.575 g | $2.00 \times 10^{-6}$ |
| $Na_2MoO_4 \cdot 2H_2O$ | — | 0.243 g | $1.00 \times 10^{-6}$ |
| $CuSO_4 \cdot 5H_2O$ | 24.97 | 1 mL | $1.00 \times 10^{-7}$ |
| $CoSO_4 \cdot 7H_2O$ | 14.06 | 1 mL | $5.00 \times 10^{-8}$ |

*Vitamins Stock Solution*

**Into 950 mL** of dH$_2$O, dissolve the thiamine · HCl, and add 1 mL of each primary stock solution. Bring to 1 liter with dH$_2$O. Filter-sterilize.

| Component | 1° Stock Solution (g · L$^{-1}$ dH$_2$O) | Quantity Used | Concentration in Final Medium (M) |
|---|---|---|---|
| Thiamine · HCl (vitamin B$_1$) | — | 500 mg | $1.48 \times 10^{-6}$ |
| Biotin (vitamin H) | 1.00 | 1 mL | $4.09 \times 10^{-9}$ |
| Cyanocobalamin (B$_{12}$) | 2.00 | 1 mL | $1.48 \times 10^{-9}$ |

## ANT Medium

### (Tompkins et al. 1995)

This medium, prepared by the Culture Centre for Algae and Protozoa (Tompkins et al. 1995), is derived from Antia's medium (Antia et al. 1969). However, it is highly modified from Antia et al. (1969), especially with the substitution of glycine for glycerol (possible typographical error; glycerine = glycerol) and Tris base for Tris · HCl. The recipe does not provide for a vitamin stock solution, and this should be prepared because it is nearly impossible to accurately weigh µg quantities. **Into 800 mL** of natural seawater, add the quantities of each component, and bring up to 1 liter with dH$_2$O. Autoclave. Final pH should be 7.6 to 7.8, and final salinity ~28.

| Component | Stock Solution (g · L$^{-1}$ dH$_2$O) | Quantity Used | Concentration in Final Medium (M) |
|---|---|---|---|
| Tris base | — | 1.0 g | $8.26 \times 10^{-3}$ |
| Glycine | — | 0.3 g | $4.00 \times 10^{-3}$ |
| KNO$_3$ | 50.0 | 1 mL | $4.95 \times 10^{-4}$ |
| NaH$_2$PO$_4$ · 2H$_2$O | 7.8 | 1 mL | $5.65 \times 10^{-5}$ |
| Trace metals solution | (See following recipe) | 2.5 mL | — |
| Thiamine · HCl (vitamin B$_1$) | — | 500.0 µg | $1.48 \times 10^{-6}$ |
| Biotin (vitamin H) | — | 1.0 µg | $4.09 \times 10^{-9}$ |
| Cyanocobalamin (vitamin B$_{12}$) | — | 2.0 µg | $1.48 \times 10^{-9}$ |

*Trace Metals Solution*

**Into 950 mL** of dH$_2$O, dissolve the EDTA and other components; bring final volume to 1 liter with dH$_2$O. Adjust the pH to 7.6–7.8.

| Component | 1° Stock Solution (g · L$^{-1}$ dH$_2$O) | Quantity Used | Concentration in Final Medium (M) |
|---|---|---|---|
| EDTA Na$_2$ · 2H$_2$O | — | 3.240 g | $2.18 \times 10^{-5}$ |
| FeCl$_3$ · 4H$_2$O | — | 1.080 g | $1.36 \times 10^{-5}$ |
| MnSO$_4$ · 4H$_2$O | — | 0.450 g | $4.56 \times 10^{-6}$ |
| ZnSO$_4$ · 7H$_2$O | — | 0.230 g | $2.00 \times 10^{-6}$ |
| Na$_2$MoO$_4$ · 2H$_2$O | 9.7 | 10 mL | $1.00 \times 10^{-8}$ |
| CuSO$_4$ · 5H$_2$O | 10.0 | 10 mL | $1.00 \times 10^{-8}$ |
| CoSO$_4$ · 7H$_2$O | 5.6 | 1 mL | $5.00 \times 10^{-8}$ |

## 2.4. Soil-Water Enriched Natural Seawater Media

### Plymouth Erd-schreiber Medium (PE)

(Tompkins et al. 1995)

This medium can be traced back to Schreiber's Medium and Erd-Schreiber's Medium (Schreiber 1927, Hämmerling 1931, Føyn 1934; see Chapter 1). The soil-water extract substitutes for micronutrients, a chelator, and trace metals; for preparation of soil-water extract, see Chapter 2. This is a good medium for maintenance culturing of many marine algae, but the undefined nature of the soil extract makes it a problematic medium for experimental research. The CCAP recipe presented here was used extensively by Mary Parke and coworkers at the Plymouth Marine Laboratory. **Combine 902.5 mL** of filtered natural seawater and 47.5 mL of dH$_2$O to produce a 95% seawater solution. Autoclave and cool. Aseptically add 1 mL of nitrate and phosphate solutions and 50 mL of soil-water extract. Final salinity should be ~32.

| Component | Stock Solution $(g \cdot L^{-1}\ dH_2O)$ | Quantity Used | Concentration in Final Medium (M) |
|---|---|---|---|
| NaNO$_3$ | 200 | 1 mL | $2.35 \times 10^{-3}$ |
| Na$_2$HPO$_4 \cdot 12H_2O$ | 20 | 1 mL | $5.58 \times 10^{-5}$ |
| Soil-water extract | — | 50 mL | — |

## General Purpose Medium (GPM), Modified

(Sweeney et al. 1959, modified by Loeblich 1975)

This is a reduced-salinity seawater medium that originated with Sweeney et al. (1959) and was then modified by Loeblich (1975). Essentially, it is Erd-Schreiber's medium supplemented with trace metals and vitamins. The salinity is reduced to ~28. It will grow many types of marine algae, but its primary application has been for growing dinoflagellates (e.g., Blackburn et al. 1989). **Into 800 mL** of filtered natural seawater, add 204 mL of dH$_2$O. Add the components as indicated. Pasteurize or autoclave (autoclaving will cause precipitation, which can be reduced by rapid cooling). Final pH should be 8.2.

| Component | Stock Solution (g · L$^{-1}$ dH$_2$O) | Quantity Used | Concentration in Final Medium (M) |
|---|---|---|---|
| KNO$_3$ | — | 0.200 g | $1.98 \times 10^{-3}$ |
| K$_2$HPO$_4$ | — | 0.035 g | $2.01 \times 10^{-4}$ |
| Soil extract | — | 15 mL | — |
| Trace metals solution | (See following recipe) | 30 ml | — |
| Vitamins stock solution | (See following recipe) | 1 mL | — |

*Trace Metals Solution*

**Into 900 mL** of dH$_2$O, dissolve the EDTA and other components. Bring final volume to 1 liter. Adjust pH to 7.5.

| Component | 1° Stock Solution (g · L$^{-1}$ dH$_2$O) | Quantity Used | Concentration in Final Medium (M) |
|---|---|---|---|
| EDTA Na$_2$ · 2H$_2$O | — | 1.2700 g | $1.02 \times 10^{-4}$ |
| FeCl$_3$ · 6H$_2$O | — | 0.0484 g | $5.37 \times 10^{-6}$ |
| H$_3$BO$_3$ | — | 1.1400 g | $5.53 \times 10^{-4}$ |
| MnCl$_2$ · 4H$_2$O | — | 0.1440 g | $2.18 \times 10^{-5}$ |
| ZnCl$_2$ | 10.4 | 1 mL | $2.29 \times 10^{-6}$ |
| CoCl$_2$ · 6H$_2$O | 4.0 | 1 mL | $5.04 \times 10^{-7}$ |

*Vitamins Stock Solution*

**Into 900 mL** of $dH_2O$, add the thiamine · HCl and 1 mL each of the two primary stocks; bring final volume to 1 liter and filter-sterilize.

| Component | 1° Stock Solution $(g \cdot L^{-1}\ dH_2O)$ | Quantity Used | Concentration in Final Medium (M) |
|---|---|---|---|
| Thiamine · HCl (vitamin $B_1$) | — | 100 mg | $2.96 \times 10^{-7}$ |
| Biotin (vitamin H) | 1 | 1 mL | $4.09 \times 10^{-9}$ |
| Cyanocobalamin (vitamin $B_{12}$) | 2 | 1 mL | $1.48 \times 10^{-9}$ |

## 2.5. Bacterial-Fungal Test Media for Seawater

### f/2 Peptone Test Medium
(Andersen et al. 1997)

This is a general marine test medium used to detect the presence of nonmethylaminotrophic bacteria when attempting to establish the axenicity of a strain. Dissolve 1 g peptone into 1 liter of f/2 medium (see previous recipe); another culture medium, marine or freshwater, may be substituted. Dispense in test tubes and autoclave. Agar can be added to produce a solid medium (see Chapter 2).

## f/2 Methylamine Test Medium

(Andersen et al. 1997)

This is a specific marine test medium to detect the presence of methylaminotrophic bacteria when attempting to establish the axenicity of a strain. Dissolve 1 g methylamine · HCl into 1 liter of f/2 medium (see previous recipe); another culture medium, marine or freshwater, may be substituted. Dispense in test tubes and autoclave. Agar can be added to produce a solid medium (see Chapter 2).

## f/2 Peptone-Methylamine Test Medium

(Andersen et al. 1997)

This is a broad-spectrum marine test medium to detect the presence of many types of marine bacteria when attempting to establish the axenicity of a strain. Dissolve 1 g peptone and 1 g methylamine · HCl into 1 liter of f/2 medium (see previous recipe); another culture medium, marine or freshwater, may be substituted. Dispense in test tubes and autoclave. Agar can be added to produce a solid medium (see Chapter 2).

## Malt Medium in Seawater

(Andersen et al. 1997)

This is a good test medium for detecting the presence of marine fungi when attempting to establish the axenicity of a strain. It uses reduced-salinity seawater as a base. Dissolve 5 g of peptone and 10 g malt extract in a mixture of 750 mL filtered natural seawater and 250 mL dH$_2$O. Final pH should be 7.0 to 7.2 after autoclaving. Dispense in test tubes and autoclave. Agar can be added to produce a solid medium (see Chapter 2). Final salinity should be ~26.

## STP Test Medium

### (Tatewaki and Provasoli 1964)

This is a complex test medium for marine bacteria from coastal habitats (e.g., phytoplankton and seaweed cultures) (Tatewaki and Provasoli 1964). The V8 vitamins solution was first published by Provasoli et al. (1957), but see also Hutner et al. (1949) and Hamilton et al. (1952) for similar compilations. Note that the cyanocobalamin in this mix results in an extremely low concentration in the final medium, and therefore this compound is added twice. For practical purposes, the cyanocobalamin could be excluded from the 8A vitamins mix without having any significant effect on the final solution. STP medium is probably too rich, as formulated, for oceanic bacteria, but it may be effective in growing some "unculturable" bacteria if extremely dilute (e.g., diluted by $10^6$ or more). A variant of this medium, STPM medium, has 1 g methylamine $\cdot$ HCl $\cdot$ L$^{-1}$ for growth of methylaminotrophic bacteria. Tatewaki and Provasoli (1964) provide a second test medium, ST$_3$ medium, that contains a rich organic mixture (C-source mix II). **To prepare,** mix 800 mL of natural seawater and 150 mL of dH$_2$O. Add the following components, making sure each is dissolved before adding the next. Autoclave. Final pH 7.5 to 7.6.

| Component | Stock Solution ($g \cdot L^{-1}$ dH$_2$O) | Quantity Used | Concentration in Final Medium (M) |
|---|---|---|---|
| NaNO$_3$ | — | 200 mg | $2.35 \times 10^{-3}$ |
| K$_2$HPO$_4$ | 10.0 | 1 mL | $5.74 \times 10^{-3}$ |
| Yeast autolysate | — | 200 mg | — |
| Sucrose | — | 1 g | $2.92 \times 10^{-3}$ |
| Na H-glutamate | — | 500 mg | $2.96 \times 10^{-3}$ |
| DL-alanine | — | 100 mg | $1.12 \times 10^{-3}$ |
| Tripticase (BBL) | — | 200 mg | — |
| Glycine | — | 100 mg | $1.33 \times 10^{-3}$ |
| 8A vitamins solution | (See following recipe) | 1 mL | — |
| Soil-water extract | — | 50 mL | — |

*8A Vitamins Solution*

(Provasoli et al. 1957)

Although not specified in the recipe, stock solutions should be prepared for those components with extremely small weights. **To 900 mL** of dH$_2$O, add the components and bring final volume to 1 liter with dH$_2$O. Filter-sterilize.

| Component | 1° Stock Solution (g · L$^{-1}$ dH$_2$O) | Quantity Used | Concentration in Final Medium (M) |
|---|---|---|---|
| Thiamine · HCl (vitamin B$_1$) | — | 200 mg | $5.39 \times 10^{-7}$ |
| Nicotinic acid (niacin) | — | 100 mg | $8.12 \times 10^{-7}$ |
| Putrecine · HCl | — | 40 mg | $2.48 \times 10^{-7}$ |
| Ca pantothenate | — | 100 mg | $4.20 \times 10^{-7}$ |
| Riboflavin (vitamin B$_2$) | — | 5 mg | $1.33 \times 10^{-8}$ |
| Pyridoxine · HCl (vitamin B$_6$) | — | 40 mg | $1.95 \times 10^{-8}$ |
| Pyridoxamine · 2HCl | — | 20 mg | $8.30 \times 10^{-8}$ |
| p-Aminobenzoic acid | — | 10 mg | $7.29 \times 10^{-8}$ |
| Biotin (vitamin H) | — | 0.5 mg | $2.23 \times 10^{-9}$ |
| Choline H$_2$ citrate | — | 500 mg | $1.69 \times 10^{-6}$ |
| Inositol | — | 1 g | $5.55 \times 10^{-6}$ |
| Thymine | — | 800 mg | $6.34 \times 10^{-6}$ |
| Orotic acid | — | 260 mg | $1.67 \times 10^{-6}$ |
| Cyanocobalamin (vitamin B$_{12}$) | — | 0.05 mg | $3.69 \times 10^{-11}$ |
| Folic acid | — | 2.5 mg | $5.67 \times 10^{-9}$ |
| Folinic acid | — | 0.2 mg | $3.91 \times 10^{-10}$ |

## References

Aghajanian, J. G. 1979. A starch grain-mitochondrion-dictyosome association in *Batrachospermum* (Rhodophyta). *J. Phycol.* 15:230–2.

Ahn, C.-Y., Chung, A.-S., and Oh, H.-M. 2002. Diel rhythm of algal phosphate uptake rates in P-limited cyclostats and simulation of its effect on growth and competition. *J. Phycol.* 38:695–704.

Aiba, S., and Ogawa, T. 1977. Assessment of growth yield of a blue-green alga: *Spirulina platensis* in axenic and continuous culture. *J. Gen. Microbiol.* 102:179–82.

Allen, M. B. 1959. Studies with *Cyanidium caldarium*, an anomalously pigmented chlorophyte. *Arch. Mikrobiol.* 32:270–7.

Allen, M. M., and Stanier, R. Y. 1968. Growth and division of some unicellular blue-green algae. *J. Gen. Microbiol.* 51:199–202.

Allen, M. M. 1968. Simple conditions for growth of unicellular blue-green algae on plates. *J. Phycol.* 4:1–4.

Andersen, R. A., Morton, S. L., and Sexton, J. P. 1997. Provasoli-Guillard National Center for Culture of Marine Phytoplankton 1997 list of strains. *J. Phycol.* 33 (suppl.):1–75.

Anderson, E. H. 1945. Nature of the growth factor for the colorless alga *Prototheca zopfi*. *J. Gen. Physiol.* 28:287–96.

Antia, N. J., Cheng, J. Y., and Taylor, F. J. R. 1969. The heterotrophic growth of a marine photosynthetic cryptomonad (*Chroomonas salina*). In: Margalef, R. ed. *Proceedings of the Sixth International Seaweed Symposium*, Subsecretaria de la Marina Mercante, Direccion General de Pesca Maritima, Madrid, pp. 17–29.

Arnon, D. I. 1938. Microelements in culture solution experiments with higher plants. *Am. J. Bot.* 25:232–5.

Beakes, G., Canter, H. M., and Jaworski, G. H. M. 1988. Zoospore ultrastructure of *Zygorhizidium affluens* Canter and *Z. planktonicum* Cantor, two chytrids parasitizing the diatom *Asterionella formosa* Hassall. *Can. J. Bot.* 66:1054–67.

Berges, J. A., Franklin, D. J., and Harrison, P. J. 2001. Evolution of an artificial seawater medium: Improvements in enriched seawater, artificial water over the past two decades. *J. Phycol.* 37:1138–45.

Bischoff, H. W., and Bold, H. C. 1963. *Phycological Studies IV. Some Soil Algae From Enchanted Rock and Related Algal Species.* University of Texas, Austin, 6318:1–95.

Blackburn, S., Hallegraeff, G., and Bolch, C. H. J. 1989. Vegetative reproduction and sexual life cycle of the toxic dinoflagellate *Gymnodinium catenatum* from Tasmania, Australia. *J. Phycol.* 25:577–90.

Bold, H. C. 1949. The morphology of *Chlamydomonas chlamydogama* sp. nov. *Bull. Torrey Bot. Club.* 76:101–8.

Carefoot, J. R. 1968. Culture and heterotrophy of the freshwater dinoflagellate, *Peridinium cinctum* fa. *ovoplanum* Lindeman. *J. Phycol.* 4:129–31.

Castenholz, R. W. 1969. Thermophilic blue-green algae and the thermal environment. *Bacteriol. Rev.* 33:476–504.

Castenholz, R. W. 1988. Culturing methods for cyanobacteria. In: Packer, L., and Glazer, A. N., eds. *Cyanobacteria. Meth. Enzymol.* 167:68–100.

Chen, Y.-B., Zehr, J. P., and Mellon, M. 1996. Growth and nitrogen fixation of the diazotrophic filamentous nonheterocystous cyanobacterium *Trichodesmium* sp. IMS 101 in defined media: evidence for a circadian rhythm. *J. Phycol.* 32:916–23.

Chu, S. P. 1942. The influence of the mineral composition of the medium on the growth of planktonic algae. Part I. Methods and culture media. *J. Ecol.* 30:284–325.

Coesel, P. F. M. 1991. Ammonium dependency in *Closterium aciculare* T. West, a planktonic desmid from alkaline, eutrophic waters. *J. Plank. Res.* 13:913–22.

Cohn, S. A., and Disparti, N. C. 1994. Environmental factors influencing diatom cell motility. *J. Phycol.* 30:818–28.

Cohn, S. A., Farrell, J. F., Munro, J. D., Ragland, R. L., Weitzell, R. E., Jr., and Wibisono, B. L. 2003. The effect of temperature and mixed species composition on diatom motility and adhesion. *Diatom Res.* 18:225–43.

Cohn, S. A., and Pickett-Heaps, J. D. 1988. The effects of colchicines and dinitrophenol on the *in vivo* rates of anaphase A and B in the diatom *Surirella*. *Eur. J. Cell Biol.* 46:523–30.

Cohn, S. A., and Weitzel, R. E. 1996. Ecological considerations of diatom cell motility: I. characterization of motility and adhesion in four diatom species. *J. Phycol.* 32:928–39.

Cohn, S. A., Spurck, T. R., and Pickett-Heaps, J. D. 1989. Perizonium and initial valve formation in the diatom *Navicula cuspidata* (Bacillariophyceae). *J. Phycol.* 25:15–26.

Cole, G. T., and Wynne, M. J. 1974. Endocytosis of *Microcystis aeruginosa* by *Ochromonas danica*. *J. Phycol.* 10:397–410.

D'Agostino, A. S., and Provasoli, L. 1968. Effects of salinity and nutrients on mono- and diaxenic cultures of two strains of *Artemia salina*. *Biol. Bull.* 134:1–14.

D'Agostino, A. S., and Provasoli, L. 1970. Dixenic culture of *Daphnia magna*, Straus. *Biol. Bull.* 139:485–94.

Darling, R. B., Friedmann, E. I., and Broady, P. A. 1987. *Heterococcus endolithicus* (Xanthophyceae) and other terrestrial *Heterococcus* species from Antarctica: Morphological changes during life history and response to temperature. *J. Phycol.* 23:598–607.

Donk, E. van, and Kilham, S. S. 1990. Temperature effects on silicon and phosphorus-limited growth and competitive interactions among three diatoms. *J. Phycol.* 26:40–50.

Elendt, B. P., and Bias, W. R. 1990. Trace nutrient deficiency in *Daphnia magna* cultured in standard medium for toxicity testing: effects of the optimization of culture conditions on life history parameters of *Daphnia magna*. *Wat. Res.* 24:1157–67.

Erata, M., and Chihara, M. 1987. Cryptomonads from the Sugadaira-Moor, Central Japan [in Japanese with English summary]. *Bull. Sugadaira Mont. Res. Center, Univ. Tsukuba,* 8:57–69.

Errécalde, O., and Campbell, P. G. C. 2000. Cadmium and zinc bioavailability to *Selenastrum capricorniutum* (Chlorophyceae): Accidental metal uptake and toxicity in the presence of citrate. *J. Phycol.* 36:473–83.

Floyd, G. L., Stewart, K. D., and Mattox, K. R. 1972. Cellular organization, mitosis, and cytokinesis in the ulotrichalean alga, *Klebsormidium*. *J. Phycol.* 8:176–84.

Foissner, I. 1989. pH-dependence of chlortetracycline (CTC)-induced plug formation in *Nitella flexilis* (Characeae). *J. Phycol.* 25:313–8.

Forsberg, C. 1965. Nutritional studies of *Chara* in axenic cultures. *Physiologia Plantarum* 18:275–90.

Fox, C. H. 1967. Studies of the cultural physiology of the lichen alga *Trebouxia*. *Physiologia Plantarum* 20: 251–62.

Føyn, B. 1934. Lebenszyklus, Cytologie und Sexualität der Chlorophycee *Cladophora suhriana* Kützing. *Arch. Protistenk.* 83:1–56.

Freire-Nordi, C. S., Vieira, A. A. H., and Nascimento, O. R. 1998. Selective permeability of the extracellular envelope of the microalga *Spondylosium panduriforme* (Chlorophyceae) as revealed by electron paramagnetic resonance. *J. Phycol.* 34:631–7.

Garcia-Pichel, F. and Castenholz, R. W. 1991. Characterization and biological implications of scytonemin, a cyanobacterial sheath pigment. *J. Phycol.* 27:395–405.

Gargiulo, G. M., Genovese, G. Morabito, M. Culoso, F., and De Masi, F. 2001. Sexual and asexual reproduction in a freshwater population of *Bangia atropurpurea* (Bangiales, Rhodophyta) from eastern Sicily (Italy). *Phycologia* 40:88–96.

Genesmer, R. W. 1990. Role of aluminum and growth rate on changes in cell size and silica content of silica-limited populations of *Asterionella ralfsii* var. *americana* (Bacillariophyceae). *J. Phycol.* 26:250–8.

Gensemer, R. W., Smith, R. E. H., and Duthie, H. C. 1993. Comparative effects of pH and aluminum on silica-limited growth and nutrient uptake in *Asterionella ralfsii* var. *americana* (Bacillariophyceae). *J. Phycol.* 29:36–44.

Gerloff, G. C., Fitzgerald, G. P., and Skoog, F. 1950. The isolation, purification and nutrient solution requirements of blue-green algae. In: Brunel, J., Prescott, G. W., and Tiffany, L. N. eds. *Proceedings of the Symposium on the Culturing of Algae.* Charles F. Kettering Foundation, Dayton, Ohio, pp. 27–44.

Gervais, F. 1997. Light-dependent growth, dark survival, and glucose uptake by cryptophytes isolated from a freshwater chemocline. *J. Phycol.* 33:18–25.

Glazer, A. N., Chan, C. F., and West, J. A. 1997. An unusual phycocyanobilin-containing phycoerythrin of several bluish-colored, acrochaetioid, freshwater red algal species. *J. Phycol.* 33:617–24.

Goldman, J. C., and McCarthy, J. J. 1978. Steady state growth and ammonium uptake of a fast growing marine diatom. *Limnol. Oceanogr.* 23:695–703.

Graham, J. M., Auer, M. T., Canale, R. P., and Hoffmann, J. P. 1982. Ecological studies and mathematical modeling of *Cladophora* in Lake Huron: 4. Photosynthesis and respiration as functions of light and temperature. *J. Great Lakes Res.* 8:100–11.

Graham, J. M., Lembi, C. A., Adrian, H. L., and Spencer, D. F. 1995. Physiological responses to temperature and irradiance in *Spirogyra* (Zygnematales, Charophyceae). *J. Phycol.* 31:531–40.

Graham, L. E., Graham, J. M., and Kranzfelder, J. A. 1986. Irradiance, daylength and temperature effects on zoosporogenesis in *Coleochaete scutata* (Charophyceae). *J. Phycol.* 22:35–9.

Guillard, R. R. L. 1975. Culture of phytoplankton for feeding marine invertebrates. In: Smith, W. L., and Chanley, M. H., eds. *Culture of Marine Invertebrate Animals.* Plenum Press, New York, pp. 26–60.

Guillard, R. R. L., and Hargraves, P. E. 1993. *Stichochrysis immobilis* is a diatom, not a chrysophyte. *Phycologia* 32:234–6.

Guillard, R. R. L., and Lorenzen, C. J. 1972. Yellow-green algae with chlorophyllide *c*. *J. Phycol.* 8:10–4.

Guillard, R. R. L., and Ryther, J. H. 1962. Studies of marine planktonic diatoms. I. *Cyclotella nana* Hustedt and *Detonula confervacea* Cleve. *Can. J. Microbiol.* 8:229–39.

Guiry, M., and Cunningham, E. 1984. Photoperiodic and temperature responses in the reproduction of the northeastern Atlantic *Gigartina acicularis* (Rhodophyta: Gigartinales). *Phycologia* 23:357–67.

Halfen, L. N. 1973. Gliding motility of *Oscillatoria*: Ultrastructural and chemical characterization of the fibrillar layer. *J. Phycol.* 9:248–53.

Hall, W. T., and Claus, G. 1967. Ultrastructural studies on the cyanelles of *Glaucocystis nostochinearum* Itizisohn. *J. Phycol.* 3:37–51.

Hamilton, L. D., Hutner, S. H., and Provasoli, L. 1952. The use of protozoa in analysis. *The Analyist* (London) 77:618–28.

Hämmerling, J. 1931. Entwicklung und Formbildungsvermögen von *Acetabularia mediterranea*. *Biol. Zentralbl.* 51: 633–47.

Harrison, P. J., Waters, R. E., and Taylor, F. J. R. 1980. A broad spectrum artificial seawater medium for coastal and open ocean phytoplankton. *J. Phycol.* 16:28–35.

Healey, F. P., and Hendzel, L. L. 1988. Competition for phosphorus between desmids. *J. Phycol.* 24:287–92.

Hepperle, D., and Krienitz, L. 1997. *Phacotus lenticularis* (Chlamydomonadales, Phacotaceae) zoospores require external supersaturation of calcium carbonate for calcification in culture. *J. Phycol.* 33:415–24.

Hoffman, J. P., and Graham, L. E. 1984. Effects of selected physiochemical factors on growth and zoosporogenesis of *Cladophora glomerata* (Chlorophyta). *J. Phycol.* 20:1–7.

Hughes, E. O., Gorham, P. R., and Zehnder, A. 1958. Toxicity of a unialgal culture of *Microcystis aeruginosa*. *Can. J. Microbiol.* 4:225–36.

Hutner, S. H., Provasoli, L., Stokstad, E. L. R., Hoffman, C. E., Belt, M., Franklin, A. L., and Jukes, T. H. 1949. Assay of anti-pernicious anemia factor with *Euglena*. *Proc. Soc. Exp. Biol. Med.* 70:118–20.

Ichimura, T. 1971. Sexual cell division and conjugation-papilla formation in sexual reproduction of *Closterium strigosum*. In: Nishizawa, K., Arasaki, S., Chihara, M., Hirose, H., Nakamura, V., Tsuchiya, Y. eds. Proceedings of the Seventh International Seaweed Symposium, Sapporo, Japan, August 8–12, 1971. *Proceedings of the Seventh International Seaweed Symposium*. University of Tokyo Press, Tokyo, pp. 208–14.

Ichimura, T. 1979. Sôrui no bunri to baiyôhô: Tansui sôrui (Isolation and culture methods of algae: Freshwater algae). In: Nishizawa, K., and Chihara, M., eds. *Sôrui Kenkyûhô (Methods in Phycological Studies)*. Kyorutsu Shuppan, Tokyo, pp. 294–305.

Ichimura, T., and Watanabe, M. 1974. The *Closterium calosporum* complex from the Ryukyu Islands: Variation and taxonomical problems (plates 13, 14). *Mem. National Sci. Mus. Tokyo* 7:89–102.

Iwasaki, H. 1961. The life-cycle of *Porphyra tenera in vitro*. *Biol. Bull.* 121:173–87.

Kato, S. 1982. Laboratory culture and morphology of *Colacium vesiculosum* Ehrb. (Euglenophyceae) [in Japanese with English summary; recipe in English]. *Jap. J. Phycol.* 30:63–7.

Kato, T., Watanabe, M. F., and Watanabe, M. 1991. Allozyme divergence in *Microcystis* (Cyanophyceae) and its taxonomic inference. *Arch. Hydrobiol. Suppl. 92: Algol. Stud.* 64:129–40.

Kaya, K., and Watanabe, M. M. 1990. Microcystin composition of an axenic clonal strain of *Microcystis viridis* and *Microcystis viridis* containing waterblooms in Japanese freshwaters. *J. Appl. Phycol.* 2:173–8.

Keller, M. D., Selvin, R. C., Claus, W., and Guillard, R. R. L. 1987. Media for the culture of oceanic ultraphytoplankton. *J. Phycol.* 23:633–8.

Kilham, S. S., Kreeger, D. A., Lynn, S. G., Goulden, C. E., and Herrera, L. 1998. COMBO: A defined freshwater culture medium for algae and zooplankton. *Hydrobiologia* 377:147–59.

Kimura, B., and Ishida, Y. 1986. Photophagotrophy in *Uroglena americana*, Chrysophyceae. *Jpn. J. Limnol.* 46:315–8.

Kirk, D. L., and Kirk, M. M. 1978. Amino acid and urea uptake in ten species of Chlorophyta. *J. Phycol.* 14:198–203.

Klaveness, D. 1982. The *Cryptomonas-Caulobacter* consortium: Facultative ectocommensalism with possible taxonomic consequences? *Nord. J. Bot.* 2:183–8.

Kobl, I., Kirk, D. L., and Schmitt, R. 1998. Quantitative PCR data falsify the chromosomal endoreduplication hypothesis for *Volvox carteri* (Volvocales, Chlorophyta). *J. Phycol.* 34:981–8.

Kugrens, P. 1984. *Octosporiella coloradoensis* gen. et sp. nov., a new tetrasporalean green alga from two Colorado Mountain lakes. *J. Phycol.* 20:88–94.

Kugrens, P., Clay, B. L., and Lee, R. E. 1999. Ultrastructure and systematics of two new freshwater red cryptomonads, *Stoeatula rhinosa*, sp. nov., and *Pyrenomonas ovalis*, sp. nov. *J. Phycol.* 35:1079–89.

Kugrens, P., Lee, R. E., and Andersen, R. A. 1987. Ultrastructural variations in cryptomonad flagella. *J. Phycol.* 23:511–8.

Kurata, A. 1986. Blooms of *Uroglena americana* in relation to concentrations of B group vitamins. In: Kristiansen, J., and Andersen, R. A., eds. *Chrysophytes: Aspects and Problems*. Cambridge University Press, Cambridge, pp. 185–96.

Kusumi, T., Ooi, T., Watanabe, M. M., Takahashi, H., and Kakisawa, H. 1987. Cyanoviridin RR, a toxin from the cyanobacterium (blue-green alga) *Microcystis viridis*. *Tetrahedron Letters* 28:4695–8.

Lehman, J. T. 1976. Ecological and nutritional studies on *Dinobryon* Ehrenb.: Seasonal periodicity and the phosphate toxicity problem. *Limnol. Oceanog.* 21:646–58.

Li, R., Yokota, A., Sugiyama, J., Watanabe, M., Hiroki, M., and Watanabe, M. M. 1998. Chemotaxonomy of planktonic cyanobacteria based on non-polar and 3-hydroxy fatty acid composition. *Phycol. Res.* 46:21–8.

Loeblich, A. 1975. A seawater medium for dinoflagellates and the nutrition of *Cachonina niei*. *J. Phycol.* 11:80–6.

Mann, D. G., Chepurnow, V. A., and Droop, S. J. M. 1999. Sexuality, incompatibility, size variation, and preferential polyandry in natural populations and clones *Selllaphora pupula* (Bacillariophyceae). *J. Phycol.* 35:152–70.

McCracken, D. A., Nadakavukaren, M. J., and Cain, J. R. 1980. A biochemical and ultrastructural evaluation of the taxonomic position of *Glaucosphaera vacuolata* Korsh. *N. Phytol.* 86:39–44.

McLachlan, J. 1964. Some considerations of the growth of marine algae in artificial media. *Can. J. Microbiol.* 10:769–82.

McLachlan, J. 1973. Growth media: marine. In: Stein, J. R., ed. *Culture Methods and Growth Measurements.* Cambridge University Press, Cambridge, pp. 25–51.

Moore, L. R., Post, A. F., Rocap, G., and Chisholm, S. W. 2002. Utilization of different nitrogen sources by the marine cyanobacteria *Prochlorococcus* and *Synechococcus. Limnol. Oceangr.* 47:989–996.

Morel, F. M. M., Rueter, J. G., Anderson, D. M., and Guillard, R. R. L. 1979. Aquil: A chemically defined phytoplankton culture medium for trace metal studies. *J. Phycol.* 15:135–41.

Morel, F. M. M., Westall, J. C., Reuter, J. G., and Chaplick, J. P. 1975. *Description of the algal growth media "Aquil" and "Fraquil." Technical Report 16.* Water Quality Laboratory, Ralph Parsons Laboratory for Water Resources and Hydrodynamics, Massachusetts Institute of Technology, Cambridge, 33 pp.

Nakahara, H. 1981. Urogurena no seikatusi (Life cycle of *Uroglena americana*). In: Kadota, G., ed. Biwako ni okeru purankuton no ijôhasseikikô ni kansuru tyôsa (Survey on red tide outbreaks in Lake Biwa) [in Japanese without English titles]. Biwako purankuton ijôhassei tyôsadan, Otsu, Japan, pp. 137–44.

Nalewajko, C., and O'Mahony, M. A. 1989. Photosynthesis of algal cultures and phytoplankton following an acid pH shock. *J. Phycol.* 25:319–25.

Nalewajko, C., Lee, K., and Olaveson, M. 1995. Responses of freshwater algae to inhibitory vanadium concentrations: The role of phosphorus. *J. Phycol.* 31:332–43.

Noël, M.-H., Kawachi, M., and Inouye, I. 2004. Induced dimorphic life cycle of a coccolithophorid, *Calyptrosphaera sphaeroidea* (Prymnesiophyceae, Haptophyta). *J. Phycol.* 40:112–29.

Ogawa, T., and Terui, G. 1970. Studies on the growth of *Spirulina platensis* (I) On the pure culture of *Spirulina platensis. J. Ferment. Technol.* 48:361–7.

Ohki, K., Zehr, J. P., and Fujita, Y. 1992. *Trichodesmium*: Establishment of culture and characteristics of N-fixation. In: Carpenter, E. J., Capone, D. G., and Rueter, J. G., eds. *Marine Pelagic Cyanobacteria: Trichodesmium and Other Diazotrophs.* Kluwer Academic Publishers, Dordrecht, The Netherlands, pp. 307–18.

Olsen, Y. 1989. Evaluation of competitive ability of *Staurastrum luetkemuellerii* (Chlorophyceae) and *Microcystis aeruginosa* (Cyanophyceae) under P limitation. *J. Phycol.* 25:486–99.

Olsen, Y., Vadstein, O., Andeerson, T., and Jensen, A. 1989. Competition between *Staurastrum luetkemuellerii* (Chlorophyceae) and *Microcystis aeruginosa* (Cyanophyceae) under varying modes of phosphorus supply. *J. Phycol.* 25:499–508.

Ooi, T., Kusumi, T., Kakisawa, H., and Watanabe, M. M. 1989. Structure of cyanoviridin RR, a toxin from the blue-green alga, *Microcystis viridis. J. Appl. Phycol.* 1:31–8.

Otsuka, S., Suda, S., Li, R., Matsumoto, S., and Watanabe, M. M. 2000. Morphological variability of colonies of *Microcystis* morphospecies in culture. *J. Gen. Appl. Microbiol.* 46:39–50.

Otsuka, S., Suda, S., Li, R., Watanabe, M., Oyaizu, H., Hiroki, M., Mahakhant, A., Liu, Y., Matsumoto, S., Watanabe, M. M. 1998a. Phycoerythrin-containing *Microcystis* isolated from P. R. China and Thailand. *Phycol. Res.* 46 (Suppl.):45–50.

Otsuka, S., Suda, S., Li, R., Watanabe, M., Oyaizu, H., Matsumoto, S., and Watanabe, M. M. 1998b. 16S rDNA sequences and phylogenetic analyses of *Microcystis* strains with and without phycoerythrin. *FEMS Microbiology Letters* 164:119–24.

Otsuka, S., Suda, S., Li, R., Watanabe, M., Oyaizu, H., Matsumoto, S., and Watanabe, M. M. 1999a. Characterization of morphospecies and strains of the genus *Microcystis* (Cyanobacteria) for a reconsideration of species classification. *Phycol. Res.* 47:189–97.

Otsuka, S., Suda, S., Li, R., Watanabe, M., Oyaizu, H., Matsumoto, S., and Watanabe, M. M. 1999b. Phylogenetic relationships between toxic and non-toxic strains of the genus *Microcystis* based on 16S to 23S internal transcribed spacer sequences. *FEMS Microbiology Letters* 172:15–21.

Otsuka, S., Suda, S., Shibata, S., Oyaizu, H., Matsumoto, S., and Watanabe, M. M. 2001. A proposal for the unification of five species of the cyanobacterial genus *Microcystis* Kutzing ex Lemmermann 1907 under the Rules of the Bacteriological Code. *Int. J. Syst. Evol. Microbiol.* 51:873–9.

Otsuki, A., Watanabe, M. M., and Sugahara, K. 1987. Chlorophyll pigments in methanol extracts from ten axenic cultured diatoms and three green algae as determined by reverse phase HPLC with fluorometric detection. *J. Phycol.* 23:406–14.

Patni, N. J., and Aaronson, S. 1974. The nutrition, resistance to antibiotics and ultrastructure of *Prototheca wickerhamii. J. Gen. Microbiol.* 83:179–82.

Price, N. M., Harrison, G. I., Hering, J. G., Hudson, R. J., Nirel, P. M. V., Palenik, B., and Morel, F. M. M. 1989. Preparation and chemistry of the artificial algal culture medium Aquil. *Biol. Oceanogr.* 6:443–61.

Pringsheim, E. G. 1946. The biphasic or soil-water culture method for growing algae and flagellata. *J. Ecol.* 33:193–204.

Pringsheim, E. G. 1955. The genus *Polytomella. J. Protozool.* 2:137–45.

Pringsheim, E. G. 1963. *Farblose Algen.* Gustav Fischer, Stuttgart, Germany, 471 pp.

Provasoli, L. 1958. Effect of plant hormones on *Ulva. Biol. Bull.* 114:375–84.

Provasoli, L. 1963. Growing marine seaweeds. In: De Virville, A. D., and Feldmann, J., eds. *Proceedings of the Fourth International Seaweed Symposium*. Pergamon Press, Oxford, pp. 9–17.

Provasoli, L. 1968. Media and prospects for the cultivation of marine algae. In: Watanabe, A., and Hattori, A., eds. *Cultures and Collections of Algae. Proceedings of the U.S.–Japan Conference, Hakone, Japan, September 1966*. Japanese Society of Plant Physiology, pp. 63–75.

Provasoli, L., and Pintner, I. J. 1953. Ecological implications of *in vitro* nutritional requirements of algal flagellates. *Ann. N. Y. Acad. Sci.* 56:839–51.

Provasoli, L., and Pintner, I. J. 1960. Artificial media for freshwater algae: Problems and suggestions. In: Tyron, C. A., Jr., and Hartman, R. T., eds. *The Ecology of Algae*. Special Publication 2. Pymatuning Laboratory of Field Biology, University of Pittsburgh, Pittsburgh, pp. 84–96.

Provasoli, L., and Pintner, I. J. 1980. Bacteria induced polymorphism in an axenic laboratory strain of *Ulva lactuca* (Chlorophyceae). *J. Phycol.* 16:196–201.

Provasoli, L., Conklin, D. E., and D'Angostino, A. S. 1970. Factors inducing fertility in aseptic Crustacea. *Helgoländer wiss. Meeresunters.* 20:443–54.

Provasoli, L., McLaughlin, J. J. A., and Droop, M. R. 1957. The development of artificial media for marine algae. *Arch. Mikrobiol.* 25:392–428.

Rippka, R. 1988. Isolation and purification of cyanobacteria. *Method Enzymol.* 167:3–27.

Rippka, R., Deruelles, J., Waterbury, J. B., Herdman, M., and Stanier, R. Y. 1979. Generic assignments, strain histories and properties of pure cultures of cyanobacteria. *J. Gen. Microbiol.* 111:1–61.

Rosowski, J. R., and Kugrens, P. 1973. Observations on the euglenoid *Colacium* with special reference to the formation and morphology of attachment material. *J. Phycol.* 9:370–83.

Rueter, J. G. 1988. Iron stimulation of photosynthesis and nitrogen fixation in *Anabaena* 7120 and *Trichodesmium* (Cyanophyceae). *J. Phycol.* 24:249–54.

Rueter, J. G., and Ades, D. R. 1987. The role of iron nutrition in photosynthesis and nitrogen assimilation in *Scenedesmus quadricauda* (Chlorophyceae). *J. Phycol.* 23:452–7.

Sako, Y., Ishida, Y., Kadota, H., and Hata, Y. 1984. Sexual reproduction and cyst formation in the freshwater dinoflagellate *Peridinium cunningtonii. Bull. Jpn. Soc. Sci. Fish.* 50:743–50.

Sako, Y., Ishida, Y., Kadota, H., and Hata, Y. 1985. Excystment in the freshwater dinoflagellate *Peridinium cunningtonii. Bull. Jpn. Soc. Sci. Fish.* 51:267–72.

Sako, Y., Ishida, Y., Nishijima, T., and Hata, Y. 1987. Sexual reproduction and cyst formation in the freshwater dinofla-gellate *Peridinium penardii. Bull. Jpn. Soc. Sci. Fish.* 53:474–8.

Schlegel, I., Krienitz, L., and Hepperle, D. 2000. Variability of calcification of *Phacotus lenticularis* (Chlorophyta, Chlamydomonadales) in nature and culture. *Phycologia* 39:318–22.

Schlösser, U. G. 1994. SAG-Sammlung von Algenkulturen at the University of Göttingen Catalogue of Strains 1994. *Bot. Acta* 107:111–86.

Schmek, C., Stadnichuk, N. I., Knaust, R., and Wehrmeyer, W. 1994. Detection of chlorophyll $C_1$ and magnesium-2,4-divinylpheoporphyrin $A_5$ monomethylester in cryptophytes. *J. Phycol.* 30:621–7.

Schreiber, E. 1927. Die Reinkultur von marinen Phytoplankton und deren Bedeutung für die Erforschung der Produktionsfähigkeit des Meerwassers. *Wiss. Meeresuntersuch., N. F.* 10:1–34.

Schultz, M. E., and Trainor, F. R. 1968. Production of male gametes and auxospores in the centric diatoms *Cyclotella meneghiniana* and *C. cryptica. J. Phycol.* 4:85–8.

Schuster, A. M., Waddle, J. A., Korth, K., and Meints, R. H. 1990. Chloroplast genome of an exsymbiotic *Chlorella*-like green alga. *Plant Mol. Biol.* 14. 859–62.

Sheridan, R. P. 1966. Photochemical and dark reduction of sulfate and thiosulfate to hydrogen sulfide in *Synechococcus lividus*. PhD. Thesis, University of Oregon, Eugene, 104 pp.

Soma, Y., Imaizumi, T., Yagi, K., and Kasuga, S. 1993. Estimation of algal succession in lake water using HPLC analysis of pigments. *Can. J. Fish. Aquat. Sci.* 50:1142–6.

Sommerfeld, M. R., and Nichols, H. W. 1970. Comparative studies in the genus *Porphyridium* Naeg. *J. Phycol.* 6:67–78.

Stanier, R. Y., and Cohen-Bazire, G. 1977. Phototrophic prokaryotes: the cyanobacteria. *Ann. Rev. Microbiol.* 31:225–74.

Starr, R. C. 1964. The culture collection of algae at Indiana University. *Am. J. Bot.* 51:1013–44.

Starr, R. C., and Zeikus, J. A. 1993. UTEX: the culture collection of algae at the University of Texas at Austin. *J. Phycol.* 29 (suppl.):1–106.

Suda, S., Watanabe, M. M., Otsuka, S., Mahakhant, A., Yongmanitchai, W., Noparatnaraporn, N., Liu, Y., and Day, J. 2002. Taxonomic revision of water-bloom-forming species of oscillatorioid cyanobacteria. *Int. J. Syst. Evol. Microbiol.* 52:1577–95.

Sweeney, B., Haxo, F., and Hastings, J. 1959. Action spectra for two effects of light on luminescence in *Gonyaulax polyedra. J. Gen. Physiol.* 43:285–99.

Takamura, N., Kasai, F., and Watanabe, M. M. 1988. The effects of Cu, Cd, and Zn on photosynthesis of freshwater benthic algae. *J. Appl. Phycol.* 1:39–52.

Takamura, N., Kasai, F., and Watanabe, M. M. 1989. Unique response of Cyanophyceae to copper. *J. Appl. Phycol.* 2:293–6.

Tatewaki, M., and Provasoli, L. 1964. Vitamin requirements of three species of *Antithamnion. Bot. Marina* 6:193–203.

Thomas, D. L., and Montes, J. G. 1978. Spectrophotometrically assayed inhibitory effects of mercuric compounds of *Anabaena flos-aquae* and *Anacystis nidulans* (Cyanophyceae). *J. Phycol.* 14:494–9.

Tilman, D., and Kilham, S. S. 1976. Phosphate and silicate growth and uptake kinetics of the diatoms *Asterionella formosa* and *Cyclotella meneghiniana* in batch and semicontinuous culture. *J. Phycol.* 12:375–83.

Tompkins, J., DeVille, M. M., Day, J. G., and Turner, M. F. 1995. *Culture Collection of Algae and Protozoa. Catalog of Strains.* Ambleside, UK, 204 pp.

von Stosch, H. 1963. Wirkungen von jod un arsenit auf meeresalgen in kultur. *In*: De Virville, D. and Feldmann, J. eds. *Proceedings of the Fourth International Seaweed Symposium*, Pergamon Press, Oxford, pp. 142–50.

Walne, P. R. 1970. Studies on food value of nineteen genera of algae to juvenile bivalves of the genera *Ostrea, Crassostrea, Mercenaria* and *Mytilus. Fish. Invest. Lond. Ser. 2.* 26(5): 1–62.

Watanabe, M. M. 1983. Growth characteristics of freshwater red tide alga *Peridinium*, based on axenic culture: Establishment of synthetic culture medium [in Japanese]. *Res. Data Natl. Inst. Environ. Stud.* 24:111–21.

Watanabe, M. M., Kawachi, M., Hiroki, M., and Kasai, F. 2000. *NIES Collection List of Strains. Sixth Edition, 2000, Microalgae and Protozoa.* Microbial Culture Collections, National Institute for Environmental Studies, Tsukuba, Japan, 159 pp.

Watanabe, M. M., Takeuchi, Y., and Takamura, N. 1987. Cu tolerance of freshwater diatom, *Achnanthes minutissima*. In: Yasuno, M., and Whitton, B. A., eds. *Biological Monitoring of Environmental Pollution.* Tokai University Press, Tokyo, pp. 171–7.

Waterbury, J. B., Watson, S. W., Valois, F. W., and Franks, D. G. 1986. Biological and ecological characterization of the marine unicellular cyanobacterium *Synechococcus*. In: Platt, T., and Li, W. K. I., eds. *Photosynthetic Picoplankton. Can. Bull. Fish. Aquatic Sci.* 214:71–120.

West, J. A., and McBride, D. L. 1999. Long-term and diurnal carpospore discharge patterns in the Ceramiaceae, Rhodomelaceae and Delesseriaceae (Rhodophyta). *Hydrobiologia* 298/299:101–13.

Westall, J. C., Zachary, J. L., and Morel, F. M. M. 1976. *MINEQL: A Computer Program for the Calculation of Chemical Equilibrium Composition of Aqueous Systems. Technical Report 18.* Water Quality Laboratory, Ralph Parsons Laboratory for Water Resources and Hydrodynamics, Massachusetts Institute of Technology, Cambridge, 91 pp.

Wolff, G., and Kück, U. 1990. The structural analysis of the mitochondrial SSU rRNA implies a close phylogenetic relationship between mitochondria from plants and from the heterotrophic alga *Prototheca wickerhamii. Curr. Genet.* 17:347–51.

Wright, R. T. 1964. Dynamics of a phytoplankton community in an ice covered lake. *Limnol. Oceanogr.* 9:163–78.

# GLOSSARY

**Acclimated growth rate** The specific growth rate of a given genotype of algae (= clone) when it is fully acclimated to a given set of growth conditions. In general, this is the maximum growth rate that the genotype can maintain under the specified growth conditions. It may take approximately 10 generations before cells fully acclimate.

**$\alpha_M$** The side reaction coefficient for inorganic complexation equal to $[M']/[M^{n+}]$.

**Anisogamous** Morphologically different gametes.

**Archeopyle** The opening in the cyst wall through which the cell exits upon germination.

**AS** Artificial seawater (= synthetic ocean water or SOW).

**Autecology** The study of the ecology of an individual species.

**Axenic culture** 1. Culture of a single algal species, free of all contaminants except viruses. 2. Culture of a single algal species, free of all contaminants, including viruses.

**Bacteriophage** A virus that infects bacteria.

**Balanced growth** Steady-state growth of a population of cells. Except for diurnal fluctuations associated with effects of the photoperiod on physiological processes, the average per cell concentration of major cell constituents remains constant, and the rate of change of these constituents in the culture (population) is constant.

**Batch culture** A closed system culture of algae in which a small inoculum is added to a vessel containing a specified volume of medium. The initial growth is not nutrient limited. Yield is controlled by the concentration of a limiting nutrient (e.g., trace metal buffer experiments)

or by light or by $CO_2/pH$ (e.g., when the culture becomes very dense).

**Binary fission** Mode of cell division in which a cell divides into two equal daughter cells.

**Capsid** The protein coat of a virus that surrounds the nucleic acid core.

**Chelate** A strong complex between an organic ligand and a metal.

**Chelation** The reaction of a metal with an organic ligand to form a chelate.

**Chelator** An organic ligand that forms stable complexes with metal ions.

**Chelex®** A chelating resin used to remove trace metals from culture media and reagent stock solutions.

**Chemostat** A type of continuous culture system in which the biomass is determined by the concentration of the limiting nutrient in the inflowing medium, and the growth rate is determined by the dilution rate (flow rate divided by the volume of the chemostat). The growth rate is equal to the dilution rate, which is set by the experimenter and held constant. When steady state is reached, cell numbers remain constant over time. The dilution rate and growth rate must be lower than the nutrient-replete growth rate at the experimental temperature and irradiance, otherwise the culture will be washed out.

**Chilling injury** Injury to a living organism that results from its cooling to a temperature below its normal growth temperature, but not caused by freezing the organism or its immediate surroundings.

**Chronic infection** The continuous production of infectious virus either by the survival of an infected host or by a situation in which a minority of cells in a population are infected and the virus spread is limited. Also known as persistent infection.

**Clonal culture** 1. Culture strain established from a single cell and maintained only by vegetative reproduction; no homothallic sexual reproduction (truly clonal). 2. Culture strain established from a single cell that may exhibit homothallic sexual reproduction (meiosis rearranges the heterozygous alleles, hence no longer truly clonal). Note that it is not always possible to tell if a cultured strain undergoes heterothallic sexual reproduction. Therefore, the assumption of true clonality for new isolates must be considered carefully.

**Clonal culture of macroalgae** A culture of a single genome set (e.g., cultures derived from a single vegetative cell or tissue or a reproductive cell) and propagated vegetatively.

**Colligative effect** An effect of dissolving a substance in solution that depends on the concentration of dissolved particles (molecules or ions), not on the chemical nature of the particles. For example, lowering of the equilibrium freezing point is a colligative effect.

**Complex** A chemical species containing one or more ligands noncovalently and reversibly bonded to a metal.

**Complexation** The reaction of a metal with one or more ligands to form a metal-ligand complex.

**Continuous culture** An open culture system. Medium flows into the culture vessel at the same rate it flows out. In steady state, the concentration of algae and the growth rate of the algae in the culture vessel are constant. The chemostat and turbidostat are types of continuous cultures.

**Coulter Counter** An instrument that can measure the volume and abundance of cells based on changes in electrical impedance.

**Crude culture of macroalgae** A culture that includes contaminants (other macroalgae, microalgae, protozoa, bacteria, and so forth).

**Cryopreservation** The process of storing at ultralow temperature (generally colder than −130°C) a living organism or portion thereof, such that it remains capable of survival upon thawing.

**Cryoprotective agent (CPA)** A chemical substance that, when exposed to a living system before cryopreservation, reduces the extent of injury during the freezing and thawing processes.

**Cryostorage** Continuous maintenance at an ultralow temperature (colder than −130°C or preferably less than −150°C).

**Cyst** Resting cell with environmentally resistant outer coating.

**Diploid** Cells with two sets of chromosomes: two copies of the basic genetic complement of the species.

**Disinfection** A process or procedure that kills or reduces the number of pathogenetic microorganisms in an environment or on a surface.

**DMSO (Me₂SO, dimethyl sulfoxide)** A frequently used cryoprotective agent.

**Dormancy** Period between resting cyst formation and germination.

**Ecdysis** Shedding of a cell wall or theca.

**EDTA** Ethylenediaminetetraacetic acid, a chelator used in metal ion buffer systems.

**EDTA\*** All EDTA species not complexed to trace metals, including protonated forms of the ligand and Ca and Mg chelates.

**[EDTA\*]** The molar concentration of EDTA\*; square brackets denote concentrations of enclosed chemical species.

**Encystment** The processes involved in cyst formation.

**Endogenous** Resulting from factors internal to an organism.

**Enrichment culture** Natural sample or rough culture with added nutrients; usually contains several algal species and contaminants.

**Entrained rhythm** An endogenous rhythm that is entrained by the environment, e.g., under a light-dark (LD) regimen (LD 12:12). During entrainment, the free-running period of the circadian rhythm is reset to exactly the 24-hour period of the LD regimen.

**Eutectic temperature** The lowest temperature at which a solution of given composition can exist in the liquid phase.

**Excystment** The processes involved in cyst germination.

**Exogenous** Resulting from causes or factors external to an organism.

**Exponential growth** A pattern of biomass (Q) or cell number (N) accretion in which the rate of increase is constant (= log phase growth). The population or biomass increase is then described by an exponential function of time, namely $Q = e^{rt}$.

**Fe′** Dissolved inorganic iron species.

**Fermentor** A closed or mostly closed vessel for heterotrophic production in which energy is supplied as organic carbon.

**Flow cytometry** A technique that can measure cell concentration and properties based on light scattering and fluorescence emission.

**Fluorescent dye** Molecule that binds to or locates in a particular region of the cell and that emits fluorescence when excited.

**Fluorescent in situ hybridization (FISH)** A technique by which fluorescently labeled oligonucleotide probes are hybridized to fixed cells. Target and nontarget cells are then discriminated by fluorescence microscopy or flow cytometry.

**Free-running rhythm** An endogenous rhythm recorded under constant conditions, e.g., in constant light.

**Gamete** Haploid reproductive cell.

**Gametolysin** The enzyme active in releasing gametes by digesting the mother cell wall.

**Glass state of water (amorphous state of solidified water)** See vitrified state of water.

**Glass transition/transformation temperature ($T_g$)** The thermal transition point at which there is a change in the physical properties of a substance, i.e., it is in a glassy (vitreous) state. Below the $T_g$ water molecules do not convert from one form to another. For example, below this temperature vitrified water does not crystallize and solidified water does not vaporize.

**Haplodiplont** The type of life cycle (e.g., *Ulva*) in which both the haploid and the diploid generations multiply by mitosis.

**Haploid** Cells with one set of chromosomes: one copy of the basic genetic complement of the species.

**Heterothallic** Requires two genetically different strains to produce compatible gametes and undergo sexual reproduction.

**Homothallic** Undergoes sexual reproduction within one strain, e.g., derived from a single cell isolation event. Compatible gametes can be produced by meiosis within a single strain; therefore, sexual reproduction can occur within a culture containing only one strain.

**Hydrous ferric oxides** Amorphous iron precipitates also commonly referred to as iron (or ferric) hydroxides.

**Hypnozygote** Dormant nonmotile zygote with a resistant outer coating.

**Induction** The act or process of inducing a latent or lysogenic virus to begin a lytic life cycle.

**Intercrosses** Two strains cultured together in conditions conducive for sexual reproduction.

**Isogamous** Morphologically identical gametes.

**Isogamy** The morphological state in which the female and male gametes cannot be distinguished by morphology.

**Isolation** Process of selecting one or more cells and establishing a culture; culture is ideally derived from a single cell without contamination but may be unialgal or mixed.

**$K^*_{MY}$** A conditional equilibrium stability constant for the complex MY defined in terms of the concentration of aquated free metal ions ($M^{n+}$).

**$K'_{MY}$** A conditional equilibrium stability constant for the complex MY, defined in terms of the concentration

of dissolved inorganic metal species ($M'$), e.g., $K'_{ZnEDTA}$.

**$K_{Ca,M}$** Equilibrium exchange constant for the reaction of Ca-EDTA or NTA chelates with trace metal M. These constants equal the thermodynamic stability constant for the formation of the trace metal chelate divided by that for the formation of the Ca chelate.

**$k_d$** The rate constant for the dissociation of a metal complex into free metal and ligand.

**$K'_d$** An equilibrium dissociation constant, equal to $1/K'_{MY}$.

**$k_f$** The rate constant for the reaction of a metal with a ligand to form a metal-ligand complex.

**$K_{hv}$** A steady state constant equal to $k_{hv}/k_f$.

**$k_{hv}$** The rate constant for the photolysis of a metal chelate (e.g., FeEDTA), resulting in the photo-dissociation of the chelate.

**Latency** Dormancy of a temperate virus.

**Ligand** A chemical constituent containing two unpaired electrons that can form a complex with a metal ion.

**Lysis** The rupture of cell membranes and loss of cytoplasm.

**Lysogen** A bacterial cell whose chromosome contains integrated viral DNA.

**Lysogeny** The ability of a virus, specifically one that infects bacteria, to survive by integrating its DNA into the genome of a host.

**Lytic virus** A virus that, as part of its life cycle, causes its host to rupture.

**$M'$** Inorganic metal species, including aquated free metal ions and complexes with inorganic ligands such as $OH^-$, $Cl^-$, and $CO_3^{2-}$.

**$[M']$** The molar concentration of inorganic metal species.

**$[M^{n+}]$** The molar concentration of free aquated metal ions, e.g., $[Zn^{2+}]$ or $[Fe^{3+}]$.

**$[M_{tot}]$** The total molar concentration of a metal; e.g., $[Cu_{tot}]$.

**Mandatory dormancy period** Period within which germination cannot occur because of internal processes or inhibition.

**Mating system** System of mating patterns, including mating type and reproductive compatibility.

**Mating type** Genetically determined physiological and biochemical differences of individuals that determine their sexual compatibility.

**Meiosis** Nuclear division that results in daughter nuclei, each containing half the number of chromosomes of the parent cell.

**Metal buffer** A stable chemical system consisting of a chelator and a lesser concentration of metal chelate, which through association and dissociation reactions maintains the free metal ion concentration at constant or near-constant levels.

**Mitosis** Nuclear division that results in daughter nuclei containing identical sets of chromosomes.

**$M^{n+}$** A free aquated metal ion, e.g., $Zn^{2+}$ or $Fe^{3+}$.

**Nanoplankton** Fraction of the plankton with size between 20 $\mu$m and 2 to 3 $\mu$m.

**NTA** Nitrilotriacetic acid, a chelator used in metal ion buffer systems.

**NTA\*** All NTA species not complexed with trace metal ions, including protonated forms and Ca and Mg chelates.

**Nucleation of ice** The initiation of an ice crystal in an aqueous solution whose temperature is at or below the equilibrium temperature for ice formation. A nucleation event may be initiated by introducing an appropriate particulate substance or perturbations to the solution.

**Oligonucleotide probes** Short DNA sequences that recognize a signature (usually on the 18S rDNA gene) specific of a taxonomic group.

**Oogamy** The morphological state in which the female gamete is immotile and generally larger than the flagellated male gamete.

**Osmotic stress** An excessively large difference in osmotic potential of the solutions contacting the two opposite sides of a cellular membrane that has the potential to cause damage to a cell.

**PAR** Photosynthetically available radiation, i.e., total irradiance over the range of 400 to 700 nm. Also called photosynthetically active radiation and sometimes extended to 350 to 700 nm.

**Parthenogenesis** Reproduction from a single gamete without fertilization by a compatible gamete.

**Pasteurization** Heating a liquid to a temperature, usually between 66° and 80°C, but always less than 100°C, for at least 30 minutes, followed by rapid cooling to less than 10°C. Note: Pasteurization of seawater is usually carried out at 95°C for 1 hour.

**Period, rhythms** The time to complete one cycle, e.g., the time distance from one maximum to the next; designated as $\tau$ (tau) in the case of a free-running rhythm (e.g., $\tau = 23$ h or $\tau = 25$ h), or T in the case of an entrained rhythm (e.g., T = 24 h).

**Persistent infection** The continuous production of infectious virus either by the survival of an infected host or by a situation in which a minority of cells in a population are infected and the virus spread is limited. Also known as chronic infection.

**Phage** A synonym for bacteriophage.

**Phase** An arbitrarily chosen reference point or part in a cycle, e.g., the day phase or the night phase, or the maximum (acrophase).

**Pheromone** Chemical substance emitted by one cell or organism into the environment that elicits a specific response from another cell/organism of the same species.

**Photobioreactor** A closed or mostly closed vessel for phototrophic algal production in which energy is supplied by light.

**Picoplankton** Fraction of the plankton smaller than 2 to 3 $\mu$m in size. Originally, this included plankton strictly between 0.2 and 2 $\mu$m; femtoplankton is sometimes used for the smaller viruses. The term *ultraplankton* has also been used for plankton between 0.2 to 5 $\mu$m.

**Planomeiocyte** Motile cell that undergoes meiosis.

**Planozygote** Motile zygote.

**Pure culture** 1. Axenic culture. 2. Unialgal culture free of protozoa, fungi, etc., but that may contain bacteria.

**Quantitative genetics** The branch of genetics concerned with the inheritance of continuously varying or meristic traits. These are traits that do not show mendelian inheritance; they are usually determined by the combined effect of many genes.

**Quiescence** Temporary cessation of development due to unfavorable environmental or internal conditions.

**Resting cyst** Dormant nonmotile hypnozygote with a resistant outer coating.

**Resting stage** Dormant nonmotile cell.

**Rough culture** Containing more than one species of algae, protozoa, or fungi.

**Safe system of work** A documented (written) method of undertaking any activity, which takes into consideration all appropriate safety issues and relevant legislation.

**Self-crosses** One strain culture in conditions conducive for sexual reproduction.

**Semicontinuous batch culture** A procedure for maintaining exponential growth without the use of continuous culture techniques. Exponential growth is maintained by careful monitoring and diluting the culture near the end of the exponential phase before the culture runs out of nutrients. A small inoculum is made into new medium and the culture is diluted (about 95% of the culture is poured out and new seawater medium is added to bring the culture to the original volume). Distinguished from another type of "semicontinuous culture" in which the culture is diluted frequently (e.g., each day) with a portion of fresh medium.

**Semicontinuous culture** A procedure for maintaining exponential growth without the use of continuous culture techniques, useful when cultures cannot be stirred continuously. Exponential growth is maintained by careful monitoring and repeated dilution of the culture while it is still in exponential phase. Biomass can be limited by the concentration of a limiting nutrient in the medium, with dilution rate (hence growth rate) determined by the experimenter (analogous to a chemostat), or biomass can be kept low by dilution of a nutrient-replete semicontinuous culture, so the average growth rate is equal to the nutrient-replete growth rate at a given irradiance and temperature (analogous to a turbidostat).

**Solution effects** Effects of solute composition and concentrations on the ability of a living system to survive cryopreservation.

**SOW** Synthetic ocean water (= artificial seawater or AW).

**Standard dormancy** The number of days from first resting cyst production to first cyst germination.

**Statocyst** Siliceous resting cell.

**Steaming** A nonspecific term for pasteurization or tyndallization (see definitions).

**Sterile technique** Working under conditions in which contamination (bacterial, fungal, viral, etc.) is avoided; = aseptic technique.

**Sterilization** A process or procedure for establishing an aseptic condition; the removal or killing of all microorganisms. There are four major types: heat sterilization, electromagnetic wave sterilization, sterilization using filtration, and chemical sterilization.

**Subjective day, subjective night** Designations for the phases corresponding to the preceding light and dark phases in an LD cycle.

**Supercooled aqueous solution** A solution that is in liquid form even though it is colder than its equilibrium temperature of freezing.

**Synchronized rhythm** Equal to an entrained rhythm; established by the environment, e.g., under a light-dark (LD) regimen (LD 12:12). During entrainment, the free-running period of the circadian rhythm is reset to exactly the 24-hour period of the LD regimen.

**Temperate virus** A virus that, upon infection of an eukaryotic host, does not necessarily cause lysis but whose genome may replicate in synchrony with that of the host for long-term dormancy. Temperate viruses may be reactivated into a lytic life cycle by stress such as ultraviolet radiation. A virus, acting similarly upon a prokaryotic host, is termed a *lysogenic virus*.

**Temporary cyst** Temporary vegetative resting cell.

**Tris** Tris(hydroxymethyl)aminomethane, a commonly used pH buffer that also chelates trace metals such as copper.

**Tyndallization** Similar to pasteurization, but repeated: heating a liquid to a temperature, usually 66° to 80°C but always less than 100°C, for at least 30 minutes, followed by rapid cooling to less than 10°C and maintained cooling until the following day, when it is heated and cooled again; the cycle is repeated once more on the third day.

**Turbidostat** A type of continuous culture system in which the dilution rate (flow rate divided by the volume of the turbidostat) is equal to the nutrient-replete growth rate at a given irradiance and temperature. Delivery is triggered by growth of the culture, as detected by an increase in a measure of biomass (e.g., attenuation or turbidity) over a threshold set by the experimenter, and halted when dilution causes the measure of biomass to drop below a second threshold set by the experimenter. Biomass oscillates in a saw tooth pattern.

**Unialgal** Culture that contains a single species of algae but may contain different phenotypes of the alga or contaminants (e.g., bacteria); usually means free of protozoa and fungi.

**Unialgal culture of macroalgae** A culture that includes only one species of alga.

**Virus** An obligate intracellular parasite of noncellular nature, consisting of DNA or RNA and a protein coat and ranging in size from 20 to 400 nm.

**Vitrified state of water** An amorphous (noncrystalline) solid state of water. It generally can be produced only by an extremely rapid rate of cooling or a very high concentration of solutes.

**VLP** An acronym for virus-like particle. Used to describe a particle resembling a virus that has not been proven to be a virus by fulfilling Koch's postulates.

**Yield** The amount of new biomass or cell number that accumulates over a period of time; the difference between the starting biomass or cell number and the amount at the end of a time interval. There is no assumption of a continuous rate of growth during the time interval.

**Zeitgeber** A synchronizing environmental factor that causes an entrained or synchronized rhythm.

**Zygote** Cell formed from the union of two gametes or reproductive cells.

# NAME INDEX

# SUBJECT INDEX

*Note:* The f or t following a page number denotes a figure or table on that page.

differential centrifugation, 120
dilution culture, 121
flow diagrams, 118f
glycerol purification, 195
goal, 117
macroalgae, 133–143
microalgae, 117–132
micropipette, 122–123
other antimicrobial agents, 126
size selective screening and
  filtration, 119–120
sonication, 120–121
testing for contaminants,
  127–129
UV radiation, 127
viral contamination, 381
vortexing, 120–121
PVP (polyvinylpyrrolidone), 174

Q

Quasi-chemostat, 299

R

Raceway ponds, 206–207, 210–213
Raft-supported long-line systems,
  230f, 231
Rate-zonal centrifugation, 381,
  381f
Reagents, 340
Recipes, 429–538. See also Culture
  media; specific recipes.
  artificial seawater media,
    487–500
  AW basal salt, 29
  bacterial-fungal test media for
    freshwater strains, 485–486
  bacterial-fungal test media for
    seawater, 527–532
  enriched natural freshwater
    media, 476–477
  enriched natural seawater +
    organics, 521–523
  enriched natural seawater
    media, 501–520
  freshwater culture media,
    431–486

freshwater media enriched with
  organics, 482–484
freshwater synthetic culture
  media, 431–475
marine culture media, 487–532
overview, 429–430
precipitate, 22, 430
soil-water enriched freshwater
  media, 478–481
soil-water enriched natural
  seawater media, 524–526
specific recipes. See Specific
  recipes.
websites, 429
Red Lists, 421
Redfield ratio, 26, 301
Reference materials (RMs), 336
Reisolation of existing cultures,
  109–110
Removal techniques. See
  Isolation/removal of specific
  organisms.
Reproductive compatibility, 409
Resting cysts, 400
Resting stages, 399–400
Restriction fragment length
  polymorphism analysis, 371
Reusable glass pipettes, 66, 69
Reverse-phase chromatogram,
  338f
RMs (reference materials), 336
RNA gel electrophoresis, 354
RNeasy Plant Mini Kit, 354
Rocephin, 161
Rope cultivation, 225, 226f
Rotifers, 215
Routine serial subculturing, 145,
  146

S

SAG, 430t
Sample freezing. See
  Cryopreservation.
Sand-dwelling organisms, 96–97
Sargasso Sea water, 113
Scale-up, 213–214
Scatter trigger, 105
Scattering errors in suspensions,
  308

Schott illuminator, 85, 86f
Scotch-Brite Dobie cleaning pad,
  160
Screening multiwell plates,
  112–114
Screw caps, 151
SD (short-day), 356
Seafoam, 97
SeaHARRE-2 intercalibration, 339
Seawater. See Marine culture
  media.
Seaweed, 219. See also Mariculture
  of seaweeds.
Sedgwick-Rafter chamber,
  245–248
Seed bank, 400
Selective culturing, 89
Selenium, 38
Semicontinuous batch culture,
  275, 277–278
Semicontinuous cultures, 297–299
Separate stock solutions, 16–17
Separation techniques. See
  Isolation/removal of specific
  organisms.
Serial dilution, 121
Settling, 94
Settling cells, 247–249
Sexual life cycle, 404–408
Sexual reproduction, 389–397
  basic sexuality in eukaryotes,
    389–390
  finding clones that can mate,
    394–395
  induction of gametogenesis,
    390–394
  sterility of clones after
    prolonged culture, 395
  terminology, 390
  zygote maturation and
    germination, 394
Shakers, 160
Sharples, 194
Sherman micromanipulator, 120
Short-day (SD), 356
Shreiber solution, 5
Shrinkable sheath tubing, 171
Siderophores, 51
Sieve, 87
Signal/noise ratio, 308
Silicate, 26
Silicate stock, 430

Printed and bound by CPI Group (UK) Ltd, Croydon, CR0 4YY

08/05/2025

01864913-0001